2nd International Conference on Hazard Mitigation in Geographic and Education Perspectives

AA007738

IOP Conference Series: Earth and Environmental Science
Volume 884

Yogyakarta, Indonesia
11-12 September 2020

ISBN: 978-1-7138-4866-0
ISSN: 1755-1307

Printed from e-media with permission by:

Curran Associates, Inc.
57 Morehouse Lane
Red Hook, NY 12571

proceedings.com

Some format issues inherent in the e-media version may also appear in this print version.

This work is licensed under a Creative Commons Attribution 3.0 International Licence.
Licence details: http://creativecommons.org/licenses/by/3.0/.

No changes have been made to the content of these proceedings. There may be changes to pagination and minor adjustments for aesthetics.

Printed with permission by Curran Associates, Inc. (2022)

For permission requests, please contact the Institute of Physics at the address below.

Institute of Physics
Dirac House, Temple Back
Bristol BS1 6BE UK

Phone: 44 1 17 929 7481
Fax: 44 1 17 920 0979

techtracking@iop.org

Additional copies of this publication are available from:

Curran Associates, Inc.
57 Morehouse Lane
Red Hook, NY 12571 USA
Phone: 845-758-0400
Fax: 845-758-2633
Email: curran@proceedings.com
Web: www.proceedings.com

TABLE OF CONTENTS

IOP Conf. Series: Earth and Environmental Science **884** (2021) 011001 doi:10.1088/1755-1315/884/1/011001

PREFACE

The Faculty of Social Sciences, Universitas Negeri Yogyakarta in collaboration with the Department of Geography of the Universiti Pendidikan Sultan Idris (UPSI) Malaysia, the Geography Department of Chulalongkorn University (CU) Thailand, the Geography Department of the University of Muhammadiyah Surakarta (UMS) Indonesia, and the Indonesian Geographical Association held The 2nd International Conference on Hazard Mitigation in Geographic and Education Perspectives (The 2nd ICHMGEP) on "Disaster Management for Enhancing Resilience, Risk Reduction and Sustainable Development Goals". This conference continues the tradition of scientific meetings (The 1st ICHMGEP) organized in 2018, which successfully gathered hundreds of participants and published conference papers in international publication outlets. The 2nd ICHMGEP 2020 provides a platform for researchers, lecturers, students, practitioners, and scientists to showcase their latest findings in areas of expertise, including geoscience, environment, science and technology innovation, mapping, and technological innovation. In addition, it aims to disseminate original research results, new ideas, innovations, practical experiments to promote scientific progress and achieve sustainable development goals.

The 2nd ICHMGEP 2020 had formerly been planned to be organized in an offline format. However, it was conducted online on September 11-12, 2020 mainly due to the Covid-19 pandemic impacts. The increasing number of Covid-19 positive cases in Yogyakarta Province since March 2020 forced the government to carry out various restrictions on activities gathering many people like a conference. This is supported by the issuance of the Covid-19 emergency response period in Yogyakarta, Indonesia from March 20 - May 29 2020. Moreover, the emergency response period was extended on May 30 - June 30 2020, 1 - 31 July 2020, 1 - 31 August 2020, and 1 - 30 September 2020. In addition, the Covid-19 pandemic offered uncertainty and was difficult to be predicted when it would end. Therefore, the organizing committee decided to change the offline format into an online format as the conference would provide great benefit for the wider community including giving solutions to cope with covid-19 related problems. Also, many supportive platforms such as zoom dan live youtube could facilitate the implementation of the online conference

The plenary session of the conference was held online on September 11, 2020, utilizing the zoom platform and broadcasted live via the official youtube channel of the Faculty of Social Sciences UNY (fis_unyofficial), with a keynote speaker Dr. Suhadi Purwantara, M.Sc. (UNY, Indonesia) and 3 invited speakers namely, Dr. Danang Sri Hadmoko, S.Si., M.Sc. (UGM, Indonesia), Assoc. Prof. Dr. Mohd Hairy Ibrahim (UPSI, Malaysia), and Dr. Bayes Ahmed (UCL, United Kingdom). The parallel session was on September 12, 2020, participated by 73 presenters via zoom. They were divided into seven groups using the zoom meeting rooms. Each presenter was given 20 minutes, consisting of 10 minutes for presentations and 10 minutes for discussions. The participants' enthusiasm during the discussion session was satisfactory, indicated by the presence of several questions addressed to the presenters. All participants were able to access the conference virtually well.

Zoom and youtube were used as they were the appropriate platforms to support the implementation of the online conference. They helped the 2nd ICHMGEP run smoothly and successfully amid the pandemic situation. To support the zoom and youtube operation, the committee utilized the facilities in the Communication Science laboratory of the Faculty of Social Sciences UNY, which was equipped with standard cameras for live streaming, mixers, and audio. Also, the committee used the FISNET laboratory, which was supported with adequate computer equipment, to host the online conference and ensure that the online conference run as expected. The conference's success was also supported by the readiness of human resources who were organizing the conference.

Overall, the implementation of The 2nd ICHMGEP was successful viewed from the number of conference participants, namely, 124 plenary session participants and 73 parallel session presenters.

Content from this work may be used under the terms of the Creative Commons Attribution 3.0 licence. Any further distribution of this work must maintain attribution to the author(s) and the title of the work, journal citation and DOI.
Published under licence by IOP Publishing Ltd

Viewed from the content, the four outstanding speakers delivered insightful and important topics via zoom during the plenary session including, "Never Stop to Think about Mitigation", "Building Multi-Dimensional Resilience to Disasters", "Resilience toward SDGs", and "Social Vulnerability Assessment in Disaster Risk Reduction (DRR)". All of these materials complement the conference theme: "Disaster Management for Enhancing Resilience, Risk Reduction and Sustainable Development Goals". To support the dissemination of research results, new inventions, insight via the conference papers, the organizing committee carried out a rigorous review of the conference paper before the conference, conducted a paper coaching clinic by inviting Prof. Dr. Sri Rum Giyarsih, M.Sc. (UGM, Indonesia) on September 21, 2020, using the zoom platform, and final review before submitting papers to the IOP publisher.

Finally, we would like to thank the Rector of Universitas Negeri Yogyakarta, who provided financial support to organize this conference. We would also like to thank all the invited speakers and participants from various cities in Indonesia, Malaysia, Thailand who have taken the time to participate in this online conference. Our gratitude also goes to IOP publishing as our partner to publish quality conference papers. We also thank all the organizers who have successfully organized this conference. Finally, I hope that this conference can contribute valuable insights for the Indonesian people and the world in general.

IOP Conf. Series: Earth and Environmental Science **884** (2021) 011001 doi:10.1088/1755-1315/884/1/011001

Lampiran

Day	Link
The 2nd ICHMGEP: Day 1	
Sep 11, 2020 02:00 PM Jakarta	Join Zoom Meeting https://uny-ac-id.zoom.us/u/abBec1EVl4
	Join You tube https://youtu.be/Us_ldyYS00Y and https://youtu.be/qHy30SY6qJA
The 2nd ICHMGEP: Day 2	
2nd ICHMGEP Room 1 Sep 12, 2020 08:30 AM Jakarta	Join Zoom Meeting https://uny-ac-id.zoom.us/j/94888826866?pwd=U1ZaanZvOHJ1VnQ5R2txN21JSEFrUT09 Meeting ID: 948 8882 6866 Passcode: ICHMGEP
2nd ICHMGEP Room 2 Sep 12, 2020 08:30 AM Jakarta	Join Zoom Meeting https://uny-ac-id.zoom.us/j/91818811030?pwd=TlNoWFBlZUw0SlByLzg2Wk9DU2xBUT09 Meeting ID: 918 1881 1030 Passcode: ICHMGEP
2nd ICHMGEP room 3 Sep 12, 2020 08:30 AM Jakarta	Join Zoom Meeting https://uny-ac-id.zoom.us/j/95203796535?pwd=YTNoOUFnbGRyckxwM1NCcHloVDE1dz09 Meeting ID: 952 0379 6535 Passcode: ICHMGEP
2nd ICHMGEP Room 4 Sep 12, 2020 08:30 AM Jakarta	Join Zoom Meeting https://uny-ac-id.zoom.us/j/94764057731?pwd=NWY1S0V1QndjVDdvTkJmeEx0VElNZz09 Meeting ID: 947 6405 7731 Passcode: ICHMGEP
2nd ICHMGEP room 5 Sep 12, 2020 08:30 AM Jakarta	Join Zoom Meeting https://uny-ac-id.zoom.us/j/97688860031?pwd=L0JtclZtWk1GZFVWSVYwYXFJN3lUUT09 Meeting ID: 976 8886 0031 Passcode: ICHMGEP
2nd ICHMGEP room 6 Sep 12, 2020 08:30 AM Jakarta	Join Zoom Meeting https://uny-ac-id.zoom.us/j/99125187476?pwd=VzRpMVZiNis2b3JUd3lBcHA0eUZMQT09 Meeting ID: 991 2518 7476 Passcode: ICHMGEP
2nd ICHMGEP room 7 Sep 12, 2020 08:30 AM Jakarta	Join Zoom Meeting https://uny-ac-id.zoom.us/j/98019514967?pwd=U0VIS3hZMG9yU3BKWnRFdFZyeWJRQT09 Meeting ID: 980 1951 4967 Passcode: ICHMGEP

IOP Conf. Series: Earth and Environmental Science **884** (2021) 011001 doi:10.1088/1755-1315/884/1/011001

The 2nd ICHMGEP: Article Coaching Clinic	
Online Article Coaching Clinic September 21st, 2020 09:00 AM Jakarta	Join Zoom Meeting https://uny-ac-id.zoom.us/j/94010441223?pwd=cGtwVzVXdHRva3B4OXdsdFZpQlVSZz09 Meeting ID: 940 1044 1223 Passcode: ICHMGEP

IOP Conf. Series: Earth and Environmental Science **884** (2021) 011001 doi:10.1088/1755-1315/884/1/011001

RUNDOWN THE 2ND INTERNATIONAL CONFERENCE ON HAZARD MITIGATION IN GEOGRAPHIC AND EDUCATION PERSPECTIVES (ICHMGEP) 2020

Platform: Zoom and Youtube

Friday, September 11, 2020		
Time	Program	PIC
03.00-03.10 p.m (Indonesian Time) 04.00-04.10 p.m (Malaysia Time) 09.00-09.10 a.m (UK Time)	Conference Opening Sing National Anthem Indonesia Raya	Committee
03.10-03.20 p.m (Indonesian Time) 04.10-04.20 p.m (Malaysia Time) 09.10-09.20 a.m (UK Time)	Welcoming Speech	Prof. Dr. Margana, M.Hum., M.A. Rector of Universitas Negeri Yogyakarta (UNY)
03.20-03.40 p.m (Indonesian Time) 04.20-04.40 p.m (Malaysia Time) 09.20-09.40 a.m (UK Time)	Plenary Session & Discussion	Dr. Suhadi Purwantara, M.Si. - Associate Professor in Faculty of Social Sciences UNY - Dean of Faculty of Social Sciences UNY
03.40-04.00 p.m (Indonesian Time) 04.40-05.00 p.m (Malaysia Time) 09.40-10.00 a.m (UK Time)		Dr. Danang Sri Hadmoko, S.Si., M.Sc. - Lecturer in Faculty of Geography UGM - Director of Partnership, Alumni and Global Initiatives of UGM
04.00-04.20 p.m (Indonesian Time) 05.00-05.20 p.m (Malaysia Time) 10.00-10.20 a.m (UK Time)		Assoc. Prof. Dr. Mohd Hairy Ibrahim - Associate Professor in Faculty of Social Sciences and Humanities UPSI Dean of Faculty of Social Sciences and Humanities UPSI
04.20-04.40 p.m (Indonesian Time) 05.20-05.40 p.m (Malaysia Time) 10.20-10.40 a.m (UK Time)		Dr. Bayes Ahmed - Lecturer in University College London, United Kingdom
04.40-05.20 p.m (Indonesian Time) 05.40-06.20 p.m (Malaysia Time) 10.40-11.20 a.m (UK Time)		Moderator
05.20-05.30 p.m (Indonesian Time) 06.20-06.30 p.m (Malaysia Time) 11.20-11.30 a.m (UK Time)	Conference Closing	Committee

Platform: Zoom

Saturday, September 12, 2020		
Paralel Session	Presenter	PIC
09.00 a.m - 01.00 p.m (Indonesian Time)	Room 1,2,3,4,5,6,7	Committee

IOP Conf. Series: Earth and Environmental Science **884** (2021) 011001 doi:10.1088/1755-1315/884/1/011001

ROOM DISTRIBUTION FOR PARALLEL SESSION
12 September 2020, Platform: Zoom

ROOM 1
Moderator: Diana Trisnawati, S.Pd., M.d
Time: 09.00 a.m – 01.00 p.m

No	Name	Paper Title
1	Dyah Respati Suryo Sumunar	Identification of Spatial Clusters of Covid-19 In Yogyakarta Using Global Moran's Index
2	Edi Nursantosa	Image Pattern Recognition In Spatial Data Using Artificial Neural Network
3	Assriyani	The Roles of Human Geography in Disaster Studies
4	Tanasiva	Tsunami Vulnerability Assessment and Its Implications for Disaster Risk Management in the coastal area of Purworejo Regency
5	Norawit Suwannakarn	Assessment of Human Impact on Flooding for Thai East Coast - Gulf Basin
6	Saegita Adi	Assessment of Soil Erosion Hazard in Prambanan District Using RUSLE (Revised Universal Soil Loss Equation)
7	Muhammad Asrori Indra Wardoyo	Geomorphological Hazard at Some Area of The Outer Dieng Volcanic Complex Post Explosive Eruption Era
8	Nungki Febriliana Wardaya	Student's Level of Understanding on Mobile Learning based Volcano Eruption
9	Husnanda Arief Adhandika	Community Preparedness of The Fire Hazard in Jetis District, Yogyakarta City

ROOM 2
Moderator: Dyah Ayu Anggraheni, M.A
Time: 09.00 a.m – 01.00 p.m

No	Name	Paper Title
1	Kharisma	The Potential of River Discharge at The Peak of the Dry Season in Some Disaster-Prone Area of Merapi Volcano
2	I Gede Tunas	Impact of Extreme Rainfall on Flood Hydrographs
3	Raras Gistha Rosardi	The Role of Village Tourism in Environmental Education and Social Disaster Management
4	Ernawati Tarmizi	Community Education Model Design in Management Household Waste in Padang City
5	Fitria Nucifera	Assessing local community resilience to tsunami in a small port (Case Study: Sadeng Port, Indonesia)
6	Pandhu Yuanjaya	The Local Wisdom of The Tenggerese People to Coexist with Disaster of The Mount Bromo
7	Antonius Ranauria Timur	Current States of Demographic Education for Indonesian Youths: Notes and Challenges
8	Dewi Febriani	Description of the dynamics of the resilience of people victims of the 2018 North Lombok earthquake
9	Eko Sulistyono	Student Satisfaction to the Communication Process During Online Learning in the Covid-19 Pandemic

10	Wahyono	Making Disaster Monument and Mural as Disaster Education to The Community of Padang
11	Irfiani Triastari	Developing Disaster Mitigation Education with Local Wisdom: Exemplified in Indonesia Schools

ROOM 3
Moderator: Danu Eko Agustinova, M.Pd
Time: 09.00 a.m – 01.00 p.m

No	Name	Paper Title
1	Indreswari Suroso	The Analysis of Dry Weather Aerial Photography in Sermo Kulonprogo Reservoir using DJI Spark Drone
2	Salsabila Aulia Putri Zakaria	Seismicity Map to Analyze the Depth and Magnitude Earthquake Zone in Kwandang Area of North Gorontalo Regency
3	Rani Muliarti	Disaster Risk Analysis of Merapi Volcano Eruption in Cangkringan District Sleman Regency
4	Wahyu Sabani	Participatory Mapping of Tsunami Evacuation Routes (Case Study of Karangbenda Village Cilacap Regency)
5	Arif Fahrizal Nugraha	Spatial Evapotranspiration Modeling Assisted With Landsat 8 Image Using Sebal And Geographically Weighted Regression Methods In Magelang District
6	Nazar Nurdin	Perceptions of Youth Religious Understanding of Natural Disaster Resilience and Preparedness
7	Kurnia Nur Fitriana	Disaster Mitigation for Elderly: The Challenges of Social Welfare Services for Elderly in The COVID-19 Pandemic
8	Saryono	Disaster Preparedness in Physical Aspect of Students in Bantul and Balikpapan Indonesia
9	Alifi Nur P N	Disaster mitigation and resiliency learning strategies of teachers during Covid-19 pandemic from several provinces in Indonesia

ROOM 4
Moderator: Benni Setiawan, S.H.I., M.S.I
Time: 09.00 a.m – 01.00 p.m

No	Name	Paper Title
1	Noor Hamidah	The Function of "WHITE BOX" Sterilization in Anticipation of Covid-19 Pandemic Prevention
2	Ratih Probosiwi	Disaster Mitigation for Women: Learn from 2018 Pandeglang Tsunami, Indonesia
3	La Ode Amaluddin	Disaster Mitigation of Dry and Clean Water Crisis in Dana Sub-District Watopute District Muna Regency
4	Hanip Hendra Setyawan	The community adaptation strategy in dealing with landslide disaster in Banaran Village Ponorogo Regency
5	Isnani Nurkholifah	Senior High School Students' Preparedness In Facing Earthquake Disaster
6	Sareta Selaby	Landslide Susceptibility in Majalengka Regency, West Java Province

IOP Conf. Series: Earth and Environmental Science **884** (2021) 011001 doi:10.1088/1755-1315/884/1/011001

7	Sasqia Fathaya Syahar	Characteristics of Landslide and Rainfall Areas in Majalengka Regency, West Java Province
8	Fitriana Yuli Saptaningtyas	Determination of the level of seismic hazard in Yogyakarta using fuzzy simple additive weighting method
9	Abdul Latif Rifa Fakhrurozi	Community Resilience Against Tsunami Disaster in the Coastal Areas of Purworejo Regency
10	Westi utami	Settelement Suitability Analysis Based on the Catastrophic Eruption Of Sinabung
11	Marita Ahdiyana	Collaborative Governance in Realizing An Inclusive Workforce for People with Disabilities to Achieve Sustainable Development Goal's

ROOM 5
Moderator: Chatia Hastasari, M.I.Kom
Time: 09.00 a.m – 01.00 p.m

No	Name	Paper Title
1	Triyanto	Strengthening Disaster Literation As An Effort To Reduce The Risk of Disaster In D.I. Yogyakarta society
2	Erry Ersani	Disaster Mitigation Snake and Ladder Game to Improve Earthquake Disaster Preparedness (A Case Study: Yogyakarta 5 Senior High School)
3	Diana Septi Purnama	Development of Pictorial Card "Kencana" (Disaster Preparation) as a Media of Group Guidance Services In SMP N 1 Wonosari
4	Ira Putri Ariyani	School Community Preparedness in Anticipation of Earthquake and Tsunami Threats in Temon Sub-District
5	Bambang Syaeful Hadi	The Effect of Google Earth-Assisted Remote Sensing Learning on Students "Spatial Thinking Ability In Solving Disaster Mitigation Problems"
6	Muhammad Nursa'ban	Spatial Representation Developing Geographical awareness for Disaster Risk Reduction
7	Nurul Khotimah	Spatial Analysis of Covid-19 Distribution: cases studies in Indonesia and Malaysia
8	Pratiwi Wahyu	Kindergarten Teacher Online Resilience in the Implementation of Online Learning in Mlati District of Sleman
9	Arif Ashari	The challenges and obstacles of teaching geography in senior high schools during the COVID-19 pandemic: Lesson learned from several provinces
10	Nurhadi	Developing Sustainable School in the Shadow of Active Stratovolcano: A Study at Disaster-Prone Area of Merapi Volcano

ROOM 6
Moderator: Aris Martiana, M.Si.
Time: 09.00 a.m – 01.00 p.m

No	Name	Paper Title

IOP Conf. Series: Earth and Environmental Science **884** (2021) 011001 doi:10.1088/1755-1315/884/1/011001

1	Sutanto Trijuni Putro	Commercial Multirotor UAV Campaign on Data Acquisition for Disaster Management
2	Akalili Zhafirah Syarafina	Physical Vulnerability Assessment of Settlements in Lahar Hazard Prone Areas along the River originated from Merapi Volcano
3	Rita Noviani	The Multi-Risk Analysis of Various Disasters in Karanganyar Regency
4	Soni Nopembri	Constructing Student's Physical Preparedness through Game-Based Activities in Disaster-Safe School
5	Amalia Gita Ayudyanti	Impact of Optical Aerosol Depth (AOD) on Light Pollution Level: a spatio-temporal analysis
6	Listyo Yudha Irawan	The Use of Machine Learning for Accessing Landslide Susceptibility Class: Study Case of Kecamatan Pacet, Kabupaten Mojokerto
7	Sukmaniar	Hazard Level of Slum Areas in Palembang City
8	Agnes Putri Devinta	The Utilization of Sentinel-2a And Aster Imagery for Monitoring The Changes Of Public Green Open Space and Oxygen Needs In Sukoharjo Regency In 2004-2019
9	Revi Hernina	Assessment of potential locations for evacuation in Depok Municipality using images from UAV
10	Thitirat Panbamrungkij	A Flood Damage and Shelter Need Assessment: A Case Study of Mueang Sing Buri, Thailand
11	Dinda Zannuba Arifah	Spatial modelling of tsunami exposure areas at Ujung Genteng coastline, Sukabumi, West Java

ROOM 7
Moderator: Gilang Jiwana Adikara, MA
Time: 09.00 a.m – 01.00 p.m

No	Name	Paper Title
1	Syarif Jamaludin	Rethinking local genius of the Merapi's community in eruption disaster risk reduction. Does it still work?
2	Muhammad Badri	The Risk Communication Disaster of COVID-19 Community Based Through "Keep the Village" Program
3	Sovia Husni Rahmia	Disaster Resilient Village: Building Community-Based Disaster Management in Indonesia
4	Herdiyan Maulana	The Role of Sense of Coherence and Social Capital on Perceived Risk: Salutogenic Model Approach on Flood Survivors in Indonesia
5	Susiana herawati	Effectiveness of Communication to Prevent Covid-19 Transmission: A Case Study of the Kare Village
6	Anggarani Berliana Surindar	Community Preparedness for Tsunami Hazard in Galur District, Kulon Progo Regency, Special Region of Yogyakarta
7	Dwi Harsono	Disaster Rehabilitation for People with Disabilities (PWDs in Facing of COVID-19 Pandemic)
8	Fauziah Latifah	The Preparedness Level of Housewives in Dealing with the Earthquake Disaster in Tempel, Sidomulyo, Bambanglipuro, Bantul

IOP Conf. Series: Earth and Environmental Science **884** (2021) 011001 doi:10.1088/1755-1315/884/1/011001

9	Listyati Setyo Palupi	Psychological preparedness for potential disaster of university students in response to covid-19
10	Cahyadi Setiawan	The Relationship Between Socioeconomic Status And Household Water Requirement In Muara Angke Inundation Area, Jakarta
11	Listyati Setyo Palupi	The Impact of Covid-19 Pandemic on the Psychological wellbeing of university students
12	Marinda Indayani	We are getting stronger as we stay longer: the influence of place attachment to urban community
13	Listyati Setyo Palupi	Psychological Preparedness for Potential Disaster of Resident around Lapindo Mud Disaster: A Descriptive Study

IOP Conf. Series: Earth and Environmental Science **884** (2021) 011001 doi:10.1088/1755-1315/884/1/011001

Documentation of The virtual conference of The 2nd ICHMGEP 2020

IOP Conf. Series: Earth and Environmental Science **884** (2021) 011001 doi:10.1088/1755-1315/884/1/011001

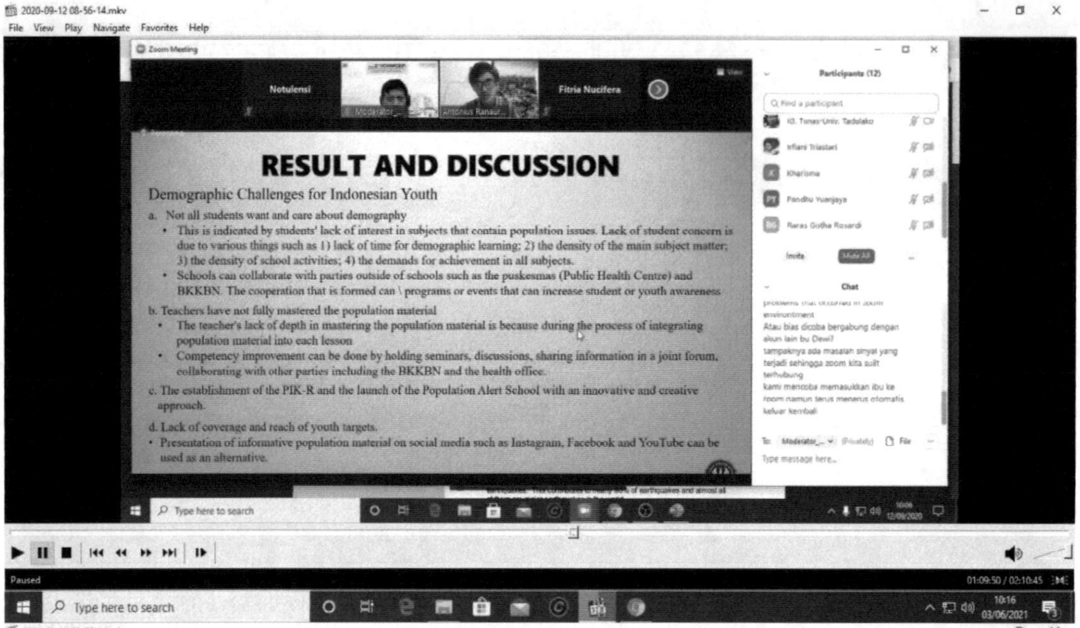

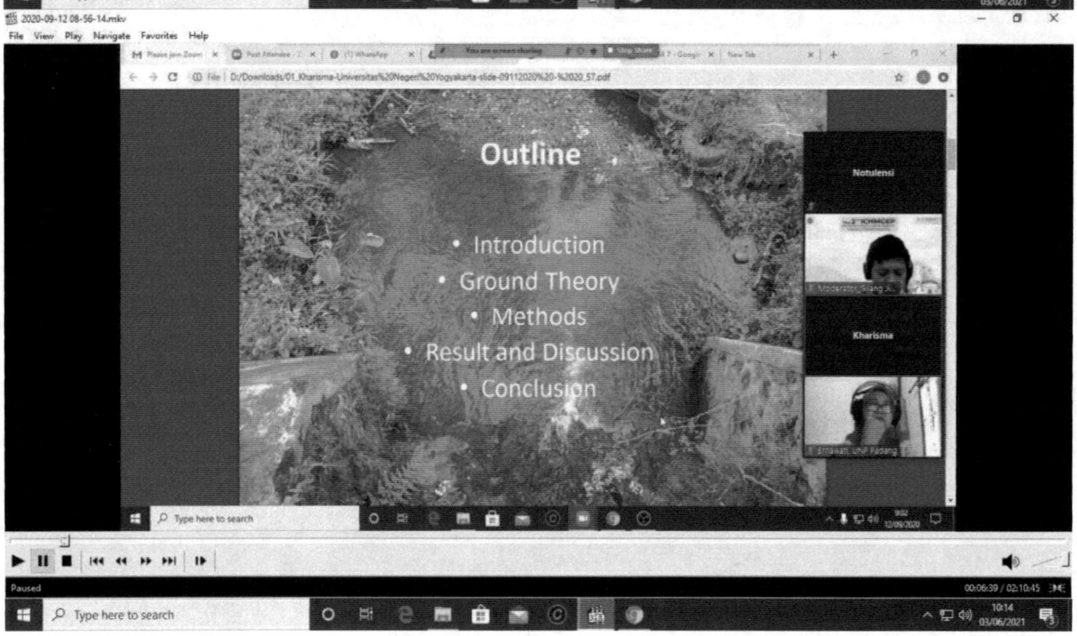

IOP Conf. Series: Earth and Environmental Science **884** (2021) 011001 doi:10.1088/1755-1315/884/1/011001

Committee of The 2nd ICHMGEP 2020

Patron
Prof. Dr. Sutrisna Wibawa, M.Pd, Rector of Universitas Negeri Yogyakarta

Steering Committee
Dr. Suhadi Purwantara, M.Si., Dean of Faculty of Social Sciences, Universitas Negeri Yogyakarta
Dr. Dyah Kumalasari, M.Pd., Vice Dean for Academic Affairs and Partnership, Faculty of Social Sciences, Universitas Negeri Yogyakarta

Conference Chair
Dr. Nurul Khotimah, M.Si., Universitas Negeri Yogyakarta

Secretary
Alifi Nur Prasetia Nugroho, M.Pd., Universitas Negeri Yogyakarta
Yuyun Farida, M.Biotech, Universitas Negeri Yogyakarta

Treasurer
Sukamto, S.Sos., Universitas Negeri Yogyakarta
Suyud, S.Pd., Universitas Negeri Yogyakarta

IOP Conf. Series: Earth and Environmental Science **884** (2021) 011001 doi:10.1088/1755-1315/884/1/011001

Editor

Dr. Azis Shafie, University, University of Malaya, Malaysia (Scopus ID: 36704187700)
Dr. Hairy bin Ibrahim, Universiti Pendidikan Sultan Idris, Malaysia (Scopus ID: 55497320300)
Dr. Nasir Nayan, Universiti Pendidikan Sultan Idris, Malaysia (Scopus ID: 55243312300)
Dr. Ekkamol Vannametee, Chulalongkorn University, Thailand (Scopus ID: 55209942200)
Dr. Pannee Cheewinsiriwat, Chulalongkorn University, Thailand (Scopus ID: 55053331900)
Dr. Yany Lopez, University of the Philippines Diliman, The Philippines (Scopus ID: 25929920200)
Prof. Dr. Mukminan, Universitas Negeri Yogyakarta, Indonesia (Scopus ID: 57204920476)
Prof. Dr. Sri Rum Giyarsih, Universitas Gadjah Mada, Indonesia (Scopus ID: 57191891077)

Scientific Committee

Dr. Kamarul Ismail, Universiti Pendidikan Sultan Idris, Malaysia (Scopus ID: 56707615400)
Dr. Nor Kalsum Mohd Isa, Universiti Pendidikan Sultan Idris, Malaysia (Scopus ID: 56800348900)
Dr. Sumayyah Aimi Mohd Najib, Universiti Sains Malaysia, Malaysia (Scopus ID: 37073006100)
Dr. Hanafi Bin Hussin, Universitas of Malaya, Malaysia. (Scopus ID: 57193666690)
Prof. Dr. Brian Stout, Western Sydney University, Australia. (Scopus ID: 54929577300)
Dr. Sonia Roitman, The University of Queensland, Australia (Scopus ID: 8366708400)
Dr. Edouard de Belizal, Universite Paris Nanterre, France (Scopus ID: 37090459600)
Dr. Brian Doucet,University of Waterloo, Canada (Scopus ID: 26534236100)
Dr. Jake Rom D. Cadag, University of the Philippines Diliman, The Philippines (Scopus ID: 24449184200)
Dr. Maria Emerita Indraningrum Srestha, Kathmandu College of Management Kathmandu, Nepal
Dr. Thitirat Panbamrungkij, Chulalongkorn University, Thailand.
Dr. Daniel L Mabazza, University of the Philippines Diliman, The Philippines
Ari Cahyono, M.Sc., Universitas Gadjah Mada, Indonesia. (Scopus ID: 55485324700)
Dr. Nur Mohammad Farda, Universitas Gadjah Mada, Indonesia. (Scopus ID: 57188869324)
Dr. Prima Widayani, Universitas Gadjah Mada, Indonesia. (Scopus ID: 57188879871)
Wirastuti Widyatmanti, S.Si., Ph.D., Universitas Gadjah Mada, Indonesia. (Scopus ID: 57188872075)
Prof. Dr. Suranto, M.Pd., M.Si., Universitas Negeri Yogyakarta, Indonesia. (Scopus ID: 57205219982)
Prof. Dr. Siti Irene Astuti Dwiningrum, Universitas Negeri Yogyakarta, Indonesia (Scopus ID: 55847799900)
Adi Cilik Pierewan, Ph.D., Universitas Negeri Yogyakarta, Indonesia. (Scopus ID: 55964444200)
Amika Wardana, Ph.D., Universitas Negeri Yogyakarta, Indonesia. (Scopus ID: 57205295162)
Dr. Hastuti, Universitas Negeri Yogyakarta, Indonesia (Scopus ID: 57210447589)
Dr. Muhammad Nursa'ban, Universitas Negeri Yogyakarta, Indonesia (Scopus ID: 57210440672)
Dr. Nurul Khotimah, M.Si., Universitas Negeri Yogyakarta, Indonesia, (Scopus ID: 57210448770)
Prof. Dr. Enok Maryani Universitas, Pendidikan Indonesia, Indonesia (Scopus ID: 57193733326)
Dr. Lili Somantri, Universitas Pendidikan Indonesia, Indonesia (Scopus ID: 57192950116)
Prof. Dr. Dewi Liesnoor Setyawati, Universitas Negeri Semarang, Indonesia (Scopus ID: 57194268597)

Prof. Dr. Eva Banowati Universitas, Negeri Semarang, Indonesia (Scopus ID: 57195296290)
Dr. Aris Munandar, S.Pd, M.Si Universitas Negeri Jakarta, Indonesia (Scopus ID: 57215382201)
Dr. Cahyadi Setiawan, M.Si Universitas Negeri Jakarta, Indonesia (Scopus ID: 57202300561)
Dr. Samadi, SPd, M.Si, Universitas Negeri Jakarta, Indonesia (Scopus ID: 57203964908)
Dr. Gatot Supangkat, M.Si. Universitas Muhammadiyah Yogyakarta, Indonesia. (Scopus ID:57188877853)
Yunus Aris Wibowo, M.Sc., Universitas Muhammadiyah Surakarta, Indonesia. (Scopus ID: 57211294558)
Siti Hadiyati Nur Hafida, M.Sc., Universitas Muhammadiyah Surakarta, Indonesia. (Scopus ID: 57204834575)
Wahyu Widiyatmoko, M.Sc., Universitas Muhammadiyah Surakarta, Indonesia. (Scopus ID: 57208673147)
Dr. Deasy Arisanty, Universitas Lambung Mangkurat, Indonesia. (Scopus ID: 57200093656)
Dr. Iya Setyasih, Universitas Mulawarman, Indonesia. (Scopus ID: 57202310086)
Dr. Pipit Wijayanti, Universitas Negeri Surakarta, Indonesia. (Scopus ID: 57200370896)
Dr. Yurni Suasti, Universitas Negeri Padang, Indonesia. (Scopus ID: 57209399944)
Dr. Dwi Wahyuni Nurwihastuti, M.Sc., Universitas Negeri Medan, Indonesia (Scopus ID: 57209454661)
Dr. Lies Indriyani, Universitas Halu Uleo, Indonesia

Local Organizing Committee
Annisa Istiqomah, M.Pd., Universitas Negeri Yogyakarta
Awanis Akalili, M.A., Universitas Negeri Yogyakarta
Kurnia Nur Fitriana, MPA., Universitas Negeri Yogyakarta
Yuhanida Milhani, M.Pd., Universitas Negeri Yogyakarta
Nainta Agustanta, S.E., Universitas Negeri Yogyakarta

Proceeding Chair
Muhammad Yuanda Zara, Ph.D., Universitas Negeri Yogyakarta
Dr. Nursida Arif, M.Sc., Universitas Negeri Yogyakarta (Scopus ID: 56236800700)
Arif Ashari, M.Sc., Universitas Negeri Yogyakarta (Scopus ID: 57203784171)
Eko Prasetyo Nugroho Saputro, M.Hum, M.I.Kom., Universitas Negeri Yogyakarta
Poerwanti Hadi Pratiwi, M.Si., Universitas Negeri Yogyakarta

IOP Conf. Series: Earth and Environmental Science **884** (2021) 011002 doi:10.1088/1755-1315/884/1/011002

Peer review declaration

All papers published in this volume of IOP Conference Series: Earth and Environmental Science have been peer reviewed through processes administered by the Editors. Reviews were conducted by expert referees to the professional and scientific standards expected of a proceedings journal published by IOP Publishing.

- **Type of peer review:**

  The 2nd ICHMGEP 2020 employed a double-blind review format, which means that both the reviewer and author identities were concealed from the reviewers, and vice versa, throughout the review process. The papers were reviewed by the internal (committee) and external reviewers (experts in the field). The internal reviewers ensured that the papers had conformed with the template, met the standard language, and showed satisfactory academic writing. Meanwhile, the external reviewers assessed the content of the papers.

- **Conference submission management system:**
  The 2nd ICHMGEP 2020 utilized a web-based management system, where a typical web was designed to organize the conference including the paper submission and the review process. The webpage can be accessed at http://ichmgep.uny.ac.id/paper-submission.

- **Number of submissions received:**

  The 2nd ICHMGEP 2020 organizers received 83 papers

- **Number of submissions sent for review:**

  The 2nd ICHMGEP 2020 organizers sent 83 papers for review

- **Number of submissions accepted:**

  There were 60 papers accepted for publication

- **Acceptance Rate (Number of Submissions Accepted / Number of Submissions Received X 100):**

  The acceptance rate was 72,28

- **Average number of reviews per paper:**

  The average number of reviews for each paper was three reviews

- **Any additional info on review process:**

Contact person for queries:
Name : Dr. Nursida Arif, S.T., M.Sc.
Affiliation: Universitas Negeri Yogyakarta
Email : nursida.arif@uny.ac.id

Content from this work may be used under the terms of the Creative Commons Attribution 3.0 licence. Any further distribution of this work must maintain attribution to the author(s) and the title of the work, journal citation and DOI.
Published under licence by IOP Publishing Ltd

Community Resilience in facing Tsunami Disaster in the Coastal Areas of Purworejo Regency

Abdul Latif RF, Chatarina Muryani, Yasin Yusup

Department of Geography Education, Sebelas Maret University

abdul.latif.rf@student.uns.ac.id

Abstract. Purworejo Regency is one of the regencies in Central Java Province, which is directly adjacent to the Indian Ocean. Because the topography of the southern coast of Purworejo Regency is relatively flat, the area's vulnerability to a tsunami is high. A resilient community is needed to reduce the number of victims in the event of an earthquake and tsunami. This study aims to analyze the level of resilience of the southern coastal community of Purworejo Regency to the tsunami disaster. This research uses a qualitative approach. Collecting data using in-depth interviews with selected respondents, observation, secondary data analysis, and questionnaires. Data analysis used the scorecard toolkit version 2 with 22 questions for each question with five answer options. With a score of 1 to 5. The result of this study indicate that of the 16 villages the coastal area of Purworejo Regency, all of them have a caution zone status.
Keyword: Tsunami Megathrust, Community Research, Purworejo

1. Introduction

Purworejo Regency is located in Central Java Province, which is in the southern part of Java Island and is directly adjacent to the Indian Ocean. Although the Meteorology, Climatology, and Geophysics Agency (2018) has never recorded a tsunami incident in Purworejo, this does not mean that the area is safe from the threat of a tsunami disaster. The Purworejo area is physically vulnerable. The Central Bureau of Statistics of Purworejo Regency noted that the altitude of the area in coastal villages is 6 - 16 mas [1], [2], [3]. There are three sub-districts in the coastal area of Purworejo district, the easternmost is Purwodadi District with an average height of 6 masl, then to the west of Ngombol and Grabag Districts which have an average altitude of 11-16 masl. It can be concluded that the morphological conditions of the Purworejo coastal area are relatively gentle.

Data shows that there were at least 28 earthquakes that impacted a tsunami in Indonesian waters between 2000 and 2018 [4]. One of them is the Sumatra megathrust tsunami on December 26, 2004 in Banda Aceh, Nangro Aceh Darusalam Province. Initially, the term megathrust earthquake was commonly used among paleoseismologists to describe a large earthquake that was very destructive [5]. The potential for megathrust or large earthquakes occurs due to the very large amount of energy released for a long time, and the plates' movement was colliding with each other [6]. In the southern region of Java, a location that may be capable of generating a large tsunami in the eastern part of the Sunda megathrust

Content from this work may be used under the terms of the Creative Commons Attribution 3.0 licence. Any further distribution of this work must maintain attribution to the author(s) and the title of the work, journal citation and DOI.
Published under licence by IOP Publishing Ltd

IOP Conf. Series: Earth and Environmental Science **884** (2021) 012001 doi:10.1088/1755-1315/884/1/012001

consisting of the Indo-Australian plate plunging below the Eurasian margin [7]. From this explanation, it can be assumed that the conditions on the southern coast of Java Island are threatened by a megathrust tsunami. Along the southern coast of Java Island is a subduction zone or a meeting zone between the Eurasian continental plate and the Indian Ocean plate. Purworejo is one of the districts that has these dangers, here is a model in the event of a megatrust tsunami in Purworejo released by gitews.org (2010)

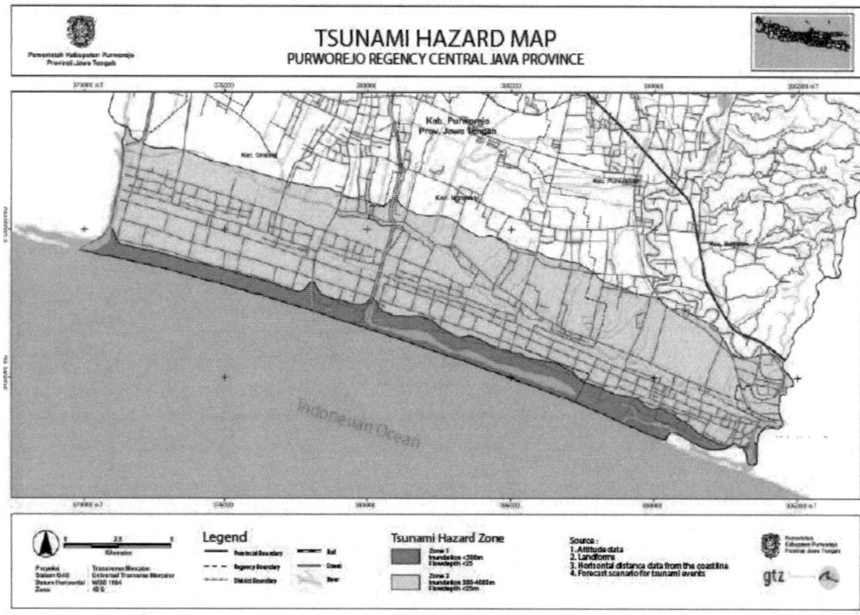

Figure1. Tsunami Hazard Map for Purworejo Regency

Natural disasters can affect life in a community. Losses arising from the occurrence of a disaster can create vulnerabilities. As Gallopin (2006) argues [8], the vulnerability will affect existing systems in the community, meaning that a system will be vulnerable when there is a disturbance. However, if a disturbance hits another system, it is not certain that the system will be affected, because each system contains a different form of vulnerability. Addressing and building resilience enables communities and households to strengthen their capacities to cope with, survive, and recover from disasters [9]. Thus, the study of community resilience is crucial. Because community resilience plays a role in dealing with a disaster or pressure. However, resilience does not have to reverse the situation as before (before the disaster). In an area that is prone to disasters such as Purworejo, ideally the community has a high level of community resilience [9].

Community resilience has a different definition in each of the disciplines that use it. The common thread or identical characteristic that defines a resilient society is that it can continue to carry out its function when under pressure, successful adaptation to new challenges, independence, and social capacity [9]. These characteristics are key to understanding how a community or community contributes to disaster exposure. Social support systems, such as the environment, family and kinship networks, social cohesion, mutual interest groups, and groups helping each other are needed to build community resilience. Various community assets should be considered when evaluating community resilience, such as community members' skills, knowledge, experience and motivation, and physical assets and connections between them [10].

Various community assets should be considered when evaluating community resilience, such as community members' skills, knowledge, experience and motivation, and physical assets and connections between them. [10]. It is important to consider internal community structures, community history and community vulnerability, and to undertake an assessment of community resources and adaptive capacity [11]. Resilience can also be considered using a systems approach by considering subsystems such as diversity, resilience, connectedness, relationships across functional scales and learning capacity [12]. The availability and resilience of critical infrastructures, such as mitigation systems, water supply, information technology and buildings must also be considered [12], [13], [14].

2. Methods

This research was conducted on the south coast of Purworejo Regency, consisting of 3 sub-districts, namely Purwodadi, Ngombol, and Grabag Districts. The qualitative research method was chosen to explore and deepen socio-environmental phenomena that cannot be obtained by quantitative methods. A qualitative research methodology is considered suitable when the researcher or researcher is either investigating a new field of study or intends to ascertain and theorize a salient problem. There are many qualitative methods developed to have a deep and broad understanding of problems through their interpretation of texts. The most common types are interviews and observations [15].

The ideal data collection uses the Focus Group Discussion (FGD) method with community leaders such as government representatives, representatives of disaster organizations, religious leaders, female leaders, youth leaders. However, because during the Covid-19 pandemic, mass gatherings were not allowed, the FGD was replaced by using in-depth interviews. In-depth interviews were carried out on a sample selected proposively representing three sub-districts, consisting of government officials (Head of Regional Disaster Management Agency of Purworejo Regency, Sub-district Head, and Village Head), religious leaders, community leaders, female leaders and representatives of community organizations and the general public. To complement the data, field observations and secondary data collection were also carried out. If done carefully and planned the interview approach can provide a rich set of data [16]. Secondary data were collected from district and district level agencies.

In-depth interview data were analyzed to understand community interaction. The qualitative research method of discourse analysis focuses more on the social context in which communication between respondents and researchers occurs. In-depth interviews are guided by Community Disaster Resilience Scorecard Toolkit Version2. According to the Scorecard consists of 4 parameters, namely community connectedness, risk and vulnerability, planning and procedure and available resources. The table contains 22 questions: 5 questions for community connectedness, seven questions for risk and vulnerability, four questions for planning and procedure and six questions for available resources. Then performed triangulation with data from direct observation and documentation. The findings from the analysis were then applied to the assessment table. The answer to each question consists of 5 options that can be included in a score of 1 to 5. From the results of the respondents' answers were analyzed to be included in the community resilience classification.

Table 1. Scorecard toolkit version 2

	Red Zone	*Caution Zone*	*Going Well*
Overall score	25% (22-33)	26-75% (34-98)	76-100% (99-110)
Community Connectedness	25% (5-10)	26-75% (11-19)	76-100% (20-25)
Risk & Vulnerability	25% (7-13)	26-75% (14-280	76-100% (29-35)
Planning & Procedures	25% (4-8)	26-75% (9-16)	76-100% (17-20)
Available Resources	25% (6-11)	26-75% (12-24)	76-100% (25-30)

Source: Torrens Resilience Institute (2015)

3. Results and Discussions

The community resilience index to the tsunami disaster is obtained by scoring from observations, documentation, and interviews. The three data with different outcomes were combined. So that complement each other and validate. These data are then used to determine the level or index on the four aspects used to determine the level of community resilience to the tsunami disaster. These aspects are community connectedness, risk and vulnerability, planning and procedures, and available resources [9].

3.1. Community Connectedness

Based on interviews with selected respondents, the results of field observations and document analysis, the conditions of the community connectedness of the southern coastal communities of Purworejo district are:

a. The proportion of populations was engaged with organizations
Every RT has a community association, whether it's an association of gentlemen, an association of PKK women, and the Youth Organization. Most of the population (about 80%) is involved in this community organization. Besides that, in coastal villages, Disaster Resilient Village has also been developed and many community members (around 50%) are actively involved in this activity. Some people are also active in religious organizations such as administrators of mosques, churches, and so on. Overall, the proportion of the population involved in the organization is good, the value is 4 (about 70 – 75%) in each sub-district.
b. Community access to various communication methods to come together and share information during an emergency
Because many people are involved in community organizations, communication access between residents for sharing communication is also good. At this time, the community has joined the WhatsApp (online chatting application) Group community between residents, there is a WhatsApp group between the Head of Regional Disaster Management Agency and the sub-district head, village head and community leaders. On the coast of Purworejo, a Tsunami Early Warning System (TEWS) has also been installed, besides that, the Regional Disaster Management Agency of Purworejo Regency has also installed a tsunami EWS along the coast. For community access, the score is 4 (between 60-75%) in each district.

Figure 2. Early warning system tool

c. The level of communication between the local governing body and population
Resilient communities are formed from the synergistic efforts of the community and the government as policymakers. The government has carried out collaborative communication with the community through government channels. The government conducts outreach through the village government and asks for and accommodates public input through village heads in forums organized by government disaster agencies. There are two forms of communication between the government and the community on the coast of Purworejo regarding disasters. The first is consultation, in which the government asks questions without any feedback. The two are related, namely the government asking questions and feedback.
d. The general relationship of the community with the larger region
The composition of the community which is more than 90% has the same belief (religion), making the people on the coast of Purworejo able to connect. There are associations with religious titles such as the grand recitation. These associations can establish and strengthen links with the wider community (between villages). However, this is an informal association. There is no specific planning or discussion regarding disaster in this association. The score on this parameter is 2.

e. The degree of connectedness across community groups
The community or community on the coast of Purworejo is relatively homogeneous. In general, there are no socially isolated communities. There were small groups formed, for example a minority faith group. This does not isolate the minority community. The community still receives attention even though there is no comprehensive forum and inventory. With the word score for this parameter is 2. The conditions for the five parameters community connectedness above can be described as follows:

Table 2. Community Connectedness

No.	*Indicator of Community Connectedness*	*Purwodadi Sub District*	*Ngombol Sub District*	*Grabag Sub District*
1	Populations was engaged organizations	4 (61 - 80%)	4 (61 - 80%)	4 (61 - 80%)
2	Community access	4 (60-75%)	4 (60-75%)	4 (60-75%)
3	Level of communication	2 (Consultation)	2 (Consultation)	3 (Engagement)
4	General relationship with a larger region	2 (Informal network)	2 (Informal network)	2 (Informal network)
5	Degree of connectedness across the group	2 (No isolated group)	2 (No isolated group)	2 (No isolated group)
	Total score	14	14	15
	Classification	Caution Zone	Caution Zone	Caution Zone

3.2. Risk and Vulnerability

Enhancing disaster risk reduction and preparedness for natural hazards requires first and foremost identification and assessment of the various vulnerabilities of the community, economy, institutional structure, and environmental resource base through tools for measuring vulnerability [17]. Regarding material assessment, it is viewed from the efforts that have been made by conducting document studies. These documents can be used to project damage and losses from a tsunami. In addition, interviews and observations were also carried out to get an in-depth study.

a. The known risks of all identified hazards of community
In Purworejo there is already a tsunami hazard map that provides information on areas that may have been affected by the tsunami, but there has been no calculation of the damage or losses exposed to the tsunami disaster. The existing map shows risks in general, there is no spatial risk specification. The map published by the government covers the entire coastal area of Purworejo. The score for this parameter is 3 (see Figure 2).

b. The trend in relative size of population permanent population and daily population
The daily population was identified from workers not residing in the study area and tourists. Researchers did not get documentation of the daily population. Based on interviews and direct observation, researchers assume that the daily population has a low quantity. This is because agriculture and aquaculture are the dominant sectors on the coast of Purworejo. The two sectors of the economy do not have great potential to attract workers from outside the region. Likewise, the tourism sector on the coast of Purworejo is also not heavily promoted and does not have a strong appeal. This condition brings advantages for community resilience in facing the tsunami. This parameter has a score of 5.

c. Rate of the resident population change in the last 5 years
Population growth in the last 5 years is calculated from population data. The data was obtained from the Central Statistics Agency of Purworejo Regency. The calculation results are as follows:

Table 3. 5 year population increase

No.	*sub-district*	***Total Population (study area)***		***Enhancement (%)***
		2013	**2018**	
1	Purwodadi	2870	3352	17
2	Ngombol	3585	4342	21
3	Grabag	10046	12168	21

The scores for Purwodadi District are 3 and Ngombol and Grabag Districts is 2.

d. Proportion of the population has the capacity to independently move to safety
To calculate the proportion of the population who can move to a safe place independently, researchers used population data from the Central Bureau of Statistics. Researchers compared the population of children under five and the elderly with a population of more than toddlers and less than the elderly. The results are as follows:

Table 4. The proportion of independent population

No.	*sub-district*	***Total population***			***Percentage***
		Independent	Toddler	Elderly	
1	Purwodadi	10043	814	1311	83
2	Ngombol	3582	287	473	82
3	Grabag	2807	240	305	84

e. Proportion of the resident population prefers communication in a language other.
The languages commonly used on the coast of Purworejo are Javanese (regional) and Indonesian. As a whole, residents on the coast of Purworejo can use these two languages or one of them. That means the use of language is not a problem. This conclusion is based on interviews from the village government along the coast of Purworejo. The score is 5 because it is less than 5%.

f. Transient population (eg, tourists, transient workers) been included in planning for response and recovery.
In the aspect of risk and vulnerability point b, it is explained that there is no documentation regarding the daily population or the working population that is not permanent. The absence of this document indicates that there is no planning for the temporary population. If there is, it cannot function without good data collection or documentation. There is only identification for tourists with inadequate procedures. The regulations that have been established by the tourism awareness group are actually quite adequate, by recording the identity of tourists. But in practice, it is not as good as planning. The score is 2.

g. The risk that your community could be isolated during an emergency event
To assess the risk of the community if required to evacuate during an emergency, the researcher conducted an assessment of the supporting infrastructure. The infrastructure in question is a path or road used for evacuation. Overall, the villages along Pesisir Purworejo already have roads connected to the arterial road (Daendels Street). The road conditions vary, some are still in the form of dirt and stone roads to asphalt or concrete roads. Ideally, roads should be built properly to facilitate the evacuation process during an emergency. In terms of quantity, villages along the coast of Purworejo have only one entry and exit access. This will hamper the evacuation process. Thus, the score is 1.

Figure 3. Some photos of the main road connecting the village by arterial line

Table 5. Risk and Vulnerability

No.	*Indicator of Risk and Vulnerability*	*Purwodadi Sub District*	*Ngombol Sub District*	*Grabag Sub District*
1	Known risks	3 (There are mapping)	3 (There are mapping)	3 (There are mapping)
2	Daily population	5 (less than 20%)	5 (less than 20%)	5 (less than 20%)
3	Population change	3 (17%)	2 (21%)	2 (21%)
4	Capacity to move Independently	5 (83%)	5 (82%)	5 (84%)
5	Population use another language	5 (<5%)	5 (<5%)	5 (<5%)
6	Transient population planning	2 (identification)	2 (identification)	2 (identification)
7	Risk of isolation	1 (there is 1 access)	1 (there is 1 access)	1 (there is 1 access)
	Total score	24	23	23
	Classification	Caution Zone	Caution Zone	Caution Zone

3.3 Planning and Procedures

Careful planning and procedures will overcome and foster resilience. Addressing and building resilience enables communities and households to strengthen their abilities to cope with, survive, and recover from disasters [9]. Planning starts at the household level. Ideally, the household should be part of the planning from pre-disaster to post-disaster.

a. Level of households resource within the community engaged in planning for disaster response and recovery.
   To measure the extent to which household resources are included in disaster planning, researchers used a questionnaire. Respondents amounted to 100 from the entire population who live in the coastal area of Purworejo. The results of the questionnaire are as follows:

IOP Conf. Series: Earth and Environmental Science **884** (2021) 012001 doi:10.1088/1755-1315/884/1/012001

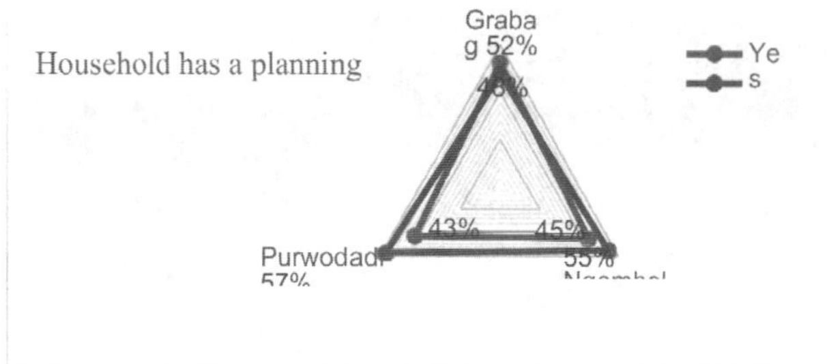

Figure 4. Household planning questionnaire result

b. Planning to reach the entire community about all-hazards resilience
Activities planned to reach all communities regarding tsunami disaster resilience need to be carried out. It is intended that knowledge about the disaster is owned collectively. These activities can refresh knowledge and abilities and instill awareness in people who need to relay to the next generation. However, at the researched location, there were no sustainable activities. The government once held socialization, but it was not carried out frequently. Planning for future outreach was not scheduled either. Based on interview data and documentation, massive socialization was carried out after the tsunami in Tasikmalaya, which impacted large waves in Purworejo. To measure a society, researchers used a questionnaire with the following results:

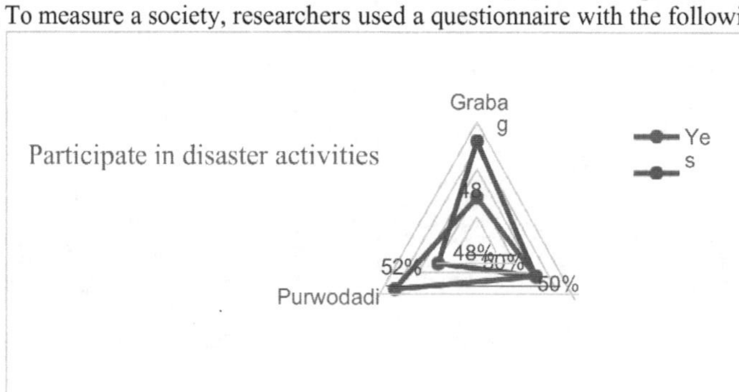

Figure 5. Disaster activities participation questionnaire result

c. The requirement for disaster readiness
The community's readiness to face the threat of a tsunami disaster is already present in the community. Some communities already have awareness of the importance of having disaster preparedness. However, the sample questionnaire shows that ready people are below 50%. The following are the results of the questionnaire:

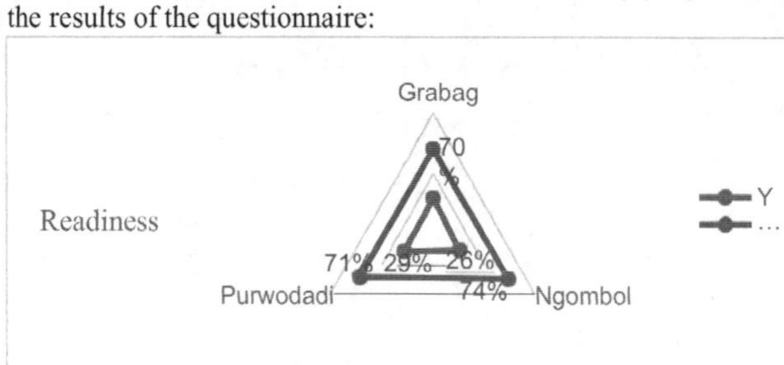

Figure 6. Readiness questionnaire result

d. Change expectations or plans in post-disaster event
In Purworejo, Regional Disaster Management Agency Purworejo recorded a catalog of earthquakes that hit Purworejo. On this note, there is no information that Purworejo had a tsunami. Thus, the post-disaster assessment was never applied to the Purworejo coast in connection with the tsunami disaster. However, the government has carried out an assessment of the tsunami disaster in other areas, such as after the tsunami in Tasikmalaya. In addition, the government also carried out assessments for other disasters that had occurred and the results were also shared publicly. This reflects that post-disaster assessments have been carried out and shared which can also be applied to a tsunami disaster should it occur.

Table 6. Planning and Procedures

No	*Indicator of Planning and Procedures*	*Purwodadi Sub District*	*Ngombol Sub District*	*Grabag Sub District*
1	Households resource level	2 (25-50%)	2 (25-50%)	3 (> 50%)
2	Planning to reach the entire community	3 (> 50%)	2 (25-50%)	2 (25-50%)
3	Requirement for disaster readiness	2 (25-50%)	2 (25-50%)	2 (25-50%)
4	Change expectations or plans	2 (Assess and share)	2 (Assess and share)	2 (Assess and share)
	Total score	9	8	9
	Classification	Caution Zone	Red Zone	Caution Zone

3.4. Available Resources

In order to assess the level of community resilience, an analysis of resources is needed. The resources in question include human and non-human resources. The availability of resources will affect disaster planning. The following are the parameters for assessing the level of available resources:

a. Local infrastructure emergency protection plan
Infrastructure is critical to the success of disaster risk reduction efforts. The infrastructure referred to includes roads, directions for evacuation, early warning system tools, gathering points, evacuation places and buildings (either residential or public facilities). In the context of disaster risk reduction and mitigation it can be needed at any time, it is necessary to determine a temporary evacuation site (TEP), a final evacuation site (FES) and an evacuation route in a tsunami-prone zone.[18]. Roads already exist in every village that connects to arterial roads (as described in the risk and vulnerability point), but there is no maintenance plan yet. Damage to roads will hamper mobilization, while in an emergency it requires mobilization in a short time. The score for this parameter is 2 (there is identification without treatment planning).

IOP Conf. Series: Earth and Environmental Science **884** (2021) 012001 doi:10.1088/1755-1315/884/1/012001

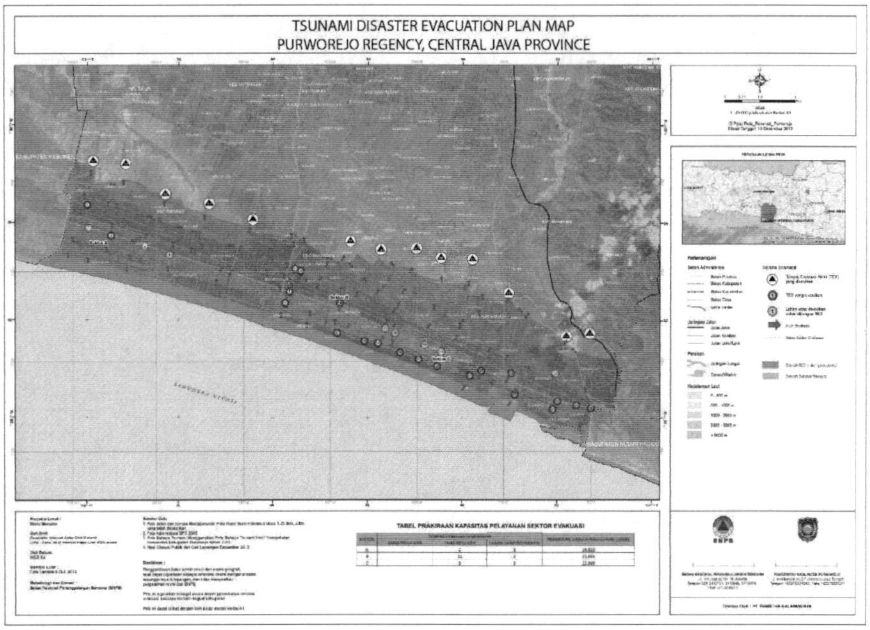

Figure 7. Evacuation plan map

b. The proportion of population with skills useful in emergency response
If the composition of the community includes people who have the skills for emergency response, it will have a good impact on risk reduction efforts. This composition can be seen from the demographic profile. Namely the proportion of people with medical and rescue skills. As a result, the people in each village still have a very low proportion. If the proportion is low, then the community only depends on assistance from outside parties such as the government and volunteers. The score for this parameter is 1.

c. Educational institutions engaged in emergency preparedness education
In Purworejo there is already an educational effort implemented by the government to prepare a generation of resilient students to face disasters. Regional Disaster Management Agency Purworejo came to schools to conduct socialization with students. However, it has not been done regularly and on a scheduled basis. The provision of disaster material is still carried out by external parties of the school (not provided by the teacher). The target of disaster education in schools is only students. At a higher level, it can be extended to all school members and parents of students. The score for this parameter is 2.

d. Available medical and public health services included in emergency planning
Regional Disaster Management Agency Purworejo in carrying out tasks facing disasters, will act as coordinator during emergency situations. As for what is being coordinated, namely the medical team, search and rescue team, logistic team, and so on. This indicates that the support services needed during an emergency response are included in emergency planning (dealing with disasters). One of them is medical services that have been included in emergency planning. The score is 2.

e. Accessible locations available as evacuation or recovery centers
The villages along the coast of Purworejo already have evacuation routes and temporary evacuation sites (see Figure 7). The community needs TES and evacuation routes so that when a disaster occurs, the community can immediately go to TES through the correct route according

to the evacuation route to reduce the risk of casualties when a tsunami occurs[19]. The planning of evacuation routes and places is carried out by Regional Disaster Management Agency Purworejo. There is no final evacuation site (FES) in this plan. Regional Disaster Management Agency explained that the final evacuation site is conditional. When a disaster occurs, the most important thing is that the community knows where to go (namely TES). Then it can become FES depending on post-disaster conditions and available resources. The problem in the field is the absence of good publications. Evacuation directions in several villages did not show the TES location. The score is 2.

f. Level of food, water or fuel readily available in the emergency situation
The supply or supply of food, water, and fuel that is prepared for an emergency has not been prepared. From the results of interviews with Regional Disaster Management Agency and document studies, there is no specific information about this plan. Thus, the score for this parameter is 1.

Table 7. Available resource

No	*Indicator of Available Resources*	*Purwodadi Sub District*	*Ngombol Sub District*	*Grabag Sub District*
1	Infrastructure emergency protection	2 (identification without a treatment plan)	2 (identification without a treatment plan)	2 (identification without a treatment plan)
2	Population with skills useful in emergency	1 (<20%)	1 (<20%)	1 (<20%)
3	Educational institutions engaged in emergency preparedness	2 (there for students)	2 (there for students)	2 (there for students)
4	Available medical and public health services	2 (rely on local services)	2 (rely on local services)	2 (rely on local services)
5	Locations available as evacuation	2 (There is inventory, but bad publicity)	2 (There is inventory, but bad publicity)	2 (There is inventory, but bad publicity)
6	Level of food, water or fuel readily availability	1 (Do not know)	1 (Do not know)	1 (Do not know)
	Total score	10	10	10
	Classification	Red Zone	Red Zone	Red Zone

From the analysis of these 4 aspects, it is then used to determine the level of community resilience to the tsunami disaster. The researcher conducted a scoring on each aspect (community connectedness, risk and vulnerability, planning and procedures and available resources). The scoring results are classified into 3 clusters. The first cluster is a low category (red zone) with an index of 0 - 25%, the moderate category (caution zone) with an index of 26 - 75% and the high category (going well) with an index of 76 - 100%.

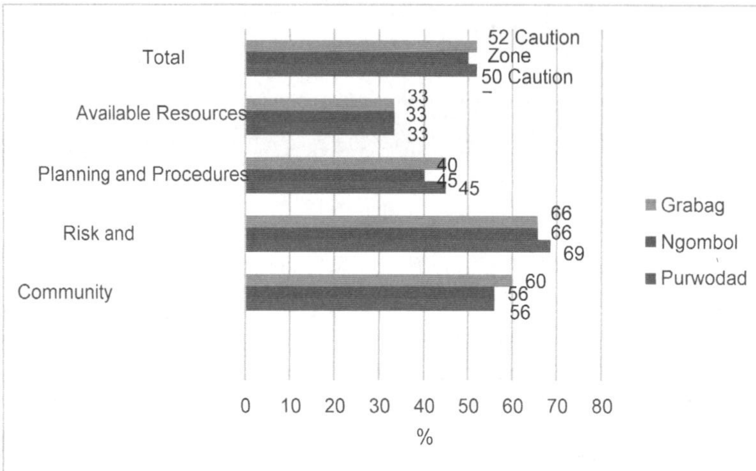

Figure 8. Final index

The phenomenon that occurs in the coastal community of Purworejo, when compared to the coastal community of Kebumen Regency, to be precise in Puring District, has a relatively high gap. In other research results, with the same parameters and calculation methods, it shows a higher index and has a good predicate. Historically, neither Kebumen District nor Purworejo District had a history of facing a tsunami disaster. This is evidenced by the absence of documents proving the existence of a tsunami incident at that place. However, with the same historical conditions, it turns out that it does not always produce the same results, as the views of the people in the two areas. In the Kebumen coastal community, there are several efforts made more frequently, this is a striking difference. Efforts made by the community to improve preparedness regarding tsunami disasters are by participating in the simulation and socialization of tsunami disasters. The efforts implemented in Kebumen district are actually no different from what has been implemented in Purworejo Regency. The difference is how it is implemented and its consistency. Generally, the more often it is done the better impact. This is evident in this case study. The skeptical attitude of the community that was formed was the impact of the low level of human resources, especially in understanding the tsunami phenomenon. Skeptic means ignoring the threat of a tsunami disaster or being ignorant. In several villages on the coast of Purworejo, the community behaves like this on the basis of the excuse that their area is still relatively far from the coast and has a high enough place to feel safe from the threat of a tsunami. Even though this is far from safe, in the event of a megathrust tsunami.

4. Conclusions

The total community resilience index in each sub-district is Purwodadi and Grabag 52% and Ngombol 50%. Of the 3 sub-districts located on the coast of Purworejo, all showed a moderate level of community resilience or a caution zone. At this level, improvements must be made to form communities that are resilient to disasters. The gap from the total index shows a very small gap, which is 2%. This is because the community's condition tends to be homogeneous, and the government's efforts are uniform along the coast of Purworejo in relation to the handling of the tsunami threat.

The government has made efforts to reduce the risk of tsunami disaster in Purworejo through the Regional Disaster Management Agency. The form of the effort is to build supporting facilities for disaster risk reduction with road infrastructure, directions, and places for evacuation and early warning system tools. These efforts still require development and maintenance planning. In some villages, there is no proper infrastructure and maintenance plans have not been established. Evaluation of the community on the coast of Purworejo shows that there is still a need for efforts to increase knowledge and skills. One of the ways is by increasing public education through socialization and publication of information. Lack of this results in low public knowledge, so that awareness of disasters is also low. In addition, people who are poor in skills need to be improved so that people can be resilient in facing disasters. Efforts to increase community resilience can be carried out by considering the approach to local wisdom values, so that the community more easily accepts them. Human and non-human resources found in Purworejo need to be framed with careful planning. This means that the existence of supporting facilities must also be accompanied by demonstrations of their use so that the infrastructure and facilities can function properly. Then, maintenance efforts should ideally also be included as part of the planning. Because in the context of facing the threat of disaster, it can be needed at any time without knowing for sure.

Acknowledgments

I would like to express my special thanks of gratitude to Prof. Dr. Chatarina Muryani, M.Si. as head of Disaster Study Center of Universitas Sebelas Maret (PSB UNS) who gave me the golden opportunity to do this wonderful project and Dr. Yasin Yusup, S.Si., M.Si. which also helped me in doing a lot of studies and i came to know about so many new things I am really thankful. Secondly i would also like to thank my parents and friends who helped me a lot in finalizing this project.

Refferences

[1] B. P. Statistik, *Kecamatan Ngombol dalam Angka 2019*. 2018.

[2] Badan Pusat Statistik, *Kecamatan Grabag dalam Angka 2019*. Purworejo, 2018.

[3] Badan Pusat Statistik, *Kecamatan Purwodadi dalam Angka 2019*. 2018.

[4] BMKG, *Katalog Tsunami Indonesia Tahun 416-2017*. 2018.

[5] J. Park *et al.*, "Performance review of the global seismographic network for the Sumatra-Andaman megathrust earthquake," *Seismol. Res. Lett.*, vol. 76, no. 3, pp. 331–343, 2005, doi: 10.1785/gssrl.76.3.331.

[6] D. Hermon, *Geografi Bencana Alam*. Jakarta: Rajawali Pers, 2015.

[7] I. E. Mulia, A. R. Gusman, A. L. Williamson, and K. Satake, "An Optimized Array Configuration of Tsunami Observation Network Off Southern Java, Indonesia," *J. Geophys. Res. Solid Earth*, vol. 124, no. 9, pp. 9622–9637, 2019, doi: 10.1029/2019JB017600.

[8] G. C. Gallopín, "Linkages between vulnerability, resilience, and adaptive capacity," *Glob. Environ. Chang.*, vol. 16, no. 3, pp. 293–303, 2006, doi: 10.1016/j.gloenvcha.2006.02.004.

[9] P. Arbon, L. Cusack, K. Gebbie, M. Steenkamp, and O. Anikeeva, "HOW DO WE MEASURE AND BUILD RESILIENCE AGAINST DISASTER IN COMMUNITIES AND HOUSEHOLDS? Prepared for the Global Assessment Report on Disaster Risk Reduction 2015," no. December 2013, 2013, [Online]. Available: http://www.torrensresilience.org/.

[10] B. Maguire and S. Cartwright, "Assessing a community's capacity to manage change : A resilience approach to social assessment," *Soc. Sci. Progr.*, no. May, pp. 1–27, 2008, doi: Yes.

[11] P. Long, N. Armstrong, K. Perrin, W. M. Parker, and M. Hidek, "Building Resilient Communities: A Preliminary Framework for Assessment," *Homel. Secur. Aff.*, vol. 6, no. 3, 2010.

[12] A. Keil, M. Zeller, A. Wida, B. Sanim, and R. Birner, "What determines farmers' resilience towards ENSO-related drought? An empirical assessment in Central Sulawesi, Indonesia," *Clim. Change*, vol. 86, no. 3–4, pp. 291–307, 2008, doi: 10.1007/s10584-007-9326-4.

[13] A. Fekete, "Common criteria for the assessment of critical infrastructures," *Int. J. Disaster Risk Sci.*, vol. 2, no. 1, pp. 15–24, 2011, doi: 10.1007/s13753-011-0002-y.

[14] R. J. T. Klein, R. J. Nicholls, and F. Thomalla, "Resilience to natural hazards: How useful is this

concept?," *Environ. Hazards*, vol. 5, no. 1, pp. 35–45, 2003, doi: 10.1016/j.hazards.2004.02.001.

[15] S. Jamshed, "Qualitative research method-interviewing and observation," *J. Basic Clin. Pharm.*, vol. 5, no. 4, p. 87, 2014, doi: 10.4103/0976-0105.141942.

[16] S. Q. Qu and J. Dumay, "The qualitative research interview," *Qual. Res. Account. Manag.*, vol. 8, no. 3, pp. 238–264, 2011, doi: 10.1108/11766091111162070.

[17] B. M. de Loyola Hummell, S. L. Cutter, and C. T. Emrich, "Social Vulnerability to Natural Hazards in Brazil," *Int. J. Disaster Risk Sci.*, vol. 7, no. 2, pp. 111–122, 2016, doi: 10.1007/s13753-016-0090-9.

[18] D. Hermon *et al.*, "A SPATIAL MODEL FOR TEP AND FES LOCATIONS AS THE CONCEPT OF TSUNAMI DISASTER MITIGATION IN PARIAMAN CITY - INDONESIA," vol. 2, no. 2, pp. 71–81, 2020.

[19] A. Putra and H. Mutmainah, "The Mapping of Temporary Evacuation Site (TES) and Tsunami Evacuation Route in North Pagai Island, Mentawai Islands Regency - Indonesia," *IOP Conf. Ser. Earth Environ. Sci.*, vol. 47, no. 1, pp. 0–16, 2016, doi: 10.1088/1755-1315/47/1/012020.

IOP Conf. Series: Earth and Environmental Science **884** (2021) 012002 doi:10.1088/1755-1315/884/1/012002

Determination of the level of seismic hazard in Yogyakarta using fuzzy simple additive weighting method

F Y Saptaningtyas, A M Abadi, Musthofa, E Y Utomo and A W Sugiyarto

Mathematics Department, Faculty of Mathematics and Natural Science
Universitas Negeri Yogyakarta, Yogyakarta, Indonesia

Email: fitrianatya@uny.ac.id

Abstract. This study aims to obtain a mapping of seismic hazard vulnerability of Special Region of Yogyakarta (DIY) using the Fuzzy Simple Additive Weighting (FSAW) method. This mapping is serves as an act of mitigation for earthquake disasters to reduce the risk of earthquake disasters and minimize fatalities or losses in earthquake-prone areas. The factors used construct the mapping of earthquake-prone areas in Yogyakarta are social, physical, and economic vulnerabilities. Each factor has several variables. The data of this research were taken from the Central Bureau of Statistics (BPS) and the National Disaster Management Agency (BNPB) of the Special Region of Yogyakarta from 2014 to 2017. The FSAW method was used to determine the level of seismic hazard for each sub-district in DIY with a high, medium, or low vulnerability category. The steps taken in the FSAW method are (1) performing fuzzification for each variable, (2) normalizing, (3) determining the weigh of each variable, (4) determining the classification of the result; whether it is categorized under high, medium, or low level. The result from this study is a map that shows areas in each district of Special Region of Yogyakarta with high, medium, and low seismic vulnerability. Based on the social, economic and physical vulnerability factors, it was found that every sub-district in Yogyakarta City were categorized under the high-level vulnerability, while the sub-districts in Bantul, Gunungkidul, Kulonprogo and Sleman Regencies had the average level of medium.

1. Introduction

Special Region of Yogyakarta is divided into 4 districts – Bantul Regency with 17 sub-districts, Gunungkidul Regency with 18 sub-districts, Kulonprogo Regency with 12 sub-districts, Sleman Regency with 17 sub-districts and a city called Yogyakarta City with 14 sub-districts. Topographically, 75,57 % of the areas in Special Region of Yogyakarta are located on plains, 23,97% are on the slope/ridge, and the remaining 0,46 percent are in the valley or watershed [1].

The total population of Special Region of Yogyakarta was recorded around 3.762.167.000 people in 2017. For each regency and city, the total population in Bantul Regency was recorded at 995.264.000 people with an average population of around 58.544.941 people per district, Gunungkidul Regency at 729.364.000 people with an average population of around 40.520.222 people per district, Kulon Progo Regency at 421.295.000 people with an average population of around 35.107.916 people per district, Sleman Regency at 1.193.512.000 people with an average population of around 70.206.588 people per district, and Yogyakarta City at 422.732.000 people with an average population of around 30.195.142 people per district [1]. By comparing the population divided by the area, the population density of an area can be obtained. The population density on Special Region of Yogyakarta is 1.180.818 people per

Content from this work may be used under the terms of the Creative Commons Attribution 3.0 licence. Any further distribution of this work must maintain attribution to the author(s) and the title of the work, journal citation and DOI.
Published under licence by IOP Publishing Ltd

square kilometer. This makes Special Region of Yogyakarta to be prone to disasters, both environmental and man-made.

Disasters are natural events or unexpected events that can cause damage to the environment and living creatures [2]. In Special Region of Yogyakarta, an earthquake measuring 5,9 on the Richter scale occurred in 2006 [3]. Seismic activity of a certain area can be determined by analyzing earthquakes that occur based on time period, frequency, power, and type of earthquake [4]. However, humans still can not predict the time for an earthquake to occur. Therefore, an earthquake is one of the disasters that can cause major damage to both the environment and human life [5]. Estimated losses resulting from earthquake disasters include: physical damage (schools, critical facilities and infrastructure), economic losses and social impacts [6].

According to the National Disaster Management Agency (BNPB), the criteria for an area to be said to be prone to earthquakes can be seen from three factors, namely social, physical, and economic vulnerability factors [7]. In the context of disaster mitigation, several studies on natural disasters have already been done. Research on potential locations for landslide evacuation using the simple additive weighting method has been conducted [8]. Fuzzy multi criteria decision making method has been used for another research to map the flood risk area [9]. Next, the mapping of earthquake risk levels around the Jayapura area has been conducted based on microtremor measurements [10]. Then, the determination of the tectonic earthquake disaster vulnerability zone in Malang district, southern region has also been performed [11]. Next, there is an analysis based on seismic data that shows an increase in earthquake power from 1973 to 2007 [12]. There is also an analysis of the earthquake disaster in the Philippines using the Twitter social media application [13]. An earthquake modeling research has also been carried out using the magnitude distribution [14].

The Simple Additive Weighting (SAW) method is one of the simplest, natural and most widely used multi-criteria evaluation methods in research [15]. The basis of the SAW method is the overall ranking index for each alternative which is calculated as the number of criteria and suitable alternative objectives. The SAW method makes an assumption for the decision results controlled by weights which express the importance of the criterion [16]. This method was developed to help make decisions on several alternative decisions to get an accurate and optimal decision [17]. If the information obtained cannot be presented completely and contains uncertainties, then the fuzzy multi attribute decision making method can be used to solve the problem [18]. Fuzzy logic is a multi-value logic derived from fuzzy set theory relating to human reasoning ranging from "almost certain" to "very unlikely" with membership values ranging between 0 and 1 [19]. The advantage of the Fuzzy Simple Additive Weighting (FSAW) method is that it can be used for decision making based on data that contains uncertainty [20], [21], [22], [23], [24]. Based on previous research, research on seismic hazard level mapping in Special Region of Yogyakarta using the FSAW method has not been performed. Therefore, in this study, a seismic vulnerability area mapping for each sub-district in Special Region of Yogyakarta will be carried out using the FSAW method.

2. Method

In this study, three aspects were used, namely social aspects, physical aspects and economic aspects. Social aspects include population density (people per hectare), gender ratio (percent), age dependency ratio (percent), poverty ratio (percent), and disabled population ratio (percent). Physical aspects include housing (million rupiah), public facilities consisting of road costs (million rupiah), public transportation (million rupiah) and waterways (million rupiah) as well as critical facilities such as hospitals (million rupiah). The economic aspect includes productive land variables consisting of forests (million rupiah), rice fields (million rupiah) and plantations (million rupiah), as well as GRDP (million rupiah). Data for each district is taken from the BPS-Statistics of D.I. Yogyakarta [1] and the Disaster Management National Agency (BNPB) [25]. The data used are the data from 2014 to 2017. The steps to determine the level of seismic hazard using the FSAW method can be seen in Figure 1.

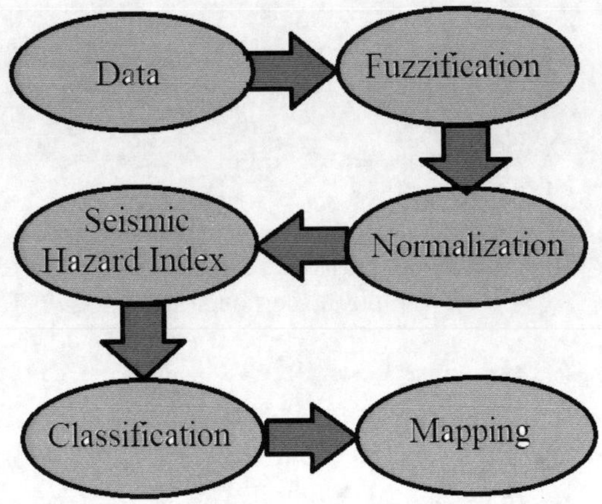

Figure 1. Method for Determining the Level of Seismic hazard Using FSAW

In more detail, the steps in Figure 1 are carried out as follows:

2.1. Fuzzy Set

The fuzzy set for each variable is constructed using an upward and downward representation [26]. The determination of the type of fuzzy set for each variable is based on the variable type, whether it is a benefit or a cost variable [27].

2.1.1. Social Aspects is divided into 5 variables, namely: population density (C1), gender ratio (C2), age dependency ratio (C3) poverty ratio (C4) and ratio of disabled population (C5). This variable is expressed by the membership function as follows.

$$\mu_{C1} = \begin{cases} 0 & x \le 0 \\ \dfrac{x}{208} & 0 \le x \le 208 \\ 1 & x \ge 208 \end{cases} \tag{1}$$

$$\mu_{C2} = \begin{cases} \dfrac{109 - x}{109} & 0 \le x \le 109 \\ 0 & x \ge 109 \end{cases} \tag{2}$$

$$\mu_{C3} = \begin{cases} 0 & x \le 0 \\ \dfrac{x}{100} & 0 \le x \le 100 \\ 1 & x \ge 100 \end{cases} \tag{3}$$

$$\mu_{C4} = \begin{cases} 0 & x \le 0 \\ \dfrac{x}{100} & 0 \le x \le 100 \\ 1 & x \ge 100 \end{cases} \tag{4}$$

$$\mu_{C5} = \begin{cases} 0 & x \le 0 \\ \dfrac{x}{100} & 0 \le x \le 100 \\ 1 & x \ge 100 \end{cases} \tag{5}$$

The graph of the membership functions C1, C2, C3, C4 and C5 can be seen in Figure 2 to Figure 6.

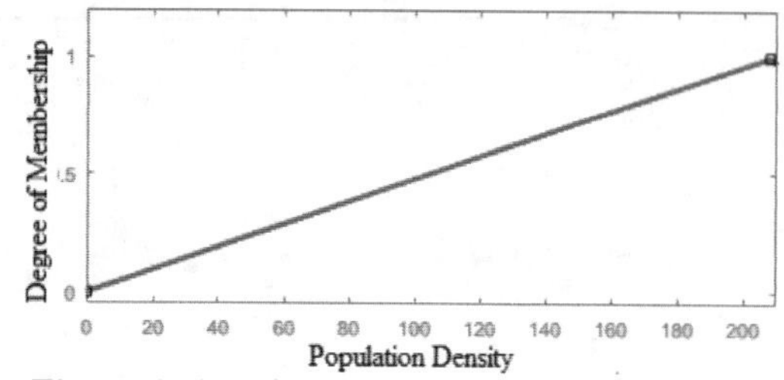

Figure 2. Graph of Membership Functions C1

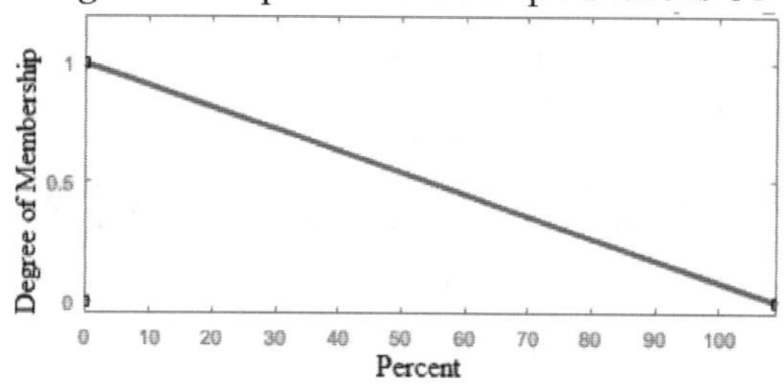

Figure 3. Graph of Membership Functions C2

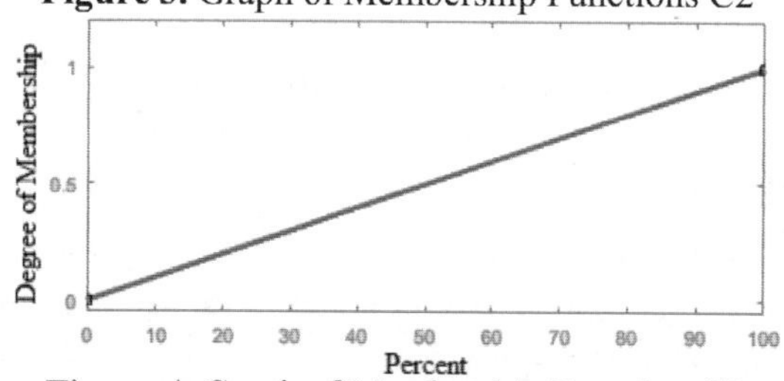

Figure 4. Graph of Membership Function C3

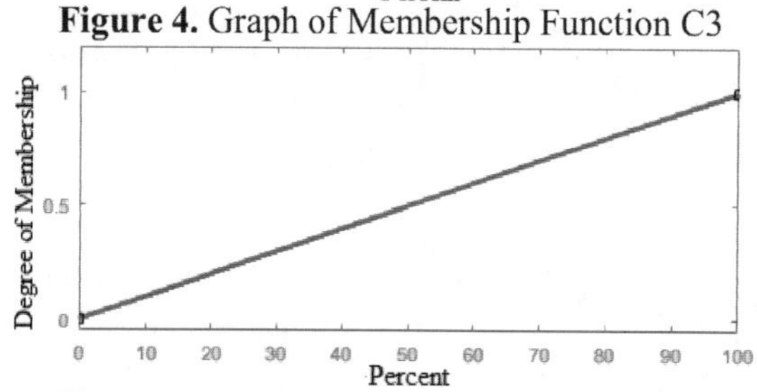

Figure 5. Graph of Membership Function C4

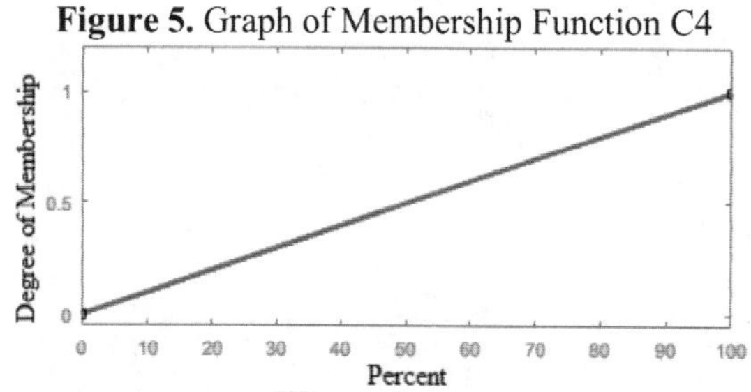

Figure 6. Graph of Membership Function C5

2.1.2. Physical Aspects is divided into 5 variables, namely: houses (P1), road costs (P2), public transportations (P3), waterways (P4) and hospitals (P5). This variable is expressed by the membership function as follows.

$$\mu_{P1} = \begin{cases} 0 & x \le 0 \\ \frac{x}{2700} & 0 \le x \le 2700 \\ 1 & x \ge 2700 \end{cases} \quad (6)$$

$$\mu_{P2} = \begin{cases} \frac{7000 - x}{7000} & 0 \le x \le 7000 \\ 0 & x \ge 7000 \end{cases} \quad (7)$$

$$\mu_{P3} = \begin{cases} 0 & x \le 0 \\ \frac{x}{152000} & 0 \le x \le 152000 \\ 1 & x \ge 152000 \end{cases} \quad (8)$$

$$\mu_{P4} = \begin{cases} 0 & x \le 0 \\ \frac{x}{3304} & 0 \le x \le 3304 \\ 1 & x \ge 3304 \end{cases} \quad (9)$$

$$\mu_{P5} = \begin{cases} \frac{552223 - x}{552223} & 0 \le x \le 552223 \\ 0 & x \ge 552223 \end{cases} \quad (10)$$

The graph of the membership functions P1, P2, P3, P4 and P5 can be seen in Figure 7 to Figure 11.

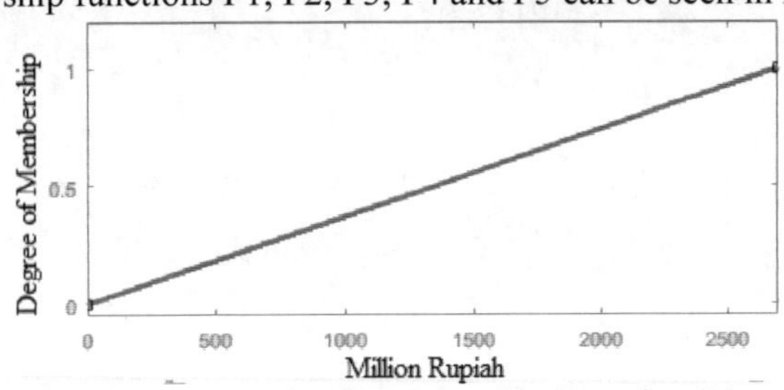

Figure 7. Graph of Membership Function P1

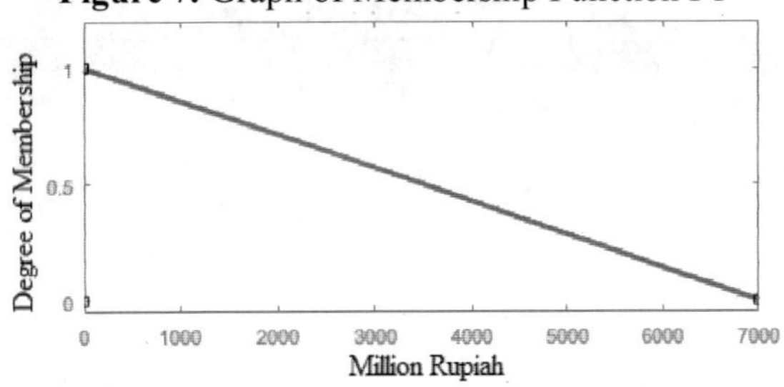

Figure 8. Graph of Membership Function P2

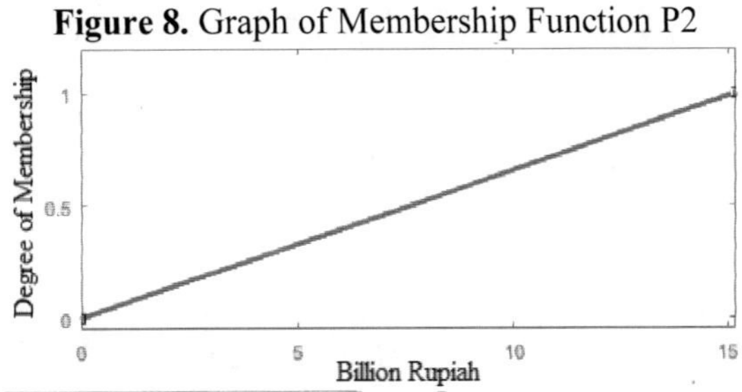

Figure 9. Graph of Membership Function P3

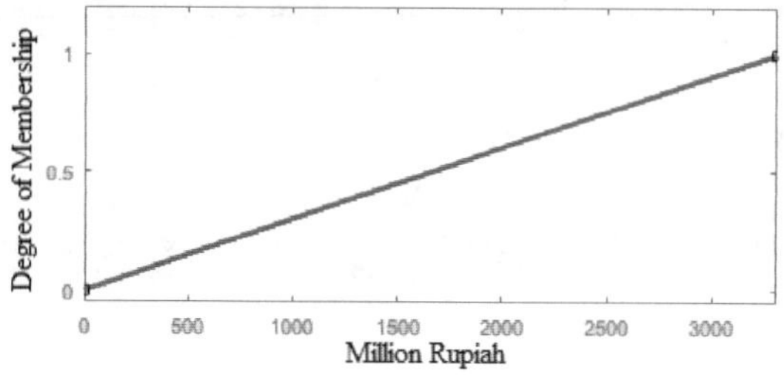

Figure 10. Graph of Membership Function P4

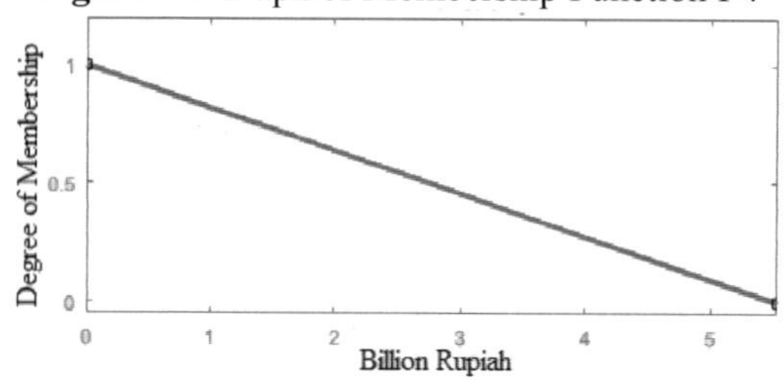

Figure 11. Graph of Membership Function P5

2.1.3. Economic Aspects is divided into 4 variables, namely: forestry (E1), rice fields (E2), plantations (E3), and GRDP (E4). This variable is expressed by the membership function as follows.

$$\mu_{E1} = \begin{cases} 0 & x \le 0 \\ \dfrac{x}{11211000} & 0 \le x \le 11211000 \\ 1 & x \ge 1121100 \end{cases} \tag{11}$$

$$\mu_{E2} = \begin{cases} 0 & x \le 0 \\ \dfrac{x}{17360000} & 0 \le x \le 1736000 \\ 1 & x \ge 1736000 \end{cases} \tag{12}$$

$$\mu_{E3} = \begin{cases} 0 & x \le 0 \\ \dfrac{x}{205322000} & 0 \le x \le 205322000 \\ 1 & x \ge 205322000 \end{cases} \tag{13}$$

$$\mu_{E4} = \begin{cases} 0 & x \le 0 \\ \dfrac{x}{2360000} & 0 \le x \le 2360000 \\ 1 & x \ge 2360000 \end{cases} \tag{14}$$

The graph of the membership functions E1, E2, E3, and E4 can be seen in Figure 12 to Figure 15.

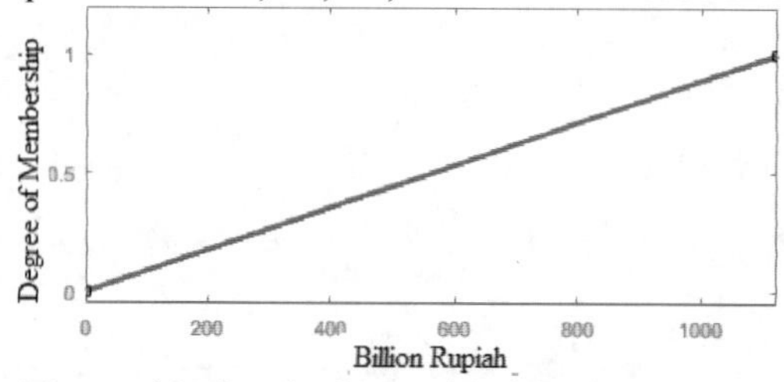

Figure 12. Graph of Membership Function E1

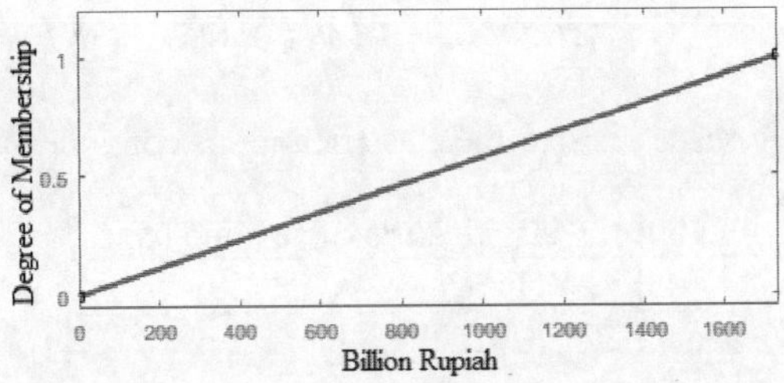

Figure 13. Graph of Membership Function E2

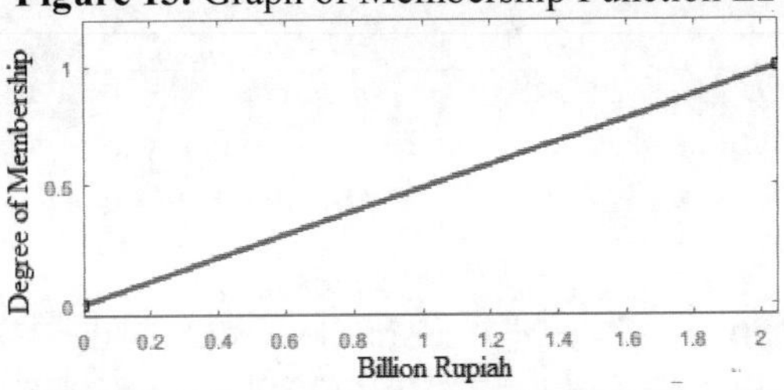

Figure 14. Graph of Membership Function E3

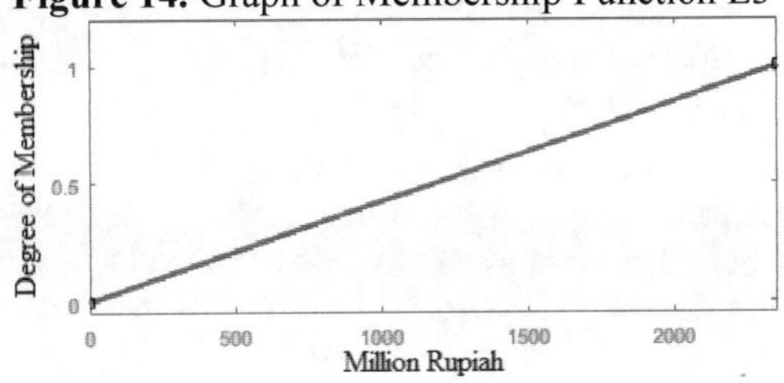

Figure 15. Graph of Membership Function E4

2.2. Data Normalization

Next, the degree of membership for each data is normalized by formula (15) [27].

$$r_{ij} = \begin{cases} \dfrac{x_i}{Max \quad x_{ij}} & Benefit \\ \dfrac{Min \quad x_{ij}}{x_i} & Cost \end{cases} \tag{15}$$

2.3. Determining the seismic hazard index

The value of the Seismic Hazard Index (IKG-Indeks Kerawanan Gempa) is calculated as follows [27].

$$V_i = \sum_{i=1}^{n} w_j r_{ij} \tag{16}$$

Where V_i is the preference value of the alternative, w_j is the weighted value of the *j* criterion and r_{ij} is the normalized value of the *i* alternative and the *j* criterion The social vulnerability index (IKS-Indeks Kerawanan Sosial), physical vulnerability index (IKF-Indeks Kerawanan Fisik) and economical vulnerability index (IKE-Indeks Kerawanan Ekonomi) are calculated with the following formula [25].

$$IKS = (0,6 \times C1) + (0,1 \times C2) + (0,1 \times C3) + (0,1 \times C4) + (0,1 \times C5) \tag{17}$$

$$IKF = (0,4 \times P1) + (0,1 \times P2) + (0,1 \times P3) + (0,1 \times P4) + (0,3 \times P5) \tag{18}$$

$$IKE = (0,2 \times E1) + (0,2 \times E2) + (0,2 \times E3) + (0,4 \times E4) \tag{19}$$

Furthermore, the IKG value is calculated by formula (16) and is obtained:

$$IKG = (IKS \times 0{,}4) + (IKF \times 0{,}3) + (IKE \times 0{,}3) \quad (20)$$

2.4. IKG Classification

After getting the IKG value for each district, the classification is continued according to Table 1 [25].

Table 1. Classification of seismic hazard

IKG Value	Level
IKG < 26	Low
26 ≤ IKG ≤ 70	Medium
IKG > 70	High

2.5. Mapping

This study uses three classifications of seismic hazard levels, namely low hazard (green), medium hazard (yellow) and high hazard (red). Each district is then mapped according to its level of vulnerability.

3. Result and Discussion

In this discussion, we will determine the level of seismic hazard using the FSAW method using the steps as shown in Figure 1. The following will give an example of calculating the level of earthquake vulnerability in Srandakan sub-district in 2017.

(i). We determine the value of Social Vulnerability Index (IKS). The values of the social vulnerability variable in Srandakan sub-district in 2017 were respectively S1 = 16.01, S2 = 98.01, S3 = 31.42, S4 = 16.76 and S5 = 0.19. Variables S1, S3, S4 and S5 are benefit variables so they use upward representation functions, while S2 is a cost variable, so it uses a downward representation function. The normalized value for the IKS variable is then calculated as follows.

$$C1 = \frac{0{,}076}{0{,}998} \times 100 = 7{,}708$$

$$C2 = \frac{0{,}003}{0{,}1008} \times 100 = 3{,}002$$

$$C3 = \frac{0{,}314}{0{,}345} \times 100 = 90{,}989$$

$$C4 = \frac{0{,}167}{0{,}189} \times 100 = 88{,}647$$

$$C5 = \frac{0{,}001}{0{,}022} \times 100 = 8{,}585$$

Next, by using formula (17), the IKS for Srandakan sub-district is obtained as

$$IKS = 23{,}747$$

(ii). We determine the value of Physical Vulnerability Index (IKF). The following is the value data from the physical vulnerability variable of Srandakan sub-district in 2017 with the values of F1, F2, F3, F4, and F5 respectively 300; 2755.01; 3773; 45.75; and 1111.4. The variables F1, F3, and F4 are benefit variables so they use upward representation functions, while the variables F2 and F5 are cost variables, so they use downward representation functions. The normalization value for the IKF variable is then calculated as follows.

$$P1 = \frac{0{,}111}{1} \times 100 = 11{,}111$$

$$P2 = \frac{0{,}004}{0{,}606} \times 100 = 0{,}726$$

$$P3 = \frac{0{,}024}{0{,}997} \times 100 = 2{,}487$$

$$P4 = \frac{0{,}013}{0{,}999} \times 100 = 1{,}385$$

$$P5 = \frac{0{,}0000003}{0{,}997} \times 100 = 0{,}00003$$

Next, by using formula (18), the IKF for Srandakan sub-district is obtained as

$$IKF = 4{,}904$$

(iii). We determine the value of Economic Vulnerability Index (IKE). The following is the value data of the economic vulnerability variable for Srandakan sub-district in 2017 with the values of E1, E2, E3, and E4, respectively 350000; 2315000; 430000; and 1331396. The variables E1, E2, E3, and E4 are all benefit variables so that the linear upward membership function is used. The normalized value for the IKE variable is then calculated as follows.

$$E1 = \frac{0{,}031}{1} \times 100 = 3{,}121$$

$$E2 = \frac{0{,}133}{0{,}846} \times 100 = 15{,}748$$

$$E3 = \frac{0{,}002}{0{,}1} \times 100 = 2{,}094$$

$$E4 = \frac{0{,}564}{0{,}999} \times 100 = 56{,}46$$

Next, by using formula (19), the IKE for Srandakan sub-district is obtained as

$$IKE = 26{,}777$$

After obtaining the IKS, IKF and IKE values from Srandakan sub-district in 2017, then the Earthquake Vulnerability Index value (20) was determined with the IKS, IKF and IKE weigh respectively 0.4; 0.3 and 0.3 and the GI values are obtained as follows.

$$IKG = 19{,}003$$

Afterwards, by using the Fuzzy SAW method, the IKG value is established as 19.003. The IKG calculation results based on the method used by BNPB do not have an upper limit so that the results can exceed 100. In the FSAW method, the data undergo normalization so that the results only range from 0 to 100. To resolve this, the equivalent value is calculated by dividing the maximum value of IKG from the BNPB method by the value the maximum IKG of the FSAW method. Then, the result will be multiplied by the IKG from the FSAW method, and the equivalent value is obtained. The maximum value of the IKG from the method used by BNPB in 2017 is 212. With this, we can now determine the equivalent value of the Srandakan sub-district.

$$EquivalentValue = \frac{212}{100} \times 19{,}003 = 40{,}287$$

Thus, the IKG value for Srandakan sub-district in 2017 is 40.287 which is categorized under the moderate hazard classification. In the same way, seismic hazard classification was performed in the Special Region of Yogyakarta from 2014 to 2017. The following graphs show the number of sub-districts for each category of seismic hazard in districts or city in the Special Region of Yogyakarta Each graph from Figure 16 to Figure 20 represents the graph for each regency or city.

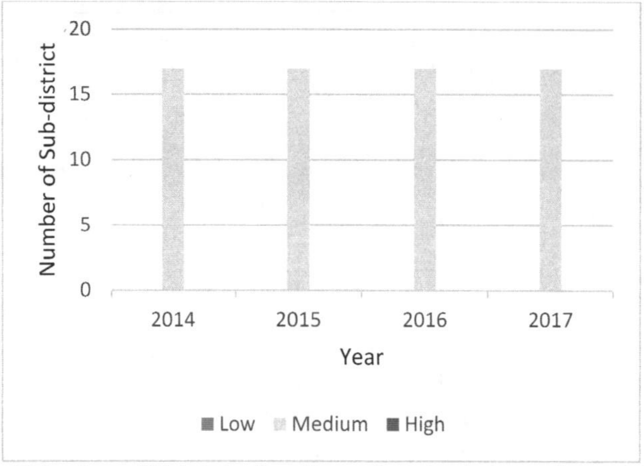

Figure 16. Graph of Number of Sub-Districts for each Seismic hazard Level in Bantul Regency

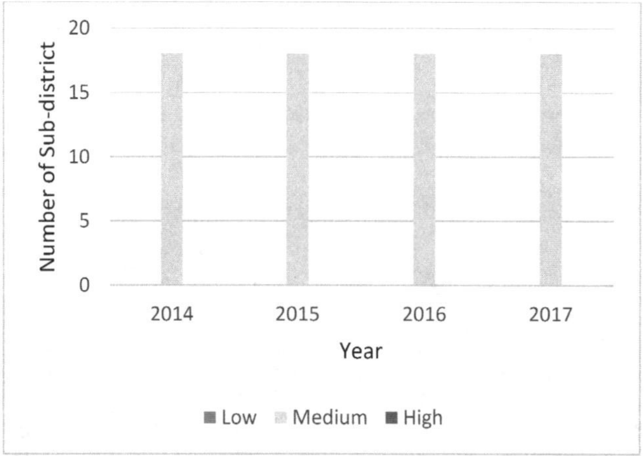

Figure 17. Graph of Number of Sub-Districts for each Seismic hazard Level in Gunungkidul Regency

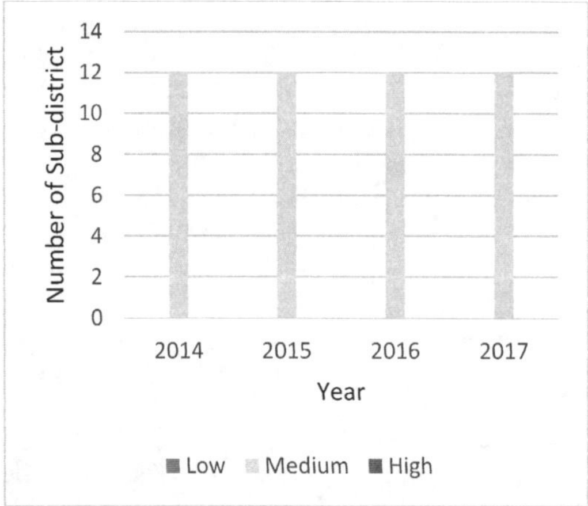

Figure 18. Graph of Number of Sub-Districts for each Seismic hazard Level in Kulonprogo Regency

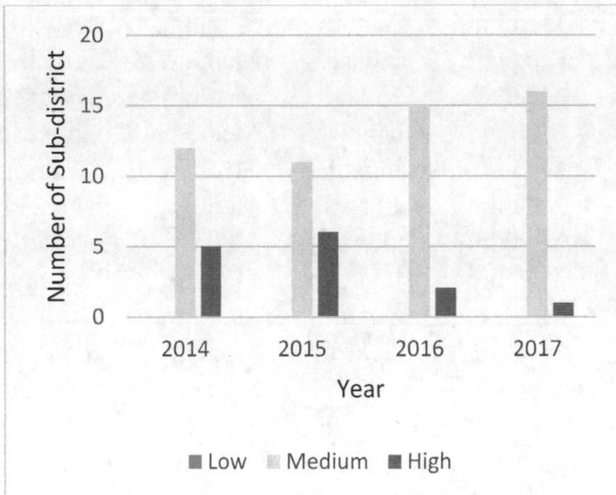

Figure 19. Graph of Number of Sub-Districts for each Seismic hazard Level in Sleman Regency

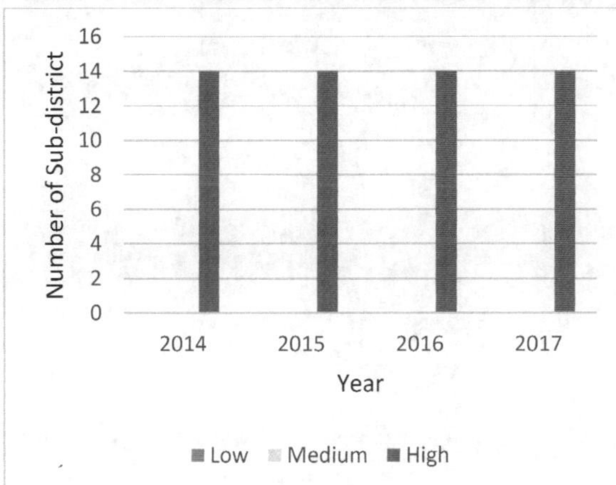

Figure 20. Graph of Number of Sub-Districts for each Seismic hazard Level in Yogyakarta City

The classification results for each district from Special Region of Yogyakarta in 2017 can be seen in table 2.

Table 2. Classification Results using the FSAW Method in 2017

Sub-District	Level	Sub-District	Level	Sub-District	Level
Srandakan	Medium	Ponjong	Medium	Mlati	High
Sandeh	Medium	Karangmojo	Medium	Depok	Medium
Kretek	Medium	Wonosari	Medium	Berbah	High
Pundong	Medium	Playen	Medium	Prambanan	Medium
Bambanglipuro	Medium	Patuk	Medium	Kalasan	Medium
Pandak	Medium	Gedangsari	Medium	Ngemplak	Medium
Bantul	Medium	Nglipar	Medium	Ngaglik	Medium
Jetis	Medium	Ngawen	Medium	Sleman	Medium
Imogiri	Medium	Semin	Medium	Tempel	Medium
Dlingo	Medium	Temon	Medium	Turi	Medium
Pleret	Medium	Wates	Medium	Pakem	Medium
Piyungan	Medium	Panjatan	Medium	Cangkringan	Medium
Banguntapan	Medium	Galur	Medium	Mantrijeron	High
Sewon	Medium	Lendah	Medium	Kraton	High

Based on

IOP Conf. Series: Earth and Environmental Science **884** (2021) 012002 doi:10.1088/1755-1315/884/1/012002

table 2, a seismic hazard level

Kasihan	Medium	Sentolo	Medium	Mergangsan	High
Pajangan	Medium	Pengasih	Medium	Umbulharjo	High
Sedayu	Medium	Kokap	Medium	Kotagede	High
Panggang	Low	Girimulyo	Medium	Gondokusuman	High
Purwosari	Medium	Nanggulan	Medium	Danurejan	High
Paliyan	Medium	Kalibawang	Medium	Pakualaman	High
Saptosari	Medium	Samigaluh	Medium	Gondomanan	High
Tepus	Medium	Moyudan	Medium	Ngampilan	High
Tanjungsari	Medium	Minggir	Medium	Wirobrajan	High
Rongkop	Medium	Seyegan	Medium	Gedongtengen	High
Girisubo	Medium	Godean	Medium	Jetis	High
Semanu	Medium	Gamping	Medium	Tegalrejo	High

mapping is constructed for each sub-district or city as shown in Figure 21 to Figure 24.

Figure 21. Mapping of Classification Results in Bantul Regency

Figure 22. Mapping of Classification Results for Gunungkidul Regency

Figure 23. Mapping of Classification Results in Kulonprogo Regency

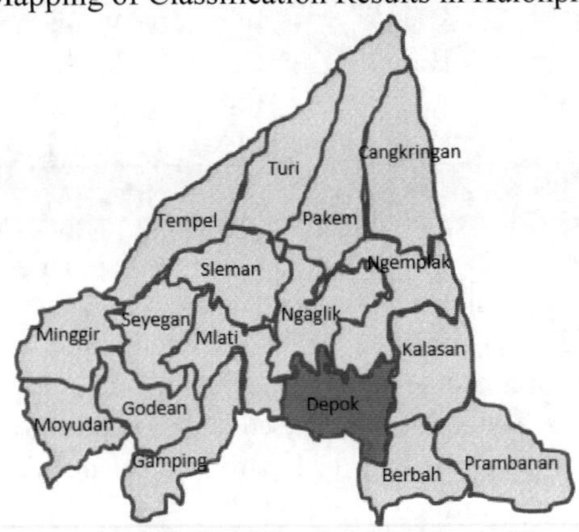

Figure 24. Mapping of Classification Results in Sleman Regency

Figure 25. Mapping of Classification Results in Yogyakarta City

Comparison of classification results with the FSAW method and the BNPB method and the SAW method can be seen in Table 3.

Table 3. Percentage of Similarity Level of Results

Year	Degree of Similarity (%)	
	FSAW with BNPB	FSAW with SAW
2014	93,58	91,02
2015	93,58	92,3
2016	98,71	92,3
2017	100	84,61

Based on table 3, the level of similarity between the FSAW and BNPB methods ranges from 93.58% to 100%, meaning that the FSAW method can be used as an alternative method of determining the level of earthquake vulnerability. Table 4 shows several sub-districts that change classification of seismic hazard levels in 2014 to 2017.

Table 4. Changes in Classification of seismic vulnerability using the FSAW Method

Sub-District	2014	2015	2016	2017
Mlati	High	High	Medium	Medium
Prambanan	High	High	Medium	Medium
Ngemplak	Medium	High	Medium	Medium
Ngaglik	High	High	High	Medium

Based on table 4, Mlati sub-district experienced a decrease in the hazard class from 2015 to 2016. In 2015 the IKG value of Mlati sub-district was 72.86 which was in the high hazard category, then in 2016 it experienced a decline of 65.15 which was included as medium hazard category. This was due to the decline in the IKF value, which was originally 19.06 then decreased to 12.69. The decrease in IKF was caused by the critical facility variable, namely the hospital. In 2016 the number of hospitals and health centers in Umbulharjo District increased, this caused a decrease in the minimum limit value of the hospital variable in this study. Prambanan Sub-district also experienced a decrease in the hazard categoty from 2015 to 2016. This was due to the decreasing IKF value. The IKF value decreases due to the decrease in the minimum limit of the hospital variable. Ngemplak sub-district experienced a decrease in the hazard category from 2015 to 2016. The cause was due to the decrease in the minimum limit of the hospital variable.

Ngaglik sub-district experienced a decrease in the hazard class from 2016 to 2017. This was due to the decrease in the IKG value in 2016 reaching 72.19 which is included as high hazard class. Then the IKG in 2017 decreased to 69.03 which was included as moderate hazard class. This was due to a decrease in the IKE and IKF values.

4. Conclusion

This research has resulted in a mapping of seismic hazard levels for each sub-district in the Special Region of Yogyakarta using the FSAW method from 2014 to 2017. According to the FSAW method from 2014 to 2017, the province of Yogyakarta Special Region is dominated by medium hazard levels with a percentage above 70%.

Acknowledgments
Researchers would like to thank Yogyakarta State University for funding this research with the Higher Education Excellence Research scheme in 2019.

References

[1] BPS 2018 *Provinsi Daerah Istimewa Yogyakarta dalam Angka 2018* (Yogyakarta: BPS) p 2
[2] Rosemarie T. Bigueras, Jocelyn O. Torio and Thelma D. Palaoag 2018 A data-driven architectural framework for LGUs in disaster preparedness and management system *Int. J. Mach. Learn. Comput.* **8** 256-261
[3] Rakhman A N and Kuswardani I 2012 Studi kasus gempa bumi yogyakarta 2006: Pemberdayaan kearifan lokal sebagai modal masyarakat tangguh menghadapi bencana *Pros. SNAST 2012*
[4] Choudhury M, Verma S and Saha P 2018 Effect of earthquake on the surrounding environment: an overview *Proceeding Int. Conf. Recent Adv. Mech. Mater.* **1** 17-23
[5] Asim K M, Idris A and Matinez-Alvarez F 2018 Earthquake prediction model using support vector regressor and hybrid neural networks *Plos one* **13(7)**
[6] Folger P 2013 Earthquakes: risk, detection, warning and research *Congressional Research Service* **7** 1-22
[7] Sunardi B, Ngadmanto D, Hardy T, Susilanto P and Nurdiyanto B 2012 Kajian kerawanan gempa bumi berbasis SIG dalam upaya mitigasi bencana studi kasus kabupaten dan kota sukabumi *Pros. Semin. Nas. Geogr. UMS Semin. Nas. Penginderaan Jauh dan Sist. Inf. Geogr. 2012* 1-11
[8] Yunarto and Sukristiyanti 2013 Pemeringkatan lokasi potensi evakuasi longsor menggunakan metode simple additive weighting (studi kasus: Kabupaten Garut) *Pros. geoteknologi* 73-86
[9] Ontah G M, Weku W C and Rindengan A J 2014 Sistem pendukung keputusan dalam memetakan wilayah risiko banjir menggunakan fuzzy multi criteria decision making *J. Mat dan Apl.* **3** 24-30
[10] Lunga S, Minarto E and Mantiri S Y 2015 Pemeteaan tingkat resiko gempa bumi di sekitar wilayah jayapura berdasarkan pengukuran mikrotremor *Spektra: J. Fis. dan Apl.* **16** 55-58
[11] Desmonda N I and Pamungkas A 2014 Penentuan zona kerentanan bencana gempa bumi tektonik di kabupaten malang wilayah selatan *J. tek. ITS* **3** 107-112
[12] Chalko T J 2008 Earthquake energy rise on earth *NU J. Discov.* **5** 1-2
[13] Lany L M, Jennifer L L and Thelma D P 2017 Corpus analysis of earthquake related tweets throug topic modelling *Int. J. Mach. Learn. Comput.* **7** 194-7
[14] Pratiwi H, Slamet I and Saputro D R S 2017 Self-exciting point process in modelling earthquake occurrences *Int. Conf. Math.: Educ. Theory Appl.* **855** 1-7
[15] ZeinEldin R A 2012 A decision support system for performance evaluation *Int. J. Comput. Appl. Spec. Issue Comput. Intell. Inf. Secur. CIIS* **2** 1-8
[16] Shakouri G H, Nabee M and Aliakbarisani S 2014 A quantitative discussion on the assessment of power supply technologies: DEA (data envelopment analysis) and SAW (simple additive weighting) as complementary methods for the "Grammar" *J. Energy* **64** 640-647
[17] Kusumadewi S and Guswaludin I 2005 Fuzzy multi-criteria decision making *Media Inform.* **3** 25-38
[18] Bima P, Isa I M and Ali M 2014 Fuzzy MADM method for decision support system based on artificial neural network to water quality assessment in surabaya river *J. Soft Comput. Decis. Support Syst.* **1** 24-29
[19] Novak V, Perfilieva I and Mockor J 1999 *Mathematical Principles of Fuzzy Logic* (Dodrecht: Kluwer Academic) p 39-46
[20] Adriyendi 2015 Multi attribute decision making using simple additive weighting and weighted product in food choice *Int. J. Inf. Eng. Electron. Bus.* **6** 8-14
[21] Pungkasanti P T 2017 Implementation of simple additive weighting (saw) method in determining high school student's interest *J. Inform. Upgris* **3** 24-27
[22] Mukodimah S, Muslihudin M and Andoyo A 2018 Fuzzy simple additive weighting and its application to toddler healthy food *Int. J. Pure Appl. Math.* **118** 1-7

[23] Nawindah 2017 Simple additive weighting (saw) mathematics method for warehouse disaster location selection in central jakarta, indonesia *Int. J. Pure Appl. Math.* **117** 795-803
[24] Priadi R and Hududillah T 2018 Risk level analysis of lightning strike with simple additive weighting method in gowa region *J. Penelit. Fis. Apl.* **8** 17-24
[25] Amri M 2016 *Risiko Bencana Indonesia* (Jakarta:Badan Nasional Penanggulangan Bencana) p 45-48
[26] Klir G J 1997 *Fuzzy Set Theory Foundations and Applications* (New York:Prentice-Hall) p 12-13
[27] Henry Wibowo S 2010 MADM-tool: aplikasi uji sensitivitas untuk model madm menggunakan metode saw dan topsis *Semin. Nas. Apl. Teknol. Inf.* E56-61

The community adaptation strategy in dealing with landslide disaster in Banaran Village Ponorogo Regency

H H Setyawan, Mukminan

Department of Geography Education, Universitas Negeri Yogyakarta, Indonesia

Haniphendra.2018@student.uny.ac.id, Mukminan@uny.ac.id

Abstract. A landslide disaster occurred in Tangkil, Banaran Village, Ponorogo Regency on April 1st, 2017 which caused many casualties. The objectives of this research were to determine the factors that caused the landslide and to determine the community adaptation strategy in dealing with landslide. This research applied descriptive qualitative method with its data collection included observation, interview, and documentation. The research results showed, 1) the factors causing landslide were steep slopes, fertile rocks and soils, land conversion, high rainfall, and human activities, 2) the community adaptation strategy involved adaptation and adjustment in dealing with landslide. The impacts of landslide have been felt by the community thus, they could change their behavior towards their environment and make adjustments to the new environment.

1. Introduction

Landslides often occur in various regions of Indonesia, with one of its factors is geographical condition. The symptoms indicating imminent landslides are the emergence of cracks in slopes occurring after rain, the presence of new springs, and the falling of various materials from the cliffs (Adiyoso and Kanegae, 2013). (Wang *et al*, 2017) stated that the occurrence of landslides is influenced by several factors, namely geology, distance from the fault, topography, vegetation, and precipitation. Landslides also often occur on natural and non-natural slopes caused by factors triggering soil displacement. Controlling changes in land use on slopes is very important as it can lead to a decrease in land stability, which in turn raises the potential of landslides during the rainy season. In addition, changes in use must be understood in detail including their effects on ecosystems, because the impacts are significant for humans (Corbett *et al,* 2017).

Landslides continue to increase from time to time, implying the various factors triggering the occurrences, the vulnerability of community increases through various development practices (Barasa, 2018). Landslides are lands that have various material components, forces, and motion mechanisms that cause complex hazards and landslides due to rainfall (Lin *et al*, 2018). The impacts caused by landslides include damages and losses that can affect people's lives. Those damages and losses can be in the form of casualties, damaged houses, and disrupted ecosystems. Therefore, it is necessary to have a good strategy and special handling measures in areas prone to landslides. Baum and Godt, (2010) found that complex landslides caused by rainfall often show characteristics of deformation, progressive failure, and threaten people's lives and property.

Content from this work may be used under the terms of the Creative Commons Attribution 3.0 licence. Any further distribution of this work must maintain attribution to the author(s) and the title of the work, journal citation and DOI.
Published under licence by IOP Publishing Ltd

Ponorogo Regency is one of the areas prone to landslides due to its morphology in form of hills, highlands, and lowlands with an altitude of 92 to 2563 masl. The potential for landslides is quite high in the area of Pulung and Sawo Districts because they have altitude of more than 1000 masl. Paimin et al. (2009) explained that the level of landslides susceptibility is measured by the parameters of natural factors that make up the formula such as 3 consecutive days of cumulative daily rain (25%), slopes (15%), rocks (10%), faults (5%), soil depth to impermeable layers (5%), and management factors including land uses (20%), infrastructures (15%), and settlements (5%). Furthermore, groundwater condition, geomorphology, and soil property regulate slope instability (Wan *et al,* 2011).

Tangkil is located in Banaran Village, Pulung District, Ponorogo Regency. On Saturday April 1st, 2017, a landslide disaster occurred on that area. The incident occurred in the morning at 07.30 WIB, when the community carried out their daily activities in their farms. At that time, many people were harvesting ginger, as the disaster ended up with 6 death, 22 buried by landslides, and 17 suffered minor injuries. The number of refugees was 178 people in 8 points provided by the government. The victims of landslides were not only the community who harvested ginger but also those who were in their houses because landslides were close to the settlement. These geological conditions are very detrimental and human activities can deteriorate due to dangerous situations (Xu *et al*, 2018).

The landslide in Tangkil was preceded by signs of cracks and the high rainfall on the night before so that the entire Tangkil villagers had started leaving their homes. But, in the next morning those people returned to their home to harvest ginger. The Ponorogo Government's had advised to the community since March 13 2017 to evacuate to a safer place, but those people decided to return home as the landslide occurred in Tangkil. Bignami *et al.* (2018) found that strategy and public awareness need to be improved, through preparedness for the risk of landslides and floods. The purposes of this research were to determine the factors causing landslides and to determine the community's adaptation strategy in dealing with landslides.

2. Methods

This research applied descriptive qualitative method. The data were collected using interview, observation, and documentation techniques. In order to get the correct sources and in accordance with the objectives, the researcher used purposive sampling technique. Interviews were conducted with the people of Tangkli, Banaran Village, Pulung District, Ponorogo Regency. Observation and documentation were carried out to determine the causes of landslides and to determine the community's adaptation strategy in dealing with landslides. This research used triangulation techniques, sources, and methods to test the validity of the data obtained to keep them in accordance with the research objectives.

3. Result and Discussions

Ponorogo located at (7°52'15.3 "S, 111 ° 27'44.5" E) has a wide variety of landscapes and has had 59 landslides occurring during 2012-2018. (Muriyatmoko *et al*, 2019) The worst landslide occurred in Tangkil, Banaran Village, Pulung District, which resulted in casualties and material losses. Banaran Village is an area with moderate class ecosystem services for regulating, protecting, and preventing disasters, which means that the land in that area has the potential for disaster when triggered by high intensity rainfall. The landscapes of Tangkil, Banaran Village, are part of the foot of Mount Liman Dorowati with a pyroclastic material, so the area is prone to landslides. (Xiao *et al*, 2020) research results showed that lower slope stability of 0.923% and failure probability of 98.5% with high rainfall caused landslides in Wanzhou County, Three Gorges Reservoir, China. Climate change significantly increased multiple hazards and tiered impacts on society, resulting in disasters such as landslides (Schiermeier, 2018).

3.1 The factors causing landslides

Slope

The geomorphology is formed by moderate to steep hills where under the slopes there are settlements and community-owned lands. 13 hectares of community lands were buried by landslides, causing considerable losses because the plants on the lands were ready to be harvested. (Naryanto *et al*, 2019) showed that the slope in Tangkil is very steep, slanting at 70-140%.

Figure 1. The direction of the landslide

The observation result obtained that the peak height of the slopes was 900-1000 masl. The distance between the top of the slope and the end point of the landslide leading to the east was about 2000 meters. The direction of the landslide was included in the slope with a slope of 7-15%, so it was used for the community settlement. Slope conditions are actually not recommended for agriculture and settlements due to the surface water flow rate cannot be controlled. There needs to be a better understanding by providing a rigorous systematic process to improve slope management, thereby reducing landslide hazards (Dai, Lee, and Ngai, 2002).

Rock and soil

The rocks in Tangkil are composed of volcanic rocks that are easily broken down, have many cracks, and overlap tertiary sedimentary rocks which can form a sliding layer. The soil properties include high organic content with a texture of sandy clay and loose. The soil characteristic on the top layer is dark and the soil solum is thick. Based on observation, the soil on the slopes of Tangkil has very good fertility, as can be seen from the soil nature and soil type. The soil types are dominated by lithosol, inceptisol, and alfisol. Therefore, human activity is very high in using the lands for agricultural cultivation of seasonal crops.

The community did not understand the impact of slope soil fertility with these seasonal crops on soil structure. Inceptisol soil has a moderate level of weathering and soil development, while alfisol soil is used by the community for agriculture. If the slope is steep, the soil can only be planted with hardy plant types that can withstand erosion. However, the community used the slopes for planting ginger, so the erosion rate is quite high due to seasonal crops. (Suprato *et al*, 2017) found that the lithosol and andosol soil have little organic elements that if planted with unsuitable plants will cause an increase in erosion rate.

Land conversion

The topography and slopes with very steep criteria are utilized by the surrounding community for agricultural cultivation activities, namely ginger, because of the large yields to the international market. The community applies a terraced planting model with a width of 3 to 5 meters. Ginger is very suitable to be planted in tropical and sub-tropical areas with temperatures ranging from 20-30 ° C, Tangkil meets the criteria for the cultivation of these plants. Land conversion causes a decrease in global biodiversity, given the importance of functions and ecosystems that exist in the hills (Newbold *et al*, 2015).

Rainfall

Rainfall infiltration increases pore pressure which reduces the effective stress in the soil and reduces the shear strength of the soil, eventually resulting in slope failure (Conte et al. 2017). One of the triggers of landslides in Tangkil was the high rainfall condition. High intensity rain flushed in all areas in Ponorogo

Regency. Before the landslide in Tangkil occurred, it rained for three days. Groundwater absorption was very high due to high rainfall intensity, so that the soil on the slopes became saturated with water resulting in landslides. Seismically induced landslide failures and the resulting ground flow are essential for appropriate design precautions. Rainwater causes an increase in soil mass thereby weakening the bonds of soil particles that lead to landslides (Susanti and Miardini, 2017).

Human activity

The human factor belongs to non-natural factors and is a key factor in the occurrence of disasters, one of which is landslides. The cultivating of plants cumulatively causes soil imbalance and will have a harmful impact. (Mandasari, Arifin, and Ali, 2016) due to the limitations and high land value, some people choose to live on the hillside and take advantage of the existing land, but as a consequence, almost every year the community is affected by landslides, as what happened in Cappagalung District.

The results of observation, documentation, and interview with the people of Tangkil, Banaran Village, showed that the people made their living as agricultural cultivation of ginger, vegetable, rice, and coffee. Some of the farmers are farming on their own land and some are farming using the slopes. Those who farm on their own land cultivate rice, coffee, and vegetables while ginger is planted on slopes. Initially, the slope area was included in protected forest and the land was classified as fertile with plants that can grip the soil, such as pine. Along with the pressure from community activities in Tangkil, many people cultivate seasonal crops and are classified as plants that cannot control/maintain the soil structure.

As a result of the activities of some people in Tangkil who use the slopes, it has an impact on other farmers who farm on their own land as the landslide hit private land and caused losses and damages to family members. Regional development must adapt to the risk of natural disasters, so that people can implement disaster mitigation and prevention (Nurohmah *et al*, 2014).

3.2 The community adaptation strategy in dealing with landslides

The landslide disaster in Tangkil in 2017 caused many losses ranging from physical and non-physical aspects. Physical losses are related to buildings and lands damage, while the non-physical aspects are anxiety and trauma experienced by the community.

Table 1. Losses from the landslide of Tangkil

No	Physical		Non physical
	Type	Number & area	
1.	Home	56	Anxiety
2.	Production forest area	2 Ha	
3.	Community land	13 Ha	

The value and usefulness of concepts, approaches, and models need to be maintained for dealing with disaster risk challenges (Okada *et al*, 2018). The community lacks knowledge of disasters, so the feeling of trauma and anxiety is so impactful. Therefore, awareness of natural disasters needs to be given since school due to the awareness of environmental sensitivity (Tuncel, 2018). Landslides also destroyed the environment in which they lived and forced people to adapt to their new environment.

The adaptation process was done by adjusting to a new environment as a place to live, so that the people of Tangkil, Banaran Village are ready for the adaptation in the new environment after a landslide. Everyone will find it difficult to re-adapt, considering that for years they have occupied their former places However, the adaptation process can be carried out in several stages, so that results can be achieved for the people affected by the landslide. The concept of community adaptation strategy that is applied by Berry (1980): 1) adaptation by reaction, 2) adaptation by adjustment, 3) adaptation by withdrawal.

The main adaptation strategy chosen to deal with landslides in Tangkil, Banaran Village is adaptation by adjustment. Adaptation is carried out by the community with the aim of being able to live in harmony and integrated with the environment (Hutcheon, 2006). The hierarchy of human needs for housing fulfillment includes: survival needs, safety and security needs, affiliation needs, estern needs, as well as cognitive and aesthetic needs (Fulliove & Fulliove 3rd, 2000). Therefore, people affected by natural disasters must meet their needs and carry out their adaptation process. Adjustment of natural and human systems to the stimulus of natural disasters is required to reduce the negative impact of disasters. An adaptation by adjustment is a strategy to adapt to changes in people's behavior to the environment for the sake of better environment.

The research result, showed that adaptation by adjustment was chosen because the land cultivation was very intensive and terracing was not in accordance with conservation principles. In addition, areas prone to landslides are those with hilly and highland areas in Ponorogo Regency (Yuniarta *et al*, 2015). The condition of Tangkil, Banaran Village after the landslide disaster made it possible to rebuild their residence, but all people must change their behavior towards their environment; not planting ginger on the slopes. The topography in Tangkil is steep and the community uses it for plantations and applies the concept of micro conservation or terracing. The characteristic of loose soil makes rainwater infiltration effective and it is added to the cultivation of ginger, so that the plant does not have a function as a soil structure and run-of reinforcement. (Hermiyanto, Winarso, and Kusumandaru, 2016) Farmers should conduct soil tests with the aim of measuring all parameters of the chemical properties of the soil, so that they can determine suitable plants for the lands.

Figure 2. Ginger plant in Tangkil

The impact of landslides has been experienced by the community. Therefore, people need to change their behavior towards their environment. Before the landslide, the slopes were planted with ginger which could not function properly to hold the soil structure. With an adaptation by adjustment strategy, the community will not plant plants that cannot function for the soil structure so that they replace plants that can strengthen the soil structure on the slopes. Hermon (2019) stated that controlling land use is very important because the land size is permanent while the need for land is getting higher. This can lead to a decrease in land stability so that in the rainy season the potential for landslides to occur is even greater. Perez *et al* (2015) mentioned that change in land use and its implications for land cover conversion rates resulting in reduced living biomass and delayed vegetation transition.

Hermon (2019) stated that every region in Indonesia has a Regional Spatial Plan (RTRW) which functions as a controlling instrument, because along with the development and progress of the area, the pressure on changes in natural spatial ecologically including in the buffer zone will also be high. Community understanding of landslides through the interpretation of the level of vulnerability that occurred in Tangkil, Banaran Village, that the level of vulnerability to landslides is high but community understanding is low. The low community understanding with a high level of landslide vulnerability can result on great losses and environmental damages. The community does not have the insight into disaster mitigation and only relies on natural phenomena in the surrounding environment. The landslide disaster in Tangkil has attracted the sympathy of various parties, so a lot of assistance has come to meet the needs of the community.

4. Conclusion

Based on the research results, it can be concluded that the factors causing landslides in Tangkil, Banaran Village are steep slopes, rocks and soil, land conversion, high rainfall, and human activities. In addition, the community adaptation strategy in dealing with landslides used is adaptation by. adjustment. Landslide destroyed the environment in which they lived and forced people adapt to their new environment. The adaptation process by adjusting to a new environment as a place to live is very important, so that the affected community must be prepared for the adaptation process in the new environment after landslide occurs.

References

[1] Adiyoso W and Kanegae H 2013 The Preliminary Study of the Role of Islamic Teaching in the Disaster Risk Reduction (A Qualitative Case Study of Banda Aceh, Indonesia). *Procedia Environmental Sciences*. **17** 918-927.

[2] Wang F, Xu P, Wang C and Jiang N 2017 Application of a GIS Based Slope Unit Method for Landslide Susceptibility Mapping Along the Longzi River, Southeastern Tibetan Plateau, China. *International Journal of Geo-Information.* **6** 172-180.

[3] Corbett D R, Walsh J D, and Zhao Y 2017 Impacts of Land-Use Change on Sedimentation in Tidal Creeks of North Carolina USA. *Journal of Geography and Earth Sciences*. **5** 1-25.

[4] Barasa B 2018 Increased Incidences, Intensity and Scope of Disasters: Manifestation of Unsustainable Development Practices. *Environment Pollution and Climate Change*. **2** 1-9.

[5] Lin F L, Wu R Q, Huang, and H Zhang 2018 Formation and characteristics of the Xiaoba landslide in Fuquan, Guizhou, China. *Landslides.* **15** 669–681.

[6] Baum R L, and J W Godt 2010 Early warning of rainfall-induced shallow landslides and debris flflows in the USA. *Landslides.* **7** 259–272.

[7] Paimin, Sukresno, dan Pramono I B 2009 *Teknik Mitigasi Banjir dan Tanah Longsor.* Balikpapan: Tropenbos Internasional Indonesia Programme.

[8] Wan S, Lei T, and Chou T 2011 A landslide expert system: image classifification through integration of data mining approaches for multi-category analysis. *International Journal of Geographical Information Science.* **4** 747-770.

[9] Xu D, L Peng S Liu, and X Wang 2018 Inflfluences of risk perception and sense of place on landslide disaster preparedness in southwestern China. *International Journal of Disaster Risk Science*. **9** 167–180.

[10] Bignami D F, Dragon A, and Menduni G 2018 Assessing and Improving Flood and Landslide Community Social Awareness and Engagement via a Web Platform: The Case of Italy. *Internasional Journal of Disaster Risk Science*. **9** 530-540.

[11] Muriyatmoko D, Utama S N, Pradhana F R, Umami J, Rozaqi A J, and Setyaningrum H 2019 Landslide prediction model of prone areas in Pulung, Ponorogo East Java. *Procedia Computer Science*. **161** 747-755.

[12] Xiao L, Wang J, Zhu Y, and Zhang J 2020 Wanzhou County, Three Gorges Reservoir, Cina. *Internasional Journal of Disaster Risk Science*. **11** 347-363.

[13] Schiermeier Q 2018 Droughts, heatwaves and flfloods: How to tell when climate change is to blame. *Nature*. **560** 20–22.

[14] Naryanto H S, Soewandita H, Ganesha D, Prawiradisastra F, and Kristijono A 2019 Analisis penyebab kejadian dan evaluasi bencana tanah longsor di Desa Banaran, Kecamatan Pulung, Kabupaten Ponorogo, Provinsi Jawa Timur tanggal 1 April 2017. *Jurnal Ilmu Lingkungan.* **17** 272-282.

[15] Dai F C, Lee C F, and Ngai Y Y 2002 Landslide risk assessment and management: An overview.

Engineering Geology. **64** 65-87.

[16] Suprapto, Nurmasari R, dan Rosyida A 2017 Analisis penyebab tanah longsor di Kabupaten Ponorogo (studi: Dusun Tangkil, Desa Banaran, Kecamatan Pulung). *Jurnal Dialog Penanggulangan Bencana*. **8** 112-119.

[17] Newbold et al 2015 Global effects of land use on local terrstrial biodevesity. *Nature*. **520** 45-70.

[18] Conte E, A Donato, and A Troncone 2017 A simplifified method for predicting rainfall-induced mobility of active landslides. *Landslides*. **14** 35-45.

[19] Susanti P D, Miardini A, and Harjadi B 2017 Analisis kerentanan tanah longsor sebagai mitigasi di Kabupaten Banjarnegara. *Jurnal Penelitian Pengelolaan Daerah Aliran Sungai*. **1** 49-59.

[20] Mandasari J, Arifin M, and Ali M 2016 Settlement Arrangement Based on Landslide Mitigation (Case Study: Capagallung, West Bacukiki District, City of Parepare). *Procedia - Social and Behavioral Sciences*. **227** 442-450.

[21] Nurohmah A, Priadmojo A, Dewi M K, Satria M R, and Saputra N 2014 Analysis of Regional Development in Connection with Multi Disaster Susceptibility in Bengkulu Province. *Procedia - Social and Behavioral Sciences*. **135** 70-75.

[22] Okada N, Chabay H, and Renn O 2018 Participatory Risk Governance for Reducing Disaster and Societal Risks: Collaborative Knowledge Production and Implementation. *Internasional Journal of Disaster Risk Science*. **9** 429-433.

[23] Tuncel G 2018 Alternative Environmental Applications In Developing "Sensitivity To Natural Environment" Values In Social Studies. *International Journal of Geography and Geography* Education. **38** 91-103.

[24] Berry J W 1980 Cultural Ecology and Individual Behavior. In: Alman I, Rapoport A, and Wohlwill J F Environment and culture. *Human Behavior and Environment*, Vol 4, Springer, Boston, MA.

[25] Hutcheon L 2006 Beginning to theorize adaptation. *A Theory of Adaptation*. **5** 1-32.

[26] Fulliove M T, and Fulliove 3rd R E 2000) what's housing got to do with it?. *American journal of public health*. **2** 183-190.

[27] Yuniarta H, Saido A P, dan Purwana Y M 2015 Kerawanan bencana tanah longsor Kabupaten Ponorogo. *E-Jurnal Matriks Teknik Sipil*. **1** 194-201.

[28] Hermiyanto B, Winarso S, and Kusumandaru W 2016 Soil Chemical Properties Index of Tobacco Plantation Land in Jember District. *Agriculture and Agricultural Science Procedia*. **9** 181-190.

[29] Hermon D 2019 Land Stability Model for Sustainable Spatial Planning in Padang City-Indonesia based on Landslide Disaster. *Journal of Geography and Earth Sciences*. **7** 19-26.

[30] Perez M E V, Hernandez M E O, and Salazar L M 2015 Analysis Of Land Use Changes And Carbon Content In Forest Covers In The State Of Mexico With Spatial And Stochastic Methods. *Journal of Geography and Earth Sciences*. **3** 33-49.

[31] Hermon D 2019 Land Stability Model for Sustainable Spatial Planning in Padang City-Indonesia based on Landslide Disaster. *Journal of Geography and Earth Sciences*. **7** 19-26.

IOP Conf. Series: Earth and Environmental Science **884** (2021) 012004 doi:10.1088/1755-1315/884/1/012004

Developing Disaster Mitigation Education with Local Wisdom: Exemplified in Indonesia Schools

Irfiani Triastari[1], Siti Irene Astuti Dwiningrum[2], Sovia Husni Rahmia[3]

[1] Student at Social Science Education Graduate Program, Yogyakarta State University, Yogyakarta, Indonesia
[2] Faculty of Science Education, Yogyakarta State University, Yogyakarta, Indonesia
[3] Student at Social Science Education Graduate Program, Yogyakarta State University, Yogyakarta, Indonesia

Irfianitriastari.2019@student.uny.ac.id

Abstract. In vulnerable disaster-geographical condition in ring of fire, schools in Indonesia have conducted innovation of disaster mitigation education recent years. A disaster mitigation model in Indonesia is the use of local wisdom. Local wisdom-based disaster education aims as grand design during learning process and it supports the formation of student resilience. Therefore, to revive the value of local wisdom, it needs reinterpretation through adaptation of local knowledge and revitalization of contemporary conditions as innovations in disaster risk reduction. Through the integration of local wisdom based on disaster mitigation curriculum in the learning process, it is expected to be able to take fast, precise and accurate steps in dealing with disasters that occur in every region in Indonesia.

1. Introduction

Indonesia is a disaster-prone country located in Southeast Asia. Based on data released by the United Nations Agency for the International Strategy for Disaster Risk Reduction (UNISDR), geographically Indonesia is an archipelago located at the junction of four tectonic plates, namely the Asian Continent plate, the Australian Continent, the Indian Ocean plate and the Pacific Ocean and it is nicknamed as the Ring of fire. In the southern and eastern parts of Indonesia there is a volcanic belt (volcanic arc) that extends from the island of Sumatra-Java-Nusa Tenggara-Sulawesi, whose sides are old volcanic mountains and lowlands which are partly dominated by swamps. This condition is very potential and prone to disasters such as volcanic eruptions, earthquakes, tsunamis, floods and landslides [1]. Based on BNPB data in the past 20 years, 28,671 disasters have occurred in Indonesia, both natural disasters and social disasters [2]. Almost every region has experienced a disaster, even some areas have a high level of risk of disaster affected. Each region also has different risks and types of disasters. Since the beginning of 2020 until the end of February 2020, 13 disasters had been recorded in 12 districts/cities in Indonesia [2]. To reduce the risk of increasing number of victims, it is necessary to increase mitigation efforts. Disaster preparedness is defined as actions that aim to improve safety in the event of a disaster. One of the efforts carried out by the government is the existence of a vision of national development in the National Long-Term Development Plan (RPJPN) that is reducing the risk of loss of life, potential

Content from this work may be used under the terms of the Creative Commons Attribution 3.0 licence. Any further distribution of this work must maintain attribution to the author(s) and the title of the work, journal citation and DOI.
Published under licence by IOP Publishing Ltd

damage and loss through increasing and understanding public awareness and building a culture of awareness and security in the people of Indonesia from disasters.

School is an effective place to instill disaster mitigation education. Based on the Education and Culture Data and Statistics Center (PDPS) in 2018/2019 the target number of disaster mitigation and preparedness education programs in the Education unit is 441,024 [3]. Students are as information delivery agents in terms of disaster mitigation. In addition, schools are a high-risk place to cause casualties in the event of a disaster. Over the past 10 years (2009-2018), more than 62,687 education units and 12 million students were affected by disaster. Considering the number of fatalities caused by the disaster, vigilance is very important. On the other hand, school buildings are also often places of evacuation and refuge after a disaster occurs. Thus, it is appropriate if the school/education unit becomes the target of mitigation education.

Indonesia's National Disaster Management Agency (BNPB) indicates that 75% of schools in Indonesia are located in disaster areas (medium or high risk) [4]. Because it is important to provide knowledge related to disaster education to students through formal school education. In addition, most children or adolescents spend more time in school. Apart from being an object of safety, especially at the secondary school level, another thing that needs to be developed is the existence of a knowledge center and facilities before, during, and after a disaster. Thus, disaster education activities in schools are considered as a concrete, effective, dynamic, and sustainable strategy in efforts to reduce disaster risk and disseminate disaster education.

In realizing these objectives, the first step is to prepare quality human resources starting from the field of Education (school). Schools are considered to be a field of growth in the quality of reliable human resources in the future. In addition to student competencies that must be global in the modern era, it is time for literacy education to involve local wisdom. In the National Education System Law a new paradigm has been born in which education based on local wisdom has the task of encouraging accelerating development in accordance with the potential of local communities in the area. This is in line with the local content in the curriculum (article 27: 1) which must prepare students to know the local area. Thus, they can work together according to their local needs, that is leading to local wisdom. In dealing with disaster risk reduction in schools, many researchers, workers in NGOs, UN agencies, and other organizations have shown that school construction with attention to building structures and disaster education is an important factor for the development of school safety, especially in the case of earthquake disasters. Earth that often happens in Indonesia [5].

Curriculum of natural disaster is important in disaster-prone countries including Indonesia to be taught from an early age. Thus, children are given knowledge about disaster. Disaster curricula are considered more effective if they are applied in primary and secondary school education curricula than practices that are easily forgotten [6]. It is realized that knowledge about disasters should be owned by all people, a concrete solution is needed in overcoming natural conditions. Education is a means that is quite effective as an effort to reduce the risk of disasters by including disaster learning materials as a compulsory lesson for students at all levels, especially schools that are located in disaster prone areas. The curriculum is integrated into local wisdom, it is expected that this effort is easily accepted and understood by students because it comes from local values of local culture [7].

This paper discusses the local wisdom-based disaster mitigation curriculum in the effort of Disaster Risk Reduction (DRR). Empowerment in the school environment is the first step to instilling student resilience in responding to a disaster. Students are expected to be able to carry out cultural values in their environment that are integrated in the learning curriculum. Thus, they are able to be precise, fast, and accurate when disasters occur in their area.

2. Methods

This research uses literature and library studies. Library research is studies related to library data collection methods, or research where the object of research is explored through a variety of library information, a list of references from all types of references such as books, journal papers, articles, dissertations, theses, theses, hand outs, laboratory manuals, and other scientific works [8]. Literature study is carried out using data which is a review of gathering various references that are appropriate to the problem and research objectives. The technique used aims to find theories that are in accordance with the problem being studied in research to be used as reference material for discussion of research results.

3. Result and Discussion

In law number 24 Year 2007 about Disaster Management, disaster is a series of events that threaten and disrupt the life and livelihood of the community caused by natural factors, non-natural factors and human factors, resulting in human casualties, environmental damage, property loss, and psychological impacts [9].Disasters referred to in this Law consists of natural disasters, non-natural disasters, and social disasters. In 2014, Indonesia received criticism from the UNISDR (United Nations Secretariat for International Strategy for Disaster Reduction) regarding policies on prevention and protection of natural disasters which were considered very weak in coordination between ministries. Coupled with a budget for countermeasures that are considered very small valued at 0.699% of the overall Indonesian national budget. Based on the classification from UNISDR (United Nations Secretariat for International Strategy for Disaster Reduction) divides disasters into two types of groups namely technological disasters and natural disasters. As a prone-disaster country, Indonesia is still struggling with the main problems of the low effort to deal with disasters, the low focus of attention on a disaster, and the weak role of schools in introducing disaster mitigation education.

Disaster mitigation is a strategic step in reduce disaster risk, both through physical development and awareness and increase ability to face the threat of disaster. The number of disasters experience by the people of Indonesia, knowledge needs to be given to the main community who reside in areas that are very risky. Disaster mitigation includes protective actions and activities that have taken place since preparation before the disaster, the disaster danger, prevention of rescue, rehabilitation and relocation efforts. Disaster mitigation includes all initiatives and actions aimed at minimizing the adverse effects of a disaster before it occurs. Within its scope, disaster preparedness encompasses the overall sweeping of activities, both structural and non-structural. Structural mitigation involves steps such as strengthening buildings and infrastructure. Nonstructural mitigation involves measures such as disaster education, disaster policy planning, forecasting, and early warning. The scope of disaster mitigation is not limited to saving lives and property, it also includes reducing the negative impact of hazards on economic development, biodiversity, and social institutions [10].

3.1. Education Disaster Mitigation Model in Indonesia

Non-stop disaster occurs in Indonesia. This should be accompanied by effort to continue disaster management such as through the education sector. Education is one of the determining factors in disaster risk reduction. Various types of integration models for disaster mitigation education are carried out in subjects. Learning model Disaster education is designed integrated in the teaching and learning process in schools without reducing the learning burden of students. Learning models are designed creatively, innovatively. Thus, the learning process does not cause rejection but instead increases resilience and objective awareness of disaster events [11].

3.1.1. Establishing Partnership Network Between Parties To Support PRB (Disaster Risk Reduction) at Schools

In supporting disaster risk reduction, disaster education conducted is through Training and Simulation methods [12]. The training is carried out by certain professional groups that have a background in community service. Besides being done in the community, it is also done in the school environment. Training activities begin with pre-disaster training, when a disaster occurs, and after a disaster. [13] Training conducted in the community is an effort to build preparedness at the pre-disaster stage. While the simulation method in disaster management is a way to be able to assess someone's readiness to face a disaster [14]. Other findings show that simulation methods that use games can lead to disaster preparedness better than not using simulation methods.

Identical school environment with the knowledge uses books. Thus, literacy becomes something appropriate in disaster management. Research on the role of geography lessons in developing disaster literacy curricula concludes that disaster literacy can play a role in building disaster literacy through setting disaster literacy curricula and the use of disaster contextual methods in classroom learning [15]. Knowledge literacy is implemented through printed books and non-printed documents. This output produces a product that can be consumed by all students and has access to the network. Books are produced by individuals, communities, certain organizations. The role of various organizations and communities supports the fulfillment of goals and needs in reducing the impact of disaster risk.

Other effort is through games. Games are widely applied in learning media. This game is interactive. Just as what has done by Rahadian created a disaster education game for autistic children that is developed using the MDA framework [16]. The games provide information on the signs of disaster, mitigation procedures and rescue bag for each individual.

In improving students' abilities to outside activities, every school provides extracurricular activities. In some disaster-prone countries such as Japan, Nepal and the Philippines, extracurricular activities are effort to reduce disaster risks as well as in Indonesia. Through extracurricular disaster mitigation education, it can provide supporting information related to safety education in schools. In the Philippines, there is a scientific science club for students to channel ideas and ideas related to the guidance given from the school [17]. Findings of other research on school-age survey preparedness to deal with flooding are concluded that school-age children can be agents of reformers to understand the importance of disaster preparedness, for parents who have children after school set a good example in maintaining environmental conditions, and children's readiness school activities for flood disasters can be taught independently in extracurricular activities [15]. Extracurricular activities can be used to facilitate students' talents and interests outside academic abilities such as sports, scouts, arts, crafts, and others. Another extracurricular can be chosen is disaster mitigation simulation training. The exercise begins with theory and simulation in the classroom and then is practiced in the field with disaster equipment. Training in schools can involve local government residents and the Local Disaster Management Agency (LDMA) [14]

3.1.2. Disaster Resilient School

In 2006, The Indonesian Institute of Sciences (LIPI) carried out research in 3 regions, such as Bengkulu City, Padang City, and Aceh Regency. From those findings, they state that school is 'public spaces' with a high level of vulnerability. On the field, there are many areas that are vulnerable to disasters. School facilities and infrastructure, school buildings are the main targets of disasters occurring mainly in school hours [18]. The description of events in China during the earthquake on May 12, 2008 had a big impact because it occurred during school hours. The disaster killed 87,000 people and 5,335 students with a magnitude 7.9 earthquake. School children who died around 6%. Chinese state media reports have launched more than 7,000 school buildings that collapsed and led to teachers and students. Due to the poor quality of building materials, many school buildings collapsed during the earthquake. That fact made Indonesia realize efforts to manage disaster preparedness in schools.

One of efforts in developing disaster preparedness in schools is through disaster preparedness school programs. Disaster prepared school is a school that is able to manage disaster risk and has the ability in disaster management planning (before and after a disaster), the availability of logistics, environmental safety Education, infrastructure, and emergency systems supported by the presence of knowledge and preparedness skills, fixed procedures (standard operational procedure), and early warning systems. The UNISDR handbook explains Safer School Construction, (Guidance Notes on Safer School Construction). It states that preparedness is the knowledge and capacity developed by the government, professional organizations that organize emergency response and post-disaster recovery, communities and individuals are expected to effectively anticipate and respond to the impact of events or conditions that can occur and will occur..

Endeavoring school disaster preparedness are a manifestation of the National Action Plan for Disaster Risk Reduction (RAN PRB) 2010-2012, namely using knowledge, innovation and education to build a culture of safety and disaster resilience.. Thus, the concept of disaster preparedness school/school of disaster alert consists of efforts to develop innovative minds to produce a culture of safety, resilience, and security for all school residents. school community preparedness and a safe learning environment are part of the main elements of disaster preparedness schools.

School of disaster alert aims to: 1) Build a culture of alertness and a culture of safety in schools by developing networks among stakeholders in the field of disaster management; 2) Increase the capacity of school institutions and individuals in realizing safer learning places for students, teachers, and all residents around the school; 3) Disseminate and develop disaster knowledge to the wider community through education in schools. The Disaster Preparedness School Program (SSB) is part of the BNPB program through the disaster safe school campaign in collaboration with the Ministry of National Education.. The disaster preparedness school (SSB) program is part of the BNPB program through a safe school campaign in collaboration with the Ministry of National Education.

3.2 Local Wisdom-based Disaster Mitigation Education

Indonesia is an archipelago with a very diverse culture. The cultures are assets in the field of tourism and the identity of a nation. However, along with the development of modernization and globalization, many of these cultures begin to fade and cultural assimilation ensued. This diversity brings up the local identity of each region. Thus, they are able to contribute to the issue of disaster development as one of the issues raised in sustainable development (Sustainable Development Goals) [19]. With the focus on disaster management, it is expected that disaster risk can be minimized in each region/area. Almost every region in Indonesia is categorized as disaster prone. Regions in Indonesia have a varied nature conditions. This is affected by geographical, geological, and climatological factors that cause each region to need adaptation to various types of disasters [20] [21] [1]. Based on research conducted by Simon Sujanjoyo in 21 regions in Indonesia, He states that the dominant disaster management efforts use technology or system applications. The second sequence is local wisdom and the rest is in the form of Education related to disaster [22]. The focus of this research is about education-based disaster mitigation local wisdom. Culturally, Indonesia is a culturally diverse country and it should be preserved as the identity of the Indonesian nation. Some studies say that a culture can survive in society if that culture is integrated into formal educational institutions[23].

Various forms of culture for disaster mitigation have begun to be applied in various regions in Indonesia, including in Central Java which is called Jogo Tonggo. *Jogo Tonggo* is a concept of disaster mitigation in Central Java. Furthermore, it becomes movements that contain local wisdom. The local wisdom side can be seen from the first, the use of phrases in Javanese that are easily understood by the wider community. Second, open space for community participation based on habits, norms, and local values in the community such as social associations at the RT, RW level, and other social communities such as the Islamic study groups, and others. As meaning in words chosen using Javanese language, this concept is applied in order to, first, the power of information and education about the dangers of Covid-19 is easily and quickly understood by local residents. Second, each member of the community in Central Java at the RW / RT level takes care of each other's neighbors and environment from the spread of the Covid-19 outbreak and the effects caused by the pandemic. [24]

Baduy people also have local wisdom. In Baduy society, there are several rules in the order of life related to nature, as follow: they are not allowed to enter the forbidden forest and take products from the forest even if they are only dry leaves. They are not allowed to change the flow of the river for some people. They are not allowed to manage the land with pesticides, use hoes and make terraces to grow rice. [25] Baduy local wisdom is related to disaster mitigation in earthquakes, landslides, floods, and fires. This is reflected in: (1) the farming tradition with the rules for selecting the location of the field when farming, the procedures for opening and burning land, and the equipment that is allowed to be used. The farming tradition avoids the danger of landslides and fires. (2) Rules and *pikukuh* is for making building houses, bridges, granaries, etc. with bamboo, fibers and kirey materials without spile. Buildings made on the ground adjust the contour of the land, erected on the base are not allowed to change the contour of the land. This is a mitigation against earthquake, landslides, floods and fires. (3) Dividing the forest zone into three areas is an effort to mitigate landslides, floods, erosion, and other disasters also as an effort to conserve the ecosystem.

In Kalimantan, the Dayak tribe also has local wisdom. As a local farmer has local wisdom in farming and opening fields. The Dayaks do not burn forests except for farming, and this is also conducted through traditional calculations that have been taught for generations. In the Dayak community as a child, they have been planted to not leave the fire alive because leaving the fire alive will have an impact on something undesirable such as a fire. [26]

In Aceh, there is a local wisdom called Smong, local wisdom precisely on the island of Simeulue in reading disasters that occur on the beach. Adopting this custom has been proven to save lives when a tsunami occurs. Smong is a simple local wisdom concept that functions as an information tool when a tsunami occurs. Shouts of a "smong" sound signal an early warning to the surrounding community that the water is receding and they must immediately run to the hills. This knowledge comes from their ancestors who learned from previous disasters. Smong is also defined as sea water that rises after an earthquake [27]

In West Java, Sundanese people also have local wisdom. Every indigenous people in Sundanese culture has a form of local wisdom that is very significant in mitigating disasters. In general, indigenous peoples have realized that if the environment is damaged, disaster will be overtaken, even though the way to maintain it is through myths and customary rules. However, this belief has been very effective in preventing disasters until now, especially erosion, floods and landslides. The forms of interpretation of local wisdom include three, as follow (1) Bamboo House Building; (2) Spatial Planning & Zoning of Land Use in Micro Scale; (3) Environmental Friendly Land Management. [28]

In Sulawesi, an indigenous Karampuang people have local wisdom in managing forests. They have their own ways and become part of their cultural system. The forest is an inseparable part of nature itself to maintain the balance of the ecosystem. There are separate rules or norms that must be obeyed by all members of the community. The Karampuang community is still very bound and obedient to its customary rules, which are full of cosmological beliefs, knowledge and views, related to environmental management and maintenance. In order to stay awake. The Karampuang Indigenous Council as a symbol of the traditional ruler, agreed to manage the existing customary forest by using knowledge sourced from the local wisdom they possessed. As it is known that these indigenous people still hold myths and ancestral messages that contain prohibitions, invitations, sanctions in managing their forests [7].

Various types of local wisdom above manifested in the form of adaptive behavior to the environment has an important role in seeking disaster risk reduction. The original knowledge that has been applied in the community has a positive impact while facing and responding to a disaster that comes. The original knowledge that people have is extraction from a variety of experiences that are hereditary from their ancestors when practicing disaster [7]. Education based on local wisdom is an education model that has high relevance for life skills by relying on the local potential of each region. This education teaches students to absorb and even manage culture that comes from within or outside their area. With local wisdom, students will adjust to the view of life around them. Through local wisdom disaster mitigation is an example of education that has a strong connection to disaster risk reduction skills [23]. The purpose of education is to be able to educate and prepare students who are able to contribute to society, the nation, and the country. Therefore every student needs to know the characteristics of their area and their needs. Geographically, disasters have different patterns and characteristics in dealing with disasters, so it requires different patterns of management. This forms the attachment of the power of local wisdom to encourage the importance of local wisdom in an area.

Table 1. The Importance of Local Wisdom for Disaster Mitigation

No	Identification of the Foundation of the Importance of Local Wisdom
1	The practices and strategies of natives contained in local wisdom, that have proven to be very valuable in dealing with natural disasters can be transferred and adapted by other communities facing similar situations
2	The integration of local wisdom into existing policies will encourage community participation to mutually empower existing capacities as the main step in disaster risk reduction activities
3	Information on local wisdom of an area can help provide knowledge regarding the cultural values of disaster management in the area.
4	Provision of policy information in direct practice can be a good example as a step in socializing disaster risk reduction

Source: Data Analyzed, 2020 [6]

In Indonesia, exploring local wisdom to minimize disaster has occurred in some regions. Each region has a variety of local knowledge with names and traditions that the use is not the same. According to local communities, local wisdom is more effective in shaping resilience about disasters compared to appeals from the government especially if education is carried out by community leaders. With various examples of forms of local wisdom in disaster mitigation above, before it is applied in classroom learning it is necessary to identify the forms of local wisdom according to the area or location of the student located. Furthermore, teachers and students together identify local wisdom to minimize disaster mitigation.

3.3 Disaster Mitigation Curriculum

Disaster mitigation is an effort to reduce disaster, both through physical development as well as awareness and capacity building to face the threat of disaster (Article 1 paragraph 6 of PP No 21 of 2008 concerning the Implementation of Disaster Management). So far, the disaster mitigation efforts undertaken by Indonesia, in terms of legislation is to have a Disaster Management Act namely Law Number 24 Year 2007 and Government Regulation Number 21 Year 2008, Regarding Disaster Management Arrangement.

Disaster mitigation in the environment Education has its own challenges. The challenge is how disaster education programs can encourage people to update information, increase risk, maintain awareness, and prepare appropriately for future disasters. As a step to develop various teaching and learning approaches, it is hoped that they can achieve the main goal of disaster risk reduction by making people have a culture of disaster preparedness [29]. To reach beyond the role of academics is also needed. Academics as one of the main stakeholders in disaster risk reduction, play a more proactive role in disaster risk reduction engagement programs and the wider community. [30]

School in Indonesia has not been fully applied disaster education besides integration with other subjects. This condition is also carried out when the material is taught. The importance of disaster education is a necessity for disaster mitigation involving educational institutions. Children who are also likely students to be the individuals who feel the most impact after a disaster. Crisis will occur in children, so that it becomes the focus in disaster management.

Disaster management education is a long-term plan for sustainable development. Sustainability Education is a lifelong learning process that leads to citizens who are informed, have creative, scientific and socially literate problem-solving skills, and a commitment to engaging in responsible individual and cooperative action. On the other hand, disaster is one of the biggest obstacles to achieving development goals. Thus, reducing disaster risk and its impact has become an important development issue. DRR (disaster risk reduction) has a natural synergy especially for the Asia-Pacific region. DRR has shown that effective education is only possible when education is able to reach all aspects including the culture of local wisdom. Learning by connecting schools and communities is considered important and effective. [31]

One of challenge in sustainable development which is currently receiving serious attention, is related to disaster management. Almost every region, especially in Indonesia, has the potential for disaster. Therefore, by incorporating subject matter into the curriculum and becoming compulsory subjects in schools, it can support the reduction of disaster risk. The curriculum designed is based on local wisdom. Thus, it is easily understood by students.

3.4 The Application of Local Wisdom Learning Through Curriculum Integration

In an effort to reduce disaster risk in schools, the Minister of National Education issued a letter with Number 70a / MPN / SE / 2010 which contains: 1) empowering the role of institutions and communities in schools, 2) integrating disaster risk reduction (DRR) into the school curriculum, 3) formed collaborations with various parties to support disaster risk reduction in schools. This potential disaster will later be included in the learning curriculum in schools (2013 Curriculum) which is called the disaster mitigation curriculum. The curriculum in Indonesia currently focuses on four core competencies, namely: 1) spiritual attitude; 2) social attitude; 3) knowledge; and 4) skills. of the four competencies, the values of local wisdom about disaster mitigation are applied..

First, The teacher maps basic competencies in each meeting and indicator and looks at potential disasters in the area students live in and local wisdom in disaster mitigation. In disaster management, it is carried out using local wisdom using the KIDA model (Knowledge, Interest, Desire, Action). The knowledge is about the hazards and risks of disaster, interest (curiosity in the event of a disaster and pre-disaster preparedness, Desire (actively participating in disaster preparedness, action (concrete steps taken when a disaster occurs) [5].

There are some different approaches to integrate disaster risk reduction. Selby and Kagawa identified that 1) an approach is driven by textbooks, 2) project approach, 3) competency-based approach centrally, 4) special subjects developed centrally, 5) a symbiotic approach and an 'special approach'[32].

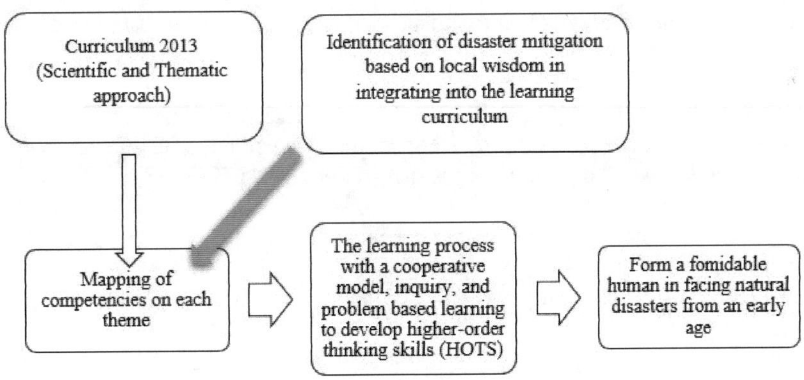

Figure 1. Disaster Mitigation Education Model Based On Local Wisdom

Based on the findings of Aiyono and Kanegae, it proves that students can instill a spiritual attitude about local wisdom in disaster mitigation. From the two schools, the results show that the role of religion in a natural phenomenon is very important. For countries like Indonesia, in their efforts to develop disaster risk reduction materials, most of them believe that disasters that occur are caused by God's punishment and focus more on self-reflection. In fact, this belief is significant in efforts to reduce disaster risk (DRR) that are effective personally so as to build preparedness when a disaster occurs suddenly. It is therefore very important to develop disaster knowledge from a religious perspective [33]. With local wisdom, social attitudes are formed, students become more aware and concerned about the importance of protecting the environment, disciplined behavior, and how to react and resilience that is built when a disaster occurs. From the value of knowledge, learning disaster mitigation through local wisdom is one form of contextual learning.

Local wisdom is a comprehensive lesson if it is integrated into the learning curriculum so that it can detect a region's disasters. In this effort, students, teachers, the community can be educated tp recognize the disasters that occur around their area. Therefore, this kind of practice needs to be developed to increase understanding in disaster management efforts. The existence of a curriculum based on local wisdom can reduce the risk of disasters and people are aware of the relationship with nature and culture. Because a community that is resilient to disasters is a society that is tolerant of nature and understands the nature it occupies. [6] Teachers must be able to design learning by integrating various appropriate topics, one of which is the community environmental education approach. Teachers can use materials related to the environment or learning methods in accordance with the theme of disaster mitigation. [34].

4. Conclusion

In this article, I argue that the development of local wisdom-based disaster mitigation education in schools in Indonesia give alternative in efforts to build awareness and skills in dealing with disasters. Local wisdom in the form of local community habits that are developed in daily life such as Smong, Jogo tonggo, Dayak tribe, Karampuang custom, Baduy community, etc. are proven to play a role in minimizing the occurrence of a disaster. Furthermore, Local wisdom is integrated into the thematic-based disaster mitigation curriculum and scientific approach through the process of identifying competency standards in the 2013 curriculum, such as spiritual attitude, social attitude, knowledge, and skills. The form of curriculum integration is continued into the learning process with cooperative models, inquiry, and problem based learning to develop higher-order thinking skills (HOTS). Learning by emphasizing cultural values provides the benefit of forming strong students. Thus, they are able to behave quickly, precise, and accurate when facing disasters.

Acknowledgments
Thank you to Yogyakarta State University and Supervising Lecturers for their support in this research.

References

[1] H. Hu, T. Lei, J. Hu, S. Zhang, and P. Kavan, "Disaster-mitigating and general innovative responses to climate disasters: Evidence from modern and historical China," *Int. J. Disaster Risk Reduct.*, vol. 28, no. August 2017, pp. 664–673, 2018, doi: 10.1016/j.ijdrr.2018.01.022

[2] D. I. B. I. Badan Nasional Penanggulangan Bencana, "Data Informasi Bencana Indonesia." [Online]. Available: https://bnpb.cloud/dibi/.

[3] Direktorat Pendidikan Khusus Ditjen Paud Dikdas Dikmen, "Sekolah Tanguh Bencana Untuk Mendukung Ketangguhan Keluarga," Jakarta, 2020.

[4] P. A. Kamil, S. Utaya, Sumarmi, and D. H. Utomo, "Strengthen Disaster Preparedness for Effective Response on Young People through Geography Education: A Case Study at School in the Tsunami Affected Area of Banda Aceh City, Indonesia," *IOP Conf. Ser. Earth Environ. Sci.*, vol. 412, no. 1, 2020, doi: 10.1088/1755-1315/412/1/012016.

[5] Y. T. Rajib Shaw, Koichi Shiwaku, 2011 *Community, Environment and Disaster Risk Management*..

[6] M. Desfandi, "Urgensi Kurikulum Pendidikan Kebencanaan Berbasis Kearifan Lokal di Indonesia [The urgency of disaster education curriculum based on local wisdom in Indonesia}," *Sosio Didakt. Sos. Sci. Educ. J.*, vol. 1, no. 2, 2014.

[7] P. Eka and E. G. U. Utama, "Pendidikan Mitigasi Bencana Di Sekolah Dasar (Sebuah Kajian Analisis Etnopedagogi)," *J. Pendidik. Dasar Indones.*, vol. Volum 2 No, no. September, pp. 18–24, 2017.

[8] N. Syaodih, 2009 *Metode Penelitian Pendidikan*. Bandung: PT Remaja Rosdakarya,.
[9] *Undang-Undang Republik Indonesia Nomor 24 Tahun 2007 Tentang Penanggulangan Bencana*. Indonesia, 2007.
[10] R. K. Bhandari, 2014 *Disaster Education and Management*. London: Springer New Delhi Heidelberg New York Dordrecht London,.
[11] S. Siti Irene Astuti D, *Jurnal Dialog Penanggulangan Bencana*, vol. 1, no. 1. 2010.
[12] Retno Wihyanti, "Analisis Inovasi Pendidikan Kebencanaan di Sekolah di Indonesia," *J. Chem. Inf. Model.*, vol. 53, no. 9, pp. 1689–1699, 2019, doi: 10.1017/CBO9781107415324.004.
[13] K. Ibrahim, E. Emaliyawati, and D. I. Yani, "Pelatihan dan Simulasi Penanggulangan Bencana Bagi Masyarakat," vol. 3, no. 1, pp. 27–38.
[14] Anwar Kurniadi; Fauzi Bahar, "The Review Of Disaster Preparedness School Program," *J. Pertahanan*, vol. 6, no. 1, pp. 46–58, 2020.
[15] A. Hakim, D. N. Inten, and D. Mulyani, "The Literation of Disaster Mitigation for Early Childhood," vol. 409, no. SoRes 2019, pp. 287–290, 2020, doi: 10.2991/assehr.k.200225.060.
[16] R. R. Rahadian Kurniawan, Affan Mahtarani, "GEMPA: Game Edukasi sebagai Media Sosialisasi Mitigasi Bencana Gempa Bumi bagi Anak Autis," *J. Pendidik. dan Pembelajaran Univ. Negeri Malang*, vol. 6, no. 1, pp. 115–120, 2013, doi: 10.12928/kesmas.v7i1.1050.
[17] T. Shaw and F. Kaneko, *International Cooperation: Interventions in Enhancing Disaster Education in Turkey*. 2016.
[18] K. P. B. INDONESIA, *Kerangka Kerja Sekolah Siaga Bencana*. Jakarta, 2011.
[19] A. Fauzy, L. Chabib, and A. S. Putra, "Tujuan Pembangunan Berkelanjutan Untuk Penanggulangan Bencana," *Asian J. Innov. Entrep.*, vol. 4, no. 3, pp. 171–180, 2019.
[20] B. Khazai, J. Anhorn, and C. G. Burton, "Resilience Performance Scorecard: Measuring urban disaster resilience at multiple levels of geography with case study application to Lalitpur, Nepal," *Int. J. Disaster Risk Reduct.*, vol. 31, pp. 604–616, 2018, doi: 10.1016/j.ijdrr.2018.06.012.
[21] Z. Wu and M. Hu, "Neotectonics, active tectonics and earthquake geology : terminology, applications and advances," *Journal of Geodynamics*, vol. 127. pp. 1–15, 2019, doi: 10.1016/j.jog.2019.01.007.
[22] S. S. Hutagalung and H. Indrajat, "Adoption of local wisdom in disaster management in Indonesia," *Int. J. Sci. Technol. Res.*, vol. 9, no. 3, pp. 48–52, 2020.
[23] R. Afrian, J. Hariadi, B. Akob, and Z. R. Islami, "Local Culture Inventory for Disaster Mitigation Learning," *IOP Conf. Ser. Earth Environ. Sci.*, vol. 412, no. 1, 2020, doi: 10.1088/1755-1315/412/1/012017.
[24] P. L. Erisandi Arditama, "Jogo Tonggo: Membangkitkan Kesadaran Dan Ketaatan Warga Berbasis Kearifan Lokal Pada Masa Pande mi Covid-19 Di Jawa Tengah," vol. 8, no. 1, pp. 35–46, 2020.
[25] I. Saleh, I. S. W. Atmaja, and R. M. Syahadat, "Prohibition in Baduy Dalam Community: Soil and Water Conservation Perspective," vol. 429, no. Icasseth 2019, pp. 180–182, 2020, doi: 10.2991/assehr.k.200402.040.
[26] N. W. & I. D. K. W. Dirhamsyah, Diar Budi Utama and Program, "Local Wisdom and Participation of the East Kalimantan Dayak Alliance in Dealing with Forest and Land Fire Disasters," *Perspektif*, vol. 9, no. 2, pp. 314–321, 2020.
[27] M. D. Siti Rahayu, Abdul Wahab Abdi, "Penerapan Metode Pembelajaran Discovery Learning Berbasis Smong Untuk Meningkatkan Hasil Belajar Mitigasi Bencana Siswa Kelas Xi Ips Sma Negeri I Simeulue Tengah," *J. Ilm. Mhs. Pendidik. Geogr. FKIP Unsyiah*, vol. 21, no. 1, pp. 1–9, 2020.
[28] E. Maryani and A. Yani, "Kearifan Lokal Masyarakat Sunda Dalam Memitigasi Bencana Dan

Aplikasinya Sebagai Sumber Pembelajaran Ips Berbasis Nilai," *J. Penelit. Pendidik.*, vol. 14, no. 2, 2016.

[29] Z. Musthofa and S. Indartono, "Disaster Mitigation Curriculum-Based on Local Wisdom to Support Sustainable Development Programs," vol. 398, no. ICoSSCE 2019, pp. 97–102, 2020, doi: 10.2991/assehr.k.200130.021.

[30] Y. F. Chen and A. Adefila, "Enhancing school safety through university engagement in DRR education," *Int. J. Disaster Risk Reduct.*, vol. 44, p. 101386, 2020, doi: 10.1016/j.ijdrr.2019.101386.

[31] R. Shaw,2014, *Disaster Risk Reduction Education: Issues and Challenges*..

[32] D. Selby and F. Kagawa, 2012, *Disaster risk reduction in school curricula: case studies from thirty countries*.

[33] H. K. Wignoyo Adiyoso, "Efektifitas Dampak Penerapan Pendidikan Kebencanaan di Sekolah terhadap Kesiapsiagaan Siswa Menghadapi Bencana Tsunami di Aceh , Indonesia," 2013.

[34] S. Ramadhan, E. Sukma, and V. Indriyani, "Environmental education and disaster mitigation through language learning," *IOP Conf. Ser. Earth Environ. Sci.*, vol. 314, no. 1, 2019, doi: 10.1088/1755-1315/314/1/012054.

The potential of river discharge at the peak of the dry season in some disaster-prone area of Merapi Volcano

Kharisma, MB Prasetyo, AF Rosa, I Fitrianingrum, MAI Wardoyo, A Ashari

Department of Geography Education, Faculty of Social Sciences,
Universitas Negeri Yogyakarta, Yogyakarta, Indonesia

kharisma.2019@student.uny.ac.id
muhammad3168fis.2019@student.uny.ac.id
amaliafadila.2019@student.uny.ac.id
isnafitrianingrum.2019@student.uny.ac.id
muhammadasrori.2017@student.uny.ac.id
arif.ashari@uny.ac.id

Abstract. With a high level of volcanic activity and many people living in the disaster-prone area, sustainable disaster management in the Merapi Volcano area is still very much needed in the future. One of the needs in building disaster management is ensuring water resources available during the pre-eruption period. So far, springs are the primary source of water that has been widely used by residents in the volcanic foothills, which are prone to eruption disasters. However, at the time of the eruption, many springs were not functioning so that alternative water sources were needed. River water can be an alternative water source, and a study of river water's potential is required to ensure its feasibility. This study was conducted on the south to the southwest side of Merapi Volcano to analyze the quantity and quality of river water as an alternative water source. Data collected through observation, literature study, and document search. Data analysis performed using descriptive analysis, matching analysis, and statistical analysis using independent sample t-test and simple linear regression. The study results show that seven rivers in the south to the southwest slope can be used as water sources because they are not lava flows. These rivers have varying discharge and water quality. There is no difference in discharge and water quality between the top and the bottom of the volcano foot. When dry season is on the peak, three rivers have no flow because some springs locations are at the volcano's bottom. Overall, based on these findings, it can be emphasized that the use of water sources is needed to do at the closest point to the evacuation barracks. However, a water treatment installation needs to be provided in the evacuation barracks so that the river water is suitable for consumption.

1. Introduction

Natural disasters due to volcanism are one of the most threatening types of disasters in Indonesia, and this is inseparable from the existence of 129 active volcanoes [1], with their activities that continuously. Data from the Center for Volcanology and Geological Disaster Mitigation [2] shows that in the first quarter of 2020, there were at least 13 volcanoes whose activity was above normal. The high potential for volcanic hazard in Indonesia is inseparable from the Indonesian archipelago geological setting, located in the subduction zone of the world's major plates [3]. The volcanic eruption disaster that has happened a lot from time to time harms a large area and affects many

Content from this work may be used under the terms of the Creative Commons Attribution 3.0 licence. Any further distribution of this work must maintain attribution to the author(s) and the title of the work, journal citation and DOI.
Published under licence by IOP Publishing Ltd

residents' lives. This condition requires adequate preparation in various aspects in order to form excellent and sustainable disaster management.

Based on the historical record of eruptions, 47% of Indonesia's eruptions occurred on the island of Java [3]. In Central Java, there is the Merapi Volcano, which is widely known as one of the most active volcanoes in the world [4]. A short return period of eruptions is a sign of Merapi's volcanic activity. Until now, Merapi Volcano has experienced eruptions more than 80 times, with a return period of eruptions occurring between 1 to 15 years, a massive eruption every 150-500 years, and a maximum interval of 12 years [5]. On the other hand, many people live in areas prone to the eruption of Merapi. After the big eruption in 2010, the Merapi hazard area's growth rate was even higher than the average national population growth [6]. Data from the Global Volcanism Program published by the Smithsonian Institution [7] shows that the population in a 30 km radius from Merapi's summit in early 2020 reached 4,348,473 people. The large number of people who live in this active volcanic area requires sustainable and innovative disaster management to reduce risks in the future.

One of the obstacles encountered in disaster management is handling refugees during the disaster emergency period, including the provision of clean water resources. Springs are the main water source that is widely used by residents in the volcanic foothills, which are prone to eruptions. Research conducted by Aurita and Purwantara [8] and Ratih et al. [9] showed that the springs in the southern to western foothills of Merapi volcano are good quality and sufficient quantity. Meanwhile, Sudarmadji et al. [10] showed that in Pakem and Turi Subdistricts alone, there are at least 37 springs used by the community to meet water needs. However, the problem is that during the crisis period, many of the springs died, or the drainage installation was unusable. The location of the springs is also mostly far from the evacuation barracks. As a result, the fulfillment of clean water sources is highly dependent on supplies from other regions. Of course, it requires a lot of money, and mobilization is not easy to provide clean water during a disaster emergency. Experience in Srumbung Subdistrict during the eruption crisis period in 2010; every household received a water reservoir that was refilled by waiting for the supply of water sources from other areas.

The water sources originating from rivers is necessary to meet water needs in an emergency of an eruption, namely by starting to carry out management during typical situations or signs of increased volcanic activity. Merapi volcano, a stratovolcano landscape, has many rivers that flow perennial, mostly the young south to the southwest side. In this area, many river channels do not originate from Merapi's volcanic cone so that it does not become a lava flow path when it erupts. This existing potential can utilize optimally. For that, of course, a study on river water sources' potential is needed, both in quality and quantity. The big eruption in 2010 and several eruptions in early 2020 occurred at the beginning of the rainy season, where the flow rate in the rivers was large enough to be used as an alternative water source. However, at the beginning of the dry season, many rivers have reduced water discharge, so it is necessary to review their availability and suitability. Further studies are necessary to do after the quality and quantity requirements of clean water are fulfilled to determine which areas can be used for optimum utilization of river water spatially.

The southern to western slopes of the Merapi volcano is heavily affected by eruptions until now. In the past, eruptions had much impact on the southwestern side of Merapi [5], shown by the deposition of lava material that often occurs in the major rivers of this region [4]. Meanwhile, the maps of eruption-prone areas and eruption-affected areas published by the Geological Agency show that the southern slopes were the areas that were directly affected by the 2010 eruptions. After the big eruption in 2020, this area is also still the part most prone to experiencing the impact of disasters. For example, during the eruption on March 3, 2020, monitoring of the Himawari Weather Satellite Image showed that volcanic ash due to the eruption was detected moving towards the Southwest - Southeast. On the other hand, many people are living in this area. Data from BPS Magelang and Sleman Regencies [11], [12], [13], [14], shows that Cangkringan, Pakem, Turi, and Srumbung Subdistricts, whose areas are part of disaster-prone areas, in 2019 reached 152,390 people. With this condition, the provision of alternative water resources to support disaster emergency management is of greater urgency for the Merapi Volcano's southern slopes.

This paper provides alternative information about the characteristics of river discharge in the disaster-prone area of Merapi Volcano, at the beginning of the dry season. Thus, this paper has two specific goals. First, this paper describes the quantity of river water and its potential to meet per capita water needs in a disaster emergency. Second, this paper describes water quality in terms of physics and its suitability for consumption, based on water quality parameters.

2. Methods

2.1. Data collection and analysis

This research is a descriptive study with a geographic approach, namely, a spatial approach that discusses and analyzes the problem carrying by emphasizing geographic themes, especially location, place, and region. The data collected is river discharge and water quality, which includes temperature, pH, dissolved oxygen (DO), total dissolved solids (TDS), electric conductivity (EC), and Total Coliform. Data collection carrying by observation, literature study, and documentation. Determining the sample's location in field observations was carried out by purposive sampling, namely the upstream, which is in the transitional area of the mountain foot and the volcanic foot plain, and the downstream part, which is entirely in the form of the volcanic foot plain. The upstream part is included in the disaster-prone area (DPA) II, while the downstream part includes non-DPA. This location selection aims to compare the potential of river water at the top and bottom and consider the recommendation of the location for the use of water resources. In the observation process, measuring the river discharge is by the velocity-area method using a buoy (eq 1 and eq 2), or a slope-area method with the manning formula (eq 3). Water quality was measured in the field using a multiparameter Hanna meter, except for Coliform, analyzed in the laboratory. Data collection is also carried by tracing documents to determine the population, land use, characteristics of springs, which are the source of discharge, and characteristics of the research area's physical environment. Table 1 shows the relationship between data types, data collection methods, and the instrument used or data sources.

The formula used in calculating the discharge using a float is as follows:

$$Q = V \; x \; A \; x \; K \text{ .... (eq 1)}$$

Where Q is discharge (m^3/ s), V is the mean flow velocity (m / s), A is wet cross-sectional area (m^2), and K is buoy coefficient. The value of K is determined by the Francis formula, namely:

$$k = 1 - 0{,}116(\sqrt{1 - \alpha} - 0{,}1) \text{ .... (eq 2)}$$

Where α is the part of the submerged float divided by the depth of the water in centimeters, or expressed in terms of the depth of the shaft (h) divided by the depth of water (d)

Calculation of discharge is also carried out using the slope -area method where the flow velocity is calculated using the manning formula as follows:

$$V = \frac{1}{n} R^{\frac{2}{5}} S^{\frac{1}{2}} \text{ .... (eq 3)}$$

Where V is flow velocity (m / s), R is hydrologic radius (m), S is hydrologic gradient, n is the manning roughness coefficient. The value of R is determined by dividing the cross-sectional area of wet / A (m^2) by the wet perimeter (m)

Table 1. The relationship between data types, data collection methods, and instrument used or data sources

Variable	Data collection methods	Instruments / data sources
River discharge	Observation	Floating method measurement equipment: Buoy, roll meter, yallon, stopwatch
		Measuring equipment for manning method: Water hose, roll meter, yallon
Water quality	Observation	Multiparameter Hanna meter
		Sample bottle to bring samples to the laboratory
Characteristics of discharge source springs	Literature Study	Ratih et al. [9] , Ratih et al. [15], Sudarmadji et al. [10]
Physical environmental conditions of the study area	Literature study	Sutikno et al. [16], Purwantara et al. [17]
Population	Documentation	Datafrom the Central Bureau of Statistics of Sleman Regency and Magelang Regency [11], [12], [13], [14]
Land use	Observation	Digital camera, compass
	Documentation	Indonesian Topographical Mapsheet Kaliurang, Pakem, Sleman and Muntilan

Analyzed data is carried out by using descriptive analysis methods, matching analysis, and statistical analysis; descriptive analysis to provide a detailed description of the discharge conditions and water quality in the study area. To support this descriptive analysis, we used statistical analysis with independent sample t-test and simple linear regression. An independent sample t-test compares the quantity and quality of river water between the top and the bottom. We are using Microsoft Excel and calculations (eq 4 and 5) to perform this analysis also for simple linear regression to determine the relationship between water quality variables related to each other. Matching analysis is used to determine whether the water from the rivers in the study area is suitable to be used as a source of clean water fulfillment. The analysis is comparing the water quality of the research results with the international water quality standards from WHO, as well as the national drinking water quality standards based on the Regulation of the Minister of Health of the Republic of Indonesia No. 492 of 2010 [18] and the Republic of Indonesia Government Regulation No. 82 of 2001 [19] (Table 2). Matching analysis is also used to determine whether river water discharge can meet the per capita needs of refugees who accommodated in refugee barracks.

$$t = \frac{[\bar{x} - \bar{y}]}{\sqrt{\frac{(\Sigma x^2/n_x) - \bar{x}^2}{n_x - 1} + \frac{(\Sigma y^2/n_y) - \bar{y}^2}{n_y - 1}}} \text{............... (eq 4)}$$

where x and y are the means of the two samples, n_x and n_y are the sizes of two samples

$$df = \frac{(SX+SY)^2}{\frac{SX^2}{n_x-1} + \frac{SY^2}{n_y-1}} \text{................................ (eq 5)}$$

where $SX = \frac{(\Sigma x^2/n_x) - x^2}{n_x-1}\frac{(\Sigma x^2/n_x) - x^2}{n_x-1}$ and $SY = \frac{(\Sigma y^2/n_y) - y^2}{n_y-1}\frac{(\Sigma y^2/n_y) - y^2}{n_y-1}$

Table 2. Standard standards of drinking water quality standards used in the matching analysis

No.	Variable	Indonesian National Standard	Standard International Standard
1.	Temperature	± 3 ° C [18]	30 ° C (WHO) [20]
2.	TDS	500 mg / l [18]	500 mg / l (WHO) [20]
3.	DHL	-	400 μS / cm (WHO) [20]
4.	DO	> 6 mg / l [19]	4.0 mg / l [21]
5.	pH	6.5 - 8.5 [18]	6.5 - 8.5 (WHO) [20]
6`.	Total coliform	0 / 100ml [18]	0 MPN / 100ml (WHO) [20]

2.2. Research Area

We do this research on the south to the southwest side of the Merapi Volcano. Administratively, most of the research areas are in Sleman Regency, Yogyakarta Special Region, which includes Cangkringan, Pakem, and Turi Subdistricts. A small number of others are in the Magelang Regency area of Central Java Province, namely in the Srumbung Subdistrict. The area of the study is 116,66 km^2 (Figure 1). Geologically, the research area is composed of volcanic material from Young Merapi and Old Merapi. Old Merapi material occupies a minimal area, namely Turgo Hill and Plawangan, which, according to Newhall et al. [22], is the oldest part of Merapi. Meanwhile, the Young Merapi material dominates in an extensive area. Based on the Geological Map of the Yogyakarta Sheet, known that the volcanic deposits of Young Merapi include Young Merapi Volcano Deposits (Qmi), Avalanche Deposits from Hot Clouds (na), and lava domes and melting (d). The abundance of this Young Merapi material is because the south to west slopes is indeed the youngest part of Merapi [22]. The youngest material of pyroclastics and lahars found in the Southwest [23].

Geomorphologically, the research area located at the volcano's foot as part of the stratovolcano geomorphology of Merapi, which includes volcanic cranes, volcanic cones, volcanic slopes, volcanic foothills, lava fields, lava fields, and volcanic foot plains [24], [25]. This geomorphological characteristic affects the amount of rain in the form of orographic rain. At the foot of the volcano, rainfall reaches 1550 mm. High rainfall affects the hydrological system, which is characterized by many rivers and groundwater with large discharge. The indicate of groundwater potential is shown by potential aquifers with discharge reaching 0.005-0.01 m^3/ s [16]. Apart from being influenced by the amount of rain, high groundwater potential is influenced by the material's characteristics in this area, which has high infiltration capability [17].

The population in the study area in 2018 reached 60,337 people spread across nine villages. The highest population was in Wukirsari Village, Cangkringan Subdistrict, with 10,291 people, while the lowest was in Kemiren Village, Srumbung Subdistrict, which was 1229 people. The highest population density of 1210 people / km^2 was found in Donokerto Village, Turi District, while the lowest population density founded in Kemiren Village at 199 people/ km^2. Land use in this area varies, consisting of settlements, mixed gardens, moor, rice fields. There are natural land uses in the form of forests and shrubs on the border with the volcanic foot area. This region is a producer of agricultural commodities, including rice, corn, vegetables, and fruits.

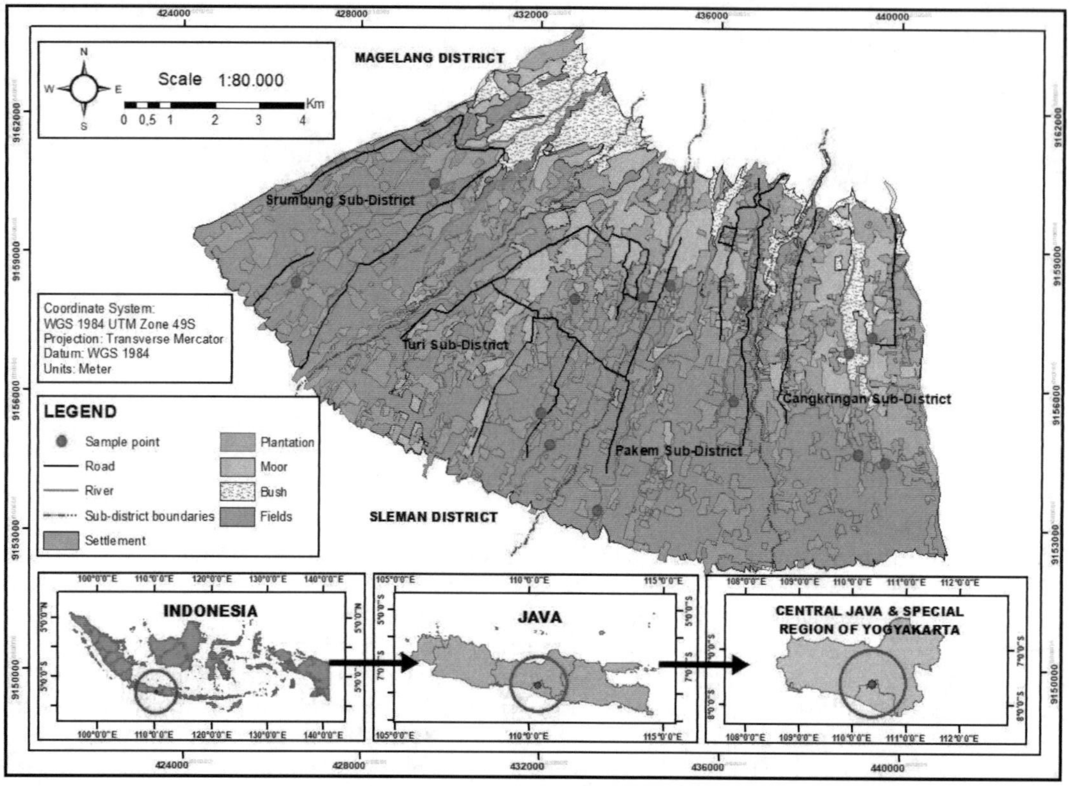

Figure 1. The study area

3. Result and Discussion

One of the characteristics of Merapi Volcano as a stratovolcano is the many river valleys that develop in a radial flow pattern. Most of the rivers are part of the Opak Watershed in Merapi's volcanic foothills, while a few are part of the Progo Watershed. There are seven rivers in this area that do not head directly to Merapi's summit, namely Pelang, Sempor, Degong, Tangkil, Opak, Tepus, and Batang. These rivers are perennial even though the amount of flow is decreasing in the dry season. Sutikno et al. [16] explained that these rivers have varying discharge and quality. River discharge and water quality are possible things as alternative water sources in an emergency of an eruption. Intake of river water for this utilization is certainly not carried out at the time of the eruption, but during regular activity, until there are signs of increased volcanic activity. During this period, river water resources can be collected and processed in the evacuation barracks to use in emergencies. The study results on the quantity and quality of river water in the study area describe as follows.

3.1. River water discharge and its potential to fulfill per capita water needs

In this study, we measured the discharge of seven rivers. The discharge of each river is measured twice, namely in the upstream and downstream areas. The upstream area is included in the disaster-prone area (DPA) II, while the downstream area includes in the non-DPA. Based on field observations in the upstream area, we know that there are three rivers which at the peak of the dry season do not have flow rates, namely the Tepus, Pelang, and Opak rivers. Meanwhile, in the downstream area, all rivers have flow rates. Based on these conditions, this study's discharge measurements were all carried out at 11 locations, four locations in the upstream area, and seven locations in the downstream area.

The results of measurements at 11 locations showed an average discharge of 63.64 liters/ second. The most massive discharge found in the downstream of the Sempor River, which is 230 liters/ second, while the smallest discharge is found in the Tangkil River downstream, which is 9 liters/ second. In the upstream area of the four rivers, the average flow rate is 36 liters/ second. Meanwhile, in the downstream part, the average discharge is 91.28 liters/ second. Overall, the river's amount of discharge in the downstream area is greater than the upstream area. In other words, rivers in the downstream area have the potential for better water quantity than upstream. However, the comparative test conducted shows that statistically, there is no significant difference between the mean river discharge in the upstream and downstream areas. Sometimes, the debit in the upstream area is more significant than the downstream area due to river water used for irrigation of agricultural land and plantations. Part of the discharge is channeled for agricultural land irrigation, thereby reducing the discharge in the downstream area. The results of the discharge measurements shown in Table 3 below.

Table 3. River discharge at 11 sample locations in the study area

River name River	discharge (liter / second)	
	Upstream	Downstream
Batang	113	61
Tangkil	61	9
Sempor	38	230
Degong	40	39
Tepus	0	170
Pelang	0	20
Opak	0	110

Source: Primary data (2020)

The size of the river discharge in the area; this is inseparable from rain, topography, geology, vegetation conditions, and humans [26]. Topographic and geological factors appear to play an essential role in determining the size of the discharge. Three rivers that are dry in the upstream area but still have flow in the downstream area, possibly because the emergence of springs is at the river's bottom. This condition cannot separate from geological and topographic factors. In general, all rivers in this area have perennial flows, although they experience reduced discharge during the dry season's peak. Characteristics of rock types in this region, which are material from the volcanic activity of Young Merapi, function as productive aquifers [16]. A large amount of groundwater stored in the aquifer can then appear as springs that supply river flow so that it still has discharge at the dry season's peak.

The volcanic foot plain's topography supported the type of rock that can store groundwater, which is not too steep. The slope of the slope, which includes in the sloping-sloping category, causes the surface flow not to flow quickly, thereby increasing the chance of infiltration, and in the end, increasing the potential for groundwater, which provides a supply base flow for the river. On the other hand, the vegetation factor is also one factor that influences the amount of discharge. Vegetation cover will cause more and more water to disappear, either by evapotranspiration or through infiltration, so that it will reduce run-off, which can affect river discharge [27]. Around the river the flow rate is measured, there is quite a lot of vegetation, especially in the upstream river, so that the river flow discharge is not so large.

An important question related to the potential of river water discharge as an alternative source of drinking water in a disaster emergency is: "Can the available river discharge meet the per capita water needs of refugees in the evacuation barracks?". Referring to the Regulation of the Minister of Health of the Republic of Indonesia Number 28 of 2019 concerning the Recommended Nutritional Adequacy Rate for the Indonesian People [28], the standard of water needs per capita varies between people. The need for water between adults and children is different, so are men and women's needs. Since the water adequacy analysis in this study was carried out to meet emergency disasters' needs, the figure for water needs per person used is a maximum figure of 2.5 liters per day.

Furthermore, the water demand figure is multiplied by the refugee barracks' capacity to find out the water needs of all the refugees in the evacuation barracks. There are refugee barracks in the study area with capacities ranging from 250, 300, to 500 people. In this calculation, maximum capacity is 500 people, so that the water need for evacuation barracks is 1,250 liters per day. The river, which has a discharge of 1,250 liters per day, means that in quantity, it has the potential to meet the needs of drinking water in the evacuation barracks closest to the river. Of course, this depends on the water storage capacity available in the evacuation barracks. By paying attention to the river discharge presented in Table 3, all rivers in the study area can be used as a good water source for the refugees in the evacuation barracks. The downstream section should be done the utilization of water by considering the yearly flow rate in the downstream area and the location of the refugee barracks in the downstream area so that it is more efficient to use in terms of distance and affordability water.

3.2. The physical quality of river water and its feasibility as an alternative to emergency drinking water

In this section, we elaborate on the results of measurements of several water quality parameters in seven rivers in the study area. Measurements were made at 11 locations, namely 4 locations upstream and 7 locations downstream. The upstream only allows for four measurement locations because, in the three rivers, namely the Tepus, Pelang, and Opak rivers, there is no flow at the dry season's peak. Meanwhile, in the downstream area, there are flows in all rivers. Water quality parameters discussed in this study are still limited to physical parameters of water quality, namely water temperature, dissolved oxygen (DO), total dissolved solids (TDS), and electric conductivity (EC), as well as one chemical parameter, namely water pH. There are variations in various water quality parameters between these rivers. Also, variations founded between the upstream rivers in the transition between the volcanic foot and the volcanic footplate and the river downstream are at the center of the volcanic footplate, mostly occupied by residents.

Water temperature is a parameter of water quality in the study area with small variations (Figure 2). Measurements made at 11 locations showed the highest temperature of 24.93°C was found in the downstream of the Opak River, while the lowest temperature was 22.42°C in the downstream of the Sempor River. The average water temperature in the study area was 23.61°C. The average water temperature difference between the upstream area (23.37°C) and the downstream area (23.74°C) was also minimal. The results of statistical tests with the independent sample t-test also showed that there was no difference between the average upstream and downstream water temperatures. The type of land use in all sample locations, which are dominated by mixed gardens, is a factor that might have a strong influence on the small variation in river water temperature in the study area. As explained by Asdak [29], clearing of vegetation along riverbanks can cause more sunlight to penetrate the surface of the water, thereby increasing the water temperature. In the study area, the characteristics of mixed gardens with relatively similar dense vegetation in all sample locations caused river water temperatures to tend to below. Concerning the presence of vegetation in riparian areas (streamside), the research results of Swartz et al. [30] showed that a wide gap between vegetation affects increasing river water temperature. However, based on research conducted in mountainous areas, the effect of the gap between this vegetation is not significant. Meanwhile, the width of the river also affects the increase in river water temperature. Referring to this information, in the study we conducted, the type of land use and vegetation density also affected river water temperature, but the effect was not significant. The water temperature in the downstream Tepus and Pelang where the land use is in rice fields, provenly the water temperature is relatively the same as other locations.

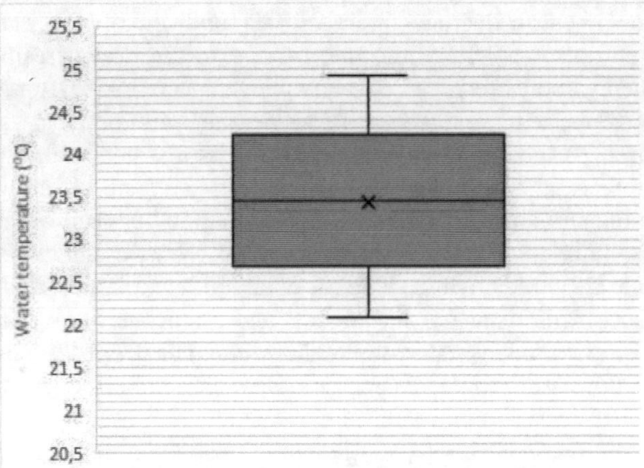

Figure 2. Boxplot for water temperature in the study area

The relatively narrow height range of only 270 meters also determines the relatively invariable river water temperature. With a narrow coverage area, relatively homogeneous land use and landform, and insignificant altitude variations, the insolation variables and albedo values in this area are also relatively uniform. Martinus et al. [31] explained that the lower the altitude, the higher the water temperature. Following this information, this study's results also indicate an increase in temperature at each decrease in altitude. However, the narrow and homogeneous research area causes small temperature variations. The influence of altitude on river water temperature is also weak, shown by R^2, which is only 0.11. Furthermore, Effendi et al. [32] explained that water temperature could vary depending on humidity and sunlight exposure. This research conduct at the peak of the dry season, where the relative humidity is generally low, and there is much sunshine. Therefore, in the rainy season, the average water temperature can be lower than these measurement results.

The results of our measurements show low mean water temperature and little variation. Compared with previous studies, this condition is relatively the same as the findings of Effendi et al. [32]. The results of this study indicate a temperature range between 22-26°C, as a result of the influence of the mountains around the river. Furthermore, water temperature is an essential variable to know. Concerning ecosystems, Davie [33] explained that water temperature needs critical attention because of its interdependence with DO and controlling the rate of chemical reactions in rivers. Meanwhile, concerning drinking water, the effect of temperature on DO will determine the amount of dissolved oxygen as consideration for consumption. Referring to the Ministry of Health of the Republic of Indonesia [18], all river water in the study area has met the requirements of drinking water based on the temperature aspect.

The results of water pH measurements at all points get the number 6.5. We assumed the volcanic landscape's characteristics with the deposition of the erupted material with acidic minerals to affect the pH of the water, which tends to be acidic. Also, land use in the catchment area in the form of mixed gardens and agricultural land assumed to produce organic matter, affecting the acidity of river water. These results are relatively the same as previous studies conducted in the dry season. Saraswati et al. [34] lowered the pH value between 5.1 and 8.9 with a mean of 7.175. Alam et al. [35] obtained pH results from 5.86 to 6.86. Meanwhile, Effendi et al. [32] obtained a dominant pH of 6.5. In previous studies, the emergence of water acidity mainly relates to population activities such as bathing, washing, and sanitation, along with water bodies due to changes in agricultural land function into residential areas. Therefore, the residue from these activities carries organic material broken down by microorganisms so that the process consumes oxygen and emits carbon dioxide. With this result, all measurement locations in the study area comply with the national water quality standards but not according to WHO's standards.

DO measurements obtained an average figure of 4.31 ppm or 54.72%. The average DO in the upstream area is 3.89 ppm or 49.48%, while the average in the downstream area is 4.55 ppm or 57.71%. Statistically, there is no difference between the average DO in the upstream and downstream areas. However, the overall range of values varied considerably, with one low value found upstream of the Batang River (Figure 3). According to Davie [33], DO is a water quality parameter that is very important for aquatic organisms' lives. High DO also affects the taste of the water, which is getting fresher. As discussed in the previous section, there is a correlation between water temperature and DO. Warm water will hold less dissolved oxygen than cold water. Thus, the higher the water temperature, the lower the DO value [29]. Meanwhile, Effendi et al. [32] explained that in addition to water temperature, DO influence salinity, water turbulence, and atmospheric pressure. The results of the analysis of the relationship between water temperature and DO from 11 samples showed that an increase in water temperature affected decreasing DO. However, the effect is very small with R^2, only 0.08. Simultaneously, the relationship between salinity DO DO indicated by a decrease in any increase in salinity with R^2 of 0.18. In comparison, the effect of atmospheric pressure on DO is an increase in DO at each increase in atmospheric pressure with R^2 of 0.25.

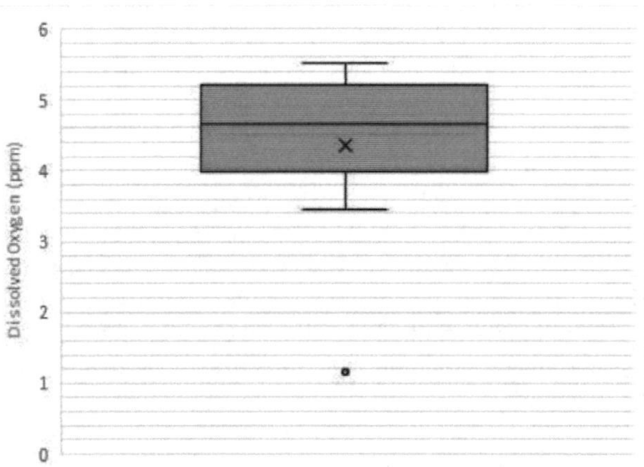

Figure 3. Boxplot for dissolved oxygen in the study area

The DO measurement results in the range of 1.15 to 5.51 ppm are small, for example, when compared to the results of measurements by Effendi et al. [32] in Cimbulawung River, which was between 7.16 up to 9.98 ppm. Meanwhile, Alam et al. [35], in their research in the dry season, obtained a minimum DO of 3.50 and a maximum of 7.20. A small DO figure is a normal condition in Merapi because the DO range in this area can be very large. Research results from Saraswati et al. [34] in three rivers, Code, Winongo, and Gajahwong, which also flow on Merapi's southern side, indicate that the DO ranges of these rivers are between 1 and 10 ppm. According to Davie [33], river water in the study area with an average temperature of 23.61°C can contain a maximum DO up to 9 ppm. Thus, the results of DO measurements in the study area, which only reached 5.51 ppm, are still far from the maximum DO. Furthermore, Asdak [29] explained that the minimum DO for human consumption is 4.0 ppm. With this threshold, the Batang upstream river water is the only one that does not qualify as drinking water.

TDS, which shows the amount of dissolved or deposited in water, is an indicator of water salinity [33]. TDS measurements carried out in 11 locations obtained an average value of 83.45 ppm. The upstream area has a smaller average TDS than the downstream area, 80 ppm compared to 85.43 ppm. Statistically, there is no difference in TDS between upstream and downstream. The seven rivers in this study are rivers that do not originate directly from the peak of Merapi. Therefore, the dissolved solid that enters the water discharge results from erosion from the land in the catchment area, in contrast to rivers that originate from the peak of Merapi, which carry the rubble of the eruption product material in the volcanic cone area. Overall, the smallest TDS founded in the downstream of the Tangkil River, which was 75 ppm, while the largest TDS founded in the downstream of the Tepus River at 104 ppm. The second-largest TDS figure is in the downstream of the Pelang River with a TDS of 93 ppm. The land in the downstream area of Sungai Tepus and Sungai Pelang is widely used for rice fields so that the potential for erosion is more significant than in other areas where land uses are mixed gardens. Therefore these two regions have the largest TDS compared to the other 9 locations. Figure 4 below shows the results of TDS measurement data in the study area.

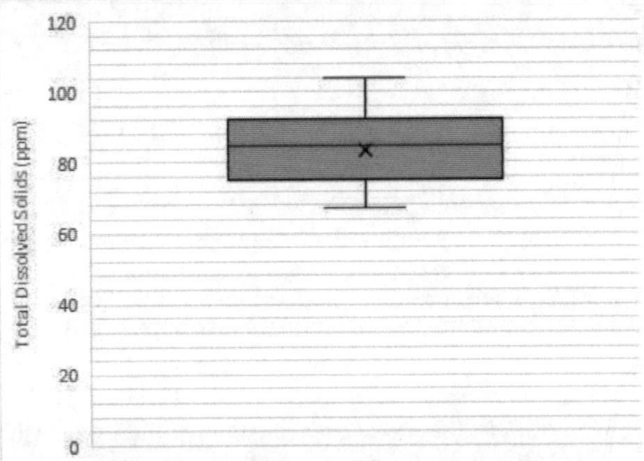

Figure 4. Boxplot for TDS in the study area

In general, the TDS in the study area is smaller than the results of the study by Saraswati et al. [34], which is a river that flows directly around Merapi's peak. The average TDS of the study was 186.54 ppm, while in this study, it was only 83.45 ppm. However, this result is still more significant than the study results by Effendi et al. [32], who obtained a TDS range between 27.40 - 53.80 ppm. The catchment area of the seven rivers in this study area, which is all in an active volcanic landscape, allows us to get many dissolved solids. Effendi et al. [32] further explained that high TDS could increase water temperature because solid materials absorb heat from sunlight. In the study area, an increase in TDS show to affect increasing temperature. However, this effect is minimal, with R^2 is only 0.03. The factors of vegetation cover and altitude assume to have more roles in influencing river water temperature. Referring to international standards from WHO and national standards, all rivers in the study area meet the criteria for drinking water quality standards in the TDS aspect.

The last aspect of the physical properties of river water that measures in this study is EC. Davie [33] explained that this EC is also an indicator of salinity. EC is the water's ability to deliver an electric current, which is a function of the concentration of the solution (water), including the total valence of dissolved ions and the level of ions that can move in the water. EC is closely related to TDS. Temperature also affects EC because ion movement in water is affected by temperature. River EC values are typically in the range of 10 to 1000 μS/ cm. The results of TDS measurements in the study area showed a mean EC of 168.55 μS/ cm. The EC range is between 135 and 208 μS/ cm (Figure 5). The lowest EC value is found in the Degong River's upstream, while the highest found in the downstream of the Tepus River. Like the TDS measurement results, in this EC measurement again, the downstream of Sungai Tepus and Sungai Pelang are the places with the highest values. The downstream area also has a higher average EC value than the upstream area, although statistically, there is no significant difference. The average EC in the downstream area was 173.14 μS/ cm, while in the upstream area, it was 160.5 μS/ cm.

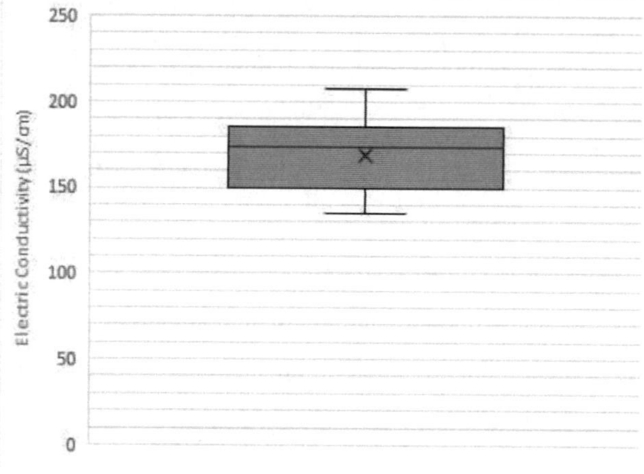

Figure 5. Boxplot for EC in the study area

TDS strongly influences EC. The more dissolved solids in river water, the greater the electrical conductivity. Both the downstream of the Tepus and Pelang Rivers has the highest EC because their TDS values are also the highest. In the study area, the increase in TDS had a powerful effect on EC's increase (Figure 6). The increase in temperature also affects the EC increase, but the effect is minimal, with R^2 only 0.16. In general, EC in the study area is smaller than the results of previous studies conducted by Saraswati et al. [34]. In this study, we found the rivers that flowed around Merapi's peak had EC between 103.3 and 966 160.5 μS/ cm, with an average of 399.16 160.5 μS/ cm. As previously discussed, the river with its head at the peak of Merapi carries much cargo from the debris of the eruption material. The summary of the results of river water quality measurements in the study area shows in Table 4.

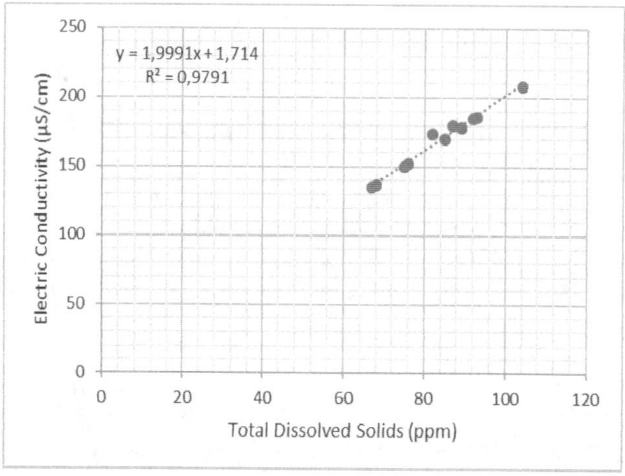

Figure 6. Relationship between TDS and EC in the study area

Table 4. Results of water quality measurements in the study area

Variable	n	Mean	Median	SD	Max	Min
Temperature (^{0}C)	11	23,61	23,66	0,80	24,93	22,42
pH	11	6,5	6,5	0	6,5	6,5

DO (ppm)	11	4,31	4,66	1,34	5,51	1,15
DO (%)	11	54,72	59,1	16,86	68,6	9,3
TDS (ppm)	11	83,45	85	11,27	104	67
EC (μS/cm)	11	173,14	174	22,77	208	135
Salinity	11	0,087	0,08	0,01	0,11	0,07

Source: Primary data (2020)

To ensure the suitability of river water as an alternative source of drinking water in an emergency of an eruption disaster, we have conducted a matching analysis of the results of water quality measurements against drinking water quality standards. The analysis showed that of the six parameters measured directly in the field, namely water temperature, pH, DO, TDS, EC, and salinity, in general, the rivers in the study area were suitable for use as a source of drinking water. In this analysis, we also added one variable analyzed in the laboratory, namely total coliform. Total coliform is an essential variable because it is the primary indicator of water's suitability for consumption [20]. The results of this variable analysis show that all rivers in the study area do not meet the total coliform standard, and this is probably due to the influence of population activities both through land use for agriculture and domestic activities. Following the results of previous research, by Sutikno et al. [16], rivers in the Merapi area are polluted by domestic waste so that if it is to as drinking water, it must be filtered and properly managed first. This research recommends that if the river water use as an alternative water source, it is necessary to provide a water treatment plant in the evacuation barracks so that the river water is suitable for consumption. Also, it is necessary to test other variables that have not been analyzed in this study to meet all the eligibility criteria according to the Regulation of the Minister of Health of the Republic of Indonesia [18]. The results of the water quality matching analysis shown in Table 5 below.

Table 5. Results of the matching analysis of several water quality parameters with water quality standards

	Parameters					
	Physical				**Chemistry**	**Biology**
	Temperature	**TDS**	**DHL**	**DO**	**pH**	**Total Coliform**
Unit	°C	Mg/l	μS/ cm	ppm	-	MPN/100ml
Tangkil (U)	24,23(**)	76(**)	152(**)	5,21(*)(**)	6,5	Not measured
Sempor (U)	22,68(**)	92(**)	185(**)	4,95(*)(**)	6,5	Not measured
Degong (U)	22,92(**)	67(**)	135(**)	4,66(*)(**)	6,5	Not measured
Batang (U)	23,66(**)	85(**)	170(**)	0,72(*)(**)	6,5	Not measured
Tangkil (D)	22,93(**)	75(**)	150(**)	3,46(*)(**)	6,5	4900(*)(**)
Sempor (D)	22,42(**)	89(**)	178(**)	5,33(*)(**)	6,5	11000(*)(**)
Degong (D)	24,08(**)	68(**)	136(**)	5,51(*)(**)	6,5	13000(*)(**)
Tepus (D)	24,41(**)	104(**)	208(**)	3,98(*)(**)	6,5	7900(*)(**)
Pelang (D)	23,96(**)	93(**)	186(**)	4,39(*)(**)	6,5	2300(*)(**)

Batang (D)	23,45$^{(**)}$	87$^{(**)}$	180$^{(**)}$	4,95$^{(*)(**)}$	6,5	4900$^{(*)(**)}$
Opak (D)	24,93$^{(**)}$	82$^{(**)}$	174$^{(**)}$	4,22$^{(*)(**)}$	6,5	11 000$^{(*)(**)}$

Note: $^{(*)}$ parameters that are not in accordance with national quality standards, $^{(**)}$ parameters that are not in accordance with international quality standards (U) = upstream. (D) = downstream.
Source: Primary data (2020)

4. Conclusion

Active volcanic areas such as Merapi needs sustainable disaster management. One aspect that needs to provides in support of realizing this goal is the availability of drinking water. Some rivers still have flow during the dry season on the south to southwest slopes of the Merapi Volcano. These rivers also do not flow from Merapi's summit so they can use as alternative drinking water sources in emergencies. In terms of quantity, all rivers can meet water needs in the evacuation barracks. Water withdrawal can carry out in the downstream area adjacent to the evacuation barracks. In terms of quality, based on limited testing on several parameters, it is known that river water in this area can use as an alternative source of drinking water. However, it is necessary to provide a water treatment plant in the evacuation barracks, so that water sourced from river water treated first so that it is suitable for consumption. It is also necessary to analyze other water quality parameters to ensure the suitability of river water as a source of drinking water.

For the evaluation process, this study's limits to 6 water quality parameters. Future research is necessary to provide a follow-up to this research's results, namely by testing other water quality parameters, both physics, chemistry, and biology. Thus, complete information can be obtained as a consideration if the river water use as an alternative water source in a disaster emergency. In future research, we recommend carrying out a study on the development of water treatment installations in refugee barracks. This study is essential to build self-reliance concerning increasing community capacity to deal with disasters and build sustainable disaster management.

Achknowledgement

This research was carried out thanks to the support of the FIS 2020 Research Grant Program with entitled The potential of river discharge at the peak of the dry season in some disaster-prone area of Merapi Volcano.

References

[1] M. A. Basyid, "Pengembangan Peta Rencana Kontijensi Bencana Gunung Api. Jurnal Rekayasa," *J. Rekayasa*, vol. 14, pp. 216–226, 2010.
[2] Pusat Vulkanologi dan Mitigasi Bencana Geologi (PVMBG), "Status Gunungapi di Atas Normal," 2020. https://vsi.esdm.go.id/ (accessed Mar. 06, 2020).
[3] H. T. Verstappen, *Garis Besar Geomorfologi Indonesia*. Yogyakarta: Gadjah Mada University Press, 2013.
[4] F. Lavigne, J. C. Thouret, B. Voight, H. Suwa, and A. Sumaryono, "Lahars at Merapi Volcano, Central Java: An Overview," *J. Volcanol. Geotherm. Res.*, vol. 100, no. 1–4, pp. 423–456, 2000, doi: 10.1016/S0377-0273(00)00150-5.
[5] S. D. Andreastuti, C. Newhall, and J. Dwiyanto, "Menelusuri kebenaran letusan Gunung Merapi 1006," *Indones. J. Geosci.*, vol. 1, no. 4, pp. 201–207, 2006, doi: 10.17014/ijog.vol1no4.20064.
[6] H. A. Sudibyakto, *Manajemen Bencana Indonesia Ke Mana?* 2011.
[7] Global Volcanism Program, "Holocene Volcano List," *Database*, 2020. https://volcano.si.edu/list_volcano_holocene.cfm (accessed Mar. 06, 2020).
[8] R. P. Aurita and S. Purwantara, "Karakteristik Mataair Kaki Lereng Gunung Merapi Dan Pemanfaatannya Di Kecamatan Dukun Kabupaten Magelang," *Geomedia Maj. Ilm. dan Inf. Kegeografian*, vol. 15, no. 1, pp. 75–85, 2017, doi: 10.21831/gm.v15i1.16239.

[9] S. Ratih, H. N. Awanda, A. C. Saputra, and A. Ashari, "Volcanic Springs, An Alternative Emergency Water Resource to Support Sustainable Disaster Management in Southern Flank of Merapi Volcano," in *IOP Conference Series: Earth and Environmental Science*, 2019, vol. 271, no. 1, doi: 10.1088/1755-1315/271/1/012012.

[10] S. Sudarmadji, D. Darmanto, M. Widyastuti, and S. Lestari, "Pengelolaan Mata Air untuk Penyediaan Air Rumahtangga Bekelanjutan di Lereng Selatan Gunungapi Merapi (Springs Management for Sustainability Domestic Water Supply in the South West of Merapi Volcano Slope)," *J. Mns. dan Lingkung.*, vol. 23, no. 1, p. 102, 2016, doi: 10.22146/jml.18779.

[11] Badan Pusat Statistik Kabupaten Magelang, *Kecamatan Srumbung dalam Angka 2019*. Magelang: BPS Kabupaten Magelang, 2019.

[12] Badan Pusat Statistik Kabupaten Sleman, *Kecamatan Cangkringan dalam Angka 2019*. Sleman: BPS Kabupaten Sleman, 2019.

[13] Badan Pusat Statistik Kabupaten Sleman, *Kecamatan Turi dalam Angka 2019*. Sleman: BPS Kabupaten Sleman, 2019.

[14] Badan Pusat Statistik Kabupaten Sleman, *Kecamatan Pakem dalam Angka 2018*. Sleman: BPS Kabupaten Sleman, 2018.

[15] S. Ratih, H. N. Awanda, A. C. Saputra, and A. Ashari, "Hidrogeomorfologi mataair kaki Vulkan Merapi bagian selatan," *Geomedia Maj. Ilm. dan Inf. Kegeografian*, vol. 16, no. 1, pp. 25–36, 2018, doi: 10.21831/gm.v16i1.20977.

[16] Sutikno, L. W. Santosa, Widiyanto, A. Kurniawan, and T. H. Purwanto, *'Kerajaan Merapi', Sumberdaya Alam dan Daya Dukungnya*. Yogyakarta: BPFG UGM, 2007.

[17] S. Purwantara, A. Ashari, and M. H. Bin Ibrahim, "The Characteristics of Infiltration on The Southern Flank of Merapi Volcanic Plain, Yogyakarta, Indonesia," *Int. J. GEOMATE*, vol. 19, no. 74, pp. 201–209, 2020, doi: 10.21660/2020.74.52941.

[18] Kementrian Kesehatan Republik Indonesia, *Peraturan Menteri Kesehatan No. 492 Tahun 2010 Tentang Persyaratan Kualitas Air Minum*. Indonesia, 2010.

[19] Pemerintah Republik Indonesia, *Peraturan Pemerintah No. 82 Tahun 2001 Tentang Pengelolaan Kualitas Air dan Pengendalian Pencemaran Air*. Indonesia, 2001.

[20] Y. Meride and B. Ayenew, "Drinking water quality assessment and its effects on residents health in Wondo genet campus, Ethiopia," *Environ. Syst. Res.*, vol. 5, no. 1, pp. 1–7, 2016, doi: 10.1186/s40068-016-0053-6.

[21] C. Asdak, *Hidrologi dan Pengelolaan Daerah Aliran Sungai*. Yogyakarta: Gadjah Mada University Press, 2010.

[22] C. G. Newhall *et al.*, "10,000 Years of explosive eruptions of Merapi Volcano, Central Java: Archaeological and modern implications," *J. Volcanol. Geotherm. Res.*, vol. 100, no. 1–4, pp. 9–50, 2000, doi: 10.1016/S0377-0273(00)00132-3.

[23] R. Gertisser, S. J. Charbonnier, J. Keller, and X. Quidelleur, "The geological evolution of Merapi volcano, Central Java, Indonesia," *Bull. Volcanol.*, vol. 74, no. 5, pp. 1213–1233, 2012, doi: 10.1007/s00445-012-0591-3.

[24] S. Setyawati and A. Ashari, "Geomorfologi Lereng Baratdaya Gunungapi Merapi Kaitannya Dengan Upaya Pengelolaan Lingkungan dan Kebencanaan," *Geomedia Maj. Ilm. dan Inf. Kegeografian*, vol. 15, no. 1, pp. 13–16, 2017, doi: 10.21831/gm.v15i1.16235.

[25] A. Ashari, "Geomorphology of the Southern Flank of Merapi Volcano in Relation to The Potential Hazards and Natural Resources: a Review," *Geomedia Maj. Ilm. dan Inf. Kegeografian*, vol. 15, no. 2, pp. 183–192, 2018, doi: 10.21831/gm.v15i2.19556.

[26] N. Abdullah and A. Muchtar, "Analisis Faktor-Faktor yang Mempengaruhi Debit Sungai Masama," *J. Hutan dan Masy.*, vol. 2, no. 1, pp. 174–187, 2007.

[27] A. W. Abd Kamal Neno, Herman Harijanto, "Hubungan Debit Air dan Tinggi Muka Air di Sungai Lambagu Kecamatan Tawaeli Kota Palu," *War. Rimba*, vol. 4, no. 2, pp. 1–8, 2016, [Online]. Available: http://jurnal.untad.ac.id/jurnal/index.php/WartaRimba/article/view/8128/6442.

[28] Pemerintah Republik Indonesia, *Peraturan Menteri Kesehatan Republik Indonesia No. 28 Tahun 2019 Tentang Angka Kecukupan Gizi yang Dianjurkan untuk Masyarakat Indonesia*. Indonesia, 2019.
[29] C. Asdak, *Hidrologi dan Pengelolaan Daerah Aliran Sungai*. Yogyakarta: Gadjah Mada University Press, 2014.
[30] A. Swartz, D. Roon, M. Reiter, and D. Warren, "Stream temperature responses to experimental riparian canopy gaps along forested headwaters in western Oregon," *For. Ecol. Manage.*, vol. 474, no. June, p. 118354, 2020, doi: 10.1016/j.foreco.2020.118354.
[31] Y. Martinus, W. Astono, and D. Hendrawan, "Water quality study of Sunter River in Jakarta, Indonesia," *IOP Conf. Ser. Earth Environ. Sci.*, vol. 106, no. 1, 2018, doi: 10.1088/1755-1315/106/1/012022.
[32] H. Effendi, Romanto, and Y. Wardiatno, "Water quality status of Ciambulawung River, Banten Province, based on pollution index and NSF-WQI," in *Procedia Environmental Sciences*, 2015, vol. 24, pp. 228–237, doi: 10.1016/j.proenv.2015.03.030.
[33] T. Davie, *Fundamentals of Hydrology*, Second Edi. London: Roudledge, 2008.
[34] S. P. Saraswati, M. V. Ardion, Y. H. Widodo, and S. Hadisusanto, "Water Quality Index Performance for River Pollution Control Based on Better Ecological Point of View (A Case Study in Code, Winongo, Gadjah Wong Streams)," *J. Civ. Eng. Forum*, vol. 5, no. 1, p. 47, 2019, doi: 10.22146/jcef.41165.
[35] M. J. B. Alam, M. R. Islam, Z. Muyen, M. Mamun, and S. Islam, "Water quality parameters along rivers," *Int. J. Environ. Sci. Technol.*, vol. 4, no. 1, pp. 159–167, 2007, doi: 10.1007/BF03325974.

The Use of Machine Learning for Accessing Landslide Susceptibility Class: Study Case of Kecamatan Pacet, Kabupaten Mojokerto

Listyo Yudha Irawan[1]*, Sumarmi[1], Syamsul Bachri[1], Damar Panoto[1], Nabila[1], Irfan Helmi Pradana[1], Rahmad Faizal[1], Melinda Meganagatha Rosbella Devy[1], Widodo Eko Prasetyo[1]
[1] Department of Geography, State University of Malang, Jl. Semarang 05, Malang 65145, Indonesia

*listyo.fis@um.ac.id

Abstract. Kecamatan Pacet, Kabupaten Mojokerto is one of an area with many landslide events in East Java Province. As a mitigation effort, this research aimed to map the landslide susceptibility class distribution of the research area. This research applied a machine learning analysis technic which combined Frequency Ratio (FR) and Logistic Regression (LR) models to assess the landslide susceptibility class distribution. FR bivariate analysis is used to normalized the data and to identify the influence significancy on each class of triggering factors. LR multivariate analysis is applied to generate the landslide probability (susceptibility) and to show the influence significancy of each triggering factor to landslide events. There are 12 triggering factors to landslide used in this research, which is: TPI, TWI, SPI, slope, aspect, elevation, profile curvature, distance to drainage, geological unit, rainfall, land use, and distance to the road. This research has 383 landslides and 383 non-landslide events as the data sample based on field survey, BPBD Kabupaten Mojokerto, and Google Earth Pro imagery interpretation. The proportion of dataset training and testing is 70% and 30%, which generated from the data inventory. This research used ROC analysis to validate the landslide susceptibility model. The result showed that the landslide susceptibility model has an AUC value of 0.91, which indicated that the model has high accuracy.

1. Introduction

Indonesia is a country with a high rate of landslide events. In 2017, landslide events were causing 163 fatalities, 185 people injured, and more than 59.641 people relocated nationally [1]. With a total of 848 occurrences in the same year, landslides are the third-highest disaster events in Indonesia after floods (979) and whirlwinds (886) [1]. Respectively, the province with the highest to lowest landslide events in Java Island is Central Java, West Java, East Java, and Special Region of Yogyakarta (D.I. Yogyakarta) Province [1], [2]. Among the four provinces, East Java Province has the highest number of casualties caused by landslides (Figure 1).

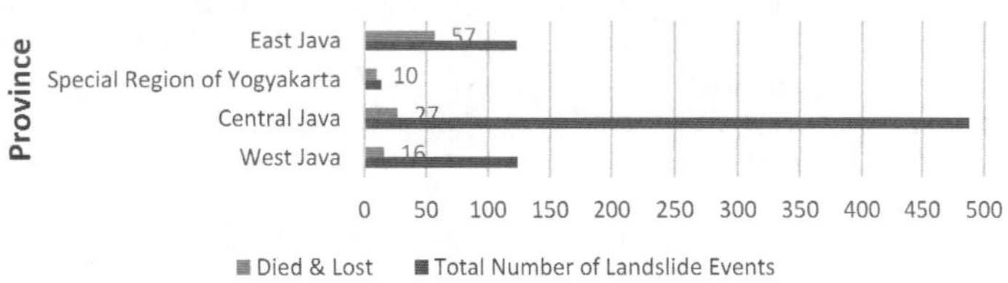

Figure 1. Diagram of Total Fatalities and People Lost Caused by Landslide
Source: BNPB (2018) Modified

Kecamatan Pacet, Kabupaten Mojokerto is one of an area with many landslide events in East Java Province. Therefore, it needs a landslide detection in Kecamatan Pacet, Kabupaten Mojokerto as a response to the high susceptibility in the area. Furthermore, the detection effort needs to be translated

Content from this work may be used under the terms of the Creative Commons Attribution 3.0 licence. Any further distribution of this work must maintain attribution to the author(s) and the title of the work, journal citation and DOI.
Published under licence by IOP Publishing Ltd

into a more communicative form for the community. Providing information about landslide susceptibility class can further increase the community preparedness to the events [3-5]. Accordingly, this research tried to map the landslide susceptibility class in Kecamatan Pacet, Kabupaten Mojokerto.

This research applied a machine learning analysis technic to assess the landslide susceptibility class. The analysis technic involved both statistical and soft computing approaches. In general, the principal work of machine learning is collecting data to create a prediction. This paper used machine learning to enhance the accuracy of landslide susceptibility class assessment. It is because of the ability of machine learning technic to analyze the complex relationship between landslide susceptibility and its triggering factors [5], [6], [7], [8]. Some of the machine learning analysis technic model to look landslide events probability (landslide susceptibility class) are; artificial neural network [9],10], support vector machine [8], [11], random forest [6], [12], logistic regression [8,13], and others.

The logistic regression model is used in this research given the high accuracy to assess a landslide susceptibility. As a consequence, this model is the most used model to assess landslide susceptibility in the world [14]. The recent trends in similar research also include the data normalization process. The process is applied to increase the prediction result [8], [11], [13], [15]. Bivariate analysis is used to normalized the data thus the influence of every triggering factor to landslide can be seen. To validate the result of landslide events probability prediction, ROC curved is used. All of the mentioned processes are used to generate appropriate landslide modeling.

2. Data Preparation and Methods

There are four phases to generate landslide events probability (susceptibility) using machine learning, which are: 1) spatial dataset preparation, 2) data normalization, 3) landslide modeling, and 4) model validation (Figure 2).

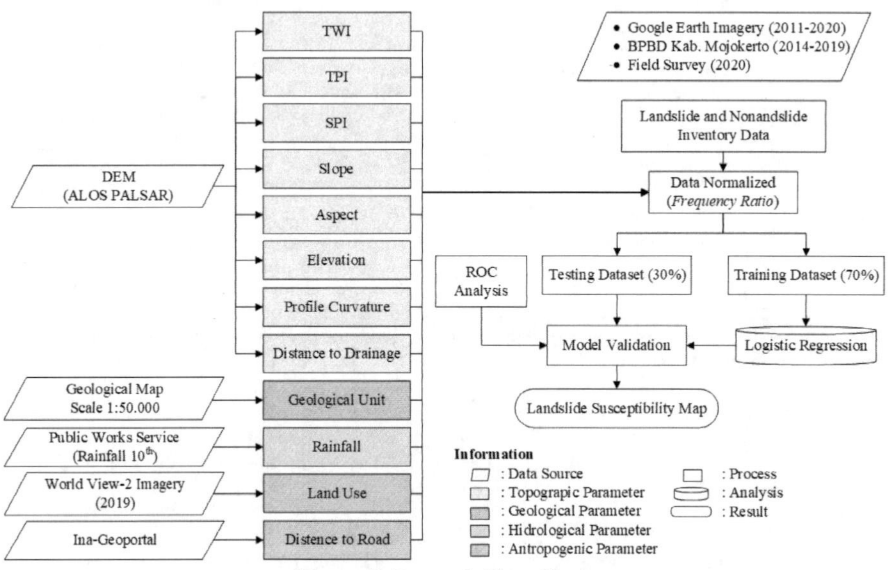

Figure 2. Research Flow Chart

2.1 *Study Area*

The study area (Kecamatan Pacet) area covers an area of 17,300 km^2, composed of volcanic landform (Figure 3). Kecamatan Pacet located within a typical tropical monsoon region with 2.518 mm/years. The BPBD of Kabupaten Mojokerto stated that there were 45 landslide events from 2015 to 2020 in the area. There were several impacts regarding the events which varied in different parts of the research area, such as; cut transportation connections, destroyed farmlands, and damaged houses.

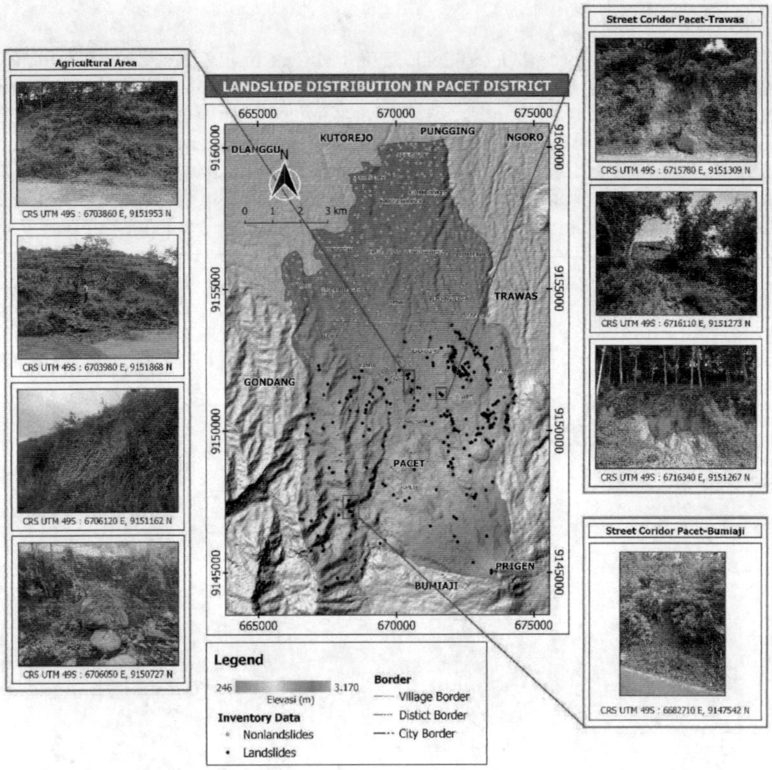

Figure 3. Landslide Distribution in Pacet, Mojokerjo
Source: Authors (2020)

2.2 Spatial Dataset Preparation

2.2.1 Landslide Data Inventory

There were three ways to do the landslide data inventory in this research, which are: 1) field survey in 2020, 2) landslide events records from BPBD Kabupaten Mojokerto data collection and 3) 2011—2020 Google Earth Pro imagery interpretation. The data collection resulted in 383 landslide events shown in Figure 3. There was a variability of landslide length in some of the landslide spots in the field—the shortest was 2 meters while the longest was 17 meters.

The use of machine learning to model the landslide probability can be considered as a typical binary classification [16], [17]. As a consequence, the exact same number of non-landslide spots are also needed [18], [19], [20], [21]. Landslide events are symbolized as "1" and non-landslide events are symbolized as "0". This research used supervised machine learning analysis to predict the landslide and non-landslide events into the dependent variable.

2.2.2 Landsilde Triggering Factor

There were twelve triggering factors to landslide used in this research, which includes topographic, geologic, hydrologic and topographic conditions. Topographic parameters were TWI (Topographic Wetness Index), TPI (Topographic Position Index), SPI (Stream Power Index), slope, elevation, profile curvature and distance to drainage [22], [23], [24]. The geologic parameter was geological unit [24], while hydrologic parameter was rainfall intensity [25]. For the anthropogenic parameter, this research accounted land use and distance from the nearest road. Distance from the nearest road was used on the account of some landslide events which are triggered by a cut slope.

The topographic parameters were generated by ALOS Palsar DEM using Q.GIS 3.21 (elevation, distance to drainage and slope) and SAGA GIS 2.3.2 (TWI, TPI, SPI, aspect and profile curvature). Factor from the lithological condition was generated by geological unit classification from 1:50.000 Geological Map of Pandaan Page, published by PVMBG (*Pusat Vulkanologi dan Mitigasi Bencana*

IOP Conf. Series: Earth and Environmental Science **884** (2021) 012006 doi:10.1088/1755-1315/884/1/012006

Geologi or Center of Volcanology and Geological Hazard Mitigation). Average rainfall intensity for the last 10 years (2010—2019) from Dinas Pekerjaan Umum Kabupaten Mojokerto (The Public Works Service of Kabupaten Mojokerto) was used as the hydrological parameter in this research. Digitation was done in order to generate the land-use factor of the anthropogenic parameter, while the road data of Kabupaten Mojokerto from Ina-Geoportal was used as the basis of distance to road factor.

Each factor should be in a raster format. Therefore, some shapefile data must be converted into a raster with the same spatial resolution of ALOS Palsar DEM which is 12 meters. Not only every raster-format spatial data must encompass the same coordinate system, but also the spatial resolution and pixel numbers [4], [9], [13], [26], [28]. If the spatial resolution and pixel numbers are inconsistent, the analysis process is rejected.

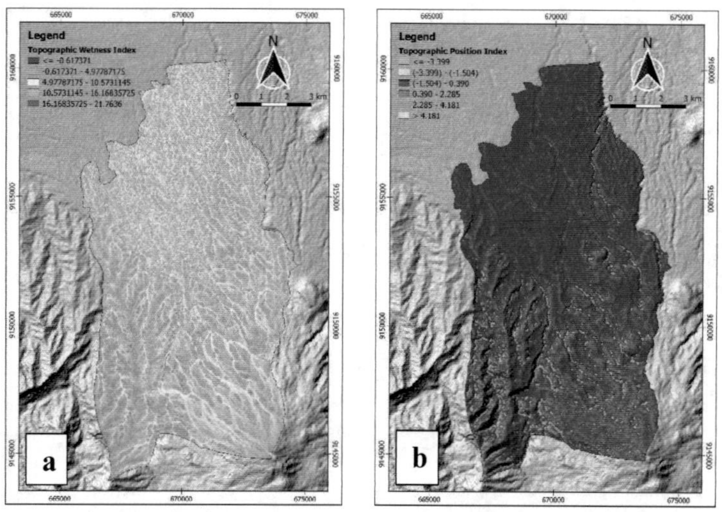

Figure 4. Landslide Triggering Factors (a. TWI; b. TPI)
Source: Authors (2020)

Topographic Wetness Index indicated the hydrological processes related to the water flow accumulation based on its slope control [28], [29], [30]. Accordingly, the water flow accumulation will increase as the TWI score is also higher—*vice versa*. The distribution of the TWI class in the research area is mostly classified as moderate to low, which can be seen in Figure 4a.

Topographic Position Index (Figure 4b) displayed the altitude difference between a certain area with the surrounding area [31]. Henceforth, the valley, slope and ridge of a terrain can be seen from the TPI value [24]. Gradually, the valley of terrain has the smallest TPI value and the ridge has the highest TPI value.

Stream Power Index (Figure 5a) indicated the potential erosion as an illustration of a geomorphologic process [24]. SPI visualized the water flow energy which affected by the potential flow erosion in a certain location [32]. As the SPI value increased, the slope erosion risk was also increased [32]. In the research area, the highest SPI value was located in the upper to lower slope of Anjasmoro and Welirang Mountain areas.

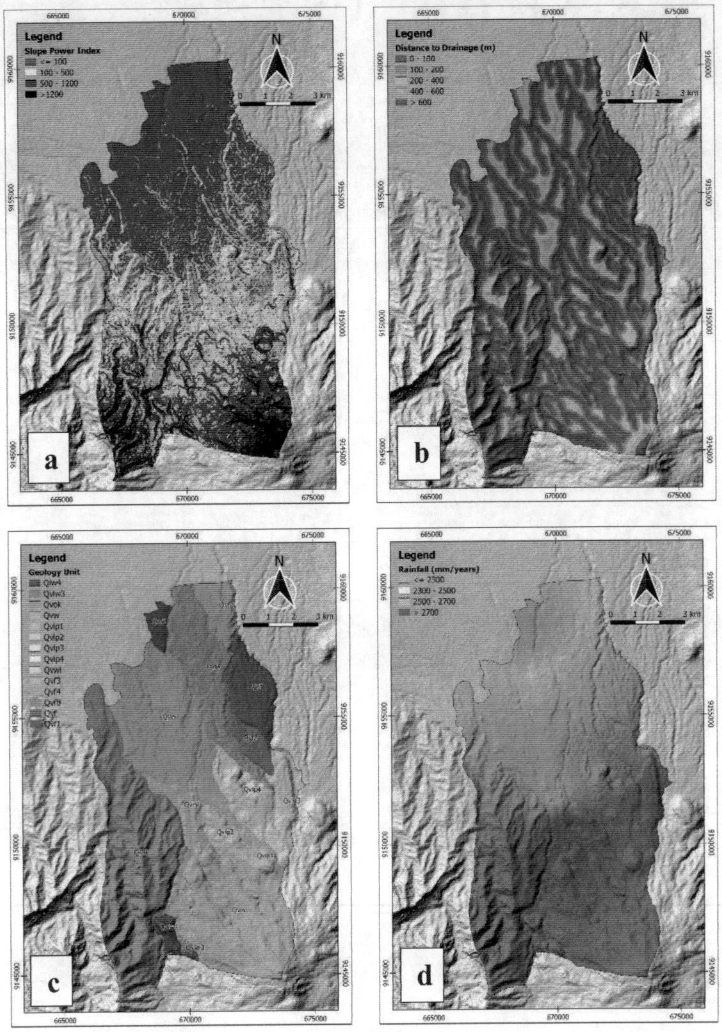

Figure 5. Landslide Triggering Factors (a. SPI; b. Slope; c. Geological Unit; d. Rainfall)
Source: Authors (2020)

Distance to drainage (Figure 5b) can affect the infiltration, runoff and soil moisture rate. As the distance to drainage is closer, the infiltration, runoff and soil moisture rate can increase. Most of the research area has a distance to drainage of 0—400 meters.

There are 14 geological units in the research area which located in the Arjuna-Welirang and Anjasmara Tua Volcanic quarter complex based on 1:50.000 Geological Map of Pandaan Page [33]. The Anjasmara Tua Volcanic complex is symbolized as Qvok and the rest of the symbol is part of Arjuno Welirang complex. The Anjasmara Tua Volcanic complex is older than Arjuno Welirang, which formed in the early Pleistocene era.

Rainfall is one of the triggering factors of landslides [21].The high intensity of rainfall can fasten the soil saturation process; thus, the slope has a greater soil mass. Consequently, it will increase the landslide hazard in the area [21]. In the research area, rain falls between 2,000 to 2,800 mm/year with the highest rainfall in the slope area of Anjasmara and Arjuna-Welirang Mountain.

IOP Conf. Series: Earth and Environmental Science **884** (2021) 012006 doi:10.1088/1755-1315/884/1/012006

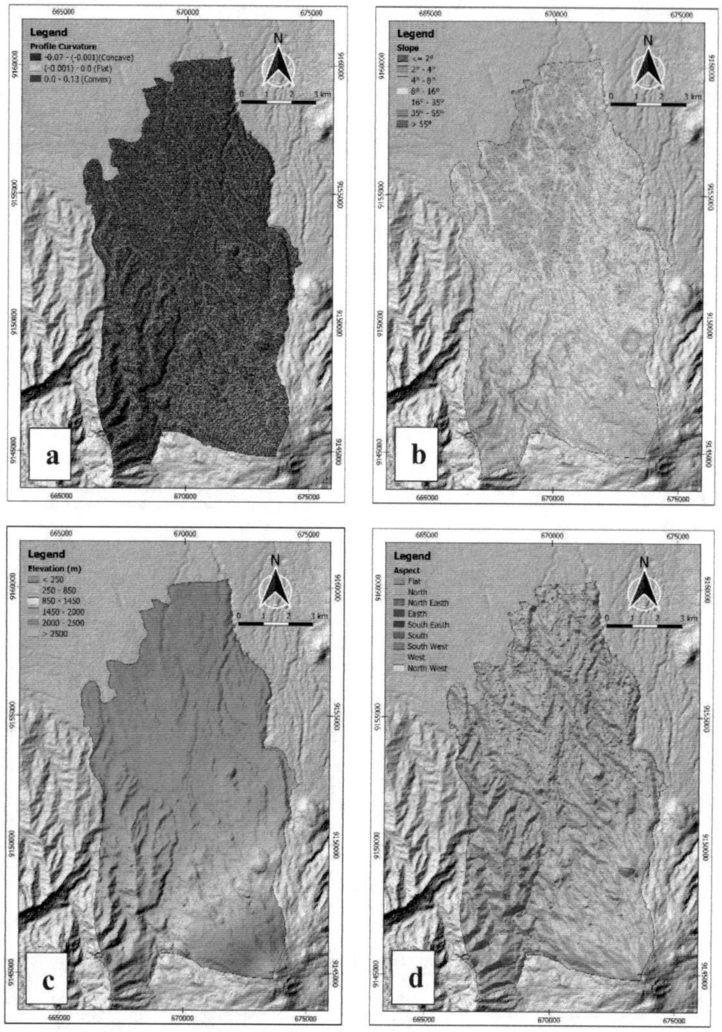

Figure 6. Landslide Triggering Factors (a. Profile Curvature; b. Slope; c. Elevasi; d. Aspect)
Source: Authors (2020)

The profile curvature in this research was classified into three classes. Profile curvature showed the gradient value of the slope, such as convex, flat and concave. The negative value represented concave, zero (0) represented flat and positive value represented convex. The extreme positive or negative value of a slope indicated an unstable slope [34].

Slope is the most important factor for slope stability analysis. A steeper slope has a higher stream number and water transport energy. It is because of the linear correlation between gravitational acceleration and slope steepness [35]. A slope can be classified into 7 classes using Van Zuidam (1973) classification, namely 0-2° (Flat), 2-4° (Gentle), 4-8° (More Gentle), 8-16° (Slightly Steep), 16-35° (Steep), 35-55° (Very Steep), dan >55° (Extreme Steep) [36]. In the research area, the lower and middle parts of Welirang and Anjasmoro Mountain are classified as Flat—Gentle and Steep—Very Steep slope respectively.

Elevation in the research area is varied between <250 masl (mainly in the northern part) to >2,500 masl (mainly in the southern part, with the range of 850 masl to 2,500 masl). Generally, landslide tends to occur in a higher elevation area [37], [38]. Such a condition happens because of the higher soil moisture and greater rainfall distribution, as well as its mountainous topographic feature.

To identify the characteristic of the erosion, runoff and sediment disposition of a slope, the aspect parameter is used in this research. Aspect value could illustrate a higher solar intensity and rainfall, which fasten the material decomposition process due to erosion and weathering. In addition, it could also affect the stability of a slope regarding its soil moisture condition.

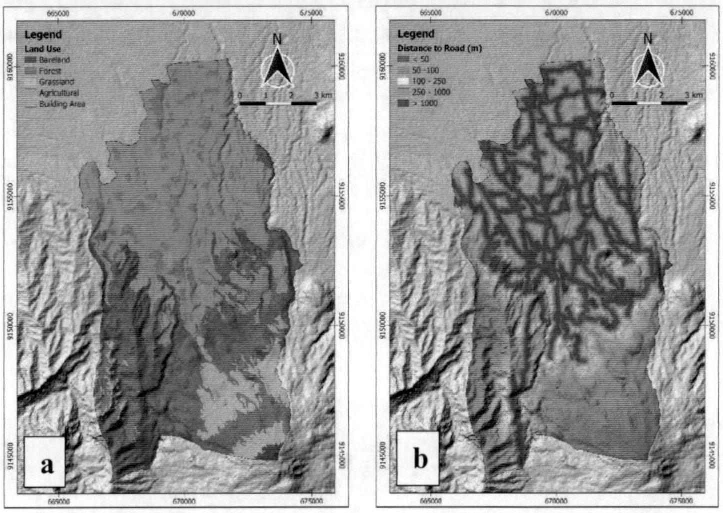

Figure 7. Landslide Triggering Factors (a. Land Use; b. Distance to Road)
Source: Authors (2020)

Land use can also affect the slope stability of an area [39]. Furthermore, it also reflects the human activities and the environmental conditions in the research area. In particular, soil stability can be improved mechanically and hydrologically by the existing vegetation characteristics. Some of the land-uses in the research area are forest, agriculture, bare lands, settlements and grasslands. The highest proportions are forest and agriculture.

Distance to road is one of the most important factors to assess a landslide susceptibility of an area. It is because the distance to road factor can see the human-related morphological changes. Some of the landslide events in the research area are triggered by cut slopes for roads.

2.3 Data Normalization

Frequency ratio (FR) computation was done using Equation 5 as shown below. From the equation, score coefficients for each class in each factor could show the correlation between landslide probability events [40].

$$FRv_1 = \frac{Lvc_i/L_s}{Ac_i/A_s} \quad (1)$$

Where Lvc_i is the pixel number of landslide events in each i class factors; L_s is the total pixel number landslide events in each class factors; Ac_i is the total pixel of the research area.

Data normalization was done after computation using Equation 5 had done. This process resulted in a value ranging from 0.01 to 0.99 after applying a min-max method [11,27,41].

$$Value_{new} = \frac{value - \min(v)}{\max(v) - \min(v)} \text{ x } (upper - lower) + lower \quad (2)$$

Where $Value_{new}$ is the data normalization result. $value$ is the true value of a class, $upper$ is a 0.99 constant and $lower$ is a 0.01 constant.

2.4 Landslide Modeling

This research used a logistic regression method for dataset training. The analysis was done using Equation 3. A landslide probability is ranging from 0 (0% landslide probability) to 1 (100% landslide probability) [42]. The correlation between landslide events with the triggering factors can be seen from the computation using Equation 4.

$$P = \frac{1}{1+e^{-z}} = \frac{e}{1+e^{z}} \quad (3)$$

Where P is landslide probability; e is exponential; z is combination score from independent variable which calculated using Equation 7.

$$Z = b_0 + b_1X_1 + b_2X_2 + ... + b_nX_n \quad (4)$$

Where b_0 is a constant from the model; b_n is score from landslide triggering factor of X_n.

2.5 Model Validation

The trend to use an ROC curve algorithm to evaluate machine learning's performance has been increasing for the last decade [43], [44]. The curve can evaluate the quality of a model based on the AUC value. In order to generate the ROC curve, true positive rate (sensitivity) and false positive rate (1-specify) values need to be plotted. The rates were computed using Equation 5 and 6 as follows:

$$sensitivity = \frac{TP}{TP+FN} \quad (5)$$

$$specificity = \frac{TN}{TN+FP} \quad (6)$$

Where TP is true positive value; FN is false negative value; TN is the total of true negative; FP is the total of false positive.

3. Results and Discussion

3.1. Frequency Analysis of Landslide Events on Each Factor Class

Frequency analysis of landslide events on each factor class can be seen in Table 1. The table includes the calculation of FR value until the normalization value (*Nv*). The total pixel of the all research area is 634,629 with the 383 pixels of them are the landslide events. As shown in the normalization value (*Nv*) column, the value of each factor varied from 0 to 1. Those numbers indicated the influence significancy of each triggering factor to the landslide events—if the numbers are close to 0, the influence is insignificant. On the other hand, if the numbers are close to 1, the factor has a significant influence on landslide.

Table 1. Correlation Between Landslide and Its Triggering Factors Using FR

Conditioning Factors	Classes	*Lvci*	*Aci*	FR	*Nv*
Elevasi	< 500	0	162085	0	0.00
	500 - 800	105	171390	1.015138	0.45
	800 - 1100	138	102166	2.238174	1.00
	1100 - 1400	95	78722	1.999625	0.89
	1400 - 1600	20	38999	0.849763	0.38
	> 1600	25	81267	0.509738	0.23
Slope	<2	0	17048	0	0.00
	2-4	0	51529	0	0.00
	4-8	15	150953	0.164653	0.05
	8-16	90	170472	0.874804	0.29
	16-35	214	209487	1.692692	0.55
	35-55	63	34183	3.053877	1.00
	>55	1	957	1.731447	0.57

Aspect	Flat	0	471	0	0.00
	North	126	144026	1.449609	0.86
	Northeast	89	100420	1.468557	0.87
	East	54	55717	1.605932	0.95
	Shoutheast	16	15729	1.685544	1.00
	South	1	6256	0.264865	0.16
	Southwest	6	34049	0.29199	0.17
	West	19	105927	0.297213	0.18
	Northwest	72	172034	0.693489	0.41
Topographic Wetness Index	< -0.617	0	17	0	0.00
	-0.617 - 4.977	170	173139	1.626954	1.00
	4.977 - 10.573	202	412158	0.812099	0.50
	10.573 - 16.168	10	44089	0.37583	0.23
	16.168 - 21.763	1	5226	0.317068	0.19
Topographic Position Index	< -3.399	0	2281	0	0.00
	(-3.399) - (-1.504)	27	31296	1.429539	0.49
	(-1.504) - 0.390	207	424473	0.808056	0.28
	0.390 - 2.285	131	165909	1.308346	0.45
	2.285 - 4.181	17	9744	2.890898	1.00
	> 4.181	1	926	1.789411	0.62
Stream Power Index	< 100	12	248978	0.079862	0.00
	100 - 500	223	235112	1.571633	0.91
	500 - 1200	123	118412	1.721197	1.00
	> 1200	25	32127	1.28941	0.74
Profile Curvature	-0.07 - (-0.001)	165	275081	0.993904	0.74
	(-0.001) - 0.0	23	87706	0.43453	0.00
	0.0 - 0.13	195	271842	1.188609	1.00
Distance to Road	< 50	44	107901	0.675691	0.10
	50 - 100	28	77803	0.596325	0.00
	100 - 250	121	157622	1.272008	0.84
	250 - 1000	138	163160	1.401479	1.00
	> 1000	52	128143	0.672403	0.09
Distance to Drainage	< 100	253	336892	1.244374	1.00
	100 - 200	98	195019	0.832665	0.67
	200 - 400	31	95873	0.53578	0.43
	400 - 600	1	4968	0.333534	0.27
	> 600	0	1877	0	0.00
Rainfall	< 2300	0	50945	0	0.00
	2300 - 2500	0	66077	0	0.00
	2500 - 2700	81	197164	0.680736	0.44
	> 2700	302	320443	1.561627	1.00
Geological Unit	Qlw4	4	6777	0.978011	0.22
	Qvlw3	0	2068	0	0.00
	Qvok	104	158469	1.087452	0.24
	Qvw	40	224506	0.295225	0.07
	Qvlp1	26	17395	2.476681	0.56
	Qvlp2	34	15646	3.600781	0.81

	Qvlp3	57	27301	3.459533	0.78
	Qvlp4	76	28369	4.439057	1.00
	Qvwl	16	18678	1.419419	0.32
	Qvf3	0	18278	0	0.00
	Qvf4	26	62796	0.686061	0.15
	Qvf8	0	13871	0	0.00
	Qvf	0	32412	0	0.00
	Qvfl	0	8063	0	0.00
Land Use	Bareland	4	3101	2.137368	1.00
	Forest	282	255264	1.830546	0.85
	Grassland	29	59042	0.813876	0.37
	Agricultural	67	276611	0.401353	0.17
	Building Area	1	40611	0.040802	0.00

Source: Authors (2020)

3.2. *LR-FR Model Construction*

Landslide probability (susceptibility) using LR-FR model were built from Equation 3 using Python's programming language. The computations resulted a value of -12.644 for the constant model (b_0). After that, a Z value (combination score of independent variables) were generated using Equation 7 as the base to estimate the landslide probability. The value of each factors (b_n) indicated the influence significancy on landslide events. As the b_n value higher, the contribution to landslide is also increased. Based on the result, slope has the highest b_n value (9.240) and distance to road has the lowest (-2.125). Accordingly, slope has the highest and distance to road has the lowest contribution to landslide events in the research area.

$$Z = -12.644 + 0.638[TWI] + 1.419[TPI] + 2.711[SPI] + 9.240[Slope] - 0.436[Rainfall] - 0.493[Profile\ Curvature] + 4.285[Laand\ Use] + 0.817[Geology] + 3.218[Elevation] - 2.125[Distance\ to\ Road] + 5.261[Distance\ to\ Drainage] + 2.170[Aspect]$$

3.3. *Model Validation*

Model validation was done in order to assure the quality of the prediction. The AUC value is ranging from 0.5 to 1.0. The accuracy of the prediction model is higher if the value approaches 1.0. On the contrary, the prediction model is less accurate as it approaches 0.5.

The result showed that the landslide susceptibility model has an AUC value of 0.91 (Figure 8). It indicated that the model has high accuracy. From the ROC curve, the true positive and false negative plotting is 0.93 and 0.1 respectively.

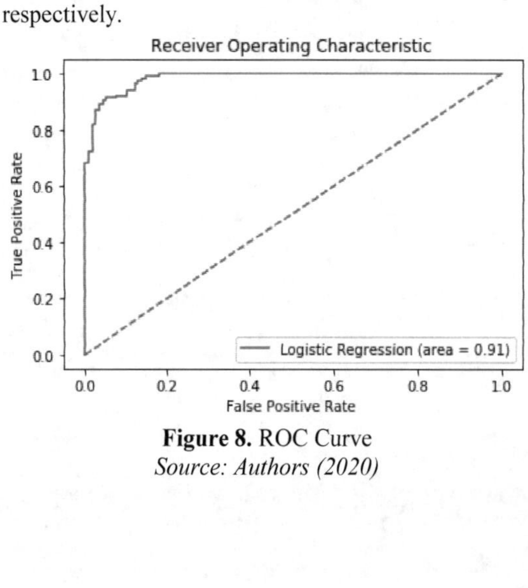

Figure 8. ROC Curve
Source: Authors (2020)

3.4. *Result Discussion*

There are 5 classes of landslide susceptibility using a natural brake method, namely very low, low, moderate, high and very high. This method is highly recommended to classify a certain spatial distribution because of its capability to optimized the differences between groups which has the same value [45], [46], [47]. The result shows that the area with very low, low, moderate, high and very high landslide susceptibility has a total percentage of 37.8%, 4.0%, 5.8%, 8.4% and 44.0% respectively (Figure 9).

The very low—low landslide susceptibility class distribution is located in the flat to More Gentle areas, such as in; Desa Kuripansari, Desa Pandanarum, Desa Sumberkembar, Desa Tanjungkenongo, Desa Warugunung, Desa Candiwatu, Desa Kesimantengah, Desa Wiyu, Desa Petak, Desa Bendunganjati, some of Desa Mojokembang, Desa Kembangbelor, Desa Nogosari, Desa Cepokolimo, Desa Pacet, Desa Sajen and Desa Kemiri. Areas with moderate landslide susceptibility class are located in undulating to the hilly area with slightly steep slope—covering the foot slope area of Anjasmoro and Welirang Mountain. Areas with high—very high landslide susceptibility class are located in hilly to a mountainous area with steep to extreme steep slope—covering the forest area in the foot slope until the upper slope of Anjasmoro and Welirang Mountain.

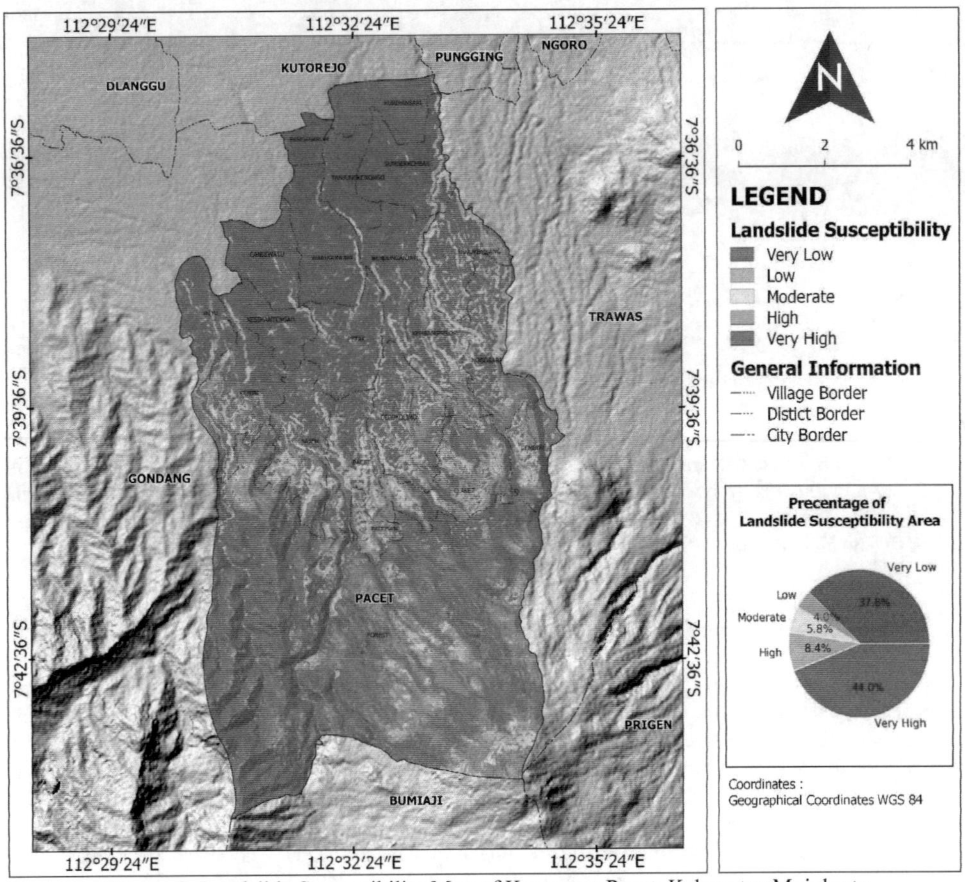

Figure 9. Landslide Susceptibility Map of Kecamtan Pacet, Kabupaten Mojokerto
Source: Authors (2020)

In Figure 10, the characteristic of an area with high to very high landslide susceptibility class can be known from the normalization value. The high to very high landslide susceptibility class characteristics are: elevation 800-1400 masl; slope 33-55°; aspect North to Southeast; TWI -0.617-

4.977; TPI 285-4.181; SPI 100-1200; profile curvature 0.0-0.13; distance to road 100-1000 m; distance to drainage <100; rainfall >2700; a geological unit of Qvlp2, Qvlp3 and Qvlp4; and land use of bare land and forest. The very low landslide susceptibility class characteristics are: elevation 500 masl; slope 2-8°; aspect flat, south, southwest and west; TWI <-0.617; TPI <-3.399; SPI <100; profile curvature -0.001 to 0.0; distance to road 50-100 m; distance to drainage >600; rainfall 2300-2500; a geological unit of Qvlw3, Qvw, Qvlp4; and land use of building area.

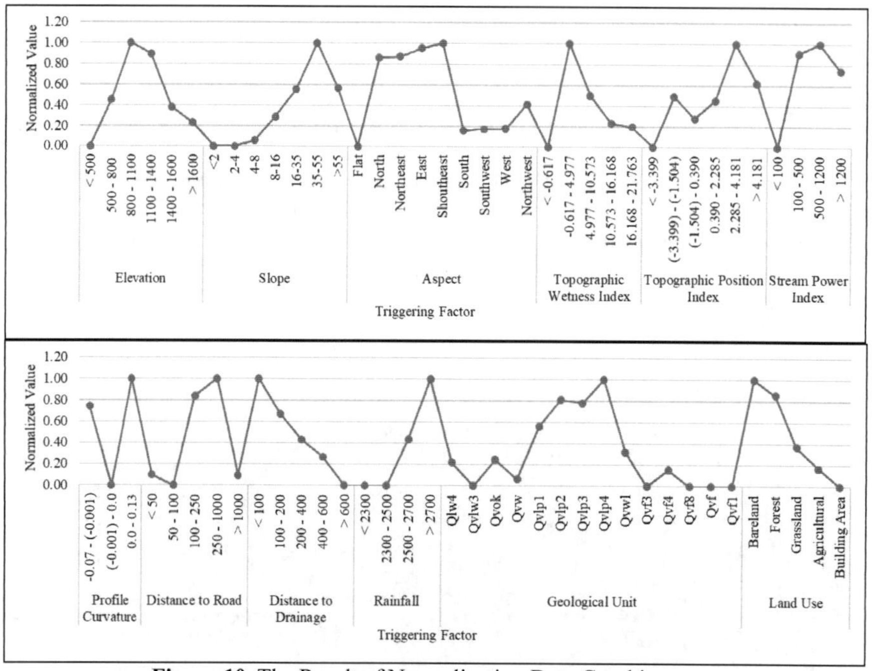

Figure 10. The Result of Normalization Data Graphic
Source: Authors (2020)

The LR-FR model using machine learning techniques resulted a high-quality result in this research, similar to previous researches. One of the case examples is landslide mapping in Mila Basin, Algeria which resulted an AUC value of 0.85 [13]. Moreover, it also resulted that slope is the most significant factor to landslide, similar to this research.

4. Conclusions

Landslide susceptibility model using LR-FR combination model of machine learning analysis has a high accuracy, which indicated from the AUC value of 0.91 from the ROC curve. There are 5 classes of landslide susceptibility using a natural brake method, namely very low, low, moderate, high and very high. The result shows that the area with very low, low, moderate, high and very high landslide susceptibility has a total percentage of 37.8%, 4.0%, 5.8%, 8.4% and 44.0% respectively. The high to very high landslide susceptibility class characteristics are: elevation 800-1400 masl; slope 33-55°; aspect North to Southeast; TWI -0.617-4.977; TPI 285-4.181; SPI 100-1200; profile curvature 0.0-0.13; distance to road 100-1000 m; distance to drainage <100; rainfall >2700; a geological unit of Qvlp2, Qvlp3 and Qvlp4; and land use of bare land and forest. Slope and distance to drainage have the highest contribution to landslide events in the research area shown the high value of score from landslide triggering.

Acknowledgments
Special thanks to Lembaga Penelian dan Pengabdian (LP2M) Universitas Negeri Malang for funding this research through PNBP UM funding scheme, thus the research, writing and publishing process of this paper can be well-completed

References*

[1] BNPB 2018 *Data Bencana Indonesia 2017* (Jakarta: Badan Nasional Penanggulangan Bencana)
[2] BNPB 2014 *Informasi Bencana: Informasi Kebencanaan Bulan Teraktual* (Jakarta: Badan Nasional Penanggulangan Bencana)
[3] Ahmed B 2015 Landslide susceptibility mapping using multi-criteria evaluation techniques in Chittagong Metropolitan Area, Bangladesh *Landslides* **12** 1077–1095
[4] Wang L-J, Guo M, Sawada K, Lin J, and Zhang J 2015 Landslide susceptibility mapping in Mizunami City, Japan: a comparison between logistic regression, bivariate statistical analysis and multivariate adaptive regression spline models *Catena* **135** 271–282
[5] Chen W, Peng J, Hong H, Shahabi H, Pradhan B, Liu J, and Duan Z 2018 Landslide susceptibility modelling using GIS-based machine learning techniques for Chongren County, Jiangxi Province, China *Sci. Total Environ* **626** 1121–1135
[6] Chen W, Pourghasemi H R, Kornejady A, and Zhang, N 2017 Landslide Spatial Modeling: Introducing new ensembles of ANN, MaxEnt, and SVM machine learning techniques *Geoderma* **305** 314-327
[7] Zhu A X, Miao Y, Wang R, Zhu T, Deng Y, Liu J, and Hong H 2018 A comparative study of an expert knowledge-based model and two data-driven models for landslide susceptibility mapping *Catena* **166** 317–327
[8] Wang Y, Wu X, Chen Z, Ren F, Feng L, and Du Q 2019 Optimizing the Predictive Ability of Machine Learning Methods for Landslide Susceptibility Mapping Using SMOTE for Lishui City in Zhejiang Province, China *International Journal of Environmental Research and Public Health* **16** 368
[9] Wang L J, Guo M, Sawada K, Lin J, and Zhang J, 2016 A comparative study of landslide susceptibility maps using logistic regression, frequency ratio, decision tree, weights of evidence and artificial neural network *Geosci. J.* **20** 117–136
[10] Bui D T, Tuan, T A, Klempe H, Pradhan B, and Revhaug I 2016 Spatial prediction models for shallow landslide hazards: a comparative assessment of the efficacy of support vector machines, artificial neural networks, kernel logistic regression, and logistic model tree *Landslides* **13** 361–378
[11] Bui D T, Tsangaratos P, Ngo P T, Pham T D, and Pham BT 2019 Flash flood susceptibility modeling using an optimized fuzzy rule based feature selection technique and tree based ensemble methods. *Science of the Total Environment* **668** 1038–1054
[12] Zhang K, Wu X, Niu R., Yang, K, and Zhao L 2017 The assessment of landslide susceptibility mapping using random forest and decision tree methods in the Three Gorges Reservoir area, China *Environ. Earth Sci.* **76** 405
[13] Merghadi A, Abderrahmane B, and Bui D T 2018 Landslide susceptibility assessmemt at Mila Basin (Algeria): a comparative assessment of prediction capability of advanced machine learning methods *ISPRS Internasional Journal of Geo-information* **7** 268
[14] Pourghasemi H R, Yansari Z T, Panagos P, and Pradhan B 2018 Analysis and evaluation of landslide susceptibility: a review on articles published during 2005–2016 (periods of 2005–2012 and 2013–2016) *Arabian Journal of Geosciences* **11** 193
[15] Pamela, Sadisun I A, Kartiko R D, and Arifianti Y 2018 Metode Kombinasi Weight of Evidence (WoE) dan Logistic Regression (LR) untuk Pemetaan Kerentanan Gerakan Tanah di Takengon, Aceh *Jurnal Lingkungan dan Bencana Geologi* **9** 77–86
[16] Achour Y and Pourghasemi H R 2020 How do machine learning techniques help in increasing accuracy of landslide susceptibility maps? *Geoscience Frontiers* **11** 871–883.
[17] Wang Y, Feng L, Li S, Ren F, and Du Q 2020 A hybrid model considering spatial heterogeneity for landslide susceptibility mapping in Zhejiang Province, China *Catena* **188** 104425
[18] Tsangaratos P and Benardos A 2014 Estimating landslide susceptibility through an artificial neural network classifier. *Nat. Hazards* **74** 1489–1516

[19] Lombardo L, Cama M, Maerker M, and Rotigliano E 2014 A test of transferability for landslides susceptibility models under extreme climatic events: application to the Messina 2009 disaster *Nat. Hazards* **74** 1951–1989

[20] Cama M, Conoscenti C, Lombardo L, and Rotigliano E 2016 Exploring relationships between grid cell size and accuracy for debris-flow susceptibility models: a test in the Giampilieri catchment (Sicily, Italy). *Environ. Earth Sci.* **75** 1–21

[21] Kornejady A, Ownegh M, and Bahremand A 2017 Landslide susceptibility assessment using maximum entropy model with two different data sampling methods *Catena* **152** 144–162

[22] Pandey V K, Sharma K K, Pourghasemi H R, and Bandooni S K 2019 Sedimentological characteristics and application of machine learning techniques for landslide susceptibility modelling along the highway corridor Nahan to Rajgarh (Himachal Pradesh), India *Catena* **182** 104150

[23] Nguyen V V, Pham B T, Vu B T, Prakash I, Jha S, Shahabi H, Shirzadi A, Ba D N, Kumar R, Chatterjee J M, Bui D T 2019 Hybrid machine learning approaches for landslide susceptibility modeling *Forests* **10** 157

[24] Bachri S, Sumarmi, Irawan L Y, Utaya S, Nurdiansyah F D, Nurjanah A E, Tyas L W N, Adillah A A, and Purnama D S 2019 Landslide Susceptibility Mapping (LSM) in Kelud Volcano Using Spatial Multi-Criteria Evaluation *IOP Conference Series: Earth and Environmental Science* **273** 1-26

[25] Hong, H., Chen, W., Xu, C., Youssef, A.M., Pradhan, B., Tien Bui, D., 2017. Rainfallinduced landslide susceptibility assessment at the Chongren area (China) using frequency ratio, certainty factor, and index of entropy. Geocarto Int. 32, 139–154.

[26] Costache R 2019 Flash-Flood Potential Assessment in The Upper and Middle Sector of Prahova River Catchment (Romania). A Comparative Approach Between Four Hybrid Models *Science of the Total Environment* ***659*** 1115–1134

[27] Bui D T, Tsangaratos P, Nguyen V T, Liem N V, and Trinh P T, 2020 Comparing the prediction performance of a Deep Learning Neural Network model with conventional machine learning models in landslide susceptibility assessment *Catena* **188** 104426

[28] Pourghasemi H R, Moradi H R, Fatemi Aghda S M, Gokceoglu C and Pradhan B 2014 GIS-based landslide susceptibility mapping with probabilistic likelihood ratio and spatial multi-criteria evaluation models (North of Tehran, Iran) *Arabian Journal of Geosciences* **7** 1857–1878

[29] Jebur M N, Pradhan B and Tehrany M S 2014 Optimization of landslide conditioning factors using very high-resolution airborne laser scanning (LiDAR) data at catchment scale *Remote Sensing of Environment* **152** 150–165

[30] Pawluszek K and Borkowski A 2017 Impact of DEM-derived factors and analytical hierarchy process on landslide susceptibility mapping in the region of Rożnów Lake, Poland ***Nat. Hazards*** **86** 919–952

[31] Jiang L, Ling D, Zhao M, Wang C, Liang Q, and Liu K 2018 Effective identification of terrain positions from gridded DEM data using multimodal classification integration *ISPRS International Journal of Geo-Information* **7** 443

[32] Pradhan A M S and Kim Y T 2015 Evaluation of a combined spatial multi-criteria evaluation model and deterministic model for landslide susceptibility mapping *Catena* **140** 125-139

[33] Setijadji L D 2010 Segmented Volcanic Arc and its Association with Geothermal Fields in Java Island, Indonesia *Proc. World Geothermal Congress(Bali, Indonesia April 2010)* pp. 25-29.

[34] Mandal S and Mondal S 2019 *Statistical Approaches fo Landslide Susceptibility Assessment and Prediction* (Switzerland : Springer International Publishing AG)

[35] Coanji T 2016 Slope analysis based on SRTM digital elevation model data: study case on Rokan 4 Koto area and surrounding International Journal of Enginering and High-End Technology **1** 71-75

[36] Van Zuidam R A 1973 *Terrain Analysis and Classification Using Aerial Photographs: A Geomorphological Approach* (Enschede : International Institute for Aerial Survey and Earth Sciences (ITC))

[37] Devkota KC, Regmi AD, Yoshida KH, Pourghasemi R, Pradhan B, Althuwaynee OF, Dhital MR 2013 Landslide susceptibility mapping using certainty factor, index of entropy and logistic

regression models in GIS and their comparison at Mugling-Narayanghat road section in Nepal Himalaya *Nat Hazards* **65** 135–165

[38] Umar Z, Pradhan B, and Ahmad A 2014 Earthquake induced landslide susceptibility mapping using an integrated ensemble frequency ratio and logistic regression models in West Sumatera Province, Indonesia *Catena* **118** 124–135

[39] Chikalamo E E, Mavrouli O C, Ettema J, van Westen C J, Muntohar A S, and Mustofa A 2020 Satellite-derived rainfall thresholds for landslide early warning in Bogowonto Catchment, Central Java, Indonesia *International Journal of Applied Earth Observation and Geoinformation* **89** 102093

[40] Cao C, Wang Q, Chen J, Ruan Y, Zheng L, Song S, and Niu C 2016 Landslide susceptibility mapping in vertical distribution law of precipitation area: Case of the Xulong Hydropower station Reservoir Southwestern China *Water(Switzerland)* **8** 270

[41] Hong H, Liu J, Tien Bui D, Pradhan B, Acharya T D, Pham B T, Zhu A.-X, Chen W, and Ahmad B B 2018 Landslide susceptibility mapping using J48 Decision Tree with AdaBoost, Bagging and Rotation Forest ensembles in the Guangchang area (China) *Catena* **163** 399–413

[42] Zhou S, Wang W, Chen G, Liu B, and Fang L 2016 A Combined Weight of Evidence and Logistic Regression Method for Susceptibility Mapping of Earthquake-induced Landslides: A Case Study of the April 20, 2013 Lushan Earthquake, China. *Acta Geologica Sinica* **90** 511–524

[43] Fawcett T 2006 An introduction to ROC analysis Pattern *Recognition Letters* **27** 861–874

[44] Majnik M and Bosnic Z 2013 ROC analysis of clissifiers in machine learning: a survey *Intelligent Data Analysis* **17** 531–558

[45] Costanzo D, Rotigliano E, Irigaray C, Jiménez-Perálvarez J D, and Chacón J 2012 Factors selection in landslide susceptibility modelling on large scale following the gis matrix method: application to the river Beiro basin (Spain) *Nat. Hazards Earth Syst. Sci.* **12** 327–340.

[46] Chalkias C, Ferentinou, Polykretis M, Polykretis C 2014 GIS-based landslide susceptibility mapping on the Peloponnese Peninsula, Greece *Geosciences* **4** 176–190

[47] Fressard M, Thiery Y, Maquaire O 2014 Which data for quantitative landslide susceptibility mapping at operational scale ? Case study of the Pays d'Auge plateau hillslopes (Normandy, France) *Nat. Hazards Earth Syst. Sci.* **14** 569–588

Geomorphological hazard at some area of the Outer Dieng Volcanic Complex post explosive eruption era

M A I Wardoyo, R Prayuda, Y Susena, F A W Baradi, A Ashari, L F Imamah

Departement of Geography Education, Faculty of Social Science
Universitas Negeri Yogyakarta, Jl. Colombo No.1 Yogyakarta, 55281, Indonesia

Email: muhammadasrori.2017@student.uny.ac.id

Abstract. Dieng area as an important and productive area needs the support of various aspects in encouraging regional growth and development. This region also has been experiences many disasters year by year. Disasters due to volcanism are events that cause many losses, especially casualties in the past. Discussions about potential disasters in the Dieng area should not only be limited to volcanism and earthquakes as hazards caused by endogenous processes. The Dieng area is an old landscape. As an old landscape that is heavily influenced by exogenous factors from a wet tropical climate, other potential hazards in the form of geomorphological hazards also need attention. In fact, there are some geomorphological hazards that occur due to the complexity of the landforms of this area, there are various processes that often occur such as mass movement and erosion and these processes usually have a negative impact on social and economic life. This paper aims to analyze the various types and levels of geomorphological hazards that exist outside of the Dieng Volcanic Complex. This region has experienced major eruptions in the past. At present, in the post-explosive era, it is necessary to know whether geomorphological hazards are limited to eruptions or if other hazards are a bigger threat. The terrain survey method or geomorphological survey is a field observation method used to achieve research objectives. The interpretation of remote sensing images, literature research, and documentation provides support in the form of additional data to complement the primary data from the observation. The results obtained show that the area of the Dieng Volcanic Complex after the explosive eruption era, which is largely composed of volcanic landforms, which physically dominates the denudation process. The denudation process is what causes mass movement and erosion to emphasize that there is a very high potential geomorphological hazard in this research location. This research area is also a productive agricultural production area. Although it cannot be separated from criticism because of its negative impact on the environment, because in the use and management of land carried out by the community it affects mass movement and erosion. Therefore, information on geomorphological hazards is very important to be provided and updated continuously in the interest of ensuring effective and sustainable disaster management.

1. Introduction

The Dieng area is one of the most important highlands in Central Java. This area is socially and culturally known as part of the development of the oldest cultural site in Java. Degroot [1] mentions it as the cradle of Javanese civilization, in its environment with the Borobudur zone in past developments in Central Java. Meanwhile, Lavigne et al [2] mentioned it as the residence of the gods. Physically this area is a volcanic complex ([2], [3], [4]). With the potential for natural and socio-cultural panoramas, the Dieng

Content from this work may be used under the terms of the Creative Commons Attribution 3.0 licence. Any further distribution of this work must maintain attribution to the author(s) and the title of the work, journal citation and DOI.
Published under licence by IOP Publishing Ltd

area is currently developing as an important tourist destination. The number of visits to this region in 2013-2017 has increased from 703,974 to 1,585,477 people [5]. Apart from being a tourist destination, the Dieng area is also a productive agricultural production area. While not ignoring criticism because of its negative impact on the environment, the use of agricultural land in the Dieng area is also large with potatoes as the main product. As a volcanic complex area, this area also has a large potential source of geothermal energy [6].

Dieng area as an important and productive area needs the support of various aspects in encouraging regional growth and development. One that needs attention is the disaster aspect. Information regarding potential hazards and disaster risks is very important to be provided and kept updated in the interests of ensuring effective and sustainable disaster management. From time to time this region also experiences many adverse disasters. Disasters due to volcanism are events that cause many losses, especially casualties in the past. The eruption incident in Sinila Crater in 1979 which triggered the release of poisonous gas in Kawah Timbang which caused the death toll of 149 people of Kepucukan Village, is an example of an eruption that caused disaster in the past. According to Aziz in Ridlo [7], there are three of the 11 craters in Dieng that are considered very dangerous because they often emit poisonous gas. The three craters are, Weigh, Sinila and Sikendang. This incident is an example of an incident that has always been a reminder and learning until now. Data from the Center for Volcanology and Geological Disaster Mitigation show that in the last five years the types of potential hazards that have become a threat to this area are volcanic eruptions, bursts of toxic gas, landslides and potential lava floods.

Discussions about potential disasters in the Dieng area should not only be limited to volcanism and earthquakes as hazards caused by endogenous processes. Dieng area is an old landscape. Harijoko et al [4] show that the oldest part is at least 3.6 million years old on Mount Prahu or 2.53 million years on Mount Bisma. Meanwhile, the youngest part alone is 90 thousand years old on Mount Pakuwaja or 70,000 years old on Mount Seroja. As an old landscape that is heavily influenced by exogenous factors from a wet tropical climate, other potential hazards in the form of geomorphological hazards also need attention. Geomorphological hazards in the form of mass movement and erosion are also common in this region. Based on data from the Center for Volcanology and Geological Disaster Mitigation during 2019, 249 disasters occurred and most of them were landslides, totaling 166 events. Meanwhile, erosion is a very massive problem with the rate of erosion in Dieng reaching 463.86 tons/ha/year. Based on the Erosion Hazard Classification according to Hammer (1981), the Erosion Danger Level (EDL) value obtained is in the very high category [8]. In the past, the Dieng area has experienced explosive eruptions. However, at this time, this region is entering a new era where geomorphological hazards are also a threat that is no less important and needs to be informed.

This study aims to analyze the various types and levels of geomorphological hazards that exist outside the Dieng Volcanic Complex. This region has experienced major eruptions in the past. At present, in the post-explosive era, it is necessary to know whether geomorphological hazards are limited to eruptions or if other hazards are a bigger threat. This study also aims to produce alternative information to support land management by land conditions so as not to cause mismanagement. Alternative information about potential hazards is also needed to support the spatial planning of the area. At this time Dieng residents mostly work in agriculture and tourism. For optimal agricultural and tourism management, detailed and actual information on geomorphological hazards is required.

2. Method

This research uses descriptive-exploratory research. Descriptive research is a research process that is more directed at revealing a problem or situation as there are facts, sometimes given interpretation and analysis. This method is used to explain the symptoms encountered in detail regarding field conditions. The approach used in this study is a geographic approach, namely a spatial approach. The spatial approach in this study is used to be able to identify and describe several geomorphological phenomena and their distribution to assess potential geomorphological hazards in the study area. This research also includes geographic themes, namely location, and place in the discussion. Geographical themes used are mainly location and place. Following previous research Ashari in 2019 [9], The theme is used to provide an overview to strengthen arguments and clarify the discussion regarding the spatial aspects of geomorphological conditions in the research area, which reflects which geomorphic phenomena are in the study area, the existence of these phenomena, and their distribution patterns?.

Primary data collection in this study employs an observation method which was carried out by directly checking the conditions in the field. At the same time, interpretation of remote sensing images, literature research and documentation is carried out to improve secondary data collection results. Observations were made using two methods, namely digital observation (image interpretation) using Geographic Information System (GIS) software and field observation. The location of the sample was determined purposively in each field unit in the study area. The use of tools, namely drones, aims to take aerial photographs of several terrains. Interpretation of remote sensing images is used to obtain additional data or secondary data regarding regional geomorphological data, regional infrastructure, and landform distribution data. In some samples in the field that have been determined, the process of observation and field measurements will be carried out. Each sample location will be selected or purposively determined in each terrain unit in the research area. Terrain units represent changes in the geomorphological conditions of an area. Geomorphological data obtained from ground surveys can also be used as a verification control for image interpretation data, because field survey results are often more realistic and accurate. Interpretation of remote sensing images is used to obtain additional data or secondary data regarding regional geomorphological data. observing in terms of terrain and relief that are difficult to reach, therefore the use of remote sensing images is very important to help complete and obtain geomorphological data from the area due to accessibility, terrain, and affordability factors. Various landform data taken using remote sensing interpretation is determined as a control to emphasize the geomorphological data in this research obtained from Landsat images. Geomorphological data obtained from direct field observation together with geomorphological data obtained using remote sensing image interpretation will complement each other and strengthen the discussion.

Other secondary data collection uses literature and documentation. The literature study was carried out to obtain secondary data used, among others, the book from Van Bemmelen and the article from Harijoko. These guidelines are used to strengthen primary data in land-based data analysis. The landform data in question is in the form of landform characteristics and potential geomorphological hazards. Further strengthening of secondary data collection is by using documentation to obtain secondary data from various documents containing maps in the research area such as RBI maps and geological maps. This method is used to provide geological information as well as existing land use in the area to be studied or observed. The entire series of field data collection in the study area is listed in table 2.1. The table 2.1 below shows the relationship between data types, data collection methods, instruments, and data sources in this study.

Table 2.1 Field data collection techniques

No	Data	Methods of collecting data	Instrument/data source
1	Landform	Observation	GPS, observation sheet, digital camera, Drone
		Imagery Interpretation	Landsat Imagaery, Google Earth
		Literature Study	Harijoko et al and Van Bemmelen
2	Slope	Observation	Observation sheet, Yallon, abney level, roll meter
		Documentation	*Peta Rupa bumi Indonesia* or RBI map (Indonesian Topographical map), Google Earth
3	Relief unit	Observation	Observation sheet, Yallon, abney level, roll meter
		Documentation	RBI map
4	Rock type	Observation	Observation sheet, geological compass, GPS
		Documentation	Geological Map

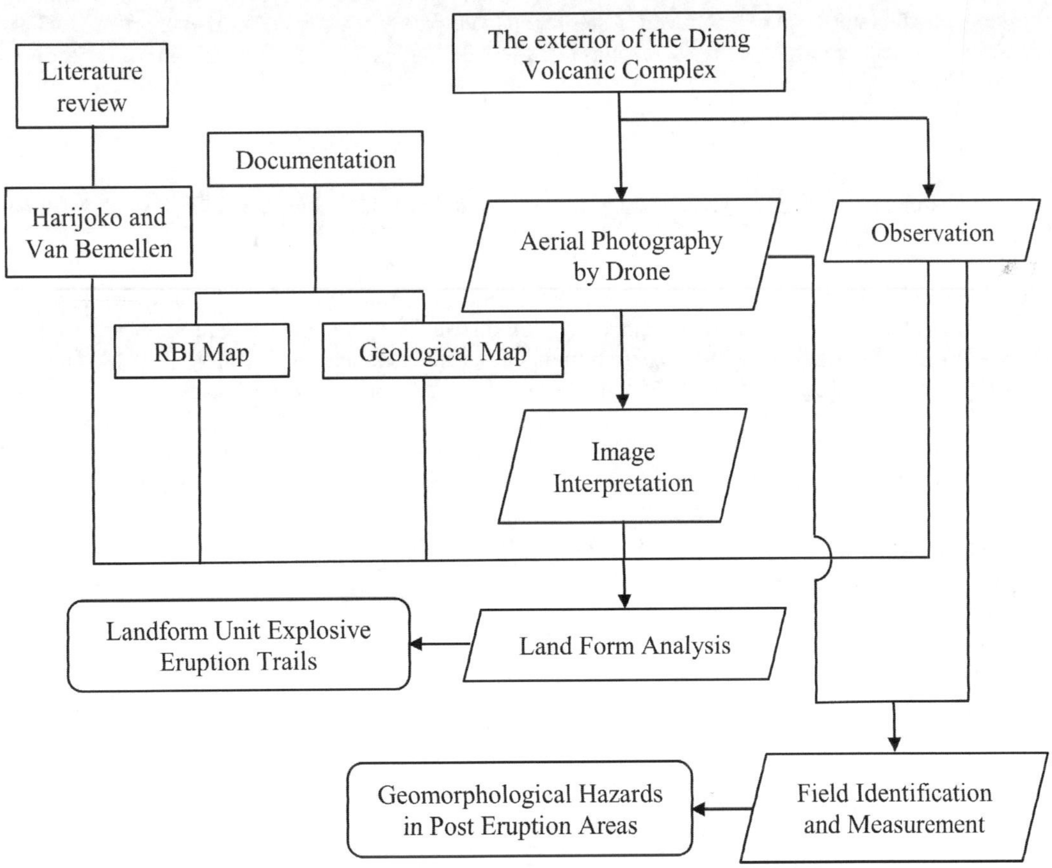

Figure 2.1 Research Procedure

In this research, the main points of emphasis are survey and geomorphological mapping. In addition, field survey activities and geomorphological mapping are carried out as the first step in conducting field analysis in order to identify and analyze the geomorphological hazards of the research area. The data analysis technique used in this research is descriptive analysis with a spatial approach to the location aspect. When using this analysis, the spatial conditions between each unit of the eruption trail part of the Dieng Volcanic Complex must be considered. The use of descriptive analysis with due observance of spatial aspects is obtained from literature studies from reference books and image interpretation which checks field conditions. The field survey was conducted by combining a comprehensive analytic geomorphological survey method and a synthetic terrain survey method. The use of descriptive analysis through the place aspect aims to identify geomorphological hazards after field identification and measurements are carried out, the use of these aspects of the area aims to determine the geomorphological hazards in each area with other areas. The analysis begins by identifying geomorphological conditions. When identifying geomorphological conditions and hazards, the emphasis of the analysis on geomorphological processes, types of landforms and land use distribution in the study area are the main steps taken to obtain relevant and detailed discussion of the surrounding terrain conditions.

This study also uses the same steps as previous research which refers to Ashari [9]. Topographic analysis and classification in the research area related to geomorphic hazard potential is carried out in several stages, namely: (1) detecting terrain related to land forms, geology, soil, hydrology, and vegetation / land use processes, to collect terrain data, (2) the second step describes and delineates the main reliefs, (3) details the terrain based on land form units, and (4) Step 4 is to select and describe the terrain features associated with the geomorphological hazard potential. For a clearer and more complete discussion, the four steps can be continued in step. Step (5) is part of the terrain assessment phase, which is a continuation of the terrain analysis and classification. This stage includes assigning terrain features to geomorphological hazards.

3. Results

3.1. Research Area

The research area covers all areas outside the eastern side of the Dieng Volcanic Complex which administratively includes the Tieng Village, Sikunang Village, Campursari Village, Sembungan Village, Maron Village, Seranggede Village, and Mlandi Village. The five villages cover two sub-sub-districs, namely Kejajar and Garung sub-districs, Wonosobo regency, Central Java. On the southwest side of the research area which is the area of Watumalang Sub-distric using secondary data sources listed in previous research from Ashari in 2019. Astronomically this area is located at 372515 mT to 384411 mT and 9191594 mU to 9203216 mU at UTM zone 49 coordinates (Figure 3.1). This research location also has a total area of 90.2 km^2. The research boundary when viewed physiographically this area is bounded to the north by volcanic landscapes and denudational mountains, namely the Dieng Volcanic Complex, in the eastern part it is bordered by the Sindoro Volcano, as well as parts of the Serayu Mountains in the south and west. Furthermore, administratively this research area is limited to the north and west by Banjarnegara Regency, the south is bordered by Watumalang and Mojotengah sub-districts, and in the east, it is limited by the central side of Kejajar Sub-distric.

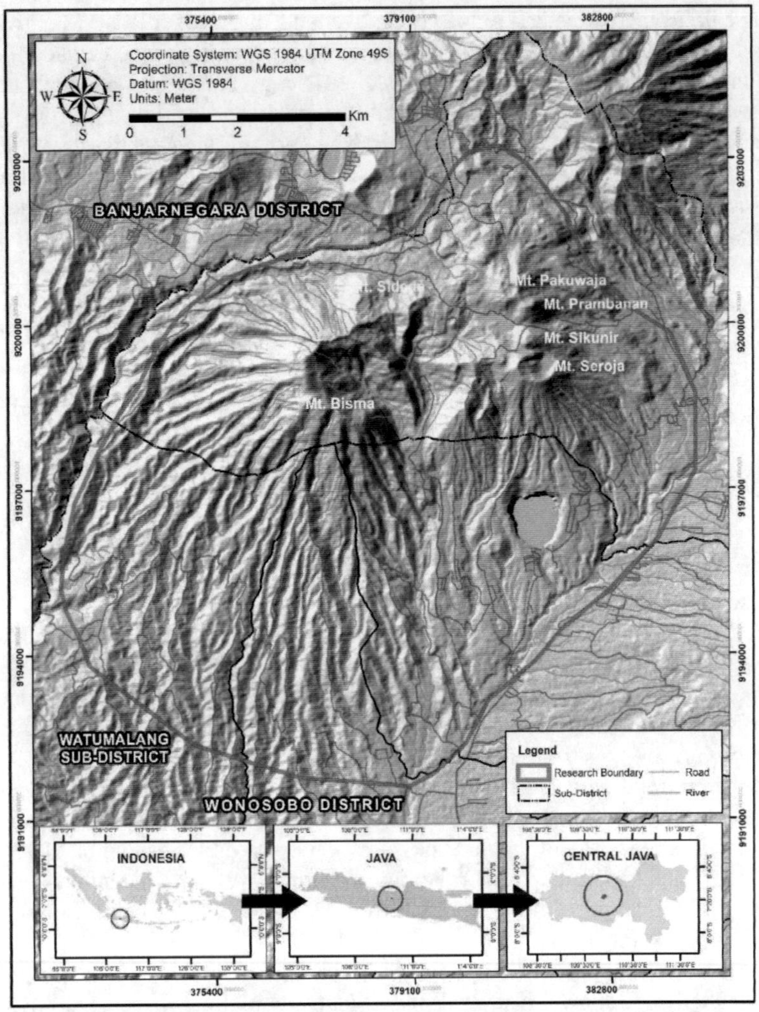

Figure 3.1 The Outer Dieng Volcanic Complex Research Area

From a geomorphological perspective, the study area includes several volcanic groups (Volcanic Complex) consisting of Mount Bisma, Mount Seroja, Mount Sidede, Mount Sikunir, Mount Prambanan, and Mount Pakuwaja, so this area is generally grouped into five sub-units of shape. The landform consist of volcanic cones, volcanic slopes, volcanic foot, volcanic footplain, and volcanic fluvial plains because the average volcano in this study is a type of stratovolcano even though it is currently in a denudational phase. However, if you look at it in more detail, this research area also includes several landforms, including volcanic cones, volcanic slopes, creaters, and caldera.

The geological conditions of the study area based on the information on the geological map of the Dieng Volcanic Complex, Wonosobo Sukhyar, et al [10] have various geological conditions, starting from young dieng surface deposits in the form of alluvium deposits in the valley area between the Volcanoes of Mount Bisma and Mount Seroja and the Cebong Lake. which consists of breccias, sand, and unconsolidated layers showing that the area has flat or sloping morphology. The sedimentary layers of the Dieng volcanic rock which are included in this area are andesite lava and andesite biotite which occupy the northern side of the study area. In general, Sikunir, Prambanan and Pakuwaja volcanic rocks are in the form of andesite and clastic lava deposits from a series of young Dieng volcanoes. There is a lava deposit on the east side of the study, namely the Sikunir lava deposit which can be assumed that an eruption occurred and there was a flow of lava movement that headed east. Meanwhile, the material composition of the Adult Dieng Volcanic Rocks, which are generally in the form of andesite and clastic lava, is found in Bisma, Seroja, and Sidede Volcanoes. Dieng volcanic rocks which are included in this area are the lower part of andesite lava and andesite quartz and volcanic clastic rocks, with a relatively lower silica content compared to the younger part. The andesite lava deposits from Bisma and Seroja volcanoes dominate the constituent materials in the study area. The climate is based on BPS data from Wonosobo Regency [11] from the three sub-sub-districs in the study (Garung, Kejajar, and Watumalang) which have high average rainfall numbers (Table 3.1).

Table 3.1 Rainfall in Garung, Kejajar, and Watumalang

No	Sub-Sub-distric	Years							
		2014		2015		2016		2017	
		Rainfall (mm)	rainy days	Rainfall (mm)	rainy days	Rainfall (mm)	rainy days	Rainfall (mm)	rainy days
1.	Kejajar	3157	212	2136	177	4141	244	3578	252
2.	Garung	3014	190	3212	159	5147	312	3433	218
3.	Watumalang	2389	245	2456	197	4043	305	2590	223

The peak of rain in the three sub-sub-districs has the same character, which occurs at the end of the year, more precisely in November and December, and the lowest rainfall occurs in the middle of the year, more precisely in July and August. Aspects of air temperature in this research area, namely the sub-districs of Kejajar, Garung, and Watumalang are following the state of the mountainous area, the temperature in this area ranges from 14° C to 27° C, but in the dry season the temperature can reach 14° C or less. In the geomorphological context, this high rainfall is a factor that affects the rate of geomorphic phenomena and an important factor that determines the development of the landscape in this area [9]. Seeing from the climatic factors of the data and also seeing from the viewpoint of the landscape of the research area which is a highland and morphological factor in several volcanic cones in the study area also encourages the formation of orographic rain so that the Dieng Volcanic Complex Area has high rainfall. many with great intensity.

Hydrological conditions are characterized by river flow systems with radial and dendritic flow patterns, radial patterns are shown in several rivers that have not been affected by high denudation, and dendritic patterns show characteristics of denudational landforms. Another hydrological condition was found in the spring belt on bending the slopes in the Outer Dieng Volcanic Complex area. In this research area, springs are the main source of water to meet daily needs such as logistics and are used to support the survival of the surrounding community. The use of these springs is by using a pipeline to the springs and distributed to the community. Springs are a source of livelihood for water needs in the study area due to several factors, one of which is the relief condition, which is hilly to mountainous, is a factor that hinders the ease of access to groundwater. The use of these springs is also based on the quality of clean and clear water rather than surface water such as rivers. Land uses include forest, rice fields, gardens, settlements, moor, and shrubs. Land use in the form of forests is still found in the volcanic cones in the Dieng Volcanic Complex. The most dominating land use in this research area is the moor, which is very unfortunate because a lot of this moor is on slopes with steep slopes, clearing land devoted to the use of moor in areas with steep slopes will increase the potential for disasters such as landslides. Settlement areas in the study area are very rare. The settlements located in the research area are in an area where the slope is not that steep and there is main accessibility. The residential area is uneven because the Wonosobo community is more centralized in urban areas. Based on data from BPS Wonosobo Regency (2019), the majority of the population in the research area, namely Kejajar, Garung, and Watumalang Sub-districs are farmers and traders, although there are other jobs such as civil servants and private employees, these jobs are only on a minor scale.

3.2. Geomorphological Study of Research Areas

Geomorphology is a complex and detailed scientific study when translating about the phenomenon of physical conditions on the earth's surface, which includes the form of rivers, hills, plains, beaches, dunes, and many others. Geomorphology examines landforms and the processes or factors that shape them. The forms, processes, and interrelationships between landscapes are at the core of understanding the origin and development of landscape forms. In geomorphology, shape or form has three aspects - composition (chemical and physical properties described by material property variables), configuration (size and shape described by geometric variables) and mass flow (by mass flow variables such as discharge rate, precipitation flow rate and evaporation). Referring to Verstappen [3], geomorphological studies are very complex and based on many aspects of landform, genesis, processes, and the environment while other scientists like Huggett explained, in geomorphological studies, there are several aspects discussed in the study, the aspects in question are structural and process aspects. [12]. The aspect of the study used to discuss a research on geomorphology in this area refers to previous research, namely the Ashari research in 2019 [9], This includes forms of dynamic geomorphological studies related to short-term processes and changes, forms of static geomorphological studies related to actual landforms, forms of environmental geomorphology studies related to landscape ecology, and forms of genetic geomorphology studies related to relief development in the long term.

Discussion about genesis, the research area is largely formed in volcanic landforms. The origin genesis of volcanoes in this area can be characterized by several volcanic cones with river flow patterns classified as radial patterns. As mentioned above in all parts of the study area, when viewed from a geomorphological perspective, there will be found parts that characterize the origin of this research area being a volcanic area, and of course, the study area will include several volcanic clusters (Volcanic Complex) consisting of Mount Bisma, Mount Seroja, Mount Sidede, Mount Sikunir, Mount Prambanan, and Mount Pakuwaja, in general, the morphology of this area is grouped into five landform units, namely volcanic cones, volcanic slopes, volcanic foothills, volcanic foot plains, and volcanic fluvial plains, but is viewed in more detail using images, there is a landform unit in the form Guungapi cone, volcanic foot slopes, creater, and caldera see Figure 3.2. This is because the average mountains in this study are Stratovolcanoes, even though they are currently in the denudational phase. Information regarding the genesis aspect of the volcanic complex in the study area is increasingly clarified in the results of image interpretation with evidence that there are several landforms such as volcanic cones that form a volcanic complex cluster in the study area.

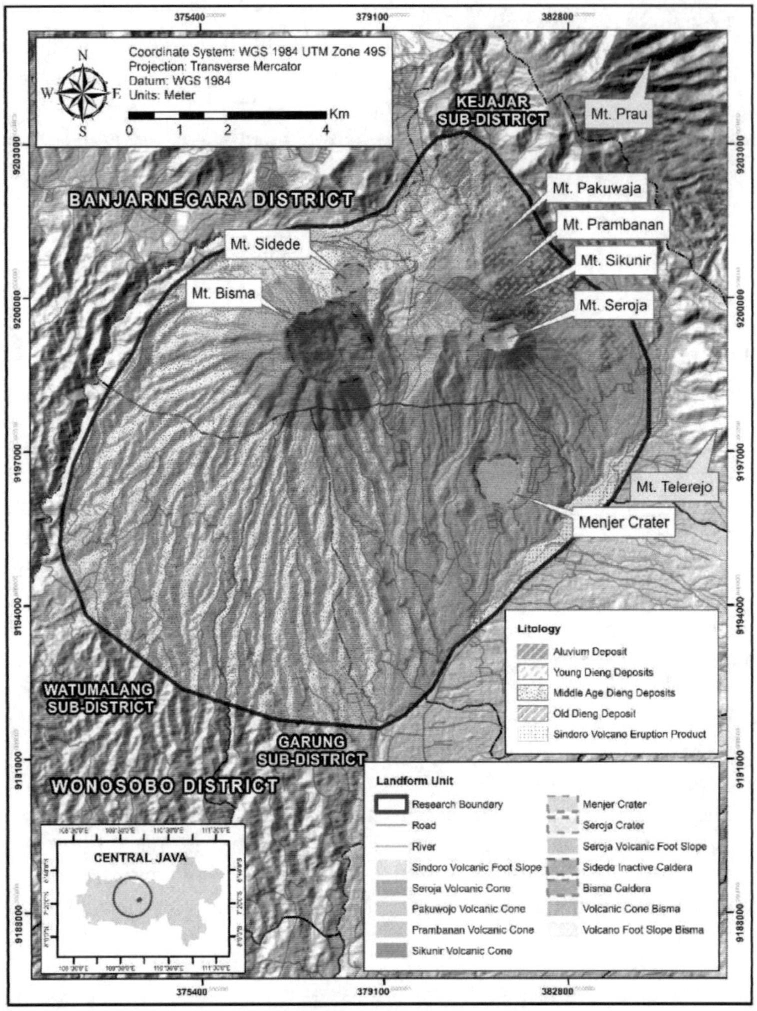

Figure 3.2 The Geomorphological Map of Outer Dieng Volcanic Complex Research Area

The process that dominates the volcanic landforms in the current research area is denudational. Indonesia's tropical climate is one reason that the denudation process has developed in mountainous areas, so that it has characteristics such as air temperature and high rainfall. Denudation is a form of a weathering process on the earth's surface that generally occurs in structural and volcanic landforms. As described by Huggett [12], weathering is the breakdown of rock by mechanical disintegration and chemical decomposition. When exposed to lower temperatures and pressures on the earth's surface and in contact with air, water, and organisms, these rocks begin to decay. Weathering tends to weaken the rock and make it more permeable so that the rock is more susceptible to erosion by erosive agents, and removal of the surface layer of the rock will speed up the weathering process.

Based on the results of the geomorphological survey regarding field conditions, various geomorphological processes have taken place in the current research area. Exogenous processes that dominate and are strong have resulted in the occurrence of several geomorphic phenomena such as erosion and mass movement processes. Geomorphological processes in the form of erosion are found in the research area in various types, namely, in the form of splash erosion, sheet erosion, furrow erosion, and trench erosion, the erosion process of some of these erosions is a factor that causes the valley floor to widen and deepen. This erosion process is also accelerated and exacerbated by human-influenced activities, especially in land management. Human activities such as land reclamation for agriculture and mooring on steep mountain slopes will significantly impact the acceleration of erosion. Soil solum that has become thin is a parameter of the fast rate of erosion. The soil erosion will continue and can be said to be massive if there is an parent soil material called regolith has appeared in several places to the surface because the topsoil has been lost.

There is a geomorphological process besides erosion that is found in the research area, the process is a mass movement consisting of various types, namely slides (rotational and translational), falls, and creeps. Fall and Slide are the most common types. Fall is the downward movement of a mass of rock and soil from which it is free falling. Fall is more common, especially in landscapes with steep rocky slopes and towering cliffs. While slides are a form of mass movement that occurs on a wide scale. This movement occurs because of the mass transfer along with the landslide (slide) plane which moves by sliding from its original place. The two subtypes are rotational and translational slides. Creep mass movement is a very slow plastic deformation of soil or rock. They occur due to pressure exerted by the weight of the soil or rock bodies and usually occur at depth, under weathered layers [12].The high intensity of rainfall, volume, and length of rainy days will affect the mass movement events. Various geomorphological processes such as erosion and mass movement is a natural cycle which when studied using the context of geomorphological hazards can be a threat or danger to living things around the area. The result of geomorphological survey and observations in several sample locations using geospatial technology and documentation as a way to assist data collection shown in Table Geomorphological survey results and observations in several sample locations Table 3.2 and Geomorphological Map of Outer Dieng Volcanic Complex Research Area Figure 3.2.

Table 3.2 Geomorphological survey results and observations in several sample locations

No	Location	Coordinate	Elevation (masl)	Erosion Type	Soil	Rock Type	Slope Shape	Slope	Slope Length	Land Use	Mass Movement Type
1	Mlandi I	49-379606 9195829	1221	Splash, sheet	dusty clay	Lahar / sediment	Concave	45% (steep)	30 m	Moor	Creep
2	Mlandi II	49-379505 9197209	1335	Splash, sheet	dusty clay	Lahar / sediment	Concave	60% (steep)	35 m	Moor	Fall
3	Mlandi III	49-379113 9197946	1501	Splash, trench	Sandy loam	Pyroclastic	Concave	90% (very steep)	55 m	Moor	Fall, slide (translational & rotational)
4	Sembungan I	49-379302 9199297	1770	Splash, trench	Sandy loam	Lahar	Concave	30% (a little steep)	7 m	Bush, Tour	Fall
5	Sembungan II	49-379246 9199685	1967	Splash, sheet	dust	Pyroclastic	Concave	38% (steep)	11 m	Moor	Fall
6	Sikunang	49-379453 9200765	2154	Splash, sheet, furrow	Sandy loam	Lahar / sediment	Concave	38 % (steep)	11 m	PLTU, Moor	Fall
7	Tieng	49-382067 9201126	1950	Splash, sheet	dusty clay	Pyroclastic	Straight	12% (inclined slopes)	30 m	Moor	Slide (rotational)
8	Seranggede	49-383461 9198128	1421	Splash, sheet, trench	dusty clay	Lahar / sediment	Concave	40% (steep)	10 m	Bush	Slide (translational & rotational)
9	Maron	49-381366 9195904	1229	Splash, sheet	Sandy loam	Lahar / sediment	Concave	34% (steep)	16 m	Tour, tea garden, tobacco	Slide (translational & rotational)

3.3. Geomorphological Hazards in the Outer Dieng Volcanic Complex

The study of geomorphological hazards in this research area includes mass movement and erosion. The strong climatic influence along with the terrain of the old outer Dieng area causes various geomorphological phenomena that can affect and threaten the safety of living things. This research area has regional characteristics that have reached the denudational phase. This is reinforced because the outer region of Dieng is an area that has been eroded by high exogenous factors so that the constructive nature of endogenous factors such as tectonic and volcanic factors is no longer dominant. The many types of geomorphological hazards in the study area indicate that the relationship between high climatic factors that affect the area. These symptoms are the main source of this research location having various kinds of geomorphological phenomena that can cause disasters to live things, especially the people around the outer Dieng, therefore it can be called a geomorphological hazard. There are several types of mass movements identified through geomorphological surveys that are often encountered in the field. These include, among others: falls, slides, and creeps.

Taking from the main point the characteristics of the landforms and the ongoing series of geomorphic processes, geomorphological hazards in the outer Dieng originate from denudational processes by the climate in the form of mass movement and erosion. The length of the rainy day, the high intensity of rainfall, and the volume, greatly influence the mass movement events. This mass movement occurred in various parts of Mount Bisma, Mount Seroja, Mount Sidede, Mount Sikunir, Mount Prambanan, and Mount Pakuwaja, both in volcanic cone units and in other volcanic units. This mass movement in outer Dieng occurs in various types. The process of mass movement found during terrain survey were of various types, namely slide (rotational and translational), fall, and creep.

This slide type mass movement occurs because of the old mountain, the high level of weathering, and the type of soil that has high aggregate as seen in the table of measurement results that the average area that has the form of slide type mass movement has clay and loamy soils. Both types of soil are classified as the last phase soil or can be said to be old and have very small soil particles so that the soil porosity is low. This type of slide occurs around the eruption of Mount Bisma because Mount Bisma is one of the oldest volcanoes in the Dieng Volcanic complex. Harijoko et al [4], state that the oldest part in the Dieng Volcanic Complex area is at least 3.6 million years old on Mount Prahu or 2.53 million years on Mount Bisma even the youngest part is 90 thousand years old on Mount Pakuwaja or 70,000 years at Mount Seroja. Of course, on Mount Bisma it is very possible to have a slide-type mass movement. Steep slope coefficient is the main factor related to the slide-type mass movement. In the field, the slides are dominated by very steep slopes, with a slope of 12% - 90%. The slide mass movements that were found consisted of rotational and translational slide types. While rotational slide mass movement with the composition of the material does not change much and is generally influenced by rotational motion, it is called a slump. This rotational slide-type has bedrock in the form of debris slump, earth slump, and rock slump. Meanwhile, translational slides are a type of mass movement that is influenced by translational motion with the composition of the material changing a lot. The basic rock types of translational slides are rapids, ice sliding, debris slide, earth slide, debris block slide, earth block slide, rock slide, and rock block slide. The adult Dieng rock formation contributes to the large mass movement of the slide type because when viewed from the type of material it is composed of, it is lava andesite.

The basic material in the Seroja Volcano, namely the Tieng, Seranggede, Maron area, which is andesite lava as a material has low porosity, make a lot of mass movement in this area especially the slide-type mass movement. The result of the terrain which has low porosity is that the infiltration in this area is not too much so that the rainfall will be responded to by the soil as runoff and cause or tend to cause slide mass movement. Even in Ashari's previous research [9], the Bisma Volcano can have several types of mass movement that are more varied, of course, this factor is caused by the terrain which is composed of Dieng volcanic material in the Bisma volcano and material from the Jembangan volcanic material in the fold belt. In this area, you can find the mass movement of the slide, creep, and fall types. Old volcanic rock material is one of the causes, this is confirmed by Hadmoko [13] who explained that based on landslide data throughout Java between 1981 and 2007, from this data it can be analyzed that landslides mainly occur on slopes of 30°-40° in weathered soil areas caused by the foundation material of an old terrain, such areas were usually in a denudational state. The steeper slope of the case on a steeper slope above 40o landslide events is reduced because the steeper location material is relatively more stable and is supported by less land use by the community. Similar discoveries have occurred in the Seroja volcanic area which is composed of several old volcanic materials such as clastic pyroclastic debris consisting of pyroclastic breccias and agglomerates with the basic material in the form of andesite lava and even experiencing many landslides. In the field, it shows that the slides in the Seroja Volcano area occur at a slope of 12% to 40% with a height ranging from 1200 masl to 1950 masl. There is also a slide event at a 90% slope with an altitude of 1500 masl in the Bisma volcano crater.

Besides the slides, there are also creep and fall type mass movements. On the other side of the Vulkan Bisma, there are many creep and fall type mass movements. The lithology factor of volcanic products which has greater permeability and porosity is very likely to have an effect on the existence of the creep and fall type mass movement coupled with the increasingly steeper slope so that creep and fall can exist. Creep is a mass movement that moves at a slow speed. Creep has a small amount of water and is often found in the fold zone [9]. In the study area, creeps were identified on upland agricultural land at the foot of the Bisma Volcano. Creep is found on a 45% slope with a height of 1221 masl. While a fall is a fall or mass of rock that moves through the air, including free-fall, jumping and rolling over boulders and scrap material without much contact with one another. Fall was identified in Vulkan Bisma, Vulkan Seroja, Vulkan Sidede, and Vulkan Pakuwaja. Fall is often found in upland agriculture with a slope of 30% to 90% and an altitude ranging from 1300 masl to 2150 masl.

Another very influential factor in mass movement is rainfall. In certain seasons, such as the rainy season, mass movement often occurs. Along with the peak of the rainy season, the intensity of mass movement also increases. In the outer area of the Dieng volcanic complex, which is based on mountainous areas and highlands, of course, has relatively high rainfall, the strato volcanic cones these areas play a very important role in the formation of orographic rainfall. Due to higher rainfall, there will be more water infiltration, causing a lot of mass movement to occur. Apart from the rainfall factor, Humans are a very influential factor in the mass movement, for example, land use. land use greatly affects the movement of the masses if it is not properly regulated and managed. From year to year, the population growth rate will increase and along with the development of life there are human needs which cause people to use more land, in the outer area of the Dieng volcanic complex or this research area, many people use the land even including lands that are prone to landslides. Land use activities by the community have also triggered landslides and various other types of mass movements.

Widespread land degradation occurs due to potato cultivation activities, landslide impacts during the rainy season in many residential points and roads along the Garung-Kejajar sub-sub-distric, flooding that occurs due to disruption of watersheds due to landslides, and drought during the dry [14]. According to Ashari [9], moor, then rice fields, and then forest are the forms of land use that occur the most landslides, especially in the moor. there is a lot of lands uses in this area which has led to mass movements in the Watumalang area. The land use includes moor and slope cutting which is used for road construction. Agriculture in the moor in this study area is found on steep slopes such as the middle slopes to the top of the mountains. the improper use of upland land resulted in an increase in the potential for large mass movements..

In the field conditions, when the geomorphological survey was carried out, there was another geomorphological hazard besides mass movement. Erosion is common in this research area and erosion is also one of the Geomorphological hazards besides mass movement. However, it is emphasized that this research only identifies the appearance or presence of erosion and observes the type of erosion in the field. Calculating the rate of erosion and determining the level of erosion hazard has not been carried out in this study. The phenomenon of erosion can also trigger various forms of threats to living things, both loss of property and damage to infrastructure and public facilities. According to Arsyad [15], an erosion event is an event of moving or transporting land or part of the land from one place to another using natural media such as water and wind. If there is erosion, soil or part of the soil from one place will be eroded and transported which is then deposited in another place. In fact, the factors that influence mass movement and erosion are almost the same and interrelated, erosion can also be caused and influenced because it is the result of interactions between climatic factors, topography, vegetation, and human intervention (management) of land [15].

There are variations in the types of erosion types found based on the results of the terrain survey in the research area. This erosion includes various types namely splash erosion, trench erosion, furrow erosion, and sheet erosion. However, splash and sheet erosion are the dominant types of erosion types or are often found in the study area. Several variations of this erosion are explained according to Nursa'ban [16], splash erosion is erosion that occurs at the beginning of rain because this erosion is caused by the influence of rainwater. The rainwater causes an impact or collision because it gives certain energy when it falls (kinetic energy), this energy can release soil particles. The release of soil particles from the soil mass due to splash erosion is very dependent on the type of soil being eroded. This spark erosion can be found in every field and research location. Splash erosion is the initiation or cause of other erosion such as sheet, furrow, and trench erosion. The erosion found in each study area shows that the outer Dieng area has a high rainfall intensity. This is supported by the rainfall data of BPS Wonosobo Regency [11] from the three sub-sub-districs in the study, namely Garung Sub-distric, Kejajar Sub-distric, and Watumalang Sub-distric which have a high average number of rainfall. The average rainfall in 2014-2017 in the three sub-sub-districs reached approximately 3,253 mm. Of course, the high rainfall is influenced by factors in the research area which is a plateau and is included in the volcanic complex. The high rainfall has a high influence coupled with the slope conditions, land use. The high intensity of rain causes an increase in the destructive power of soil particles, namely soil aggregates so that there is a lot of splash erosion in each location that has been identified, this is in accordance with research [9].

Of course, splash erosion will stop if the raindrops are no longer able to penetrate the thickness of the water layer, this is when the sheet erosion process begins. Sheet erosion is a continuation of splash erosion and can be found clearly over a relatively uniform surface area. Next is a continuation of sheet erosion. Sheet erosion is the research area can be found in the mean sample locations of the outer Dieng volcanic complex. Sheet erosion is continued erosion of the predominant splash erosion in this area, the average sheet erosion occurs on several slopes from sloping to steep. The next erosion is groove erosion starting from the presence of surface runoff concentration. This erosion often occurs on land on the slopes of mountains to form grooves. Erosion also greatly affects the activities of increasing the socio-economic needs of a place. The more the population continues to increase and coupled with the use of land to meet economic needs, the negative impacts on the environment often occur. The natural vegetation change in the research area is the use of moor. As has been explained above, the moor is a land use that is very vulnerable to mass movement and erosion. Planting plant species in the moor also greatly influences the type and magnitude of erosion. The average types of crops cultivated in this region include corn, chilies, potatoes, tobacco, and various other types of crops. As only in the research of Khadija et al [17], in the Kalaya River Basin, Morocco. Degradation of vegetation cover is still a very common problem in many locations in Morocco, in fact, it directly affects the annual amount of land loss, this is due to side effects and negatives such as population growth, technological advances, and an increase in the forest-economy. The degradation of natural vegetation has reached surprising levels and has even caused various natural disasters such as erosion. The impact of forest and land degradation has made the Dieng Plateau area save a lot of potential disasters [14] including toxic gases caused by the volcanic activity of the Dieng volcanic complex [18].

According to Nursa'ban [16], if the distance reaches hundreds of meters, what happens is not channel erosion but surface runoff erosion. When the size of the groove is very large, it cannot be removed just by regular plowing, or the channel is directly connected to the main drain, then the erosion that occurs has met the category of trench erosion. Subsequent erosion is trench erosion which is found in several locations with 30-90% slopes. This is certainly relevant to Ashari's research [9] at Watumalang that the results at a slope of 30° or more show that the erosion rate of the channel and trench erosion type is getting bigger and growing. It can be said that the steeper a slope will increase the result of erosion that occurs in an area. The effect of a steep slope will accelerate the runoff rate. When the surface runoff rate is high, it will certainly increase the strength of the runoff to erode the land that is passed. This is confirmed according to Guerra et al [19], which explains that the dominant hillside processes are associated with gravity and flowing water. Other influences such as human activities also play an important role in the hillside process, due to land-use change and logging of natural vegetation. Especially in tropical areas, where rainstorms and other extreme weather are very likely, the signs of erosion are clear, erosion in tropical areas is usually higher and filled with sediment due to deposition.

Although mass movement and erosion are natural phenomena, human activities often accelerate the erosion process. Erosion can occur naturally, due to the angle of inclination and rainfall, and other factors as described above. Humans are a major and important role in this geomorphological process, they present various modes of events and causes and effects as a very influential role in nature today. According to Andriana [8], Dieng is an area that must be protected from production activities and other human activities that can damage its protective function. According to the Wonosobo, Level II Sub-distric Regional Regulation Number 1 of 1996 concerning the Regional Spatial Plan for the Wonosobo Level II Distric, the Dieng Plateau is part of a protected area. In Regional Regulation No. 1 of 2004 [20] concerning Regency Spatial Planning. Banjarnegara, Dieng Plateau Area is also a Protected Function Area. The Dieng Plateau area is a protected area that should be protected from production activities and other human activities that can damage its protection function. However, in fact, this area is used by humans to meet their needs by utilizing large-scale land for planting seasonal crops, namely potatoes, because potatoes are a superior commodity for farmers in the Dieng Plateau. Various natural events caused by the use of human land in this area, among others, cause geomorphological hazards that can cause losses to life around the outer area of the Dieng volcanic complex, the potential geomorphological hazards, both erosion, and mass movement in the outer area of the Dieng volcanic complex, are certainly reinforced by the denudation process by climatic influences. The outer Dieng area, which is part of the Dieng volcanic complex, has various variations in the age of the landform unit, which makes the phenomena of mass movement and erosion in this area very varied. The outer area of the Dieng volcanic complex, if viewed from genesis, has a lot to do with volcanic events, but these events played an active role in the past, but this volcanism does not play a role as a trigger for geomorphological hazard at this time but the results of the denudation process such as mass movement and erosion.

4. Conclusion

The Outer area of the Dieng Volcanic Complex, is one of the most vulnerable areas and currently facing geomorphological hazards caused by the denudation process in the form of erosion and mass movement. The Outer area of the Dieng Volcanic Complex, if viewed from genesis, has a lot to do with volcanic events, but these events played an active role in the past there will be found parts that characterize the origin of this research area being a volcanic area like a geological system in this place. However, volcanic activity as in the past does not pose a geomorphological hazard in the present era, but the results of the denudation process such as mass movement and erosion. Due to the sharp climatic influence and slow endogenous processes that create new problems, namely the denudation process, this kind of process is the main cause of damage or can be called a geomorphological hazard. The varying ages of volcanoes and the strong weathering of the rocks exacerbate this situation and encourage geomorphological disasters to become more severe and life-threatening. Based on the research results as described above, the outer area of the Dieng Volcanic Complex is faced with environmental degradation problems which are getting worse over the years. The main cause besides the denudation and relief process is the conversion of forest land to cultivation land and in some areas, it has turned into settlements. Conservation behavior that is built by the community and the government is very much needed, this is because the natural vegetation of the study area if reforestation or reforestation is not carried out, will continue to be eroded by the factor of land clearing as cultivation land or residential land. Humans should be aware to protect and provide positive treatment to their surroundings because of the environment human can survive.

5. Acknowledgment

We are very grateful to the Faculty of Social Sciences, Yogyakarta State University who has funded its research through the 2020 Student Research Funding Program and to Mr. Arif Ashari who has guided in writing this article, and not to forget to all employees and the community in Kejajar and Garung Sub-districs, Wonosobo Regency which gives permission for the research to be carried out.

References

[1] V. M. Y. Degroot, "Candi, Space and Landscape: A Study on the Distribution, Orientation and Spatial Organization of Central Javanese Temple Remains," *Sidestone Press*, pp. 1–385, 2009, [Online]. Available: http://www.amazon.com/dp/9088900396.

[2] F. Lavigne *et al.*, "People's behaviour in the face of volcanic hazards: Perspectives from Javanese communities, Indonesia," *J. Volcanol. Geotherm. Res.*, vol. 172, no. 3–4, pp. 273–287, 2008, doi: 10.1016/j.jvolgeores.2007.12.013.

[3] H. T. Verstappen, *Garis Besar Geomorfologi Indonesia*. Yogyakarta: Gadjah Mada University Press, 2013.

[4] A. Harijoko *et al.*, "Geochronology and magmatic evolution of the Dieng Volcanic Complex, Central Java, Indonesia and their relationships to geothermal resources," *J. Volcanol. Geotherm. Res.*, vol. 310, pp. 209–224, 2016, doi: 10.1016/j.jvolgeores.2015.12.010.

[5] Department of Tourism and Culture, "Statistik Kunjungan 2013 – 2017," *Pemerintah Kabupaten Banjarnegara*, 2018. https://budparbanjarnegara.com/about/statistik-kunjungan/.

[6] A. Harijoko, R. Uruma, H. E. Wibowo, L. D. Setijadji, A. Imai, and K. Watanabe, "Long-Term Volcanic Evolution Surrounding Dieng Geothermal Area, Indonesia," *Proc. World Geotherm. Congr. 2010 Bali, Indones. 25-29 April 2010*, no. 2, pp. 25–29, 2010.

[7] R. Muhamad, "40 Tahun Horor Kawah Sinila Tewaskan 149 Warga Dieng," *Liputan 6*, 2019. https://www.liputan6.com/regional/read/4112169/40-tahun-horor-kawah-sinila-tewaskan-149-warga-dieng (accessed Mar. 16, 2020).

[8] R. Andriana, "Evaluasi Kawasan Lindung Dataran Tinggi Dieng Kabupaten Wonosobo," pp. 1–135, 2007, [Online]. Available: eprints.undip.ac.id/18791/1/RENI_ANDRIANA.pd.

[9] A. Ashari, "Understanding geomorphological hazard in Watumalang Area post Bisma volcanic era. Denudational or multihazard ?," 2019, doi: 10.1088/1755-1315/271/1/012006.

[10] R. Sukhyar et al., "Geologic Map of Dieng Volcano Complex Central Java," *Volkanological Surv. Indones. Direktur*, 1986.

[11] BPS Kabupaten Wonosobo, *Kabupaten Wonosobo dalam Angka 2019*. Wonosobo: BPS Kabupaten Wonosobo, 2019.

[12] R. J. Huggett, *Fundamentals of Geomorphology*. 2011.

[13] D. S. Hadmoko, F. Lavigne, J. Sartohadi, C. Gomez, and D. Daryono, "Spatio-Temporal Distribution of Landslides in Java and the Triggering Factors," *Forum Geogr.*, vol. 31, no. 1, pp. 1–15, 2017, doi: 10.23917/forgeo.v31i1.3790.

[14] S. Mutmainah, "Rehabilitasi Lahan Kritis Sebagai Praktik Disaster Governance Berbasis Komunitas di Kawasan Dieng Plateau Kabupaten Wonosobo," *J. Polit. Gov. Stud.*, vol. 7, no. 3, pp. 171–180, 2018.

[15] S. Arsyad, *Konservasi Tanah dan Air*, 2nd edition. Bogor: IPB Press, 2012.

[16] M. Nursa'ban, "Pengendalian Erosi Tanah Sebagai Upaya Melestarikan Kemampuan Fungsi Lingkungan," *Geomedia*, vol. 4, no. November, pp. 93–116, 2006.

[17] K. B. H. Lech-hab, L. K. Issa, A. Raissouni, A. El Arrim, A. A. Tribak, and R. Moussadek, "Effects of Vegetation Cover and Land Use Changes on Soil Erosion in Kalaya Watershed

(North Western Morocco)," *Int. J. Geosci.*, vol. 06, no. 12, pp. 1353–1366, 2015, doi: 10.4236/ijg.2015.612107.

[18] H. Nugraha et al., "Volcanic Hazard Management in Dieng Plateau Area. Proceedings Ecosystem Based Disaster Risk Reduction," no. December 2012, 2013.

[19] A. J. T. Guerra, M. A. Fullen, M. DO C. O. Jorge, J. F. R. Bezerra, And M. S. Shokr, "Slope Processes, Mass Movement and Soil Erosion: A Review," *Pedosphere*, vol. 27, no. 1, pp. 27–41, 2017, doi: 10.1016/S1002-0160(17)60294-7.

[20] Peraturan Daerah, "Tata Ruang Wilayah Kabupaten Banjarnegara." Pemerintah Daerah Kabupaten Banjarnegara, Banjarnegara, 2004.

IOP Conf. Series: Earth and Environmental Science **884** (2021) 012008 doi:10.1088/1755-1315/884/1/012008

Developing Sustainable Schools in the Shadow of Active Stratovolcano: A Study at Disaster-Prone Areas of Merapi Volcano

Nurhadi, A Ashari, S A Sutrisnowati

Department of Geography Education, Universitas Negeri Yogyakarta

Email: nurhadi_fise@uny.ac.id

Abstract. Education is a right for people that must be given to children, even in an emergency of eruption disaster. To ensure the sustainable education during the crisis situations, ideas on how to develop a sustainable school system in the stratovolcano area which is prone to eruption disaster are needed. This research was conducted at schools located in the western side of Merapi Volcano, attempting to provide alternative information about the potential and role of schools in providing education during emergency situations of eruption disaster. Data in the study were collected through interviews, observations, remote sensing image interpretations, and documentations. Data were analysed by scoring and spatial analysis using geographic information systems. The results of the study are as follows. First, schools have the potential to provide education during the disaster emergency situations. School potential consists of physical infrastructure and human resource potential in the form of the role of the principal and teachers. Second, most of the level of school preparedness in facing disasters is still in the low category. This needs to be strengthened to ensure that the schools continue to function during the crisis period. Last, a division of roles in the sister school system is needed based on the location of the school toward the eruption centre. The schools in the safe zone act as a buffer for schools in vulnerable areas. Overall, schools have the potential to keep running the learning process during the crisis periods. This requires good management through the sister school system within the framework of sustainable school initiatives.

1. Introduction

The study of improve in the field of eruption disaster risk reduction is still very relevant in Indonesia because Indonesia is one of the countries with the highest volcanic disaster potential in the world [1]. Data from the Indonesian Centre for Volcanology and Geological Hazard Mitigation shows that from January to the beginning of April 2010, there were at least 20 volcanoes that experienced an above normal activity phase, where some volcanoes erupted. Many problems arose then because the areas were included in the Disaster-Prone Areas (DPA) eruption and was occupied by many residents. Lavigne et al [2] explain that the slope of 130 active volcanoes in Indonesia has indeed been inhabited by high density population for thousands of years. Among the various regions in Indonesia, Java Island is the region which has the highest level of volcanism. Historical eruption records show that almost half of the eruptions in Indonesia occurred on Java [1]. On the other hand, Java Island which is only 6.9% of the country's surface is occupied by 60% of Indonesia's population [3].

Merapi Volcano in Central Java is one of the most active volcanoes in the world [4] characterized by a short eruption range from one to fifteen years [5]. After the VEI 4 scale eruption in 2010, volcanic

Content from this work may be used under the terms of the Creative Commons Attribution 3.0 licence. Any further distribution of this work must maintain attribution to the author(s) and the title of the work, journal citation and DOI.
Published under licence by IOP Publishing Ltd

activity has increased again in Merapi Volcano in 2018. The increase in volcanic activity is characterized by small-scale eruptions that have occured repeatedly and continuously from May 2018 to April 2020. The high risk of eruptions disaster is not only caused by the level of volcanic activity but also the large number of residents living in those disaster-prone areas. Based on the data from the Global Volcanism Program [6], in the beginning of 2020, the population of the areas around 30 km from Merapi Volcano is not less than 4,348,473 inhabitants. Meanwhile, the Indonesian National Disaster Management Agency [7] explain that the population in the disaster-prone areas of Merapi reaches 226,618 people, with a population density of 720.1 people/km^2.

Dukun Sub-district in Magelang District, which lies on the western flank of Merapi Volcano, is an area that is greatly affected by eruptions. This is because many eruptions have struck this area since the past time [5]. Repeated eruption disasters offer a great impact on people's lives and livelihoods. In the 2010 eruption, for example, many aspects of life could not run normally when residents had to evacuate for more than one month. The education is among the most affected sector by eruptions. Many schools do not carry out education because they are located in disaster-prone areas and residents are displaced to other places. The disruption of the educational process, as recorded by the Indonesian National Disaster Management Agency [7], has resulted in a loss of 57% which is greater than the loss caused by the damage of the school's infrastructure.

Learning from the experience of the 2010 eruption, it is necessary to formulate strategies for dealing with future crisis conditions so that education can still be run. Based on the data from the Ministry of Education and Culture of the Republic of Indonesia (2020), there are 40 schools from elementary to secondary levels with a total of 4510 students in this region. Some of these schools are included in DPA and some are not. In a disaster emergency situation, schools which are not included in DPA can help schools in the DPA area organize education. Various schools need to work together in developing sustainable school initiatives. As an initial step, it is necessary to identify the potential of the schools, the role of the schools in handling disasters that have been going on so far, and the level of school's preparedness in dealing with disasters.

The concepts of sustainable and safe schools have developed in the field of disaster risk reduction in various countries. In this concept, schools continue to function properly during the crisis situations [8], even they help manage disaster emergencies [9]. In general, a comprehensive school safety policy has been applied in most countries in the world [10]. Based on this experience, in the disaster-prone area of Merapi Volcano, a sustainable and safe school system needs to be developed to ensure the sustainability of education in crisis and post-disaster recovery situations. At present, in the Dukun Sub-district, a sister village system has been developed between the refugee-contributing villages and refugee-receiving villages. It is better to enhance this sister village with the sister school system in order to realize a sustainable education in those areas.

This research attempts to provide alternative information about the development of sustainable schools in disaster-prone areas. Thus, this research has three goals. First, we illustrate the potential of schools in handling emergency eruption disaster. Second, we describe the level of school preparedness in facing eruption disasters. Third, we discuss the role of schools in disaster prone areas in developing sustainable schools. The rest of this paper is organized as follows. Section 2 describes the research methods, section 3 elaborates the results, section 4 presents the discussions, and section 5 explains the conclusion of this work.

2. Methods

2.1. The Study Area

This research was conducted in the Dukun Sub-district, Magelang Regency, Central Java, which lies on the western flank of Merapi Volcano. Geomorphologically, the study areas cover all parts of the stratovolcano landform from the peak to the volcanic footplain, however the areas which are occupied by residents only spread in the volcanic foot and volcanic footplain. Viewed from its position against the threat of Merapi Volcano eruption, the area occupied by residents is divided into Disaster-Prone Area (DPA) III, DPA II, DPA I, and non-DPA (Figure 1). This DPA categorization is carried out based on the DPA map published by the Indonesian Center for Volcanology and Geological Hazard Mitigation. DPA III is the area with the highest level of hazard.

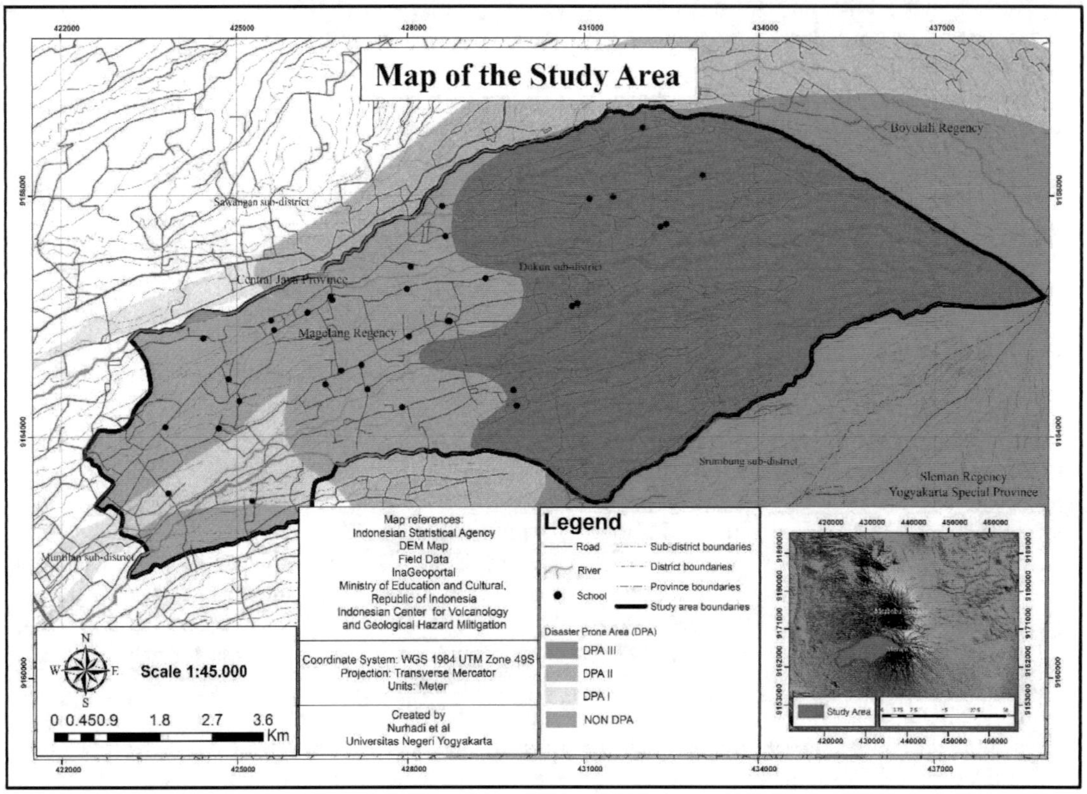

Figure 1. The Study Area

The study area has annual rainfall that varies between 295 to 1734 mm. Hydrologically, in the Dukun Sub-district, there are perennial rivers such as Pabelan, Trising, Senowo, Lamat, and Blongkeng [11]. The rivers are perennial because they get a huge water supply from the springs around this area as part of the volcanic spring belt in the Merapi Volcano [12] or Merbabu Volcano area [13]. The Dukun sub-district has varied aquifer conditions, namely aquifers with moderate productivity in volcanic foot to aquifers with high productivity in volcanic footplain [11]. The potential of this good aquifer is due to the types of material in this region. According to the Geology Map Yogyakarta Sheet 1993 [14], the types of material in the study area consist of the Young Merapi Volcano Deposits, the Old Merapi Volcano Deposits, as well as the Lava Dome and Lava Melt. Pyroclastic and lava materials, both old and young deposits, dominate the types of material in this area. This condition influences the existence of widely distributed productive aquifers. The types of land use based on the Indonesian Rupabumi Map of Kaliurang and Muntilan Sheets (2000) consist of settlements, rice fields, dry fields, mixed gardens, forests and shrubs. Rice fields are the most common type of land use whose coverage reach 47% of the area while settlements are only 9%.

Specifically, this research was conducted at all schools in the Dukun Sub-district from elementary to secondary levels. Based on data from the Ministry of Education and Culture of the Republic of Indonesia, the number of schools in this region in 2019 reaches 35 schools, consisting of 28 elementary schools (ES), 5 junior high schools (JHS), and 2 senior high schools (SHS). Ten schools are included in

DPA III, 14 schools belong to in DPA II, two schools are included in DPA I, and 10 other schools are outside of DPA (see Figure 1). The distance from the eruption center ranges from 7 to 16 km (Figure 2). The schools included in DPA III is closer to the eruption center. The total number of students in 2019 was 4850. In DPA III, there were 1,047 students (22%) that spread in 10 schools; in DPA II, there were 2,425 students (50%) that spread in 14 schools; and in non-DPA and DPA I areas, there were 1,378 students (28%) that spread in 11 schools (Figure 3). There are many students in DPA II because there is one junior high school and one senior high school in that region which has a very large number of students since the schools are supported with adequate facilities.

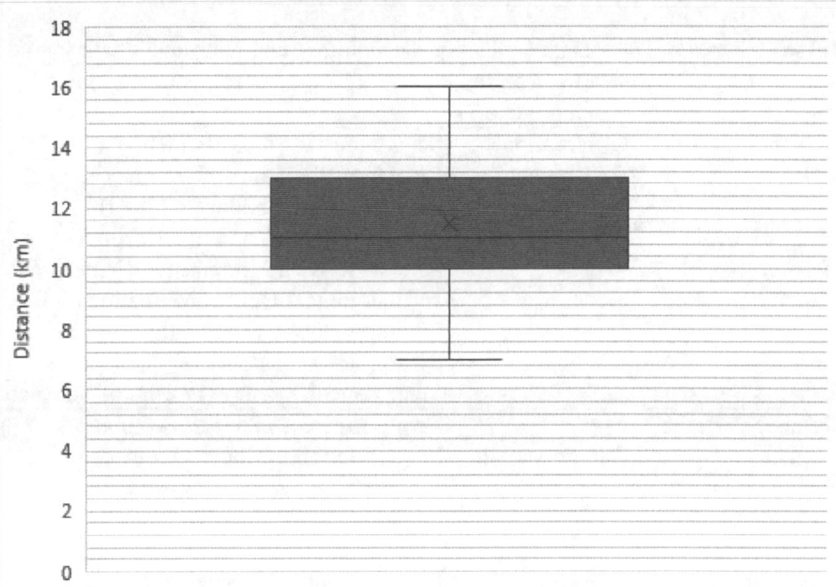

Figure 2. The boxplot for school distance from the eruption center of Merapi Volcano

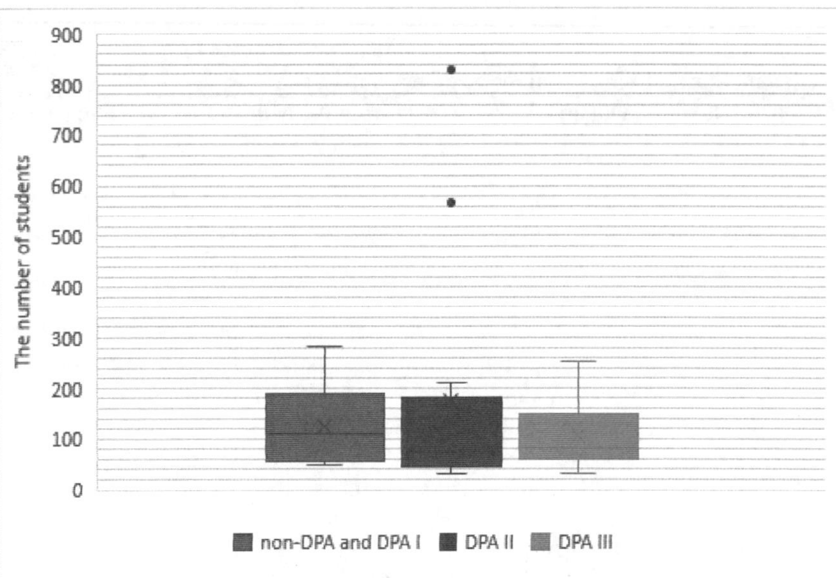

Figure 3. The boxplot for the number and distribution of students in the study area

IOP Conf. Series: Earth and Environmental Science **884** (2021) 012008 doi:10.1088/1755-1315/884/1/012008

2.2. Data collection and analysis

The data collected in this study include the school's conditions and school's preparedness in facing disasters. The data were collected using some techniques such as field observations, interviews, and document searches. In the process of data collection, those three methods mutually complement, support, and validate the data obtained from each method. In collecting data through interviews, 12 schools were taken as samples. The samples were determined purposively by considering the variations in the distribution of schools in DPA III, DPA II, and DPA I or non-DPA, as well as variations in the educational level. The relationship between data types, data collection methods, and instruments used or data sources is presented in Table 1 below.

Table 1. Types of Data, Data Collection Techniques, and instrument/data sources

No	Type of Data	Data Collection Techniques	Instrument/data source
1	School locations	Observation	School location measurement using GPS
		Documentation	Data from the Ministry of Education and Culture of the Republic of Indonesia (2019)
2	Eruption hazard level at school location	Documentation	Map of Merapi Vulnerable Eruption Eruption Areas. Published by the Indonesian Center for Volcanology and Geological Hazard Mitigation (2002)
		Observation	Measurement by GPS on the boundary markers of the hazard area installed by the Regional Disaster Management Agency (BPBD) of Magelang Regency
3	School distance from the eruption center	Documentation	*Peta Rupabumi Indonesia* (Indonesian Topographical Map) Muntilan Sheet. Published by the National Coordinator for Survey and Mapping Agency of Indonesia (2000)
4	School distance from main river channel		
5	School actions in the previous disasters	Interview	Interview with the principal or vice principal using the interview guide instrument
6	School involvement in disaster management		
7	School property (including land area, classrooms, sanitation, clean water facilities, electricity facilities, internet facilities, decent classrooms)	Interview	Interview with the school principal or vice principal of facilities and infrastructure
		Observation	Observation sheet
		Documentation	Data from the Ministry of Education and Culture of the Republic of Indonesia (2019)
8	Size of settlements around the school areas	Documentation	*Peta Rupabumi Indonesia* Muntilan Sheet. The calculation is done with ArcGIS

In data analysis, there are three stages of analysis carried out based on the three research objectives. First, to investigate the potential of schools in handling emergency eruption disasters, a review and scoring of school infrastructure were carried out. Scoring was performed by considering school infrastructure standards refferring to the regulations of the Indonesian Ministry of Education. Second, to examine the level of school preparedness, five-parameters scoring of disaster preparedness schools in Indonesia is conducted which include: (1) knowledge and attitudes, (2) policies and guidelines, (3) emergency response plans, (4) disaster warning systems, and (5) disaster mobilization. Third, to analyze the recommended role of schools in building sustainable schools, an analysis of geographical information systems is carried out by employing buffering, clip, and network techniques.

This analysis used 13 sample schools which were determined purposively representing the schools in DPA III, II, I, and non-DPA as well as the ES, JHS, and SHS categories.

Buffering analysis was performed to determine the position of the school toward the source of danger, both the eruption center and the main river channel. This analysis was supported by plotting the school location of the DPA zone division. The position of this school becomes a consideration whether the school is an affected school or a supporting school. The schools in DPA III definitely must be evacuated. Schools in DPA II can become supporting schools because DPA II is not evacuated except a large scale eruption of VEI 4 occurs as in 2010. Since the range of Merapi eruptions is relatively short but large eruptions occur in a long time span between 150-500 years [15], [16], the schools in DPA II can be categorized as supporting schools. After the eruption in 2010, BPBD Magelang Regency expanded the eruption hazard area where part of the area that was previously included in DPA II was included in DPA III. For this reason, we conducted a re-selection for schools referring to the DPA map in the DPA II area. The schools in DPA II which are located more than 11 km from the eruption center can function as supporting schools. However, the facilities and level of school preparedness must be considered. The percentage of settlements and infrastructure around the school also become a consideration in determining the eligibility criteria of the supporting schools. The percentage of settlements around the school was analyzed utilizing GIS using the clip method. The final step is performing a network analysis. This analysis is used to determine the closest route from affected schools to the supporting schools. All GIS analysis processes are carried out using ArcGIS and QGIS.

3. Results

3.1. The potential of schools in handling emergency eruption disaster

In this section, we illustrate the potential of schools in the Dukun sub-district in handling emergency eruption disasters. School potential is determined based on the infrastructure of the schools as well as human resources including the principal and teachers. There are several aspects of school infrastructure that need to be considered, including the area of school, the ratio of school area to students, classrooms, clean water sources, sanitation, electricity, internet, libraries, and laboratories.

The land area and number of classes are varied, especially at different levels of education such as ES, JHS, and SHS. ES with a higher number of units generally has fewer students because the coverage of student residence is also very limited. In one village area, there is at least one ES. A densely populated village can have several ESs. Meanwhile, the higher the level of education is, the farther the range of student residence will be. Also, the number of schools are increasingly limited along with the higher of the educational level, but the schools has greater area and classrooms in accordance with the number of students in the school.

ES area among 28 schools in the study area varies in the range of 200 m^2 to 6000 m^2. Moreover, the number of students varies from 31 to 281 students. Most schools at the ES level has a number of students and area that is relatively similar to one another. The different conditions were found in JHS. Public schools have a larger capacity and area compared to private schools, as indicated by the gap between two public schools and three private schools in the region. The same thing was also found in SHS level. Public schools have more students and areas. But in SHS level, a school type also influences the number of students. There are fewer students in vocational SHS than other SHSs. Overall, based on standards set by the Indonesian Ministry of Education, 77% of schools meet the standards regrading the ratio of school area to student while the aspect of total area only reaches 62%.

In addition, the number of classrooms is not significantly different in ES, but it is quite different in JHS and SHS. This condition influenced by the area and the number of students as explained above. The schools with large area have more capacity so that the number of students is high and the number of classrooms is also large to accommodate the students. According to the Indonesian Ministry of Education standards, all schools should have a minimum number of classrooms which are equal to or more than the number of classes (study groups) in the school. With regard to the classroom quality, 86% of all schools in the study area have feasible classrooms. The schools which have feasible classrooms in

ES category are 86%, JHS is 100%, and SHS is 50%. In the study area, there are two SHSs, one of which only has 40% of feasible classrooms and another one has 100%.

In regard to the ratio of classrooms to students, the range of ratio for ES is between 1:5 to 1:25 with an average 1:15. The ratio range for JHS is between 1:8 to 1:32 with an average of 1:18. In this JHS category, there is a gap on the ratio of classrooms to students for instance the gap regarding the number of classrooms. Meanwhile for the SHS category, there are differences between the two schools, namely 1:33 to 1:57. The number of classrooms and the ratio of classrooms to students shows that the classroom facilities and their use are not too different at the ES level, whereas in they are different in JHS and SHS level. The data distribution of land area, number of students, and classrooms is presented in Table 2, while the correlation between school area and number of classrooms to the number of students is shown in Figure 4.

Table 2. The school's facility and user

		School Level		
		ES	JHS	SHS
The ratio of school area to students (m^2/students)	N	21	5	2
	Mean	20	32	12
	median	17	24	12
	SD	12	18	5
	Max	43	53	15
	Min	3	13	8
School area (m^2)	N	21	5	2
	Mean	1798	3887	7150
	median	1610	3300	7150
	SD	1236	2929	7566
	Max	6000	7515	12500
	Min	200	100	1800
The number of classroom (unit)	N	21	5	2
	Mean	7	9	15
	median	6	6	15
	SD	2	6	15
	Max	13	18	25
	Min	4	3	4
The ratio of classroom to students (units/students)	N	21	5	2
	Mean	1:15	1:18	1:45
	median	1:14	1:17	1:45
	SD	1:6	1:9	1:17
	Max	1:25	1:32	1:57
	Min	1:5	1:8	1:33

The data on the area and classrooms indicate that the schools in the study area have facilities that meet the needs of their students. In other words, the schools recruit students based on the quota. One school to another at the same level and type show relatively the same condition regarding the area of land and classrooms. To keep running the educational processes during the disaster emergency situations, the schools located outside DPA (maximum in DPA II) can assist the schools located in the most dangerous areas (DPA III) by sharing facilities to keep learning going on. The sister school system can be designed by involving schools that are close together but in different DPA areas. The schools in non-DPA areas, for example, can facilitate schools from DPA III where the number of students, area, and number of classrooms is relatively similar. This condition will certainly get better if the targeted supporting schools have more areas and classrooms than that of the affected schools.

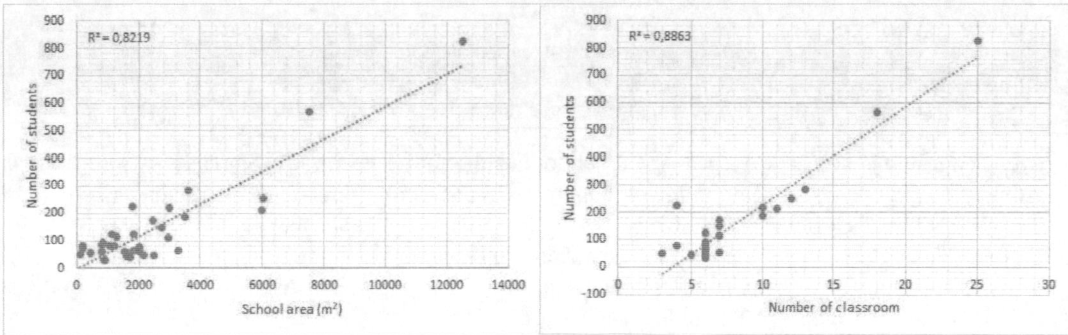

Figure 4. The relationship between the school areas and the number of students (A), the number of classrooms and the number of students (B)

To support the ideal learning activities, many supporting facilities are needed, not only dealing with the area of the school and the number of classrooms. Other facilities such as clean water, sanitation, electricity and internet are also required. In terms of clean water sources, all schools have clean water sources and can fulfil their own needs. The source of clean water mostly comes from the springs (46%), especially in schools that are located in the volcanic foot. In this region, there are indeed many springs [12] related to the position of buckling slope which is the boundary between landforms [11]. Another source of clean water that is widely used is wells (34%), especially in schools that are located in the volcanic footplain. In the sanitation aspect, 9% of schools do not meet the standards set by the Indonesian Ministry of Education. In regard to the electrical energy, there are 23% of schools that do not meet standards. Meanwhile for internet access, 34% of the schools are not connected to the internet network. In relation to utilizing the potential of schools for disaster management, some problems arise because the schools located in safe areas have nonstandard electricity, sanitation and internet facilities. The results of our scoring also showed that 3 of 35 schools in the study area had a nonstandard infrastructure. The three schools are located in a safe zone that should be able to function as an evacuation site for the affected schools.

3.2. The level of school preparedness in facing eruption disaster

Another aspect that needs to be considered in developing a sustainable school in the DPA of Merapi eruption is the degree of school preparedness in facing disasters. The better school preparedness will improve the resilience to face disasters and encourage the successful implementation of schools during the crisis period. The results of the scoring analysis indicate that the level of school preparedness is included in the low and medium categories. 77% of schools are in the low category, 23% are in the medium category, and none is in the high category. There is no variation in preparedness level between ES, JHS, and SHS. The level of preparedness is low and is found in ES, JHS, and SHS. In addition, there is no variation viewed from the location of the schools in the DPA. Some schools in DPA III have low preparedness and some others belong to the medium category in terms of the preparedness. Likewise, in DPA II, DPA I, and non-DPA, some schools have low and medium degree of preparedness.

The reason for the absence of schools with high preparedness is because some elements of preparedness have not yet been fulfilled which include (1) there has not been good coordination with National Disaster Management Agency (BNPB) and Regional Disaster Management Agency (BPBD) as the institutions which are in charge of disaster management, (2) there is no SOP for disaster management in schools, (3) there is no special fund allocated for disaster management, (4) there are no special teachers assigned as coordinators in disaster management, (5) there is no coordination with student's parents related to disaster management, and (6) there is no decree which strengthen the tasks and roles in disaster management. The aspects that the schools currently own are presented in Table 3 below.

Table 3. The preparedness aspects owned by the schools

Preparedness aspects	Level of preparedness			
	Low		Medium	
Knowledge and attitude	3 of 8 sub-aspects	Activating the school health unit / junior doctor, Junior Red Cross and Scouts	3 of 8 sub-aspects	Same as low preparedness aspects
		Increasing the capacity of teachers relevant to lessons and preparedness activities		
		Developing cooperation with disaster and education observer institutions		
Policies and guidelines	0 of 4 sub-aspects	None	0 of 4 sub-aspects	None
Emergency response plan	3 of 6 sub-aspects	Make copies of important school documents and keep them in a safe place	4 of 6 sub-aspects	Same as low preparedness aspects, added with the following aspects
		Prepare medicines and first aid kits		Make an evacuation plan
		Conduct first aid training guided by relevant agencies		
Disaster warning system	2 of 6 sub-aspects	Prepare communication equipment	4 of 6 sub-aspects	Same as low preparedness aspects, added with the following aspects:
		Prepare disaster warning equipment		Open communication access to official disaster warning information
				Disseminate the school's disaster warning system to the entire school community
Disaster mobilization	1 of 6 sub-aspects	Conduct training in schools or facilitate school representatives to attend preparedness training	2 of 6 sub-aspects	Same as low preparedness, plus added with the following aspects:
				Develop cooperation with local government agencies related to disaster management

The aspects of policy and guideline with four sub aspects have not yet been possessed by the schools in the study area due to various administrative and technical constraints faced by schools. In regard to the policy and guidelines, there are currently instructions and directions related to disaster learning in schools from the Magelang Regency Government. However, technically there is no integration of preparedness materials in learning, there is no disaster preparedness group, there is no special funding budget for disaster preparedness activities in schools, and there is no continuous disaster training or simulation. Teachers still need assistance for the integration of disaster materials such as determining the competency standards and developing a lesson plan for learning in class. Because of this obstacle, the teachers are not able to implement disaster learning and still focus on their duty to teach compulsory subjects. In addition, routine disaster evacuation simulations have also not been conducted. The schools that have conducted simulations after the 2010 Merapi eruption disaster were still limited, accompanied by disaster volunteers. This activity is not continously carried out.

To develop a sustainable school, there are some urgents aspects of preparedness to fulfil, including: (1) making policies related to the implementation of sister schools to realize a sustainable learning, (2) establishing disaster preparedness clusters, not only for handling emergency situations but also for coordinating the organization of learning in crisis situations and coordinating the schools that need assistance, (3) formulating evacuation plans which are equipped with signs and procedures, including the procedure of how one school must evacuate to another school, (4) performing rescue simulation and implementing a learning in crisis situations between two schools within the sister school system.

3.3. Sister Schools: the role of schools in developing sustainable schools

To establish a sustainable school, a collaboration between schools in one region is needed. Among the 35 schools in this study area, 18 of them are in danger areas so that it is not possible to conduct normal learning in these schools during the crisis. Even in the emergency status of alert (level 4), residents in the danger area have to evacuate to other places. Moreover, there are 17 schools that are outside of the danger area. These schools can still organize the learning process. Utilizing the potential of the infrastructure, the schools in safer areas can assist other schools in hazard areas so that learning activities in these schools can still be carried out. Sister school systems in a network of sustainable school initiatives need to be encouraged and developed to facilitate this need.

The sister school system at DPA Merapi Volcano is not something new. In Sleman Regency, which is on the southern side of Merapi Volcano, a sister school has been developed involving 20 schools. Based on this experience, the sister school system can also be developed on the western side of Merapi Volcano by involving schools in the region. Sister schools can be establised between schools by paying attention to three important aspects, namely, their proximity, the number of students, and the similar condition of the infrastructure, and the same level of education. Two schools in the danger zone and the safe zone which are located next to one another will enable the process of mobilization between schools easier and more efficient. The relatively same number of students and infrastructure between the two schools will guarantee that supporting schools can meet the needs of the affected schools to organize learning. The same level of education, for example ES and ES, will make it possible to access facilities more effectively. For example, the use of libraries, laboratories, and others.

Based on the scoring analysis, all 18 schools in the danger zone have good facilities. However, schools in this area must totally evacuate to other places. 18 of these schools became affected schools which had to be assisted by supporting schools. The level of preparedness in dealing with disasters is crucial in the mobilization process. However, because the level of preparedness still ranges from low to moderate, this aspect needs to be improved in the future. Furthermore, among the 17 schools in the safe zone, three schools have nonstandard infrastructure. We carried out the re-selection process so that 14 schools were recommended to be supporting schools in the sister school system. Taking into account the distance, infrastructure, and student numbers, the division of school partnerships along with the best route between schools is presented in Figure 5. The learning process during the crisis period is also different from the normal period. During this crisis period, a good coordination between teachers and

IOP Conf. Series: Earth and Environmental Science **884** (2021) 012008 doi:10.1088/1755-1315/884/1/012008

parents of students or between the school and the surrounding community is needed. This condition is supported by easy access to reach the settlements around the schools.

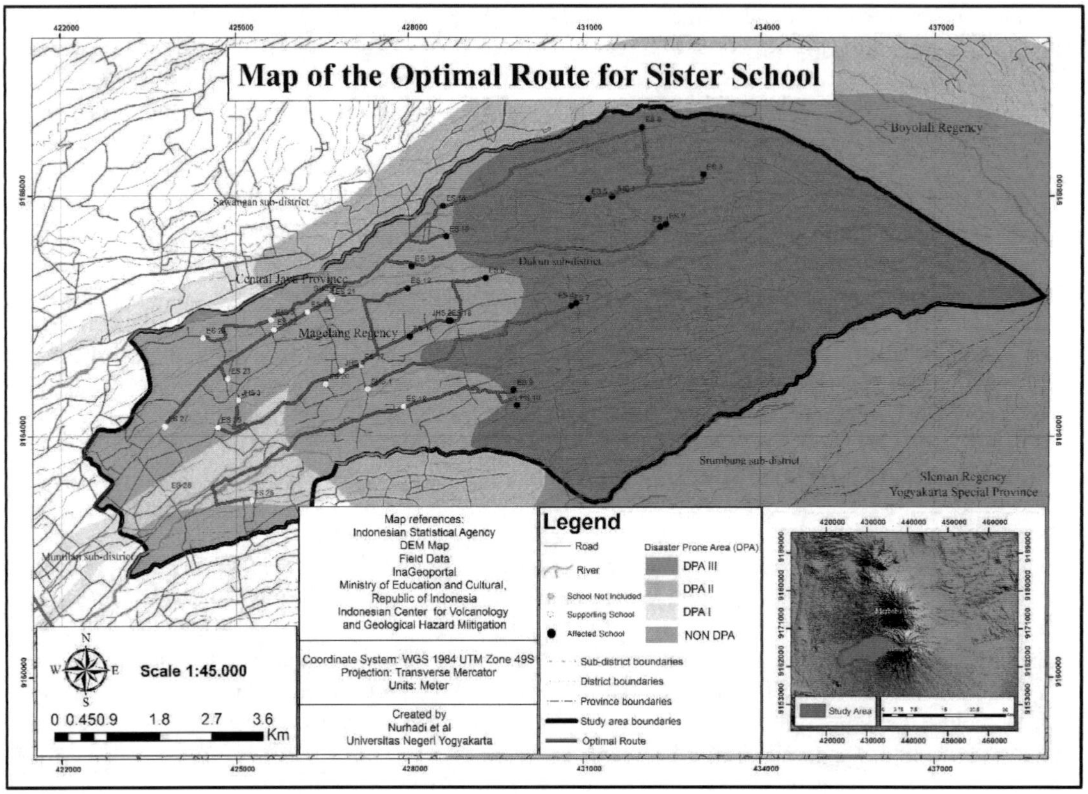

Figure 5. The sister school system for Dukun Sub-district

4. Discussions

A sustainable school is one of the efforts that can be realized to ensure that education can still be performed during the crisis period in an emergency situation of eruption disaster. The efforts to develop a sustainable educational environment have been carried out in various countries. For example, the Sustainable Campus Initiative developed in Shonan Fujisawa Campus Keio University. To build a sustainable campus, various innovations have been carried out within the campus environment in order to keep the campus function normally during the crisis period [8]. The development of sustainable schools in Merapi eruption prone areas also has the same goal, but it's carried out in a different way. In developing this sustainable school, the approach used is the regional complex. The collaboration between schools in one region through a network of sister schools is a feature of the development of sustainable schools in the Merapi area, instead of the campus independence initiative as explained by Bai et al [8].

The collaboration between schools is an important factor in determining the realization of a sustainable schools. This is due to the different characteristics of the eruption disaster that pose a threat in this region. The characteristics of the earthquake hazard underlying the development of sustainable campus as explained by Bai et al [8] are different from the eruption disaster. Thus, the approach used in building a sustainable school is also different. In eruption disasters, there is zoning of danger areas that are constantly under threat during the eruption period. In other words, a school cannot function for a long time so it needs to be assisted by other schools so that the learning process can still be running. Based on the experience from the Merapi eruption in 2010, residents of DPA III had to evacuate for several months. In this condition, sustainable learning can only be realized if it joins with other schools, utilizing the existing infrastructure at the school, with adjusted learning hours.

The sister school system is actually not a new concept at DPA Merapi. On the southern side of Merapi, the other part most affected by eruption in 2010, has implemented a sister school system. Mei et al [17] explain that sister school plays an important role in maintaining the student's safety during the learning process and ensuring a sustainable education during the eruption. The BPBD of Magelang Regency acts as a liaison between the two schools in this sister school system. In addition, a coordination is also carried out with local government, especially with regard to evacuation efforts. Furthermore, Pambudi and Ashari [18] explain that school principals and teachers play a crucial role in organizing disaster preparedness schools. We found that teachers in the study area had also received training on disaster, even at the national level. A disaster training which has ever been experienced by teachers is part of school preparedness. Even schools with a low level of preparedness generally meet this aspect. However, further coordination is still needed so that the teacher's role can function more optimally. Principals and teachers also play an important role in bridging communication between schools and parents of students, even building cooperation with parents of students [9]. In order to realize this role effectively and efficiently, it is necessary to divide the tasks in the disaster preparedness group. Until now, the formal distribution of tasks between school principals and teachers in disaster preparedness groups needs to be improved. In addition, the principal and teachers are also responsible for providing knowledge about disaster and safety. The teacher can provide knowledge with various methods during the learning process [19].

Spatial analysis that shows the location of schools toward the source of danger and the location and distance between schools is an important part that needs to be considered in building a sister school network in the Merapi eruption disaster area. The location of the school towards the source of danger will ensure that the school is safe enough to continue to carry out education during the eruption period and assist the affected schools. The existence of good school facilities increasingly supports the establishment of safe schools. Green et al [20] emphasizes the need to build safe schools. The experience in various countries also shows that the community can contribute in creating safe schools through community-based school development. Schools that are built together with the community will further enhance the linkages between the school and the surrounding community. The selection of school location must be well considered in a truly safe area. In the Merapi eruption DPA, many schools are in the safe zone. Instead of building new schools, schools in safe zones can be improved both in the quality of their buildings and supporting facilities. At present, many schools in the safe zone lack the supporting facilities especially electricity and internet networks.

Furthermore, Green et al [10] state that the location of schools is an important part of the safe school facility which is included in one of the three pillars in developing safe schools. Beside the location of the school, assessment and retrofitting of the existing schools and their care are also an important part of the pillars of the safe school facility. In DPA Merapi, many schools are located in safe zones but the facilities are inadequate. Therefore, retrofitting schools is needed. In the future, improving facilities and the role of principals and teachers is a very important to be improved. The lack of facilities and the role of the principal and teachers, not only hamper the development of the sister school system but also weakens school preparedness to deal with disasters [21].

5. Conclusion

The eruption disaster of Merapi Volcano can occur in a long period of time. Experience from a major eruption that occurred in 2010 showed that residents in the DPA had to evacuate for several months. This condition causes disruption of the learning process in schools located in DPA. For this reason, the development of sustainable schools is needed. The nature of the eruption disaster that continues to pose threats to DPA has prevented the schools in DPA from functioning normally during this period. For this reason, sustainable schools must be developed through a collaboration between schools within the sister school network. During the crisis period, schools in DPA III and part of DPA II became the affected schools that can continue the learning process in the supporting schools located in safe zones. In this paper, we propose the possibility of developing sister school networks by considering location, distance, and educational infrastructure factors in schools. Sister school networks can be established but infrastructure improvements and school preparedness are factors that must be seriously managed so that the goals of sustainable school development can be achieved.

For the evaluation process, this research is still limited to the examination on the potential of infrastructure owned by schools for disaster management and disaster preparedness. There are several things that need to be further investigated including school readiness to implement the sister school system, sister school implementation scenarios, and stakeholder support in developing sustainable schools through sister school networks.

6. Acknowledgement

The authors would like to thank the Faculty of Social Sciences Universitas Negeri Yogyakarta that provides a financial support for this research through Research Group Grant Programme on 2019. We also thank the lecturers and laboratory staffs at the Geography Education Department, and research assistants who helped collect and analyse the data.

References

[1] H. T. Verstappen, *Garis Besar Geomorfologi Indonesia*, 1st ed. Yogyakarta: Gadjah Mada University Press, 2013.
[2] F. Lavigne *et al.*, "People's behaviour in the face of volcanic hazards: Perspectives from Javanese communities, Indonesia," *J. Volcanol. Geotherm. Res.*, vol. 172, no. 3–4, pp. 273–287, 2008.
[3] D. S. Hadmoko, F. Lavigne, J. Sartohadi, and C. Gomez, "Spatio-Temporal Distribution of Landslides in Java and The Trigerring Factors," *Forum Geogr.*, vol. 31, no. 1, pp. 1–15, 2017.
[4] F. Lavigne, J. C. Thouret, B. Voight, H. Suwa, and A. Sumaryono, "Lahars at Merapi volcano , Central Java : an overview," *J. Volcanol. Geotherm. Res.*, vol. 100, pp. 423–456, 2000.
[5] S. D. Andreastuti, C. Newhall, and J. Dwiyatno, "Menelusuri kebenaran letusan Gunung Merapi 1006," *J. Geol. Indones.*, vol. 1, no. 4, pp. 201–207, 2006.
[6] Global Volcanism Program, "Volcanoes of the World - Merapi 263250," *Holocene volcano list*, 2013. [Online]. Available: https://volcano.si.edu/volcano.cfm?vn=263250. [Accessed: 29-Jun-2019].
[7] Badan Nasional Penanggulangan Bencana, *Rencana aksi rehabilitasi dan rekonstruksi wilayah pasca bencana erupsi Gunung Merapi di Provinsi DI Yogyakarta dan Provinsi Jawa Tengah Tahun 2011-2013*. Jakarta: Badan Nasional Penanggulangan Bencana, 2011.
[8] Y. Bai, Y. Ikeda, S. Ota, and H. Kobayashi, "Sustainable Campus Initiative at Keio University after the Great East Japan Earthquake Disaster," *Int. J. Disaster Risk Sci.*, vol. 3, no. 2, pp. 123–130, 2012.
[9] H. Kawasaki, S. Yamasaki, M. M. Rahman, Y. Murata, M. Iwasa, and C. Teramoto, "Teachers-parents cooperation in disaster preparation when schools become as evacuation centers," *Int. J. Disaster Risk Reduct.*, vol. 44, no. December 2019, pp. 101–445, 2020.
[10] R. Paci-Green, A. Varchetta, K. Mcfarlane, P. Iyer, and M. Goyeneche, "Comprehensive school safety policy: A global baseline survey," *Int. J. Disaster Risk Reduct.*, vol. 44, no. May 2019, pp. 101–399, 2020.
[11] Sutikno, L. W. Santosa, Widiyanto, A. Kurniawan, and T. H. Purwanto, *"Kerajaan Merapi", Sumberdaya Alam dan Daya Dukungnya*. Yogyakarta: BPFG UGM, 2007.
[12] R. P. Aurita and S. Purwantara, "Karakteristik Mataair Kaki Lereng Gunung Merapi dan Pemanfaatannya di Kecamatan Dukun Kabupaten Magelang," *Geomedia*, vol. 15, no. 1, pp. 75–85, 2017.
[13] A. Ashari and E. Widodo, "Hidrogeomorfologi dan Potensi Mataair Lereng Baratdaya Gunung Merbabu," *Maj. Geogr. Indones.*, vol. 33, no. 1, pp. 48–56, 2019.
[14] W. Raharjo, Sukandarrumidi, and H. M. D. Rosidi, "Peta Geologi Lembar Yogyakarta, Jawa," Bandung, 1995.

[15] S. D. Andreastuti, B. V. Alloway, and I. E. M. Smith, "A detailed tephrostratigraphic framework at Merapi Volcano, Central Java, Indonesia: implications for eruption predictions and hazard assessment," *J. Volcanol. Geotherm. Res.*, vol. 100, pp. 51–67, 2000.

[16] R. Gertisser, S. J. Charbonnier, J. Keller, and X. Quidelleur, "The geological evolution of Merapi volcano, Central Java, Indonesia," *Bull. Volcanol.*, vol. 74, pp. 1213–1233, 2012.

[17] E. T. Mei *et al.*, "Sister School for Merapi Volcano Disaster Risk Reduction," in *International Conference on Environmental Resources Management in Global Region*, 2019, pp. 1–9.

[18] D. I. Pambudi and A. Ashari, "Enhancing role of elementary school in developing sustainable disaster preparedness: a review with some examples from disaster-prone areas of Merapi," in *International Conference on Hazard Mitigation in Geographic and Educational Perspectives*, 2019, pp. 1–13.

[19] I.-M. Andersson, K. Gunnarsson, and G. Rosèn, "Role of Headmasters, Teachers, and Supervisors in Knowledge Transfer about Occupational Health and Safety to Pupils in Vocational Education," *Saf. Health Work*, vol. 6, pp. 317–323, 2015.

[20] R. Paci-green, B. Pandey, H. Gryc, N. Ireland, J. Torres, and M. Young, "Challenges and benefits of community-based safer school construction," *Int. J. Disaster Risk Reduct.*, vol. 43, pp. 101–384, 2020.

[21] A. A. Shah *et al.*, "Schools' Flood Emergency Preparedness in Khyber Pakhtunkhwa Province, Pakistan," *Int. J. Disaster Risk Sci.*, vol. 9, pp. 181–194, 2018.

IOP Conf. Series: Earth and Environmental Science **884** (2021) 012009 doi:10.1088/1755-1315/884/1/012009

Assessment of potential locations for evacuation in Depok Municipality using images from UAV

Amrih Halil, Revi Hernina*, Rokhmatuloh

Department of Geography, Universitas Indonesia, Kampus UI Depok 16424, Depok

*Corresponding author's email: revie12@yahoo.com

Abstract. This study aims to perform assessment of public facilities in Perumnas Depok Utara, Depok Municipality, Indonesia for evacuation locations using images from UAV. A UAV device, a DJI Phantom 4 was flown at 100 metres height to capture seven locations of public facilities located within Perumnas Depok Utara residential area. Raw images from DJI Phantom 4 underwent image processing and georeferencing process to generate smooth aerial images. Public facilities locations were identified from acquired images and later digitized to generate layers of assessment parameters. Each locations were scored based on rating of all criteria's parameters. Seven criteria serve as foundation layers for evacuation assessment namely size, ownership, availability, capacity, infrastructure, transportation and safety. Study result shows that Jawa Soccer Field is the first rank of potential evacuation location while Balai Rakyat Meeting Place is the least potential evacuation location. Total area for potential evacuation locations around 43,000 m^2 and can accommodate 28,961 people or equal to 60.5% of Beji Village's population.

1. Introduction

The growing attention to plan and prepare the shelter in urban areas for evacuation during natural disaster events such as earthquakes and floods has been discussed in various recent publications. This is reflected by incorporating quickest time to reach the evacuation and emergency shelters [1], and the assessment of evacuation location based on several parameters [2]. Consequently, the planning of emergency shelters and evacuation locations should consider strategic location where it can be reached timely and accommodate more people conveniently. Such considerations are not easy to fulfil especially in dense urban environment where not much open spaces or public facilities are strategically located. One of the first public facilities developed within residential areas in Indonesia was Perumnas housing projects in Depok Municipality. Built around 1974 [3], Perumnas Depok I and II were equipped with strategically public facilities such as playgrounds, soccer fields, badminton fields and green open spaces. These facilities still can be seen today in Perumnas Depok I, II and Utara where mostly located in the centre of residential units. However, these public facilities have not been assessed for their availability as emergency shelters and evacuation locations. This assessment is really important, since emergency readiness in the wake of disasters can save many lives. In addition, detail assessment of public facilities needs holistic mapping which can be provided by remote sensing technologies. When [4] performed a comprehensive literature review to provide guidelines for the use of Unmanned Aerial Systems (UAS) in flood emergency response, it was identified that UAS can be used in all three phases of the emergency response namely; pre, during and post. One of the UAS application in pre-emergency response phase is to conduct flood risk assessment such as determining terrain elevations, flood extent modelling, evacuation's routes identification and flood warning [4]. Unmanned aerial vehicle (UAV) is a device of

Content from this work may be used under the terms of the Creative Commons Attribution 3.0 licence. Any further distribution of this work must maintain attribution to the author(s) and the title of the work, journal citation and DOI.
Published under licence by IOP Publishing Ltd

remote sensing which can capture images with very high spatial resolution and rapid time. Previously, UAV aerial mapping has been conducted by [5] in Batuhiu and Pangandaran Beach, West Java Province in order to plot post-tsunami buildings' spatial distribution in 2006. The spatial distribution of buildings developed after 2006 tsunami which was captured by UAV then compared with buildings before 2006 tsunami event generated from satellite imagery. The information come from differences of building spatial distribution before and after 2006 tsunami event were formulated to analyze tsunami risk, and use it for proposing proper mitigation to reduce the impact. By observing those previously demonstrated capabilities of UAV in gathering spatial data to be used in disaster risk assessment, therefore this study employs UAV aerial mapping to capture spatial characteristics of assessed public facilities in Perumnas Depok Utara. Hence, the process of capturing spatial data using UAV will be described in this paper as well as the study objective which is to perform assessment of public facilities in Perumnas Depok Utara for evacuation locations.

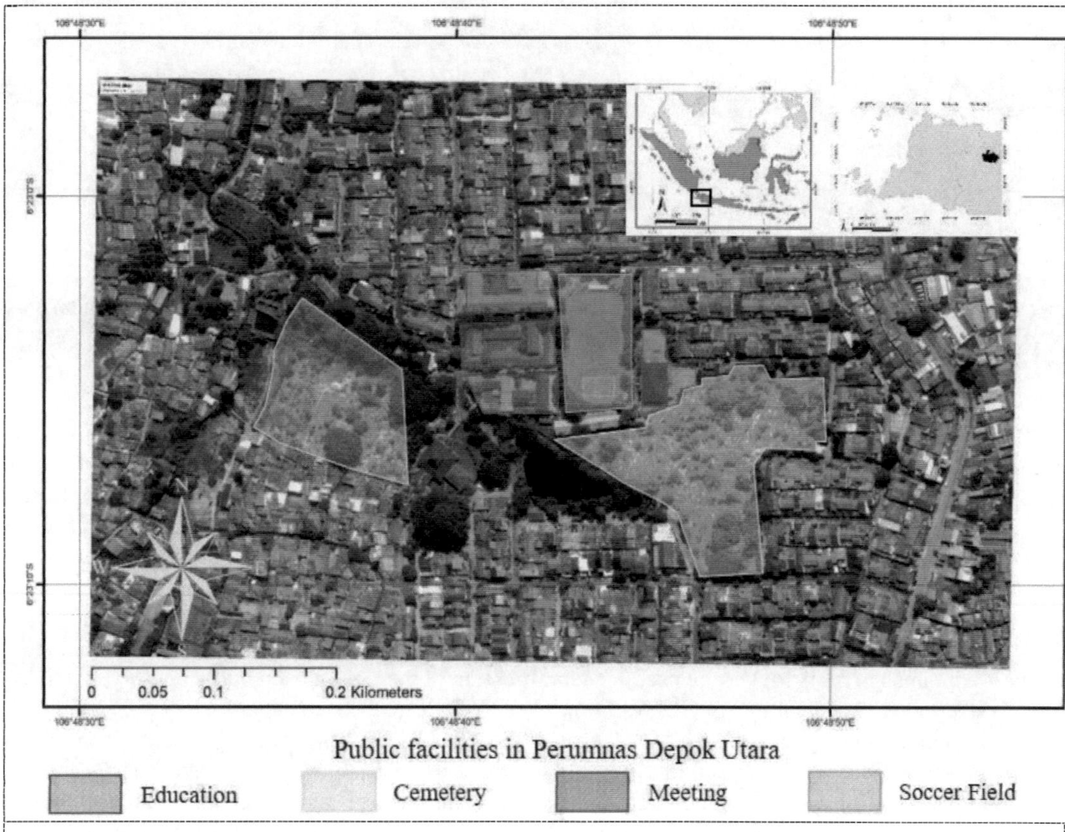

Figure 1. Study location (Data processing using image downloaded from Google Earth).

2. Methods

The study flowchart can be observed in Figure 2. Firstly, a UAV device namely DJI Phantom 4 was flown at 100 metres height at public utilities locations within Perumnas Depok Utara residential area. Six locations were identified as potential locations for emergency evacuation to be evaluated based on assessment parameters. After image processing and georeferencing, very high resolution images from UAV were acquired to compute the size of each public facility location. The data derived from UAV images were classified further to be assessment parameters based on previous study [2] namely size, ownership, availability, capacity, infrastructure, transportation and safety. The resulted assessment of every parameters indicated the suitable location for evacuation in Perumnas Depok Utara.

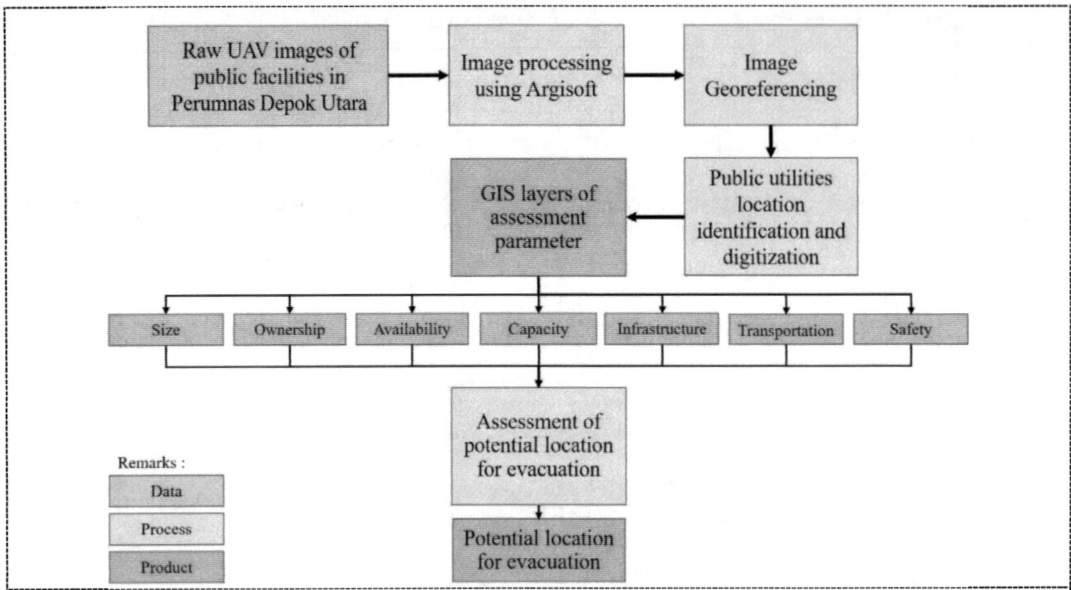

Figure 2. Study flowchart.

2.1. Study location
The study location is known as Perumnas Depok Utara. It was built in 1975 and commissioned later in 1976 for residential use [6]. In this study, the public facilities are Balai Rakyat Meeting Place, Jawa Soccer Field, Cemetery I and II, schools of SDN Beji 6, SDN Beji 4 and SMP Muqorobin (Figure 1).

2.2. Image processing
Acquired UAV images from DJI Phantom 4 were processed using Agisoft Photoscan. Imported photos were input in a single project name. Stored photos underwent alignment process to generate 3D model. For georeferencing, acquired photos were referenced with high spatial resolution images which were downloaded from Google Earth. Resulted RMSE was observed to see the accuracy of georeferencing process. Georeferencing process was performed using ArcGIS software.

2.3. Public utilities location identification and digitization
In this study, images from UAV were used for identification of the size of each public facility locations. Area measurement menu within ArcGIS was employed to estimate the size of each facility. During field survey for aerial photo, interviews with local around those public facilities were conducted primarily to seek information whether those facilities are public or private. In addition, the existence of pedestrian, supply water, electricity and sewage system were surveyed around the target locations. Hence, once all information were complete, they input into spatial attribute table of each location using ArcGIS. Digitization of all acquired images from UAV were performed in ArcGIS to delineate all public facilities potentially as evacuation locations.

2.4. Assessment of potential location for evacuation
After all potential places of evacuation were digitized in form of vector map, rating of each parameter from all criteria were input into spatial attribute table within ArcGIS (Figure 3). In this paper, score of 1 indicates that the acquired parameter value for each location less potential than those with score of 2. Therefore, the bigger score means that input value is the better or more potential to become a location for evacuation. Consequently, it is better not to use 0 to represent less potential because the parameter is not empty but rather indicate more potential or not. However, for capacity criterion because the values were calculated from size of each location divided by 1.5 m^2/person therefore ranking method for each

location was applied to indicate which one was the biggest. Hence, for capacity criterion input for spatial attribute table was converted from rank using following equation [7]:

$$w_j = \frac{n - r_j + 1}{\sum_{j=1}^{n} n + r_j + 1} \tag{1}$$

where w_j=the weight of *j-th* attribute, r_j=rank of *j-th* criteria, dan n=number of criteria with circumstance that rank 1 is given to the most important criteria. Moreover, all six criteria and its parameters for potential rating can be observed in Table 1. Especially for capacity criterion, 1-2 scale is replaced with ranking method. Rank 1 is given for the biggest value of capacity criterion while the smaller

Table 1. Criteria for assessment and its parameters (modified from [2])

Criteria	Parameters	Scoring (the bigger the better)
Size	500- <2,000 m^2	1
	2,000- 10,000 m^2	2
Ownership	Private	1
	Public	2
Availability	School, sport facility, green areas, playgrounds, outdoor sport areas, other open areas	2
	Others	1
Transportation	Pedestrian access	2
	No pedestrian access	1
Capacity	Calculated from size divided by 1.5 m^2/person	Rank 1 for the biggest value followed by the smaller capacity
Safety	Building risks	1
	No Building Risks	2
Infrastructure	Water supply, sewage system, electricity	2
	None above	1

In detail, the process of value scoring for every potential location using spatial attribute table in ArcGIS and final summed values can be observed in Figure 3.

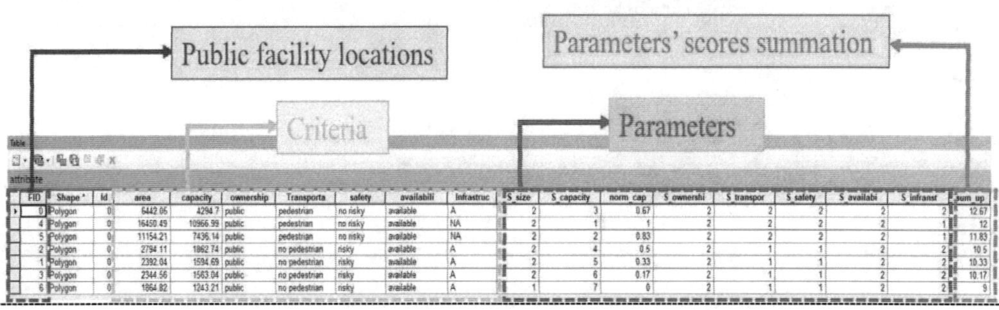

FID	Shape *	Id	area	capacity	ownership	Transporta	safety	availabili	Infrastruc	S_size	S_capacity	norm_cap	S_ownershi	S_transpor	S_safety	S_availabi	S_infranst	sum_up
0	Polygon	0	6442.06	4294.7	public	pedestrian	no risky	available	A	2	3	0.67	2	2	2	2	2	12.67
4	Polygon	0	16450.49	10966.99	public	pedestrian	no risky	available	NA	2	1	1	2	2	2	2	1	12
5	Polygon	0	11154.21	7436.14	public	pedestrian	no risky	available	NA	2	2	0.83	2	2	2	2	1	11.83
2	Polygon	0	2794.11	1862.74	public	no pedestrian	risky	available	A	2	4	0.5	2	1	1	2	2	10.5
1	Polygon	0	2392.04	1594.69	public	no pedestrian	risky	available	A	2	5	0.33	2	1	1	2	2	10.33
3	Polygon	0	2344.56	1563.04	public	no pedestrian	risky	available	A	2	6	0.17	2	1	1	2	2	10.17
6	Polygon	0	1864.82	1243.21	public	no pedestrian	risky	available	A	1	7	0	2	1	1	2	2	9

Figure 3. Parameters' score computation using spatial attribute table in ArcGIS.

IOP Conf. Series: Earth and Environmental Science **884** (2021) 012009 doi:10.1088/1755-1315/884/1/012009

3. Results and Discussion

3.1. UAV images

After the raw images from UAV underwent image processing and georeferencing, ready to be analysed image is shown in Figure 4. Three potential locations for evacuation namely Jawa Soccer Field, Balai Rakyat Meeting Place and Cemetery were identified and later be digitized. When combined together all of these three locations have approximately area around 24,000 m^2.

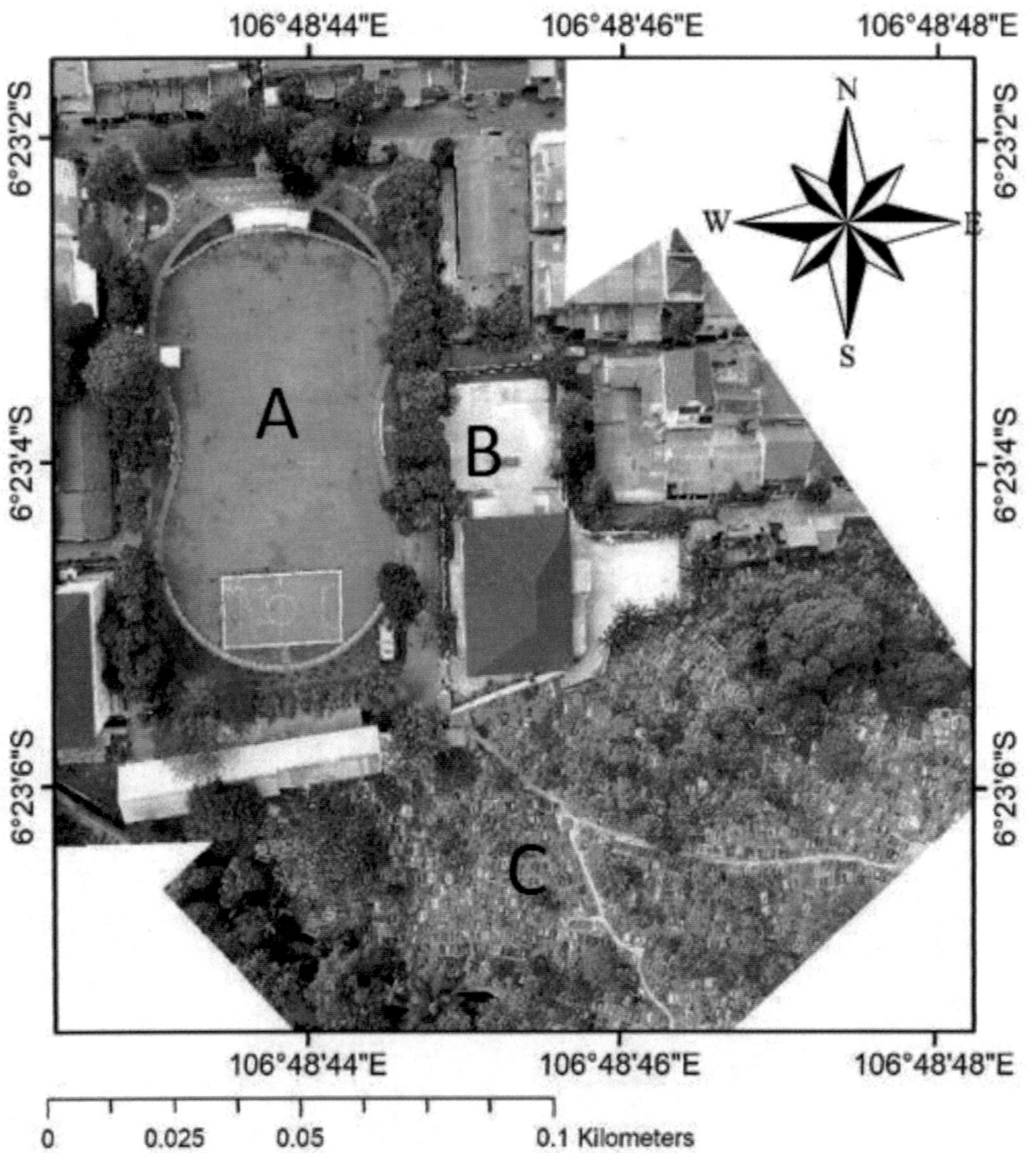

Figure 4. UAV image of the study area: (a) Jawa Soccer Field, (b) Balai Rakyat Meeting Place, (c) Public Cemetery

3.2. Potential location identification and digitization

Once potential locations for evacuation are identified, the process of digitization from images to become vector maps begin. It can be seen from Figure 4, the Jawa Soccer Field and Cemetery have no secondary hazard risks caused from the building unlike the Balai Rakyat Meeting Place. On the contrary, cemetery has no water supply, electricity and sewage system thus make it low rating score in infrastructure criterion when compared to schools and meeting place. Vector maps in Figure 5 resulted from digitization process can be measured to provide size information and later capacity. Interestingly, the biggest size is cemetery with an area around 16,000 m^2 followed by the second cemetery around 11,000 m^2.

3.3. Potential location for evacuation

Calculation result shows Jawa Soccer Field is the first rank for potential evacuation while cemetery is the second. Though cemetery has the bigger size than Jawa Soccer Field but it lacks of water supply, electricity and sewage system. In addition, it can be observed in Figure 5 that all school buildings are ranked lower than open spaces. It because of school buildings have risks during earthquake while open spaces are relatively safer. Furthermore, all of these seven locations are closely located to each other it offers better strategic locations during emergency because all supports can be alocated in one single location. Not to mention it is located in the center of Perumnas Depok Utara, where all residents from all directions can reach this site relatively fast. This strategic location of public facilities in Perumnas Depok Utara has fulfilled the first step towards urban resilience planning, which according to [1], reflected by quickest evacuation to assembly points and shelters in the wake of the disasters.

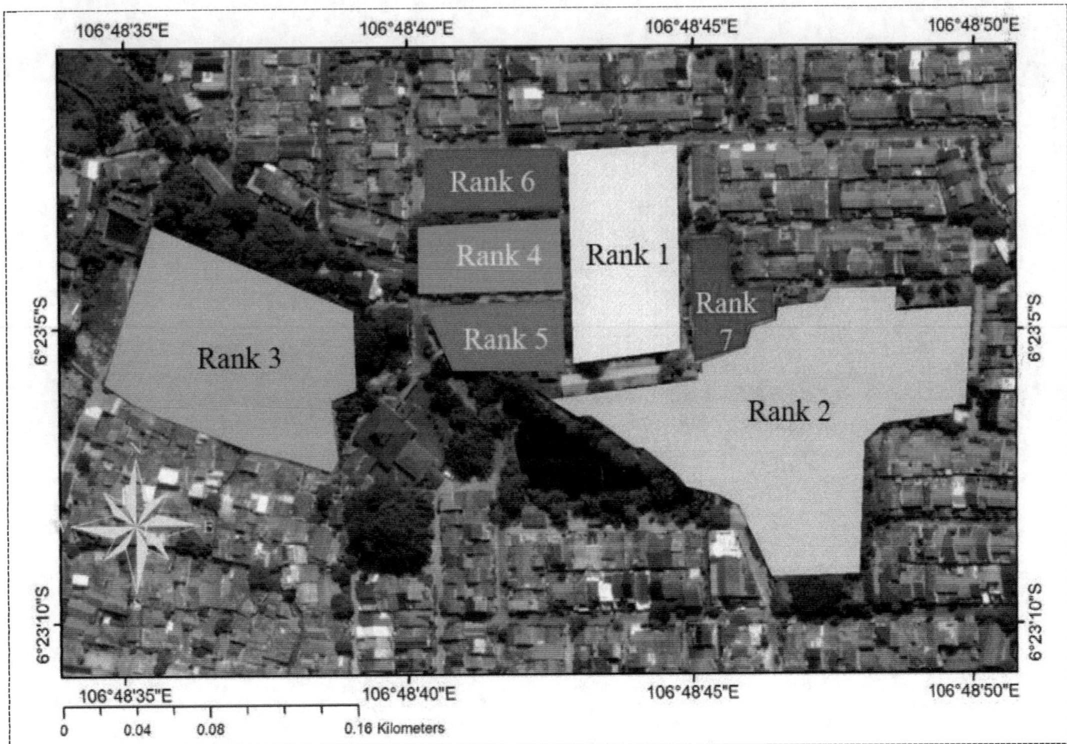

Figure 5. Rank of potential evacuation location based on parameters' assessment (Data processing using image downloaded from Google Earth).

With the total capacity around 28,961 people, these all seven potential locations during earthquake can support 60.5% of total population in Beji Village [8]. However, certain places like cemeteries need to be equipped with water supply, electricity and sewage system. Though schools offer multiple floors to accommodate more people inside the building, they also have secondary hazard risks especially during earthquake. For further study, these all seven locations need further assessment whether they are close to water source to prevent from flooding risks during evacuation process. In prior study related to evacuation shelter planning [9], existing shelter locations in Pangandaran area prepared for tsunami hazards consist of public facilities such as mosques, fields, regent offices, and halls. In addition, [9] added potential shelter locations for tsunami mitigation included various types of buildings i.e. mosques, schools, public health centers, government buildings, and shopping centers. Presumably, public facilities have the advantages of large space to accommodate refugees in disaster events, not to mention the ownership of those facilities belong to the government which can be used instantly compare to private-owned facilities.

4. Conclusion

Using seven criteria and its parameters, the assessment of public facilities in this study has found that Jawa Soccer Field is the most potential evacuation location while Balai Rakyat Meeting place is the least potential for evacuation location. Though Cemetery I and II are very potential for evacuation location, they need to be equipped with more facilities such as water supply, electricity and sewage system. For further research, these potential locations need more assessment particularly in flood risks.

Acknowledgments

Authors would like to express sincerely gratitude to Universitas Indonesia for funding this study under Hibah PITTA grants.

References

[1] Zhao L, Li H, Sun Y, Huang R, Hu Q, Wang J and Gao F 2017 *Sustainability* **9** 2098
[2] Yucel G 2018 *Disaster Science and Engineering* **4**
[3] Perumnas 2020 https://perumnas.co.id/tentang-perumnas/#sejarah-perumnas
[4] Salmoral G, Rivas Casado M, Muthusamy M, Butler D, Menon P and Leinster P 2020 *Water* **12** 521
[5] Mardiatno D, Malawani MN and Nisaa RM 2020 *International Journal of Disaster Risk Reduction*, **46**, 101523
[6] Yulianti R 2020 https://www.radardepok.com/2017/01/perumnas-diresmika-soeharto-lapangan-jawa-jadi-primadona/
[7] Malczewski J and Rinner C 2015 *Multicriteria Decision Analysis in Geographic Information Science* (New York: Springer)
[8] BPS 2019 *Beji subdistrict in figures* (Depok: BPS)
[9] Husa N and Damayanti A 2019 *IOP Conf. Ser.: Earth Environ. Sci.*, **311**, 012023

Assessment of Soil Erosion Hazard in Prambanan District Using RUSLE (Revised Universal Soil Loss Equation)

S. A Mulya [1], N. Khotimah [2]

[1,2]Department of Geography Education, Faculty of Social Sciences, Universitas Negeri Yogyakarta, Yogyakarta, Indonesia

Email: [1]saegita.adi2016@student.uny.ac.id; [2]nurulkhotimah@uny.ac.id

Abstract. Prambanan District which located in Daerah Istimewa Yogyakarta Province has the potential for land degradation due to erosion processes. With the characteristics of annual rainfall more than 2000 mm / year, topography with a slope of more than 20% in upland areas, as well as the conversion of upland to dryland agriculture are factors that can trigger the erosion process more quickly. If the rate of erosion speed exceeds the ability of the soil to regenerate the soil body, its productivity will be disrupted and accelerate the formation of critical soil. Therefore, it is necessary to know the estimated rate of erosion, tolerable distribution of erosion, and the potential danger of erosion that occurs. The purpose of this study was to (1) predict the rate of erosion, (2) calculate the permissible erosion value, (3) identify the rate & index of erosion hazard. Data were collected using field surveys and soil sampling using stratified random sampling techniques with land units as the unit of analysis. The value of erosion was predicted using the Revised Universal Soil Loss Equation (RUSLE) method. The RUSLE method is described by the following equation, $A=R*K*L*S*C*P$, where; A as estimated averages annual loss of soil, R is the rainfall erosivity factor, K is the soil erodibility factor, LS is the slope length factor, C is the cover management factor, & P is the conservation practice factor. The results showed that the erosion value ranged from 0.39 - 268.55 tons/ha/year. Permissible erosion ranges from 8.4 – 15 tons/ha/year for Latosol and 27.4 ton/ha/year for Regosol. The Rate of Erosion Hazard is dominated by moderate erosion, covering an area of 1330.7 ha or 31.8% of the total area. The Erosion Hazard Index is dominated by the low class (<1.0) which is covered over 2703.1 ha or 64.61% of the total area.

1. Introduction

Erosion is one of the exogenic processes that form the morphology in the land. Issues related to erosion occur due to the increasing demand for agricultural land, and the activities in managing this land potentially accelerate the natural erosion process that previously happens at a normal rate are now accelerated. If the rate of erosion exceeds the ability of the soil to regenerate its body, soil genesis will be disrupted and shrinking land productivity [1]. Erosion has become the major cause of the establishment of critical soils in Java Island. The use of land without followed by proper land management can trigger the natural erosion process to be accelerated so that it slowly degrades or becomes critical soil. There is a huge presence of critical land in Indonesia reached 14 million ha in which the largest are the regions with the topography of hills [2]. 36% of the eastern part of Java Central (including the Daerah Istimewa Yogyakarta Province) suffered from erosion in the rate of heavy/severe, moderate erosion rate occurs an area of 10.5%, and light erosion rate spread over 4.5% of its area.

Content from this work may be used under the terms of the Creative Commons Attribution 3.0 licence. Any further distribution of this work must maintain attribution to the author(s) and the title of the work, journal citation and DOI.
Published under licence by IOP Publishing Ltd

Erosion occurs most easily in the management of upland slopes as dryland agriculture, rainfed fields, and shrubland. Dryland agriculture on the upland areas is the most vulnerable component of ecosystems that are susceptible to the degradation of land.

Several studies have shown that the Prambanan District has factors that indicate a rapid rate of erosion at several places. The prevalent erosion of landslides has made Prambanan one of the districts with a high level of landslide hazard in the Sleman Regency [3]. Landslide erosion is a type of erosion, which moves massive amounts of mass simultaneously in a relatively short time [4]. Prambanan District has a high susceptibility to landslide erosion. The distribution of landslide risk ranges from moderate to very high hazard, was located in villages that vary in hills with slope variations of more than 15%, namely Sambirejo, Gayamharjo, and Wukiharjo[5]. Other physiographic characteristics that trigger erosion in this area are the topography of the area and the annual rainfall. Prambanan District is dominated by the topography with a slope of more than 15% and it is widely used for dryland agriculture [6]. The average annual rainfall that occurs is 2231 mm/year [7]. It can be assumed that the Prambanan District high rainfall has a strong correlation with its soil erosivity index. The rate of erosion will affect soil productivity. The further human intervention on land, its productivity will change to the minimum capacity [4]. Improper land management not only increasing the rate of erosion but also establishing a critical land. The critical land in Prambanan District discovered spread over 1,022 ha or 25% of its area, which is identified in upland areas with slopes ranges 15 to 70% [8].

Soil loss tolerance (T) is defined as the rate of erosion which should not exceed the rate of soil development [9]. Soil loss tolerance is known as a widely used concept for assessing potential risks of soil erosion and is a measure for assessing the effectiveness of soil and water conservation projects [10]. Physical degradation of soil can occur due to the intensity of erosion that quickly removes soil layers, where the process of soil loss is more durable than the process of forming soil solum. To prevent a decrease in soil productivity in Prambanan District, it is necessary to calculate the soil loss tolerance value. Research conducted in the southern region of the Prambanan District, namely in the Petir River Basin, showed an erosion value ranges of 8.73 - 47.60 tons/ha/year.

The Petir watershed has an area of 1,762.75 ha, of the entire area only 235.03 ha (13.05%) has an erosion rate below the standard of soil loss tolerance value. Meanwhile, most of the areas (84.8%) had an erosion rate that exceeded the soil loss tolerance rate. These can indicate that critical land can become a dangerous warning to the upland areas in the Prambanan District. The RUSLE (Revised Universal Soil Loss Equation) equation is a mathematical model that is widely used to estimate the amount of soil loss due to erosion [11]. The result of the RUSLE erosion model is an estimate of the amount of soil lost over one year in tons per hectare (tons/ha/year).

Erosion could potentially be a threat to the preservation of natural land resources in the Prambanan District. Therefore, it is necessary to predict the value of the rate of erosion so that land management practice can refer to the limit of soil capacity while maintaining the sustainability of soil resources. To discover the frequency of environmental damage in Prambanan District due to erosion, it is necessary to conduct a study of erosion hazard assessment in Prambanan District, therefore the research was carried out with the aim of (1) predicting the rate of erosion, (2) calculating the value of soil loss tolerance, (3) ascertaining the rate & index of erosion hazard.

2. Theoretical Framework

Soil erosion is defined as the result of moving or transporting soil or parts of land from one place to another by natural mechanisms, specifically water or wind [12]. Erosion is a geomorphological process that runs in the development of landforms [13]. In Tropical climates like Indonesia regions, most of the erosion occurrence is caused by the force of water [12]. The mechanism of erosion by water involves the process of breaking soil aggregates, discharging particles from the aggregates, transporting materials by surface runoff, and deposition of soil materials which are controlled by the erosive forces of raindrops and surface runoff [14]. The factors controlling soil erosion are the erosivity of the eroding agent, the erodibility of the soil, the slope of the land, and the nature of the plant cover [15]. Surface erosion classified into 4 types, specifically, 1) splash erosion, 2) sheet erosion, 3) rill erosion, 4) gully erosion [15].

The RUSLE equation model is widely used to determine the weight of soil loss or estimate the rate of erosion. According to Renard et al., (1997) RUSLE equation can be described as follows [11]; $A=R*K*L*S*C*P$; where; ***A*** (total amount of erosion/estimated soil erosion) is the weight of soil lost per hectare for a certain period of time. ***R*** (soil erosivity factor) is defined as the ability of rain to cause erosion, the impact of which is determined by the duration, number, and size of raindrops falling to the soil. ***K*** (Soil erodibility factor) is defined as the rate of soil susceptibility to detachment and transport of soil particles. ***L*** (slope length factor) is the ratio between the amount of soil erosion and the length of a certain slope under identical conditions. ***S*** (slope steepness factor) is the ratio between the amount of erosion that occurs in a plot of land with a certain steepness under identical conditions. ***C*** (plant & cover management factor) is the effect of cropping and management practices on the runoff and soil erosion rate. ***P*** (soil conservation action factor) is the ratio of soil loss using a specific support practice to the corresponding soil loss after up and down cultivation.

Soil loss tolerance or soil permissible erosion (T), is a concept to measure soil loss to a tolerable limit, hence the soil depth can be maintained for plant growth and allows sustainable productivity of the soil [10]. The soil in the Indonesia region is estimated to have a T value of 15-33 tons/ha/year or 1.25-2.5 mm / year, due to high rainfall and temperature [9]. Thompson (1957) formulated a method of calculating the T value using indicators 1) soil depth; 2) permeability of the subsoil, 3) substratum conditions, & 4) soil volume weight [4]. Indicators 1, 2 & 3 are used to determine the value (t), then multiplied by the volume weight to make the value in tons/ha/year [9].

Erosion hazard is an estimate of the maximum amount of soil lost on land if soil management has not changed. The Rate of erosion hazard (REH) is the ratio between the amount of eroded soil and the effectiveness of the soil which does not take into account the expected sustainability period, the amount of erosion requested, or the process of ordering land orders [16]. The Index of erosion hazard (IEH) value is the ratio of the erosion value (A) to the indicated erosion rate (T). The erosion hazard index can be used to determine the level of erosion hazard that occurs in a plot of land [17].

3. Methods

The design of this research is quantitative descriptive mixed with explanatory research. the descriptive method was to describe the occurrences of erosion symptoms and collect information related to erosion indicators in Prambanan District. Data was collected by field observations using land units that combine parameters of the slope, soil type, and land use. Soil sampling was done by using a proportionate stratified random sampling technique.

The data collected in this research included primary and secondary data in the forms of a description of 10 years period amount of rainfall obtained from the nearest climatology station soil properties, length and incline of the slope, plant management system from the field. Furthermore, samples of whole soil and nonwhole soil were retrieved based on the ring sample in order to measure the texture, permeability, organic contents in the soil, and bulk density.

Erosion rate was calculated and analyzed using RUSLE method which according to Renard et al., (1997) describe as follows: $A=R*K*L*S*C*P$;

(***R***) Rain erosivity factor is determined using the Abdurachman (1989) equation to measure the erosivity index which has high accuracy (81.13%) in providing estimates of soil erosivity values in Indonesia [18]. The equation is stated as follows:

$$R= (Rain^{2,263}*MaxP^{0,678})/(40,056*(Days^{0,349})$$

where; (***R***) is erosivity of monthly rainfall (tons/ha/year); *Rain* is the average monthly rainfall (cm); *Days* is the average number of rainy days per month; *MaxP* is the maximum amount of rainfall for 24 hours in the month concerned.

(***K***) Soil erodibility factor value in RUSLE model is obtained by the following equation [17];

$$100K=[2,1*M^{1,14}* (10^{-4}) *(12-a) +3,25*(b-2)+2,5*(c-3)]$$

where; ***K*** is the value of soil erodibility factor; ***M*** determined as the calculation of *[(% dust + very fine sand) ∗ (100-% clay)];* ***a*** is the percentage of organic material; ***b*** determined as the level of soil structure; ***c*** determined as the level of soil permeability.

(***L***) The slope length factor value is calculated using the following equation;

$$L=(X/22)^m$$

Values of ***m*** will be determined according to steepness of the slope as follows; (m=0,2 for slope <1%), (m=0,3 for slope 1-3%), (m=0,4 for slope 3,5-5%), (m=0,5 for slope >5%).

(***S***) The slope steepnes factor value is calculated using the following equation;

$$S=6,0065+0,00454*s+0,0065*s2 \text{ [for (s}^1\text{) slope values <12\%]}$$

$$S=(slope/9)^{1,35} \text{ [for (s}^1\text{) slope values >12\%]}$$

(s^1) slope was determined in advances by dividing altitude differences between two elevation (*d*) with length of the slope (*l*); [$s^1 = d/l$; in percent]

The ***C&P*** factor value is determined by conducting field observations on each land unit. The types of crop management & conservation practice refer to the classification as shown in Table 1.

Table 1. Values of C&P factor for different land uses.

Type of vegetation	*C factor*	*Type of land conservation*	*P factor*
Irrigated fields	0.561	Lack of conservation	1.000
Rainfed fields	0.210	Traditional bench	0.400
Settlement	0.128	Proper built-up terrace	0.040
Shrubland	0.250	Fair built-up terrace	0.150
Dryland agriculture	0.400	Plant on slopes (>20%)	0.900
Community forest with more plant litter	0.005	Plant on slopes (9-20%)	0.750
Community forest with less plant litter	0.001	retaining stone wall	0.100

Source: Secondary data & field survey, 2020

Soill loss tolerance (T) is calculated using the following equation;

$$B=t*Soil\ Bulk\ Density*10$$

Erosion Hazard is calculated by comparing the amount of erosion that occurs with the thickness of soil solum. The erosion hazard index was calculated using the Hammer (1981) equation [19]; IEH = A / T. The erosion hazard index is classified as follows; (<1.0 = Low; 1.0 - 4.0 = Medium; 4.01 - 10.0 = High;> 10.01 = Very High).

4. Results & Discussion

4.1. Study Area

Prambanan District is located at 7° 44' 19"- 7° 49' 51" N & 110° 27' 56"- 110° 32' 44"E, which is one of the districts in Sleman Regency, Daerah Istimewa Yogyakarta Province. Prambanan District has an area of 4185 ha (41.85 km) which is divided into 6 village administrative areas namely Bokoharjo, Madurejo, Sumberharjo, Wukirharjo, Gayamharjo, and Sambirejo. The population is 48.734 people with a population density of 1.179 / km^2. The area of Prambanan District is located at an altitude of 87 - 413 meters above sea level. The topography of the area is categorized into 3 slope classes, namely Plains (0 - 15%) covering an area of 2,082 ha or 51.2, Undulating hills (15 - 40%) covering an area of 1,795 ha or 44.1% & Hills with steep slopes (> 40%) covering an area of 191 ha or 4.7%. The Schmidth & Ferguson classification shows that Prambanan has a type C climate (slightly wet) with average rainfall for the last 10 years of 2355 mm/year (2010-2019). The geological conditions in the upland area formed by the *Semilir Lapilli Unit* & the *Kebobutak Sandstone Unit*, while the plains are composed of the *Fine Sand Unit of Fluvial Loose Material.* The geomorphological condition is formed of 42% Alluvial Plains, 7.3% Escarpment hills & 50.7% Denudational slopes & hills. Types of soil are Regosol 22.7%, Latosol 68.1% & Kambisol 9.2%.

IOP Conf. Series: Earth and Environmental Science **884** (2021) 012010 doi:10.1088/1755-1315/884/1/012010

In this study, land units were used to determine the physical characteristics of the Prambanan District based on topographic conditions, soil types, and land use. The selection of observation points and research sampling based on the results of overlapping three maps, namely slope map, soil type map, and land use map. The overlapping results of the three maps produce 19 land units. Table 2 shows the land units and the value of each erosion factor.

4.2. Estimated Soil Erosion (A)
The total amount of surface erosion in the Prambanan District will be determined by the quantity of rainfall, soil characteristics, length & steepness slope, land management & conservation measures taken. The RUSLE method for calculating the rate of erosion (A) uses multiplication of factors, namely rain erosivity (R), soil erodibility (K), slope length & slope (LS), crop & cover management (C) & conservation practice (P), the result of multiplying these factors become an equation to predict the amount of soil loss expressed in units of tons/ha/year. Table 2 shows the prediction of soil erosion from the multiplication of each factor. The explanation for each factor is as follows:

Table 2. Estimated values of soil los on each Land Unit

Land unit	*K*	*LS*	*C*	*P*	*A* (ton/ha/year)
I-Re-Si	0.59	0.19	0.01	0.40	0.39
I-Kb-Si	0.59	0.22	0.01	0.40	0.45
I-Kb-Tk	0.59	0.18	0.50	1.00	45.93
I-Re-St	0.63	0.19	0.20	1.00	20.95
I-Re-Tk	0.59	0.21	0.20	1.00	21.92
I-Re-Bp	0.63	0.21	0.13	1.00	15.03
I-Kb-Bp	0.59	0.22	0.13	1.00	14.47
II-Lt-St	0.34	5.78	0.21	0.04	14.64
I-Lt-Si	0.34	0.25	0.56	0.04	1.68
I-Lt-St	0.34	2.18	0.21	0.04	5.53
II-Lt-Bp	0.30	2.72	0.13	1.00	92.20
I-Lt-Bp	0.30	0.91	0.13	1.00	30.81
II-Lt-Hm	0.31	6.34	0.01	0.90	7.71
I-Lt-Hm	0.31	2.38	0.01	0.75	2.41
II-Lt-Tk	0.44	6.94	0.80	0.10	212.92
I-Lt-Tk	0.44	1.56	0.80	0.15	71.84
III-Lt-Hm	0.46	18.67	0.00	0.90	6.69
III-Lt-Tk	0.66	11.64	0.40	0.10	268.55
III-Lt-St	0.66	9.49	0.21	0.15	172.50
				Total	907.14

Source: Field Survery & Analysis of Primary Data, 2020

4.2.1. Rainfall Erosivity Factor(R)
The rainfall erosivity in the Prambanan District for 10 years ranged from 500-1800 cm/year. The average erosivity value during the last ten years was 875.1 cm/year. The variation in the erosivity of rain is determined by the amount of rainfall, the number of rainy days, and the maximum amount of rain [18]. A large amount of rainfall does not always create a high erosion rate if the intensity of the rainfall is low, on the other hand, heavy rain in a short time can cause erosion. The higher the rain erosivity value, the higher the rainwater's ability to break the soil aggregate getting stronger. The falling rain will

fill the macro pore space as a result, the infiltration rate will be hampered, and the surface runoff will increase [19].

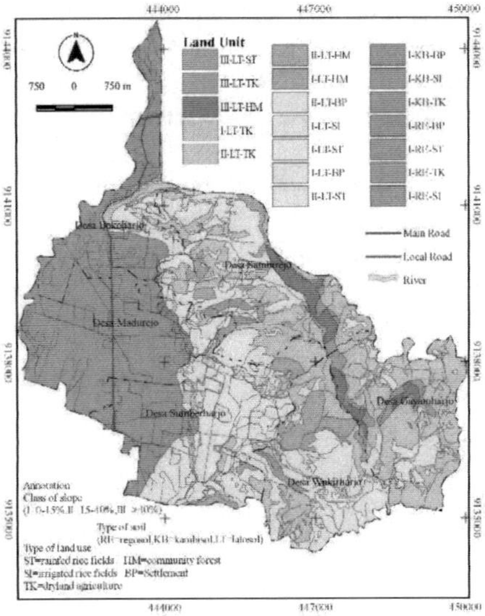

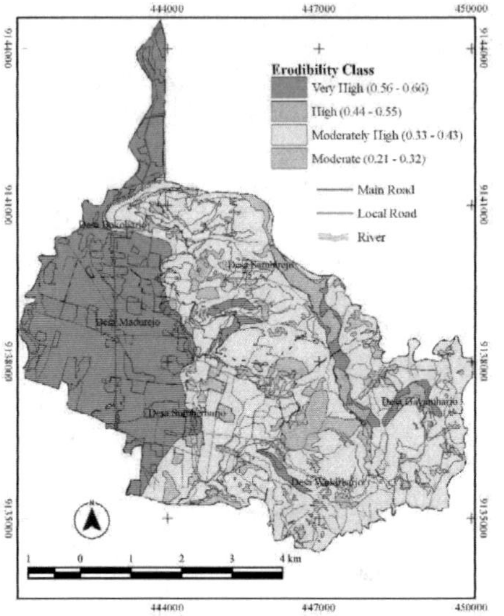

(*left*) **Figure 1.** Land unit in the study area as an analysis unit, (*right*) **Figure 2.** The distribution of soil erodibility factor values in Prambanan District.

4.2.2. Soil Erodibility Factor (K)
Soil erodibility is determined by soil resistance to the destructive force of rainfall and soil ability to absorb water [19]. The erodibility of soil in the Prambanan District varied from 0.30 - 0.66 (moderate - very high). The higher the percentage of sand and silt in the soil, the more susceptible to the soil to be eroded, while soils with a higher percentage of clay tend to be stronger against erosion [13]. Regosol soil type has very high erodibility (32% areas) & Latosol on slopes >40% has also high erodibility due to the percentage of soil texture with the presence of sand particles that are greater than silt & clay.

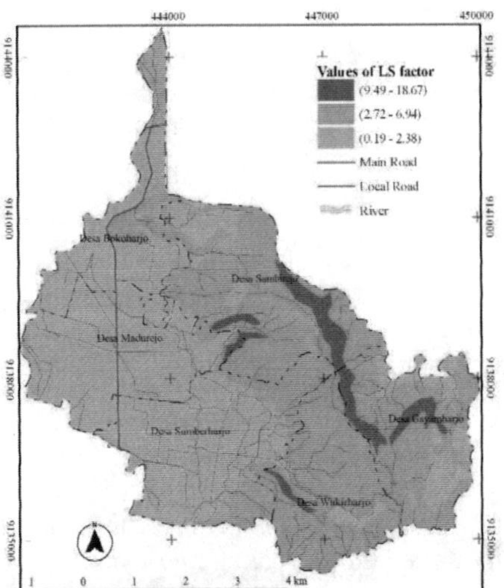

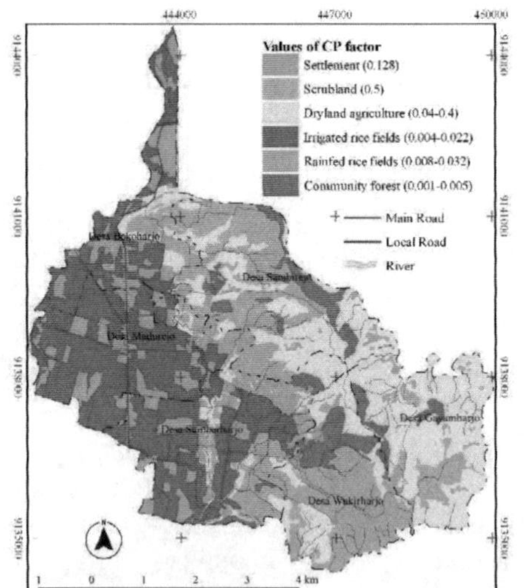

(*left*) **Figure 3.** The distribution of LS factor values in Prambanan District, (*right*) **Figure 4.** The distribution of crop management & conservation practice values in Prambanan District

4.2.3. Length & Steepness of Slope Factor (LS)

The longer the slope length (L) will increase the accumulation of surface runoff which results in an increase in the amount of eroded soil under the slope. Increasing the slope of a slope (S) will increase the velocity and erosivity of the runoff [20]. The length of the slopes in the study area varies from 22 - 87 m, with slopes varying from 1.1% - 14.8 on Alluvial plains, 16.7 - 33% on Denudational hills & 44 - 73% on the Escarpment hills. The value of the LS factor varies from 0.19 - 16.8. the higher the value of the LS factor will affect the amount of surface runoff accumulation.

4.2.4. Land Cover Management & Conservation Practice Factor (CP)

Cover and management factors describe the impact of crop management on runoff and erosion rates. Vegetation cover generally reduces the kinetic energy of raindrops before finally touching the ground [20]. The conservation action factor is the ratio of the amount of soil lost to the specific support measures appropriate to reduce the rate of erosion as the type of crop management on the land changes. The goal of land conservation practice is to reduce surface runoff velocity, as well as reduce erosion rates [20]. Rainfed fields and dryland agriculture are the largest land use in hilly areas, with conservation measures being taken are the construction of bench terraces & stone retaining walls. The plains area is dominated by irrigated rice fields with conservation measures in the form of traditional terraces. Scrubland has the highest CP value while Forest Community has the lowest CP value. The CP value will affect the size of the resulting erosion value, land use with a high CP value indicates that the vegetation and conservation practices that have been taken on the land are not sufficient to control the rate of erosion.

4.3. Soil Loss Tolerance (T)

Soil loss tolerance (T) is the rate of erosion that is still allowed or tolerated to maintain sufficient soil thickness for plant growth [17]. The calculation of erosion is allowed using the Thompson (1957) method which uses parameters of soil depth, subsoil permeability, substrate conditions, and soil bulk density, which results are expressed in tons/ha/year [17]. Regosol soil has a T value of 27.4 tons/ha/year, while Latosol soil has a T value ranging from 8.8 - 15 tons/ha/year. Soil thickness is a factor that greatly affects the T value [17]. If the actual erosion rate exceeds the soil loss tolerance rate it will decrease the soil productivity.

4.4. Rate & Index of Erosion Hazzard

Table 3. Rate & Index of Erosion Hazzard in Prambanan's Disctrict.

REH	Area		A (ton/ha/yr)	IEH	Area	
	ha	(%)			ha	(%)
Very Heavy	964.6	23	106 – 267	Very High	105.8	2.5
Heavy	585.7	14	7.7 – 92.2	High	1197.2	28.6
Moderate	1330.7	31.8	1.7 – 30.8	Moderate	179	4.3
Light	362.6	8.7	15 – 46	Low	2703.1	64.61
Very Light	941.4	22.5	0.4 – 14.5			

Source: Analysis of Primary Data, 2020

The rate of erosion hazard is determined using a ratio of soil solum depth and the amount of the erosion rate [19]. The rate of erosion hazard in the Prambanan District is dominated by the moderate category, which occurs in 31.8% of the total area. A moderate rate of erosion hazard was found on the land unit with a slope of 16-29% which held the amount of erosion ranges from 1.7 - 30.8 tons/ha/year. Very heavy erosion hazard rate occurs on land units with utilization as dryland agriculture and rainfed fields on slopes> 32%. Moderate erosion hazard rate occurs on land units with a slope of <14% with the amount of erosion 15-46 tons/ha/year. Very light erosion hazard rate occurs on land units with a slope of <3% with the land use as an irrigated rice fields.

A large amount of erosion does not necessarily have a heavy erosion hazard rate, because it also needs to consider the depth of soil solum. If a high amount of erosion occurs on soils that have decent dept of soil solum (> 90cm), the rate of erosion does not necessarily degrade the soil fertility. on the contrary, if the soil solum is too shallow (<30cm), even a small amount of erosion can threaten the fertility of the soil.

The Erosion Hazard Index (IEH) was determined using the Hammer method (1981) based on the ratio of the total amount of erosion (A) and soil loss tolerance (T) [12]. The Erosion Hazard Index in the Prambanan District is dominated by the light category (64.61%) mostly spread over on land with a slope of <14%. The erosion hazard index value can diverge from the erosion hazard rate because to determine the erosion hazard index does not involve the soil depth factor.

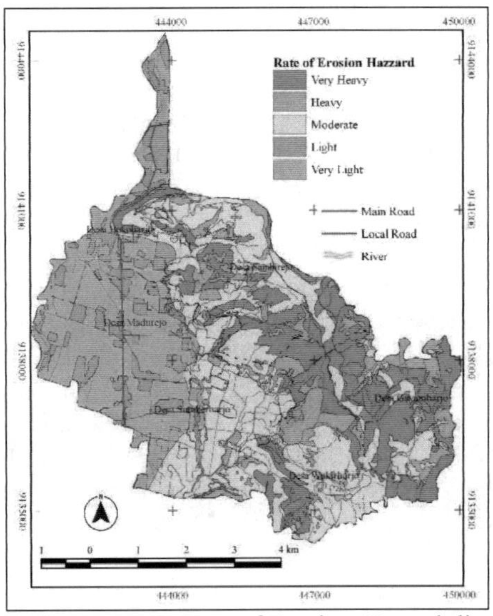

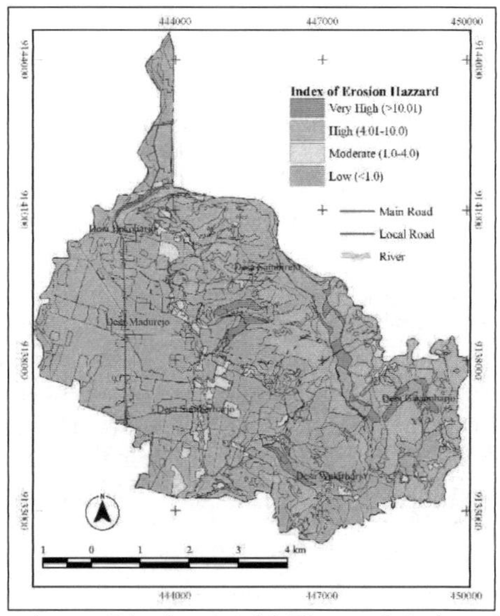

(*left*) **Figure 5.** Rate of Erosion Hazard distribution in Prambanan District, (*right*) **Figure 6.** Index of Erosion Hazard distribution in Prambanan District

5. Conclusion

Soil erosion prediction using the RUSLE method estimates the soil erosion rate (A) in Prambanan District to be 907.14 tons/ha/year. The land unit that has the largest erosion value occurs on land with latosol soil, slope> 30%, and land use as rainfed fields or dryland agriculture. The land unit that has the smallest erosion value occurs on land with Regosol soil, slope <3%, and land use as irrigated rice fields. The soil loss tolerance value (T) in Prambanan District is strongly influenced by soil thickness, hence Regosol soil (> 90cm) has a T value of 27.4 tons/ha/year, while Latosol soil (30-75cm) has a T value ranging from 8.8 - 15 tons/ha/year. Erosion Hazard Level in Prambanan District is dominated by moderate erosion hazard level, which is spread over 1330.7 ha or 31.8% of the district area. The Erosion Hazard Index in the Prambanan District shows that 64.61% of its area (2703.1 ha) is land unit with a total erosion rate (A) which is still below the permissible erosion value (T) so that the land has a low erosion hazard index.

Acknowledgments

Gratitude was expressed to the lecturers of geography education department for their many helpful suggestions which improved the original manuscript.

References

[1] Haile G and Fetene M 2011 Assesment of Soil Erosion Hazard in Kilie Catchment, East Shoa, Ethiopia *Land Degrad. Develop*. Published online in Wiley Online Library
[2] Rafi M 2019 Pendugaan Erosi Dan Tingkat Bahaya Erosi Di Areal PT ANTAM UBPE Pongkor Jawa Barat *Insitut Pertanian Bogor Scientific Repository*
[3] Badan Pusat Statistik 2020 Kabupaten Sleman Dalam Angka 2019 (Sleman: BPS Kabupaten Sleman)
[4] Irwan B 2013 Erosi (Jakarta: Prenadamedia Group)
[5] Rachma 2018 Analisis Kerawanan Bencana Longsorlahan Di Kecamatan Prambanan Kabupaten Sleman *Universitas Muhammadiyah Surakarta Online Journals*
[6] Nani H, Budi K and Hendri S 2014 Pemetaan Potensi Air Tanah untuk Mendukung Pengembangan Pertanian Lahan Kering *Jurnal Sumberdaya Lahan* 8 2 95-106
[7] Suhadi P, Slamet S, Pramono H and Setyawan P 2019 The Impact of Land Use Change on Water Resources in The Southern Foot Plain of Merapi *IOP Conf. Series: Earth and Environmental Science 338(2019) 012031*
[8] Dessy I 2017 Analisis Tingkat Kekritisan Lahan Kawasan Budidaya Pertanian Kabupaten Sleman Tahun 2016 *Universitas Muhammadiyah Surakarta Online Journals*
[9] Suhadi P and Muhammad N 2012 Pengukuran Tingkat Bahaya Bencana Erosi Di Kecamatan Kokap *Jurnal Geomedia* 10 1 111-128
[10] Xingwu D, Xiaoning S, Yanbo L, Li R and Detai F 2016 A new method to calculate soill loss tolerance for sustainable soil productivity *Agron. Sustain. Dev.* INRA and Springer-Verlag France
[11] Rudiarto I and Doppler W 2013 Impact of land use change in accelerating soil erosion in Indonesian upland area: A case of Dieng Plateau, Central Java – Indonesia *International Journal of AgriScience* 3 7 558-576
[12] Sitanala A 2012 *Konservasi Tanah & Air* (Bogor: Penerbit IPB Press)
[13] Arif A 2013 Kajian Tingkat Erodibilitas Beberapa Jenis Tanah Di Pegunungan Baturagung Desa Putat Dan Nglanggeran Kecamatan Patuk Kabupaten Gunungkidul *Jurnal Informasi* 39 2 15-31
[14] Shi Z, Fang N, Wu F, Wang I, Yue B and Wu G 2012 Soil erosion processes and sediment sorting associated with transport mechanisms on steep slopes *Journal of Hydrology* 454-455 123-130
[15] Morgan R 2005 *Soil Erosion and Conservation Third Edition* (London: Longman)
[16] Novia I 2012 Indeks dan Tingkat Bahaya Erosi Kawasan Hutan Pendidikan Gunung Walat Kabupaten Sukabumi *Insitut Pertanian Bogor Scientific Repository*
[17] Cahyu N and Suprapto D 2015 Pemetaan Tingkat Bahaya Erosi Menggunakan Model Revised Universal Soil Loss Equation (RUSLE) Di Daerah Aliran Sungai Petir Daerah Istimewa Yogyakarta *Jurnal Bumi Indonesia* 4 1 71-82
[18] Bambang S 2011 The Effect of Rain Erosivity Generated from Different Formulae on A Raster-Based Erotion Modeling (Case Study in Merawu Watershed, Banjarnegara, Central Java) *Agritech* 31 3 250-259
[19] Rhoshandhayani K T, Marga M and Indarto I 2019 Prediksi Erosi di Wilayah Jawa Timur *Jurnal Ilmu Lingkungan* 17 2 323-332
[20] Yahya F and Samer N 2015 Spatial assessment of soil erosion risk using RUSLE and GIS techniques *Eniron Earth Sci* Springer-Verlag Berlin Heidelberg

Rethinking local genius of the Merapi's community in eruption disaster risk reduction. Does it still work?

S Jamaludin[1], A Ashari[2], E P N Saputro[3]

[1]Department of Social Sciences, Master Program, Universitas Negeri Yogyakarta
[2]Department of Geography Education, Universitas Negeri Yogyakarta
[3]Department of Communication Science, Universitas Negeri Yogyakarta

Email: syarif0667pasca.2019@student.uny.ac.id

Abstract. Merapi Volcano in Central Java which has a high level of activity still holds hazard potential in the future. To reduce the disaster risks, a good disaster management is required. People who live in hazardous areas possess important social capital to enhance disaster management, in the form of local genius to cope with eruption. However, a question arises whether local genius still exists and is accepted by members of the younger generation. This paper presents a discussion about the existence of local genius on the younger generation in the southwest region of Merapi Volcano. The data were collected using some techniques which include interviews, focus group discussions, and literature study. The data were analyzed descriptively supported with scoring and statistical analysis. The results of the study show that local genius is not widely recognized and understood by the younger generation. There are no differences on the understanding of local genius among young members of the society who differ in age. The lack of teaching about local genius between generations causes a low understanding of young generation on local genius. Despite having a high motivation for preserving local genius, the younger generation perceive the local genius as a socio-cultural assets rather than as a source which enhances a disaster management. Overall, this paper presents new insights to understand the existence of local genius among the younger generation.

1. Introduction

The natural disasters caused by volcanic eruptions still pose a threat to the lives and livelihoods of the population in Indonesia till the present time. A study conducted by the Indonesian National Disaster Management Agency [1] shows that the risk of a volcanic eruption disaster in Indonesia is exposed to no less than 2,396,761 residents that spread across all provinces in Indonesia, which offers a potential loss of up to Rp 13.6 trillion. In the historical time range, volcanic eruptions are included in the natural disasters which cause great losses and fatalities. This condition correlates with the number of active volcanoes in Indonesia which account for 13% of the active volcanoes in the world [2], while the high density population occupancy in volcanic areas has been running for thousands of years [3]. With regard to this, attention and studies pertaining future eruption disaster risk are still relevant to be carried out.

Merapi Volcano located in the Central Java is one of the most active volcanoes in the world [4], indicated by a short eruption period of between 1 and 7 years and a large eruption of every 15 years [5]. With this characteristic, it poses great negative impacts for the population. Based on the data from the Indonesian National Disaster Management Agency [6], Merapi Volcano eruption at a scale of VEI-4 in 2010 caused a total loss of Rp 3.62 trillion and 386 people's deaths. Therefore, to reduce the disaster risks caused by the Merapi Volcano eruption, a good disaster management is required. Local genius that

Content from this work may be used under the terms of the Creative Commons Attribution 3.0 licence. Any further distribution of this work must maintain attribution to the author(s) and the title of the work, journal citation and DOI.
Published under licence by IOP Publishing Ltd

grow as a result of the interaction of the community with its environment for a long time [7] becomes an important social capital to enhance the disaster management. Previous studies show that there are various forms of local genius pertaining to the introduction and management of disasters in the area of Merapi [3], [7] - [9].

Local genius to deal with the eruption disaster can be a social cultural asset of the community and support the implementation of disaster management. In this regard, the question that arises is how the existence of local genius among the current millennial generation is. In the modern era today, the existence of local genius is likely to be displaced by the advancement of information technology. Local genius may still exist, but its effectiveness in handling eruption disasters needs to be reviewed. In addition, it is also necessary to know how the role of local genius as one of the references to deal with Merapi eruption amid the information disclosure from various sources that can be easily accessed by people in disaster prone areas.

This paper attempts to provide alternative information about the existence of local genius to cope with disasters among the younger generation. More specifically, the research area is limited to the southwestern part of Vulkan Merapi to further explore previous studies conducted by Setyawati et al [7] and Septiana et al [9]. There are two more specific aspects discussed in this paper. First, we illustrate the understanding and perception of the young generation on the local genius of the community in dealing with the eruption disaster. Second, we discuss how the position of disaster local genius at this time.

2. Methods

This research builds on previous research which shows the existence of local genius to cope with the eruption disaster in the southwest slope area of Merapi Volcano. In this research, a thorough discussion on how the existence of local genius among the younger generation is presented. The collected data includes the understanding and perception of the younger generation on local genius. Data collection was carried out by employing some techniques namely, focus group discussions, interviews, and literature studies. The sampling was conducted randomly on members of the community aged 18 to 40 years from the seven villages in the District of Srumbung. The selected seven villages are included in the eruption-prone areas, namely Kaliurang, Kemiren, Ngablak, Srumbung, Mranggen, Ngargasoka, and Tegalrandu village. In the previous study, Setyawati et al. [7] identified young generation's understanding on local genius in three villages namely Mranggen, Ngargasoka, and Tegalrandu. Some of these research results also become sources of data to support this research.

The collected data were then analyzed descriptively which was supported with scoring and statistical analysis. The descriptive analysis was carried out to provide detail descriptions about various types of local genius in dealing with catastrophic eruptions and the understanding and perception among the younger generation. Scoring analysis was performed to determine the extent of understanding and acceptance of local genius (Table 1). The higher the obtained score is, the better the understanding and the acceptance of local genius will be. In addition, a statistical analysis was employed as a follow up of the scoring analysis which intended to get more information about the variations on the existence of local genius in different age group. In the statistical analysis, we tested whether there are differences of mean score that represent the existence of local genius, between respondents at the age of 20s and 30s. The analysis was conducted by employing t test using SPSS and calculated with eq 1 and eq 2 [10]. In addition, we also tested whether age influences the understanding and acceptance of local genius. The analysis was performed by utilizing simple linear regression using Microsoft Excel.

Table 1. Aspects assessed in the scoring analysis

No	Aspects assessed	Indicators and scores		
		Score 1	Score 2	Score 3
1	Knowledge on the existence of various forms of local genius in dealing with disasters	Don't know	Know little bit	Know thouroughly

2	Understanding on the meaning of various types of local genius	Don't understand	Understand little bit	Thouroughly understand
3	The teaching on local genius in dealing with disasters	Never get a teaching	Know the local genius from the secondary sources	Get direct teaching
4	The benefits offered by the existence of local genius in dealing with disasters	Productive counter	Never get benefits	Get benefits
5	Motivation in preserving	Not interested	Just for understanding	Very interested
6	Involvement in the preservation of local genius	Never	Little involvement	Actively involve in the preservation

$$t = \frac{[\bar{x} - \bar{y}]}{\sqrt{\frac{(\Sigma x^2/n_x) - \bar{x}^2}{n_x - 1} + \frac{(\Sigma y^2/n_y) - \bar{y}^2}{n_y - 1}}} \text{ .... (eq 1)}$$

where $\bar{x}$ and $\bar{y}$ are the means of the two samples, n_x and n_y are the sizes of two samples

$$df = \frac{(SX+SY)^2}{\frac{SX^2}{n_x - 1} + \frac{SY^2}{n_y - 1}} \text{ .... (eq 2)}$$

where $SX = \frac{(\Sigma x^2/n_x) - x^2}{n_x - 1}$ and $SY = \frac{(\Sigma y^2/n_y) - y^2}{n_y - 1}$

3. Results and Discussions

3.1. Interviews and focus group discussions with respondents

The interviews were conducted in six villages which geomorphologically are located in the south-western flank of Merapi Volcano. The selection of research location is based on the research objectives, namely to follow-up the previous research findings conducted by Setyawati et al [7] and Septiana et al [9] who investigated the local genius in dealing with eruptions in the region. The six villages consist of Kaliurang, Kemiren, Ngablak, Srumbung, Ngargasoka, and Tegalrandu, which are administratively included in Srumbung District, Magelang Regency, Central Java. The history of Merapi eruption indicates that this area was greatly affected by eruptions in the past [4], [5], [11, [12], [13]. The interaction between community life and the eruption that repeatedly occurred in this region also contibutes to the development of local genius in dealing with eruption disaster.

Sixty interviews were already conducted. The respondents interviewed in this research include community members aged 18 to 40 years (Figure 1) residing in the six villages. In the interview process, the questions have been prepared covering several aspects related to the introduction and understanding of local genius, the teaching of local genius that has been experienced, and the motivation to participate in the preservation of local genius (Table 2). The questions in the interview are open, so that respondents can freely and broadly give answers, instead of only depending on the guiding question. Interviews were conducted in Bahasa Indonesia and the results were filled in the interviews guide.

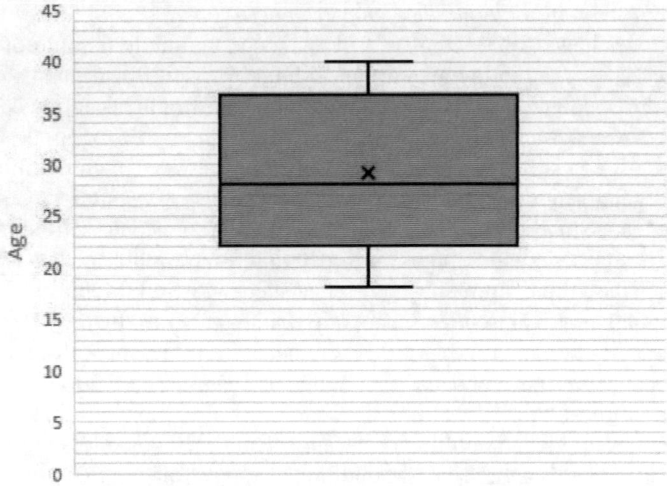

Figure 1. Boxplot for respondent's age

Table 2. Questions asked during semi-sctructured interviews

Variable	Questions
Socio-demographic	Question about personal details which include name, address, age, occupation, and education
Knowledge about local genius	What do you know about the advices, teachings, habits, or myths of the predecessors in dealing with the eruption of Merapi?
Understanding on the meaning of local genius	What is the meaning of advices, teachings, habits, or myths about the Merapi eruption disaster?
	What are the objectives and expected actions of the various forms of local genius?
The teaching on local genius which has ever been experienced	How do you know about these various forms of local genius? are they taught directly or do you obtain information from daily life in the community?
	Who usually shares information about local genius in dealing with disasters?
The benefits offered by the local genius in disaster management	What are the benefits of local genius for dealing with the Merapi eruption disaster?
Motivation in preserving local genius	What roles will you play as a young generation to preserve local genius?
Involvement in the preservation of local genius	What are your activities and how have you carried out your activities in preserving the local genius?

Beside interviews, we also conducted focus group discussions which invlove community members of 18-40 years old from the six villages. In the focus group discussion, various aspects pertaining to the existence of local genius among young people, as asked in the interview, were discussed. The focus group discussion allowed participants to perform broader and deeper discussion, where participants carry out mutually supportive interaction, mutually correcting, and compare experiences with each other. The information obtained from focus group discussions becomes another reference to discuss issues other than information obtained from the interviews. In addition, information from interviews, focus group discussions, and literature reviews is complementary and corrective.

3.2. *The existence of local genius among the younger generation*

Various earlier studies have explained that local genius pertaining to the introduction and efforts to deal with eruption disasters exist in the area around Mount Merapi. Local genius consists of spiritual beliefs, myths, including people's behavior in dealing with catastrophic eruptions [3], [8]. Specifically in the southwest slope area of Merapi Volcano, Setyawati et al [7] in her research explain that there are various forms of local genius conveyed in communication between communities. They consist of cultural semiotics, vegetal semiotics, faunal semiotics and physical semiotics. Cultural semiotics are teachings, advice, guidance, and even myths related to Merapi's volcanic activity and efforts to avoid its dangers. Vegetal semiotics include signs of vegetation which indicate that eruptions will occur. Faunal semiotics are in the form of animal behavior that stay away from Merapi Volcano as an indicator of eruption. Meanwhile, physical semiotics are physical symptoms that exist before the eruption including high air temperatures and rumbling that originates from Merapi Volcano.

Many different local genius have become a guideline in the society for a long time, especially in the past when an access to information was limited. Septiana et al [9] explain that the teaching on local genius is also carried out in the community. There are several methods of teaching local genius to the younger generations, namely through behavior, traditional activity, oral stories, community associations, and *gethok tular* or word of mouth between community members. However, it seems that the inheritance of local genius to young generation today is not very effective. This is proven by the research of Setyawati et al [7] carried out in the villages of Mranggen, Ngargosoka, and Tegalrandu which show that local genius is only understood by one third of the young generation in the region. The next third only hears information about local genius without having understanding, while another third never even hears about the existence of local genius.

In a further investigation that we conducted in the six villages on the southwest slopes of Merapi Volcano, we found that there were no significant variations on the understanding of local genius in the younger generation. This is based on the scoring of the six aspects related to the understanding of local genius (Table 1) where the results show a fairly large grouping of mean score between 10 and 14. The scoring results show a range of score between 7 to 17 without low scores i.e scores between zero up to six (Figure 2). The aspects that contribute to a lot of high score is the benefits of local genius. The younger generation generally preceive that local genius is useful as sources of information in the introduction of the eruption disaster and its handling. However, the younger generation keeps referring to the official information delivered by government agencies and preceive local genius as an additional reference in the context of socio-cultural assets. In the focus group discussions, some even expressed their opinion that the use of local genius was considered a setback in the context of information seeking, since accurate sources of information could be easily accessed at this time.

The aspect that contribute to many low scores is an understanding of local genius. Many young people have heard various information about different forms of local genius, but they do not understand the meaning behind the information. Local genius is often implicitly taught through stories, advice, or myths. The lack of systematic learning between generations becomes a factor that influences the low understanding of these various types of local genius. In this regard, some received direct teaching on local genius through advice and exemplary behavior, habits, and culture. Meanwhile, many younger generation know local genius from *gethok tular* (word of mouth).

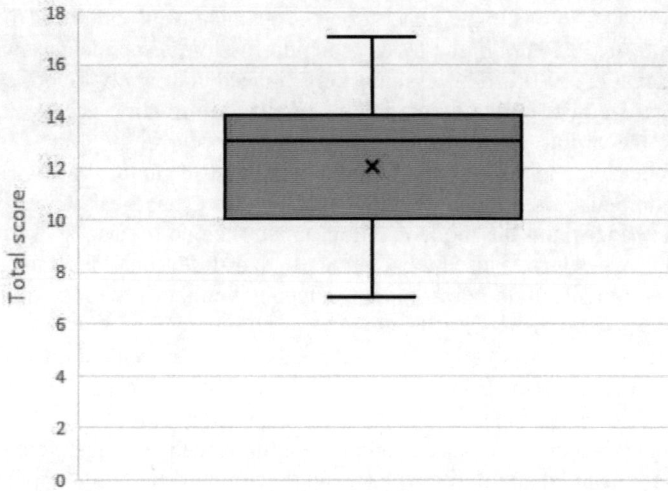

Figure 2. Boxplot of representation score of the understanding on local genius

An interesting question to further explore is: is the acquisition of local genius in the younger generation influenced by one's age? The higher of one's age should enable them to gain more teaching and experience related to local genius. In this research, we involved respondents aged 18 to 40 years. These respondents can simply be grouped into two different generations, 20s and 30s generation. To examine the differences on the acquisition of local genius between generations, we conducted a different test on the mean of scoring results for the aspects of local genius acquisition in the two groups of generations. Since the median age of the respondent is 28 years, the first age group is 18 to 28 years and the second group is 29 to 40 years. The results of the scoring in each age group are presented in Table 3.

Table 3. Scores of local genius representation in different age groups

Age Group	N	Mean	Median	SD	Max	Min
Under 28	31	11,61	13	2,70	16	7
Upper 28	29	12,45	13	2,16	17	9
Total	60	12,02	13	2,47	17	7

The results of the different test analysis show that there was no significant difference between the mean scores of the generation in their 20s and 30s. The calculation using eq 1 and eq 2 get a t score of 1.33 with a degree of freedom 20. The t value is at a significance level of 0.05 greater than the critical value. Meanwhile, the results of the independent t-test analysis using SPSS showed a sig value of 0.194 which was greater than 0.05. Both of these results indicate that there is no difference in the mean score that represents the acquisition of local genius of the 20s to 30s age group.

A simple linear regression analysis shows that age has a positive effect on the acquisition of local genius by young people. But, this effect is very small as indicated by the R2 value of 0.07 with the equation $y = 0.0906x + 9.3637$. This shows that age hardly ever influences the one's acquisition of local genius. This is also in accordance with the results of the independent sample t test which indicates that there is no difference in the acquisition of local genius by young community members of different age. The lack of local genius teaching from the previous generation and the ease of information seeking have driven the low acquisition of local genius by young people aged 18 to 40 years. The higher age does not have an effect because the values of local genius have long been ignored. For young generation, local genius is often preceived as a myth that cannot be used as an accurate reference for dealing with disasters.

The low understanding of the younger generation on local genius is partly due to the asumption that local genius is ancient. As we found in the focus group discussion, respondents say that local genius is a back step in the current era of life. This assumption is indeed commonly found, as explained by Lan [14] that local genius is considered backward. The ease of information access, the advancement in education levels, and the various forms of modernization that occur in the southwest slopes of Merapi become factors that contribute to the lack of understanding and trust in local genius among the younger generation. This condition is in accordance Ford et al [15] stating that local genius is influenced by the rapid environmental changes and the modernization which include the access to formal schools. The rapid environmental changes have caused local genius such as traditional observations and forecasting become innacurate. Meanwhile, modernization may change livelihoods, weaken traditional institutions, and reduce communication between generations.

The delivery of information about local genius to the younger generation can be performed by teaching, explaining, and transferring knowledge and experience. This can be optimized by utilizing social media such as Facebook, Line, YouTube video clips [21]. The use of social media to convey information to the younger generation has a strategic role because the younger generation in the current era uses more social media than traditional media to communicate, seek information, exchange experiences, learn and so on.

The use of social media to convey information to the younger generation offers various advantages such as social media enable quick information delivery, allow users to modify messages to be more interesting for example by combining text and images, and overcome barriers of space and time. In this regard, the older generation needs to have media literacy so that they can optimally pass on the values of local genius to the younger generation.

3.3. *The current function of local genius: disaster management and / or socio-cultural assets?*

The question that needs to be addressed about the local genius existed in the southwest slope area of Merapi Volcano at this time is: Is the local genius still relevant to be used as a reference to deal with disasters? When viewed from the content and meaning of various types of local genius, it is still relevant as one of the references in recognizing and facing eruption disasters. But, the problem now is that the knowledge of the younger generation and intergenerational teaching on local genius is inadequate. Parties that have sufficient knowledge and understanding often do not position local genius formally. Talks related to local genius are only limited to oral stories, which is slightly delivered, for example when there is an increase in Merapi volcanic activity.

Although many young people are motivated to learn and preserve (in Javanese: *nguri-nguri*) the local genius, but many do not trust information or actions that originate from local genius in today's era. Moreover, local genius is often preceived as myths (part of cultural semiotics), causing a lack of attention from the younger generation. In fact, in addition to cultural semiotics, there are faunal semiotics, vegetal semiotics and physical semiotics in the form of introducing natural signs related to the volcanic activity of Merapi. In our focus group discussion, young people at the 20s and 30s agreed that local genius can be an additional source of information for local people along with official information and direction given by the government. For example, the government shares information about the increased volcanic activity of Merapi, on the other hand the community indentify the physical symptoms of eruption such as an increase of air temperature or rumbling (physical semiotics) and wild animals which move to the settlements (faunal semiotics). This information is expected to increase people's alert on the disaster that may occur. In line with this idea, an experience in Bangladesh [16] proves that local genius combined with western-euro-centric scientific knowledge is proven to improve the community resilience to cope with flood. Based on this experience, the community resilience can be realized by combining the benefits of indigenous knowledge with modern technological solutions.

The efforts to integrate indigenous knowledge with official government policies in disaster management has become a global thought. In the Philippines [17], for example, indigenous people participate in managing the Typhoon Haiyan disaster by employing innovative methods that integrate local genius in the programs they carried out. Here, the ederly even dedicates special activities to ensure the transmission and preservation of their local genius to the younger member of their community. However, on a national scale, formal efforts have not yet been made to realize such integration, for example in a product of law. Compared to the experience in the Philippines, the implementation of local genius in the southwest region of Merapi is lower. Meanwhile in Japan [18], studies were carried out on the architecture of ancient cities which from time to time survived various catastrophic events. Strong traditional knowledge is used to support the application of disaster mitigation methods in modern architectural designs. Furthermore, this knowledge is passed on to future generations. Learning from the experience in Japan, the local genius of the Merapi community also needs to be integrated with modern disaster management to function in local scale disaster management in the region. Meanwhile in China, the exploration of ancient wisdom and the use of ideas from ancestors is useful in helping to find more effective and efficient strategies for managing stormwater in urban areas [19].

The idea of combining local genius with modern science has been widely practiced. Local genius which is combined with scientific knowledge can be a great resource to navigate the changing world of people's lives [14]. For example, in a climate change adaptation strategy, the integration of unique and specific indigenous knowledge can be one of the best ways to implement a more effective and sustainable climate change adaptation strategy in the community [20]. Why is local genius important to support scientific knowledge? Ford et al [15] say that local genius is proven to be a major source of resilience and become a repository of the accumulated experience.

Apart from its role in supporting disaster management that is still debatable among the younger generation, the local genius is basically a socio-cultural asset and part of the culture of the Javanese people. The content delivered is also the result of community interaction with the physical environment where they live for a long time. As a socio cultural asset, further teaching and efforts to preserve various aspects of local genius are very important to be realized. Research related to local genius has also been carried out. If the introduction and preservation of local genius is performed, an effective communication strategy is required to convey information to the younger generation

4. Conclusions

Local genius in the form of introduction and adaptation to the potential eruption hazards has long been developing in the southwest slope communities of Merapi Volcano. In the past, this local genius became an important reference for the community in disaster risk reduction. However, in the modern era, the knowledge and understanding of the younger generation on local genius is quite low. This is found in community members aged 18 to 40 years. In this case, the higher age doesn't guarantee of better understanding of local genius. This condition occurs due to the lack of teaching from members of the old community to the younger generation. The lack of teaching certainly affects the perception of the younger generation of the local genius as a reference in disaster management in the region. Information sourced from the local genius is often preceived as inaccurate. However, the younger generation actually has a high motivation in preserving the local genius as one of the socio-cultural assets of the people in the region. This allows the integration of local genius with the official disaster management policies from the government sourced from scientific studies.

For evaluation process, this research is limited to the understanding of local genius of the community members based on age groups. There are several things that need to be carried out in further research, including the understanding of local genius based on the level of education, gender, and the perception of the younger generation on the function of local genius in disaster management. Future research is recommended to develop a model of integration between local genius and official disaster management from the government.

Acknowledgement

The authors would like to thank the teaching staffs at the Department of Geography Education of Universitas Negeri Yogyakarta for conducting a preliminary study of local genius and community-based disaster risk reduction on the southwest slopes of Merapi Volcano. Special thank goes to Assoc Prof.

Suparmini, Assoc Prof. Nurhadi, Assoc Prof. Sriadi Setyawati, and Assoc Prof. Heru Pramono who have provided direction for performing further exploration of studies conducted in this region since 2014.

References

[1] Badan Nasional Penanggulangan Bencana, *Rencana Nasional Penanggulangan Bencana 2015-2019*. Jakarta: Badan Nasional Penanggulangan Bencana, 2014.
[2] A. Zaennudin, "The characteristic of eruption of Indonesian active volcanoes in the last four decades," *J. Lingkung. dan Bencana Geol.*, vol. 1, no. 2, pp. 113–129, 2010.
[3] F. Lavigne *et al.*, "People's behaviour in the face of volcanic hazards: Perspectives from Javanese communities, Indonesia," *J. Volcanol. Geotherm. Res.*, vol. 172, no. 3–4, pp. 273–287, 2008.
[4] F. Lavigne, J. C. Thouret, B. Voight, H. Suwa, and A. Sumaryono, "Lahars at Merapi volcano , Central Java : an overview," *J. Volcanol. Geotherm. Res.*, vol. 100, pp. 423–456, 2000.
[5] S. D. Andreastuti, C. Newhall, and J. Dwiyatno, "Menelusuri kebenaran letusan Gunung Merapi 1006," *J. Geol. Indones.*, vol. 1, no. 4, pp. 201–207, 2006.
[6] Badan Nasional Penanggulangan Bencana, *Rencana aksi rehabilitasi dan rekonstruksi wilayah pasca bencana erupsi Gunung Merapi di Provinsi DI Yogyakarta dan Provinsi Jawa Tengah Tahun 2011-2013*. Jakarta: Badan Nasional Penanggulangan Bencana, 2011.
[7] S. Setyawati, H. Pramono, and A. Ashari, "Kecerdasan Tradisional dalam Mitigasi Bencana Erupsi pada Masyarakat Lereng Baratdaya Gunungapi Merapi," *Socia*, vol. 12, no. 2, pp. 100–110, 2015.
[8] K. Donovan, "Doing social volcanology: exploring volcanic culture in Indonesia," *Area*, vol. 42, no. 1, pp. 117–126, 2010.
[9] M. E. Septiana *et al.*, "Disaster Education Through Local Knowledge in Some Area of Merapi Volcano," in *IOP Conference Series: Earth and Environmental Science 271 (2019) 012011*, 2019, pp. 1–15.
[10] D. Ebdon, *Statistics in Geography*, Second Edi. Oxford: Blackwell Publishers, 1985.
[11] S. D. Andreastuti, B. V. Alloway, and I. E. M. Smith, "A detailed tephrostratigraphic framework at Merapi Volcano, Central Java, Indonesia: implications for eruption predictions and hazard assessment," *J. Volcanol. Geotherm. Res.*, vol. 100, pp. 51–67, 2000.
[12] R. Gertisser, S. J. Charbonnier, J. Keller, and X. Quidelleur, "The geological evolution of Merapi volcano, Central Java, Indonesia," *Bull. Volcanol.*, vol. 74, pp. 1213–1233, 2012.
[13] S. Setyawati and A. Ashari, "Geomorfologi Lereng Baratdaya Gunungapi Merapi Kaitannya dengan Upaya Pengelolaan Lingkungan dan Kebencanaan," *Geomedia*, vol. 15, no. 1, pp. 45–60, 2017.
[14] N. T. P. Lan, "Changing approaches to indigenous knowledge and Its relationship with scientific knowledge," *Humaniora*, vol. 27, no. 2, pp. 186–194, 2015.
[15] J. D. Ford, N. King, E. K. Galappaththi, T. Pearce, G. Mcdowell, and S. L. Harper, "The Resilience of Indigenous Peoples to Environmental Change," *One Earth*, vol. 2, no. 6, pp. 532–543, 2020.
[16] I. Chowdooree, "Indigenous knowledge for enhancing community resilience: An experience from the south-western coastal region of Bangladesh," *Int. J. Disaster Risk Reduct.*, vol. 40, no. 101259, 2019.
[17] G. P. Cuaton and Y. Su, "Local-indigenous knowledge on disaster risk reduction: Insights from the Mamanwa indigenous peoples in Basey, Samar after Typhoon Haiyan in the Philippines," *Int. J. Disaster Risk Reduct.*, vol. 48, no. 101596, 2020.
[18] T. Okubo, "Traditional wisdom for disaster mitigation in history of Japanese Architectures and historic cities," *J. Cult. Herit.*, vol. 20, pp. 715–724, 2016.
[19] L. Zhang, Z. Yang, A. Voinov, and S. Gao, "Nature-inspired stormwater management practice: The ecological wisdom underlying the Tuanchen drainage system in Beijing, China and its contemporary relevance," *Landsc. Urban Plan.*, vol. 155, pp. 11–20, 2016.

[20] C. C. Makondo and D. S. G. Thomas, "Climate change adaptation: Linking indigenous knowledge with western science for effective adaptation," *Environ. Sci. Policy*, vol. 88, pp. 83–91, 2018.

[21] B. Chunngam and T. Worasesthaphong, "A procedure of local wisdom: learning, transferring, and utilizing social media, " *Rmutsb Academic Journal (Humanities And Social Sciences)*, vol. 3, pp. 215-228, 2018.

Tsunami Vulnerability Assessment and Its Implications for Disaster Risk Management in the coastal area of Purworejo Regency

Tanasiva[1], **Chatarina Muryani**[1] **and Pipit Wijiyanti**[1]

[1]Department of Geography Education, Universitas Sebelas Maret, Surakarta, Indonesia

Email: tanasiva15@gmail.com

Abstract. Tsunamis are disasters with unpredictable events, but the occurrence of tsunamis in Indonesia always has a significant impact on every sector of life, especially the economy and society. The position of the Indonesian State, which is located in the subduction zones as well as the increasing human activity in coastal areas, are the factors that trigger the tsunami, which is accompanied by losses and damages. One of them is Purworejo Regency, which is directly adjacent to the Indian Ocean. Vulnerability assessment is the focus of this paper. This study uses a quantitative analysis approach with analytical methods in the form of scoring analysis. Weighted cell-based data processing is the main cog in vulnerability assessment. The combination of economic and social parameters creates a vulnerability. Financial vulnerability is measured based on GRDP and productive land, while social vulnerability is estimated based on population density and vulnerable communities. The vulnerability analysis results are in the form of total vulnerability level in the medium class, the level of economic vulnerability in the low, quality, and social vulnerability in the medium class associated with the implication of disaster risk management.
Keywords: Tsunami, Purworejo, Vulnerability, Risk Management.

1. Introduction

Tsunamis are rare catastrophic events, not all seismic activities on the seabed can cause tsunamis[1]. The level of occurrence of a tsunami is challenging to predict, but if it has occurred, it can cause enormous damage if prevention is not prepared. The destructive power of a tsunami disaster can be significant if it has entered the coastal area and can reach a reasonably wide area from the coastline. The distance of the tsunami inundation depends on the coastal zone [2]. The impact of tsunami disasters, such as human casualties and economic losses, has increased more than 100 times over the past twenty years compared to the previous two decades[3].

Indonesia is surrounded by converging plate boundaries so that it has the opportunity to display patterns of repeated earthquakes, tsunamis, and tsunamis from volcanic eruptions. For example, in the last two years (2017-2018), there have been 15 fatal seismic events and two disastrous tsunamis with a death toll of> 5,500 deaths.[4]. The impact of earthquakes continues to grow due to increased vulnerability and exposure in earthquake-prone areas. Human vulnerability to earthquake hazards increases due to the proliferation of high-risk objects increased population in earthquake-prone regions and significant

Content from this work may be used under the terms of the Creative Commons Attribution 3.0 licence. Any further distribution of this work must maintain attribution to the author(s) and the title of the work, journal citation and DOI.
Published under licence by IOP Publishing Ltd

destabilization of cities[5]. The most massive earthquake and tsunami events that have occurred in Indonesia include Aceh and Palu-Donggala with a total death toll of 128,645 for the tsunami that occurred in Aceh[6] and 2,037 for the tsunami that occurred in Palu-Donggala[7].

Java Island is located in the southeast of Sundaland, which has two plate motion systems, namely the Eurasian Plate in the north and the India-Australia Plate in the south. Recent records of significant earthquakes (> magnitude 6) by the USGS show that Java experienced three quiescent intervals of earthquakes followed by increased seismic activity. The seismic pattern indicates that the Java Trench plate boundary accumulates potential movement at a rate of 7 cm/year for 30-40 years, followed by 30-40 years of more significant and more frequent earthquake events. However, only a small part and the movements accumulated over the last 40 years of seismic calm have been released, which means it is very likely that the Java region will experience a significant earthquake generating a tsunami shortly [8]. There are about 4.35 million people threatened by the tsunami along the southern coast of Sumatra, Java, and Bali. The population boom is high, along with the increasing level of risk along the south coast of Java in particular.

Tsunami risk in Central Java Province is classified as medium to high, written in [9] with physical vulnerability reaching 2,677,841,000, economic vulnerability reaching 359,862,000, and social vulnerability reaching 130,608,000. One of the districts in Central Java Province, which is directly adjacent to the coastline is Purworejo Regency. Three sub-districts, including Grabag, Ngombol and Purwodadi are located on the coast. Purworejo Regency is one of the areas adjacent to the Eurasian subduction zone and the Indo-Australian plate. This fact causes Purworejo Regency to have the potential for a tsunami that comes from a tectonic earthquake. Large-scale tectonic earthquakes in shallow water have the potential to generate tsunamis with high destructive force. Geomorphologically, Purworejo Regency has a mountainous area, namely Menoreh Hill. Tectonically it represents the volcanic arc of the central part of Java Island from the tertiary era. The area has a stratigraphy that varies from the oldest to the youngest. The geological structures found in Menoreh Hill are in the form of joints and lateral slip faults. The volcanic rock found in Menoreh Hill has weathered extensively and extends to a thickness of up to 5 meters.

Purworejo Regency is in an area with a flat slope category. In this regard, a risk assessment must be carried out in the context of disaster management development in the Purworejo Regency area. Purworejo Regency has three sub-districts bordering the coastline, which have tectonically active status. This condition is a strong basis for the occurrence of a tsunami disaster in this area. Therefore preventive measures are needed to reduce the value of the risks that may occur. The coastal area of Purworejo Regency is indicated by the phenomenon of settlement development for 15 years, which happened as a result of land conversion from non-settlement to residential land as a result of social housing construction.[10]. The development of human activities in Purworejo Regency is also influenced by two main axes, namely Jalan Lintas Selatan (JLS) Daendels and New Yogyakarta Airport (NYIA) in Kulon Progo. The coastal area of Purworejo Regency, which is directly adjacent to the Indian Ocean, is not the right zoning area for human activities.

Tsunami readiness is an essential key in reducing tsunami risk. One important information to improve the preparedness of coastal communities to the tsunami is an accurate tsunami vulnerability assessment. Many risk assessments have been carried out to calculate the potential losses caused by a tsunami. Integrated geospatial data analysis is critical in the tsunami risk assessment phase. Tsunami risk is formed based on the hazard posed as well as the vulnerability that occurs. Tsunami susceptibility is analyzed after the evaluation of tsunami potential and probability. It depends on how close the community is to the source of the hazard and its social and economic characteristics[11].

The incidence of the tsunami in Indonesia greatly affected almost every sector of the economy, including agriculture, fisheries, tourism, transportation, housing, and health [12]. In order to build a better tsunami disaster mitigation, it is necessary to have proper analysis related to tsunami vulnerability and risk assessmentMany simulators and geospatial analysis applications for tsunami susceptibility and risk assessment have been developed. For example, Arikawa and Tomita [13] combine Storm and Tsunami simulators in the ocean and Pacific coastal areas. Hazard mapping to display the characteristics of these hazards both from the nature and type of hazard that occurred, the time of the disaster, and the duration of the impact and the area of influence as hazard zones that are useful for disaster mitigation measures. This study examines the potential for tsunami-affected regions, where the tsunami vulnerability area will be calculated using economic and social vulnerability parameters, which will later be used in tsunami disaster risk management planning.

2. Methods

2.1. Research area

In this study, tsunami vulnerability mapping uses social and economic vulnerability parameters applied in the coastal area of Purworejo Regency, Central Java (Fig.1). This area is bordered by Kebumen Regency and Kulonprogo Regency in the west and east. Of course, it is also directly adjacent to the Indian Ocean in the south. Purworejo Regency is based on its astronomical location at 07^0 32 "- 7^0 54" South Latitude and 109^0 47 "28" - 110^0 8 "20" East Longitude. Purworejo Regency is divided into 16 districts with an area of 103,481 ha and a 22 km coastline.

Determining the group or population as a description of the study [14] is used in determining the research sample in this study. 16 Villages directly adjacent to the coastline were selected using the cluster random sampling method as samples in this study. The selection of the number of samples in this study used the results of the calculation of the Slovin formula on the total population of the study, which obtained a sample size of 104 people spread across 16 villages.

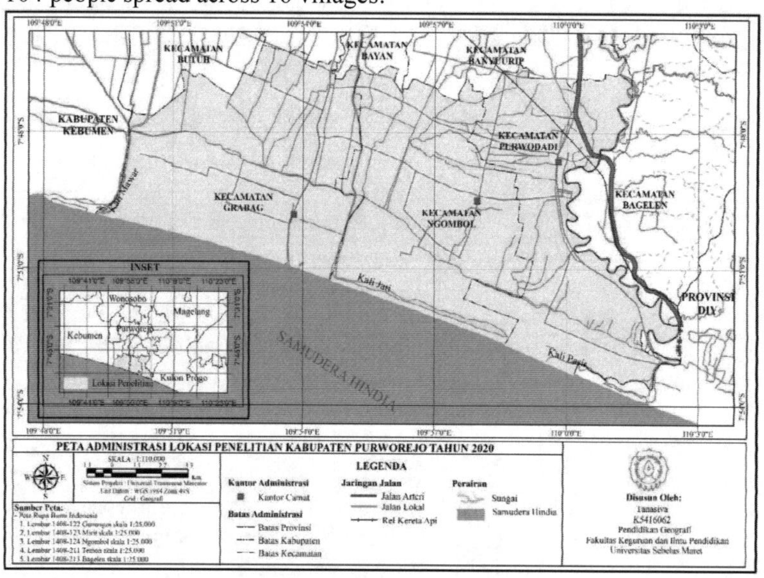

Figure 1. Research area

2.2. Dataset

Parameters of economic vulnerability and social vulnerability are needed to make a map of tsunami vulnerability in the study area. The parameters of economic vulnerability include Gross Regional Domestic Product (PDRB) and productive land in rupiah. Meanwhile, the parameters of social vulnerability consist of population density, poor people, gender, age, and disability. Productive land area data is obtained from land-use maps and district or sub-district books in numbers converted into rupiah, while GRDP is obtained from sector or district reports in figures. Social vulnerability data is collected from the Indonesian Central Bureau of Statistics.

2.3. Geospatial Analysis

Data analysis was carried out through geospatial analysis. They are starting with the determination of parameters, pre-processing, geospatial analysis, and results. Data consists of two types of data, excel, and vector. IKONOS imagery as raster data is used to create land use maps. All parameters are classified into three classes based on the economic vulnerability of the tsunami. Social vulnerability data is taken from the Indonesian Central Bureau of Statistics. Information is classified and scored based on social tsunami vulnerability criteria. It consists of population density, gender, age, poor and disabled people.

Table 1. Parameters for the Preparation and Scoring of Economic Vulnerability.

Parameters	Weight (%)	Score	Class		
			Low	Medium	Hight
Productive Land	60	Class / Max Grade Value	<50 million	50 – 200 million	>200 million
PDRB	40		<100 million	100 – 300 million	>300 million

Economic Vulnerability = (0.6*productive land) + (0.4*PDRB)

Table 2. Parameters for the Preparation and Scoring of Social Vulnerability.

Parameters	Weight (%)	Score	Class		
			Low	Medium	Hight
Population density	60		<500 people/km^2	500-1000 people/ km^2	>1000 people/ km^2
Gender Ratio (10%)					
The ratio of Vulnerable Age Groups (10%)		Class / Max Grade Value			
Poor Population Ratio (10%)	40		<20%	20 – 40%	>40%
The ratio of Disabled People (10%)					

Social Vulnerability = (0.6*log(population density /0.01)/ log(100/0.01)) + (0.1* gender ratio)+ (0.1* ratio of vulnerable age groups)+ (0.1* poor population ratio)+ (0.1* ratio of disabled people)

Data analysis for creating a tsunami vulnerability map as a combination of economic and social tsunami vulnerability was carried out using cell-based modeling about geographic information systems. All parameters applied in the model are converted to a raster dataset. This data consists of a matrix of cells and is classified under different weights. A weighted overlay is a type of conformity analysis that helps analyze geographic data based on several criteria. Weighted overlays allow users to combine the pressures of several kinds of information and visualize them, where several factors can be evaluated at once[15].

3. Results and Discussion

3.1. Economic vulnerability:
Two parameters of economic vulnerability are processed in the tabular form then converted into spatial data, which is then converted into raster cell types and analyzed using a weighted overlay to create a tsunami vulnerability map. Raster cells of all parameters are classified based on their value into three classes of vulnerability representing low, medium, and high vulnerability. Vulnerability classes and their weights are described in Table 1, and tsunami vulnerability maps are described in Figure 2.

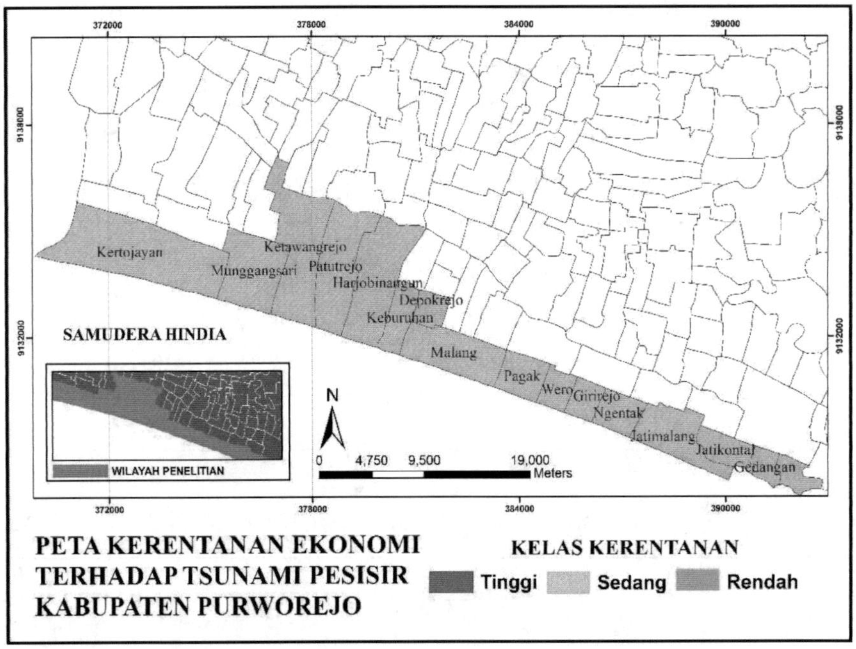

Figure 2. Economic Vulnerability Map

Figure 2 depicts a tsunami vulnerability map in the coastal area of Purworejo Regency based on economic parameters. The entire Purworejo area is described as low vulnerability due to the economic conditions of the coastal community of Purworejo Regency when faced with a tsunami disaster, and they can still take precautions. It can be seen from the productive land, which is scattered along the coastal area, which is still relatively large.

3.2. Social Vulnerability

The social vulnerability can be interpreted as a form of limitation of a community group or individual to the impact of an event, in this case, a tsunami disaster, which will affect the ability to prevent and recover from these impacts. The social vulnerability map is created using four parameters and is weighted. The calculations are based on criteria as described in Table 2. Social vulnerability map as described in Figure 3.

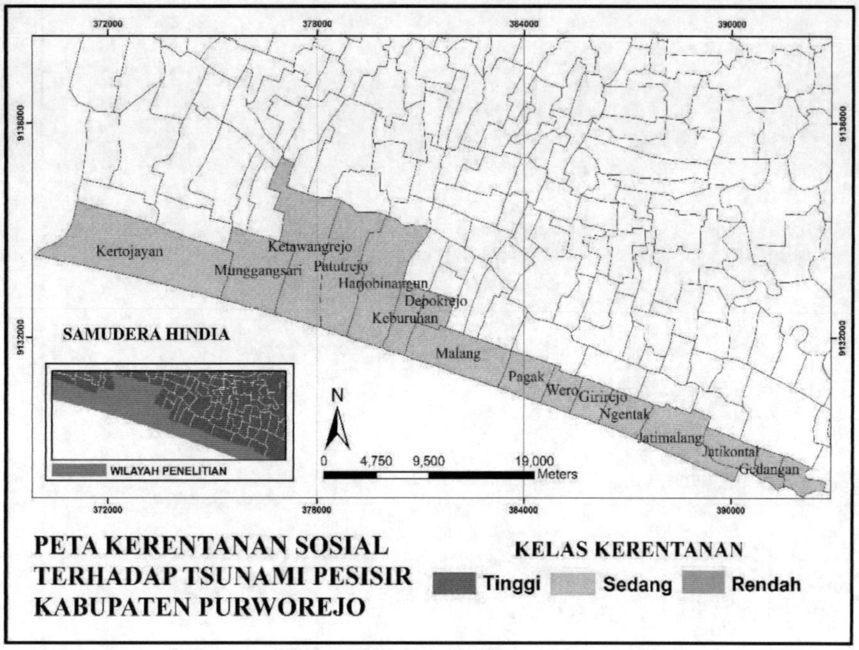

Figure 3. *Social Vulnerability Map*

Figure 3 illustrates that the entire Purworejo area is described as moderate vulnerability. The population explosion is very pronounced in Purworejo Regency, especially in its coastal area. This becomes a special note in the event of a tsunami disaster. There needs to be a direction for the people living in the area, especially those classified as vulnerable communities.

3.3. Vulnerability

The combination of economic and social parameters for assessing tsunami-prone areas using cell-based modeling is one approach to support risk analysis. Cell-based modeling calculates each parameter pixel based on its score and weight. The analysis follows the arithmetic logic below

$$\textit{Total vulnerability} = \Sigma\,(\textit{physical vulnerability} * \textit{weight}) + (\textit{social vulnerability} * \textit{weight}) \quad (1)$$

The total vulnerability describes the area of vulnerability due to the tsunami in which two parameters of economic vulnerability and four parameters of social vulnerability are combined as described in Figure 4

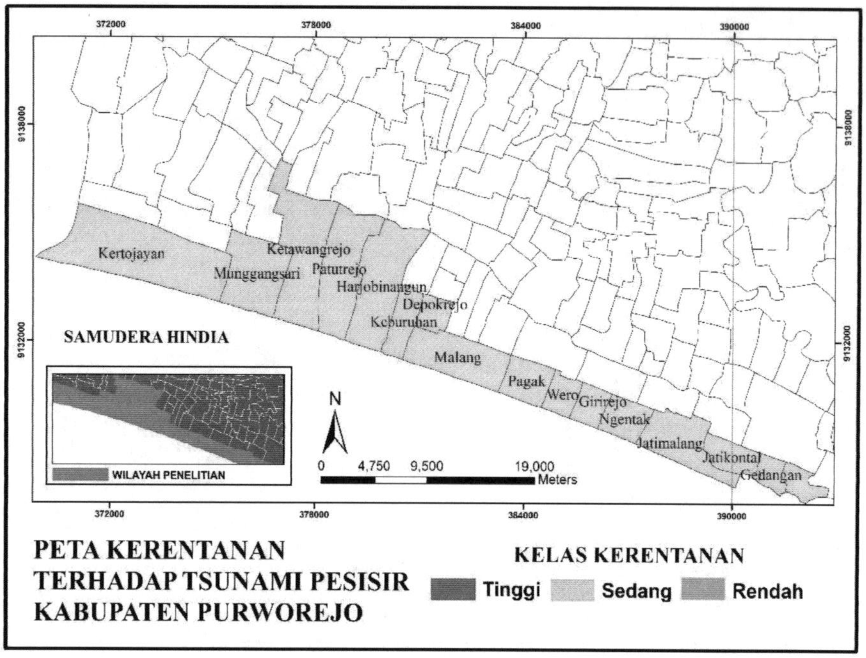

Figure 4. Vulnerability Map

Table 3. Total Vulnerability Areas

Vulnerability Class		Areas (ha)
Low		0
Medium		277.490,682
High		0

The vulnerability map, as described in Figure 4 includes information on the total area per vulnerability class (Table 3). The information contained in the map above plays an important role in establishing an effective early warning system. Social vulnerability maps provide information about villages or sub-districts with relatively high vulnerability. By integrating economic and social vulnerability in the early warning system, coastal communities will be aware of the risks. As a result, tsunami preparedness will increase.

3.4. Disaster risk management

Based on the previous calculation of vulnerability, it was found that people living in the coastal area of Purworejo Regency have a moderate level of vulnerability to tsunami disasters. The involvement of economic and social factors in disaster risk reduction is very important because it is felt that these two factors overlap directly in daily life. The ability of the coastal community of Purworejo Regency to rebuild the economy after the tsunami disaster is still considered capable. People who live in the coastal area of Purworejo Regency still depend on nature for their lives, where it is proven that the area of productive land is larger than the area of non-productive land. That reason is the benchmark that the economic vulnerability of the coastal community in Purworejo Regency is low.

One of the points that must be considered is the social skills possessed by the coastal community of Purworejo Regency. The population density and gender ratio, which are still not equal, cause social vulnerability to be classified as moderate. There needs to be a real movement so that there are not too many casualties during the tsunami disaster. Cultivating an understanding of disaster risk management in itself is one of the efforts that are felt to reduce the number of casualties who fall after a disaster. Disaster risk reduction can be made through risk management carried out individually or in groups. One of the forms of risk management that can be carried out personally is self-management, which includes management actions in the way of developing individual skills that support disaster activities. It is hoped that this will reduce the number of victims in the event of a tsunami disaster.

4. Conclusions

Weighted cell-based data processing represents a good outcome in assessing vulnerability. The combination of the economic and social vulnerability parameters in this assessment also represents a good approach for creating a tsunami vulnerability map. The map is important for determining the area that is likely to be affected by a tsunami wave. The results illustrate critical information in many applications to support disaster risk management. Furthermore, to obtain a more accurate vulnerability map and tsunami risk map, this method can be applied by adding other vulnerability parameters such as building density, mangrove forest area, distance from the coast, presence of barrier islands, beach type, and tsunami direction.

5. References

[1] D. Wicaksono, F. Ardiansyah, G. A. Nugroho, and C. A. T. S, "*ANALISIS MULTI-SKENARIO DAMPAK TSUNAMI DI KAWASAN PESISIR KABUPATEN KULON PROGO* ," vol. **1**, 2014.

[2] T. Mikami *et al.*, "*Field Survey of the 2018 Sulawesi Tsunami: Inundation and Run-up Heights and Damage to Coastal Communities*," *Pure Appl. Geophys.*, vol. **176**, no. 8, pp. 3291–3304, 2019.

[3] United Nations Office for Disaster Risk Reduction (UNISDR), "**Tsunamis account for $280 billion in economic losses over last twenty years**," *UNISDR*, 2018.

[4] R. Harris, W. Meservy, K. Stuart, M. Deng, C. Emmett, and H. Suliaman, "*Seismic and Tsunami Risk of the Java Trench and Implementation of Risk Reduction Strategies* Downloaded 08 / 22 / 19 to 91 . 212 . 53 . 253 . Redistribution subject to SEG license or copyright ; see Terms of Use at http://library.seg.org/ Seismic and Tsuna," pp. 4786–4789, 2019.

[5] A. Ismail-Zadeh, A. Soloviev, V. Sokolov, I. Vorobieva, B. Müller, and F. Schilling, "*Quantitative modeling of the lithosphere dynamics, earthquakes and seismic hazard*," *Tectonophysics*, vol. **746**, pp. 624–647, 2018.

[6] A. Rofi, M. A. Mercy, C. Indonesia, and S. Doocy, "*Tsunami mortality and displacement in Aceh province , Indonesia*," vol. **30**, no. February 2005, pp. 340–350, 2006.

[7] R. Triyono *et al.*, "*KATALOG TSUNAMI INDONESIA TAHUN 416-2018," Badan Meteorologi Klimatologi dan Geofisika*, 2019.

[8] S. Hall *et al.*, "*Awareness of tsunami natural warning signs and intended evacuation behaviors in Java, Indonesia*," *Nat. Hazards*, vol. **89**, no. 1, pp. 473–496, 2017.

[9] Mohd. Robi Amri *et al.*, *Risiko Bencana Indonesia*. Jakarta: BNBP, 2016.

[10] A. P. Damayanti *et al.*, "*Faktor-faktor yang Mempengaruhi Perkembangan Permukiman di*

Wilayah Pesisir Kabupaten Purworejo The Factors Influencing The Development of Settlements in The Coastal Area," *J. Pembang. Wil. dan Perenc. Partisipatif*, vol. **14**, no. 2, pp. 154–172, 2019.

[11] S. L. Cutter, J. T. Mitchell, and M. S. Scott, "*Revealing the vulnerability of people and places: A case study of Georgetown county, South carolina*," *Hazards, Vulnerability Environ. Justice*, vol. **90**, no. 4, pp. 83–114, 2012.

[12] J. K. Levy and C. Gopalakrishnan, "*Promoting disaster-resilient communities: The great Sumatra - Andaman Earthquake of 26 December 2004 and the resulting Indian Ocean Tsunami*," *Int. J. Water Resour. Dev.*, vol. **21**, no. 4, pp. 543–559, 2005.

[13] T. Arikawa and T. Tomita, "*Development of high precision tsunami runup calculation method based on a hierarchical simulation*," *J. Disaster Res.*, vol. **11**, no. 4, pp. 639–646, 2016.

[14] P. A. Asra and A. Prasetyo, *Pengambilan Sampel Dalam Penelitian Survei*. Jakarta: Rajawali Pers, 2015.

[15] Guntur, A. B. Sambah, F. Miura, Fuad, and D. M. Arisandi, "*Assessing tsunami vulnerability areas using satellite imagery and weighted cell-based analysis*," *Int. J. GEOMATE*, vol. **12**, no. 34, pp. 115–122, 2017.

The Effect of Google Earth-Assisted Remote Sensing Learning on Students 'Spatial Thinking Ability in Solving Disaster Mitigation Problems

Bambang Syaeful Hadi[1], Mukminan[2], Muhsinatun Siasah M[3], Kimpul E. Sariyono[4]
[1,2,3,4] Department of Geography Education, Faculty of Social Sciences, UNY, Yogyakarta, Indonesia

email: bambang_saefulhadi@uny.ac.id

Abstract. Spatial thinking ability (STA) have an important role in the study of geography which is currently supported by many geospatial technologies. Remote sensing learning has a strategic position to support the formation of student STA. This study aims to (1) test the effectiveness of Google Earth-assisted remote sensing learning on students' spatial thinking skills to solve the disaster mitigation problems, and (2) examine the relationship between STA students and remote sensing learning achievements. This study uses a quasi-experimental design. The subjects in this study were students of the Department of Geography Education. Subjects were treated as remote sensing learning with the help of dynamic imagery in Google Earth. The experimental and control classes used are geography education students who are taking remote sensing courses. Data collection is done by the test method. The test instrument was in the form of multiple-choice questions developed based on the STA concept proposed by Gresmehl & Gresmehl. Data analysis techniques to test hypotheses are t-test and Pearson product-moment correlation. The expected results of the research are Google Earth-assisted remote sensing learning is effective for improving student STA in solving disaster mitigation problems. This can be seen from the test results that show the coefficient t = 30.187 with degrees sig = 0,000. There is a positive and significant relationship between STA students with remote sensing learning achievement. This can be seen from the high significance coefficient.
Keywords: spatial thinking ability, remote sensing, google earth

1. Introduction

Remote sensing is a strategic subject in the Department of Geography Education which is under the auspices of the Higher Education Institutions for Education Personnel or Lembaga Pendidikan Tenaga Kependidikan (LPTK). It is said to be strategic because this course is just no a science and a tool for data collection by almost all geographic branches and an analytical tool that can produce fast accurate multipurpose findings, it can also function as a learning media in schools that enable goals. Learning can be achieved optimally if it used learning media[1]. The media, in addition to explaining learning material, can also increase student motivation and achievment [2]. Remote sensing gives a very big role in the development of geography and various fields related to the spatial aspects of a particular phenomenon.

Remote sensing learning at the university level often experiences problems in terms of image availability and image processing applications. This learning requires a lot of images, both as lecture material and as a learning medium. For learning purposes, images of various spatial resolutions, temporal resolutions, spectral resolutions, multi-sensor, and multi-vehicle images should be available. The price of digital images, which is still expensive, is an obstacle to procurement, so the need for images for learning cannot be adequately fulfilled. There are currently only Landsat 8 and Sent Continental imagery available freely or unpaid, provided by NASA. Landsat 8 imagery is only one type of medium spatial resolution image, so learning that requires high and low spatial resolution images cannot be fulfilled. The powerful image processing application needed also has an expensive price, so learning only uses open-source, free applications, such as ILWIS and QGIS.

The implementation of remote sensing courses is not by the minimum requirements due to the lack of supporting facilities and infrastructure. Learning facilities in the form of images and software are not available adequately. The supporting infrastructure is in the form of a representative

Content from this work may be used under the terms of the Creative Commons Attribution 3.0 licence. Any further distribution of this work must maintain attribution to the author(s) and the title of the work, journal citation and DOI.
Published under licence by IOP Publishing Ltd

laboratory, the lack of fulfillment of the number of computer units needed for training in digital image processing, the number is not proportional to the number of students, and the specifications of the computer units are not by the software currently available in remote sensing circles, which generally require high specifications.

A very important component of remote sensing learning is the image. Lack of image causes the implementation of learning not to be carried out optimally. In remote sensing learning, including practicum learning, the Geography Education Study Program requires aerial photographs of various scales, but these aerial photographs are not sufficiently available. Satellite images with various spatial resolutions and multiple channels required for learning are also not available adequately. The Geospatial Laboratory of this department does not have sufficient data to carry out adequate learning of geospatial technology. In general, in some LPTK colleges, there is no budget to purchase a new image with various resolutions, which of course costs tens or even hundreds of millions of rupiah. This constraint causes remote sensing learning to be unable to keep up with newness through technological developments. In reality, it only uses less varied images with relatively long image life. The impact is that the accuracy of the results of image interpretation is not as expected, because the objects depicted in the image and the existing relations do not match.

Remote sensing has a wide potential for various applications to solve problems [3]. In addition to applied needs, remote sensing can also be used as a medium in learning, especially to develop students' spatial thinking ability (STA). In the industrial era 4.0, which is accompanied by the proliferation of various information (big data era) today requires learning in various educational institutions to educate students to be able to think and solve problems. One of the important problems that always befall this nation is the problem of natural disasters. STA which is supported by the availability of imagery on Google Earth allows it to be used by education to train students to solve spatial problems around disaster mitigation problems. Google Earth has considerable potential to enhance methods for teaching geography and helping students develop other capabilities [3]. Several problems of disaster mitigation include determining the location of evacuation, determining safe evacuation routes, determining points for assisting, and determining the location of EWS (Early Warning Systems) tools

Google was a powerful application to get spatial information and the features on earth [4]. Therefore many researchers are interested in utilizing this application for various purposes. Some research on the use of Google Erath for mitigation learning [5] [6] [7], is not specific to spatial thinking development. Research only emphasizes how to use Google Earth for mitigation. In other research, Google Earth is used as a source of imagery for mitigation purposes. Some studies use Google Earth only to develop STA, but these were not used to develop disaster mitigation capabilities. Even though the development of this STA can trigger students to solve spatial problems independently. The development of STA is emphasized in this study with the aim that students can solve problems related to Google Earth assisted mitigation.

The main competence of utilizing remote sensing imagery for mitigating natural disasters that are urgently needed is the ability to sort information, analyze, and draw logical conclusions with supporting tools. STA is a type of thinking aimed at the ability to solve spatial problems [8]. Remote sensing learning has not been oriented towards the achievement of increasing students' spatial thinking skills in solving problems. The use of Google Earth, which is very prevalent by various groups, including in learning, has not reached a wider audience. Spatial thinking requires that users ask "what is where" and "why"; the most common virtual globe tasks only include the "what" [9]. What, where, and why are urgent question that related STA. Google Earth could to answer these question and to develop STA[10].

2. Methods

2.1 Research Design

This type of research is experimental, specifically quasi-experiments whose subjects cannot be completely controlled. In this study, the experimental class that was determined was 1 class, and the control class was also 1 class. The experimental group was not randomly selected. Therefore, the experimental design used was a non-randomized pretest-posttest control group design. This design model uses an experimental group and a control group as samples that are not randomly selected, but requires a minimum of two groups provided that both groups are homogeneous. This design was chosen because in the world of education, especially in learning, the implementation of research does not always allow random selection of subjects. After all, subjects are naturally formed in one whole group (whole groups that are formed naturally), namely class.

2.2. Variables

The variables of this study include the independent and dependent variables. The independent variable in this study is a qualitative variable (fix factor), namely the use of Google Earth. The dependent variable includes 2 variables, namely the variable spatial thinking ability and remote sensing learning achievement. Spatial thinking ability variables include location conditions, connections, comparisons, aura, regions, hierarchies, transitions, analogies, patterns, spatial associations [11].

The ability to think spatial is the ability to think inclusively as an amalgamation of declarative forms, perception of knowledge, and several cognitive operations that can be used to change, combine or operate on this knowledge, where this ability is measured using a set of questions, and the results are expressed with a certain score, from 50 to 100.

Remote sensing learning achievement is the score obtained by students after learning disaster mitigation materials assisted by Google Earth. The Scores were measure through a series of test and expressed as a number from 0-100.

2.3. Population and Sample

The population of this study was students of the Department of Geography Education, Faculty of Social Sciences, Yogyakarta State University. According to the records of the student affairs subdivision, the number of students in this department is 415 students. The sample of this study were students who were selected to serve as the experimental and control classes, which numbered 87 people (class A and B), respectively (41 and 46 people). The sample was not chosen randomly because if it was chosen randomly it would change the nature of the class.

Sampling was carried out by non-random methods by a quasi-experimental research design or also known as purposive technique (purposive sampling). The sample taken was not an individual student but the sample taken was a class, using quota sampling. Sampling is not done individually, but in a per class manner, it is intended that the experiment can run naturally.

2.4 Data Collection Techniques

There are 2 data collection techniques used in this study, namely tests and observations. The test technique was used to collect data on students' special thinking skills and remote sensing learning achievement. The test was carried out twice, namely pre-test and post-test. The pretest was carried out before the treatment (treatment) of the research subject was carried out. Post-test was carried out after treatment. Both tests were conducted to measure the effectiveness of the treatment in improving students' spatial thinking skills. The results of this test are then used for testing with inferential statistics, especially the t-test. Observations are used to collect student activity data while carrying out remote sensing learning with treatment using Google Earth.

2.5 Research Instruments

There are 2 research instruments used for data collection purposes, namely STA test questions and learning achievement test questions. To measure student STA, multiple-choice questions were developed based on the imagery provided by Google Earth. The STA component developed for the preparation of questions is based on a concept developed by Gresmehl & Gresmehl [12]. This instrument has been developed and used for research purposes whose validity and reliability have been tested [13]. The number of questions from this instrument is 35 items.

The questions used to measure remote sensing learning achievement are based on the basic competencies stated in the syllabus, namely the basic competencies in interpreting satellite images and aerial photographs. The number of questions is 40 multiple-choice questions. The number of questions from the two sets of questions is 75 items. The test is carried out in stages, namely in the first stage,

STA test questions are given, then in the second stage questions are given to measure geography learning achievement.

2.6. Data Analysis Techniques

The data analysis techniques used to test the hypothesis are the t-test and Pearson Product Moment Analysis. The t-test is used to (a) determine the differences in the spatial thinking abilities of the experimental class and the control class, (b) to test the differences of the student's spatial thinking abilities of the pre-test and post-test results, (c) to test the differences in spatial thinking abilities between male and female students. Pearson Product Moment Analysis is used to examine the relationship between spatial thinking skills and student remote sensing learning achievement.

3. Result and Discussion

3.1 Ability to Think Spatially

3.1.1 Experiment Class

To find out the score of the spatial thinking ability of the experimental group or students who carry out learning activities using Google Erath media, with a sample of 36 students, a descriptive analysis was carried out. The analysis showed that the average STA score for this group was 66.22, with a standard deviation of 4.776, a median of 69.00, and a mode of 65. The minimum score was 60 and the maximum score was 78, the range of scores obtained was from 60-78. The theoretical range of scores of the instrument is 0-100.

The frequency distribution of students' spatial thinking ability scores treated with Google Earth media is by the following table where the score 76-78 is the score with the highest frequency (27.78%) and the score> 78 is the highest score with the lowest frequency, namely 1 or 2.78%. Based on the classification (very low class = 0-20, low = 21-40, medium = 41-60, high = 61-80, and very high = 81-100) it is known that all students in this group have high grade STA, none students who belong to the low and very high class (Fig 1).

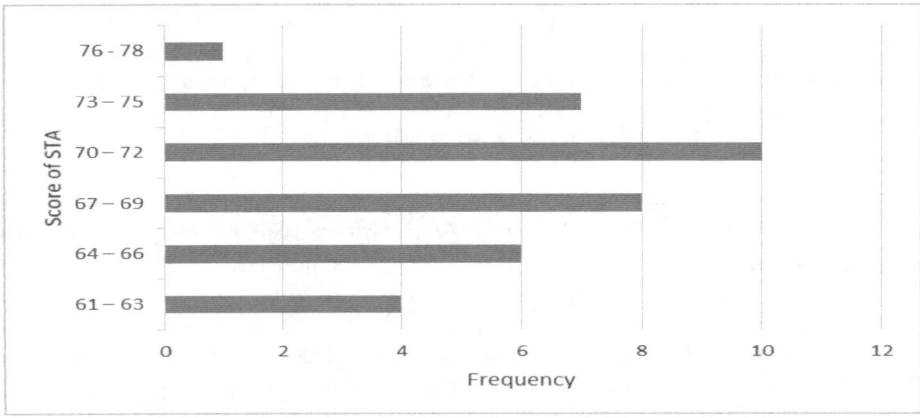

Figure 1. Line graph of STA Score in the Experimental Group

3.1.2 STA of Control Class

The control class or group of students learning to improve spatial thinking skills aided by conventional remote sensing or not using Google Earth media was represented by 42 samples of students. The average score was 63.36 with a standard deviation of 15.693. The minimum score is 45 and the maximum is 75, the median is 65, and the mode is 65 with a variance of 43,388. The achievement of this score shows that in this control class as much as 39.54% of students have STA abilities at the middle to lower level (Fig 2)

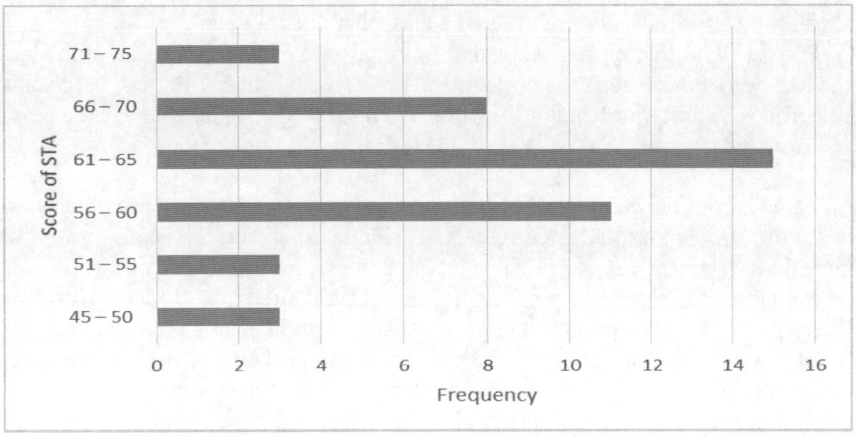

Figure 2. Frequency Distribution Graph of STA Score in Control Group

3.2. Remote Sensing Learning Achievement
3.2.1. Experimental Class Learning Achievement

Remote sensing learning achievement referred to in this study is limited to learning achievement in the cognitive aspects. Achievements are obtained from the average midterm exam (nd course assignments that have been obtained by the participants. Information regarding learning achievement scores from the experimental group is needed to provide a statistical picture, so that it can understand the logical relationship between the STA position and determine the role of remote sensing learning achievement as an independent variable. Learning achievement (knowledge) is an independent variable that can affect the ability to solve spatial problems, as in the next section of this study which conducts inferential analysis. The discussion of learning achievement includes both groups, namely the achievement of the experimental group in general and according to spatial resolution.

Before discussing the results of the descriptive analysis of the achievement variables according to spatial resolution, first discuss the distribution of the experimental class achievement scores in general. The results of the descriptive analysis of remote sensing learning achievement in general from the experimental group were learning achievement scores ranging from 48-90, average score = 70.20, median and mode = 70, SD = 9,124, and variance = 84,899. Judging from the central distribution, it can be seen that the student achievement scores are relatively even, this can also be seen from the SD score which shows a deviation from the average score of 9,124. The learning achievement score of the experimental group was included in the good category.

3.2.2 Control Class Learning Achievement

The remote sensing learning achievement of the control group showed a varied distribution, namely scores ranging from 48-81. The central tendency scores included the mean = 65.81, median = 65, mode = 65, SD = 9.214, and variance = 31.274. From the graph, it appears that learning achievement is more concentrated in the intervals of 61-65 and 66-70. The largest group is in the interval of 61-65. This learning achievement is classified as a medium class. When compared with the success criteria set by the instructor of this course, namely the achievement of KKM (minimum competency) with a score of 70, the control class average score can be stated as not meeting the standard of success.

3.3 The Effect of Google Earth Assisted Remote Sensing Learning

The use of Google Earth as remote sensing learning media with problem-based learning methods is used to train students to use imagery as a medium and a geospatial technology tool that can help develop spatial thinking skills. Remote sensing learning through the study of geography themes/problems requires students to use images available on Google Erath to solve problems so that their STA can be trained. To determine the effect of Google Earth assisted remote sensing learning on the experimental class, it requires the control class as a class that does not carry out Google Earth assisted learning. The problems that must be solved in this study are the determination of the location of settlements, the location of the evacuation of people affected by volcanic eruptions, the location of flood prevention facilities, priority for handling hazards and disaster risks, and where is the best route for evacuation and drop aids.

The data obtained from the STA pre-test and post-test results of the experimental group that received treatment with Google Earth-assisted remote sensing learning and the control group who did not receive Google Earth-assisted remote sensing learning were then analyzed inferential statistics. The analysis model used to test this hypothesis is the t-test. Paired t-test analysis was used to determine the difference between the pre-test and post-test scores of the paired groups, namely the pre-test and post-test of the experimental group and the pre-test and post-test of the control group. The independent t-test model was used to test the differences in STA scores between the experimental group and the control group, both between pre-test and pre-test and between post-test and post-test scores.

The hypothesis proposed to test the effect of remote sensing methods assisted by Google Earth on the spatial thinking ability of geography education students is H_0 = there is no difference in the average STA score between the experimental group and the control group. H_1 = there is a difference in the average STA score between the experimental group and the control group.

To answer the above hypotheses, the analysis was carried out with an independent t-test. Before deciding on the hypothesis, first look at the variance equality in Levene's test as presented in the appendix. In the Equal variances assumed Levene's column, it is known that the coefficient of F=2.854 with a value of sig.=0.092. The sig number is still>0.05, so it can be declared insignificant. Therefore, equal variances are not assumed, from which the value of t = 9.069 is obtained with sig.= 0.000.

3.4. Relationship between STA and Learning Achievement

Based on the results of the Pearson Correlation analysis as presented in Table 1, it is known that the coefficient of the relationship between the STA variable and remote sensing learning achievement is 0.489 with a significant degree of 2 tails of 0.002. These results justify the hypothesis that nil. which states that there is no relationship between STA and remote sensing learning achievement is rejected. This is based on the results of the analysis which shows that the significant coefficient is below 0.005. Therefore it can be stated that there is a positive and significant relationship between STA and remote sensing learning achievement.

Table 1. Results of Analysis of Relationship between STA and Learning Achievement

		STA	Learning Ach of Remote Sensing
STA	Pearson Correlation	1	.489**
	Sig. (2-tailed)		.002
	N	36	36
PresBel_PJ	Pearson Correlation	.489**	1
	Sig. (2-tailed)	.002	
	N	36	36

**. Correlation is significant at the 0.01 level (2-tailed).

4. Discussion

STA is an important factor that must be taught to students majoring in geography education. Geography students have an important task to solve spatial problems based on geospatial technology. Google earth is a form of geospatial technology that can help students improve STA in solving disaster mitigation problems. The results of this study are in line with what Schwartz have explored regarding the use of big data, including with Google Earth for disaster mapping, particularly for flood mapping and flood vulnerability assessments [14]. But the research [14] did not reveal STA as part of learning outcomes using Google Earth which is useful for solving problems related to mitigation. His research were only for mapping floods and assessing flood vulnerability for disaster decision-making.

Another study using imagery that has been done shows that spatial resolution images can improve students' abilities in solving spatial problems related to disasters. This study strengthens the conclusions from the results of this study which show that Google Earth can be used effectively as a learning medium to train students in solving disaster mitigation problems. This evidence confirms that the nature of Google Earth which has dynamic spatial resolution is in line with the image properties used by Hadi[13].

Disaster mitigation is a part of disaster management that requires in-depth and comprehensive studies. Google earth technology is able to provide the necessary variables for disaster management purposes. Many variables are spatial in the broad study area. For this purpose, Google earth technology alone is not sufficient. He needs someone who has an STA. studies conducted by several researchers [8], [5] for mitigation purposes require a variety of geospatial technologies. This research only uses one geospatial technology, but the human factor is more developed, so that one technology becomes more effective.

The conclusion of this study is also relevant to what has been done by several researchers [15], [8], [11] who use geospatial technology to improve students' spatial abilities and learning achievement. This research is also able to show that students' sensing learning achievement can be improved by utilizing this dynamic image. Research on the use of integrated geospatial technology was carried out by Hyunh [8] which showed that the use of geospatial technology was very effective in improving student achievement and competence. There is an interesting finding in this study, which shows that a Google Earth assisted remote sensing learning does not only improve student STA. But students have increased geographical knowledge to solve problems related to disaster mitigation. Geographical knowledge and the use of Google Earth media in the study and study of remote sensing have an important role in developing spatial thinking skills [16]. Another interesting finding in this study is that the use of dynamic Google Earth to increase spatial thinking encourages increased student interaction with media and interactions between students. This finding is in line with other studies that use virtual globes to hone STA, especially in relation to spatial logic ((the emphasis is on the ability to answer "why" a feature lies there) [9].

5. Conclusion

From the description that has been stated above, it can be concluded that: (1) Remote sensing learning assisted by Google Earth is effective in improving student STA for solving hazards mitigation problems. It can be seen from the test results which show the coefficient t = 30.187 with degrees sig = 0.000. The use of Google Earth can increase student STA, because of its dynamic nature (zoom in and zoom out), provides complete features of the earth's surface, and support attribute information to spatial features. (2) There is a positive and significant relationship between student STA and remote sensing learning achievement. It was evidenced by the high correlation coefficient and the degree of significance = 0.000. STA helps students in image interpretation activities, which require understanding the spatial features of the image. Thus, in the remote sensing learning assessment, students' high STA can obtain high learning achievement. Google Earth is a powerful application for learning geography and remote sensing, including developing spatial thinking skills in disaster mitigation. So, geography teachers must apply this application in learning to train students in discovering various spatial phenomena and attributes.

6. Acknowledgement

The researchers would like to thank LPPM Yogyakarta State University, especially the head of Department of Geography Education, Faculty of Social Sciences for their good support of our research. In addition, this publication has been supported by Vice Rector of academic who provides research opportunities and support to be published in international scientific forums. Thanks also go to IGI, the Indonesian Geographical Association through the State University of Padang, which has organized an international Geography conference.

References

[1] I. Shah and M. Khan, "Impact of Multimedia-aided Teaching on Students' Academic Achievement and Attitude at Elementary Level," US-China Educ. Rev. A, 2015.

[2] J. Lee, L. Lin, and T. Robertson, "The impact of media multitasking on learning," Learn. Media Technol., 2012, doi: 10.1080/17439884.2010.537664.

[3] T. C. Patterson, "Google earth as a (not just) geography education tool," J. Geog., 2007, doi: 10.1080/00221340701678032.

[4] O. Mutanga and L. Kumar, "Google earth engine applications," Remote Sensing. 2019, doi: 10.3390/rs11050591.

[5] J. N. Hird, E. R. DeLancey, G. J. McDermid, and J. Kariyeva, "Google earth engine, open-access satellite data, and machine learning in support of large-area probabilisticwetland mapping," Remote Sens., 2017, doi: 10.3390/rs9121315.

[6] M. A. Brovelli, Y. Sun, and V. Yordanov, "Monitoring forest change in the amazon using multi-temporal remote sensing data and machine learning classification on Google Earth Engine," ISPRS Int. J. Geo-Information, 2020, doi: 10.3390/ijgi9100580.

[7] M. Yu, Y. Huang, Q. Xu, P. Guo, and Z. Dai, "Application of virtual earth in 3D terrain modeling to visual analysis of large-scale geological disasters in mountainous areas," Environ. Earth Sci., 2016, doi: 10.1007/s12665-015-5161-5.

[8] N. T. Huynh and B. Sharpe, "The role of geospatial reasoning in effective gis problem solving: K-16 education levels," Geomatica, 2009.

[9] J. Schöming, M. Raubal, M. Marsh, B. Hecht, A. Krüger, and M. Rohs, "Improving interaction with virtual globes through spatial thinking: Helping users ask 'Why?,'" in International Conference on Intelligent User Interfaces, Proceedings IUI, 2008, doi: 10.1145/1378773.1378790.

[10] G. (UNY) Ardyodyanto, "Pemanfaatan Google Earth Dalam Pembelajaran Geografi Untuk Meningkatkan Hasil Belajar Siswa Kelas X Sma Widya Kutoarjo," Yogyakarta State University, 2014.

[11] R. S. Bednarz and J. Lee, "The components of spatial thinking: Empirical evidence," in Procedia - Social and Behavioral Sciences, 2011, doi: 10.1016/j.sbspro.2011.07.048.

[12] B. S. Hadi, Penginderaan Jauh Pengantar ke Arah Pembelajaran Berpikir Spasial, Fisrt. Yogyakarta: UNY Press, 2019.

[13] B. S. Hadi, "Pengaruh Penggunaan Citra Multiresolusi Spasial Dalam Pembelajaran Penginderaan Jauh Terhadap Kemampuan Berpikir Spasial Mahasiswa Calon Guru Geografi," 2017.

[14] B. Schwarz et al., "Mapping Floods and Assessing Flood Vulnerability for Disaster Decision-Making: A Case Study Remote Sensing Application in Senegal," in Earth Observation Open Science and Innovation, 2018.

[15] J. Lee and R. Bednarz, "Effect of GIS learning on spatial thinking," J. Geogr. High. Educ., 2009, doi: 10.1080/03098260802276714.

[16] Wakabayashi Y, "Role of Geographic Knowledge and Spatial Abilities in Map Reading Process: Impilcations for Geospatial Thinking," *Geogr. reports Tokyo Metrop. Univ.*, 2013.

IOP Conf. Series: Earth and Environmental Science **884** (2021) 012014 doi:10.1088/1755-1315/884/1/012014

The Relationship Between Socioeconomic Status and Household Water Requirement in Muara Angke Inundation Area, Jakarta

Cahyadi Setiawan, Muzani, Warnadi

Department of Geography Education, Faculty of Social Science, Universitas Negeri Jakarta

e-mail: cahyadi-setiawan@unj.ac.id

Abstract. Jakarta has approximately 40 percent of the area below sea level at high tide, one of which is located at Muara Angke. The area of Muara Angke experiences inundation and has groundwater contaminated by urban community activities and saltwater intrusion. As for the purpose in this research is two calculate the relationship between socioeconomic status and household water requirement. This research was conducted for 5 months from March to July during the Covid-19 pandemic. Respondents were 40 households which were determined by simple random sampling. Multiple regression is used to determine the relationship between the independent variables and dependent variable. The results showed that there was a strong relationship between socioeconomic status and household water requirement because the value was above 0.8. The value R Square obtained is 78.9% which can be interpreted that the independent variables contribute 78.9% to the dependent variable and remaining 21.1% is influenced by other factors. The most influential variable is the number of household members.

Keywords: Socioeconomic Status; Household Water Requirement

1. Introduction

Jakarta has 40% of the area which is below sea level at high tide and is in the form of a coastal swamp, making it very possible for flooding to occur. The sharply increased use of urban space also affected the return period of flood-stricken areas. The topography of the area, which is unsuitable for a built-up area and designated as a wetland used to collect water during floods since the 1960s, has been built for various urban purposes. The canal system was unsuccessful because of the flat topography of Jakarta. It made the water couldn't flow by gravity. The existence of mud sedimentation also causes the water flow to be not smooth, so that the construction of canals only acts to cope with flooding for a moment [1], [2].

Jakarta is experiencing a rapid development in land use change from a vegetation area to a built-up area. The concentration of development of residential areas, industrial and office areas, infrastructure, and infrastructure facilities to support activities is expanding and consuming an area. The concentration of population density results in unavoidable overexploitation of groundwater. The development of built-up areas and exploitation of groundwater causes more intensive land subsidence due to not only natural factors but also human factors [3].

Content from this work may be used under the terms of the Creative Commons Attribution 3.0 licence. Any further distribution of this work must maintain attribution to the author(s) and the title of the work, journal citation and DOI.
Published under licence by IOP Publishing Ltd

The city of Jakarta has grown to become the largest and most important metropolitan city in Indonesia because it has become the center of state administration and business. The city of Jakarta also has an important role in national and international trade, being a center of investment for foreign investors, not only in manufacturing but also in construction and services. This has boosted the economy of the city of Jakarta and made Jakarta one of the favorite cities for urbanization destinations. Economic and social changes have an impact on changing land use patterns, for example the city center of Jakarta has changed significantly in recent decades. Settlements in downtown Jakarta have turned into commercial areas, offices, and elite settlements such as apartments. Development in the city center of Jakarta affects the development of satellite cities in the vicinity such as Depok, Bogor, Tangerang, and Bekasi areas. These satellite cities grow in the service, residential and industrial sectors to support the city of Jakarta [4].

The availability of water resources and demand for water are subject to many uncertainties, so it is necessary to conserve water resources to ensure the sustainability of life in the future by taking into account the factors that affect water consumption [5]. Most of the urban population takes groundwater for their daily needs from shallow groundwater by making wells which can be in the form of dug wells, PAM water and deep groundwater (boreholes). The use of groundwater will continue to grow and increase in line with urban development. In line with urban development, water problems also increase both in terms of quantity and quality. This situation has also happened in Jakarta that facing serious water pollution problems. It due to economic development, the rapid growth of population, and urbanization [6].

The increasing population causes an increase in groundwater use which has an impact on decreasing the quantity of groundwater. The existence of a deficit of groundwater causes water vacancies in the soil pores, resulting in land subsidence and continuous intrusion in coastal areas. Apart from decreasing the quantity of groundwater reserves, the community also faces the problem of decreasing groundwater quality [7], [8], [9]. Groundwater is not the only source to fulfil household water demand in Jakarta, but it is necessary to utilise other sources of water [10]. There are many problems in this area, so that the aim of this research is two calculate the relationship between socioeconomic status and water quantity needed for domestic uses.

2. Methods

This research was conducted for 5 months from March to July 2020 during the Covid-19 pandemic. Respondents were 40 households which were determined by simple random sampling. Multiple regression is used to determine the relationship between the independent variables and dependent variable. The research implementation procedure was carried out in four stages. Stage 1, literature studies to build a theoretical foundation. Stage 2, the preparation of research instruments in the form of observation guidelines and questionnaires. The observation guideline is in the form of a basic map in the form of a PDF map which can be used during field observations with the application on a cellphone, while the questionnaire is used as a guide when the interview is conducted. Stage 3, collecting field data and secondary data from related agencies. Stage 4, analysis of research data according to research objectives. Multiple regression analysis with SPSS version 23 to determine the socio-economic condition variables that affect the amount of household clean water needs. The sample was determined by simple random sampling as much as 10% of the 369 population, namely 40 households.

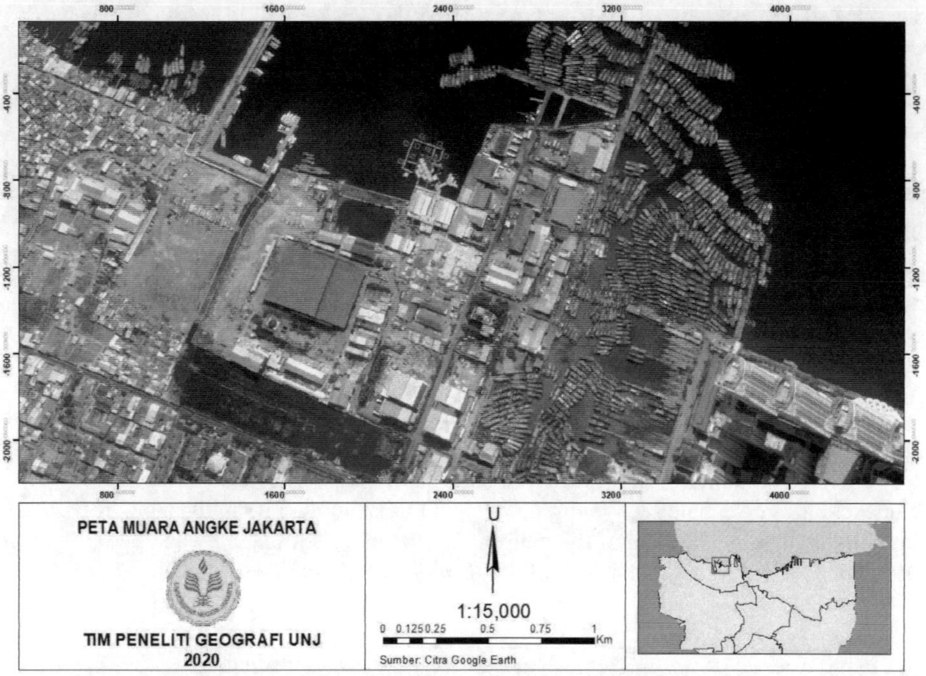

Figure 1. Research Location

3. Result and Discussion

3.1. Number of Household Members

The number of household members can be used as the basis for calculating the amount of household water needs each day or every month. Most of the number of household members is less than or equal to 4 people who are the nuclear family, which consists of father, mother, and their children. The number of household members consisting of 2 people is usually a young couple who has not been blessed with children or an elderly couple whose children are already married. The number of household members that is more than 4 people is dominated by the extended family. This type of household usually consists of several families, namely consisting of father, mother, children, grandchildren, and their relatives. The management of household finance with nuclear family is certainly different from household with extended family. According to Noh [11] that the number of household member is one of the factors that affected stress level.

Table 1. The Number of Household Members

Household Members	Amount	Percentage
2	6	15.0
3	10	25.0
4	12	30.0
5	6	15.0
6	6	15.0
Total	40	100.0

3.2. Education of The Head of The Household

It is important to know the education of head of household because it is related to the level of education completed by the head of the household. The higher the level of education a person is expected to have more knowledge as well as experience, especially related to the work they do. Most of the head of household respondents have low education, namely primary and junior high school education. Noh [11] stated that the educational level also affected stress level.

Table 2. The Education of Head of Household

Education	Amount	Percentage
Primary School	12	30.0
Junior Hight School	14	35.0
Senior High School	11	27.5
Bachelor's degree	3	7.5
Total	40	100.0

3.3. Age of the Head of the Household
Age is related to the productive age in the field of employment. People who are at the productive age are expected to be able to do their job well. Most of the respondents were <50 years old. This shows that the head of the household is dominated by young people who fall into the productive age group of the workforce. With a mature age of the head of the household, it is expected that they can better manage stress in their life. Noh [11] stated that age is one of the factors that affected stress level.

Table 3. Age of Head of Household

Age	Amount	Percentage
< 30	8	20.0
30 – 39	11	27.5
40 – 49	14	35.0
50 – 59	3	7.5
60^{+}	4	10.0
Total	40	100.0

3.4. Household Total Income
Household total income is the total income of all working household members. Most of the household total income is below three million rupiah. The greater the household total income, the more household needs that can be met, including household water needs. There is a small proportion of household whose household members work alone with an income below the minimum wage of Jakarta Special Province. It is this group of households with low income that still depend on a large part of their water needs from groundwater. his is in line with Kooy [12] that stated the significance of shallow sub-surface groundwater services for the poorest household and increasing services between income groups. The household total income is proportional to household water demands but it is inversely proportional to the share of groundwater utilisation [10]. Income is one of the parameters used to determine the socioeconomic conditions of a household. Increasing the economic capacity of the community will increase the need for water in line with the facilities owned and the higher demands of life. According to van Gent [13] household's suitability with its social environment affects its likelihood of leaving.

Table 4. Household Total Income

Total Income (Rp/month)	Amount	Percentage
< 3	27	67.5
3 – 4,5	2	5.0
4,5 – 6	6	15.0
> 6	5	12.5
Total	40	100.0

IOP Conf. Series: Earth and Environmental Science **884** (2021) 012014 doi:10.1088/1755-1315/884/1/012014

3.5. House Area

The area of the house is calculated based on the length, width, and number of stories of the respondent's house. The limited land owned makes people tend to build their houses upwards, so to determine the area of the building it is necessary to consider the dimensions of the length, width, and number of stories of the house. In the research location, there are many houses with an area of less than 36 m^2 and a small part of which is more than 54 m^2. The higher the socioeconomic conditions of the household, the more likely the household has a wider area of house building.

Table 5. House Area

Area (m^2)	Amount	Percentage
≤ 36	28	70.0
36 – 45	4	10.0
45 – 54	5	12.5
> 54	3	7.5
Total	40	100.0

3.6. Household Water Requirements

The amount of water needed in the dry season is greater than that of the rainy season. The amount of water needed during the dry and rainy seasons ranges from 134 - 172 liters/person/day with an average of 155 liters/person/day in the dry season and 151 liters/person/day in the rainy season. In general, it is known that there is more water demand during the dry season than during the rainy season. Most of the total household water needs are less than 450 liters/day to meet various household water requirement, including for cooking, bathing, and washing. Using sufficient water is essential for ecological sustainability. This is similar to what conveyed by Ebersbach that acting ecologically sustainably and not exhausting natural resources is becoming more and more important [14].

Table 6. Total Household Water Requirements

Water Requirement (liters/day)	Amount	Percentage
< 300	12	30.0
300 – 450	14	35.0
450 – 600	7	17.5
> 600	7	17.5
Total	40	100.0

3.7. Relationship between Socioeconomic Status and Household Water Requirement

The analysis of the needs for the relationship between socioeconomic status and household water requirements was carried out using multiple regression using SPSS version 23. The Y variable is the household water requirement, while the x variable consists of number of household members (X1), education of the head of the household (X2), age of the head of the household (X3), household's income total (X4), and house area (X5). The full regression results can be seen in Appendix 2. The unknown value is 0.824. This value indicates that there is a strong relationship between socio-economic conditions and household water needs because the value is above 0.8. Through this table also obtained the value of R Square which shows the regression model formed by the interaction of the independent and dependent variables. The value obtained is 78.9% which can be understood that the independent variable has a contribution of 78.9% to the y variable and 21.1% is obtained from factors outside the variables X1, X2, X3, X4, and X5.

The criteria for the significance level or linearity of the regression can be determined by the F test. The provisions of the f test if the sig value <0.05, then the regression model is linear, and vice versa.

The regression results show that the sig value is 0.004 so that it is <0.05. This means that the research data is significant, so the linear regression model meets the linearity criteria. The regression equation model obtained with the constant coefficients and variable coefficients in the unstandardized coefficients B column, the regression equation model is obtained as follows:

$$Y = 4.235 + 115.583\,X1 + 4.078\,X2 + 0.418\,X3 - 2.487\,X4 - 0.530\,X5$$

This means that the number of household members is the variable that most influences the amount of household water requirement. The more household members, the greater the need for water. The education and age of the head of the household also contribute to the increase in household water requirement. Socioeconomic status is affected household water requirement. This is in line with Gomes [15] who found a relationship between socio-economic factors related to water access in rural area. Therefore, socio-economic factors influence the fulfillment of household water requirement in urban and rural area. This is also in line with the findings of Reniko [16] which states that economic factors greatly influence household water fulfillment, in addition to demographic and social factors.

4. Conclusion

The results showed that there was a strong relationship between socioeconomic status and household water requirement because the value was above 0.8. The value R Square obtained is 78.9% which can be interpreted that the independent variables contribute 78.9% to the dependent variable and remaining 21.1% is influenced by other factors. The most influential variable is the number of household members.

5. Acknowledgment

We would like to express our deepest gratitude to various parties, including: The Government of Jakarta Special Province, all respondents and all research assistants who have supported this research.

References

[1] M. S. Team and P. T. Mirah Sakethi, *Why are There Floods in Jakarta?: Flood Control by the Government of the Province of DKI Jakarta*. Mirah Sakethi, 2010.

[2] I. B. Pramono, "Distribution of Water Infiltration Rate in Ciliwung Watershed," *Forum Geogr. Vol 30, No 1 July 2016*, 2016.

[3] H. Z. Abidin, H. Andreas, I. Gumilar, Y. Fukuda, Y. E. Pohan, and T. Deguchi, "Land subsidence of Jakarta (Indonesia) and its relation with urban development," *Nat. Hazards*, vol. 59, no. 3, p. 1753, 2011.

[4] A. K. Ali, "Cities of the World: World Regional Urban Development," *J. Geog.*, vol. 109, no. 4, pp. 166–167, Jul. 2010.

[5] N. Bich-Ngoc and J. Teller, "A Review of Residential Water Consumption Determinants BT - Computational Science and Its Applications – ICCSA 2018," 2018, pp. 685–696.

[6] P. Luo *et al.*, "Water quality trend assessment in Jakarta: A rapidly growing Asian megacity," *PLoS One*, vol. 14, no. 7, p. e0219009, Jul. 2019.

[7] S. ichi Onodera *et al.*, "Effects of intensive urbanization on the intrusion of shallow groundwater into deep groundwater: Examples from Bangkok and Jakarta," *Sci. Total Environ.*, vol. 404, no. 2–3, pp. 401–410, 2008.

[8] M. A. Marfai, F. Yulianto, D. R. Hizbaron, P. Ward, and J. Aerts, "Preliminary assessment and modeling the effects of climate change on potential coastal flood damage in Jakarta," *Yogyakarta Amsterdam Gadjah Mada Univ. vu Univ.*, 2009.

[9] J. Aerts, D. C. Major, M. J. Bowman, P. Dircke, and M. Aris Marfai, *Connecting Delta Cities: Coastal Cities, Flood Risk Management and Adaptation to Climate Change*. Amsterdam: VU University Press, 2009.

[10] C. Setiawan, S. Suratman, and M. A. Marfai, "The Relationship between Total Income and

Groundwater Utilization on Fluviomarine Landform Area in Jakarta," *Forum Geogr. Vol 31, No 1 July 2017*, 2017.

[11] J.-W. Noh, K.-B. Kim, J. Park, J. Hong, and Y. D. Kwon, "Relationship between the number of family members and stress by gender: Cross-sectional analysis of the fifth Korea National Health and Nutrition Examination Survey," *PLoS One*, vol. 12, no. 9, pp. e0184235–e0184235, Sep. 2017.

[12] M. Kooy, C. T. Walter, and I. Prabaharyaka, "Inclusive development of urban water services in Jakarta: The role of groundwater," *Habitat Int.*, vol. 73, pp. 109–118, 2018.

[13] W. van Gent, M. Das, and S. Musterd, "Sociocultural, economic and ethnic homogeneity in residential mobility and spatial sorting among couples," *Environ. Plan. A Econ. Sp.*, vol. 51, no. 4, pp. 891–912, Jan. 2019.

[14] M. Ebersbach, D. Malkus, and A. Ernst, "Factors that affect primary school children's sustainable behavior in a resource dilemma," *J. Exp. Child Psychol.*, vol. 184, pp. 18–33, 2019.

[15] M. Gomez, J. Perdiguero, and A. Sanz, "Socioeconomic Factors Affecting Water Access in Rural Areas of Low and Middle Income Countries," *Water* , vol. 11, no. 2. 2019.

[16] G. Reniko, N. D. Phonchi-Tshekiso, P. K. Mogomotsi, G. E. J. Mogomotsi, and H. Chirefu, "Analysis of Factors Influencing Domestic Water Consumption in Karoi Town, Zimbabwe," *J. Sustain. Dev.*, vol. 13, no. 2, p. 82, 2020.

Development of Pictorial Card "Kencana" (Disaster Preparation) as a Media of Group Guidance Services in SMP N 1 Wonosari

Diana Septi Purnama, Naafi' Fajar Oktaviana

Department of Educational Psychology and Counselling, Faculty of Science Education, Universitas Negeri Yogyakarta

Email : dianaseptipurnama@uny.ac.id

Abstract. This study is aimed to develop a product called the Kartu Kesiapsiagaan Bencana "KENCANA" (Disaster Preparedness) pictorial card which can be used as a media for group guidance services in junior high schools. The material theme of this card is physical, psychological, social, and spiritual disaster preparedness. This research is an R&D (Research and Development) research by adopting the Borg and Gall development model. The stages of this research consisted of: 1) the data collection stage, 2) the product planning stage, 3) the product development stage and 4) the validation and testing stages. The research subjects were 8th grade students of SMP N 1 Wonosari, Gunungkidul. The results showed that the pictorial cards KENCANA (Disaster Preparedness) were suitable for being used by the counseling teacher as a media for group guidance services and also used as a media for evaluating student understanding of the preparedness material that had been previously presented with good criteria based on all aspects that have been validated and tested.
Keywords: pictorial cards, group guidance services

1. Introduction

Indonesia is a country that has a lot of cultural, ethnic, racial and religious diversity. Indonesia is known as a beautiful country for its natural beauty. But in reality, Indonesia also has a high risk of natural disasters. Based on its geographical location, Indonesia is located between the junction of three active tectonic plates, namely the Indo-Australian Plate in the southern region, the Eurasian Plate in the northern region and the Pacific Plate in the eastern region. The three plates move and collide with each other, so that the Indo-Australian Plate plunges under the Eurasian plate and causes earthquakes, volcanic routes, and faults or faults. The subduction of the Indo-Australian Plate which moves relative to the north and the Eurasian Plate which moves relative to the south creates an earthquake route and a series of active volcanoes along the islands of Sumatra, Java, Bali and Nusa Tenggara. (Kurniawan, et al 2011: 1). There are approximately 5,590 watersheds (DAS) in Indonesia. This has resulted in Indonesia becoming one of the countries with a high risk of natural disasters, such as earthquakes, tsunamis, volcanic eruptions (129 active volcanoes), and ground movements. (Supartini, et al 2017: 10)

Everyone has a risk of potential disasters, so disaster management and management is the business of all parties (everybody's business). Therefore, it is necessary to carry out various roles and responsibilities (shared responsibility) in increasing disaster preparedness for all levels, including children, adolescents, adults and the elderly. (Supartini, et al 2017: 11).

Content from this work may be used under the terms of the Creative Commons Attribution 3.0 licence. Any further distribution of this work must maintain attribution to the author(s) and the title of the work, journal citation and DOI.
Published under licence by IOP Publishing Ltd

IOP Conf. Series: Earth and Environmental Science **884** (2021) 012015 doi:10.1088/1755-1315/884/1/012015

Victims of natural disasters experience very complex situations and conditions in dealing with natural disasters both physically, psychologically and socially (Rusmiyati & Hikmawati 2012: 98). Likewise for children who very clearly have a negative impact on the psychological and social fields of disasters in their lives (Mondal, et al., 2013; Uemoto, Asakawa, Takamiya, Asakawa, & Inui, 2012; Kilic, Ozguven, & Sayil, 2003. ; Kar, 2009; Yonekura, Ueno, & Iwanaka, 2013), it requires an adequate spiritual level for psychological strengthening caused by disasters (Dew et al., 2010; Hernandez, 2011). According to the disaster-prone index of Indonesia, the disaster threat in the Special Region of Yogyakarta includes natural disasters in the form of floods, earthquakes, tsunamis, droughts, extreme weather, landslides, volcanoes, abrasion, social conflicts, epidemics and disease outbreaks, with all districts included in the class. disaster-prone (Kurniawan, et al 2011: 86). According to the National Disaster Management Agency (BNPB), one of the districts in the Special Region of Yogyakarta, Gunungkidul Regency, has a high potential for disaster, this is evidenced by a score of 76 which is included in the disaster-prone class according to data from the disaster-prone index in the Special Region of Yogyakarta (Kurniawan, et al 2011: 86). With these disaster-prone conditions, Gunungkidul is one of the districts that has often experienced natural disasters. In fact, almost part of the villages in Gunungkidul are prone to potential disasters. For this reason, it is deemed necessary to raise awareness of the dangers of natural disasters, so that people can anticipate and be aware of these occurrences. In an effort to overcome various possible disasters that are prone to occurring, it is necessary to have knowledge and awareness of facing disaster preparedness. This can be given to all components of society, including school students. Moreover, in the Education System Law number 33 of 2019 concerning the Implementation of the Disaster Safe Education Unit Program, it is stated:

a. That in order to provide protection and safety to students, educators, and education personnel from disaster risk, it is necessary to improve disaster preparedness and mitigation in education units;
b. That in order to ensure the continuity of education services in education units affected by disasters, it is necessary to handle emergency and post-disaster situations;
c. Whereas based on the considerations as referred to in letters a and b, it is necessary to stipulate a Regulation of the Minister of Education and Culture concerning the Implementation of the Disaster Safe Education Unit Program.

Based on the law, teachers or educators can provide material related to disaster preparedness to students, by means of socialization or by simulation or can also use media that can cover various ways of learning students. This can be provided with group guidance services to students using existing media, namely by using picture cards to provide convenience in understanding school children about disaster preparedness awareness both physically, psychologically, socially, and spiritually.

Various ways can be done as a form of disaster preparedness easily and efficiently. One of the efforts that can be done is by developing learning media. The development of instructional media in group guidance services is determined by the ability of the teacher to analyze the needs of students with the media to be used. Group guidance services will be easy to run if the supervising teacher uses media or methods by utilizing tools, materials and facilities in schools that can support group guidance services.

Picture cards as a media for group guidance services because they are guided by the percentage of human absorption ability, which according to Daryanto (2010: 14) that the sense of sight has a big role to absorb information. The description of the percentage of human absorption capacity is; taste 2.5%, touch 3.5%, smell 1%, hearing 11%, and vision 82%. The large percentage of human absorption capacity on sight is one of the reasons for the development of this picture card as a media for group guidance services. Because the accuracy of media selection will greatly affect the learning outcomes of students (Daryanto 2010: 13). The ability to receive visual messages includes reading visuals correctly, understanding the meaning contained therein, connecting elements of the content of visual messages with verbal messages or vice versa (Nana 2010: 11). The visual messages on picture cards are thought to enhance the teaching and learning process, because picture cards can maximize the sense of sight.

The appropriate picture card is given to seventh grade junior high school students because at the level of cognitive development they are 12 years old on average in the formal operational phase, which is the phase where the child can think of something that might happen (hypothesis) and something abstract, this phase children can think critically and think at a high level (Bujuri: 2018). Picture cards can also train students to analyze the pictures given. The importance of understanding each level of children's cognitive abilities is also a priority guide in the process of education delivery. Given that BKan services will be effective and children will get maximum information if the materials and service methods used are in accordance with the child's cognitive abilities, starting from the concrete thinking stage to the formal thinking stage (Bujuri: 2018).

While the method used by most teachers is the lecture method, where the lecture method according to Widayati (2004: 68) tends to educate children to memorize a lot, children who have a visual learning style will have difficulty accepting the material given, for its use the cramah method tends to be boring when used For activities that are relatively long, the teacher also finds it difficult to conclude that students understand and are interested in what is being conveyed, and students become more passive because of the lack of interaction.

This research is aimed for developing picture card media to provide information related to disaster preparedness and to find out the feasibility of the picture card media being developed. Picture cards are expected to provide practical benefits for students, which can help provide information related to disaster preparedness through picture cards, then for teachers it can be used as a model for activities in providing material related to disaster preparedness for students, and for schools as a material for consideration of policy making. implementation of BK services in an effort to improve the quality of educational services for students in order to know or be aware of disaster preparedness.

2. Method

This research used Research and Development (R&D) by adopting Borg and Gall's development model. The final result of this research is the KENCANA (Disaster Preparedness) picture card suitable for use as a media for group guidance services at SMP N 1 Wonosari. This development research was conducted at SMP N 1 Wonosari which is located on Jl. Colonel Sugiyono No.35 B, Wonosari Gunungkidul Yogyakarta Special Region 55801. Limited trial data collection for students was carried out for 6 months in 2020. The limited trial subjects in this study were 23 students of class VII SMP N 1 Wonosari.

The Borg & Gall development model was chosen because it is in accordance with the research that will be developed where the researcher wants to produce a certain product and it takes research that is needs analysis and can test the effectiveness of a product. Due to the limitations of researchers, the development design used in this study was simplified into only four stages. The stages of development carried out by researchers, namely; 1) the search and data collection stage, 2) the product planning stage, 3) the product development stage, 4) the validation and testing phase.

Data collection techniques in this study used interviews with teachers and students, observation, and questionnaires to material experts, media experts. The instrument used in this study was a validation questionnaire for material and media experts and a user questionnaire.

The data analysis used in the picture card development process as a media for group guidance services is descriptive analysis in accordance with the development procedures that have been carried out. Descriptive analysis is a form of data analysis to test the generalization of research results based on one sample. This descriptive analysis was carried out through descriptive hypothesis testing. The data analysis used is the quantitative descriptive analysis technique which will determine the tendency of the research findings to be in the very good, good, enough, less or very poor category.

The KENCANA (Disaster Preparedness) pictorial card developed in this study is applied one level of feasibility of group guidance service media with a minimum assessment criterion when it has a mean score> 2.6 - 3.4 from the results of the assessment by material experts, media experts, teachers guidance and counseling, and class VII junior high school students.

3. Result and Discussion

Based on the results of information and literature studies that have been carried out, it is necessary to provide information media related to disaster preparedness for students. One alternative media that can be used is a card with the picture KENCANA (Disaster Preparedness) as a media for group guidance services for grade VII junior high school students.

The initial product design of the KENCANA (Disaster Preparedness) pictorial card is a two-dimensional media included in printed media. This picture card creation uses the inkscape application, where the pictures taken before editing are from the freepick.com page. The initial product of pictorial cards is in the form of rectangular cards measuring 200.5 x 300.5 in the size section of the application, but for a print size of 11 x 14 with the assumption of the researcher so that it can be seen clearly for grade VII junior high school students. The typeface used as an explanation for the bottom part is Poor Richard with a size of 36, while for the typeface used as an explanation on the back of the card is sans-serif with a size of 60.5. The KENCANA (Disaster Preparedness) picture card is also equipped with a manual for how to use it, and is equipped with a container for storing picture cards.

The picture card contains material on disaster preparedness physically, psychologically, socially and spiritually. This picture card product is validated by material experts and media experts who are competent in their respective fields. The validator of this material expert is a Counseling Guidance Lecturer who is competent in the field of group guidance services and material on disasters and disaster preparedness. Validators are media experts, namely Educational Technology Lecturers who are competent in the field of media and learning resources. While the product assessment lember was carried out by counseling teachers, and seventh grade students of SMP N 1 Wonosari.

Planning for the manufacture of card products with the KENCANA (Disaster Preparedness) picture requires several stages, namely; determine the objectives of group guidance services to be achieved by students using the media, reviewing material on the subject of disaster preparedness, determining design equipment and making designs from cards with the picture KENCANA (Disaster Preparedness)

3.1. Material Expert Validation

Material expert lecturers conduct assessments related to the KENCANA picture card material. Material expert validation was carried out twice by material expert validators. Assessment data by material experts can be seen in Figure 1 below.

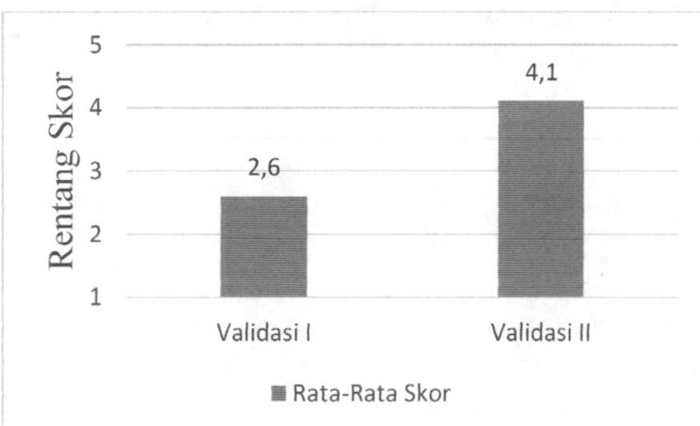

Figure 1. Bar chart of average material validation scores

Based on the results of the data from the material expert validation stage II, it can be obtained that the mean score is 4.10 which is in the score range> 3,4 - 4,2 in the "good" category. So based on the validation of phase II material experts, cards with the picture KENCANA (Disaster Preparedness) can be categorized as a good and feasible medium for field trials without revision.

3.2. Media Expert Validation

The media expert lecturer conducted an assessment related to the appearance aspects, the quality of the materials, the KENCANA card packaging, and the suitability of the media used. Media expert validation was carried out three times by media expert validators. Assessment data by media experts can be seen in Figure 2 below.

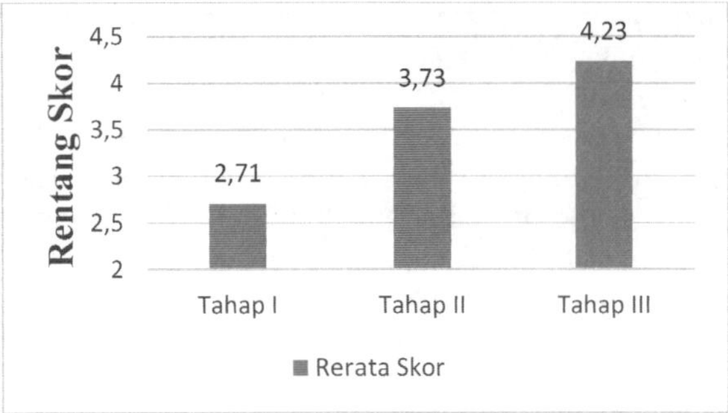

Figure 2. Chart of mean media validation scores

Based on the results of the data from the stage III media expert validation, it can be obtained a mean score of 4.3 which is in the score range> 4.2 in the "Very Good" category. So based on the validation of stage III media experts, the KENCANA (Disaster Preparedness) pictorial card can be categorized as a good and feasible medium for field trials without revision.

After the product developed was declared feasible for the trial without revision by material experts and media experts, then the product was assessed by the counseling teacher as users and grade VII students of SMP N 1 Wonosari as limited trial subjects.

Based on the results of data from users (BK teachers), it was obtained a mean score of 4.3 which was in the score range> 4.2 in the "Very Good" category and was suitable for use as a trial without revision.

Limited trial implementation was carried out on 23 students. Subject selection was carried out randomly with the school's considerations. The implementation of this limited trial was carried out online by making a video demonstration of the KENCANA picture card, this was carried out because learning activities were still being carried out in their respective homes, so the researchers changed the research plan that had been carried out face-to-face to online. Then later the students will evaluate the KENCANA picture card demonstration video through the google form that has been made by the researcher.

Of the 23 students who rated the KENCANA card, the mean score was 3.99 which was in the score range> 3.4 - 4.2 with the "Good" category.

4. Discussion

4.1 Development Procedure

The development procedure used in this research is a procedure adapted from the Borg & Gall development model with the following research procedures:

4.1.1 Search and Data Collection Stage

In this stage, the researcher measures and collects relevant data that will be used as material for media development planning through observation and literature study.

4.1.2 Product Planning Stage

The product to be made is a card with the picture KENCANA (Disaster Preparedness) as a media for group guidance services. The material on the card with the KENCANA picture is disaster preparedness physically, psychologically, socially, and spiritually. The software used in the manufacture of this product is Inkscape and the image support software used is freepick.com.

4.1.3 Product Development Stage

In order to produce the desired product in this study, the researcher made several stages, namely:

1) Determining the Material

The material used as material for making the product is based on the needs of the students, then the researchers then collect material on disaster preparedness physically, psychologically, socially and spiritually.

2) Determining the Media
This study used software which is used in the manufacture of this product and supporting the pictures taken is using freepick.com. Additional software in this research is Adobe Premier Pro which is used to create a video demonstration of the KENCANA (Disaster Preparedness) Picture Card, as well as Google Drive and Google Forms as storage places that can be accessed by students for product assessment.

4.1.4 Validation and Trial Phase

The researcher made several stages, namely:

1) Creating a product assessment grid for material expert validation, media expert validation, counseling teacher validation as a user, and limited trials for students.
2) Making expert validation instruments and limited trials based on a grid with very good (SB), good (B), adequate (C), poor (K), and very lacking (SK) criteria for assessing the feasibility of group guidance service media .
3) Validation of material experts, media experts, and counseling teachers as users as well as limited trials for students

4.2 Feasibility of the KENCANA (Disaster Preparedness) Picture Card as a Media for Group Guidance Services is as follows:

1) The material on the KENCANA picture card was deemed feasible to be tested without revision by the material expert obtaining a mean score of 4.10 in the good category.
2) Based on the validation of media experts, the KENCANA (Disaster Preparedness) pictorial card can be categorized as a good and feasible medium for field trials without revision obtaining a mean score of 4.3 which is very good.
3) Based on the results of the validation by the BK teacher on the KENCANA picture card as a media for group guidance services, the average score was 4.3 which was categorized as very good and suitable for use as a media for group guidance services.
4) The limited trial was given to class VII SMP N 1 Wonosari as many as 23 students and the mean score was 3.99 in the good category, so that the KENCANA picture card was suitable for use as a media for group guidance services.
5) Based on the assessment of material experts, media experts, counseling teachers and students, the card with the picture KENCANA (Disaster Preparedness) as a media for group guidance services gets a mean score of 4.1725 which is in the "good" category. So that the card with the picture KENCANA (Disaster Preparedness) is declared appropriate as a media for group guidance services for class VIII students.

Conclusion

This research resulted in a product in the form of a card with the picture KENCANA (Disaster Preparedness) as a media for group guidance services. The development procedure used in this study is a procedure adapted from the Borg & Gall development model with research procedures; 1) The stage of searching and collecting data, 2) The stage of product planning, 3) The stage of product development, 4) The stage of validation and testing.

Based on the assessment of material experts, media experts, counseling teachers and students, the card with the picture KENCANA (Disaster Preparedness) as a media for group guidance services gets a mean score of 4.1725 which is in the "good" category. So that the card with the picture KENCANA (Disaster Preparedness) is declared appropriate as a media for group guidance services for class VIII students.

The KENCANA (Disaster Preparedness) pictorial word card can be developed again with a wider scope of material and can be added by adding a larger number of picture cards, so that students do not easily memorize the pictures that have been seen.

References

[1] Bujuri, Dian A. 2018 Analisis Perkembangan Kognitif Anak Usia Dasar dan Implikasinya dalam Kegiatan Belajar Mengajar. *Literasi*, Volume IX, No. 1

[2] Daryanto. (2010). *Media Pembelajaran.* Yogyakarta: Gava Media.

[3] Dew, R. E., Daniel, S. S., Goldston, D. B., McCall, W. V., Kuchibhatla, M., Schleifer, C., Triplett, M.F., & Koenig, H. G. (2010). A prospective study of religion/spirituality and depressive symptoms among adolescent psychiatric patients. *Journal of Affective Disorders*, 120 (1-3), 149-157. doi:10.1016/j.jad.2009.04.029

[4] Kilic, E. Z., Ozguven, H. D., & Sayil, I. (2003). The Psychological effects of parental mental health on children experiencing disaster: *The jurnal of experience Bolu earthquake in Turkey. Family Process,* 42, 485-495. http://dx.doi.org/10.1111/j.1545-5300.2003.00485.x.

[5] Kurniawan, L, dkk. 2011. *Indeks Rawan Bencana Indonesia.* Jakarta : BNPB (Badan Nasional Penanggulangan Bencana).

[6] Mondal, R., et al. (2013). Acute stress-related psychological impact in children following devastating natural disaster, the Sikkim earthquake (2011) India. *Journal of Neuorsciences in Rural Practice,* 4, 19-23. http://dx.doi.org/10.4103/0976-3147.116434.

[7] Nana Sudjana . (2010). *Media Pengajaran.* Bandung: Sinar Baru Algensindo

[8] Rusmiyati, C., Hikmawati, E. (2012). Penanganan Dampak Sosial Psikologis Korban Bencana Merapi. *Jurnal Informasi*, Vol. 17, No. 02

[9] Supartini, E, dkk. 2017. *Latihan Kesiapsiagaan Bencana membangun kesadaran, kewaspadaan, dan kesiapsiagaan dalam menghadapi bencana.* Jakarta : BNPB (Badan Nasional Penanggulangan Bencana).

[10] Uemoto, M., Asakawa, A., Takamiya, S., Asakawa, K., & Inui, A. (2012). Kobe earthquake and post-traumatic stress in school-aged children. *Journal International of Behavior and Medicine,* 19, 243-251. http://dx.doi.org/10.1007/s12529-011-9184-3

[11] Widayati, A. 2004. *Jurnal Pendidikan Akuntansi Indonesia.* Vol. III No. 1 – Tahun 2004 Hal. 66 - 70

[12] Yonekura, T., Ueno, S., & Iwanaka, T. (2013). Care of children in a natural disaster: Lesson learned from the great East Japan earthquake and tsunami. *Jurnal of Pediatric Surgery International,* 29, 1047-1051. http://dx.doi.org/10.1007/s00383-013-3405-6

Community Education Model Design In Management Household Waste In Padang City

Ernawati[1], Muslim Suardi[2], Afdhal[1], Affandri Jasrio[1], Dilla Angraina[1]

[1]Department of Geography, Faculty of Social Sciences, State University of Padang
[2]Mohammad Natsir Bukittinggi University Pharmacy Study Program

ernawati@fis.unp.ac.id

Abstract. This study aims to design a community education model in waste management in the city of Padang. This type of research is qualitative research using data collection techniques, namely literature study, observation and interviews. The results showed that the paradigm of housewives regarding waste management that relies on downstream must be replaced with a new paradigm, namely waste management from upstream to downstream which covers the entire waste cycle. The obstacle in waste management is the perception of the community that waste management is the responsibility of the local government, not the responsibility of each individual. Furthermore, the sanctions given to dumping garbage in any place have not provided a deterrent effect. Therefore, based on the results of the field and literature studies carried out, obtained a community education model design based on the recycle waste management model by synergizing the government, community and company / private sector and emphasizing adult education. That is, the design of the resulting model encourages people's knowledge and self-confidence, so that it can also encourage positive change, both physically and mentally in a real, comprehensive, and sustainable manner.

Keywords: Society, Educational Model, Waste Management

1. Introduction

The acquisition of the Adipura trophy in 2017 for the City of Padang is inversely proportional to the fact that many hidden places that are far from the city center are still piles of rubbish. The results of research by [1] state that: (1) The industrial waste generation of Padang City is 5.057 kg / person / day (0.164 kg / m2 / day) in weight units or 6.569 liters / person / day (0.441 liter / m2 / day) in volume units; (2) The industrial waste generation of Padang City is greater than other cities in West Sumatra; (3) The composition of industrial waste in Padang City is dominated by organic waste 79.31% consisting of 9.90% food waste, 20.06% paper, 18.86% plastics, 8.69% textiles, 1.62% rubber, garbage yard 0.54%, wood 14.18%, and bark 5.47%; and (3) The composition of 20.69% inorganic waste consists of 0.28% glass, 1.36% cans, 0.12% metal, and others 18.93%. The amount of waste generated shows that the current waste management is not optimal, and is limited to conventional management.

[2] states that only 4 sub-districts out of 11 sub-districts that have just been served well by DKP or DKP are only able to provide services of 62.98% of the entire community. Furthermore, from the 540 tonnes of waste per day generated in the city of Padang, only 44%

Content from this work may be used under the terms of the Creative Commons Attribution 3.0 licence. Any further distribution of this work must maintain attribution to the author(s) and the title of the work, journal citation and DOI.
Published under licence by IOP Publishing Ltd

of the waste per day can be managed at the TPA [3]. The city of Padang only has 63 fleets to transport garbage which is not proportional to the amount of garbage [4]. In addition, [5] in her research stated that the people in Koto Tangah District, Padang City have low knowledge of the dangers of waste and think waste is a material that has no economic value and is the responsibility of the janitor.

Impactif the waste is not managed properly according to [6] as follows: (1) Source of disease, (2) air pollution, (3) unpleasant odors and dangerous to health, and (4) silting rivers which trigger flooding. The city of Padang is regularly flooded when there is heavy rain.

Solutions that have been done by the local government is strengthening the discipline of upstream waste through the formation of a team of cleaning volunteers, setting the time for disposing of garbage at 17,000 WIB, strengthening solid waste infrastructure (use of sweeping cars), and regulations for disposing of garbage (implementing a Hand Catching Operation pattern) in the field against waste throwers) [7]. However, this program has not been able to make the community more concerned about carrying out independent waste management. Therefore, an educational model is offered with the advantages of producing increased knowledge [8] and improving outcome expectations, self efficacy, self reward, self instruction, self evaluation, self reinforcement, maximizing social support, and independent application [9]. This study aims to develop an educational model that can increase the active role of the community in managing waste into products of economic value. These objectives are in accordance with the strategic issues contained in the 2016-2020 UNP Research Strategic Plan in the field of Education and Education, the sub-topic of Non-Formal Education Development (Attitude and Behavior Development), and it is very important to do this, because there has been no research in Indonesia in managing waste using a model. education.

2. Method

To achieve the goal of designing a community education model in waste management in the city of Padang, the type of research chosen is qualitative. This type of research is conducted to obtain reasonable and natural and holistic data. Qualitative research seeks to see, observe, appreciate, and describe the problem to be studied as a complex phenomenon that must be studied holistically or thoroughly [10]. The data that the researchers obtained from the results of observations, interviews and documentation were presented clearly and supported by relevant theories. In data collection, the researcher is the main instrument, the presentation of the results of the analysis begins with specific findings, which are then drawn to general conclusions.

2.1. Research Locations

This research was conducted in Bungoasang Village, Koto Tangah District, Padang City, Padang City Environmental Service, Bungoasang Village Head Office, Padang Mayor Office.

2.2. Research Informants

The informants in this study are all people who are considered to have the competence and information about the research data that researchers need to provide income, thoughts, research and reinforcement about environmental behavior. Key informants in this study were the Padang City Government, the Bungoasang Village Head, the Padang City Environmental Service, the Bungo Padang Village Community consisting of the head of RT 01 RW 05 (Mr. Afrizal), the Manager of the Waste Bank, Lidah In-Law (Mrs. Rika), and about 15 other community people. The limitation on the number of informants was based on the saturation of the data.

2.3. Data Collection Techniques and Tools

The technique of collecting data is by conducting observations, interviews, through interview guides, and data collection tools in the form of available records [11]. In line with Faisal, [12] states that data collection is basically all forms of data acceptance that are carried out by measuring and calculating it. Likewise, [13] states that data collection in qualitative research can be done through observation, interviews, questionnaires and documentation studies. In this study, data were collected by means of: observation, interviews, and in-depth interviews.

2.4. Data analysis technique

Researchers conducted data analysis starting when data collection took place and after completing data collection. At the time of the interview with the informant the researcher had conducted an analysis of the answers to the information being interviewed, after the answers to the information were analyzed, if the researcher did not feel the information was credible. In accordance with the opinion of Miles and [14] states that "activities in qualitative analysis are carried out interactively and take place continuously and thoroughly, so that the data is saturated, activities in data analysis are data reduction, data display (data display). , and Conclusion (conclusion drawing / verification). "

The following is a flow diagram and a fishbone of the research that will be carried out:

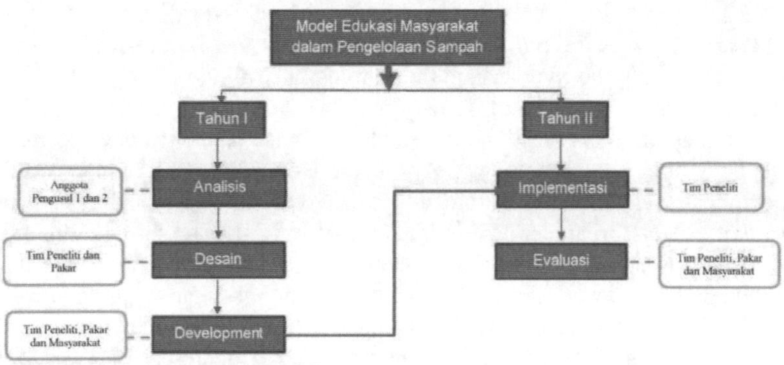

Figure 1. Research Flowchart

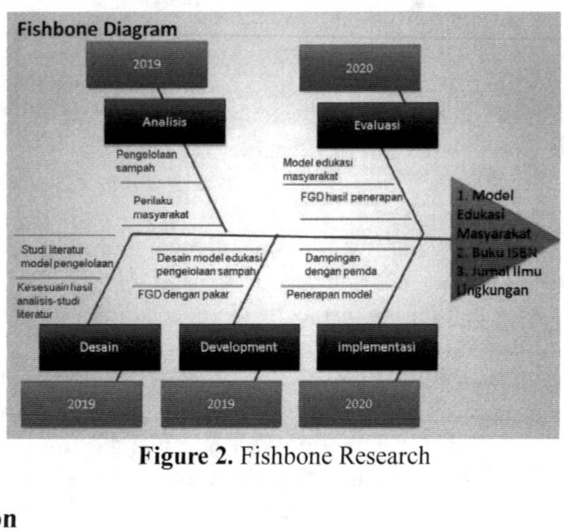

Figure 2. Fishbone Research

3. Results and Discussion

3.1 Needs Analysis

Garbage is a consequence of human activity and every human being must produce waste or garbage. Perms that occur due to waste will increase along with the increase in population. The population that is increasing has resulted in the amount of waste increasing. The amount of waste generation is very closely related to the population, because the larger the population, the greater the waste produced [15].

Therefore, it is necessary to carry out effective waste management so that the waste has economic value.

Waste management that has been carried out to date is seeing waste as a resource that has no benefit and relies on a downstream approach. It is time for waste management that relies on the downstream to be abandoned and replaced with a new paradigm, namely waste management from upstream to downstream. Waste management with a new paradigm views waste as a resource that has benefits, while its management rests on a source approach (upstream-downstream approach). The new waste management paradigm covers the entire life-cycle of waste from upstream (before it is produced) to downstream (in the phase where the product has been used and becomes waste) which is then sent to the TPA.

This is in accordance with the results of the needs analysis carried out through interviews and discussions with the people of Bungoasang Village, Koto Tangah District, Padang City.Residents in Kelurahan Bungo Posting can sort the waste according to its category and the sorted waste is placed in front of their house and then picked up by the garbage officer once every two days or once in three days at the most. This garbage officer does not come from the Padang city office, but from the agreement of all residents of RT 01 to make him a garbage officer, all residents pay a monthly garbage collection fee of Rp. 20,000 - Rp. 30,000 to the garbage officer, and residents who do not pay money. These transports carry their garbage directly to the TPS near the Tabing Railway and they are not allowed to throw garbage into the river around the Bungoasang village.

Types of waste generated by households are organic waste such as vegetable waste, kitchen waste and inorganic waste that can be processed and recycled such as paper waste, cans, etc. The amount of waste generated in 3 days for organic waste is 4 kg. However, this organic waste has not been processed into compost because it is constrained that the community does not have knowledge of how to process it so that all this organic waste is transported by trash rickshaws. While paper, cans, oil plastics, and electronic waste are sorted and placed in their respective bins in the house, the sorted waste will be channeled to the waste bank. This waste sorting has been going on since early 2018 and the community has started to think and act by sorting waste which has economic value.Some of the things that become obstacles in waste management are the inadequate use of waste, the perception in the community that waste management is the responsibility of the local government, the lack of community participation in waste management, limited landfill locations (Final Processing Sites) and limited land for technical processing.

From the perspective of implementing regulations, it turns out that the implementation of regional regulations is in fact 21/2021 concerning the regulation of waste disposal has been partially realized, and there are still many community attitudes in disposing of waste that are not in accordance with the regulations. In addition, the implementation of penalties for violators of regional regulations has not been optimal. Apart from the rules for waste management, it has not yet been touchedsolid waste management at the household level, such as sorting waste.

The low level of public knowledge about waste management and utilization so that it has economic value, results in a low public view of the value of up to, and conditions also affect the quality and aesthetics of the environment. Based on the results of the needs analysis, the design of an educational model in waste management that is carried out will be more optimal if the government, society and companies work together so that waste management can be carried out optimally. Through a household waste management approach that involves the role of the government, companies and the community, it is hoped that an innovative solution will be obtained to familiarize the community in managing waste, which previously caused waste to turn into waste into a blessing by involving related systems. The educational model design is compiled based on the principlesthe main principle of adult education: "what students learn, not what the teacher teaches" [16]. This means that the final result that is assessed is what the adult obtained from a meeting (in this case a community discussion), and not what the teacher did in that meeting. According to [17], adult education is not enough just to provide additional knowledge or skills, but must be equipped with self-confidence. Increased knowledge equipped with sufficient self-confidence can encourage positive change, both physically and mentally in a real, comprehensive and sustainable manner.

3.2 Interim Model Design

In this model design, changes that occur as a result of learning occur in each participant and learning citizens through interaction with other individuals to learn together with confidence. Changes in behavior in terms of cooperation in various activities are the result of changes in behavior after the learning process. The following is an educational model design compiled in this study based on the results of existing research and theoretical studies:

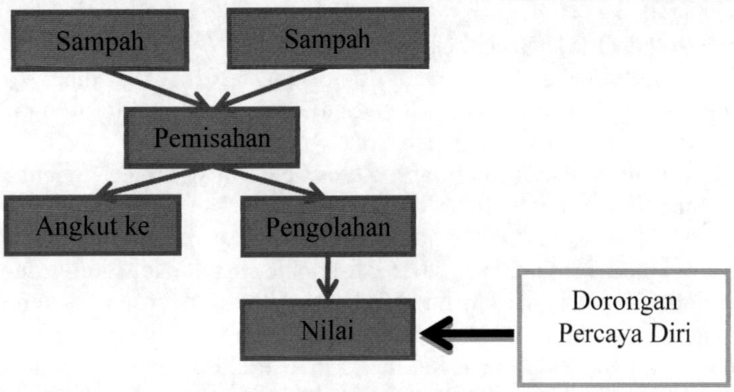

Figure 3. Diagram of the Educational Model Design for Household Waste Management Models in the City of Padang

4. Conclusions and Suggestions

Waste management that has been carried out to date is based on the view that waste is a resource that has no benefit and relies on a downstream approach. Thus an educational model in waste management is needed.What is an obstacle in waste management in this study is the use of waste is not optimal, there is an assumption in the community that waste management is the responsibility of the local government, lack of community participation in waste management, limited landfill locations (final processing sites) and land for technical processing. , the local regulation has not been optimized 21/2021 concerning fines for littering, there is no Regarding solid waste management at the household level, there are groups of people who do not know about the use of waste with the model of reusing waste into economic value products, the support of community leaders and the government has not yet been maximized to make the environmental cleanliness program a success. Therefore, it is necessary to design a model that synergizes government, society and companies through the approach of familiarizing the community in managing waste, with an approach adult education.

Reference

[1]. Alfath and Hadiguna. 2015.Proposed Waste Management Logistics Policy in Padang City:Lessons Learn Locating and Routing Garbage Transportation. Journal of Science and *Technology* UIN Suska Riau. Vol. 12 (2): 146-154.

[2]. Ministry of Education. 2001. Big Indonesian Dictionary Language Center 3rd Edition. Balai Pustaka: Jakarta.

[3]. Dwiyanto, Bambang. 2012. Model of Increasing Community Participation and Strengthening Synergy in Urban Waste Management. Journal of Development Economics. Vol. 12 (2): 239-256.

[4]. Ernawati. 2018.Waste Management Model Based on Community Education in Koto Tangah Subdistrict, Padang City, Indonesia. Sumatra Journal of Disaster, Geography and *Geography Education*. Vol. 2 (1): 118-123.

[5]. Ikhsandri, 2014. Study of Waste Management Infrastructure in the Jakabaring Developing

Area, 15 Ulu Urban Village, Palembang City. Journal of Civil and Environmental Engineering. Vol 2.

[6]. Knowles, Malcola, 1973. The Adult Leaner: Negleted Species. American Society For Training and Development, Madison, Wis.

[7]. Kushartanti, Sumaryanti, and Sriwahyuniati. 2015. Development of the Education Model – Supervision–*Integrated Physical Activity Evaluation for Metabolic Syndrome Patients in the Hospital Wirosaban Yogyakarta.* Higher Education Leading Research Reports. state University Yogyakarta: Yogyakarta

[8]. Lunandi, AG 1980. Adult Education. Gramedia.

[9]. Miles, MB, & Huberman, AM (1994), Qualitative data analysis: Asourcebook of new *methods*(2nd ed.). Thousand Oasks CA: Sage.

[10]. Nabila, S., Ramelan, AH., Himawanto, DW. 2016. The Inner Community Empowerment Model Waste Management in Slum Settlement Areas in Mojosongo Village. Seminar *National Chemistry and Chemical Education* VIII "Increasing the Professionalism of Educators and Researchers in Chemical Science in the Era of the Asean Economic Community (MEA)". UNS: Surakarta.

[11]. Nasution, 2002, Research Methods, Scientific Research, Earth Literacy, Jakarta.

[12]. Setyawati and Herlambang, 2015. E-booklet-based Nutrition Education Model to Improve Nutritional Knowledge of Mother Toddlers. UPGRIS Informatics Journal. Vol. 1: 86-94.

[13]. Surjandari I., Hidayatno A., Supriatna A. 2009. A dynamic model for waste management Reducing the Stacking Load. Journal of Industrial Engineering. Vol. 11 (2): 134-147.

[14]. Suryanto, Ari, et al. 2005. Study of Economic Potential with the Application of the 3Rs in Management *Household Waste in Depok City.* Depok.

[15]. Suwerda, Bambang. 2012.*Garbage Bank (Theory and Application Study).* Rihama Library: Yogyakarta.

[16]. Suyanto. 2017. Building Sodaqoh Garbage Awareness as a Model of Empowerment Public. Journal of Community Empowerment: Media for Thought and Da'wah *Development.* Vol. 1 (2): 245-256

[17]. Tallei, Trina E., Iskandar, J., Runtuwene, S., & Filho, Walter L. 2013. Local Community-based Initiatives of Waste Management Activities on Bunaken Island in North Sulawesi, Indonesia. Research Journal of Environmental and Earth Sciences. Vol. 5 (12): 737-743

The Role of Sense of Coherence and Social Capital on Perceived Risk: The Salutogenic Model Approach on Flood Survivors in Indonesia

H Maulana[1], **G Gumelar**[2], **G Irianda**[3]
[1,2,3] Faculty of Psychology, Universitas Negeri Jakarta

Email: herdiyan-maulana@unj.ac.id

Abstract. The study of flood risk perception has been received growing attention in multi-disciplinary research and practice. Indonesia's government approach on managing the impact of flood is highly dependence on structural engineering solutions. Few empirical attempts have systematically established to understand whether positive psychological capacity strategy helps flood survivors to overcome the negative impact of flood. This study aims to investigate the effectiveness of salutogenic and social capital models on flood risk perception. A national survey was used to collect the data of flood survivors across Indonesia. A battery comprising socio-demographic information, measures of salutogenic variable (sense of coherence), social capital (sense of community and social trust), and individual risk perception on flood was administered to the Indonesian adult (N = 194). This study findings showed that the overall model successfully predicted the perceived risk dimensions. However, different routes of correlation across variables were identified. Discussion and future recommendation are presented with regard to the study finding.

Keywords: Disaster preparedness, Flooding, Indonesia, Salutogenic, Social capital

1. Introduction

Natural disaster affects large number of people across the world. The World Health Organization (WHO) data shows a dramatic increase in the number of natural disasters in the past five decades. Scientific reports have evidences confirming that the increases of trend are mostly related to human activities. The high number of urbanizations causes excessive population growth (Cohen, 2006). As a result, sprawling cities and excessive consumption of fossil fuels increased in agricultural production leading to massive deforestation and mining activity (Satterthwaite, 2009). Without these environmental protections, there is a higher threat of flooding once the rainy season comes.

Indonesia listed as the third most vulnerable country to flooding in Asia, just below China and India, and ranked seventh in the world in terms of number of people affected by the surge (Indonesian Weather Bureau, 2016). Indonesia is a massive archipelago, a home of 268 million people who lived in across 17,000 islands. It is located at the equator with a tropical climate that often encounter with natural disasters. Wet season in Indonesia spans around October to April every year, causing heavy rainfall comes in the intense tropical downpours across the region. Indonesian Weather Bureau (2016) indicated 30% of natural disaster occurred in Indonesia is flooding. People lived in big cities (e.g. Jakarta, Surabaya, Bandung, Semarang, Medan) are faced with recurring severe flooding every year due to high density, massive construction and poor drainage system.

On early January 2020, the overnight heavy downpours caused riverbanks across Jakarta overload with floodwaters, inundating thousands of houses and buildings, submerged hospitals, local markets, airport and bus terminals, including the presidential palace complex. It forced the local authorities to cut the electricity and telecommunication lines as the waters reached as high as two meters in numerous commercial and residential areas. At least 53 people die, and more than 400,000 people were evacuated

Content from this work may be used under the terms of the Creative Commons Attribution 3.0 licence. Any further distribution of this work must maintain attribution to the author(s) and the title of the work, journal citation and DOI.
Published under licence by IOP Publishing Ltd

from their homes. Estimated financial loss due to the floods have reached around 5.2 trillion rupiah. Flood was also hit numerous cities in Java, Sumatera, and Kalimantan regions, swept thousands of buildings, farms, and rice fields, destroying people's livelihood and transportation systems. Death tool raised and millions of people in total had affected by this incident.
In response to this recurring problem, the government develops a national flood mitigation initiative (Murniningtyas, 2011). There are two strategies used in the flood prevention plan; the structural and non-structural approaches. The structural approach emphasised in technical mitigation using conventional engineering approaches (e.g. builds large scale embankments, dam, diversion canals and river normalization). While the non-structural approach focused on developing regulations for land use management and flood early warning system. Both of these strategies were criticized for put too much emphasis on technical-oriented indicators that might not sufficient to prevent losses from recurring hazards. The structural strategy may also not offer full protection against flood due to unpredictable rainfall pattern, which often creates a false sense of protection among population at risk and lead them to take less precautionary steps to adjust with the upcoming floods. The government report itself acknowledge that the initiative has placed small attention to people and community participations might do to deal with the risk (Murniningtyas, 2011). Therefore, it is important to consider alternative, non-structural measure of disaster impact from the point of view of the survivors.

Scholars have identified that understanding the individual perception of risk is an important step to design better disaster mitigation plan (Wachinger, Renn, Begg, & Kuhlicke, 2013). Perception of risk refers to individual efforts to collect, select, and interpret cues of uncertain impact of the disaster (Renn, 1989). These cues may refer to individual subjective experiences about the incident (Vanneuville & Kellens, 2011), whether they are direct (e.g. witnessing flood) or indirect experiences such as acquiring information from news or others. Morgan, Fischhoff, Bostrom, and Atman (2002) suggest that perceived risk is a form of cognitive schema that people use to evaluate the potential hazards. Xu et al. (2019) explained that risk perception consists of four dimensions; 1) Possibility, 2) Worry, 3) Controllability, and 4) Resilience. Probability refers to the likelihood of whether a disaster will happen again in the future time. Worry measures the emotional state when a disaster occurs. While controllability denotes to the degree of individual capacity to control their negative belief about the disaster that occurred, and Resilience represents people's perceptions about the impact of a disaster, whether it will have a positive or negative impact on them. This mental model is developed from the social learning process, personality attributes, and prior success coping with previous stressor that constantly maintained through social influence mechanism (media & peer influences). Gainey, Alper, and Chappell (2011) suggest that perception of risk is determined by a combination of personal and community level characteristic.

Personal exposure to risk during the flood could be understood using the salutogenic model of health and stressor by Antonovsky (1988). This model denotes to people way of thinking that focus on a desire for positive and optimal experiences built upon sense of coherence (Mittelmark et al., 2017). Antonovsky (1988) defines Sense of coherence (SOC) as a combination between optimism and sense of control. It is used to evaluate the ability to handle stressor based on the available psychological and social resources (George, 1996; Mittelmark et al., 2017). People with high SOC have better chance to seek meaning in life experiences and are more likely to appraise stressor as a nonthreatening stimulus than persons with lower SOC. According to Mittelmark et al. (2017) SOC encourages people to think that; 1) the risk they receive, whether they are from internal or external sources, can be predicted and controlled; 2) they have resources that available for them to use to manage the risk; and 3) the risk as a challenge and worthy for time investment and involvement.

The link between sense of coherence on perceived risk could be also explained through Protection Motivation Theory (PMT) by Rogers (1985). The PMT posits that individual perception of risk is developed from a fear-drive model which indicated that fear motivate people to do a trial and error act to reduce the adverse impact of risk. If the risk mitigation plan is oriented to managing the fear, then people are likely to build their protective plan based upon negative emotional state, which in turn may lead to maladaptive psychological responses, such as denial or anger. In the contrary, if the mitigation plan emphasises positive capacity advice, then people would be reinforced to perform constructive behaviour in responding the threat (Norman, Boer, & Seydel, 2005).

Furthermore, researchers have been interested in the link between community level of characteristic and perceived risk. In the present study, the authors examine the community level characteristic of risk through sense of community and social trust variables. Earlier studies have recognized on both variables' roles in perceived risk (Ahadzie, Dinye, Dinye, & Proverbs, 2016; Barberi, Davis, Isaia, Nave, & Ricci, 2008; Bronfman, Cisternas, López-Vázquez, & Cifuentes, 2016; Paton, Millar, & Johnston, 2001). Social capital variables, such as trust and sense of community, promote individual adaptive capacity in the aftermath of the natural hazards (Ahadzie et al., 2016; Babcicky & Seebauer, 2017). Antonovsky (1988) suggested that the presence of social capital (sense of community and social trust) can play a key role in providing social resource during the disasters. In the management of flood mitigation, people collective action based on these social variables appear to be a key aspect to promote individual resiliency during the flood (Noviekayati, Utomo, Pratikto, & Rumambo, 2019; Siegrist, Cvetkovich, & Roth, 2000). It means that when people have a high level of sense of community and trust to others, it may help them to control the risk perception because they know they have a reliable source to overcome the stress.

Gainey et al. (2011) indicated that low perceived risk may bring out a positive impact upon mitigation plan when sense of social cohesion and supports are present. In flood risk management cycle, social trust and sense of community can accelerate decision making process in responding imminent threat of the flood when backed by a strong sense of coherence. Both factors can complement in disaster risk management and increase people's efficacy when responding the current and in the aftermath of the flood. By understanding the association between sense of coherence and social capital towards risk perception, the flood mitigation plans would make more sense, practically realistic, and engaging as they developed based on personal and social level perspectives.

1.1. Research aim

A few studies in Indonesia have been conducted to investigate the impact of flood towards a higher risk of mental health problems. Previous findings indicated that flood may induced traumatic stress disorder (Taufiq, Susanty, Titi, & Nurlina, 2014). Bradford et al. (2012) suggest that it is important to understand the way people developed their perception of risk and what factors underlying it. This study hypothesised that salutogenic and social capital variables would predict perceived risk among flood survivors in Indonesia. This model offers a brand-new psychological and non-structural approaches to enhance the quality of flood risk mitigation in Indonesia. In practical context, this study finding would be expected to serve as an alternative source of information for the stake holders (government, environmental researcher, health practioners) to improve the effectiveness of flood management delivery.

2. Method

2.1 Sample

The sample were recruited from across Indonesia using an online survey. A total of 194 participants who have experienced flooding were voluntary participated in this study. Participants personal information (e.g. name, address, phone) were not provided to ensure the confidentiality. The participants were also asked for their consent prior to the survey. The following table 1 describes sample characteristic of this study.

Table 1. Respondents characteristics

Characteristic	Distribution N (%)
Gender	
Female	123 (64 %)
Male	71 (36 %)
Age group	
Young adult (18-28 years-old)	122 (71%)
Middle adult (29 - 40 years-old)	41 (27%)
Late adult (41- 65 years-old)	31 (2%)
Highest education	
Junior high school	4 (2%)
Senior high school	70 (36%)
Undergraduate	85 (44%)
Post-graduate	32 (17%)
Not answer	3 (1%)

2.2 Measures

2.2.1 Demographic information. Participants completed a demographic survey asking about their age, gender, highest education background, and place of living. A question asking whether they have experienced flooding in the previous year was asked to ensure participant meet with the inclusion criteria.

2.2.2 Perceived Risk. Perceived risk of flood was measured by the modified risk perception scale (Xu et al., 2019). The scale consists of 13 items asking participant's perception of risk of flood. The scale comprises four sub-scales measuring each dimensions of risk; possibility (item 1 to item 6), worry (item 7 to item 9), controllability (item 10 & item 11) and resilience (item 12 & Item 13). The participants used a 5-point Likert scale (1 = strongly not agree to 5 = strongly agree) to indicate the degree of perception of risk. The original scale was translated into Indonesians. The scale has been reported to have an acceptable internal consistency (Cronbach's alpha = .73) (Xu et al., 2019). Analysis of reliability based on the present study indicate the Cronbach's alpha of the scale to be .76, and its dimensions were .75, .85, .57, and .66 respectively indicating moderate to high reliability level (Hinton, McMurray, & Brownlow, 2014).

2.2.3. Sense of coherence. The Sense of coherence was measured using the 13 item version of SOC Scale, comprising three main dimensions; comprehensibility, manageability, and meaningfulness (Antonovsky, 1979). A semantic scale of 1 to 7 points (1 = never have the feeling to 7 = always have the feeling) was used to indicate respondent's feeling about the questions/statements. A total SOC score was obtained by summed all the score answers ranging from 13 to 91. A higher score indicates participant's higher sense of coherence. The scale has been initially tested across cultures and available in 49 different languages. The Indonesian version of the scale was developed by the author using translation and back-translation protocol (Brislin, 1970). The authors using Cronbach's alpha to examine scale validity and further checking the scale factorial validity using the Exploratory Factor Analysis (EFA). Both analysis results indicated that item # 7 ("*Doing the things, you do every day is: a source of deep pleasure and satisfaction—a source of pain and boredom*") was troublesome due to poor item fit, and therefore the authors decide to exclude this item. As a result, the 12 item of Indonesian version of SOC scale would be best used in the study (Cronbach's alpha = .906) rather than the original 13 item version. Similar issue was also found in others studies with different combination of items to exclude in the SOC scale. For example, Drageset and Haugan (2016) reported that the item # 2 was problematic due to error covariance in their model. While another study by Lerdal, Fagermoen,

Bonsaksen, Gay, and Kottorp (2014) indicated that item # 1 has failed to demonstrate acceptable goodness of fit using the Rasch model.

2.2.4 Sense of community. The sense of community (SOCM) was measured using the Brief Sense of Community Scale (BSCS) developed by Peterson, Speer, and McMillan (2008). This scale consists of 8 items using a 5-point Likert-type response ranging from 1 = strongly disagree to 5 = strongly agree. Respondent was asked to respond a question about their community/neighbourhood (e.g. "*I can get what I need in this neighbourhood*"). The scale has satisfactory psychometric properties in the original version (Cronbach' alpha = .92) (Peterson et al., 2008) as well as in a different language (e.g. German) (CFI = .994; TLI = 988; RMSEA = .042) (Wombacher, Tagg, Bürgi, & MacBryde, 2010). Reliability analysis based on the present study indicated the Cronbach's alpha for the Indonesian version of the scale to be .85 indicating satisfactory internal consistency.

2.2.5 Trust. Social trust was measured using General Trust Scale (GTS) developed by Jovanović (2016). The GTS comprises two different trust orientations: the interpersonal trust and institutional trust. All items were rated on an 11-point Likert scale, ranging from 0 (not trust at all) to 10 (trust completely). This scale has been translated and adapted into the Indonesian language and cultural context. Reliability analysis based on this study showing adequate psychometric properties of the scale (Cronbach'α = .88).

2.3 Analysis

Total scores in the scales were used to examine how Sense of coherence, Sense of community, and Trust associated with each dimensions of Perceived risk. The authors using an open-source statistical software program JASP (JASP Team, 2020) to conduct multi regression analyses on the four separate dimensions of perceived risk (Possibility, Worry, Controllability, and Resilience). A set of dummy variables of gender (1 = male, 0 = female), senior high school (1 = yes, 0 = no), graduate (1 = yes, 0 = no), post-graduate (1 = yes, 0 = no), middle adult (1 = yes, 0 = no), and late adulthood (1 = yes, 0 = no) were prepared to test whether different demographic factors have an impact on perceived risk dimensions. Junior high school category and young adulthood category was used as an anchor for education and age group variables. Only independent variables with $P < 0.05$ in the univariate analyses were put into multivariate models and likelihood ratio using $P < 0.10$ and $P < 0.15$ as entry and removal criteria.

3. Result

3.1 Descriptive analysis

There were no missing values in the data. All scales were checked for validity and reliability by using the Cronbach alpha coefficient. Data were also subjected to standard statistical assumption checks, including normality, linearity, and homogeneity. The inter-correlations among all variables were tested using the Bivariate Pearson correlation which are provided in the Table 2. The correlation analysis revealed there were no significant correlations between demographic factors (gender, education background, and late adulthood) and perceived risk dimensions, except for the middle adulthood group that negatively associated with the worry. As seen in the Table 2, the predictor variables (sense of coherence, sense of community, and trust) were indicating mild to low correlation to independent variables (possibility, worry, controllability, and resilience).

IOP Conf. Series: Earth and Environmental Science **884** (2021) 012017 doi:10.1088/1755-1315/884/1/012017

Table 2. Descriptive and Bivariate Pearson correlation matrix

Variables	*M*	*SD*	Pearson's Correlations										
			1	2	3	4	5	6	7	8	9	10	11
Gender		.483	—										
SMA		.481	-.192*	—									
D3S1		.497	.062	-.663*	—								
S2S3		.165	.124	-.334*	-.392*	—							
Middle adult		.409	.052	-.336*	.179*	.212*	—						
Late adult		.367	.194*	-.269*	-.102	.299*	-.226*	—					
Sense of coherence	45.397	15.326	-.086	.274*	-.151*	-.299*	-.184*	-.089	—				
Sense of community	29.392	6.059	-.012	-.056	.012	-.013	.073	.081	.012	—			
Interpersonal trust	6.727	1.749	.144*	-.110	-.046	.062	.038	.189	.092	.452*	—		
Institutional trust	14.747	5.003	.182*	.055	-.132	-.033	-.070	.270*	.118	.386*	.516*	—	
Possibility	12.284	4.323	-.067	-.030	.027	-.074	-.039	.012	.455*	-.014	.160*	.016	—
Worry	9.541	3.501	-.005	.073	-.059	-.134	-.209*	.140	.485*	.128	.284*	.342*	.563*
Controllability	6.902	1.884	-.064	-.058	.018	.094	.054	.023	.065	.350*	.243*	.319*	.061
Resilience	7.129	1.934	.016	.005	-.091	.078	.005	.066	.045	.367*	.228*	.222*	-.222*

3.2 Multi regression analysis

The proposed model of this study is based on the earlier literature indicating the perceived risk as a multi-dimensional construct (Xu et al, 2019). Hence, the perceived risk variable was treated as four separate sub-scales (possibility, worry, controllability, and resilience) and therefore were analysed through four separate multi-regression analyses (Table 3).

Table 3. Multiple regression analyses of perceived risk dimensions

	Possibility		**Worry**		**Controllability**		**Resilience**	
	B	**ββ**	**B**	**β**	**B**	**β**	**B**	**β**
Constant	18.684		17.973		3.023		3.214	
Self-coherence	-0.127	-0.450**	-0.102	-0.446**	0.005	0.039	0.004	0.033
Sense of community	0.044	0.062	0.012	0.021	0.082	0.264**	0.102	0.319
Interpersonal Trust	-0.518	-0.210*	-0.269	-0.134	0.016	0.015	0.048	0.043
Institutional Trust	0.104	0.121	-0.160	-0.229*	0.077	0.205	0.028	0.073
R^2	.237		.330		.164		.144	
Adjusted R^2	.221		.316		.146		.126	

** $P = < .001$; * $P = < .01$

In the first analysis, a multiple regression was performed to predict the Possibility based on Sense of coherence, Sense of community, Interpersonal trust, and Institutional trust. The overall model of independence variables was explained 23% of the variance and significantly predicted the Possibility of risk (F (4,189) = 213.986, $p < .001$). The analysis results show that Sense of Coherence (B = -.127, p = < .001) and Interpersonal trust (B = -.518, p = .008) did negatively significantly predict the Possibility, however, Sense of community (B = .044, p = .394) and Institutional trust (B = .104, p = .113) failed to significantly predict the Possibility.

Second analysis was determined to predict Worry based on Sense of coherence, Sense of community, Interpersonal trust, and Institutional trust. All four variables added significantly to the prediction (F (14,190) = 238.175, $p < .001$) and was explained 33 % of the variance of Worry. The analysis results show that Sense of Coherence (B = -.102, p = < .001) and Institutional trust (B = -.160, p = .002) did negatively significantly predict the Worry. In the contrary, Sense of community (B = .012, p = .757) and Interpersonal trust (B = -.269, p = .069) failed to significantly predict Worry.

The third analysis results showing that Sense of coherence, Sense of community, Interpersonal trust, and Institutional trust explained 14% of the variance and significantly associated with the Controllability (F (14,180) = 9,266, $p < .001$). However, only Sense of community (B = .082, p = < .001) that significantly predicts the Controllability, while Sense of coherence (B = .005, p = .558), Interpersonal trust (B = .016, p = .859), and Institutional trust (B = .077, p = .011) did not significantly predicted the Controllability.

Final multiple regression analysis was carried out to test between Sense of coherence, Sense of community, Interpersonal trust, and Intrapersonal trust to the Resilience. The analysis indicated the overall model explained the variance of resilience by 12% and significantly predicted it (F (14,189) = 7.980, $p < .001$). The model shows only Sense of community (B = .102, p = < .001) that significantly predicted the Resilience, while Sense of Coherence (B = .004, p = .633), Interpersonal trust (B = .048, p = .605) and Institutional trust (B = .028, p = .365) did not significantly predict the Resilience.

4. Discussion

Indonesia has experienced frequent and recurrent flooding every year. However, limited studies were found in the literature on the question whether the salutogenic and social capital factors are associated with the perceived risk. The present study aims to investigate the roles of Sense of coherence, Sense of community and Trust on Flood perceived risk among Indonesians. The current findings indicated that the hypothesis was supported as the overall independence model (Sense of coherence, Sense of community, and Interpersonal and Institutional Trust) significantly associated with all perceived risk dimensions. Nevertheless, the findings also indicate that different patterns of relationships emerge between these independent variables to each dimension of perceived risk.

The findings suggest that only sense of coherence and interpersonal trust that significantly predict the possibility of risk. This means that when people are optimistic and at the same time trusting one another, they will likely to belief that the risk of current flooding will not affect them in the future. We found that higher sense of coherence indicates that people appraisal of risk as more predictable and controllable. Such appraisal might lead to effective coping strategy to handle the negative symptons of risk, such as feelings of worry. While people's trust towards the social institutions (government, law, and media) would help them to dispose positive emotional response to the risk compare to those who have less trust. Our multiple regression model also indicates that sense of community plays a significant role as a factor influencing both the controllability of risk and resilience.

This finding further supports the idea that having control in a stressful situation and trusting in one another reinforces one's self-confidence in managing the future risks (Ahadzie et al. 2016; Mittelmark et al. 2017). Compare to people with less sense of coherence, people with stronger sense of coherence reported a lower frequency of negative emotions. This situation might due to their emotional response to the risk were less intense, which may help them to appraise the stressor as a manageable stimulus (Mittlemark et al., 2017). Our findings show only a sense of coherence that predicts the negative dimensions of risk (possibility and worry). This relationship could help to illustrate why the salutogenic model may best to explain the advantage of positive response over negative response to flood. The salutogenic model emphasize people's positive regard to risk that drives seeking for constructive and productive solutions (Norman, Boer, & Seydel, 2005).

Further, it seems that a community to some extent helps people to manage risk by providing conducive and supportive environment. Our finding parallel to Noviekayati et al., (2019) study indicating people's sense of community facilitates people positive response to risk and further prevent the mental health issue. Positive social relations with the community often promote attachment and satisfaction with the residential neighbourhood, which may facilitate one's sense of well-being. Further, this condition is not only enhancing one's happiness, but it also positively associated with social connectedness which in turns also increased one's sense of social support among one another, which extremely beneficial in the times of disaster (Yanuarti, 2006; Gumelar, Akbar, Suryaratri, Erchanis, & Wahyuni, 2020). Although the effect sizes of these social capital variables vary across all dimensions of risk, this study suggesting that sense of community and trust played important roles in determining the perception of risk. Particularly for those lived in an area vulnerable to flood (e.g. Jakarta), both factors are important to help them tackle the adverse impact of the disaster (Babcicky & Seebauer, 2017).

Improving sense of coherence and social capital variables has been recognized as an effective prevention strategy against the risk and mental health issues. A sense of coherence may be improved through an intervention involving mental health experts. Therefore, this relevant intervention should be included in the risk mitigation plan. While creating strong community in Indonesia should be more feasible given its strong collectivistic culture value.

5. Limitation and future direction

We identify a number of limitations for further refinement in the possible future study. Firstly, the participants sample size was considerably low. Such sample size might affect the prediction power of the analysis and might not fully represent the situation in other regions. Therefore, larger and more diverse sample from across Indonesia is warranted for further study. Next to that, this current study was employed a cross-sectional analysis, which could not provide explanation on the nature of causality across variables. Advance analysis using experimental or longitudinal approaches may provide comprehensive perspective about the complex relationship among variables.

6. Conclusion

This study aimed to examine the salutogenic and social capital model of the perceived flood risk. First, we found that both salutogenic and social capital aspects are important to understand how people manage their perceived risk of flood. Second, while the overall model was reported significantly associated to the perceived risk, different patterns emerged on the association variance between the salutogenic and social capital variables to each perceived risk dimensions. High level of sense of coherence was identified as key factors that contribute to lower perception of risk and low anxiety level. While the sense of community seems to be especially important to enhance people's sense of control and resilience in the aftermath of the flood. Finally, the results of this current study may assist policy makers, health practitioners, and community leaders in developing intervention and delivering flood mitigation plan based on the needs of flood survivors.

References

[1] Ahadzie, D., Dinye, I., Dinye, R., & Proverbs, D. (2016). Flood risk perception, coping and management in two vulnerable communities in Kumasi, Ghana. *Flood Risk Management and Response*, 84.

[2] Babcicky, P., & Seebauer, S. (2017). The two faces of social capital in private flood mitigation: Opposing effects on risk perception, self-efficacy and coping capacity. *Journal of Risk Research, 20*(8), 1017-1037.

[3] Barberi, F., Davis, M. S., Isaia, R., Nave, R., & Ricci, T. (2008). Volcanic risk perception in the Vesuvius population. *Journal of volcanology and geothermal research, 172*(3-4), 244-258.

[4] Bradford, R., O'Sullivan, J. J., Van der Craats, I., Krywkow, J., Rotko, P., Aaltonen, J., . . . Schelfaut, K. (2012). Risk perception--issues for flood management in Europe. *Natural Hazards & Earth System Sciences, 12*(7).

[5] Bronfman, N. C., Cisternas, P. C., López-Vázquez, E., & Cifuentes, L. A. (2016). Trust and risk perception of natural hazards: implications for risk preparedness in Chile. *Natural hazards, 81*(1), 307-327.

[6] Cohen, B. (2006). Urbanization in developing countries: Current trends, future projections, and key challenges for sustainability. *Technology in society, 28*(1-2), 63-80.

[7] Drageset, J., & Haugan, G. (2016). Psychometric properties of the Orientation to Life Questionnaire in nursing home residents. *Scandinavian journal of caring sciences, 30*(3), 623-630.

[8] Gumelar, G., Akbar, Z., Suryaratri, R. D., Erchanis, H., & Wahyuni, L. D. (2020). The Effect of Family Resilience towards Household Disaster Preparedness in Coastal Coast District of Sumur, Banten. *E&ES, 448*(1), 012085.

[9] Gainey, R., Alper, M., & Chappell, A. T. (2011). Fear of crime revisited: Examining the direct and indirect effects of disorder, risk perception, and social capital. *American Journal of Criminal Justice, 36*(2), 120-137.

[10] George, V. D. (1996). Field-workers' sense of coherence and perception of risk when making home visits. *Public Health Nursing, 13*(4), 244-252.

[11] Hinton, P. R., McMurray, I., & Brownlow, C. (2014). *SPSS explained*: Routledge.

[12] Lerdal, A., Fagermoen, M. S., Bonsaksen, T., Gay, C. L., & Kottorp, A. (2014). Rasch analysis of the sense of coherence scale in a sample of people with morbid obesity–a cross-sectional study. *BMC psychology, 2*(1), 1.
[13] Mittelmark, M. B., Sagy, S., Eriksson, M., Bauer, G. F., Pelikan, J. M., Lindström, B., & Espnes, G. A. (2017). *The handbook of salutogenesis*: Springer.
[14] Morgan, M. G., Fischhoff, B., Bostrom, A., & Atman, C. J. (2002). *Risk communication: A mental models approach*: Cambridge University Press.
[15] Murniningtyas, E. (2011). Kebijakan nasional mitigasi dan adaptasi perubahan iklim. *no. November*, 1-19.
[16] Norman, P., Boer, H., & Seydel, E. R. (2005). Protection motivation theory. *Predicting health behaviour, 81*, 126.
[17] Noviekayati, I., Utomo, D. B., Pratikto, H., & Rumambo, M. G. (2019). Rational Decisions in Communities Affected by Annual Floods in East Java, Indonesia.
[18] Paton, D., Millar, M., & Johnston, D. (2001). Community resilience to volcanic hazard consequences. *Natural hazards, 24*(2), 157-169.
[19] Renn, O. (1989). Risk perception and risk management. 14th Congress of the World Energy Conference.
[20] Satterthwaite, D. (2009). The implications of population growth and urbanization for climate change. *Environment and urbanization, 21*(2), 545-567.
[21] Siegrist, M., Cvetkovich, G., & Roth, C. (2000). Salient value similarity, social trust, and risk/benefit perception. *Risk analysis, 20*(3), 353-362.
[22] Wachinger, G., Renn, O., Begg, C., & Kuhlicke, C. (2013). The risk perception paradox—implications for governance and communication of natural hazards. *Risk analysis, 33*(6), 1049-1065.
[23] Xu, D., Liu, Y., Deng, X., Qing, C., Zhuang, L., Yong, Z., & Huang, K. (2019). Earthquake Disaster Risk Perception Process Model for Rural Households: A Pilot Study from Southwestern China. *International journal of environmental research and public health, 16*(22), 4512.

IOP Conf. Series: Earth and Environmental Science **884** (2021) 012018 doi:10.1088/1755-1315/884/1/012018

Impact of Extreme Rainfall on Flood Hydrographs

I G Tunas[1]*, H Azikin[1] and G M Oka[1]

[1]Department of Civil Engineering, Universitas Tadulako, Jalan Soekarno-Hatta Km. 9 Palu, Central Sulawesi, 94117, Indonesia

*Email: tunasw@yahoo.com

Abstract. Extreme rainfall is the main factor triggering flooding in various regions of the world including Indonesia. The increase in intensity and duration of current extreme rainfall is predicted as a result of global climate change. This paper aims to analyze the impact of extreme rainfall to the peak discharge of flood hydrographs at a watershed outlet in Palu, Sulawesi, Indonesia. Maximum daily rainfall data for the period 1990-1999 recorded at the Palu Meteorological Station, Central Sulawesi were selected using the Annual Maximum Series Method, and grouped into two types. Type I is the maximum daily rainfall data with extreme events and Type II is the maximum daily rainfall data without extreme events. Frequency analysis was applied to the two data groups using the best distribution method of: Normal, Normal Log, Pearson III Log, and Gumbel to obtain the design rainfall of each data group. In the next stage, the design rainfall transformation into a flood hydrograph is performed using the Nakayasu Synthetic Unit Hydrograph based on a number of return periods in one of the rivers flowing into Palu Bay, namely the Poboya River. The analysis results show that the design rainfall graphs with both extreme rainfall and without extreme rainfall are identical at the low return period and divergent at the high return period with a difference of up to 21.6% at the 1000-year return period. Correspondingly, extreme rainfall has a greater impact at the peak of the flood hydrograph with increasing return periods ranging from -1.28% to 26.81% over the entire return period.

1. Introduction

The increasing intensity and frequency of floods in various regions of the world are currently considered as the impact of extreme rainfall [1], [2], [3]. In Indonesia, in the last ten years, almost all areas from Sabang to Merauke have been affected by flooding due to changes in rainfall characteristics [4], [5]. Changing the nature of the rainfall leading to extreme rainfall is one of the characteristics of climate change, which is marked by an increase in the intensity and duration of rainfall [6], [7], [8].

Extreme rainfall is rainfall with a depth of more than 100 mm/day. In general, daily rainfall in Indonesia is less than 100 mm in depth. Although there are several areas with high annual rainfall above 3000 mm/year, in fact the depth of daily rainfall is still included in the normal rainfall category. In line with the widespread impact of global climate change, currently there are many extreme rainfall events in Indonesia [9] .

Studies on extreme rainfall and its impact on flood hydrographs have not been widely carried out, especially in Indonesia with the characteristics of a tropical climate. The studies that have been conducted are generally limited to prediction and projection of future extreme rainfall using a variety

Content from this work may be used under the terms of the Creative Commons Attribution 3.0 licence. Any further distribution of this work must maintain attribution to the author(s) and the title of the work, journal citation and DOI.
Published under licence by IOP Publishing Ltd

IOP Conf. Series: Earth and Environmental Science **884** (2021) 012018 doi:10.1088/1755-1315/884/1/012018

of approaches, as conducted by Rinaldi (2006, 2017), Saumi (2017), Rahmayani and Sutikno (2019), Simanjuntak and Safril (2020), and Simanjuntak *et al.* (2020) [10], [11], [12], [13], [14], [15]. None of these publications has examined the impact of extreme rainfall on floods, especially based on hydrographs.

In connection with the widespread impact of global climate change on extreme rainfall as a trigger for flooding, this research is very important to do with the aim of predicting the impact of extreme rainfall on peak floods. As a flood-prone area, this study will contribute to disaster mitigation programs, especially as a preliminary study and as a reference for identifying priority areas for disaster management and determining hydrological quantities for flood control.

2. Material and Methods

2.1. Data

This analysis is based on rainfall data from the Mutiara Meteorological Station, Palu, Central Sulawesi. This station is an observation station for daily climate data including: precipitation, wind speed, humidity, temperature, and sun exposure with an automatic recording system. The rainfall data used in this study are the maximum daily rainfall data for the period 1990-2019 with an observation length of 30 years.

Other data required besides rainfall data is the catchment area. The catchment area functions as a medium for transforming rainfall into discharge through the river network system in a catchment [16]. In this study, the catchment studied was the Poboya catchment, a river that passes through Palu City and has an outlet in Palu Bay. This river is interesting to study because part of the drainage system in Palu City, especially East Palu, has an outlet in the Poboya River. The river with a catchment area of around 69.82 km^2 is one of the rivers that is prone to flooding and plays an important role as a primary drainage channel in Palu City (Figure 1). The watershed with an elongated morphometric shape and tends to narrow downstream is located in the topography with an elevation of 5 m-1460 m.

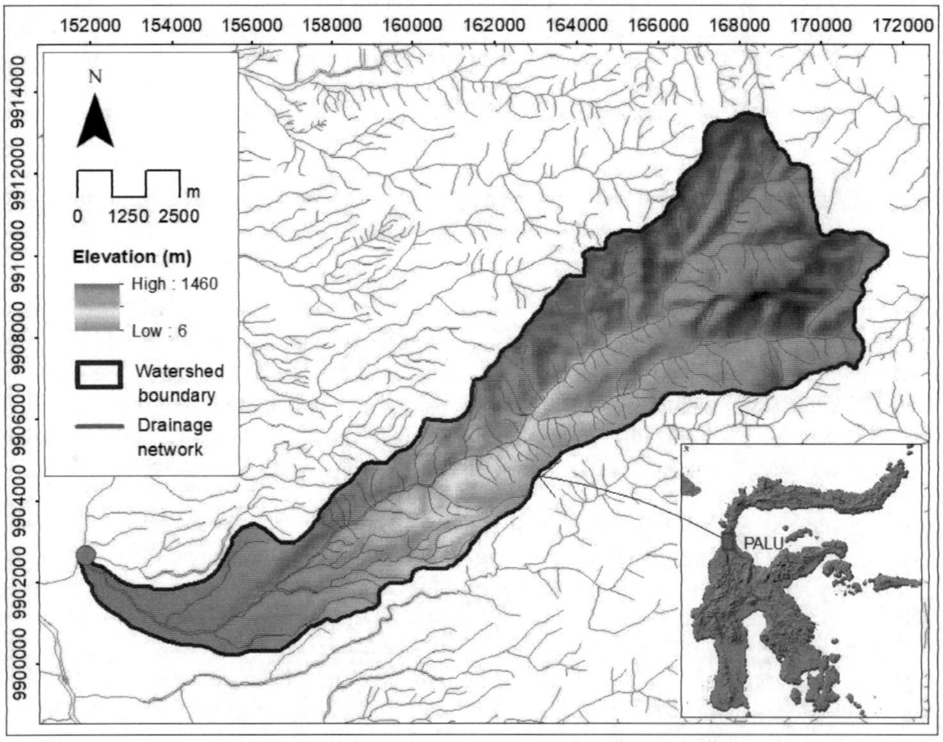

Figure 1. Morphometry characteristics of site area

2.2. Methods

Maximum daily rainfall data for the period 1990-2019 were selected using the Annual Maximum Series approach. Frequency analysis was applied to 30 maximum daily rainfall data with two categories: rainfall with extreme events and without extreme events. The rainfall depth of more than 100 mm/day categorized as extreme rainfall. Four frequency distribution methods: Normal, Normal Log, Gumbel and Log Person III, are performed to determine the design rainfall using general equation:

$$X_T = \bar{X} + K_T S \tag{1}$$

X_T refers to design rainfall (mm), $\bar{X}$ = average maximum daily rainfall (mm), K_T = frequency factor, and S = standard deviation.

Furthermore, the design rainfall is transformed into discharge using the Nakayasu Synthetic Unit Hydrograph (SUH) approach. The SUH method uses the following equations:

$$Q_p = \frac{AR_0}{3.6(0.3T_p + \alpha T_g)} \tag{2}$$

$$T_p = T_g + 0.8\,T_r \tag{3}$$

$$Q_a = \left(\frac{T}{T_p}\right)^{2.4} Q_p \qquad 0 \le T \le T_p \tag{4}$$

$$Q_{d1} = 0.3^{\left(\frac{T-T_p}{T_{0.3}}\right)} Q_p \qquad T_p < T \le (T_p + T_{0.3}) \tag{5}$$

$$Q_{d2} = 0.3^{\left(\frac{T-T_p+0.5T_{0.3}}{1.5T_{0.3}}\right)} Q_p \qquad (T_p + T_{0.3}) < T \le (T_p + T_{0.3} + 1.5T_{0.3}) \tag{6}$$

$$Q_{d3} = 0.3^{\left(\frac{T-T_p+1.5T_{0.3}}{2T_{0.3}}\right)} Q_p \qquad T_p > (T_p + T_{0.3} + 1.5T_{0.3}) \tag{7}$$

where: Q_p=peak flow (m^3/s), A=watershed area (km^2), R_0=unit rainfall (mm), T_p=peak time (hour), α=run-off coefficient, T_g=time of concentration (hour), and T_r=rainfall duration (hour). $T_{0.3}$ is time required to reach 30% of peak discharge from the hydrograph peak (hour). Q_a= discharge on the rising side of the hydrograph (m^3/s). Q_{d1}, Q_{d2}, and Q_{d3} are the discharge corresponding to $T_{0.3}$, $1.5T_{0.3}$, and $2T_{0.3}$, respectively (m^3/s).

The effect of extreme rainfall to the peak of the flood hydrograph is calculated by comparing the difference between the hydrograph peak discharge from the two data groups at each return period with the hydrograph peak discharge without extreme events. The equation applied is as follows:

$$\Delta = \left(\frac{Q_e - Q_0}{Q_0}\right) 100\% \tag{8}$$

where Δ is the impact of extreme rainfall to the peak of the flood hydrograph (%), Q_e and Q_0 are the peak of the hydrograph due to both with extrem rainfall and without extrem rainfall, respectively (m^3/s).

3. Result and Discussion

3.1. Extreme Rainfall

Identification of maximum daily rainfall data in the study area for the period 1990-2009, 30 data were selected. This data includes the maximum daily rainfall which represents each year of the data. Evaluation of all of these data indicates that during that period there was extreme rainfall. The maximum rainfall as shown in Figure 2, actually shows a regular sinusoidal graph over the observation period. The extreme rainfall with a frequency of one event in 2009 tends to influence the trend of the graph. In fact, in 1996 there was also high intensity rain approaching the extreme rainfall, which also affected the maximum rainfall pattern.

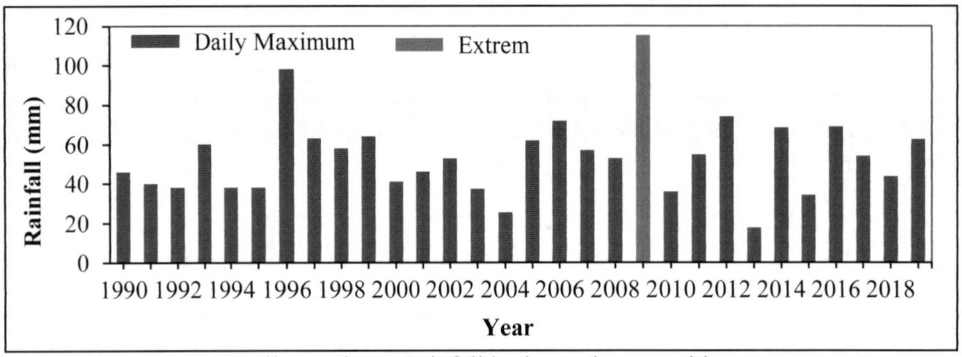

Figure 2. Daily maximum rainfall in the study area with extreme event

The frequency analysis of the 30 rainfall data which is clustered into two groups results in a design rainfall as illustrated in Figure 3. The pattern of increasing rainfall of the two data groups shows a similar trend, especially in the low return period. Trend changes enlarge with increasing return periods. Another thing that can be observed in Figure 3 is the difference in design rainfall based on extreme events and without extreme events. Rainfall extremes affect the increase in design rainfall by up to 21.6% in the 1000-year return period. This indicates that the extreme event has a major impact on design rainfall.

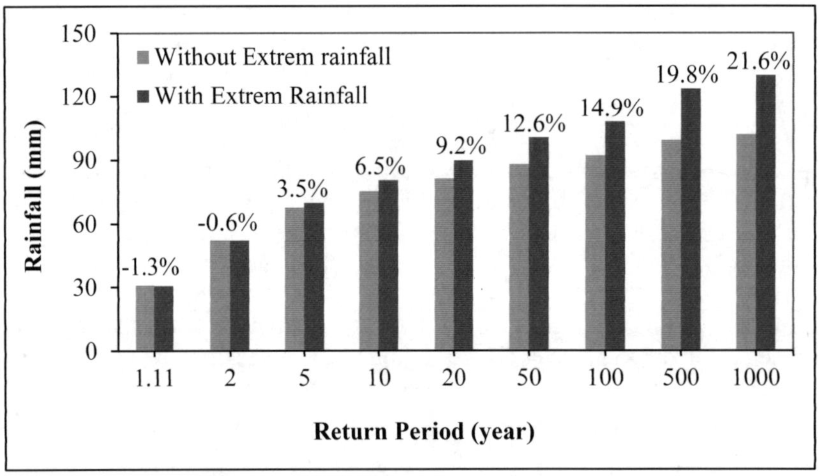

Figure 3. Design rainfall predicted with and without extreme event

3.2. Flood Hydrograph

Based on the two design rainfall groups as shown in Figure 3, then using the Nakayasu approach can be determined the flood hydrograph in the Poboya River. The flood hydrograph was determined based on the current land cover condition of the Poboya catchment, where the land cover in the middle and downstream segments had been degraded. Traditional gold mining in the middle segment of this catchment has contributed greatly to the decline in catchment carrying capacity in response to rainfall. Likewise, the development of urban settlements in the downstream segment is also predicted to trigger an increase in discharge in rivers. Changes in land cover in the two catchment segments have been deemed to have triggered flooding in this river apart from being a result of extreme rainfall.

The flood hydrograph resulting from the hydrograph based transformation is shown in Figure 4. Two groups of graphs with significant differences between Figure 4a and Figure 4b. The graph in Figure 4b is a hydrograph as a result of the rainfall extreme. The peak of the flood hydrograph without extreme rain reached 145.67 m^3/s in the 1000 year return period. In the same return period, the hydrograph peak discharge with extreme rain reached 184.60 m^3/s. This proves that the peak of the flood hydrograph is very much influenced by the depth of the rainfall, especially the rainfall extremes.

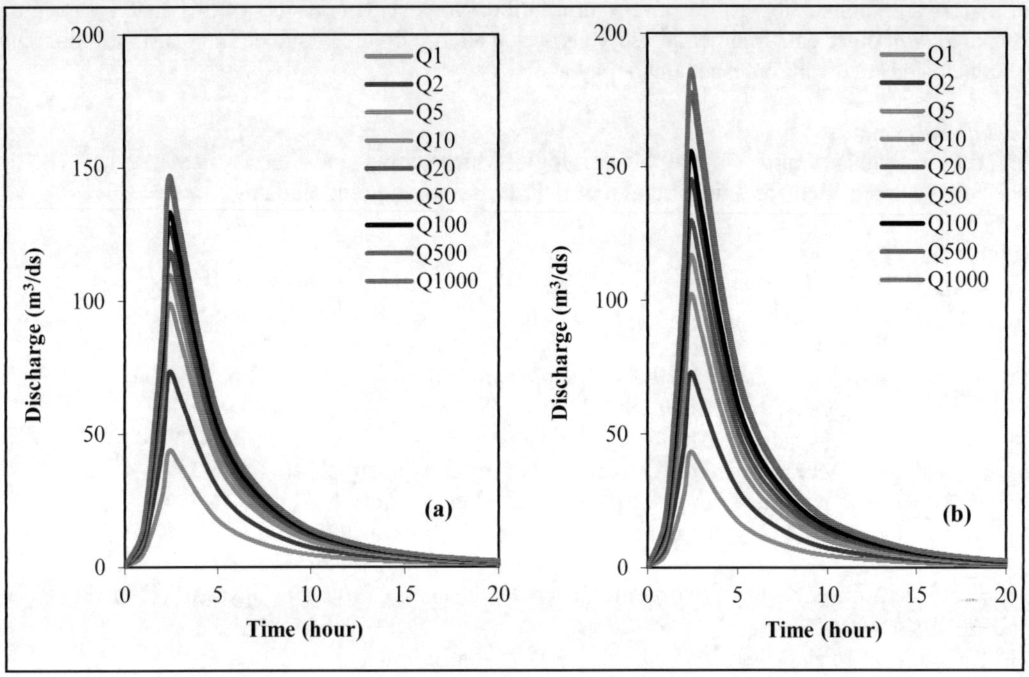

Figure 4. Flood hydrograph transformed from design rainfall. (a) without extreme rainfall, (b) with extreme rainfall

Table 1. Increasing of peak flood as an impact of extreme rainfall

Return Period (year)	Peak Flood (m^3/s)		Change (%)
	Without Extreme Data	With Extreme Data	
1	38.17	37.68	-1.28
2	63.89	63.53	-0.56
5	85.60	88.60	3.50
10	95.07	101.40	6.67
20	102.57	112.63	9.81
50	110.52	125.88	13.90
100	115.45	135.03	16.95
500	124.43	154.14	23.89
1000	127.48	161.65	26.81

Quantitatively, the influence of extreme rainfall on the hydrograph peak can be expressed as a percentage. Table 1 shows the change in the hydrograph peak discharge in the two data groups based on the return period. The change in peak discharge increases with increasing return period. Based on Table 1, the change in discharge at the largest return period reached 26.81%.

4. Conclusion

The effect of extreme rainfall at the peak of the flood hydrograph in the Poboya River, Palu, Central Sulawesi based on two groups of maximum daily rainfall data for the period 1990-1999 has been evaluated. Two data groups were classified based on the presence or absence of extreme events in the data group.

The results indicated that extreme rainfall significantly affected the hydrograph peaks. Extreme rainfall has a greater impact at the peak of the flood hydrograph with increasing return periods ranging from -1.28% to 26.81% over the entire return period. However, the results of this analysis need to be compared with other data in other areas by involving a more diverse number and distribution of data to get conclusions that can represent a wider area.

Acknowledgements
The authors highly appreciate the Meteorology, Climatology, and Geophysical Agency (BMKG) especially Mutiara Meteorological Station of Palu for supplying rainfall data

References
[1] Borga B, Stoffel M, Marchi L, Marra F, and Jakob M 2014 *J. Hydrol.* **518** 194–205
[2] Luo P, Mu D, Xue H, Ngo-Duc T, Dang-Dinh K, Takara K, Nover D, and Schladow G 2018 *Sci. Rep.* **8** 12623
[3] Ali H, Mishra V, and Pai D S 2014 *J. Geophys. Res. Atmos.* **119(22)** 12,621–12,641
[4] Tunas I G, Tanga A, and Oktavia S R 2020 *J. Ecol. Eng.* **21(2)** 190–200
[5] Tunas I G, and Maadji R 2018 *Int. J. Adv. Sci. Eng. Inform. Technol* **8(6)** 2413–2420
[6] Hamada A Y, Murayama, and Takayabu Y N 2014 *J. Climate* **27** 8151–8169
[7] Valverde M C, and Marengo J A 2014 *Open J. Modern Hydrol.* **4(1) 10–26**
[8] Martha T R, Roy P, Govindharaj K B, Kumar K V, Diwakar P G, Dadhwal V K 2015 *Landslides* **12** 135–145
[9] Brotowiryatmo S H 2000 *Hydrology: Teory, Problem and Solving* (Yogyakarta: Nafiri Offset) [in Indonesian]
[10] Rinaldi A 2016 *Al-Jabar: J. Pend. Mat.* **7(1)** 75–84 [in Indonesian]
[11] Rinaldi A 2017 *Proc. Sem. Nas. Mat. & Pend. Mat.* **2(1)** 21–28 [in Indonesian]
[12] Saumi T F 2018 *Logika* **8(2)** 145–150 [in Indonesian]
[13] Rahmayani D, and Sutikno 2019 *J. Sains & Seni ITS* **8(2)** 2337–3520 [in Indonesian]
[14] Simanjuntak P P, and Safril A 2020 *J. Ilmu & Inovasi Fisika* **4(1)** 1–11 [in Indonesian]
[15] Simanjuntak P P, Nopiyanti A D, and Safril A 2020 *Jukung: J. Tek. Lingk.* **6(1)** 45–53 [in Indonesian]
[16] Tunas I G 2019 *J. Ecol. Eng.* **20(7)** 112–125

Disaster Mitigation of Dry and Clean Water Crisis in Dana Sub-District Watopute District Muna Regency

La Ode Amaluddin[1], Rahma Musyawarah[2], La Harudu[3], La Ode Nursalam[4], Laode Utusafar Sufiandi[5]

[1,2,3,4,5] Department of Geography Education, Halu Oleo University, Kendari

Laode.amaluddin@uho.ac.id

Abstract. In the Southeast Sulawesi region, the area most frequently experiencing drought and clean water crises is Dana Sub-District, Watopute District, Muna Regency. This disaster was caused by physical and social factors. This study aims to describe the mitigation efforts of drought and clean water crisis in the Dana Sub-District, Watopute District. This research uses descriptive qualitative research methods. The population in this study were all the heads of families in Dana Sub-District, totaling 604 people. The sample in this study was selected using simple random sampling technique where the sample was randomly selected as much as 10% of the total population, so the sample in this study was 60 people. The results showed that the drought and water crisis in the Dana Sub-District, Watopute District occurred due to climate change and geomorphological conditions of the area as well as the lack of cooperation in the community in paying water bills. The peak of drought and clean water crisis in this region often occurs from July to November. To overcome this, various efforts have been made by the community, for example taking water from rivers and springs, buying clean water, and collecting water in storage tanks. Furthermore, the local government has also issued several policies, namely the development of the District Capital City Drinking Water Supply System and procurement of water pumps from springs, but according to the local community these government policies have not been effective. Furthermore, water dropping is also not maximally done because the community is less cooperative in paying water bills.

1. Introduction

Climate change is one of the issues that is of concern to the global community today. Indicators of climate change are marked by the increasing number of extreme climatic events in both intensity and distribution[1]. Based on global data, about 75-80% of the increasing tendency of natural disasters is related to the phenomenon of climate change[2]. These natural disasters include floods, flash floods, landslides, tropical storms, hurricanes, rainstorms, blizzards, sandstorms, tornadoes, orography (strong winds), extreme temperatures, heat waves, cold waves, frost, fire forests, drought, and etc[3].

One of the most serious natural disasters in Indonesia is drought. Undang-Undang No. 24 of 2007 concerning Disaster Management explains that drought is a condition of water availability that is far below the need for water for life, agriculture, economic activities and the environment. He further explained that drought is a dry period that is longer than normal and causes water availability to be far below water demand[4].

Content from this work may be used under the terms of the Creative Commons Attribution 3.0 licence. Any further distribution of this work must maintain attribution to the author(s) and the title of the work, journal citation and DOI.
Published under licence by IOP Publishing Ltd

Drought natural disasters are commonly called creeping disasters. Drought often gets less attention even though it has a serious impact on food security, forest fires, and can even cause death. This is because the formation of these natural disasters tends to be slow, has indirect consequences, often goes undetected, and can occur over a long period of time from months to years[5], [6], [7].Furthermore, drought is classified into 4 categories, namely meteorological drought, agricultural drought, hydrological drought, and socio-economic drought [8], [9], [10].

In general, drought in Indonesia occurs as a result of the El Niño-Southern Oscillation (ENSO) which occurs as a result of changes in sea surface temperature (SST) in the Pacific Ocean[11]. The effect of ENSO in Indonesia is rainfall that is higher than normal (La Niña) and rainfall below normal (El Niño)[7],[12].

The National Disaster Management Agency (BNPB) noted that from January to August 2019 there had been 60 drought incidents in Indonesia. Furthermore, the Meteorology, Climatology and Geophysics Agency predicts that the 2019/2020 rainy season will be delayed in 254 season zones in Indonesia. There are 7 areas affected by drought, namely West Java, Central Java, Special Region of Yogyakarta , East Java, Bali, West Nusa Tenggara and East Nusa Tenggara covering 101 cities, 781 districts and 2,731 sub-district/villages[13].

Particularly in the Southeast Sulawesi region, the area most frequently experiencing drought disasters is the Dana Sub-district, Watupute District. This is as a result of the occurrence of a long dry season or days without rain for more than 60 days from September to November 2019. This drought has made it very difficult for the people of Dana Sub-district to get clean water, so that people must be able to protect and save. in water use. Another effort made by the community is to buy water from the Regional Drinking Water Company in the local area, considering that water is a basic and vital need for all living things that cannot be replaced with other resources.

2. Methods

2.1. Time and Location of Research
This research was conducted in October 2019 in Dana Sub-district, Watopute District, Muna Regency.

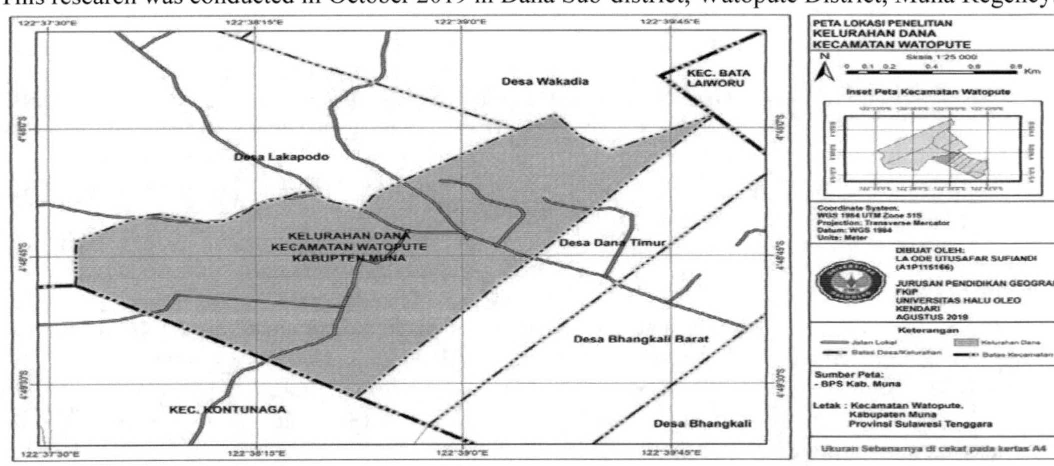

Figure 1. Map Of Research Location (Muna District Central Statistics Agency)

2.2. Types of research
This research is a qualitative descriptive study in which this research will describe the efforts to mitigate drought and clean water crisis in the Dana Sub-district, Watopute District, Muna Regency.

2.3. Population and Sample
The population in this study were all heads of households in Dana Sub-district, Watopute District, Muna Regency, amounting to 604 people. The sample in this study was selected using simple random sampling technique where the sample was randomly selected as much as 10% of the total population, so the sample in this study was 60 people.

2.4. Data Source
The data sources in this study are:

a. Primary data, sourced from the results of questionnaires and interviews with the community regarding efforts to mitigate drought and clean water crisis in the Dana Sub-district, Watopute District, Muna Regency.
b. Secondary data, sourced from the National Disaster Management Agency, The Central Statistics Agency, population data from the Watupute District Office, national journals and international journals.

2.5. Data Analysis Technique
The data in this study were analyzed descriptively and in percentage (%). To find out the description of each variable is done as follows:

a. Creating a frequency distribution table, which is meant to make the research data easy to understand.
b. Based on the frequency distribution table, the% (percentage) of each item in the questionnaire was determined using the formula:

$$P = \frac{F}{N} X100$$

Information:
P = Percentage of choices
F = Number of answer frequencies
N = Number of respondents
100 = percentage (%)

3. Methods

3.1. Difficulty Level of Getting Clean Water During the Dry Season
Community responses regarding the level of difficulty in obtaining clean water during the dry season in Dana Sub-District can be seen in Table 1 below.

Table 1. Community Responses About The Level Of Difficulty Getting Clean Water When Dry Season in Dana Sub-District (Source: Primary Data, 2019).

No.	Variables	Frequency (f)	Percentage (%)
1.	Very difficult	38	63,33
2.	Difficult	22	36,67
3.	Easy	-	-
	Total	**60**	**100**

Table 1 above shows that most of the community, namely 38 people (63,33%), consider it difficult to get clean water during the dry season and no community thinks it is easy.

3.2. Clean Water Dry Period
The period of dry clean water in Dana Sub-District can be seen in Table 2 below.

Table 2. Period of Dry Water in Dana Sub-District (Source: Primary Data, 2019).

No.	Variables	Frequency (f)	Percentage (%)
1.	July - November 2019	20	33,33
2.	July - October 2019	15	25
3.	July - September 2019	13	21,67

4.	July - August 2019	12	20
	Total	**60**	**100**

Table 2 above shows that as many as 20 people (33,33%) considered that the dryness of clean water occurred in July-November 2019, while 12 people (20%) thought that the dryness of clean water occurred in the July-August 2019 period.

3.3. The Importance of Mitigation Efforts to Drought and Clean Water Crisis

The response of the community regarding the need for mitigating efforts for clean water drought in Dana Sub-District can be seen in Table 3 below.

Table 3. The Importance of Mitigation Efforts to Drought and Clean Water Crisis in Dana Sub-District (Source: Primary Data, 2019).

No.	Variables	Frequency (f)	Percentage (%)
1.	Yes	56	93,33
2.	No	4	6,67
	Total	**60**	**100**

Table 3 above shows that as many as 56 people (93,33%) consider that it is necessary to mitigate clean water drought disasters while 4 people (6,67%) think that disaster mitigation is unnecessary.

3.4. Participation in Drought and Clean Water Crisis Mitigation Activities

Community responses regarding participation in drought and clean water crisis mitigation socialization activities in Dana Sub-District can be seen in Table 4 below.

Table 4. Participation in Drought and Clean Water Crisis Mitigation Socialization Activities In Dana Sub-District (Source: Primary Data, 2019).

No.	Variables	Frequency (f)	Percentage (%)
1.	Followed	24	40
2.	Never Followed	36	60
	Total	**60**	**100**

Table 4 above shows that most of the community, namely 36 people (60%) have never participated in any socialization activities for drought and clean water crisis mitigation and only 24 people (40%) have participated in these activities.

3.5. Government Efforts to Mitigate Drought and Clean Water Crisis

The public response to the government's efforts to mitigate drought and clean water crises in Dana Sub-District can be seen in Table 5 below.

Table 5. Mitigation Efforts Against Drought and Clean Water Crisis in Dana Sub-District (Source: Primary Data, 2019).

No.	Mitigation Efforts	Frequency (f)	Percentage (%)
1.	Construction of reservoirs and rainwater collection tanks	23	38,33
2.	Droping clean water with a Regional Drinking Water Company's water tanker	13	21,67
3.	Reforestation	11	18,33
4.	Education about clean water	13	21,67
	Total	**60**	**100**

Table 5 above shows that as many as 23 people (38.33%) think that the mitigation efforts carried out by the government should be the construction of reservoirs and rainwater storage tanks (PAH) and 11 people (18.33%) want reforestation activities.

3.6. Community Efforts to Mitigate Drought and Clean Water Crisis

Community efforts in mitigating drought and clean water crises in Dana Sub-District can be seen in Table 6 below.

Table 6. Community Efforts to Mitigate Drought and Clean Water Crisis in Dana Sub-District (Source: Primary Data, 2019).

No.	Mitigation Efforts	Frequency (f)	Percentage (%)
1.	Take water from the river and springs	10	16,67
2.	Buy clean water	27	45
3.	Hold water in the reservoir	23	38,33
	Total	**60**	**100**

Table 6 above shows that at the time of the drought, 27 people (45%) chose to buy clean water while only 10 people (16.67%) took water from rivers and springs.

3.7. Estimated Cost of Purchasing Clean Water

The costs incurred by the community to buy clean water in Dana Sub-District can be seen in Table 7 below.

Table 7. The Estimated Cost Spent by The Community to Buy Clean Water in a Week in Dana Sub-District (Source: Primary Data, 2019).

No.	Mitigation Efforts	Frequency (f)	Percentage (%)
1.	Rp.5000- Rp.50.000	28	46,67
2.	Rp.50.000 - Rp.100.000	18	30
3.	> Rp.100.000	14	23,33
	Total	**60**	**100**

Table 7 above shows that as many as 28 people (46.67%) spent around Rp. 5,000-Rp. 50,000 while 14 people (23.33%) spent about> Rp. 100,000 to buy clean water.

4. Disscussion

Drought and clean water crisis are global challenges for today's modern society. In general, drought can be defined as a temporary reduction in water supply or humidity that is significantly below the normal or expected volume for a specific period of time. Drought can also be interpreted as a condition where there is a lack of water, in this case it is usually connoted with a lack of rain water[14]. Natural disasters, drought and water shortages, are not only caused by physical environmental factors but can also be caused by socio-cultural factors.

Based on the results of monitoring in 2019 by the Kendari Meteorology, Climatology and Geophysics Agency, it shows that the Southeast Sulawesi region generally experiences days without rain in the long category, very long to [15]. This phenomenon has an impact on drought and clean water crises in several regions in Indonesia, including Dana Sub-District, Watopute District, Muna Regency.

Watopute District is very vulnerable to drought and difficulty in getting clean water. Apart from being caused by climate change and atmospheric phenomena, disasters in the region are caused by geomorphological conditions. Watopute Subdistrict is a karst topographic area, where so far the karst is seen as an arid, rocky area, and always faces problems with water resources, especially during the dry season[16],[17].

Groundwater flow in the karst region, does not follow the groundwater flow equation in general. The water flow system on the karst is unique, the water flow is controlled by a conduit (cavity) that forms an underground "river". Further explained, karst areas are formed by solid but easily soluble rocks such as limestone, dolomite, gypsum, anhydrite, and some other rocks that are impermeable (can not escape water), so without cracks and cavities, karst is difficult to store and drain air. Karst stores water for only a short time (low storage) and its porosity is low[18].

According to the people of Watopute District, drought and the peak of clean water shortages in their area often occur from July to November. To overcome this, various efforts have been made by the community, for example taking water from rivers and springs, buying clean water, and collecting water in storage tanks.

In the area of Watopute District, there is a river and a spring at Tumbulaha Tombi, but the distance is about 2-3 meters from community settlements, the terrain is winding, the roads are landslide and steep, so that the dominant community buys clean water to meet their needs. Most people spend around Rp. 5,000,-50,000, - per day to buy water. This budget is included in the high category if it is accumulated in a month's water usage. Furthermore, the community also holds water in a storage tank, especially when the rainy season arrives and at the time of dropping clean water by Regional Drinking Water Company.

Rain water harvesting is defined as a method of collecting or storing rainwater or surface runoff during high rainfall for further use when rainwater is low. There are 2 categories of rain water harvesting techniques, namely: 1) roof top rain water harvesting; and 2. techniques for harvesting rainwater (and surface runoff) using reservoir structures, such as trench dams, reservoirs, ponds, reservoirs, and so on. Rain water harvesting is an important agenda for Global Environmental Water Resources Management in the context of overcoming water imbalances during the rainy and dry seasons (lack of water), lack of clean water supply for the world's population, and overcoming floods and droughts[19].

The drought disaster and clean water crisis in the Watopute District area make the people in the area really want and urge the government to make efforts to mitigate the disasters they experience because water is a vital human need and therefore must be available in order to survive[20]. But on the other hand, drought has not become a priority for BPBD and Muna Government because there are no victims. So far, local governments have focused more on emergency measures and less on disaster anticipation and mitigation efforts. Dry clean water is still considered a less dangerous event than other natural disasters such as floods and fires[21].

Actually, there have been several policies that have been carried out by the Watopute District government and the Muna Regency government, including the construction of the District Capital City Drinking Water Supply System which is planned to drain all residents' houses in one village and the provision of water pumps from springs, but according to the local community this government policy not yet effective. Furthermore, water dropping is also not maximally carried out because according to the President Director of the Tirta Dharma Muna Drinking Water Company, the community is less cooperative in paying water bills so that the company temporarily stops services[22].

Drought disasters and clean water crisis are complex problems, so it needs cooperation between stakeholders and the local community. In addition to policies issued by the government, other efforts can be made to mitigate drought and clean water crises in Karst topographic areas such as in Watopute District, namely conservation of karst hills by identifying types of karst in watershed catchments; engineering by controlling groundwater extraction; maintaining infiltration or increasing groundwater infiltration in areas disturbed by human activities through infiltration wells, injection wells, reservoirs, absorption lakes, and mining engineering so that groundwater infiltration increases. Furthermore, reforestation can also be carried out as an effort to restore hydrological, geomorphological and biochemical functions. With forests, it is expected that the geomorphological process in the form of limestone dissolution will be able to take place naturally again. Forests produce CO_2 in the soil which when reacting with H_2O will dissolve limestone ($CaCO_3$). With the karstification process as in natural conditions, it is hoped that cracks will form near the surface that can store groundwater[16].

5. Acknowledgments

We agree to thank the leader of the Faculty of Teacher Training and Education for supporting and giving us permission to conduct this research. Lastly, thank you for the editorial team dan reviewer of the 2nd International Conference on Hazard Mitigation in Geographic and Education Perspective (ICHMGEP).

References

[1] W. Estiningtyas, R. Boer, I. Las, dan A. Buono, "Identifikasi dan Delineasi Wilayah Endemik Kekeringan Untuk Pengelolaan Risiko Iklim di Kabupaten Indramayu," *JMG*, Vol. 13, No. 1, April 2012, doi: 10.31172/jmg.v13i1.114.

[2] Sultonulhuda, Herdiansyah, H., dan Chrisandini, "Mengintegrasikan Kemampuan Adaptif Masyarakat dalam Adaptasi Perubahan Iklim dan Pengurangan Risiko Bencana," presented on Pelatihan Adaptasi Perubahan Iklim dan Pengurangan Risiko Bencana, Jakarta, 2013, [in the network]. Available on: https://adoc.tips/mengintegrasikan-kemampuan-masyarakat-dalam-adaptasi-perubah.html. (Accessed on July 2020).

[3] J. Hidalgo dan A. A. Baez, "Natural Disasters," *Critical Care Clinics*, Vol. 35, No. 4, hlm. 591–607, October 2019, doi: 10.1016/j.ccc.2019.05.001.

[4] Nagarajan, R, *Drought Assessment*. Dordrecht-Netherlands: Springer, 2009.

[5] Dai, A, "Drought Under Global Warming: A Review," *Wiley Interdisciplinary Reviews: Climate Change*, Vol. 2, No. 1, hlm. 45–65, 2010, doi: 10.1002/wcc.81.

[6] A. F. Van Loon, "Hydrological Drought Explained: Hydrological Drought Explained," *WIREs Water*, Vol. 2, No. 4, hlm. 359–392, July 2015, doi: 10.1002/wat2.1085.

[7] S. J. Sutanto, "Wawasan Mengenai Sistem Peringatan Dini Kekeringan Di Indonesia" *Jurnal Sumber Daya Air*, Vol. 13, No. 1, hlm. 53, December 2017, doi: 10.31028/jsda.v13.i1.53-68.

[8] I. W. Nuarsa, I. W. S. Adnyana, dan A. R. As-syakur, "Pemetaan Daerah Rawan Kekeringan di Bali-Nusa Tenggara dan Hubungannya Dengan ENSO Menggunakan Aplikasi Data Penginderaan Jauh," Vol. 15, No. 1, hlm. 12, 2015.

[9] Boken, V.K., Arthur P. Cracknell, dan Ronald L. Heathcote, *Agricultural Drought and Its Monitoring and Prediction: Some Concepts. In Monitoring and Predicting Agricultural Drought: A Global Study*. New York, United State of America: Oxford University Press, 2005.

[10] Wang, A, D. P. Lettenmaier, dan J. Sheffield, "Characteristics of Agriculture Drought in China During 1950-2006.," presented on The 91st American Meteorological Society Annual Meeting, Washington DC-United State of America, 2011.

[11] Sutanto S, J., Hoffmann G., Adidarma W., dan Röckmann T., "Correlation of drought related to ENSO and water isotopes in Indonesia" presented on 4th International Seminar of HATHI, Yogyakarta, 2013.

[12] Jun-Ichi H., Yamanaka M. D., Matsumoto J., Fukao S., Winarso P. A., dan Sribimawati T, "Spatial and temporal Variations of The Rainy Season Over Indonesia and Their Link to ENSO," *Journal of the Meteorological Society of Japan*, Vol. 80, No. 2, hlm. 285–310, 2002.

[13] Ari Supriyanti Rikin, "1,73 Juta Jiwa Terdampak Kekeringan di 7 Provinsi," *BERITA SATU*, 2019. Available on: https://www.beritasatu.com/nasional/573650-173-juta-jiwa-terdampak-kekeringan-di-7-provinsi#:~:text=Bahkan%20dari%20data%20yang%20dihimpun,73%20juta%20warga%20terdampak%20kekeringan. (Accessed on July 2020).

[14] P. D. Raharjo, "Teknik Penginderaan Jauh dan Sistem Informasi Geografis Untuk Identifikasi Potensi Kekeringan," *MST*, Vol. 14, No. 2, Mei 2011, doi: 10.7454/mst.v14i2.700.

[15] Inilahsultra.com, "BMKG Ingatkan Warga Sultra Waspada Potensi Bencana Kekeringan," *inilahsultra.com Berpijak di Jalur yang Tepat*, 2019. Available on: https://inilahsultra.com/2019/09/22/bmkg-ingatkan-warga-sultra-waspada-potensi-bencana-kekeringan/.(Accessed on July 2020).

[16] A. Ashari, "Konservasi Bukit Karst Sebagai Tindakan Mitigasi Kekeringan di Daerah Tangkapan Hujan Sub Sistem Geohidrologi Bribin-Baron-Seropan Karst Gunungsewu," *GM*, Vol. 10, No. 1, March 2015, doi: 10.21831/gm.v10i1.3599.

[17] Haryono, Eko, "Nilai Hidrologis Bukit Karst," presented on Seminar Nasional Eko-Hidrolik Teknik Sipil Universitas Gadjah Mada 28-29 March 2001., Yogyakarta, 2001.

[18] Pusat Penelitian Biologi LIPI, "Penataan Kawasan Karst Untuk Pemanfaatan dan Perlindungan Kawasan," *Pusat Penelitian Biologi Lembaga Ilmu Pengetahuan Indonesia*, 2017. Available on: http://www.biologi.lipi.go.id/index.php/2017-01-04-03-52-26/2017-01-04-03-57-30/staf-adm/9-

yt-sample-data/category1/308-penataan-kawasan-karst-untuk-pemanfaatan-dan-perlindungan-kawasan. (Accessed on July 2020).

[19] B. Harsoyo, "Teknik Pemanenan Air Hujan (*Rain Water Harvesting*) Sebagai Alternatif Upaya Penyelamatan Sumberdaya Air di Wilayah DKI Jakarta," *JSTMC*, Vol. 11, No. 2, hlm. 29, December 2010, doi: 10.29122/jstmc.v11i2.2183.

[20] D. Hidayati, "Memudarnya Nilai Kearifan Lokal Masyarakat Dalam Pengelolaan Sumber Daya Air," *JKI*, Vol. 11, No. 1, hlm. 39, January 2017, doi: 10.14203/jki.v11i1.36.

[21] F. Falah, "http://ejournal.forda-mof.org/ejournal-litbang/index.php/JPPDAS/article/view/4363," *JPPDAS*, Vol. 2, No. 2, hlm. 151–172, October 2018, doi: 10.20886/jppdas.2018.2.2.151-172.

[22] Rani, "Digarap Sejak 2016, SPAM di Muna Tak Kunjung Mengalir," *DETIK SULTRA Media Informasi Masyarakat Sulawesi Tenggara*, 2018. Available on: https://detiksultra.com/digarap-sejak-2016-spam-di-muna-tak-kunjung-mengalir/. (Accessed on July 2020).

Collaborative Governance in Realizing An Inclusive Workforce for People with Disabilities to Achieve Sustainable Development Goal's

M. Ahdiyana[1], S. Andari[2], A. Listyawati[3], A. Mardiyati[4]
[1] Public Administration Departement, Universitas Negeri Yogyakarta
[234] Center for Research Development and Social Welfare Services of Indonesian Ministry of Social Affairs

marita_ahdiyana@uny.ac.id[1], soetjiandari@gmail.com[2], andayani307@yahoo.com[3], animardiyati35@gmail.com[4]

Abstract. The number of workers with disabilities in public and private institutions continues to increase. However, this number has not met the quota target as stipulated in the law on people with disabilities. Therefore, attention from government agencies, the private sector, and other stakeholders is needed. Collaborative governance is often expected to be able to solve complex policy problems and support the creation of sustainable development. This paper aims to explain the stages of collaborative governance in realizing an inclusive workforce for people with disabilities to achieve sustainable development goals in the Special Region of Yogyakarta. This study used a qualitative descriptive design. Data collection was carried out by observation, in-depth interviews, and literature study. The results showed that the collaboration of stakeholders involved in realizing an inclusive workforce was needed. Collaboration has been carried out with the manpower office as the main actor although in its implementation it is still not aimed at achieving the same interests. At the initiation stage, the collaborative process carried out is not yet based on a clear cooperation framework and action plan from each stakeholder. Although there have been deliberation efforts made, they have not succeeded in creating an effective working group and building consensus to achieve inclusive workforce. In addition, the partnership that has been formed has not encouraged stakeholders to jointly evaluate the progress that has been made in cooperation. Thus it can be concluded that collaborative governance in realizing an inclusive workforce for people with disabilities has been implemented but has not yet reached the stage of being able to build a strong framework of cooperation and create broader support for this cooperation.

Keywords: Collaborative governance, inclusive workforce, people with disabilities, sustainable development goals

1. Introduction

Every country must carry out the global development agenda in the context of sustainable development. In Indonesia, the goals of Sustainable Development Goals (SDGs) are contained in Presidential Regulation Number 59 of 2017 concerning the Implementation of Achieving the Sustainable Development Goals, namely in order to sustainably improve the economic welfare of the community, maintain the sustainability of social life, maintain environmental quality and develop

Content from this work may be used under the terms of the Creative Commons Attribution 3.0 licence. Any further distribution of this work must maintain attribution to the author(s) and the title of the work, journal citation and DOI.
Published under licence by IOP Publishing Ltd

inclusive and implementing governance that is able to maintain the improvement of the quality of life from one generation to the next. Development is said to be sustainable when economic, social, environmental and government objectives are in harmony. High economic growth must be accompanied by even income society and not by exploiting natural resources and destroying the environment.

Persons with disabilities (PWDs) have the same rights in state life. This has been regulated in Law Number 8 of 2016 concerning Persons with Disabilities. One of the rights listed in Article 11, namely the Right to Work, Entrepreneurship, and Cooperatives. Article 53 paragraph (1) of the Law obliges the Government, Regional Government, State-Owned Enterprises and Regional-Owned Enterprises to employ at least 2 percent of PWDs from the number of employees or workers. Article 53 paragraph (2) requires private companies to employ at least 1 percent of PWDs from the number of employees or workers. In addition, Article 145 also contains criminal sanctions and fines for those who obstruct and / or prohibit PWDs from obtaining the right to work. The maximum imprisonment is two years and a maximum fine of 13.465 USD.

Data from the Manpower Office for 2019, shows that the number of workers with disabilities in government agencies, state-owned / regional-owned enterprises, and the private sector continues to increase. Recorded in 2017 the number reached 4,286 people. This amount accounts for 0.90 percent of the total workforce in Indonesia. Meanwhile, in 2018 there was an increase to 4,537 people or 0.96 percent of the total workforce in Indonesia [1]. The number has indeed increased, but the number of workers with disabilities has not met the quota target for disabled workers as regulated in Law No.8 Article 53 of 2016 Article 53. Whereas according to the findings of the International Labor Organization, if PWDs are involved in the development process, they have the potential contributing to gross domestic product (GDP) by 3-7 percent [2]. Therefore, serious efforts are needed to improve the quality of PWDs, so that they have the potential in the formal and informal sectors. The attention of all government agencies, the private sector, and other stakeholders is needed so that these needs can be met.

Based on the 2018 Basic Health Research, Indonesian Ministry of Health, the Special Region of Yogyakarta ranks third as one of the provinces with the highest proportion of people with adult disabilities in Indonesia. The top five provinces in Indonesia with the proportion of adults with disabilities aged 18-59 years are the highest in Central Sulawesi (40.6%), second in South Sulawesi (33.6%), the third in Special Region of Yogyakarta (33.2%), fourth is West Sulawesi (32.4%) and fifth is Southeast Sulawesi (28.9%) [3].

Since 2012, the Special Region of Yogyakarta Government has issued a policy to support the fulfillment of the rights of PWDs, especially in obtaining a job. This is stated in the Special Region of Yogyakarta Regional Regulation Number 4 of 2012 concerning Protection and Fulfillment of the rights of PWDs. In order to fulfill the rights of PWDs, the Special Region of Yogyakarta Government through the Special Region of Yogyakarta Manpower and Transmigration Office has made various efforts. One of the efforts made is providing skills training for PWDs so that they can enter the job market or do business independently [4]. Economic empowerment for PWDs is carried out in two ways, namely intrapreneurship and entrepreneurship. Through intrapreneurship, it means preparing workers with disabilities to have the ability and skills to enter the labor market. Meanwhile, entrepreneurship empowerment is to prepare PWDs through business education or training so that they are able to establish independent businesses.

The regulation emphasizes the need for the involvement of all stakeholders in fulfilling the rights and needs of PWDs. Collaborative governance is often expected to be able to solve complex policy problems [5]. However, based on several studies it has been found that realizing 'collaborative advantage' is often very difficult given the many actors, organizations and interests that must be managed. Collaborative governance is increasingly seen as a proactive policy instrument, whereby collaborative strategies can be implemented on a larger scale and extended from one local context to another [6]. Nevertheless the collaboration process is less than optimal, cause the lack of private companies and state agencies participation in employing PWDs though there are quota policies (2%

for public and 1% for private industries) [7]. Based on the background of the above problems, the authors are interested in researching the collaborative governance process in achieving an inclusive workforce for PWDs to achieve the goal of sustainable development in the Special Region of Yogyakarta. Several studies related to inclusive workplaces have been more related to the management of PWDs workers [8]. In addition, it is also about how digital technology is able to break down traditional barriers in communication, interaction, and access to information for PWDs [9]. Collaborative governance can be drawn from a wide variety of practice and research areas in public administration [10]. So this research is important to add insight, especially about the collaborative governance process, especially in realizing an inclusive workforce for PWDs.

2. Method

This research used qualitative methods with a descriptive approach. Determination of research subjects was done purposively. The data collection methods used were in-depth interviews, observation and documentation [11]. Data validity check were source and method triangulation The data analysis technique used in this research was qualitative data analysis interactives model [12].

3. Finding and Discussion

Discussions about how Collaborative Governance in Realizing An Inclusive Workforce for PWDs to Achieve SDGs in the Special Region of Yogyakarta will be discussed using collaborative governance process as proposed by Morse & Stephens [13], which consists of: assessment, initiation, deliberation, and implementation.

3.1. Assessment Stage

The need for collaboration has been stipulated in the Special Region of Yogyakarta Regional Regulation Number 4 of 2012 concerning the Protection and Fulfillment of the rights of PWDS. The regulation also emphasizes the importance of involvement of all stakeholders in fulfilling the rights and needs of PWDs. To achieve this goal, so far the Special Region of Yogyakarta Manpower and Transmigration Service has made efforts to collaborate with related stakeholders. The Special Region of Yogyakarta Manpower and Transmigration Office acts as a leading actor while other stakeholders are the Social Service, Integrated Rehabilitation Center for Persons with Disabilities, private companies, non-governmental organizations such as the Center for Improving Qualified Activity in Life of People with Disabilities (CIQAL), welfare workers Social Districts, and the PWDs group.

Based on this explanation, the assessment of the initial conditions that will determine the success of the collaboration and the reasons for the need for collaboration is very strong. This is supported by the existence of regulations at the regional level [14]. To achieve the fulfillment of the rights and needs of PWDs, collaboration of related stakeholders is very necessary. Based on the history and experience of the cooperation that has been carried out by the Special Region of Yogyakarta's Manpower and Transmigration Service, it is the main actor. One of the dynamic discussions about collaborative governance focuses on the importance of the role of the leader in collaborative governance [15]. The main actor in this case can be said to be the leader. Meanwhile, other actors participating in the collaboration include non-governmental organizations, private companies, and the PWDs group. Although the cooperation of related stakeholders has been carried out, however in its implementation this cooperation has not been based on an agreement to achieve common goals. Each stakeholder is still guided by their respective goals. Based on the results of interviews conducted, the encouragement of private companies to employ employees with disabilities is based more on meeting the quota regulations for workers with disabilities, not aimed at fulfilling the rights and needs of disabilities.

So it can be concluded that some focus in the assessment process at this stage, namely: is necessary collaboration, are preconditions in place, who are the stakeholder, and who might fill key roles (sponsor, convener, and facilitator) have been evident in this stage. The main prerequisite needed for collaboration, namely building a network, has been found in the collaborations that have been carried out by the labor agency with private companies, NGOs, Integrated Rehabilitation Centers for Persons with Disabilities, District Social Welfare Workers, & PWDs groups. However, it was found in the field that this was not based on the achievement of the common goal of fulfilling the rights and needs of PWDs.

3.2. Initiation Stage

The Special Region of Yogyakarta Manpower and Transmigration Office has initiated cooperation, which is carried out through routine skills training activities for PWDs and non-disabled. This activity is included in the Operational Plan for the Implementation of Physical Activities. For financing comes from the Regional Expenditure Budget which is adjusted to the needs. The Operational Plan for the Implementation of Activities contains the stages of implementation and action at each stage of the implementation of the activities of the Regional Apparatus, which is equipped with the weight and schedule for the implementation of each action and the stages of the activity.

The training was conducted on a group basis. In several districts in the Special Region of Yogyakarta to the sub-district level there are already PWDs groups. This makes it easier for the group mechanism to be given training. The training provision mechanism is based on the existing needs of PWDs. Requests for training came from proposals for disability groups in each region. The disability group produced a proposal that was endorsed by the Village Head. The proposal is submitted to the Special Region of Yogyakarta Manpower and Transmigration Office for further selection process. Selection is carried out because of the large number of proposals submitted each year, so it needs a priority scale.

After the disability group that will be given training passes the selection, a location check will be carried out to see the conditions of the PWDs, whether they are still able to be trained to be productive or not. In addition, the assessment of the feasibility of business premises for groups with disabilities after being given training. In July 2019, a skills training for persons with physical disabilities was carried out in the field of food processing (see Table 1). Activities were held for 6 groups with 5 members for each group. The group came from Bantul Regency and Sleman Regency. The training was held for 6 days in the form of debriefing both theory and practice. The training materials include business motivation, business planning and simple bookkeeping. In addition, there are also materials on product innovation and marketing from instructors and practitioners.

The Special Region of Yogyakarta Manpower and Transmigration Office has been collaborating with various parties in skills training for PWDs. Practitioners and observers of disabilities who are included in this empowerment program both as resource persons and training mentors include training instructors from the Special Region of Yogyakarta Manpower and Transmigration Service, and the Job Training and Productivity Development Center for the Special Region of Yogyakarta [16].

Table 1. List of People with Disabilities in Economic Empowerment Training 2019

No.	Name	Address
1	Wahyu Mulat Widodo	Sedayu, RT. 053, Argosari, Sedayu, Bantul
2	Yusup Cahyo Dhadhari	Jaten, RT. 040 Argosari, Sedayu, Bantul
3	Waljiyo	Surobayan RT. 009 Argomulyo, Sedayu, Bantul
4	Wahyu Slamet	Tapen RT. 016, Argosari, Sedayu, Bantul
5	Sareh Hartono Nosareh	Sungapan Dukuh RT. 064 Argodadi Sedayu, Bantul
6	Jumariyam	Jogonalan Kidul, RT.04 Tirtonirmolo, Kasihan, Bantul
7	Sri Rahayu	Gonjen RT. 02 Tamantirto, Kasihan, Bantul
8	Suliyem	Gonjen RT. 02 Tamantirto, Kasihan, Bantul
9	Tumijan	Jogonalan Kidul, RT.04 Tirtonirmolo, Kasihan, Bantul
10	Yuman Sutaniaji	Glondong, RT. 01 Tirtonirmolo, Kasihan, Bantul
11	Wignyo Sudarto	Bulus Wetan RT.002 Sumberagung, Jetis, Bantul

IOP Conf. Series: Earth and Environmental Science **884** (2021) 012020 doi:10.1088/1755-1315/884/1/012020

No.	Name	Address
12	Sutijono	Bulu RT.003 Trimulyo, Jetis, Bantul
13	Darini	Nogosari RT 006, Sumberagung, Jetis, Bantul
14	Supomo	Pule RT.004, Sumberagung, Jetis, Bantul
15	Jayadi	Bulus Wetan RT.002, Sumberagung, Jetis,Bantul
16	Istirochah	Turgo RT/RW 006/003 Purwobinangun, Pakem, Sleman
17	Hanif Adhi Pratama	Watuadeg, RT 001 RW 019 Purwobinangun, Pakem, Sleman
18	Purwanto	Candi RT/RW 004/013 Purwobinangun, Pakem, Sleman
19	Eka Tias Safitri	Tritis RT/RW 001/001 Purwobinangun, Pakem, Sleman
20	Agus Eko Suyadi	Tegal Sembung RT 03 RW 35, Purwobinangun, Pakem, Sleman
21	Siti Badriyah	Gondang RT/RW 002/025 Banyurejo, Tempel, Sleman
22	Rofingah	Nglengis RT/RW 005/026 Banyurejo, Tempel, Sleman
23	Mariyem	Tangisan RT/RW 004/010 Banyurejo, Tempel, Sleman
24	Tiyah Kamsiah	Tangisan RT/RW 004/010 Banyurejo, Tempel, Sleman
25	Sariyanti	Karang RT/RW 003/024 Banyurejo, Tempel, Sleman
26	Maryanti	Kemban RT/RW 003/019,Wonokerto, Turi, Sleman
27	Retna Ningsih	Manggungsari RT/RW 004/012, Wonokerto, Turi, Sleman
28	Tri Endriyani	Kopen, Becici RT/RW 004/023, Wonokerto, Turi, Sleman
29	Sudarmono	Banyurip, Jambusari RT 004 RW.015, Wonokerto, Turi, Sleman
30	Purnama	Garongan RT/RW 001/018, Wonokerto, Turi, Sleman

Source: Office of Man Power and Transmigration Yogyakarta Province, 2019

Based on data obtained in the field, the initiation stage has been carried out by the Special Region of Yogyakarta Manpower and Transmigration Office after the clarity of regulations that require collaboration and hope for successful collaboration. In this stage, the Special Region of Yogyakarta Manpower and Transmigration Office plays the role of organizer and sponsor of skills training for PWDs. As noted by Torfing, collaborative innovation exploration can be chosen as a way for public and private stakeholders to solve cooperation problems [17]. This network-based collaboration can be expected to increase skills, ideas, energy and resources between government and its partners across institutional boundaries and in the non-profit and private sector. This is done by involving relevant stakeholders such as the Integrated Rehabilitation Center for PWDs, Social Services, District Social Welfare Workers, Non-Governmental Organizations and PWDs groups. The design of the skills-giving process also has a mechanism as defined. So at this stage, questions such as how to frame the issue, how to engage stakeholders, who or what else is needed, and what kind of process have appeared in the cooperation of related stakeholders.

One of the efforts made to fulfill the rights and needs of PWDs is through the provision of skills so that PWDs are able to achieve independence or increase their competence so that they can be absorbed in the workforce. In this case, the Special Region of Yogyakarta Manpower and Transmigration Office has involved relevant stakeholders. Each stakeholder has a role to play in fulfilling the rights and needs of persons with disabilities in work. However, in this initiation stage, the cooperation process carried out by related stakeholders is not yet based on a clear framework of action and action plans from each stakeholders.

3.3. Deliberation Stage

The Special Region of Yogyakarta Manpower and Transmigration Office has conducted dialogue and cooperation with related stakeholders. One of them is the Integrated Rehabilitation Center PWDs related to the handling of PWDs. The Integrated Rehabilitation Center for PWDs has a duty as a technical implementer in social protection, social rehabilitation and medical rehabilitation for persons with sensory disabilities (blind and deaf speech), intellectual, physical, and disabled people. As a Technical Implementing Unit, its main task is to provide social, mental and psychological rehabilitation services for PWDs. These institutions carry out vocational training or skills training as an additional program.

However, the cooperation between the Rehabilitation Center and the Special Region of Yogyakarta Manpower and Transmigration Office is still very limited. Rehabilitation centers carry out rehabilitation programs for PWDs through vocational programs. Meanwhile, the skills training program is only an additional program. Thus the cooperation carried out is limited to providing training material. This is because each institution has the authority related to the disability empowerment program. The rehabilitation center conducts various skills training such as: massage, plantation for the blind, embroidery, batik, making brooms, mats for intellectual disabilities, processing snacks, computers, graphic design, sewing, electrical engineering for physical and deaf disabilities. Vocational training was carried out by inviting trainers from various fields such as disabled entrepreneurs and Community Empowerment Institutions. For sewing and massage training in collaboration with course institutions that issue certificates for PWDs who take training lessons.

Sub-district Social Welfare Workers have a role in monitoring training participants who have been provided with business capital in the form of production equipment. They are assigned to the sub-district level to the village level by monitoring and regularly reporting on the progress of the business run by the PWDs group. Kecamatan Social Welfare Workers have an important role in bridging information and communication with PWDs. This institution was formed through the Minister of Social Affairs Regulation Number 28 of 2018 to realize community participation and as the spearhead in the implementation of social welfare at the sub-district level.

At the deliberation stage, the stakeholders conduct a deliberation in which they will formulate how they will work together and what program criteria they will run. Competencies required at this stage are group facilitation, listening, consensus building, and interest-based negotiation. As stated by Morse and Stephen [13], several questions such as: how to develop effective working groups, what ground rules, how to invest options and decide, and how to facilitate mutual learning can be used to discuss this phase of deliberation. The Special Region of Yogyakarta Manpower and Transmigration Office has made a number of efforts by conducting cooperation and dialogue with stakeholders regarding their respective roles in skills training. In addition, it also provides guidance and socialization to companies in the Special Region of Yogyakarta regarding the quota for disabled workers. In this stage a meeting forum has been held between stakeholders who are involved in the cooperation although it is not routine in nature, only at the beginning of the activity implementation. Whereas a good collaborative process can be built through dialogue or face-to-face communication between stakeholders [18]. The deliberations held are also still limited in nature, have not led to how to create effective cooperation signs to achieve the goal of fulfilling the rights and needs of PWDs.

3.4. Implementation Stage

Economic independence for PWDs is important because job vacancies for PWDs are very limited. As stated by the Head of Placement and Job Expansion, the Special Region of Yogyakarta Manpower and Transmigration Office, skills training is aimed at encouraging the growth and development of an entrepreneurial spirit at PWDs [19]. Training in sewing, carpentry, animal husbandry and fisheries can be done at the Job Training Center. Apart from being encouraged to become independent entrepreneurs, PWDs are also prepared to become the professional workforce needed by companies.

IOP Conf. Series: Earth and Environmental Science **884** (2021) 012020 doi:10.1088/1755-1315/884/1/012020

For that it is necessary to prepare the competence so that it is easier to absorb employment opportunities. However, it turns out that the Vocational Training Center has not implemented any disability-friendly infrastructure.

The Yogyakarta Special Region Manpower and Transmigration Service has also encouraged a number of companies to comply with Law number 8 of 2016 concerning PWDs. At the beginning of 2019, there were only 20 companies that employed people with disabilities. Head of the Special Region of Yogyakarta Manpower and Transmigration Service, Andung Prihadi, argued that the provisions of Article 53 paragraph 2 concerning the 1 percent quota for employees with disabilities in reality have not met expectations [20]. So that on December 13 2018 the Manpower and Transmigration Office invited 16 private and industrial business organizations to sign a moral commitment from the business world and the Special Region of Yogyakarta to implement the provisions of the regulation [7]. The raising and signing of commitments from the business world and the industrial world were carried out by the Indonesian Chamber of Commerce and Industry, the Indonesian Entrepreneurs Association, the Association of Indonesian Hotels and Restaurants, the Association of Indonesian Women Entrepreneurs, the Association of The Indonesian Tours and Travel Agencies, the National Association of Simple Healthy Housing Developers, the Association Indonesian entrepreneur representatives from Yogyakarta. This signing is intended to increase awareness, understanding and commitment to increase productivity and implementation of Law Number 8 of 2016.

Based on data from early 2019, in the Yogyakarta Special Region there were only 20 companies that employed people with disabilities, including PT Mega Andalan Kalasan, PT Lezak Nesia Jaya, PT Adi Satria Abadi, Grand Quality Hotel, PT Sogan Jaya Abadi, PT Mataram Tunggal Garment , PT Eagle Glove Indonesia, PT Sport Glove Indonesia, Sheraton Mustika Ratu, PT Arib Westa Kusuma, PT Ide Studio, PT Sungchang Indonesia, PT Budi Makmur Jaya Murni, Kenji, PT IGP Internasional, PT Busana Remaja Agracipta, Mandiri Craft Foundation, PT Ameya Living Style, PT Komitrando Emporio, and PT Kyara Beauty Med. Chairman of the Indonesian Employers with Disabilities Association, Bambang Susilo, admitted that there are still few companies in the Special Region of Yogyakarta that employ people with disabilities [joglosemarnews.com]. This is reinforced by Masykur Isnan's statement, May Bank Indonesia's Industrial Relations, that the awareness of companies in the Special Region of Yogyakarta to absorb workers from people with disabilities is still low. According to him, this was caused by a number of factors, including minimal socialization from related agencies and companies that were considered not ready to implement the regulation. The reason why companies have not implemented Law No.8 of 2016 is due to difficulties in recruiting workers with disabilities, lack of information regarding institutions that supply workers with disabilities, and also factors from PWDs themselves who feel uncomfortable working in companies. Apart from this, based on interviews with several PWDs, their families do not support them in their work because they are worried about their safety. Sanctions for violating institutions have not been enforced, because the supporting infrastructure for employing people with disabilities is not widely owned by institutions.

The Sleman Regency Manpower Office and Sanata Dharma University held a Job Fair The Path of Future 2019 on September 7 - 8, 2019 (see figure 1). It was recorded that there were 2. 215 job vacancies from 40 companies including vacancies for PWDs. The Job Fair with the Vocational School of Gadjah Mada University was also held on October 8-10, 2019. In this Job Fair, one of the criteria for a company that is needed is a company that has vacancies for PWDs. There are 12 companies that open vacancies for PWDs with 170 job vacancies. At the Job Fair, the role of Kerjabilitas as an intermediary greatly helped PWDs to find jobs. Kerjabilitas is a support institution in the form of an online platform and connects PWDs with companies or inclusive job providers. Through Kerjabilitas, PWDs will get information on job vacancies, how to make a good CV, tips in interviews, and they will even be delivered to companies that directly ask to find workers with disabilities. Kerjabilitas will also monitor whether one is accepted and where it is placed. This is as stated by the Kerjabilitas Account Executive, Sella Monika, that many PWDs who have higher education backgrounds even up to S2 are capable and deserving of work so they should be given opportunities [21].

Figure 1. Job Fair The Path of Future 2019
Source: harianjogja.com

Government Regulation Number 70 of 2019 concerning Planning, Implementation and Evaluation of the Respect, Protection and Fulfillment of the Rights of PWDs. This Government Regulation is an implementing regulation of Article 27 paragraph (3) of Law Number 8 of 2016 concerning PWDs. Planning, Implementation and Evaluation are carried out at the national and regional levels. At the regional level, the implementation is based on the Provincial or Regency / City Medium-Term Development Plan, or Regional Action Plan for PWDs. The Provincial Action Plan for PWDs is a development planning and budgeting document related to the Respect, Protection and Fulfillment of the rights of PWDs, which is the elaboration of the Master Plan for Development of Disabilities at the regional level.

Focus Group Discussion on the preparation of Regional Action Plans in the context of protecting and fulfilling the rights of PWDs was held on Tuesday, 23 October 2018 at the Hall of the Special Region of Yogyakarta Social Service. The FGD was carried out based on the suggestions and initiatives of the Committee for the Protection and Fulfillment of the Rights of PWDs [22]. This is in accordance with one of the duties of the committee, namely to provide suggestions, considerations and recommendations to local governments, district / city governments, Regency / City Regional People's Representative Council in making policies related to the protection and fulfillment of the rights of persons with disabilities. Parties invited to the FGD were program / activity planning elements from 16 Regional Work Units in accordance with the mandate of the Regional Regulation. However, not all were present, along with the functional social workers of the Yogyakarta Social Service and members of the Committee for the Protection and Fulfillment of PWDs' Rights. The objective of the FGD implementation is to encourage regional work units to be more effective and optimal in implementing the mandate of the regional regulation by preparing a program / activity plan and budgeting that is responsive to PWDs.

On 2 November 2018, a Regional Action Plan Development Meeting was also held at the Office of the Disability Committee on Protection and Fulfillment of PWDs Rights [23]. The meeting was attended by representatives of disability committees in districts / cities in the Yogyakarta Special Region and other institutions concerned with disabilities. As stated by Setia Adi Purwanta, Chairman of the Committee, this meeting was held because PWDs have not felt any significant changes after the passing of the Law [7]. So it is hoped that the Regional Action Plan can be completed immediately and bring positive changes to the fulfillment of PWDs rights.

CIQAL together with the Muhammadiyah Community Empowerment Council, and the Yogyakarta Independent Legal Ad Institute have initiated a Seminar on the Draft Regional Action Plan as the Implementation of the District / City Level Disability Perspective SDGs on September 17, 2019 [24]. The activity held by the Forum for the Strengthening of the Rights of Persons with Disabilities was not only attended by PWDs but also human rights activists. This aims to encourage local governments to have a strong commitment to inclusive development to realize the SDGs. This can be achieved primarily through the provision of jobs, transportation infrastructure, roads and access to information.

The implementation stage has been carried out after an agreement between stakeholders has been made in carrying out the collaboration process. In this stage of implementation, various parties have involved supporting each other, including designing governance, monitoring the progress of agreements and evaluating results, as well as other efforts related to the establishment of a partnership relationship. Several questions related to this stage, including: who will do what, how to broaden support, what kind of governance structure, and how to monitor progress have been answered. However, the cooperation and partnerships that have been well established among stakeholders have not been able to create a strong synergy to evaluate the results together, because the cooperation carried out is often incidental.

4. Conclusion

Collaborative governance in realizing an inclusive workforce for people with disabilities has been implemented but has not yet reached the stage of being able to build a strong framework of cooperation and create broader support for this cooperation. At the assesment stage, collaboration has been carried out with the Office of Manpower and Transmigration Special Region of Yogyakarta as the main actor although in its implementation it is still not aimed at achieving the same interests. At the initiation stage, the collaborative process carried out is not yet based on a clear cooperation framework and action plan from each stakeholder. Although there have been deliberation efforts made, they have not succeeded in creating an effective working group and building consensus to achieve inclusive workforce. The partnership that has been formed has not encouraged stakeholders to jointly evaluate the progress that has been made in cooperation.

Based on the results of these conclusions, several recommendations are given as follows: collaboration among stakeholders should be encouraged to achieve an inclusive workforce for PWDs; It is necessary to formulate a clear framework and action plann for establishing cooperation; guidance and outreach to companies in Special Region of Yogyakarta regarding the quota for workers with disabilities should be increased; and enforcement of sanctions against companies that do not apply a quota for workers with disabilities.

Acknowledgement

We express our gratitude to the Ministry of Social Affairs through the Yogyakarta Social Welfare Service Research and Development Center for providing financial assistance so that collaborative research with the Department of Public Administration Universitas Negeri Yogyakarta can be carried out.

References

[1] H. Sovia, "Kesempatan Kerja Bagi Penyandang Disabilitas," hukumonline.com, Oct- 2019

[2] Putra, Pamungkas Satya. 2018. *Aksesibilitas Perlindungan Hukum Bagi Tenaga Kerja Penyandang Disabilitas* di Kabupaten Karawang: Universitas Singaperbangsa.

[3] Kementrian Kesehatan RI. 2018. Infodatin Pusat Data dan Informasi Kementerian Kesehatan RI. Nov 2019

[4] Medcom.id. 2019. Pentingnya Memberikan Pemahaman Kewirausahaan bagi Penyandang Disabilitas.

[5] Bradford, N. 2016. Ideas and Collaborative Governance: A Discursive Localism Approach. *Urban Affairs Review*. https://doi.org/10.1177/1078087415610011

[6] Ansell, C., & Gash, A. 2018. Collaborative Platforms as a Governance Strategy. *Journal of Public Administration Research and Theory*, *28*(1), 16–32. https://doi.org/10.1093/jopart/mux030

[7] Undang-Undang Republik Indonesia Nomor 8 Tahun 2016 Tentang Penyandang Disabilitas.

[8] Poerwanti, S. D. 2017. Pengelolaan Tenaga Kerja Difabel untuk Mewujudkan Workplace Inclusion. *INKLUSI*. https://doi.org/10.14421/ijds.040101

[9] Raja, D. S. 2016. Bridging the disability divide through digital technologies. In *World Development Report*.

[10] Emerson, K., Nabatchi, T., & Balogh, S. 2012. An integrative framework for collaborative governance. *Journal of Public Administration Research and Theory*. https://doi.org/10.1093/jopart/mur011

[11] Bowen, G. A. 2009. Document analysis as a qualitative research method. *Qualitative Research Journal*. https://doi.org/10.3316/QRJ0902027

[12] Milles, M. B., Saldana, J., & Huberman, M. A. 1994. SAGE: Qualitative Data Analysis: A Methods Sourcebook: Third Edition. In *Qualitative Data Analysis:*

[13] Morse, R. S., & Stephens, J. B. 2012. Teaching Collaborative Governance: Phases, Competencies, and Case-Based Learning. *Journal of Public Affairs Education*. https://doi.org/10.1080/15236803.2012.12001700

[14] Peraturan Daerah Istimewa Yogyakarta Nomor 4 Tahun 2012 Tentang Perlindungan dan Pemenuhan Hak-Hak Penyandang Disabilitas.

[15] Scott, T. A., & Thomas, C. W. 2017. Unpacking the Collaborative Toolbox: Why and When Do Public Managers Choose Collaborative Governance Strategies? *Policy Studies Journal*. https://doi.org/10.1111/psj.12162

[16] jogja.antaranews.com. Disnakertrans memfasilitasi penyandang disabilitas menciptakan lapangan kerja. Dec- 2018

[17] Torfing, J. 2016. Collaborative innovation in the public sector. In *Collaborative Innovation in the Public Sector*. https://doi.org/10.4337/9781849809740.00032

[18] Ansell, C., & Gash, A. 2008. Collaborative governance in theory and practice. *Journal of Public Administration Research and Theory*. https://doi.org/10.1093/jopart/mum032

[19] jogja.antaranews.com. Disnakertrans memfasilitasi penyandang disabilitas menciptakan lapangan kerja. Dec- 2018

[20] jogja.tribunnews.com. Pemda DIY sebut baru ada 20 perusahaan pekerjakan disabilitas. Dec-2018

[21] Jogjapolitan.harianjogja.com. Disnaker Sleman Sanata Dharma gelar job fair 2. 215 lowongan kerja disediakan. Sept – 2019

[22] dinsos.jogjaprov.go.id. Focus group discusion penyusunan rencana aksi daerah (rad) dalam rangka perlindungan dan pemenuhan hak penyandang disabilitas. Nov-2018

[23] usd.ac.id. Rencana aksi daerah hak disabilitas. Nov - 2018

[24] krjogja.com. SDGs tak boleh abaikan difabel. Aug - 2019

The Risk Communication of COVID-19 Disaster Community Based Through "Keep the Village" Program

Muhammad Badri

Department of Communication Science, Faculty of Da'wah and Communication, Universitas Islam Negeri Sultan Syarif Kasim Riau, Indonesia

Email: muhammad.badri@uin-suska.ac.id

Abstract. The spread of COVID-19 is increasingly massive and it involves local transmission. Facing these issues, community-based risk reduction is truly needed. This study aims to describe the society-based risk communication for COVID-19 disaster prevention through "Keep the Village" program, called *Jaga Kampung*. This study used a qualitative approach by collecting data through interviews with COVID-19 response volunteers in the *Jaga Kampung* program in Riau Province, Indonesia. This study finds out that the *Jaga Kampung* program has the aim to prevent the spread of COVID-19, maintain security stability, and protect people's food needs. The implementation of these activities involves the collaboration of communication actors from the police and the army who works in the village, village leader, village health worker, society leader, and volunteer. The officers of *Jaga Kampung* convey education and socialization of health risks humanistic and persuasively to villagers and newcomers. Communication is done through face-to-face, outdoor media, and social media such as WhatsApp, Facebook, YouTube, and Instagram. This community-based risk communication can increase society's participation in the prevention of COVID-19 so that the health and socio-economy of society is maintained.

Keywords: Risk communication, COVID-19, Community, Keep the village.

1. Introduction

The Law of the Republic of Indonesia Number 24 of 2007 concerning Disaster Management places the epidemic of disease as non-natural disasters, namely the disasters caused by non-natural events or series of events [1]. The Coronavirus disease (COVID-19) pandemic is a non-natural disaster that has recently damaged people's lives. Moreover, the spread of COVID-19 is no longer between countries and regions, however, it has involved the local transmission. The risk of transmission has also begun to threaten rural societies. Facing this phenomenon, the reduction of COVID-19 risk at the society level is very essential.

Considering this pandemic situation, strong risk communication is needed especially in the affected countries [2]. Understanding risk is a systematic way of dealing with handling the hazards and insecurities [3]. Awareness of risks is the best way to prevent and slow down the transmission of the COVID-19 pandemic. Risk awareness is achieved through risk assessment communication [4]. Knowledge about risk exists in a social context and is connected to the actor's activities [5].

One way to reduce risk at the village level is through community-based prevention of COVID-19. The program to protect villages from COVID-19 is widely implemented in various regions in Indonesia, one of which is in Riau Province, with "Keep the Village" program (called *Jaga Kampung*).

Content from this work may be used under the terms of the Creative Commons Attribution 3.0 licence. Any further distribution of this work must maintain attribution to the author(s) and the title of the work, journal citation and DOI.
Published under licence by IOP Publishing Ltd

Jaga Kampung is a socialization and education activity aimed at preventing the transmission of COVID-19 in village areas. This concept emphasizes the importance of preventing COVID-19 through communication and educational activities by communication actors from government elements and volunteers at the village level. The *Jaga Kampung* activity is also expanded through a food security program.

Covid-19 risk communication at the village level is needed to address information gaps and misinformation. It is due to the era of social media, information related to COVID-19 can be disseminated to the wider community through various media to create private information spaces for the public [6]. As a result, the public gains information from many informal and anonymous sources whose accuracy and the fact cannot be guaranteed. Through society-based communication, the residents discover who the communicator is, can measure their credibility, so that communication on COVID-19 prevention is expected to be more effective.

Disease knowledge and awareness, in addition to risk perception, are determining factors in how people respond to and engage in preventive behavior [7]. In general, there are four objectives of risk communication: (1) Ensure that all message recipients are able to understand its content and increase their knowledge of the risks; (2) Build a trusting relationship between senders and recipients of risk messages; (3) Persuade recipients of messages to change their attitudes or behavior regarding the causes of risk; and (4) Provide space for the involvement of stakeholder which is effective on risk issues [8].

One of the important factors in disaster management is communication between the government and the affected communities. Communication should not only happen in an emergency. Indeed before a disaster, contextual communication must be needed by educating the public about the threats faced. There are three things that must be precise in communicating with the affected society, namely: (1) the content of the message (information), (2) the communication media used, and (3) the speed of sending information [9].

Effective risk communication is very important in risk management to build public confidence in the government's ability to face risks [10]. In the study of disaster risk communication in rural areas, communication is positioned as a dialogue and knowledge exchange to facilitate risk understanding and decision making. Thus, it is recommended to build existing strengths in the society, collaborative partnerships with local organizations, local media usage, empower local organizations, and accommodate the needs of rural communities [11].

Several previous studies related to COVID-19 risk communication explained that a good level of knowledge of COVID-19 indicates a high-risk perception of the disease [7]. Efficient strategies that rely on socio-economic factors in which it should be further enhanced in countries, regions, and cities with serious epidemics [15]. To develop a local-level response strategy, democratic access to information, and participatory decision-making is needed [4]. Considering the differences in response capabilities between urban and rural areas, it is imperative to develop action plans, response scenarios, and strategies to optimize the use of the equipment and human resources in the fight against COVID-19 [16]. The publicity service of Village society in preventing COVID-19 has a major impact on society [6].

Many research-related to COVID-19 prevention has not been discussed yet about society-based risk communication approaches at the village level. This study aims to describe society-based risk communication for COVID-19 prevention through the *Jaga Kampung* activity. This research is conducted to cover the research gap on COVID-19 prevention communication at the village level. The *Jaga Kampung* program is interesting to be discussed because the model of risk reduction for COVID-19 is not limited to disease transmission factors, but it also food security factors to maintain the socio-economic immunity of the society.

2. Method

This study used a qualitative approach by collecting data through online interviews with COVID-19 response volunteers in the *Jaga Kampung* program in Riau Province, Indonesia. The research subjects were those involved in the *Jaga Kampung* program in the border areas. The assumption is the border area is the vulnerable area to the transmission of COVID-19. The data that had been collected, it was analyzed and presented descriptively together with secondary data from the news of mass dealing with *Jaga Kampung*.

3. Result and Discussion

The *Jaga Kampung* program is an implementation of the archipelago tough village program (called *Kampung Tangguh Nusantara*) launched by the Chief of the Indonesian Police and the Commander of the Indonesian National Army, in the context of handling COVID-19 in the health, security and socio-economic fields. The name of the program at the regional level is adjusted to the socio-cultural community. The aim of the program is to prevent the spread of COVID-19, maintain the stability of society security, and protect food needs by empowering the land. The Indonesian Police Chief stated:

> "We hope that the whole society to the RT/ RW level (neighborhood and hamlet level) will participate in preventing the transmission of COVID-19. We are optimistic that if the RT level is already strong then Indonesia will be strong." [17]

The social movement of *Jaga Kampung* is carried out in a structured manner through the command line down to police personnel who are tasked with fostering community security and order at the village level (called *Bhabinkamtibmas*) and soldiers tasked with fostering village societies (called *Babinsa*). The implementation of these activities involves village leaders, community leaders, health workers, and village volunteers. The *Jaga Kampung* team routinely educates the public to comply with health protocols and socializes the importance of providing handwashing facilities in every home, maintaining distance, and wearing masks when going outside [18]. Apart from involving society, the *Jaga Kampung* program also links to the company's Corporate Social Responsibility (CSR) activities with society, especially to assist the production of food crops [19].

3.1. Jaga Kampung as an Educational Arena for COVID-19

Jaga Kampung is said to be the front line for COVID-19 prevention at the village level. To support the implementation of Keep Village, a guard post has been established at the village entrance. The post is guarded by Bhabinkamtibmas, Babinsa, village leaders, health workers, and volunteers, alternately. The post of *Jaga Kampung* is an arena for educational activities to the public about all things related to COVID-19.

The *Jaga Kampung* team conducts educational activities on the use of masks, washing hands using soap, maintaining distance, not crowding, and other COVID-19 prevention health protocols. Education is carried out directly (face to face) with the society, through outdoor media, and through social media. Direct education through outreach to family, neighbors, and residents. Outdoor media through banners posted at various strategic points in the village. Education on social media using WhatsApp and creating creative content on Facebook, YouTube, and Instagram which is followed by the residents.

The COVID-19 educational material is provided by the Indonesian government through the COVID-19 Handling Task Force (called the COVID-19 Task Force) and can be accessed on its official website (https://covid19.go.id). The COVID-19 Task Force provides more than 300 educational materials aimed at groups of people at risk, the general public, parents, health workers, religious and community leaders, teachers and students, mothers and children, travelers, and business people. Educational materials in the form of narration, images, and audiovisuals, with content tailored to the target group.

Official educational materials from the COVID-19 Task Force can be downloaded and distributed to the public. In addition, parties with an interest in educating COVID-19 can modify and develop creativity in reproducing educational content. The delivery of educational messages, both face-to-face and through the media, is adjusted to the socio-cultural aspects of the community, such as the use of local languages, local social norms, local culture, and local wisdom.

The education that is most often carried out is reminding the public about the importance of washing hands and using masks. People who are not used to wearing masks often neglect not to wear

IOP Conf. Series: Earth and Environmental Science **884** (2021) 012021 doi:10.1088/1755-1315/884/1/012021

masks when they go outside. It is important that messages like this are continuously conveyed so that the public has awareness of the importance of masks to prevent transmission of COVID-19. Moreover, the research also found that wearing masks throughout the community can contribute to the control of COVID-19 by reducing the emission of infected saliva [20]. Meanwhile, increasing the awareness of the population about handwashing will provide the greatest benefit for reducing the pandemic [21]. The research informant said:

> "Public awareness grows because we often urge it to prevent COVID-19. The message that we often convey is that the public will be more disciplined and orderly in the rules of the new normal era by still paying attention to health protocols, such as washing hands frequently and using masks."

Figure 1. The Village Guard Post at the village gate

Figure 2. Educational messages for washing hands with soap

Covid-19 education at the village level, such as that carried out by the *Jaga Kampung* team, is relatively effective in preventing the spread of the virus. At the time this article was written, the village that was the research location had not experienced a positive case of COVID-19. The COVID-19 cluster, which continues to increase in Riau Province from June to August 2020, generally clusters of transmission in offices, companies, and urban areas. Dealing with this phenomenon, an inclusive and community-based approach to public health resilience and preparedness is urgently needed [25].

3.2. The Model of Risk Communication for the Prevention of COVID-19

Risk communication is one way to protect and reduce risks faced by society [12]. With risk communication, there will be an exchange of assessments, estimates, and opinions about hazards and risks among the various stakeholders involved [13]. Professionals and risk researchers consider public participation in risk assessment and management important [14]. People-centered risk communication is more effective than top-down government communication approaches. Risk communication must utilize the social networks that already exist in society because it has a significant effect on taking protective action [22].

Risk communication in *Jaga Kampung* is carried out through a persuasive approach. Persuasive communication puts forward a strategy of inviting, not coercing. Persuasive communication is in accordance with the characteristics of the people of Riau, who generally have a Malay culture, namely avoiding conflict, being soft, tolerant, and peaceful. The *Jaga Kampung* team delivers messages on the risk of COVID-19 to the community passing the *Jaga Kampung* post.

In addition, the team on duty also visit the houses and stalls to convey health protocols to residents. If there are residents from outside the village who come to the village, they are required to report to the *Jaga Kampung* post to undergo temperature checks and health checks. The society has also not been allowed to hold activities that invite crowds, such as weddings, because they have the potential to trigger transmission. This research informant said:

"When conducting socialization and educating the public to always pay attention to health protocols, we use a humanist communication approach, including when warning residents not to get together."

This humanist and persuasive approach make communication to be more effective because there is no resistance from residents. The society receives the Jaga Kampung team well and participates in the success of preventing COVID-19. In fact, many residents and shop owners have the awareness to build an independent handwashing place in front of their house or place of business. When in early August 2020, Riau Province experienced a spike in COVID-19 cases, *Jaga Kampung* officers intensified their communication activities and sprayed disinfectants.

Reducing the risk of COVID-19 through *Jaga Kampung* also targets food security. The activities include planting food crops on village-owned land and unused society lands. The goal is to create food security at the community level and revive the society's economy. Planting activities are carried out in mutual cooperation with society. Mutual Assistance is a distinctive culture of the Indonesian people and is an important social asset to overcome COVID-19 [23]. The crops are then distributed to people who need them to reduce the cost of spending on food.

The issue of economic risk needs to be considered because the economic recovery is expected to be long-lasting [24]. Facing this condition, Indonesia needs to strengthen economic resilience [25]. Moreover, COVID-19 pandemic is likely to affect global food supplies [26]. It is hoped that this model of maintaining food supply at the village level will strengthen the food security of society. If the majority of villages in Indonesia, even globally, are able to do this, it can reduce the burden on countries that are threatened with the economic recession.

The construction of a risk communication model for COVID-19 prevention through *Jaga Kampung* can be seen as a collaborative model involving the police (*Bhabinkamtibmas*), soldiers (*Babinsa*), village leaders, village health workers, community leaders, and volunteers. The message conveyed was in the form of health risks and socio-economic risks due to COVID-19. The communication channel, the *Jaga Kampung* post as the main arena, is implemented through interpersonal communication and social media. The message delivery strategy is persuasive and humanist. The target audience is the villagers and outsiders who come. Communication feedback is in the form of community participation who are volunteering to be involved in preventing COVID-19. The expected effect is a reduction in the risk of COVID-19 and socio-economic immunity.

This village-level communication model is similar to that of rural communities in China, in which both of them involve volunteers and the police, through face-to-face communication, persuasion, with the aim of prevention. Communication channels are also interpersonal and multimedia. This type of communication has strong direction, interaction frequency, and high feedback. This model is considered to be the most effective for COVID-19 communication [6]. The difference is, *Jaga Kampung's* communication does not involve party members and it does not use coercive strategies.

4. Conclusion

The "Keep the Village" program model is one of the best practices in society-based risk communication for COVID-19 disaster prevention. This is in accordance with the program's mission, to create the resilience of society against COVID-19 which will have implications for resilience at the national level. The success of risk communication at the village level is determined by a humanist and persuasive communication strategy, through education, not through implementation. This approach model can create public awareness to comply with health protocols, as well as participate in COVID-19 prevention activities. The idea of society-based COVID-19 disaster risk reduction is interesting to disseminate because the context of its activities is not only on health issues but it also concerns the security and socio-economic issues. In the future, security and socio-economic issues are important to pay attention to because the effects of the COVID-19 pandemic are not only on public health but also on the security and economic resilience of a country.

References

[1] Undang-Undang Republik Indonesia Nomor 24 Tahun 2007 tentang Penanggulangan Bencana (The Law of the Republic of Indonesia Number 24 of 2007 concerning Disaster Management)

[2] Husnayain A, Fuad A, Su EC. Applications of google search trends for risk communication in infectious disease management: A case study of COVID-19 outbreak in Taiwan. *International Journal of Infectious Diseases*. 2020 Mar 12.

[3] Beck U. *Masyarakat Risiko: Menuju Modernitas Baru*. Jogja: Kreasi Wacana; 2015.

[4] Chatterjee R, Bajwa S, Dwivedi D, Kanji R, Ahammed M, Shaw R. COVID-19 Risk Assessment Tool: Dual application of risk communication and risk governance. *Progress in Disaster Science*. 2020 Jun 2:100109.

[5] Lidskog R and Sundqvist G. Sociology of risk. In *Essentials of risk theory* 2013 (pp. 75-105). Springer, Dordrecht.

[6] Wang K, Wang S, Wang J. Shaping and Publicization of Personal Information Spacec of COVID-19 Information Dissemination. In *Journal of Physics: Conference Series* 2020 Jun 1 (Vol. 1574, No. 1, p. 012169). IOP Publishing.

[7] Karasneh R, Al-Azzam S, Muflih S, Soudah O, Hawamdeh S, Khader Y. Media's effect on shaping knowledge, awareness risk perceptions and communication practices of pandemic COVID-19 among pharmacists. *Research in Social and Administrative Pharmacy*. 2020 Apr 23.

[8] Renn O. *Risk governance: coping with uncertainty in a complex world*. London: Earthscan; 2008.

[9] BNPB. *Rencana Induk Sistem Komunikasi Bencana 2015-2019*. Jakarta: BNPB; 2015.

[10] Covello VT. Strategies for overcoming challenges to effective risk communication. In *Handbook of risk and crisis communication* 2010 Sep 28 (pp. 155-179). Routledge.

[11] Cole JM, Murphy BL. Rural hazard risk communication and public education: Strategic and tactical best practices. *International journal of disaster risk reduction*. 2014 Dec 1;10:292-304.

[12] Verroen S, Gutteling JM, De Vries PW. Enhancing self-protective behavior: efficacy beliefs and peer feedback in risk communication. *Risk analysis*. 2013 Jul;33(7):1252-64.

[13] Scheer D, Benighaus C, Benighaus L, Renn O, Gold S, Röder B, Böl GF. The distinction between risk and hazard: understanding and use in stakeholder communication. *Risk Analysis*. 2014 Jul;34(7):1270-85.

[14] McComas KA, Arvai J, Besley JC. Linking public participation and decision making through risk communication. In *Handbook of risk and crisis communication*. 2009:364-85.

[15] Ren H, Zhao L, Zhang A, Song L, Liao Y, Lu W, Cui C. Early forecasting of the potential risk zones of COVID-19 in China's megacities. *Science of The Total Environment*. 2020 Aug 10;729:138995.

[16] Tran BX, Nguyen HT, Pham HQ, Le HT, Vu GT, Latkin CA, Ho CS, Ho RC. Capacity of local authority and community on epidemic response in Vietnam: Implication for COVID-19 preparedness. *Safety Science*. 2020 Jun 5:104867.

[17] Antara. Panglima dan Kapolri luncurkan Kampung Tangguh Nusantara di Lombok. Antara News [Internet]. 2020 Jul 9 [cited 2020 Aug 6]. Available from: https://www.antaranews.com/berita/1600238/panglima-dan-kapolri-luncurkan-kampung-tangguh-nusantara-di-lombok

[18] Riauonline. Polda Riau Luncurkan Program Jaga Kampung, Ini Tujuan Mulianya. [Internet]. 2020 Jun 27 [cited 2020 Aug 6]. Available from: https://www.riauonline.co.id/riau/read/2020/06/27/polda-riau-luncurkan-program-jaga-kampung-ini-tujuan-mulianya

[19] Jawapos. Pilot Project Jaga Kampung Polda Riau Tanam Bahan Pangan. [Internet]. 2020 Jul 26 [cited 2020 Aug 6]. Available from: https://www.jawapos.com/jpg-today/26/07/2020/pilot-

project-jaga-kampung-polda-riau-tanam-bahan-pangan/

[20] Cheng VC, Wong SC, Chuang VW, So SY, Chen JH, Sridhar S, To KK, Chan JF, Hung IF, Ho PL, Yuen KY. The role of community-wide wearing of face mask for control of coronavirus disease 2019 (COVID-19) epidemic due to SARS-CoV-2. *Journal of Infection*. 2020 Apr 23.

[21] Lin YH, Liu CH, Chiu YC. Google searches for the keywords of "wash hands" predict the speed of national spread of COVID-19 outbreak among 21 countries. *Brain, Behavior, and Immunity*. 2020 Apr 10.

[22] Haer T, Botzen WW, Aerts JC. The effectiveness of flood risk communication strategies and the influence of social networks—Insights from an agent-based model. *Environmental Science & Policy*. 2016 Jun 1;60:44-52.

[23] Badri M. Mobilizing social capital to overcome corona disasters. *Zenodo*. 2020 Jul 27. Available from: http://doi.org/10.5281/zenodo.3961901

[24] Nicola M, Alsafi Z, Sohrabi C, Kerwan A, Al-Jabir A, Iosifidis C, Agha M, Agha R. The socio-economic implications of the coronavirus pandemic (COVID-19): A review. *International journal of surgery* (London, England). 2020 Jun;78:185.

[25] Djalante R, Lassa J, Setiamarga D, Mahfud C, Sudjatma A, Indrawan M, Haryanto B, Sinapoy MS, Rafliana I, Djalante S, Gunawan LA. Review and analysis of current responses to COVID-19 in Indonesia: Period of January to March 2020. *Progress in Disaster Science*. 2020 Apr 4:100091.

[26] Udmale P, Pal I, Szabo S, Pramanik M, Large A. Global food security in the context of COVID-19: A scenario-based exploratory analysis. *Progress in Disaster Science*. 2020 Jul 22:100120.

Spatial Representation Developing Geographical awareness for Disaster Risk Reduction

Muhammad Nursa'ban[1*], Mukminan[2], Muhsinatun Siasah Masruri[3], Iis Kurnia Wati[4], Rias Murti Lestari[5], Endah Kusumastuty Hartoyo[6]

[1,2,3,4,5,6]Geography Education, Universitas Negeri Yogyakarta

*m_nusaban@uny.ac.id

Abstract. Reducing the risk of various natural disasters makes us learn more about the meaning of safe and sustainable living space. Geographical awareness which is formed by understanding spatial representation should be owned as a capital for disaster mitigation. This study aims to describe the level of geographic awareness with the spatial representation of high school students in Yogyakarta. The study used a quantitative descriptive research design with a questionnaire technique by stratified random sampling on 180 samples of high school students in Yogyakarta. Data are presented as a percentage in tables, charts, and graphs. The results showed that the level of geographic awareness through understanding the spatial representation of high school students' disaster risk reduction varied from low to high. A high understanding of spatial representation is shown by around 30 percent of students, 50 percent in the medium category, and less than 15 percent in the low category. Three factors of geographical awareness, namely: caring, polite, responsive, and tolerant attitude towards the surrounding environment; trying to be part of the solution to environmental problems; love the homeland. All respondents agreed with geographic awareness through spatial representation as an alternative to disaster risk reduction.

Keywords: spatial representation, geographic awareness, disaster, risk reduction

Content from this work may be used under the terms of the Creative Commons Attribution 3.0 licence. Any further distribution of this work must maintain attribution to the author(s) and the title of the work, journal citation and DOI.
Published under licence by IOP Publishing Ltd

1. Introduction

The National Disaster Management Agency released a report that at least natural disasters such as earthquakes, forest fires, eruptions, floods, and tsunamis that occurred in several parts of Indonesia from 2015 to early this year left very many victims. More than 503,874 people have been recorded suffering from acute respiratory infections (ISPA) and dozens of them have died worldwide due to forest fires in Sumatra. The earthquakes in several places where the dissertation of the tsunami claimed many lives and damaged infrastructure and other losses. Various natural disasters have made people learn more about the meaning of a safe and sustainable living space. The community should respond to this case phenomenon by understanding the potential disaster conditions in its surroundings. Efforts that are systematic and structured within the framework of understanding potential disasters along with mitigation and risk reduction efforts should be taken as mitigation capital. They could spatially think by analysing these events through an integrated spatial relationship between geographical phenomena such as climate, soil conditions, astronomical location, population dynamics, social phenomena and so on in the surrounding area. Every level of society can interpret that a potential disaster can occur anytime and anywhere, so it requires preparedness.

One of the alternatives for disaster risk reduction is through formal education based on curriculum. This material can be integrated in a subject or stand alone. In the Indonesian SMA national curriculum, Geography is a subject that seeks to find and understand spatial phenomena or symptoms through aspects and processes of the earth, the causal relationship between spatial factors, humans, and the environment. In the aspect of knowledge and skills, students will be able to think critically in determining the underlying geographical concept / theory, behave, act smartly, wisely, and be responsible in dealing with social, economic, ecological, and disaster problems. This subject is part of fostering knowledge, attitudes and behaviour of disaster awareness and awareness of loving the country, instilling pride as a nation, and being responsible for the integrity of the State. The statement indicates that awareness of potential, mitigation, and disaster management as well as nationalism is a form of geographical awareness. Awareness of the condition of the geographical environment around it is a form of spatial ability (Jo & Bednarz, 2014; Heffron, 2012a).

Dimensions and factors of attitude to build geographical awareness are interpreted as: "as a form of living and practicing the teachings of the religion they adhere to and living and practicing honest behavior, discipline, responsibility, care (mutual cooperation, cooperation, peace), polite, responsive and proactive and show an attitude as part of the solution to various problems in interacting effectively with the social and natural environment and in placing oneself as a reflection of the nation in the world's interactions ". This statement indicates that the attitude of building geographical awareness is equivalent to awareness of environmental conditions. Baiquni (2013) has stated that the elementary environment includes Abiotic, Biotic, and Cultural. Attitude aspects and indicators of building geography awareness as a result of learning geography with a spatial thinking perspective are based on the assessment of attitudes in the Indonesian national curriculum of geography syllabus.

The attitude of building geographical awareness will have various perspectives and focuses related to environmental problems that are influenced, including different cultures, places and situations (Heffron, 2012). The same opinion is that awareness of geography will shape a person to understand their environmental problems both physically and man-made wisely and wisely (Alderman, 2017). Indonesia's spatial context, which presents the diversity of the earth's surface space as a place for human life to live with its natural and social aspects as well as interactions, the interrelation between these aspects is an object of study that must be understood by the community, especially in the world of education at every level. The plurality of Indonesia, physical and social phenomena should be described in the curriculum with a multidisciplinary approach and study of spatial perspectives. Studying this aspect will produce geographic areas that characterize the similarities and differences of objects, phenomena, patterns, problems, potentials that exist in the earth's surface.

Geographical perspective on the dynamics of the physical and social environment will be seen from the aspects of spatial integration and spatial interdependence both between places and between scales (Bonnet, 2008). This perspective can be realized in real and abstract forms (and / or representations) both visually, verbally, mathematically, digitally, and in a (cognitive) mindset (Kemdikbud, 2016). The five spatial representations have a constructive conception that is supported by the concept of space, means of representation, and the process of reasoning. The description of the concept of representation shows a spatial understanding of the standard intellectual structure of geography. Visual representation, describes a concept or phenomenon through the sense of sight. Verbal representation is interpreted as a way of looking at geographic phenomena expressed verbally or abstractly so that environmental dynamics can be seen seen from the aspects of spatial integration and interdependence. Mathematical representation, interpreted by definite and precise calculations. Digital representation, describes a quantity that is not proportional to its value, but expressed using a symbol called a digit. The activity of spatial representation in geography encourages students to understand the variety of objects, phenomena, potentials, problems that exist in the earth's surface along with their attributes, characters and character colors regarding physical and social conditions. In describing, understanding, identifying, evaluating, and overcoming these geographic problems. Students can be assisted by a number of technologies such as remote sensing, maps, and Geographic Information Systems (GIS) as a vehicle for spatial thinking.

Geography learning contains spatial representations equipping students by placing spatial thinking as the main characteristic in learning (Nursa'ban et al, 2020). The characteristics of spatial thinking show the ability that involves knowledge and understanding of spatial concepts, using knowledge skills flexibly, thinking skills and habits to utilize tools, and providing reasons for solving problems and making decisions (Jo and Berdnarz, 2010: 50). The focus of geography as a science that studies the location and organization of human activity space on the earth (Lambert and Morgan (2010: 73). Spatial depiction can be done through cognitive reflexes that affect symbol systems as in geography embodied in the form of maps (Uttal (2000: 285).

Spatial representation should be applied in schools. However, the results of the initial study indicate that spatial representation is still not understood and is still a matter of debate between teachers as practitioners and developers of geography education as experts (Nursa'ban, 2019). As many as 41.45% of geography teachers do not understand practical spatial representations in learning. On average, 16% do not know the context of spatial representations in geographic material. This description implies that the ongoing geography learning still emphasizes the conceptual and procedural abilities of the initial knowledge that has not led to factual and metacognitive aspects related to geography material taught using spatial representation. This study will describe the level of geographic awareness with the spatial representation of high school students in Yogyakarta as a barometer of Indonesian education.

2. Method

The study used a descriptive study through a quantitative approach, namely Pretest-Postest Control Group Design, to examine the effect of geography learning treatment containing spatial representations on learning outcomes of geographical awareness. In this study, the geographic material studied is about food security, industrial materials and renewable energy.

The attitude dimension of building geographical awareness in this study is developed into three variables, namely: Caring and responsiveness to environmental problems, being part of the solution to environmental problems (social and natural), and loving and proud of the country. The study was conducted on 180 students at three high schools in Yogyakarta, which were obtained based on stratified random sampling. They consist of students who have an initial understanding of low, medium, and high spatial representations proportionally. Two groups, namely experimental and

IOP Conf. Series: Earth and Environmental Science **884** (2021) 012022 doi:10.1088/1755-1315/884/1/012022

control, were chosen randomly and were given a pretest to determine the initial conditions that were not significantly different.
The data collection instrument used a questionnaire and observation sheet. Questionnaires are used to measure the level of understanding of geographical awareness, while observation ensures that the learning process uses the same method. Validation of the contents and constructs of the instrument through expert judgment and limited legibility testing. The effect of change is using the Product moment correlation calculation with the reliability estimation using the KR-20 technique. The data were analysed using inferential statistical analysis. Simple regression analysis is used to predict how far the value of the dependent variable will change, if the value of the independent variable is manipulated

3. Result and Discussion

The results of the pretest and posttest regarding the ability of spatial representation are presented in Figure 1.

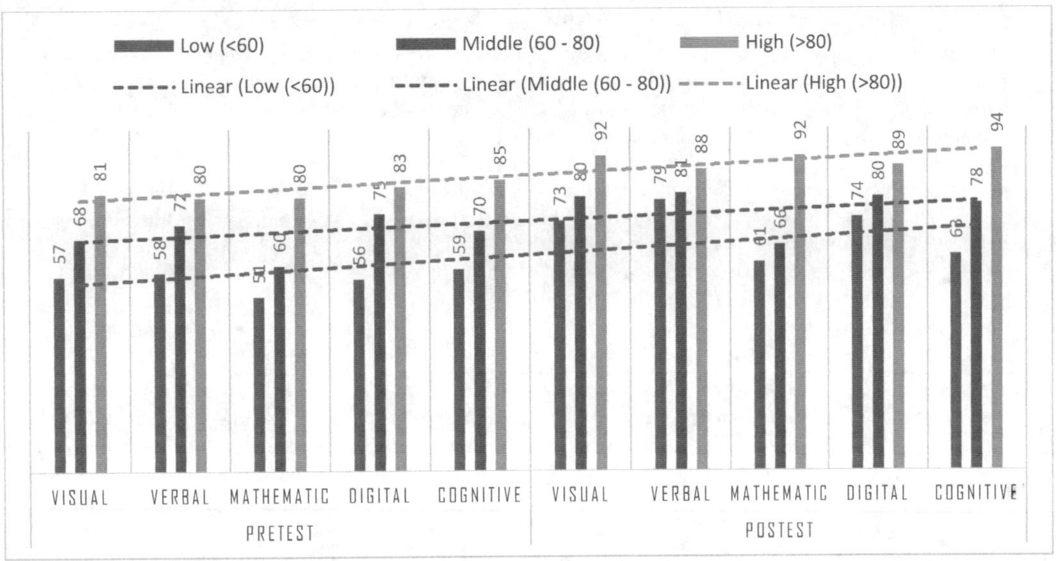

Figure 1. Pretest and Posttest results of spatial representation ability

Table 1. Describing the trendline of students' understanding of spatial representations in general has increased. The lowest condition for understanding of spatial representation lies in mathematical ability with the mean of all levels at the time of the pretest, namely at a score of 63 out of a total of 100.After the post-test, the mathematical representation increases to 73. The change in the highest increase to low average successively occurs at the low, high, and middle. The highest difference in change at the low level is in the verbal representation, on the other hand, the lowest change at this level occurs in cognitive representations. Meanwhile, visual representations experience the highest changes at the middle level, and mathematical representations at the high level.

An overview of the level of geographic awareness through understanding the spatial representation of high school students' disaster risk reduction in the material of food security, industrial materials and renewable energy is presented in Figure 2. Indicators of geographic awareness presented through spatial representations are presented in Table 1.

IOP Conf. Series: Earth and Environmental Science **884** (2021) 012022 doi:10.1088/1755-1315/884/1/012022

Table 1. Indicators of geographic awareness through learning with spatial representations

Variable	Indicators/Activities				
	Visual	Verbal	Mathematic	Digital	Cognitive
1. Caring and responsive to environmental problems	Designing a mind map for the preservation of renewable energy sources	State the forms of conservation of renewable energy sources appropriately	Calculating the availability of food with the population in an area	Looking for references via the internet environmental conservation efforts in areas of the world with various natural characteristics (climate and topography)	Understand the benefits of natural resources to support sustainable development
2. Become part of the solution to environmental problems (social and natural)	Creating environmental preservation posters for renewable energy	State various efforts to protect the environment for food availability and renewable energy sources	Calculating the magnitude of the risk of using renewable energy with fossil energy	Creating chain messages through social media about an invitation to preserve the environment	Understand the effect of energy availability for sustainable development through environmenta l conservation efforts
3. Love and proud of your homeland	Presenting data on maps about national food availability	State the names of areas with weak and strong food and energy availability	Make number coding for several regions based on the amount of energy absorption with the energy produced	Design diagrams and tables for presentations on energy and food availability that are downloaded from the central bureau of statistics	Understandin g national natural resources as energy and food potential in realizing sustainable development

Indicators of spatial representation presented in Table 1, show an overview of the activities and understanding carried out by students during the learning process. These indicators are invoices for the learning process carried out individually or in groups.

The results of the assessment of geographical awareness through the content of spatial representations in learning presented in Table 1 indicate that most students fall into the middle category. The level of understanding mostly uses visual representations through the process of describing phenomena according to the indicators in table 1. During learning, students enthusiastically participate in these activities and a warm and pleasant classroom climate appears to be formed.

On the other hand, the number of students who fall into the "high" category is more than the "low" category. This indicates that the accumulative spatial representation can target the "geographic

IOP Conf. Series: Earth and Environmental Science **884** (2021) 012022 doi:10.1088/1755-1315/884/1/012022

awareness" of most students (85%) who are subjected to research in the "middle" and "high" categories.

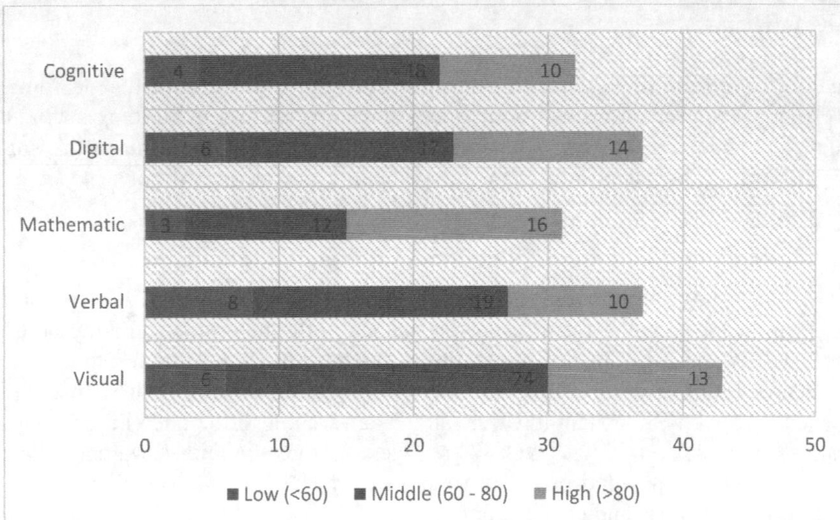

Figure 2. The level of geographic awareness through understanding the spatial representation of high school students' disaster risk reduction

The results of the study explain briefly that geographical awareness as a reflection of the competence of knowledge and attitudes can show a caring attitude and responsiveness to environmental problems so that students can be part of the solution to problems in the environment. This understanding can then form students' awareness to feel love, love, and pride for their homeland. Knowledge of environmental conditions and the potential of the country's territory. An understanding of geography awareness for students can support identifying the location and phenomena of potential resources in Indonesia as a form of superiority in the surrounding area, such as energy sources, foodstuffs, industrial raw materials, and large-scale needs and other physical phenomena.

Activities in interpreting geographical awareness through learning with spatial representations are a form of geography literacy for the younger generation in analysing interactions, interconnections, and implications between regions and with other countries. separatism. It is hoped that students of geography awareness are able to understand, among other things, spatial patterns in the national and global context so that they display caring behaviour towards the environment and utilize natural resources wisely and have tolerance for the diversity of national cultures. Of course, it also displays the behaviour of loving the country, being proud of being a nation, and being responsible for the integrity of the State.

Students can respond to this geographical awareness as an integrated spatial relationship between geographical phenomena such as climate, soil conditions, astronomical location, and population dynamics in an area. The general description of the physical condition of an area in the form of a stretch of land, water, and air as well as social, in which there are living and inanimate objects as well as symptoms that interact with one another is a manifestation of the national identity space This spatial (spatial) condition should be known and understood by students as part of society and the nation's next generation and should also be able to understand themselves as part of their national identity.

The plurality of physical and social phenomena should be a description of learning through spatial representation content. This perspective will produce geographic areas that characterize the similarities and differences of objects, phenomena, patterns, problems, potentials that exist in the earth's surface. This capability is an asset in carrying out the process of disaster risk reduction through attitudes, smart, wise and responsible actions in dealing with social, economic, ecological and disaster problems.

Conceptually, the results of this study indicate a systematic investment that learning programs containing spatial representations can build geographical awareness in building spatial thinking in understanding variations in objects, phenomena, potentials, problems in the earth's surface along with their attributes, characters and character colours, especially conditions physical and social interactions in it.

4. Conclusion

The results of the study found that geographic awareness formed by understanding spatial representations can become capital for disaster risk reduction through understanding the life space of students. The level of understanding of geography awareness through learning containing students' spatial representations consists of three levels, namely "low", "middle", and "high". Most of the students (85%) experienced increased understanding of geography awareness through learning containing spatial representations at the middle and high levels. Three factors of geographical awareness, namely: Care and responsiveness to environmental problems, being part of the solution to environmental problems (social and natural) and being proud of the country.

5. Acknowledgment

The researcher expressed his gratitude to the Faculty of Social Sciences, YSU for funding this research.

6. References

[1] S. Heffron, "GFL2! The updated geography for life: National geography standards, second edition," *The Geography Teacher,* p. 9 (2): 43–48, 2012.

[2] Uttal, D.H., Miller, D.I. and Newcombe, N.S. , "Exploring and enhancing spatial thinking: Links to achievement in science, technology, engineering, and mathematics?," *Current Directions in Psychological Science,* pp. 22 (5), 367–373, 2013.

[3] M. Ministry of National Education of Indonesia, *Regulation of ministry of national education number 23/2016,* Jakarta, 2016.

[4] A. Bonnet, What is geography?, Los Angeles: Sage Publication, 2008.

[5] I. Jo and S. W. Bednarz, "Dispositions toward teaching spatial thinking through geography: conceptualization and an exemplar assessment," *Journal of Geography,* pp. 113: 5. 198-207, 2014.

[6] M. Nursa'ban and R. Abe, "Pedagogical Content of Spatial Thinking Geography of Prospective Teachers in AUE and YSU," *Pedagogika / Pedagogy,* vol. 135, no. 3, pp. 185-199, 2019.

[7] M. Nursa'ban, K. Kumaidi and M. Mukminan, "Factors of Critical Spatial Thinking for a Geography Metacognition Assessment in Indonesian Senior High Schools," *Review of International Geographical Education (RIGEO),* pp. 10 (2), 186-204, 2020.

[8] D. Alderman, "Making every week about geography awareness and advocacy," AAG Newletter, 2017.

[9] Baiquni, "Menggugat (h) pemikiran & praktek pembangunan Berkelanjutan," in *Filsafat sains geografi*, Yogyakarta, Deepublish., 2013, pp. 160-186.

The Multi-Risk Analysis of Various Disasters in Karanganyar Regency

R Noviani[1], P Wijayanti[2], C Muryani[3], Ahmad[4], Sarwono[5], Sugiyanto[6] dan S Prihadi[7]

[1,2,3,4,5,6,7]Geography Education Study Program, Faculty of Teacher Training and Education, Universitas Sebelas Maret Surakarta, Indonesia

Email : ritanoviani@staff.uns.ac.id

Abstract. Karanganyar Regency is one of the districts in Central Java which prone of the natural disasters. Based on data Posdes 2018, it is known that the types of disasters that often occurred in Karanganyar Regency during the last three years, from 2015 to 2017 were landslides, floods, and drought. Therefore, it is important to know the multi-risk level of the three types of disasters. The method used in this research is scoring and weighting, as well as overlay between constituent parameters. Based on the results of research using the VCA matrix multiplication method, it is known that the high risk level covers 3,589.67 Ha or 4.46% areas that spread in 15 districts (except Jatipuro and Jumapolo); moderate risk level covers 34,624.94 Ha or 43.04% areas that spread in 16 districts (except Tasikmadu); The low risk level covers 42,254.98 Ha or 52.53% areas that spread in 7 districts (Colomadu, Gondangrejo, Tasikmadu, Matesih, Tawangmangu, Jumapolo, Jatiyoso).
Keywords: Landslide, Flood, Drought, Multi-Risk Disaster

1. Introduction

Karanganyar Regency is prone to disasters based on data from BNBP Central Java in 2015, Karanganyar Regency has a high potential for the threat of landslides, drought and technological failures from 12 disaster threats in Central Java Province. This is influenced by geological, geomorphological, climatological and social characteristics. The configuration of relief which varies greatly in wet tropical climate conditions is the reason why Karanganyar Regency prone to various kinds of both endogenous and exogenous disasters. These conditions are supported by the high population size and density, coburnplus low awareness and knowledge of disasters in Karanganyar Regency. This shows that the risk of natural disasters in Karanganyar Regency is quite high.

UNDP (2010) defines risk as the likelihood of dangerous consequences - loss of life, damage to property, loss of livelihoods, disruption of economic activity, and environmental damage - resulting from interactions between natural or human-induced hazards and vulnerable conditions. Risk assessment is the process of determining the nature and level of that risk, by analyzing hazards and evaluating existing conditions of vulnerability that together have the potential to harm people, property, services, livelihoods and the environment on which they depend.

It can be seen that the high risk of disasters that occur in an area, including Karanganyar Regency, is not only influenced by the many types of hazards that threaten, but also due to the increasing number of

Content from this work may be used under the terms of the Creative Commons Attribution 3.0 licence. Any further distribution of this work must maintain attribution to the author(s) and the title of the work, journal citation and DOI.
Published under licence by IOP Publishing Ltd

people who are vulnerable to the threat of disasters and the low community capacity in dealing with disasters [6], [13].

This condition requires Karanganyar Regency to carry out disaster risk reduction management which is more targeted, integrated, and comprehensive. It can be supported by a disaster risk assessment in Karanganyar Regency, because disasters cannot be avoided, but their impact can be minimized using the right information available at the right time. One source of information can be used is a map of disaster-prone areas and levels of disaster risk. It can be the basic basis for supporting the implementation of disaster management [15].

Disaster management is taking immediate and appropriate actions to deal with disasters and their components, their consequences under pressure and time due to hazards that are not clearly defined and lack of adequate and accurate information and other resources. Therefore, disaster management needs to be well planned. Disaster management and prevention differ according to the type and size of the disaster and its severity. Planning must include strategic planning, determining specific objectives, environmental strategic analysis, using prediction tools related to crisis types and hazard analysis, coordination of efforts and design of effective organizational structures [17].

Through a risk assessment and capacity to deal with disasters, including hazard and vulnerability assessments, disaster management directives can be developed, especially disaster mitigation and preparedness [2], [3], [5]. Disaster mitigation is a term used to refer to all actions to reduce the impact of a disaster that can be taken before it occurs, including long-term preparedness and risk reduction measures. This includes planning and implementing measures to reduce risks associated with known natural and man-made hazards, and planning processes for effective responses to disasters that do occur (Coburn et al, 1994; Mohamed, 2019).

Considering the importance of this in reducing the impact of disasters, carrying out disaster risk analysis is an important part in disaster management activities. Disaster risk analysis is a research activity and a study of activities that allow disasters to occur [14]. As for the Regulation of the Minister of Energy and Mineral Resources No. 15 of 2011 concerning Guidelines for Volcanic Disaster Mitigation, Land Movement, Earthquakes and Tsunamis, states that one of the considerations in disaster risk assessment is the result of an analysis of disaster-prone areas [11].

However, disaster risk assessment has only covered one type of risk from one disaster. There have not been many studies that combine the various risks from many disaster threats or what can be called the term multicaster risk, namely the combination of various types of disaster risks generated by various types of hazards (multihazard). The assessment of various disaster risks in an integrated manner is known as a multirisk assessment [7].

Based on the potential for several types of disasters that have occurred in Karanganyar Regency, research on the multiple risks of various disasters in Karanganyar Regency is important considering that Karanganyar Regency really needs this information as one of the important steps in disaster risk reduction to be applied in district development planning. The information on multi-risks can be useful for reducing risks not only from one type of disaster, but also for another increasing risk of other hazards. Therefore, this research was conducted by focusing on several problems, the characteristics and levels of vulnerability to various disasters in Karanganyar Regency, the capacity of the people of Karangnyar Regency in facing disaster-prone, and the multi-risk distribution pattern of various disasters in Karanganyar Regency.

2. Method

This research are using descriptive quantitative. Data analysis technique used to measure the level of disaster risk by scoring to determine the high, medium and low risk of a disaster and calculated using the equation:

$$R = HxV/C$$

with the equation identity:
R = Risk

H = Hazard
V = Vulnerability
C = Capacity

Hazard Mapping (H)
The disaster threat map is created based on data and records of events that have occurred in Karangnyar Regency during the last three years, It is 2015, 2016 and 2016, and were obtained from the Village Potential data of Central Java Province in 2018.

Vulnerability Mapping (V)
There are three disaster vulnerability analyzed in this study:

Landslide Vulnerability. The parameters to determine the level of landslide vulnerability can be seen in Table 1.

Table 1. Landslide hazard parameters

No	Parameter	Weight
1	Slope	0,30
2	Rainfall	0,20
3	Geology	0,30
4	Land Use	0,20

Source: Paripurno et al, 2006 *in* [4]

Flood Vulnerability. The parameters to determine the level of flood vulnerability can be seen in Table 2.

Table 2. Flood hazard parameters

No	Parameter	Weight
1	General Flood Area	0,25
2	Rainfall	0,25
3	Elevation	0,25
4	Land Use	0,25

Source: Paripurno et al, 2006 *in* [4]

Drought Vulnerability. The parameters to determine the level of drought vulnerability can be seen in Table 3.

Table 3. Drought hazard parameters

No	Parameter	Weight
1	Rainfall	0,33
2	Land Use	0,27
3	Water Sources	0,20
4	Types of Soil	0,13
5	Elevation/Slope	0,7

Source: [9]

Capacity Mapping (C)
The parameters used to determine disaster capacity which based on PERKA BNPB No. 2 of 2012 which consists of several parameters as listed in Table 4 [1], the data of which are obtained from data Posdes (the Village Potential) of Central Java Province in 2018.

Table 4. Disaster capacity parameters

Capacity Parameters	Weight (%)	Capacity Class					
		Low	Value	Medium	Value	High	Value
Number of Health Workers	25	<10 person	1	10-20 person	3	>20 person	5
Number of Health Facilities	25	<10 unit	1	10-20 unit	3	>20 unit	5
Disaster Socialization	25	None	0	-	-	Exist	3
Disaster Anticipation Effort	25	None	0	-	-	Exist	3

Source: [1]

Disaster Risk Mapping
The calculation of the score and classification of disaster risk from the parameters of hazard, vulnerability and capacity using the VCA matrix multiplication method is carried out according to Table 5.

Table 5. VCA Multiplication Matrix

V/C		Capacity (C)		
		High	Medium	Low
Vulnerability (V)	Low	Low	Low	Medium
	Medium	Low	Medium	High
	High	Medium	High	High
Risk H*(V/C)		V/C		
		Low	Medium	High
Hazard (H)	Low	Low	Low	Medium
	Medium	Low	Medium	High
	High	Medium	High	High

Source: [4]

Multi-Risk Disaster Mapping
After generating the risk map for disasters, landslide, floods, and drought, the merging process is carried out by multiplying the matrix according to Tables 6 and 7.

Table 6. Multi-Risk Floods and Landslide VCA Matrix

Multi-Risk Disaster		Landslide Risk		
		Low	Medium	High
Flood Risk	Low	Low	Low	Medium
	Medium	Low	Medium	High
	High	Medium	High	High

Source: [4]

Table 7. Multi-Risk Flood, Landslide and Drought VCA Matriks

Multi-Risk Disaster		Drought Risk		
		Low	Medium	High
Flood and Landslide Risk	Low	Low	Low	Medium
	Medium	Low	Medium	High
	High	Medium	High	High

Source: Researcher analysis, 2020

3. Result and Discussion

Karanganyar Regency is one of the districts in Central Java that astronomically located at coordinates 110⁰43'38"-111⁰11'24" East Longitude and 7⁰6'17 "-7⁰46'07" South Latitude, while geographically, Karanganyar Regency located at the west of Mount Lawu slope.

The risk level of landslide, floods, and drought, and also the multi-risks of the three disasters in Karanganyar Regency can be explained as follows:

Hazard Map

The disaster threat map is created by using data and records of disaster events that have occurred in Karanganyar Regency during the last three years, It is 2015, 2016 and 2016, Based on the disaster event data, it is known that the type of disaster that most often occurs in Karanganyar Regency is landslides, hemorrhoid, and drought.

Landslide Hazard Map. During the last three years the intensity of landslide hazards that occurred in most areas of Karangnyar Regency was classified as low, with a percentage covering 94.91% of the area, while those classified as moderate were 3.39% of the area and those classified as high were 1.69% of the area, as shown at Figure 1.

IOP Conf. Series: Earth and Environmental Science **884** (2021) 012023 doi:10.1088/1755-1315/884/1/012023

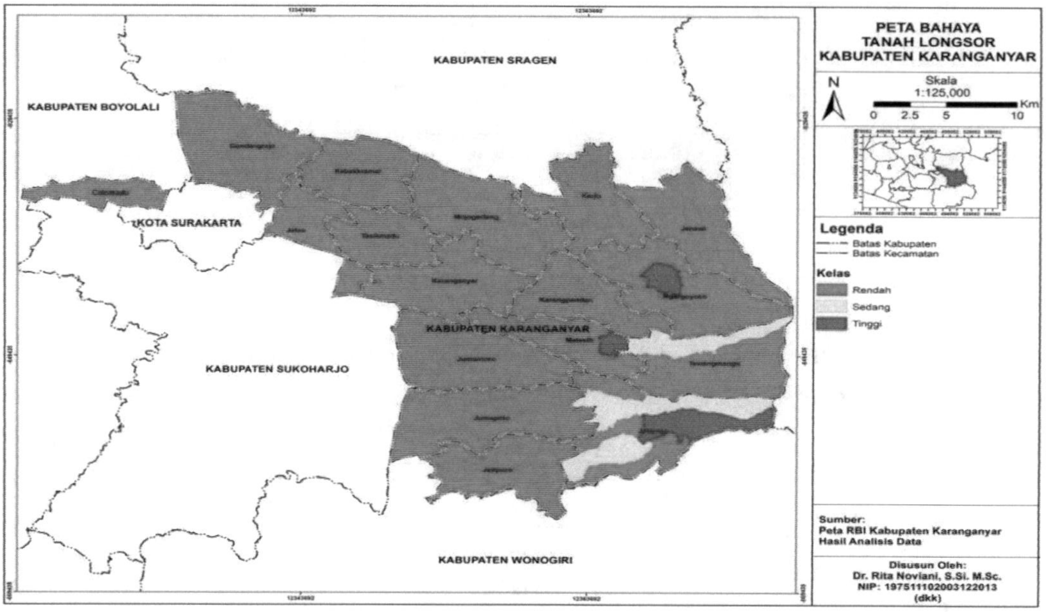

Figure 1. Landslide Hazard Map

Flood Hazard Map. During the last three years, the intensity of the flood hazard that occurred in most areas of Karangnyar Regency was classified as low, with a percentage covering 93.22% of the area, while those classified as moderate were 5.65% of the area and those classified as high were 2.25% of the area, as shown at Figure 2.

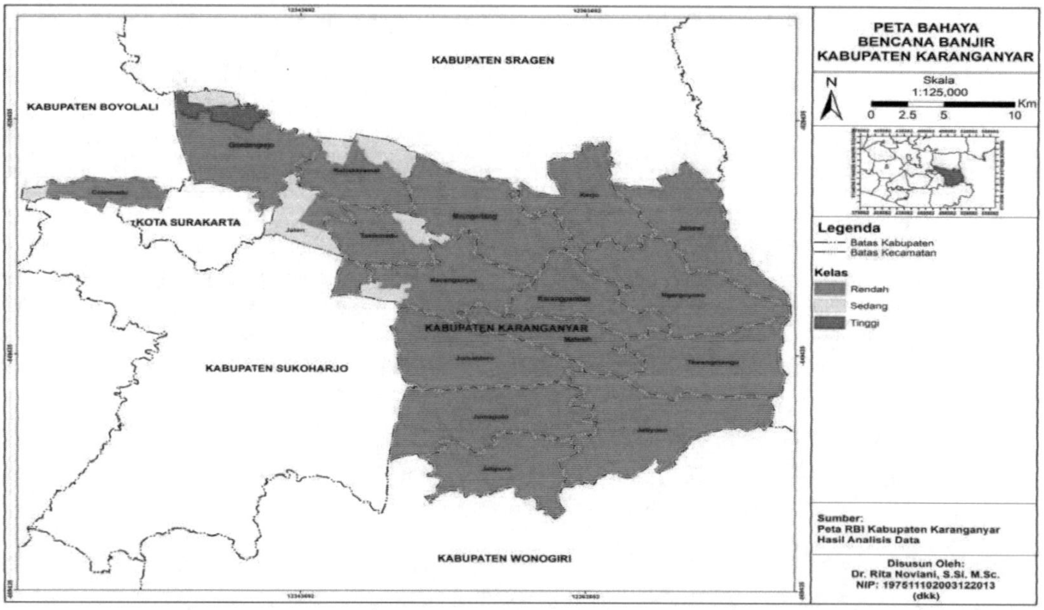

Figure 2. Flood Hazard Map

Flood Hazard Map. During the last three years, the intensity of drought in most areas of Karangnyar Regency is low, with a percentage covering 97.74% of the area and 2.26% of the area being classified as high, as shown at Figure 3.

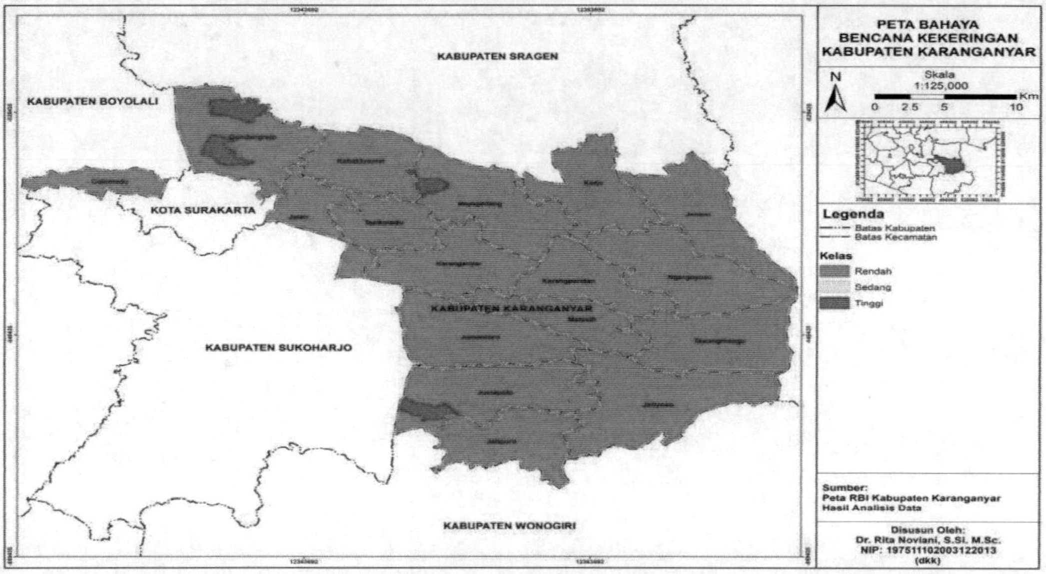

Figure 3. Drought Hazard Map

Vulnerability Map

The vulnerability of landslides, floods, and drought disaster in Karanganyar Regency are as follows:

3.2.1 Landslide Vulnerability Map. Results of weighting and overlapping landslide parameters consisting of slope, rainfall, geology, and land use shows that 8.04% of Karangnyar Regency is in the high hazard category, 71.84% is included in the hazard category moderate, and 19.11% is included in the low hazard category, as seen in Figure 4. The factor that mostly influenced this condition is slope, which 47.00% of the area in Karangnyar Regency has a slope of 8-15% (gentle) and conditions. geology, most of which are included in the Lahar Lawu (Qlla) formation, with andesite, basalt and a few pumice components of various sizes mixed with volcanic sand. This formation is indeed scattered, especially filling the plain area at the foot of the volcano or forming several low hills (Samodra, 1997 in [8]).

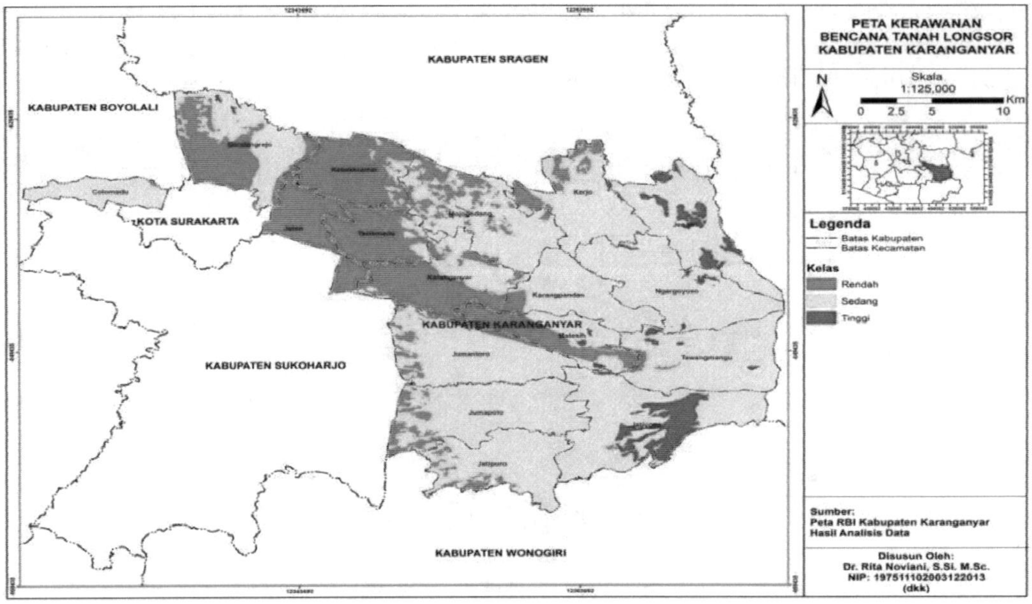

Figure 4. Landslide Vulnerability Map

3.2.2 Flood Vulnerability Map. Based on the weighting and overlaying results of flood parameters consisting of general flood zones, rainfall, altitude, and land use, it can be seen that 27.15% of Karangnyar Regency is in the high hazard category, 64.16% is included in the category moderate vulnerability, and 7.67% are included in the low hazard category, as seen in Figure 5. This condition is most influenced by the high rainfall in Karanganyar Regency (2000-2500 mm / year), as well as 37.92% land use. its territory is in the form of irrigated rice fields.

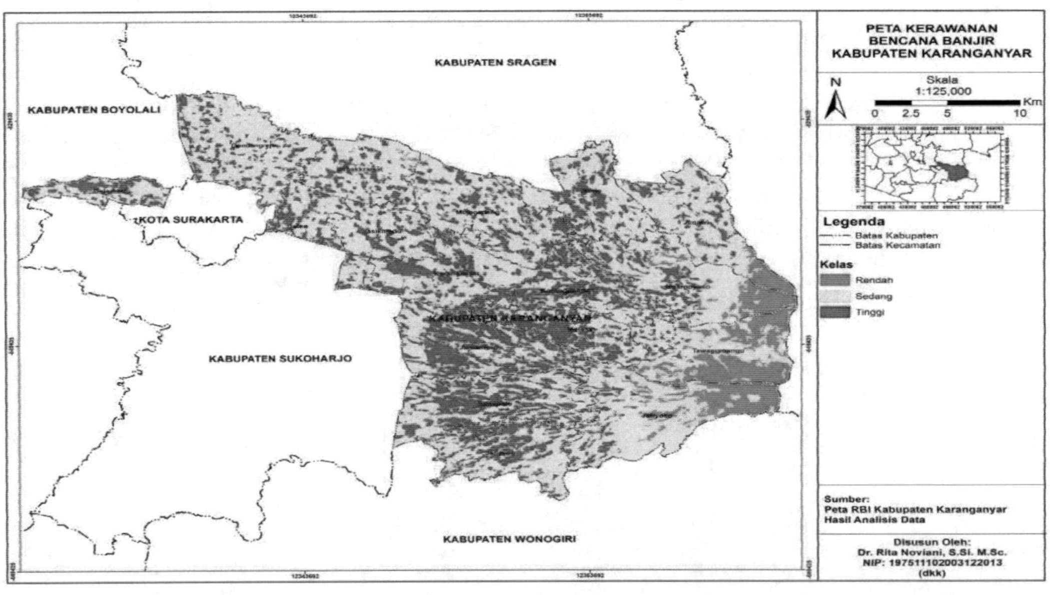

Figure 5. Flood Vulnerability Map

3.2.3 Drought Vulnerability Map. Based on the weighting and overlaying results of drought parameters which consist of rainfall, land use, water sources, soil type, and slope, it can be seen that 20.99% of Karanganyar Regency is in the high vulnerability category, 52.37% is included in moderate hazard category, and 26.59% included in the low hazard category, as seen in Figure 6. This condition is most influenced by the high rainfall in Karanganyar Regency (2000-2500 mm / year), the land use is 37.92%. Its territory is in the form of irrigated rice fields, and is also influenced by the existence of water sources. Karanganyar Regency is included in Lahar Lawu formation, one of which is characterized by the discovery of many water sources (Samodra, 1997 in [8]).

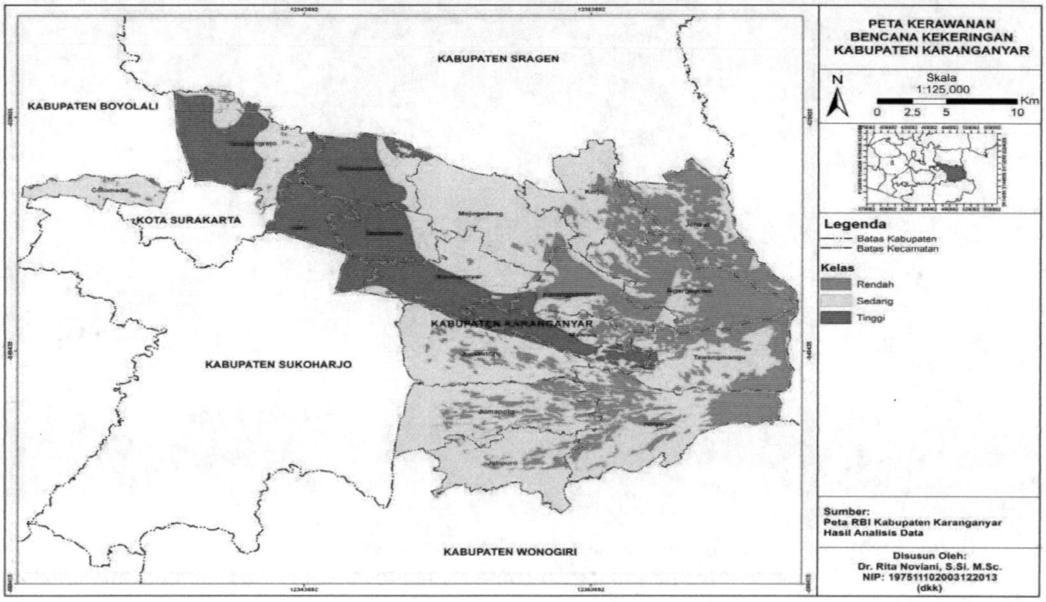

Figure 6. Drought Vulnerability Map

Disaster Capacity Map

Based on the weighting and overlaying results of disaster capacity parameters consisting of the number of health workers, the number of health facilities, socialization and disaster anticipation efforts, it is known that 75.71% of the area covering 134 villages is in the low capacity category, 19.21 % of areas covering 34 villages are in the medium capacity category, and 5.09 & areas covering 9 villages are included in the high capacity category, as seen in Figure 7. The factor that most influences village areas with high disaster capacity is the number of health facilities (> 20 people) and health workers (> 20 people) who are also classified as high. Meanwhile, there are no factors for socialization of disasters and efforts to anticipate disasters in most areas.

IOP Conf. Series: Earth and Environmental Science **884** (2021) 012023 doi:10.1088/1755-1315/884/1/012023

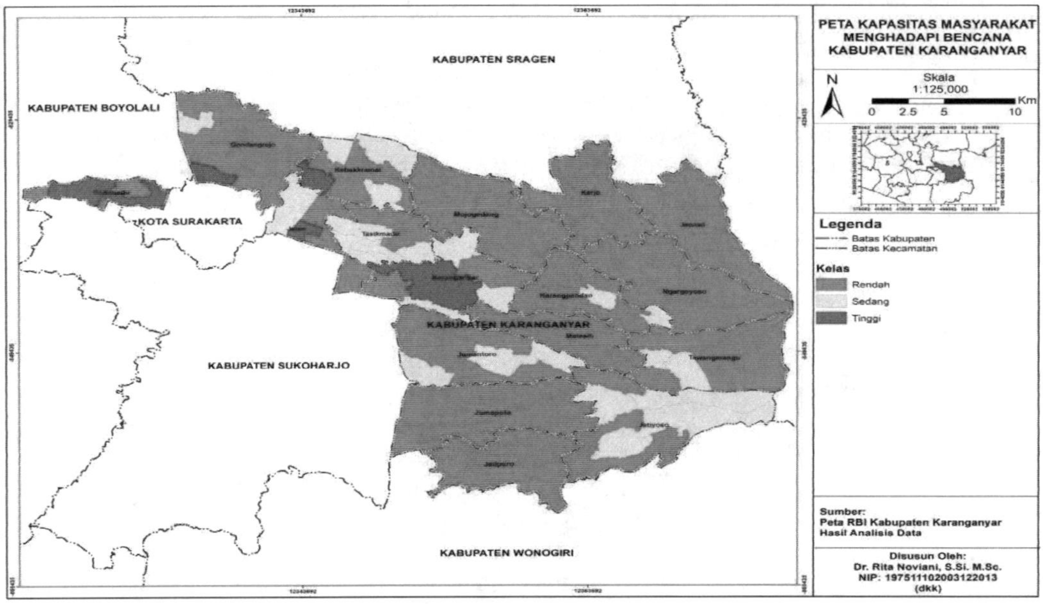

Figure 7. Community Capacity Map of Karanganyar Regency in Facing Disaster

Disaster Risk Map

Landslide Risk Map. Karanganyar Regency Landslide risk map which was generated from the analysis of hazard, vulnerability and capacity mapping using the VCA matrix multiplication method and the results were as listed in Table 8.

Table 8. Results of Landslide Risk Map Area

Risk	Matrix VCA	
	Large (ha)	Persentage (%)
High	6.456,47	8,03%
Medium	51.649,55	64,21%
Low	22,363,56	27,80%

Source: Researcher analysis, 2020

Based on the percentages listed in Table 8, and visualized in Figure 8, It shows that 64.21% of the area in Karangnyar Regency is at moderate risk of landslides. Areas with a high risk of landslides with a percentage of 8.03% are spread over four sub-districts, Jatiyoso and a small part of Tawangmangu, Matesih, and Ngargoyoso. This is because the threat of landslides in most areas of Jatiyoso Subdistrict is classified as moderate and high, with moderate and high levels of vulnerability, but the capacities are classified as medium and low, It means that theoretically it proves the equation R = H x V / C, that with high danger and vulnerability, while the capacity is low, the disaster risk of an area is high.

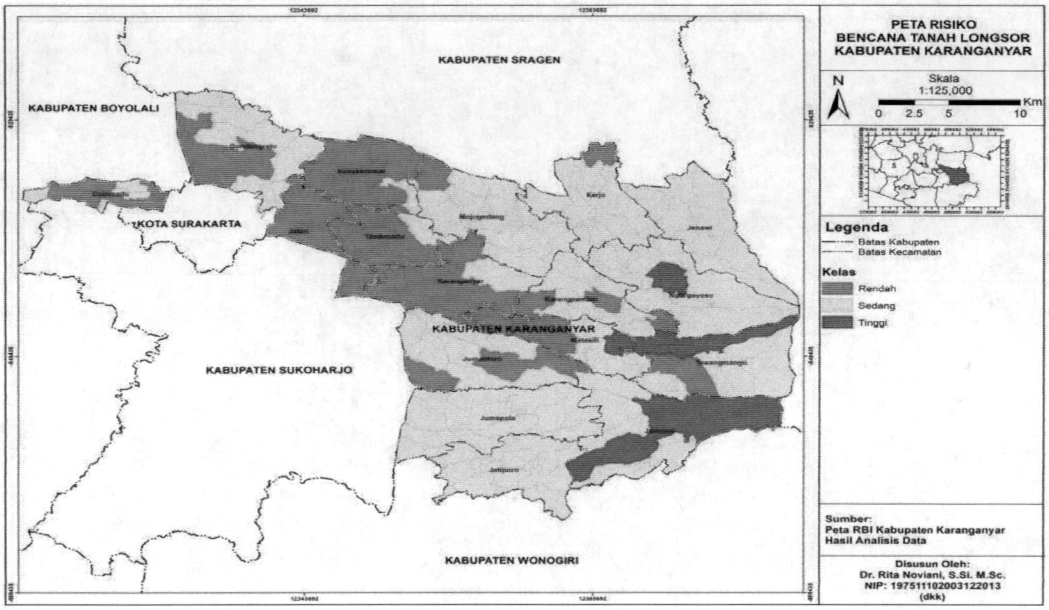

Figure 8. Landslide Risk Map

Flood Risk Map. The flood disaster risk map in Karanganyar Regency is generated from the analysis of the hazard, vulnerability and capacity mapping using the VCA matrix multiplication method and the results are as listed in Table 9.

Table 9. Result of Flood Risk Map Area

Risk	Matrix VCA	
	Large (ha)	Large (ha)
High	3.130,86	3,89%
Medium	5.4237,70	67,43%
Low	2.3101,02	28,72%

Source: Researcher analysis, 2020

Based on the percentages listed in Table 9, it can be visualized in Figure 9, which shows that 67.43% of the area in Karangnyar Regency has a moderate flood risk. Areas with high flood risk with a percentage of 3.89% are spread over a small area of the five sub-districts, namely Gondangrejo, Colomadu, Jaten, Karanganyar, and Tasikmadu. Of the five sub-districts, most Vast are scattered in the northern part of Gondangrejo District with the threat of flooding that is classified as Medium and High, with the level of vulnerability classified as Medium and High as well, but the capacity level is classified as Low, resulting in a High flood level.

IOP Conf. Series: Earth and Environmental Science **884** (2021) 012023 doi:10.1088/1755-1315/884/1/012023

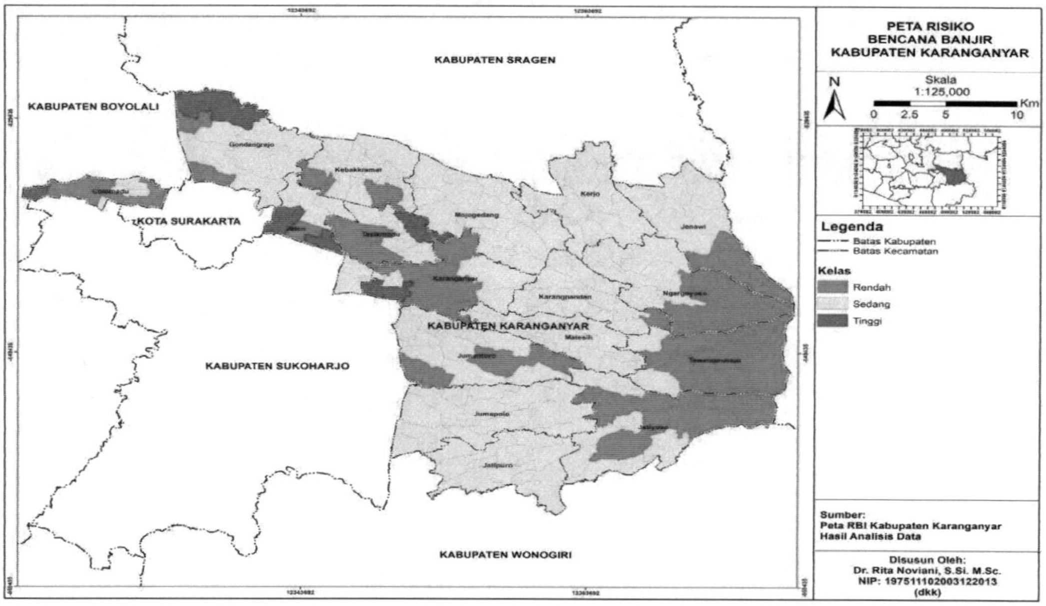

Figure 9. Flood Risk Map

Drought Risk Map. The drought disaster risk map in Karanganyar Regency is generated from the analysis of hazard, vulnerability and capacity mapping using the VCA matrix multiplication method and the results are as listed in Table 10.

Table 10. Results of Drought Risk Map Area

Risk	Matrix VCA	
	Large (ha)	Large (ha)
High	1.554,77	1,93%
Medium	50.232,88	62,45%
Low	28.681,92	35,66%

Source: Researcher analysis, 2020

Based on the percentages listed in Table 10, it can be visualized in Figure 10, which shows that 62.45% of the area in Karanganyar Regency has a moderate risk of drought. Region with a high risk of dryness with a percentage of 1.93% is spread across two districts, Gondangrejo and Jumapolo. Gondangrejo, with threats classified as Low and High, and the level of vulnerability that is classified as Medium and High as well, but the capacity level is classified as Low, resulting in a High flood level.

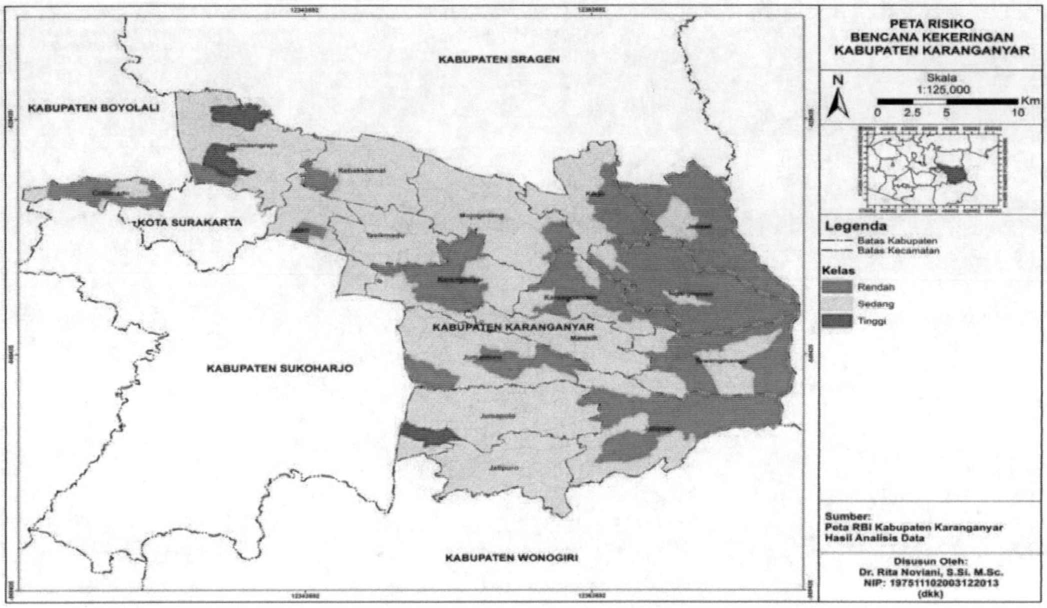

Figure 10. Drought Risk Map

Disaster Multi-Risk Map

The results of making a risk map of landslides, floods and drought were then overlaid again using the multiplication matrix method listed in table 7, it is known that the areas in Karanganyar Regency with a High Risk level are multi risk of landslides, floods, and drought. 3,589.67 Ha, or 4.46% spread over 15 districts (except Jatipuro and Jumapolo) with the distribution in 9 villages. Areas in Karanganyar Regency with a Risk Medium level of risk of landslides, floods, and drought 34,624.94 Ha, or 43.04% spread over 16 sub-districts (except Tasikmadu) with a distribution to 78 villages. As for the rest, the areas in Karanganyar Regency with a Low Risk level for multiple risks of landslides, floods, and drought at 42,254.98 Ha, or 52.53% spread over 7 districts (Colomadu, Gondangrejo, Tasikmadu, Matesih, Tawangmangu, Jumapolo, Jatiyoso) with distribution in 90 villages. A multi risk map of landslides, floods and drought in Karanganyar Regency can be seen in Figure 11.

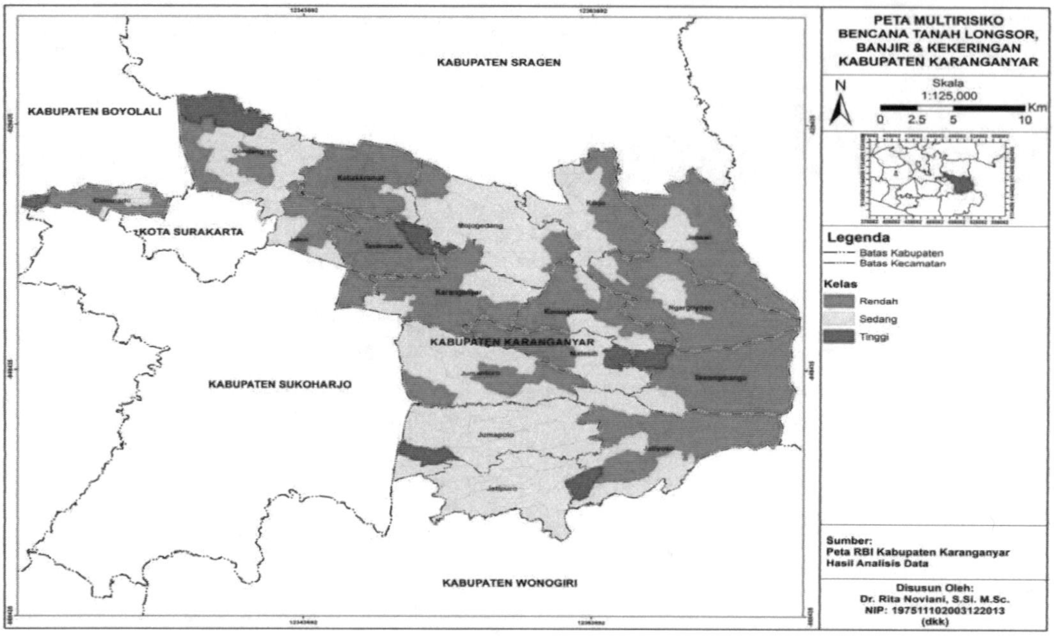

Figure 11. Multi-Risk Map of Landslides, Floods, and Drought Disasters in Karanganyar Regency

4. Conclusion

The conclusion of this research are:

1. The hazards of landslides, floods, and drought in most areas of Karanganyar Regency are classified as low with 94.91% each; 93.22%; and 97.74%
2. The vulnerability of landslides, floods, and drought in most areas of Karanganyar Regency is classified as moderate with 71.84% each; 64.16%; 52.37%.
3. Community capacity of Karanganyar Regency in facing disasters in most areas is classified as low with a percentage of 75.71% covering 134 villages.
4. Disaster risk mapping in Karanganyar Regency obtained from overlaying hazard maps, hazard maps, and capacity maps shows that the risk of landslides, floods and drought in most areas of Karanganyar Regency is moderate with a percentage of 64 , 21%; 67.43%; 62.45%.
5. The Multi-risk disaster map obtained from overlaying risk maps of landslides, floods and drought, it is known that:
   a. The high risk level covers an area of 3,589.67 hectares or 4.46%, which is spread over 15 sub-districts (except Jatipuro and Jumapolo) with a distribution in 9 villages.
   b. The moderate risk level covers an area of 34,624.94 Ha or 43.04%, spread over 16 sub-districts (except Tasikmadu) with the distribution to 78 villages.
   c. The low risk level covers an area of 42,254.98 Ha, or 52.53% spread over 7 districts (Colomadu, Gondangrejo, Tasikmadu, Matesih, Tawangmangu, Jumapolo, Jatiyoso) with distribution in 90 villages.

5. Acknowledgments

Thanks to Sebelas Maret University for providing research funding assistance, and also thanks to all parties who have provided technical assistance, so that this research can run well.

References

[1] BNPB 2012 *Peraturan Kepala Badan Nasional Penanggulangan Bencana Nomor 02 Tahun 2012*

Tentang Pedoman Umum Pengkajian Risiko Bencana (Jakarta: BNPB)

[2] Eiser J R, Bostrom A, Burton I, Johnston D M, McClure J, Paton D, Pligt J V D, White M P 2012 Risk Interpretation and Action: A Conceptual Framework for Responses to Natural Hazards *Inter. J. of Disaster Risk Reduction.* **1** 5-16

[3] Flanagan B E, Gregory E W, Halisey E J, Heitgerd J L, Lewis B 2011 A Social Vulnerability Index for Disaster Management *J. of Homeland Security and Emergency Management* **8 (1)** 1-22

[4] Gunadi B A J, Nugraha A L, Suprayogi A 2015 Aplikasi Pemetaan Multirisiko Bencana Di Kabupaten Banyumas Menggunakan Open Source Software GIS *J. Geodesi UNDIP* **4 (4)** 287-96

[5] Kaku K and Held A 2013 Sentinel Asia: Space-Based Disaster Management Support System In The Asia-Pacific Region *Inter. J. of Disaster Risk Reduction* **6** 1-17

[6] Lavigne F 2010 Penaksiran Multirisiko Bencana di Wilayah Parangtritis: Suatu Analisis Serbacakup untuk Membangun Kepedulian Masyarakat Terhadap Berbagai Kejadian Bencana *Ulasan Publikasi* ed Sunarto, Marfai M A, and Mardiatno D (Yogyakarta: PSBA UGM)

[7] Mardiatno D 2014 Penaksiran Resiko Multi Bencana di Wilayah Kepesisiran Parangtritis, Suatu Analisis Serbacakup Untuk Membangun Kepedulian Masyarakat Terhadap Berbagai Kejadian Bencana (Yogyakarta: PSBA UGM)

[8] Pambudi D, Sakur M, Ismail K, Dwiyono, Setiadji L D 2014 Delineasi Daerah Prospek Panas Bumi Berdasarkan Kelurusan Citra Landsat Dan Digital Elevation Model (Dem) Daerah Gunung Lawu, Provinsi Jawa Tengah Dan Jawa Timur *Proc. Semnas Kebumian Ke-7* (Yogjakarta: Jurusan Teknik Geologi Fakultas Teknik UGM) M2P-04 438-46

[9] Permadi M G and Adiputra A 2019 Kajian Risiko Kekeringan Di Kabupaten Cianjur *JGEL* **3 (1)** 34-44

[10] Peraturan Pemerintah Republik Indonesia Nomor 21 Tahun 2008 Tentang Penyelenggaraan Penanggulangan Bencana 2008 *LNRI Tahun 2008* Nomor 42

[11] Sagala H dan Yasaditama H I 2012 Analisis Bahaya dan Resiko Bencana Gunungapi Papandayan, Studi Kasus Kecamatan Cisurupan Kabupaten Garut *Forum Geografi* **26 (1)** 1-16

[12] Sudibyakto 1997 Manajemen Bencana Alam Dengan Pendekatan Multidisiplin: Studi Kasus Bencana Gunung Merapi *Majalah Geografi Indonesia* **12 (22)** 31-41

[13] Sudibyakto 2007 Potensi Bencana Alam Dan Kesiapan Masyarakat Menghadapi Bencana (Preparedness for Vulnerable Communities) *Pengantar Diskusi Bulanan* (Yogjakarta: PSPK UGM)

[14] Sunarto 2011 Standard Operating Procedure (SOP) Mitigasi Bencana *Proc. Semiloka Nasional Urgensi Pendidikan Mitigasi Bencana* (Yogjakarta: Fakultas Ilmu Sosial dan Ekonomi UNY) 15-16

[15] Sunarto and Rahayu L 2006 Fenomena Bencana Alam di Indonesia *J. Kebencanaan Indonesia* **1 (1)** 1-5

[16] Sunarto, Marfai, Aris M, dan Mardiatno, Djati 2010 Penaksiran Resiko Multi Bencana di Wilayah Kepesisiran Parangtritis, Suatu Analisis Serbacakup Untuk Membangun Kepedulian Masyarakat Terhadap Berbagai Kejadian Bencana (Yogyakarta: PSBA UGM)

[17] Vitalii S, Khorram-Manesh A, and Nyberg L 2017 *Handbook of Disaster and Emergency Management* ed Khorram-Manesh A (Sweden: Kompendiet) Chapter 2 p 23-29

Constructing Student's Physical Preparedness through Game-Based Activities in Disaster-Safe School

S Nopembri[1], S T Maryana[1], Saryono[1], and D S Purnama[2]
[1]Department of Physical Education, Faculty of Sport Sciences, Universitas Negeri Yogyakarta, Colombo Road Number 1, Yogyakarta, Indonesia
[2]Department of Guidance and Conselling, Faculty of Education, Universitas Negeri Yogyakarta, Colombo Road Number 1, Yogyakarta, Indonesia.

Correspondence: soni_nopembri@uny.ac.id

Abstract. Indonesia is a country that has great potential for disaster. A major earthquake hit the Special Region of Yogyakarta, especially Bantul Regency, in 2006. This has demanded the establishment of a disaster-safe school that is ready to anticipate disasters. Disaster-safe school students need to be fully prepared in various aspects, including physical aspect. Therefore, this study aims to determine the increase in physical fitness before and after carrying out game-based physical activities for upper grade students in a Disaster-Safe School. This study adopted an experimental method that involved one group pre-test-post-test design. Participants were upper grade students (grade 4 to 5) in Madrasah Ibtidaiyah Negeri 2 Bantul, totaling 74 people under such criteria as aged 10 to 12 years and following the complete treatment. The instrument used was the Indonesian Physical Fitness Test (IPFT) for children aged 10-12 years. Data analysis using paired sample t-test. The results showed that there was an increase in the physical fitness of students after being given treatment in the form of game-based physical activity for 12 meetings. Good physical fitness is a form of physical preparedness in facing various emergencies including disasters. The better physical fitness level of students, the more physically prepared they will be when facing disasters.

1. Introduction

Indonesia is a disaster-prone country. Likewise, Special Region of Yogyakarta, Bantul Regency in particular, is among the most-disaster prone regions. A tectonic earthquake of 5.8 – 6.2 SR that once hit the area on May 27, 2006 is a highly risky disaster [1]. One of preventive measures taken by the government to mitigate the risks of disasters includes implementation of disaster-safe school program. Disaster-safe schools are schools that are committed to implementing the culture of safety and health, aware of risks, and fully equipped with pre-, during, and post-disaster plans as well as prepared to respond to emergencies and disasters [2]. Nevertheless, in fact, disaster-safe school has merely been a status due to the fact that schools solely focus on development of facilities while preparedness of students has remained untouched.

Children are often the most prone to disaster impacts as their lack of understanding of the risks around them often leads to absence of preparedness in coping with disasters. Meanwhile, it is obvious that great potential of disaster comes with greater needs for anticipative measures. Education has been seen as the most effective way to provide understanding about disaster in an effort to nurture disaster preparedness [3]. To build schools' disaster preparedness and raise awareness of all elements, disaster risk reduction should be made part of education in either individual or collective level in school setting and integrated into both intracurricular and extracurricular curriculum in formal education institution [4].

Integration of disaster preparedness into school subject can be done through, for example, Physical Education subject that aims at improving quality of human lives, both physically and spiritually. Physical Education can serve as a tool to nurture disaster preparedness that includes physical preparedness (through development of motor skill and physical ability, knowledge, and reasoning as well as formation of healthy living habit to stimulate balance growth and development through physical

Content from this work may be used under the terms of the Creative Commons Attribution 3.0 licence. Any further distribution of this work must maintain attribution to the author(s) and the title of the work, journal citation and DOI.
Published under licence by IOP Publishing Ltd

activities and sports [5]. Sound physical preparedness manifests itself in physical fitness. However, as a matter of fact, the level of physical fitness in school remains low. A study reported that physical fitness of grade IV and V students in disaster prone zones was poor [5]. It is apparent that such condition is far from ideal for mitigating disaster risks through physical preparedness.

Physical preparedness can be improved through physical education learning. However, as physical fitness at school remains poor, improvement efforts through learning management and improvement of physical fitness of students as the main goal are necessary [7]. There were obstacles in the implementation of physical education at school such as the lack of time allotted for learning, the lack of facilities and infrastructure available, teachers' lack of skills in managing classroom as well as monotonous materials given in the class. Facing such conditions, students do not take PE subject seriously because of serious boredom they experience, making it difficult to improve their physical fitness.

Physical fitness is a manifestation of physical preparedness that requires model and learning variation based on needs for handling emergency situations. Game-based physical activities implemented in physical education can be an alternative that suits the characteristics of the students. Activities that involve games and movements bring relatively permanent effects. Playing games will improve physical, psychological, and social qualities of children besides increasing their physical fitness and motor skill while developing their attitude and behavior to build an active lifestyle. Providing students with some game-based physical activities to train their physical and motor skills by incorporating movements that demonstrate physical fitness components will affect children's physical fitness. The games should focus on children's physical endurance training in handling emergency situations so that they are physically prepared to mitigate disaster risks. Therefore, this research aims to examine improvement of students' physical fitness through game-based physical activities in disaster-safe school setting.

2. Method

2.1. Design

This research was an experimental research that adopted One Group Pre-test-Post-test Design. In a One-Group-Pre-test-Post-test experiment, a single group of participants is pretested before receiving particular treatment. After certain period of time, this group of people will undergo another test (post-test). Treatment results are considered more accurate as conditions before and after treatment are compared [8]. Steps in this research included determination of research subject (a total of 74 from grade 4 to 6 students), pre-test using Indonesian Physical Fitness Test (*Tes Kesegaran Jasmani Indonesia*) instrument for children aged 10-12 years, provision of treatment in form of game-based physical activities through PE subject every once a week (according to classroom schedule) for 12 meetings from July to November 2019, and a post-test using the same instrument.

2.2. Participants

Table 1. Details of the Number of Participants based on Grades and Gender

No	Grade	Male	Female	Total
1.	4	15	16	31
2.	5	9	12	21
3.	6	17	5	22
	Total	41	33	74

This research took place in a disaster-safe school namely State Madrasah Ibtida'iyah 2 Bantul (commonly known as MIN Kebonagung) from July to November 2019. Research participants were 74 grade 5-6 students selected based on criteria such as age (10 to 12 years old), full participation in pre-test and post-test as well as full participation in 12 treatment sessions. The characteristic of participants can be seen in table 1.

2.3. Data Collection

Data were collected using test and measurement techniques. Instrument used in this research was Indonesian Physical Fitness Test (IPFT) for children aged 10 to 12 years old [9]. Test validity for male and female participants was 0.884 and 0.897 respectively. Meanwhile, reliability for male and female participants was 0.911 and 0.942 respectively. IPFT comprises of 40-meter sprint, 60-second pull up, 30-second sit up, vertical jump, and 600-meter running. The 40-meter sprint aims at measuring speed using time unit (second). Pull up, on the other hand, is to measure strength and endurance of arm and shoulder muscles in seconds. Vertical jump is to measure explosive power in centimeters (cm). Meanwhile, sit-up measures strength and endurance of abdominal muscles by counting the number of movements within 60 seconds. Lastly, the 600-meter running aims to measure heart and lung endurance, blood circulation and breath in time unit (minutes and seconds). Students' achievement from each item was converted into scores ranging from one to five before being totaled and classified into test norms as presented in Table 2.

Table 2. Norms for Indonesian Physical Fitness Test (IPFT) for Children Aged 10 – 12 Years [9]

Total Score	Classification
22 – 25	Very Good
18 – 21	Good
14 – 17	Fair
10 – 13	Poor
05 – 09	Very Poor

2.4. Data Analysis

Data that compared physical fitness in both pre-test and post-test were analyzed using paired sample t-test. As data analysis required the use of normally distributed data, normality and homogeneity tests were performed [10]. Performance of normality and homogeneity tests aimed to generate better analysis results. Data analysis was performed using SPSS 20 for Windows.

3. Results

Table 3. Descriptive Statistics for Pre-Test and Post-Test for Students' Physical Fitness

Statistics	Pre-test	Post-test
N	74	74
Mean	12.80	14.34
Median	13.00	14.00
Mode	11.00	13.00
Std. Deviation	3.53	1.75
Minimum	6.00	11.00
Maximum	22.00	19.00
Sum	947.00	1061.00

Results of descriptive statistics for pre-test and post-test data on students' physical fitness are presented in Table 3. From results of normality test using Kolmogorov-Smirnov formula, it is known that both pre-test and post-test data were normally distributed ($p > 0.05$). Meanwhile, results of homogeneity test indicated homogeneity of pre-test and post-test data ($p > 0.05$). As results of assumption tests for both pre-test and post-test data indicated normal distribution and homogeneity of data, data analysis could be proceeded. Results of paired sample t-test demonstrate significant differences in students' physical fitness between pre-test and post-test or after performing game-based physical activities within 12 meetings ($t=4.224$, $p=0.000$). Students demonstrated increase in physical fitness after receiving treatment as indicated by 12.03% increase ($M_{pre}=12.80 < M_{post}=14.34$). It can be seen in table 4.

Table 4. Summary Analysis of Students' Physical Fitness in Pre-Test and Post-Test

Physical Fitness	Mean	Paired Sample t-test		
		t	*df*	*p*
Pre-test	12.80	4.224	73	0.000
Post-test	14.34			

Students' physical fitness improved after receiving treatment during the 12 game-based physical activity training sessions. Referring to the given data, differences in physical fitness of upper grade students at MIN 2 Bantul Year 2019/2020 can be presented in figure 1.

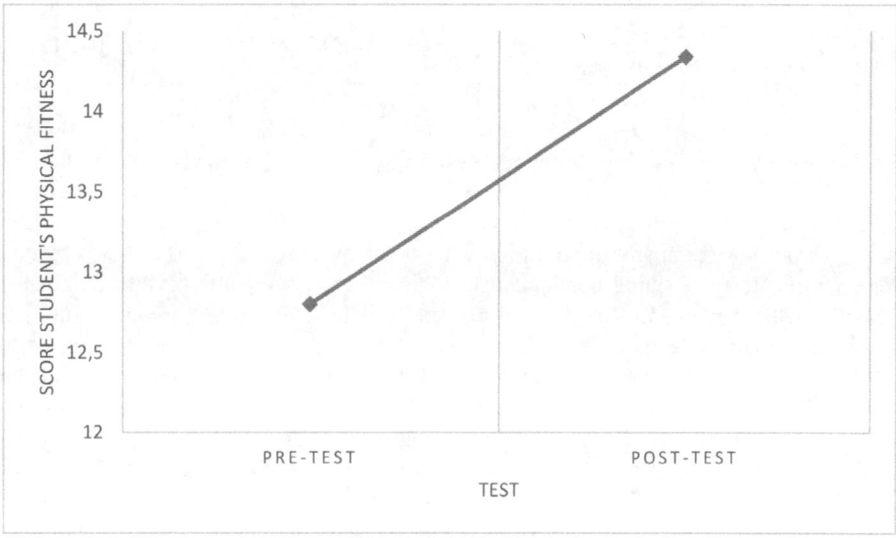

Figure 1. Differences in Mean of Students' Physical Fitness in Pre-Test and Post-Test

Results of paired sample t-test on Indonesian Physical Fitness Test (IPFT) demonstrate significant differences in several components including speed, strength/endurance of arm/shoulder and abdominal muscles, and cardiovascular endurance. On the other hand, no significant difference is demonstrated in explosive power despite small increase in post-test score. This can be seen in Table 5.

IOP Conf. Series: Earth and Environmental Science **884** (2021) 012024 doi:10.1088/1755-1315/884/1/012024

Table 5. Summary Analysis of Students' Physical Fitness in Pre-Test and Post-Test based-on IPFT Components

Components	Test	Mean	t	df	*P*
Speed	Pre-test	8.53	4.047		0.000
	Post-test	7.97			
Strength/endurance of arm/shoulder muscles	Pre-test	18.14	3.422		0.001
	Post-test	12.74			
Strength/endurance of Abdominal muscles	Pre-test	13.27	-7.294	73	0.000
	Post-test	18.82			
Explosive Power	Pre-test	26.81	-0.063		0.950
	Post-test	26.85			
Cardiovascular Endurance	Pre-test	3.90	4.714		0.000
	Post-test	3.09			

4. Discussion

Results of this research demonstrate that game-based physical activities influence students' physical fitness positively. This fact is in line with the statement saying that a regular training for six to eight weeks will generate specific results in which our bodies adapt to the given training [11]. Systematic, progressive and repetitive training will repair human's organ system and will eventually make physical performance optimal [12]. Training conducted three times a week is suitable for beginners and will produce significant improvement. Regular and measurable physical training in sufficient amount and time will change the ability to produce greater amount of energy and improve physical performance [13][14]. Motions performed repeatedly during training will cause conditioned reflex development, motion learning, and motion memorization process [12]. Results of this research are also supported by results of the study that physical activity learning program (fundamental movement) integrated into school subject significantly improves children's cognitive, psychomotor, and affective abilities [15]. Physical activities comprise of all types of movement produced by the body when performing daily activities such as working, exercising, or doing chores (doing laundry, sweeping), travelling (walking, cycling, riding motorcycle), and enjoying recreation (sports, outbound) [15].

Results suggest that performing game-based physical activities within 12 meetings can improve physical fitness up to 12.03%. It demonstrates that physical activities in form of games can influence students' physical fitness. Playing activities are means of socialization that allow students to explore, discover, express their feelings, get creative, and learn in fun ways [17]. Play significantly influences a child's development [18]. Meanwhile, play possesses certain values (the value of play), which include development of social, emotional, and cognitive skills [17]. In addition, play provides children with many benefits that can support their development. If children are given opportunities to engage in activities that require body movements, their bodies will be healthy and their muscles will grow strong [19]. Besides, being engaged in such activities will allow other body parts to move while enabling children to channel their extra energy and prevent them from getting anxious. For children, sitting for hours is not only boring but also uncomfortable and full of pressure. It is apparent that young children often exhibit active behavior, move a lot, and possess limited attention span [20].

Improved physical fitness among students found in this research was directed towards development of physical preparedness in coping with emergency situations such as disasters. Similarly, physical fitness is a manifestation of body's power and ability to adapt to the given physical release without experiencing excessive fatigue [21]. Preparedness is an effort to anticipate certain condition through organization of effective and efficient steps. Disaster preparedness is part of disaster management process as well as critical element in proactive disaster risk reduction control activity prior to occurrence of disasters. In this regard, the concept of preparedness being implemented focused on the ability to prepare for emergency situation related to disaster effectively and efficiently. One of the ways to develop physical preparedness is maintaining physical fitness. Physical activities that vary are not only the most effective but also the safest alternative to attain physical fitness as they can increase fitness as well as improve stress tolerance and concentration level [22]. Regular, continuous, gradual physical activities that involve the right movement are highly recommended and able to improve the functions of body organs.

5. Conclusion

Game-based physical activities done in 12 meetings of physical education (PE) subject in disaster-safe schools can significantly improve students' physical fitness. Treatment given by PE teachers based on researchers' intervention has contributed to students' physical preparedness. It implies that PE teachers need to create activities that vary and include game-based activities in their learning to improve physical fitness of their students. It is recommended that future researchers impose stricter control over factors that influence physical fitness such as age, gender, genetics, food intake, environment, and sleeping habit. Future researches should also consider adding control group into research subjects to control other factors.

6. Acknowledgement

The authors would like to thank the Ministry of Research, Technology, and Higher Education for the provision of PDUPT research grant in 2019. This writing is also part of second author's undergraduate thesis.

7. References

[1] Badan Perencanaan Daerah Provinsi DIY 2008 *Profil Kebencanaan DIY* (Yogyakarta: BAPEDA) pp 5-9
[2] Badan Nasional Penanggulangan Bencana (BNPB) 2014 *Peraturan Kepala Badan Nasional Penanggulangan Bencana Tentang Penerapan Sekolah/Madrasah Aman Bencana* (Jakarta; BNPB)
[3] Al-Nashr M S 2018 Integrasi Pendidikan Siaga Bencana Dalam Kurikulum Madrasah Ibtidaiyah *Magistra: Media Pengembangan Ilmu Pendidikan Dasar dan Keislaman* **6(2)** 82-101.
[4] Rahma A 2018 Implementasi program Pengurangan Risiko Bencana (PRB) melalui pendidikan formal *Jurnal VARIDIKA* **30(1)** 1-11.
[5] Sesfao A 2019 Upaya Meningkatkan Kesegaran Jasmani Melalui Pendekatan Bermain Dalam Pembelajaran Pendidikan Jasmani (PenelitianTindakan Kelas di kelas VII-G SMP Negeri 14 Tasikmalaya) *Jurnal Wahana Pendidikan* **5(3)** 31-6.
[6] Triyono A 2018 *Tingkat Kesegaran Jasmani Siswa Kelas IV dan V Daerah Rawan Bencana Gunung Merapi di Kabupaten Sleman* (Yogyakarta: Universitas Negeri Yogyakarta)
[7] Darmawan I 2017 Upaya Meningkatkan Kebugaran Jasmani Siswa Melalui Penjas *Jurnal Inspirasi Pendidikan* **7(2)** 143-54.
[8] Sugiyono 2016 *Metode Penelitian Kuantitatif, Kualitatif Dan R &D* (Bandung: Alfabeta) p 74
[9] Departemen Pendidikan Nasional 1999 *Tes kesegaran Jasmani Indonesia* (Jakarta: Pusat Kesegaran Jasmani dan Rekreasi Depdiknas).
[10] Arikunto S 2005 *Manajemen Penelitian* (Jakarta: Rineka Cipta) p 299
[11] Bompa T O and Harf G G 2009 *Periodization Training for Sports: Theory and Methodelogy of Training Fifth Edition* (United State of America: Human Kinetics) p 51
[12] Nala N 2011 *Prinsip pelatihan fisik olahraga* (Denpasar: Komite Olahraga Nasional Indonesia Daerah Bali) pp 37-9
[13] Meirizal U 2014 Aplikasi Aktivitas Fisik Siswa Sekolah Dasar sebagai Upaya Peningkatan Kesegaran Jasmani *Inovasi Sekolah Dasar* **1(1)** 157-65.
[14] Burhaein E 2017 Aktivitas fisik olahraga untuk pertumbuhan dan perkembangan siswa SD *Indonesian Journal of Primary Education* **1(1)** 51-8.
[15] Martin M and Hands B P 2003 Implementing a fundamental movement skill program in an early childhood setting: The teachers' perspectives *Australian Journal of Early Childhood* **28(4)** 40-6.

[16] Thomas J R, Nelson J K, and Silverman S J 2005 *Reseach Methods in Physical Activity* (United States of America: Human Kinetics) p 305
[17] Yuliani dan Sujiono B 2010 *Bermain Kreatif Berbasis Kecerdasan Jamak* (Jakarta: Indeks) pp 34-6
[18] Khasanah I, Prasetyo A, Rakhmawati E 2011 Permainan Tradisional sebagai Media Stimulasi Aspek Perkembangan Anak Usia Dini *Jurnal Penelitian PAUD* **1 (1)** 91-105
[19] Triharso A 2013 *Permainan kreatif & edukatif untuk anak usia dini* (Yogyakarta: Penerbit ANDI) p 10-3
[20] Stork S and Sanders S W 2008 Physical Education in Early *Chilhood The Elementary School Journal* **108(3)** 197-206
[21] Wiarto G 2015) *Panduan Berolahraga Untuk Kesehatan dan Kesegaran* (Yogyakarta: Graha Ilmu) p 55
[22] Nopembri S, Saryono, and Rhitaudin A 2019 *Aktivitas-aktivitas Jasmani Berbasis Psikososial.* (Yogyakarta: Istana Agency)

IOP Conf. Series: Earth and Environmental Science **884** (2021) 012025 doi:10.1088/1755-1315/884/1/012025

A Flood Damage and Shelter Need Assessment: A Case Study of Mueang Sing Buri, Thailand

Pattaramone Manawongcharoen[1] and Thitirat Panbamrungkij[1*]

[1] Geography and Geoinformation Research Unit, Department of Geography, Faculty of Arts, Chulalongkorn University, Thailand

[*]Corresponding author: Thitirat.pa@chula.ac.th

Abstract. Flooding is one of the main disasters in Thailand and Mueang Sing Buri is among those areas hit. Located on the Chao Phraya River Basin, in the central region of Thailand, the area receives a large amount of runoff during monsoon seasons which causes frequent flood disasters. The aims of this research are to create a flood hazard map and to estimate the number of people that may need shelter after the occurrence of a flood, and to evaluate whether the shelter capacity is adequate in Mueang Sing Buri. To explore the potential locations of emergency shelters, the relevant information related to flooding was initially recorded, such as building detail, flood depth, elevation map, and flood risk map. The available space of each building varies by the characteristics of building types. The calculation of shelter capacity thus depends on characteristics of the buildings, accessibility, and percent of vacant area. The emergency shelter assessment benefits many sectors in the design of preparation plans for hazard management.

1. Introduction

Flooding naturally occurs every year in Thailand, around the Chao Phraya River Basin in particular. This basin is the most important basin in Thailand, covering 30% of Thailand's land area and about 40% of the country's population [16]. It combines the economic and historical heartlands of Thai settlements, including the capital of Thailand, Bangkok. However, this basin receives a large amount of runoff during monsoon seasons that causes frequent flood disasters. Many communities have suffered from extensive severe flooding, especially in 2011 and 2013. These disasters resulted in a detrimental impact on life and properties. These severe floods have been aggravated by major factors such as rapid urbanisation, reduction of flood retention areas and the intensification of agricultural practices. The Thai government has controlled floods through the construction of multi-purpose reservoirs, dikes (diversions) and other flood control infrastructures. These methods have been successful in reducing the volume of floodwater, however they can still fail when floodwater rises quickly in the areas with no flood-protection infrastructure [3]. Apipattanavis et.al 2018 [1] mentioned that Thailand developed strategic plans for better water management for the next 20 years regarding the Committee on Water Resources Policy and Management, established in 2015. However, the plans do not focus extensively on emergency shelter or early warning.

Emergency shelter is likely to be one of the most important preparations for many hazard management systems. After a devastating flood, the local government should prepare efficient shelters

Content from this work may be used under the terms of the Creative Commons Attribution 3.0 licence. Any further distribution of this work must maintain attribution to the author(s) and the title of the work, journal citation and DOI.
Published under licence by IOP Publishing Ltd

for victims. Whitehead et al. (2000) claimed that households with higher incomes tend to evacuate to motels/hotels and households with higher education and/or with pets and children are more likely to stay with family. On the other hand, low-income and minority families with children or elderly members are likely to evacuate to shelters. Flood shelters in three locations along three major rivers were visited, and communities were consulted to address selected issues regarding shelters. To explore the potential locations of emergency shelters, information related to flood management was initially recorded such as building details, flood depth, elevation, and flood risk areas. The available space of each building varies by the characteristics of building types. For example, hospitals and stadiums have more floors and provide a larger floor space than single residential and religious buildings. In Bangladesh, there are 3 major types of flood shelters which are community shelters, school-cum-shelters, and individual homesteads. Each type has different functions and purposes to serve refugees [11]. In terms of mass shelter design framework proposed by London Resilience Group (2018) [8], emergency shelters have been divided into 2 groups; 1) emergency evacuation centres designed for short-term living and large numbers of evacuees (5000+) and 2) emergency rest centres designed for no more than 200 evacuees providing basic facilities for longer stays. The suggested building types for the first group are large public and private spaces such as large exhibition halls, large night club venues, large sports centres, shopping centres and large multiplex cinemas. The possible building types for the second group is dependent on the arrangements of local authorities.

However, some building types for supporting evacuees are occupied by more daytime occupants than night-time occupants. The calculation of shelter capacity thus depends on the use of the buildings, the population occupying the building in the daytime and percent of vacant area which can be used for the construction of temporary camps or tents outside the buildings. Some land uses have large vacant spaces such as recreation areas or parks and public religious zones, while others have few vacant spaces such as hospitals.

The aims of this research were to assess flood-risk buildings and populations affected in the 2011 floods. The second is the finding of shelter availability for evacuation. Ultimately, the results of this research can provide local authorities and policy decision-makers with a baseline flood hazard assessment to design urgent mitigation measures and support local communities through education and preparedness for future flooding.

2. Study area

Sing Buri province is located in the Chao Phraya River Basin, a major rice production area in the central region of Thailand (Figure 1). The province has been regularly subjected to large flooding events during monsoon seasons in Thailand. In addition, most areas of the province are flood plains, which have elevations ranging approximately from 3.5 to 15 metres [14]. In 2011 there was severe flooding which caused extensive damage to life and property of communities. It ranged from late July 2011 to mid-January 2012, triggered by the landfall of tropical storm Nock-ten. Damaged areas accounted for 9.1% of total land area of the country covering 69 provinces [7]. Many flood victims lost their homes and needed to evacuate to temporary shelter. Particularly, Mueang Sing Buri district was inundated by water up to two metres deep in some residential and commercial areas in 2011. Residents had to move their belongings to higher ground without prior planning, and some families evacuated to emergency shelters.

The total area of the province is 822.48 km^2. Most land use is agriculture (74.22%), which is mostly paddy rice, accounting for 66.87% of total agricultural areas, followed by built-up areas accounting for 18%. The population of the province was 208,446 with 77,432 households in 2019 [13]. This paper will evaluate the capacity of possible emergency shelters to support the number of residents who may be affected by floods in future disasters.

IOP Conf. Series: Earth and Environmental Science **884** (2021) 012025 doi:10.1088/1755-1315/884/1/012025

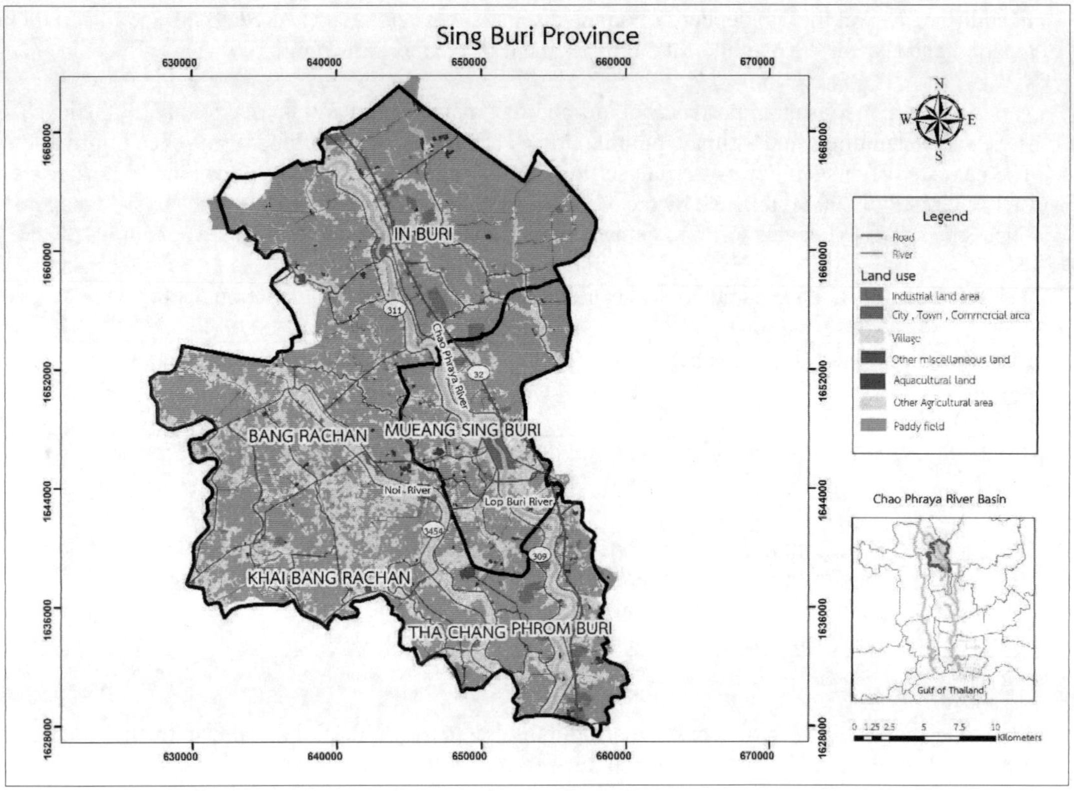

Figure 1. Sing Buri province

3. Methodology

The study started with flood inundation mapping by using a set of flood depth points from a field survey. Next, the Kriging interpolation technique was applied to create a flood hazard map from flood inundation data of Sing Buri province though the Geographic Information System (GIS). Kriging is a spatial interpolation method that generates an estimated surface from a scattered set of points with z-values. The principles of Kriging consider both the distance and the degree of variation between surrounding known data points when estimating values in unknown areas [10]. Generally, Kriging has 2 types: point Kriging and block Kriging. The point Kriging estimates parameter at a point and suits smaller amounts of data, while the block Kriging estimates parameter at the centre of a block and suits larger amounts of data [12]. Kriging has been widely applied to interpolate various geographic data such as rainfall, soil and geology operating on GIS environment [2, 10]. Its general formula is calculated using Eq [10]:

$$\mathrm{Z}(s_0) = \sum_{i=1}^{n} W_i \, \mathrm{Z}(s_i)$$

where:

$Z(s_0)$ = the measured value at the *i*th location

W = an unknown weight for the measured value at the *i*th location

s = the prediction location

N = the number of measured values

In addition, the weight (*W*) depends upon the distance between the measured points and the prediction location and also on the overall spatial arrangement of the measured points.

A flood hazard map is an important tool in flood risk management and is used to qualify the levels of flood risk to buildings and estimate building losses using the relevant flood occurrence information. The next step of this study was to analyse flood damage. Regarding the Atlas of Flood Maps proposed by EXICMAP in 2007 published by the Ministry of Transport, Public Works and Water Management, The Netherlands [5], levels of flood damage classified by flood depth classes are as follows (Table 1):

Table 1. The characteristics of flood damage in effect of inundation depth levels

Flood depth (metre)	**Flood damage characteristics**
0-0.5	Most houses will stay dry and it is still possible to walk through the water
0.5-1	There will be at least 0.5 m. of water on the ground floor and electricity will have failed
1-2	The ground floor of houses will be flooded, and the inhabitants have either to move to the first floor or evacuate
2-5 and >5	Both the first floor and often also the roof will be covered by water. Consequently, evacuation is the logical choice of action now

Then, the study integrated the consequence of building losses with the hazard map to analyse the total number of people in affected buildings. In the meantime, the land use map, flood map and flood risk building map were overlaid to extract the possible places for offering shelters. The final step was to calculate the capacity of these shelters to support the total number of people suffering from the flood and needing to go to these shelters. Thus, the shelter availability was presented in a map for evacuation preparation.

4. Results

4.1. Flood hazard map

In this study, LiDAR derived Digital Surface Model (DSM) and the set of flood depth data from a field survey of the 2011 flood were analysed to create the flood inundation map using interpolation (Kriging technique) to interpolate flood depth of Sing Buri province (Figure 3)

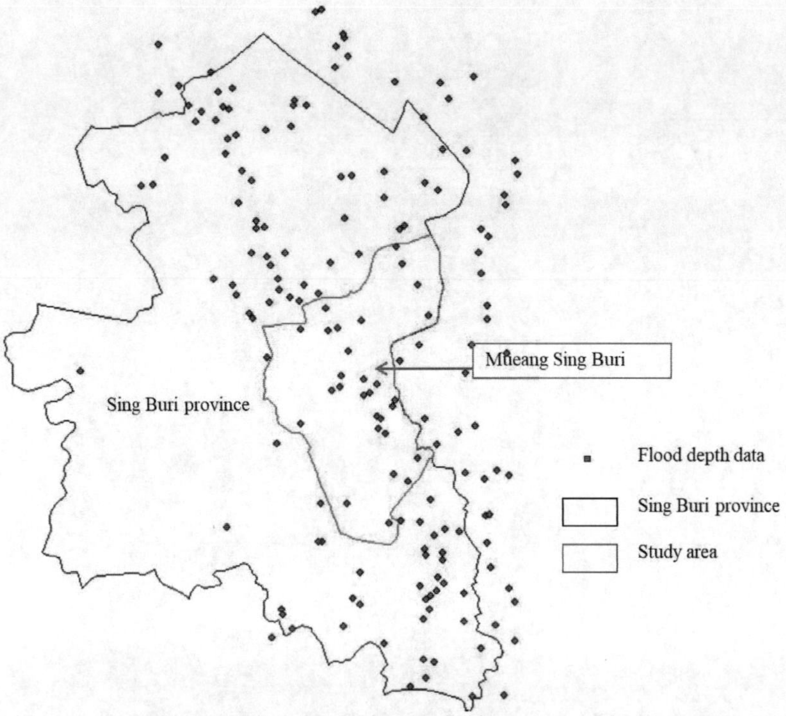

Figure 2. Flood depth data from Sing Buri (data source: the Department of Mineral Resources)

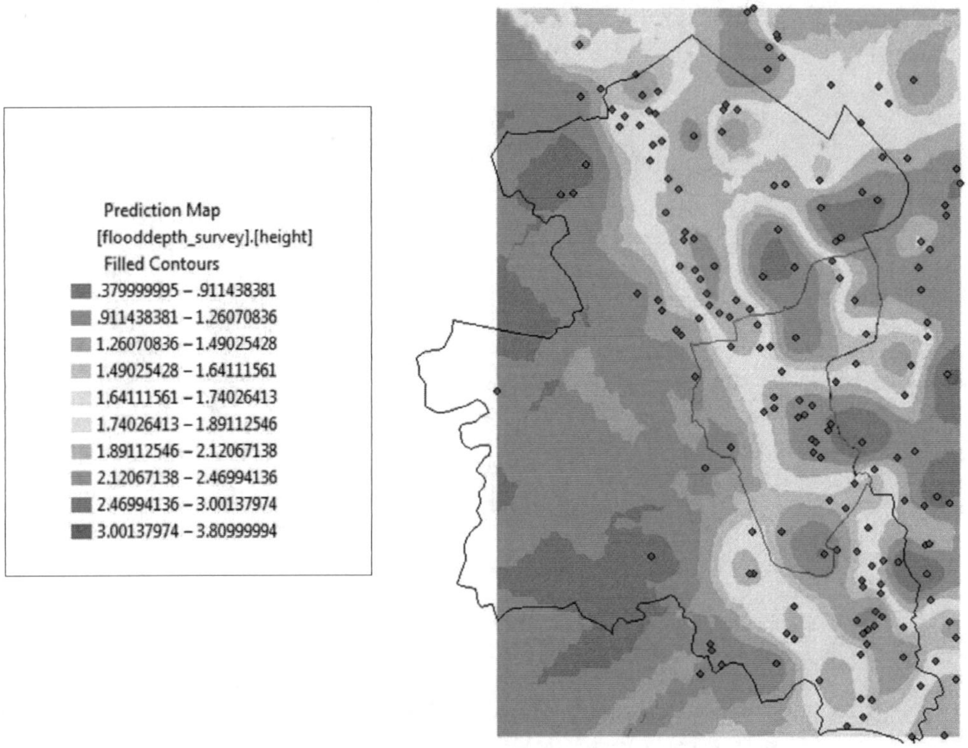

Figure 3. Flood inundation model computed by the Kriging interpolation of flood depth

In regard to the field survey after the flood, there were some areas and main streets that had not been affected by flooding due to dikes and flood control infrastructures. Therefore, it was necessary to eliminate those areas from the flood inundation map and to recheck flood magnitude with a specific return period between 50 years and 100 years derived from secondary data calculated from Nays2D Flood program (data source: the Department of Mineral Resources). A flood hazard map was then developed (Figure 4) which was subsequently classified to better depict probability of flood between 0 and 3 corresponding to flood depth interpolation.

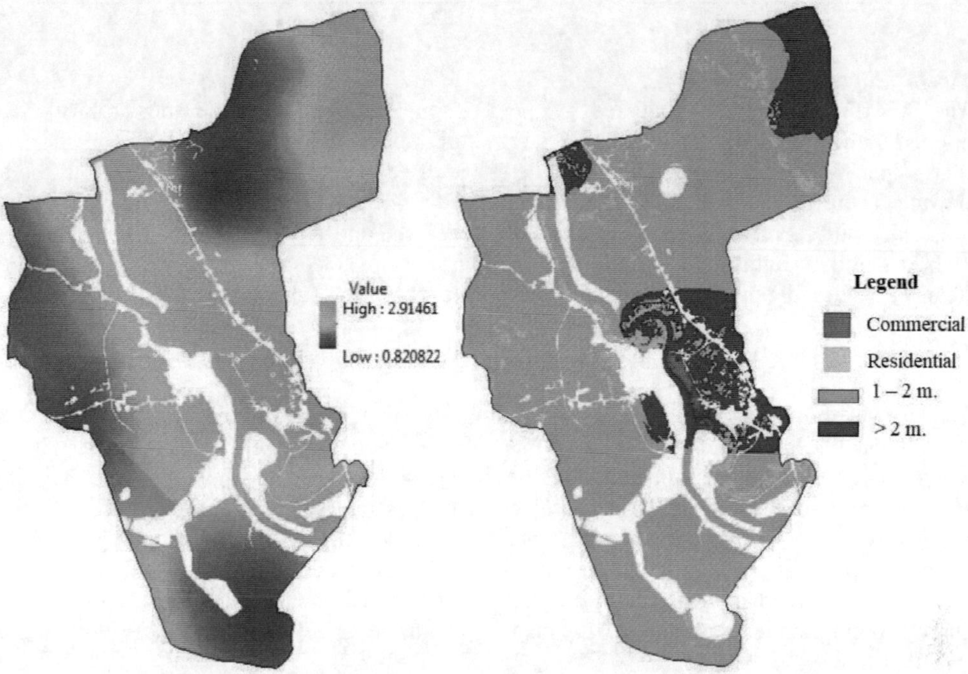

Figure 4. Flood hazard map for Mueang Sing Buri

4.2 Calculation of building damage and losses

The flood damaged area and elements at risk in this study were developed from the overlay of the built-up area/land use map and flood hazard map. The percentage of buildings affected by flooding of 1 metre or more was calculated. With at least 1-metre flood depth, the ground floor of houses will be flooded, and the inhabitants have either to move to the first floor or evacuate. The main land use type of Sing Buri is agriculture followed by residential area and commercial area respectively. Generally, most people affected by flooding live in the residential and commercial built-up areas. Thus, this study selected damaged commercial and residential buildings and then estimated the number of flood victims. The total number of buildings affected by flooding is shown in Table 2, however records of the amount people who live in each building/house was not available. Therefore, the average number of occupants per household in Sing Buri was calculated. In 2011 the total population was 52,867 and total households was 15,289 [13]. The average number of people living in each dwelling was 3 people/dwelling. The number of flood victims that required emergency shelter was estimated to be around 33,132 people (Table 2).

Table 2. Number of affected buildings and people in Mueang Sing Buri

Building type	Number of affected buildings	Number of people affected by flood (homeless)
Commercial area	1265	5060
Residential area	7018	28072

Total	8283	33132

4.3 Shelter capacity

The Sphere Project [15] conducted by the London School of Hygiene and Tropical Medicine recommends that residents have at least 3.5 m2 of indoor living space per person as minimum standards in a disaster response. The available space of each building was obtained from the area of floorspace. Therefore, shelter capacity was calculated from the floorspace, the number of occupants who use the buildings daily, and vacant area. However, the entirety of this floorspace cannot be used to support shelters. In this study, it was assumed that 60% of total could function as vacant areas for shelters. Therefore, the maximum capacity of each shelter was calculated as:

Maximum capacity (number of people) = ((Building area* 0.6) / 3.5 m^2)

Shelter availability was initially analysed through the integration of land use map, elevation map, and flood risk building map as the base maps for extraction of non-flooded areas. Then, these extracted possible areas were linked with the flood damage map to show possible buildings for shelters on the building map. Then, the selection of possible shelters was overlaid with the non-affected hazard map to find potential shelters that were not affected by flooding. The basic criteria were concentrated on the high land and non-flooding areas which focused on the land-use types of public places and recreation parks. In addition, the potential shelters should be close to the road access. According to the criteria above, the map result was generated using on GIS (Figure 5).

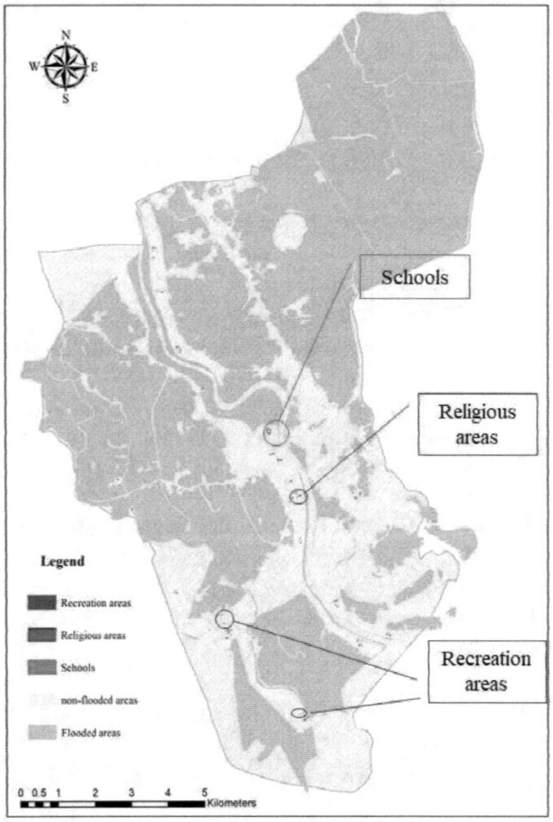

Figure 5. Selected candidate buildings for emergency shelter

Next, these non-damaged buildings were selected in built-up areas to apply as shelters, which included schools, recreation parks, and stadium/multi-purpose buildings. Areas that were located on high land at an altitude more than 6 metres (about 2 floors) were prioritised for selection as available shelters. The results showed that 37 buildings located in unaffected-flood areas were selected as candidates for setting emergency shelters which combined could accommodate around 18,807 people (Table 3).

Table 3. Available buildings and their capacities to support flood-affected people

No.	Schools	Capacity	No.	Temples	Capacity	No.	Multi-Purpose Buildings	Capacity
1	Wat Phrom Sakhon School	353	19	Wat Phrom Sakhon	1,164	36	Chak Si Village Multi-Purpose Building	23
2	Wat Khlong Sam Yaek School	159	20	Wat Khlong Sam Yaek	360	37	Phra Non Village Multi-Purpose Building	36
3	Wat Phu Langka Mittraphap 22 School	325	21	Wat Chaeng Phrom Nakhon	761			
4	Wat Chaeng Phrom Nakhon School	376	22	Wat Klang Phrom Nakhon	798			
5	Wat Rat Prasit School	162	23	Wat Chaksi	573			
6	Wat Srisakorn School	476	24	Wat Daowadueng	819			
7	Wat Sawang Arom School	182	25	Wat Buddha	403			
8	Wat Sao Thong Thong School	268	26	Wat Phranon Chaksi Worawihan	2,134			
9	Thetsaban 3 Kindergarten	200	27	Wat Pho Kaeo Nopphakhu	432			
10	Pho Rat Waranuson Kindergarten	46	28	Wat Yuat	440			
11	Sing Buri Kindergarten	2,156	29	Wat Rat Prasit	665			
12	Saeng Thien Kindergarten	264	30	Wat Si Sakhon	431			
13	Sing Buri Vocational Education College	921	31	Wat Sala Bap	415			
14	Sing Buri Non-Formal Education Center	466	32	Wat Sathian Watthanadit	31			

15	Chaksi School	544	33	Wat Sao Thong Thong	869			
16	Chom Chon Wat Phra Non Chaksi School	304	34	Wat Na Phra That	145			
17	Matthayom Tessaban 4 School	204	35	Wat Sawang Arom	488			
18	Pho Rat Waranuson Commercial School	414						
Capacity/building use		7,820			10,928			59
Total capacity		**18,807**						

According to the calculations of the maximum capacity of the 37 available shelters, shelter no.11, a school, had the largest capacity to support the victims, followed by the religious buildings no.26 and no.19 respectively. Most schools had lower capacity in the daytime as they are full of students. However, students may not attend school during a flood in reality, so these schools could support affected people to their maximum capacities without subtraction of occupants.

The capacities of available shelters (no.1-37) were compared to the number of affected people who need shelters after flooding. According to the previous results of vulnerable people made homeless in this scenario, they amount to 33,132 people. Therefore, the proposed shelters showed not enough capacity to support evacuees at this time. It is necessary to consider other potential shelters including temporary emergency shelter such as tents in public open spaces.

5. Conclusion

Emergency shelter is likely to be one of the most important preparations for many hazard management systems because it is the basic need of the homeless after an emergency event. An application of GIS is its use as a tool to perform not only the hazard risk assessment but also identify shelter availability. The results revealed that there were some public places with the potential to serve as flood shelters such as public recreation spaces, schools, and temples. In addition, the temple had the largest capacity to support flood victims. As the results of this study show, the capacity of emergency shelters analysed was insufficient for the number of victims. However, according to the information from the Disaster Prevention and Mitigation Centre of Sing Buri province, 18 evacuation centres for flood victims were established in 2011. Among these, only two places, Sing Buri Vocational Education College and Wat Sala Bap, were located in our proposed shelters. Therefore, the list of unaffected buildings from this study could be taken into consideration for establishing more emergency shelters within the Mueang Sing Buri area so that in the case of future flood occurrence, there would be enough space for those who need it. Furthermore, suitability shelter assessments may be used with the existing shelters for more effective management. For example, the assessment of the suitability of a location, and whether it is located in a high risk area for landslides due to a high volume of water in the soil [9], accessibility of the shelter within a short period of time during flood occurrence, nearest route from the shelter to other vital facilities such as hospitals or command centres [6]. A Thiessen Polygon could be used to analyse coverage of the shelter [4] because these shelters can be used not only to provide temporary accommodation but also serve as centres for providing essential needs within the area as well.

In addition, further research is suggested to consider available daytime and night-time shelter capacity separately. Most daily occupants at schools use these spaces during the daytime only. The night-time occupants are the priority for shelters because they are residents and need emergency shelters for sleeping. However, this study presented maximum capacity of shelters in general without considering the number of temporary daytime occupants at schools in particular. Therefore, these available shelters with fluctuating occupancy can support as many night-time occupants as the floorspace allows without the subtraction of daytime occupation. In addition, the study considered the outdoor shelter through the vacant space of each shelter building. These places can be used for constructing mobile shelters or erecting tents. In reality, several factors should be considered when calculating the specific space of living per person in the shelter. It may vary based on the type of land use and financial support from the government to prepare proper shelters.

Acknowledgements
This research paper was supported by the Chulalongkorn University-Faculty of Arts research fund. We would like to thank Mr Surasak Boonlue at the Department of Mineral Resources, Thailand for providing the set of flood depth data from a field survey of the 2011 flood and other information related to flood events.

References

[1] Apipattanavis S, Ketpratoom S and Kladkempetch P 2018 Water Management in Thailand *Irrigation and Drainage* **67** 113-7
[2] Arabameri A, Saha S, Chen W, Roy J, Pradhan B and Bui D T 2020 Flash flood susceptibility modelling using functional tree and hybrid ensemble techniques *Journal of Hydrology* **587** 125007
[3] Bond H, Cross K, Glotzbach R and Richaud B 2018 Flood and Drought Management Tools Case Study.
[4] Dhamodaran R and Suresh V M 2020 Nearest Emergency Shelter location and Management using Thiessen Polygon: A Case Study of Greater Chennai Corporation *Journal of Information and Computational Science* **10**
[5] EXCIMAP 2007 Atlast of Flood Maps: Examples from 19 European countries, USA and Japan. Ministry of Transport, Public Works and Water Management, The Netherlands)
[6] Huang Y, Yin Z and Chu H 2019 Suitability Assessment of Emergency Shelters Based on GIS: A Case Study in Urban Function Optimization Area of Shanghai(vol 234)
[7] Kotera A, Nagano T, Hanittinan P and Koontanakulvong S 2016 Assessing the degree of flood damage to rice crops in the Chao Phraya delta, Thailand, using MODIS satellite imaging *Paddy Water Environ* **14** 271-80
[8] London Resilience Group 2018 Mass Shelter Framework: London Resilience Partnership Mass Shelter Framework. LONDON RESILIENCE GROUP)
[9] Melgarejo L-F and Lakes T 2014 Urban adaptation planning and climate-related disasters: An integrated assessment of public infrastructure serving as temporary shelter during river floods in Colombia *International Journal of Disaster Risk Reduction* **9** 147-58
[10] Paramasivam C R and Venkatramanan S 2019 *GIS and Geostatistical Techniques for Groundwater Science,* ed S Venkatramanan, *et al.*: Elsevier) pp 23-30

[11] Rahman M A, Mallick F H, Mondal M S and Rahman M R 2015 *Hazards, Risks and Disasters in Society,* ed A E C John F. Shroder, Samantha Jones, Bernard Manyena, Janaka Jayawickrama, pp 145-59
[12] Shi G 2014 *Data Mining and Knowledge Discovery for Geoscientists,* ed G Shi (Oxford: Elsevier) pp 238-74
[13] Singburi Provincial Statistical Office 2020 Social Database. Singburi Provincial Statistical Office
[14] Takaya Y 1969 Topographical Analysis of the Southern Basin of the Central Plain, Thailand *The Southeast Asian Studies* **7** 293-300
[15] The Sphere Project 2004 The Humanitarian Charter and Minimum Standards in Disaster Response. London School of Hygiene and Tropical Medicine)
[16] The UN World Water Assessment Programme 2003 Chao Phraya River Basin, Thailand: Pilot case study by Office of the Natural Water Resources Committee (ONWRC) of Thailand. In: 1st UN World Water Development Report: Water for People, Water for Life. (the United Nations Educational, Scientific and Cultural Organization (UNESCO): the United Nations Educational, Scientific and Cultural Organization (UNESCO)) pp 387-400

IOP Conf. Series: Earth and Environmental Science **884** (2021) 012026 doi:10.1088/1755-1315/884/1/012026

Spatial Evapotranspiration Modeling Assisted With Landsat 8 Image Using Sebal And Geographically Weighted Regression Methods In Magelang District

AF Nugraha[1], BS Hadi[2]

[1] Geography Education Student, Yogyakarta State University, Sleman, Colombo Street No.1, Indonesia
[2] Teacher of Geography Education, Yogyakarta State University, Sleman, Colombo Street No.1, Indonesia

Arif.fahrizal2016@student.uny.ac.id, bambang_saefulhadi@uny.ac.id

Abstract. Information about evapotranspiration is very important in relation to vegetation because it can be used for planning both in urban planning and agriculture. Magelang Regency has a lot of vegetated green land, both agricultural and non-agricultural and has no information about evapotranspiration. The calculation of evapotranspiration uses the SEBAL (Surface Energy Balance Algorithm for Land) method and modeling uses the GWR (Geographiccaly Weighted Regression) model. Calculation and modeling assisted by QGIS 2.14, QGIS 3.6, SPSS 20, and GWR 4.09 applications. The results showed that (1) GWR evapotranspiration model with significance (sig.) 5% is divided into 3 sub-district groups according to the significant variables in the sub-district (2) NDVI and Surface Albedo variables have a small effect on a global scale and have a large effect on a local scale.

1. Introduction

Water is a fundamental element for every living thing on the surface of the earth and water is also a need that must be met by living things. Water is very important for living things, so water management and development is needed so that water resources can be available at any time. Water resource management is closely related to climate. Climate has a role in the process of the hydrological cycle. The origin of water has several sources, one of which comes from the hydrological cycle. The hydrological cycle has parts that support each other so that rain occurs. The part of the hydrological cycle that has a role in supplying water vapor to become rainwater is evaporation and evapotranspiration.

Soil surface evapotranspiration (ET) is the process of transferring large volumes of water from soil evaporation and transpiration of vegetation to the atmosphere [1]. Evapotranspiration is not only important for the global biogeochemical cycle, evapotranspiration also plays a major role in the energy balance of the earth's surface [2]. About 75% of the total precipitation is returned (evapotranspiration) by plants and soil [3] so that the calculation of evapotranspiration is very important, especially for agriculture because it can calculate water supply, the amount of water that becomes steam into the atmosphere and urban planning. Information regarding evapotranspiration is needed by people who are in the agricultural sector to have a plan regarding the development and maintenance of their agriculture. Information that can be mapped and very important for agriculture is evaporation mapping

Content from this work may be used under the terms of the Creative Commons Attribution 3.0 licence. Any further distribution of this work must maintain attribution to the author(s) and the title of the work, journal citation and DOI.
Published under licence by IOP Publishing Ltd

(evapotranspiration) because the amount of water that evaporates will have an impact on the amount of water available and needed by plants to grow. One of the regions in Indonesia that has quite extensive agricultural land is Magelang Regency. The agricultural area in Magelang Regency consists of wet and dry land, the distribution covers all districts except Ngablak and Pakis districts. The districts with the largest agricultural land area are Salaman, Mungkid, Mertoyudan, Secang, Grabag, Dukun, Bandongan and Kajoran Districts. Wet and dry agricultural land in Magelang Regency is managed to support the protection of sustainable food agriculture land covering an area of approximately 42,070 hectares spread across 21 districts. Agricultural land in Magelang Regency experienced land conversion with the percentage of land conversion into buildings based on the results of the research, namely 412.65 ha (66.91%) irrigated rice fields, 47.07 ha (7.54%) rainfed rice fields, 24.00 ha moor. (3.84%), mixed gardens 114.47 ha (18.34%), grasslands and shrubs 21.96 ha (3.36%) [4].

Evapotranspiration cannot be measured directly, but it is concluded through the calculation method [5]. There are still evpotranspiration data processing which is manual calculation from several climate stations such as using the evaporation pan. The calculation of evapotranpiration can be done using the Lysimeter Calculation tool, the Eddy Correlation Technique, and the Bowen Ratio Technique. This method is accurate but in a wider area it will be inaccurate. This limitation is what motivates the calculation to be developed using remote sensing data from satellites in a wider area coverage [6]. Remote Sensing has advantages apart from its wide area coverage but also because it is very effective in measuring evapotranspiration. Remote Sensing is also low cost, fast and accurate timing. In addition, Remote Sensing data are available globally [7] making it easier to obtain data. Remote Sensing and GIS can be integrated in processing spatial and attribute data in order to obtain appropriate and accurate results. The interactions between factors are so complex that it is difficult to calculate and predict them. GIS is able to simplify the complexity of the spatial data so that it can be analyzed quickly, precisely, and accurately by using modeling [8] One popular type of modeling, namely GWR (Geographically Weighted Regression) spatial modeling is used to determine regression models by paying attention to spatial aspects in the form of relationship between one region and another. The GWR model is a local form of regression. The GWR model allows the researcher to assess the possible spatial variations in the relationship between the dependent and independent variables across the observation area. The basic idea of the GWR model is to consider elements of geography or location as a weight in estimating the model parameters. This model is a locally linear regression model which produces a model parameter estimator that is local for each point or location where the data is collected. The dependent variable is predicted by the independent variable, each of which has a regression coefficient depending on the location where the data is observed [9]. This modeling can be used to model the results of the calculation of evapotranspiration using the SEBAL (Surface Energy Balance Algorithm for Land) method.

The surface energy balance model simulates the micro-scale energy exchange process between the soil surface and the near-ground atmosphere layer. These processes include radiation, sensible heat, latent heat, and subsurface heat exchange processes. The calculation of a spatially distributed energy budget will require spatial data from sources such as satellite imagery, models that measure and use different model parameters in the simulation. [10] The data required in the SEBAL method are routine weather data such as wind speed, humidity, radiation. sun, and air temperature. No data on land use, soil type, or hydrological conditions are required to apply SEBAL [11]. Image data that can support infrared-based evapotranspiration calculations include Landsat 8 OLI / TIRS imagery. The capacity of Landsat 8 imagery to estimate evapotranspiration using Remote Sensing is quite satisfactory [12].

2. Research Methods

2.1. Data Sources and Research Variables

This study used daily weather data on January 19, 2020, which was recorded online at Mlati Weather Station, Sleman. The spatial data used in this study is a map database of Magelang Regency and Landsat 8 OLI / TIRS imagery on January 19, 2020 recording date with wrs path 120 and wrs row 65. The variables used in this study are Evapotranspiration (Y), NDVI Vegetation Index (X1) and Surface Albedo (X2).

2.2. Research methods
The stages of research carried out to obtain model results that can explain the relationship between variables that affect evapotranspiration in Magelang Regency are as follows:

2.2.1. NDVI Vegetation Index Calculations
Vegetation Index is an index used to predict the variation in the characteristics of vegetation. Vegetation index information to distinguish vegetation and other land uses can be done with the NDVI (Normalized Difference Vegetation Index) formula as follows:

$$NDVI = \frac{NIR-RED}{NIR+RED} \quad (1)$$

Where: NIR is the near infrared channel and RED is the red band [13]

2.2.2. Calculation of Surface Albedo
Surface albedo is the ratio of the returned radiation flux to the radiation flux that occurs in the solar spectrum. Surface albedo has a role in the occurrence of soil heat. The formula for calculating the surface albedo is as follows:

$$\alpha = \frac{\alpha_{toa}-\alpha_{path\_Radiance}}{{\tau_{sw}}^2} \quad (2)$$

Where: αpath_radiance is the average portion of incoming solar radiation in all bands that are scattered back to the satellite before it reaches Earth's surface and τsw is the transmissivity of the atmosphere

2.2.3. Calculation of Plant Reference Evapotranspiration Value (ETo)
The plant reference evapotranspiration (ETo) can be determined either from weather data or from the evaporation pan. ETo can also be estimated from the loss of evaporation from the water surface. ETo calculations use the FAO Penman-Monteith method which has a standard for calculating the ETo value [14].

2.2.4. Calculation of Evapotranspiration by SEBAL (Surface Energy Balance Algorithm for Land) Method
The calculation of evapotranspiration can use the SEBAL (Surface Energy Balance Alghoritm for Land) method. The SEBAL model evapotranspiration calculation is done with the help of the spatial data processing application QGIS 2.18, 3.6, Grass 7.2 and the SEBAL module python coding script. The equation for the SEBAL method is as follows:

$$\lambda ET = R_n - G - H \quad (3)$$

Where: λET is the latent heat flux (W / m2), Rn is the net radiation flux at the surface (W / m2), G is the ground heat flux (W / m2) and H is the sensible heat flux to the atmosphere (W / m2) .

2.2.5. Data Normality Testing, Local Multicollinearity, Autocorrelation and Heteroscedasticity
Before doing data modeling, several tests were conducted to be able to do modeling. The tests are in the form of normality test, multicollinearity test, autocorrelation test, heterocedacity test. In this study, the Kolgomorov-Smirnov data normality test method was used with a significance level of 5%.

2.2.6. Spatial Evapotranspiration Modeling GWR (Geographically Weighted Regression) Model
Mathematically, the GWR (Geographically Weighted Regression) model can be written as follows:

$$y_i = \beta_0(u_i, v_i) + \sum_{k=1}^{p} \beta_k(u_i, v_i)x_{ik} + \varepsilon_i \quad (4)$$

Where: Yi is the observed value of the-i location response variable, X_{ik} is the value of the-k predictor variable at the-i observation location, (ui,vi) are the coordinates of the (longitude, latitude) observation location, $\beta 0(ui,vi)$is the coordinate / intercept of GWR, $\beta k(ui,vi)$ is the regression coefficient k at the i-th observation location and εi is the error at location which is assumed to be identical, independent, and normally distributed with the mean zero and constant variance (IIDN (0, σ2)) [15]

2.2.7. Calculation of the GWR Weighting Function

One of the weighting methods commonly used is the gaussian kernel. The Gaussian weighting function will give zero weight when location j is at or outside radius b from location i, whereas if location j is within radius b, it will get a weight that follows the Gaussian function, as follows:

$$w_{ij} = \mathrm{e}\left[-1/2\left(\frac{d_{ij}}{b}\right)^2\right] \quad (5)$$

$$d_{ij} = \sqrt{(u_i - u_j)^2 + (v_i - \mathrm{v})^2} \quad (6)$$

Where: wij is the weighting function, b is the radius of comparison, j is Euclid's distance and (ui, vj) is the coordinates of location i and j.

2.2.8. Bandwidth Selection

Bandwidth can be regarded as the radius of interaction in the area. A wide bandwidth value can cause the model to be biased and smooth, but if the bandwidth value is narrow, the model depends on the characteristics in the area. Cross validation is a method that can be used as a criterion for obtaining the optimum bandwidth. The optimum bandwidth is used to produce the minimum cross validation coefficient. with the coefficient formula as follows:

$$CV = \sum_{i=1}^{n}[y_i - \widehat{y_{\neq i}}(b)]^2 \quad (7)$$

Where: y ≠ i is the prediction value yi (fitting value) with the observation at location i omitted from the prediction process and b is the radius of comparison.

2.2.9. GWR Model Suitability Testing

The suitability test (F test) of the GWR model was conducted to determine whether the GWR model was significantly better at modeling data (the relationship between the independent and dependent variables) than the OLS model. The formulation of the hypothesis is:

H0 : GWR and OLS models are the same in modeling data

H1 : The GWR model is significantly better at modeling data

The null hypothesis can be rejected if the value of Fcount is greater than Ftable (Fcount> Ftable). Fcount is the statistical value of the F test and Ftable is the value of the F distribution of degrees of freedom (DF1; DF2).

2.2.10. Partial Testing of the GWR Model Variables

The variable partial test (t test) was conducted to determine the significant variables of the GWR model in each observation area. The t test is done by comparing the t-count value with the t-table value on the basis of decision making, if the t-count value is greater than the t-table value ($t_{count} > t_{table}$) then the j-variable has a significant effect in the observation area.

2.2.11. Testing the Spatial Model of Evapotranspiration

The GWR spatial model can be known better than the OLS model by comparing four alternative measures that consider three indicators such as R-square, AIC, RSS, and BIC / MDL. The GWR model will be better than the OLS model if the R-square value of the GWR model is greater than the R-square model of the OLS model (R-square model GWR> R-square model OLS). This comparison illustrates

IOP Conf. Series: Earth and Environmental Science **884** (2021) 012026 doi:10.1088/1755-1315/884/1/012026

that the GWR model is much better than the OLS model which does not pay attention to spatial elements or the diversity of observed areas. The results of the comparison are strengthened by the comparison of the AIC, RSS, and BIC / MDL values which are a function of the weighting in the model. If the AIC, RSS, and BIC / MDL values of the GWR model are lower than the AIC, RSS, and BIC / MDL values of the OLS model (AIC, RSS, and BIC / MDL of the GWR model> AIC, RSS, and BIC / MDL of the GWR model),

3. Result and Discussion

The initial stage carried out in this study was to process the weather and spatial data into the NDVI vegetation index value and surface albedo to find the evapotranspiration value. In this study, the calculation of the NDVI vegetation index, surface albedo and evapotranspiration used the QGIS 2.14 and 3.6 application with the annoying python coding script developed by Wagner Wolff [16] and Rafael Tieppo [17] in order to obtain the NDVI, Albedo, and Evapotranspiration index results per district. which is presented in the following table:

Table 1. NDVI Vegetation Index, Albedo and Evapotranspiration Index data by district

sub-district	Coordinate		NDVI	Albedo	Evapotranspiration
	X	Y			
Ngluwar	420446	9152419	0.601955679	0.136324366	4,204425223
Ngablak	430656	9182129	0.570099042	0.16124729	5,195991043
Borobudur	411157	9157190	0.645782624	0.150673376	4,840524585
Srumbung	429819	9160370	0.664102298	0.140696603	5,450545428
Salam	423710	9156353	0.617036863	0.137734977	4,371749121
Dukun	431744	9165558	0.638230267	0.139919195	5.857745794
Muntilan	419023	9160872	0.552787339	0.143852451	3,718333839
Mungkid	417433	9164052	0.583036024	0.14262657	4,355209526
Sawangan	430405	9170998	0.639543365	0.137100596	5.864547974
Mertoyudan	411826	9166312	0.593292432	0.15088806	4,006644167
Candimulyo	423543	9170915	0.719016705	0.132270444	5.882813565
Pakis	431577	9176438	0.651394023	0.134941896	6,153909743
Grabag	425049	9183301	0.680835285	0.130137763	5,989527199
Windusari	408395	9180455	0.668718496	0.139679443	5,853360041
Tegalrejo	420028	9175601	0.679406893	0.137624452	5,536195754
Bandongan	408897	9175685	0.647570246	0.140646337	5,3090004
Secang	420613	9181794	0.591159131	0.134939556	4,997323644
Kaliangkrik	404043	9175099	0.658707779	0.145282494	5.897064158
Tempuran	407140	9166814	0.682836967	0.139054727	5,256529557
Salaman	405633	9160956	0.664382526	0.142026074	5,112738942
Kajoran	400528	9170831	0.699023464	0.135280007	6,074401564

The next stage in this research is to conduct some data testing before modeling. Some of the results of the data testing are as follows:

Table 2. Normality Test Data

	Evapotranspiration Residue

IOP Conf. Series: Earth and Environmental Science **884** (2021) 012026 doi:10.1088/1755-1315/884/1/012026

N	21
Kolmogorov-Smirnov Z	0.626
Asymp, sig. (2-tailed)	0.829

Based on the output in Table 41, it is known that the Asymp.Sig (2-tailed) significance value of 0.829 is greater than 0.050 (0.829> 0.050), so according to the basis for decision making in the One Sample Kolmogorov-Smirnov normality test, it can be concluded that the data is normally distributed. and further analysis can be carried out, namely multicollinearity analysis. The multicollinearity test results are presented in table 3 as follows:

Table 3. Tolerance and VIF values Variable Free

No.	Variable	Tolerance	Decision	VIF	Decision
1.	NDVI Vegetation Index (X1)	0.723	Tolerance value> 0.10 means no multicollinearity	1,383	VIF value <10 means there is no multicollinearity
2.	Surface Albedo (X2)	0.723		1,383	
(Data does not contain multicollinearity)					

Based on the results of Table 42, it is known that the Tolerance value of the vegetation index variable (X1) and surface albedo (X2) is 0.723. Tolerance value of 0.723 is greater than 0.10 (0.723> 0.10), according to the basis for decision making, the independent variable data does not have multicollinearity. This conclusion is reinforced by the results of the VIF value of the independent variables (X1 and X2) of 1.383, according to the basis for decision making, the independent variable does not have multicollinearity. The next test, namely the autocorrelation test between data. The results of the autocorrelation test are presented in table 4 as follows:

Table 4. Independent Variable Autocorrelation Test

No.	Variable	DW	T	K	DL	DU	4-DW	4-DL	4-DU
1.	Vegetation Index (X1)	2,590	21	2	1,1246	1,5385	1.41	2,8754	2,4615
2.	Surface Albedo (X2)								

Table 4 shows the values of the Durbin-Watson autocorrelation test. The DW (Durbin-Watson) value is 2.590 which comes from 21 samples with 2 independent variables. The value of DL (Durbin Lower) is 1.1246, DU (Durbin Upper) is 1.5385, (4-DW) is 1.41, (4-DL) is 2.8754, and (4-DU) is 2.4615 . If the DW value is greater than DU (2.590> 1.5385), the conclusion is that there is no positive and negative autocorrelation. These results indicate that the DW value is greater than the DU value. This basis is what researchers use to continue research. The last test is the heteroscedasticity test. The test results are presented in Table 5 as follows:

Table 5. Heteroscedasticity Test (Glejser)

Model	Sig.
Constant	0.862
NDVI Vegetation Index (X1)	0.309
Surface Albedo (X2)	0.406

Based on Table 44, it is known that the significance value (sig.) Of the NDVI vegetation index variable (X1) is 0.309 and the significance value (sig.) Of the surface albedo variable (X2) is 0.406. In accordance with the basis of heteroscedasticity test decision making, if the significance value is greater than 0.05, the data does not have heteroscedasticity. The significance value (X1) 0.309 is greater than 0.05 (0.309> 0.05) and the significance value (X2) 0.406 is greater than 0.05 (0.406> 0.05), it can be concluded that the research data does not have any heterocedasticity symptoms. . After passing several tests, modeling can be continued.

The next step is to find the optimum bandwidth by looking at the smallest CV value. This study uses a Gaussian adaptive kernel type. The search results are as follows:

Table 6. Bandwidth Search Results

No.	Name	Result
1.	Best Bandwidth Size	3
2.	Minimum CV	0.185

Table 6 shows that the bandwidth value in this modeling is 3. The bandwidth value of 3 means the 3 closest neighbors (sub-districts) which significantly affect a district, for example in the GWR model the evapotranspiration value in Muntilan District is influenced by 3 other districts such as Mungkid, Salam, Dukun. The bandwidth value of 3 can be said to be narrow (small), which means that the GWR model will depend on the characteristics of the observation area. The determination of the bandwidth is in accordance with the minimum CV value of 0.185. The next step of GWR modeling is testing the GWR model whether the GWR model is better than the OLS model in modeling data or what is commonly called the F-Test. The F-test is carried out by looking at the results of the F-test ANOVA Table which is presented in Table 7 below:

Table 7. ANOVA Value of GWR Model

No.	Source	SS	DF	MS	Fcount	FTabel
1.	*Global Residual*	4,665	18			
2.	GWR Improvement	4,162	11,017	0.378		
3.	Residual GWR	0.503	6,983	0.072	5,241	3.55

Based on Table 7, it is known that the value of Fcount is greater than FTabel (5,241> 3.55), so that the null hypothesis can be rejected, which means that the GWR model can explain the relationship between the predictor variables and the response variable better than the global regression model. After the F-test (suitability) of the model, a partial t-test is then performed for each variable. The t-test is done by comparing the value of t count is greater than t table (t count> t table) with a significance of 5%. Partial testing in Ngluwar District as a sample for the GWR model as shown in Table 8 below:

Table 8. Partial Test Parameter Model GWR Ngluwar District

No.	Variable	Coefficient	thitung	t table
1.	*Intercept*	-4.57253	-1.362075	
2.	NDVI (X1)	14,265941	6,605984	0.68836
3.	Albedo (X2)	2,325665	0.120826	
4.	Local R2	0.938075		

Table 8 shows that the NDVI variable has a t-count value of 6.605984 and albedo has a t-count value of 0.120826 with a t-table value in this GWR model is 0.68836. Based on the basis of decision making, the NDVI variable has a significant effect on evapotranspiration in Ngluwar District (6.605984> 0.68836) while the surface albedo variable has no significant effect on evapotranspiration in Ngluwar District (0.120826> 0.68836). NDVI and surface albedo variables have an R-square value of 0.938, which means that these variables simultaneously (together) have an effect on evapotranspiration in Ngluwar District by 94% (93.8%). The results of the partial test then form 3 groups of districts, namely districts that are influenced by 2 independent variables, sub-districts that are influenced by the NDVI vegetation index variable and sub-districts that are not influenced by 2 variables. These results are presented in table 9 below:

Table 9. Subdistrict Groups according to Significant Variables

No.	Significant Variables	sub-district
1.	NDVI and Albedo	Serumbung, Salam, Bandongan and Kaliangkrik
2.	NDVI	Ngluwar, Ngablak, Borobudur, Dukun, Muntilan, Mertoyudan, Candimulyo, Grabag, Windusari, Tegalrejo, Secang, Tempuran, Salaman and Kajoran
3.	There are no significant variables	Sawangan and Pakis

Based on Table 9, 3 groups of sub-districts were formed which have the same significant variables that affect evapotransopiration in Magelang Regency. The advantage of the GWR model is that the data obtained can be used and displayed in the form of a thematic map as shown in the following figure:

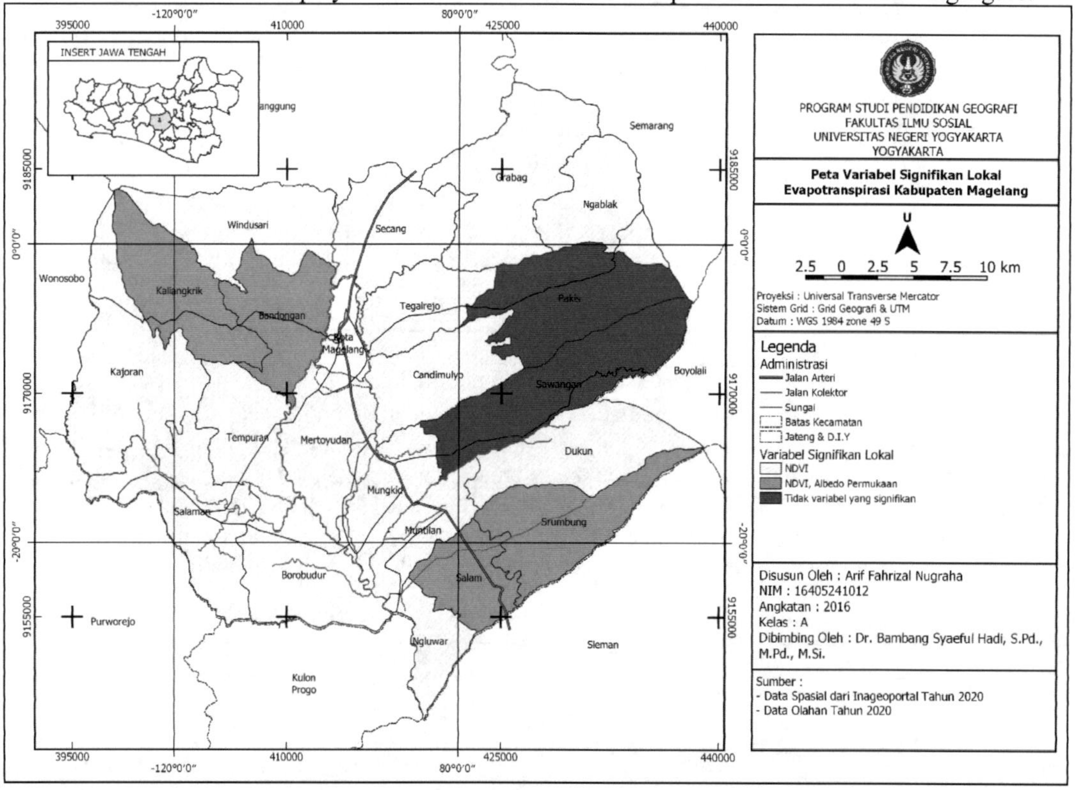

Figure 1. Map of Local Significant Variables of Evapotranspiration in Magelang Regency

Writing the GWR model in Ngluwar District based on the results of the variable partial test can be written as follows:

$$y_{Ngluwar} = -4{,}57253 + 14{,}26594X1_{Ngluwar} \quad (8)$$

The GWR model for Ngluwar District above explains that evapotranspiration will increase by 14.26594 percent if the percentage of the NDVI vegetation density index increases by one percent. The writing of the GWR model for each district is presented in Table 10 as follows:

Table 10. Similarities of the GWR Model for Each District in Magelang Regency

No.	sub-district	The GWR Evapotranspiration Model Equation
1.	Ngluwar	$y_{Ngluwar} = -4{,}57253 + 14{,}26594X1_{Ngluwar}$

2.	Ngablak	$Y_{Ngablak} = 0{,}460992 + 8{,}040957X1_{Ngablak}$
3.	Borobudur	$Y_{Borobudur} = -2{,}179644 + 13{,}127731X1_{Borobudur}$
4.	Srumbung	$Y_{Srumbung} = -8{,}094273 + 16{,}317505X1_{Srumbung} + 20{,}671994X2_{Srumbung}$
5.	Salam	$Y_{Salam} = -7{,}799461 + 15{,}754234X1_{Salam} + 19{,}149443X2_{Salam}$
6.	Dukun	$Y_{Dukun} = 6{,}477171 + 5{,}4005X1_{Dukun}$
7.	Muntilan	$Y_{Muntilan} = -2{,}426599 + 13{,}116907X1_{Muntilan}$
8.	Mungkid	$Y_{Mungkid} = -2{,}426599 + 12{,}052746X1_{Mungkid}$
9.	Sawangan	$Y_{Sawangan} = 9{,}083786$
10.	Mertoyudan	$Y_{Mertoyudan} = 1{,}027299 + 11{,}888648X1_{Mertoyudan}$
11.	Candimulyo	$Y_{Candimulyo} = 0{,}881438 + 9{,}623253X1_{Mungkid}$
12.	Pakis	$Y_{Pakis} = 7{,}540954$
13.	Grabag	$Y_{Grabag} = -1{,}535918 + 9{,}244204X1_{Grbag}$
14.	Windusari	$Y_{Windusari} = -2{,}959963 + 12{,}816733X1_{Windusari}$
15.	Tegalrejo	$Y_{Tegalrejo} = 2{,}292194 + 7{,}821521X1_{Tegalrejo}$
16.	Bandongan	$Y_{Bandongan} = -11{,}023869 + 19{,}116307X1_{Bandongan} + 128{,}216185X2_{Bandongan}$
17.	Secang	$Y_{Secang} = -0{,}10842 + 8{,}039527X1_{Secang}$
18.	Kaliangkrik	$Y_{Kaliangkrik} = -7{,}799461 + 15{,}754234X1_{Kaliangkrik} + 19{,}149443X2_{Kaliangkrik}$
19.	Tempuran	$Y_{Tempurn} = -0{,}696998 + 13{,}148116X1_{Tempuran}$
20.	Salaman	$Y_{Salaman} = -0{,}808486 + 12{,}76885X1_{Salaman}$
21	Kajoran	$Y_{Kajoran} = -7{,}875864 + 17{,}238589X1_{Kajoran}$

Acknowledgments
Gratitude was expressed to the lecturers of geography education department for their many helpful suggestions which improved the original manuscript.

References

[1] Allen R G, Tasumi M and Trezza R 2007 Satellite-Based Energy Balance for Mapping Evapotranspiration with Internalized Calibration (METRIC)-Model J. Irrigation and Drainage Engineering **133** 380

[2] Sousa D and Small C 2018 Spectral Mixture Analysis as a Unified Framework for The Remote Sensing of Evapotranspiration Remote Sensing 10 2

[3] Immerzeel W, Droogers P, and Gieske A 2006 *Remote Sensing and Evapotranspiration Mapping : State of The Art* (Netherlands: NIVR) p 7

[4] Handari, MF A W, Bambang A N and Purnaweni H 2012 *Prosiding Seminar Nasional Pengelolaan Sumberdaya Alam dan Lingkungan* Strategi Perlindungan Lahan Pertanian Berkelanjutan di Kabupaten Magelang (Semarang) p 77

[5] Westerhoff R S 2015 Using Uncertainly of Penman and Penman-Monteith Methods in Combined Satellite and Ground-based Evapotranspiration Estimates *Remote Sensing of Environment* **169** 102

[6] Allen R G, Tasumi M and Trezza R 2002 *SEBAL (Surface Energy Balance Algorithms for Land) Advanced Training and User's Manual – Idaho Implimentation, Version 1.0* (USA : The Idaho Department of Water Resources) p 7

[7] Gamage N, SmakhtinV and Perera B J C 2011 *The 19th International Congress on Modelling and Simulations* Estimation of Actual Evapotranspiration Using Remote Sensing Data (Perth: The Modelling and Simulation Society of Australia and New Zealand Inc) p 3361
[8] Purwanto, Astina I K and Yusuf S 2015 Pemanfaatan Sistem Infromasi Geografi untuk Pemodelan Spasial Pengembangan Wisata Pantai Di Kabupaten Tulungagung. *J. Pendidikan Geografi* **20** 15
[9] Fadli M R, Goejantoro R and Wasono 2018 Pemodelan Geographically Weighted Regression (GWR) dengan Fungsi Pembobot Tricube Terhadap Angka Kematian Ibu (AKI) Di Kabupaten Kutai Kartanegara Tahun 2015 *J. Eksponensial* **9** 13
[10] Abtew W and Melesse A 2013 *Evaporation and Evapotranspiration Measurement and Estimations* (London: Springer) p 141
[11] Immerzeel W, Droogers P, and Gieske A *2006 Remote Sensing and Evapotranspiration Mapping : State of The Art* (Netherlands: NIVR) p 30
[12] Semmens K A, Anderson M C, Kustas W P, Gao F, Alfieri J G, Mckee L, Prueger J H, Hain, C R, Cammalleri, C, Yang Y, Xia T, Sanchez L, Alsina M M and Velez M 2016 Monitoring Daily Evapotranspiration Over Two California Vineyards using Landsat 8 in A Multi-Sensor Data Fusion Approach *Remote Sensing for Environment* **185** 168
[13] Susantoro T M and Wikantika K 2015 Pengolahan Data Landsat 8 Untuk Ekstraksi Objek di Permukaan Laut *J. Ilmiah Geomatika* **21** 100
[14] Food Agriculture Organization 1990 FAO Irrigation and Drainage Paper No. 24 (FAO-24). http://www.fao.org/3/X0490E/x0490e02.htm accessed on 7 July 2020
[15] Lutfiani N, Sugiman dan Mariani S 2019 Pemodelan Geographically Weighted Regression (GWR) dengan Fungsi Pembobot Kernel Gaussian dan Bi-Square *UNNES Journal of Mathematics* **8** 85
[16] Wolff W 2016 wwolff7/SEBAL_GRASS doi: 10.5281/zenodo 167350
[17] Tieppo R C, Dallacort R, and Andrea M C S 2018 SEBAL model by means GRASS and PYTHON doi: 10.5281/zenodo 303413

IOP Conf. Series: Earth and Environmental Science **884** (2021) 012027 doi:10.1088/1755-1315/884/1/012027

Community Preparedness of the Fire Hazard in Jetis District, Yogyakarta City

H A Adhandika, Hastuti

Departement of Geography Education, Faculty of Social Science
Universitas Negeri Yogyakarta, Jl. Colombo No.1 Yogyakarta, 55281, Indonesia

Email: husnanda.arief2016@student.uny.ac.id

Abstract. This study aims: (1) to determine the level of community preparedness in facing fire hazards in the Jetis District of Yogyakarta City and (2) to understand the efforts to increase community preparedness to deal with fire hazards in the Jetis District of Yogyakarta City. This research uses descriptive methods with quantitative and qualitative approaches. In this research, we utilize the environmental approach as a tool. The indicator that we use here is based on the 2006 UNESCO / ISDR LIPI to measure the level of community preparedness. This research was conducted at Jetis District of The City of Yogyakarta. The population of this study was all family heads in Jetis Subdistrict, at least we came to 9282 families. The number of samples was determined using the Slovin formula and obtained 99 respondents. We chose the Proportional Random Sampling in three villages in Jetis Subdistrict as the sampling technic. The data was gained from the process of observations, structured interviews with standardized guidelines, in-depth interviews with three main resource persons, and documentation. The analysis technique that we used is the frequency table and descriptive analysis. The results show that: (1) the level of community preparedness in the Jetis Subdistrict in facing the danger of fire was in the "Less Ready" category (55.56%). The level of community preparedness on the knowledge and attitudes of the majority was in the category of "Ready" with 63.63%. The majority of emergency plan variables are in the unprepared category as much as 56.57%. The majority of disaster warning system variables are in the ready category that is 56.57%. Resource mobilization ability is a variable with respondents who are at the lowest level of preparedness, the majority are in the category of not ready as much as 76.77%. (2) the efforts to increase fire hazard preparedness in the community in Jetis District have been carried out, namely: conducting training and simulation of fire prevention, procurement of supporting facilities for fire prevention, making fire engines, and establishing disaster relief posts in Jetis District. Efforts that must be increased are: educating the public about the risks of fire hazards, increasing synergy between community organizations and local governments, and adding firefighting infrastructure.

1. Introduction

Disaster is one of the events as well as a natural process that we cannot avoid, but we can reduce the impact caused by a disaster. Disasters always threaten every human life or what are known as risks. Disasters are divided into three types according to Law number 24 of 2007 natural disasters, disasters and social disasters. One of the non-natural disasters is fires, namely residential and forest / land fires. Residential fires rarely get attention even though during 2018 according to BNPB data there were around 422 fire incidents. Fires can be caused by human negligence or occur naturally, for example due to high temperatures (BNPB: 2018).

Content from this work may be used under the terms of the Creative Commons Attribution 3.0 licence. Any further distribution of this work must maintain attribution to the author(s) and the title of the work, journal citation and DOI.
Published under licence by IOP Publishing Ltd

An urban area is a zone or area that is the center of economic activity, the center of government and population concentration with a heterogeneous way of life. The area as the center of economic and government activity has an attraction consisting of the many facilities, facilities and infrastructure which are represented by the dominance of built-up land in urban areas to meet the living needs of the population in it (Adiarto, 2013: 2). The problem that often arises in urban areas in Indonesia as a result of the attractiveness of urban areas is high population growth and it is estimated that by 2020 the population living in urban areas will reach more than 60% of the population (Fire Prevention and Management of the Yogyakarta Fire Service: 2005).

The attractiveness of urban areas results in population concentration which encourages urban change and development. There are several things that come up frequently. One of them is the problem of population growth. The population that is always increasing will cause problems in urban life because it is to meet the growing population development, while the land used never increases. Urban land is occupied by a large number of residents, as a result urban areas experience a fairly complex development process of built-in areas, which can include the wider area of built-up land and densification (compaction). This situation will cause a decrease in environmental quality in urban areas.

Administratively, Jetis District is part of Yogyakarta City. The Jetis District area has several large buildings such as hotels, markets to universities which are one of the main factors of attraction to support the development of the area towards urban areas. Jetis sub-district is built by 2 rivers, one of which is Jetis District and the other Sub-district, Code River in the east and Winongo River in the west, that is, it is not surprising that the growth of settlements in Jetis District is high, namely with a longitudinal pattern along the river. Jetis District has an area of 1,703 or 5.23% of the area of Yogyakarta City.

The results of the population registration of Jetis District which was carried out by BPS in 2019 had a population of 27,312 people with their livelihoods mostly in the service and trade sectors because they were supported by the large number of offices, educational facilities and trade / market places in Jetis District.

Almost the entire area of Jetis District is a permanent building, covering an area of 93.59% of the total area of the Jetis District. The large area resulted in the development of settlements in this area so dense. Recorded from the 2018 census, the population density of Jetis District reached 16,066 (BPS Kota Yogyakarta 2019) so that it can be categorized that the population density in Jetis District is quite high because the population density level is more than the area of Jetis District. The population density map shows that Jetis District is included in the category of rather high population density (Fire Service 2019).

High population growth in an area is of course directly proportional to the growth of buildings in that region. The large number of buildings and residents will certainly cause both physical and social hazards, one of which is fire hazard. Fire is a danger that often occurs in big cities in Indonesia, including Yogyakarta City.

The majority of land use in Jetis District is in the form of permanent buildings and semi-permanent buildings, which is one of the factors that influence the handling of fire emergencies. This is due to the condition of the building materials which are relatively susceptible to fire in the event of a fire. Other factors that influence are human error / lack of community vigilance, short-circuit electricity, and the condition of Jetis District itself, where the majority of permanent buildings are built, both for offices, settlements, and markets.

Impact of the fire hazard can of course be minimized with disaster management efforts. Disaster management itself consists of several stages, one of which is preparedness. Preparedness is a series of activities carried out to anticipate disasters through effective and efficient measures (Law No.24 of 2007). Increasing community preparedness for fire disasters aims to train people not to panic when a disaster occurs and also to reduce the level of losses when a disaster occurs. The level of preparedness of a community can decrease over time and with the occurrence of socio-cultural, political, and economic changes of a society. It is necessary to always monitor and know the condition of community preparedness and make efforts to always maintain and increase the level of preparedness (Hidayati, et al, 2009: 7).

2. Research Methods

The data analysis used in this research is quantitative and qualitative Research is a descriptive study with quantitative and qualitative approaches, namely research whose results are expected to reveal facts from things that are observed and measured so that it can provide an overview and analysis of the level of community participation in fire hazard preparedness. Geography used is the environmental approach because it deals with the interactions between organisms and their natural environment. Indicators used to measure the level of community preparedness and efforts to increase community preparedness based on LIPI / UNESCO 2006.

The population in this study were people in Jetis District. Based on bps data in 2018, there were 27,312 people with 9282 families in 3 villages. The number of samples was determined using the Slovin formula and obtained 99 respondents. The sampling technique used was proportional random sampling in three sub-districts in Jetis sub-district. Data obtained from interviews, in-depth interviews, and documentation. The data analysis technique used is quantitative and qualitative analysis.

3. Results

3.1. Decription area

Jetis District is located in the western part of Yogyakarta City. Jetis District is located at 7,773219 ° -7,788496 ° South Latitude and 110.368873 ° - 110.369218 ° East Longitude. Jetis District has an area of 1,703 consisting of 3 villages, namely Cokrodiningratan, Jetis, and Gowongan. The population composition of Jetis District is 27,312 people, consisting of 13,251 male residents (48.51%) and 14,061 female residents (51.49%).

3.2. Community Preparedness Level

There are 5 variables to measure the level of preparedness to face a disaster according to LIPI / ISDR, namely knowledge and attitudes, emergency response plans, policies, disaster warning systems, and resource mobilization capabilities. This study uses 4 of the 5 variables, namely knowledge and attitudes, emergency response plans, disaster warning systems, and resource mobilization capabilities because the main respondents in this study are people who do not have the authority to determine policies related to hazard preparedness, especially fire hazards.

3.2.1 Knowledge and Attitude

Knowledge becomes the basis for carrying out the right activities in anticipating disasters.

Table 1. Level of Community Preparedness for Knowledge and Attitude variables

No	Preparedness Level	Frequency (Soul)	Percentage (%)
1	Ready	63	63,63
2	Less Ready	34	34,34
3	Not Ready	2	2,02
Total		99	100

Source: Primary Data (2020)

The ready category can be illustrated that the respondent has a level of knowledge about fire starting from the causes of fire, the impact of the fire, the signs of fire, knowledge of buildings with a low risk of fire, relatively safe house building conditions, and discussions within the family to support their attitude. fire emergency response. The majority of respondents who are in the unprepared category have weaknesses in the condition of the building which has a small risk when a fire occurs. Factors that cause are economic limitations that have an impact on the choice of building materials and the location of the house which must be in a densely-built village. The majority of respondents who are in the unprepared category have low scores on almost every indicator due to the relatively low level of knowledge (SD and SMP), age, and economic level.

3.2.2. Emergency Response Plan

Emergency planning is an important part of activities that need to be carried out in the context of community preparedness to anticipate fires.

Table 2. Level of Community Preparedness, Emergency Response Plan variable

No	Preparedness Level	Frequency (Soul)	Percentage (%)
1	Ready	41	41,41
2	Less Ready	56	56,57
3	Not Ready	2	2,02
Total		99	100

Source: Primary Data (2020)

The level of community preparedness in the emergency response plan variable is mostly at the unprepared level. There are many things that affect it, such as the relatively low level of knowledge (SD and SMP), age, to economic factors. Environmental conditions also affect the level of community preparedness. There are still locations that cannot be passed by fire engines and dry hydrants have not been installed, although in some hard-to-reach locations there are already dry hydrants. The problem is that the distribution of dry hydrants is still uneven.

The ready category indicates the condition of the family and its environment that is ready to face the danger of fire, such as the readiness of evacuation, starting from items for emergencies both in the family or RW environment and participation in training and fire simulations. This condition will facilitate the evacuation process in a fire emergency so as to minimize the loss of life and property.

3.2.3. Disaster Warning System

The disaster warning system is an important part of community preparedness to face hazards, including fire hazards. Signs that are given from the warning system will be conveyed to the wider community either directly or indirectly, then the public can respond to the warning.

Table 3. Levels of Community Preparedness for Disaster Warning System variables

No	Preparedness Level	Frequency (Soul)	Percentage (%)
1	Ready	56	56,57
2	Less Ready	39	39,39
3	Not Ready	4	4,04
Total		99	100

Source: Primary Data (2020)

Ready category indicates that the community is ready. The community is considered to have known the existence and function of a disaster warning system in their environment, such as there are sources of information about fire hazards, there are tools for traditional and modern disaster warnings, and respondents are aware of the kentongan codes which the majority are still used as a warning tool. In contrast to the level of lack of readiness, the majority of which are filled by respondents with education levels below SMA, as well as jobs that take up a full day so that they do not have time to understand the disaster warning system. The lowest level is not ready, indicating that the respondent does not know the form of disaster early warning, from traditional to technology-based. Respondents do not know in their surroundings whether there are bells, sirens, or other electronic media within the RT or RW as a tool for disaster warning. The most influencing factor is age because some family heads are elderly.

The second indicator measures technology-based disaster warning systems. Entering the era of globalization, the technology that can be used as a danger warning is increasingly sophisticated and varied. Examples are TV, radio, newspapers, magazines, to the internet. The majority of respondents agreed that in the neighborhood where they live there is already a technology-based disaster warning tool because the condition of the respondent's residence is in an urban area, thus facilitating the flow of information.

3.2.4. Resource Mobilization Capability

The ability to mobilize resources is a training program (resources), community action, funds, community social capital.

Table 4. Community Preparedness Level of Resource Mobilization Ability variable

No	Preparedness Level	Frequency (Soul)	Percentage (%)
1	Ready	13	13,13
2	Less Ready	76	76,77
3	Not Ready	10	10,1
Total		99	100

Source: Primary Data (2020)

As many as 78% of respondents, based on the results of filling out the questionnaire, said that apart from the head of the family, no one had ever attended fire preparedness training, even though fire emergency conditions could occur at any time, including when the family head was not at home.

The level of knowledge regarding fire hazard preparedness which is still quite low is strengthened from the interview data on the indicator of family agreement to participate in training and the frequency of participating in disaster preparedness or hazard training, especially fire. As many as 73% of respondents stated that they did not have a family agreement to participate in the training because they would only participate in training spontaneously, or only family representatives who were usually followed by the head of the family. Many factors influence this, such as the activity of each family member to the low level of community awareness of attending training. This is corroborated by a statement from a resource person who comes from the Bokomi Community management who also assesses that the level of public awareness of disaster preparedness is still low. So far, the participation of family heads has been good enough. As many as 70% of household head respondents stated that they had attended disaster preparedness training, although not everything they had attended specifically discussed fire. The ability to mobilize resources also includes the financial readiness of the community to deal with emergencies, especially fires. Public awareness to support preparedness from a financial perspective is still quite low due to the economic factors of the community, most of whom are still in the middle to lower class. 66% said they had no plans for funding in an emergency. They only rely on modest funds and assistance from the surrounding community.

3.2.5. Community Preparedness Level

Preparedness is a part of the disaster management process and in the concept of disaster management that is currently developing. Increased preparedness is one of the important elements of pro-active disaster risk reduction activities, before a disaster occurs.

Table 5. Levels of Community Preparedness in Facing Fire Danger

No	Preparedness Level	Frequency (Soul)	Percentage (%)
1	Ready	41	41,41
2	Less Ready	55	55,56
3	Not Ready	3	3,03
Total		99	100

Source: Primary Data (2020)

Referring to the assessment of the calculation results using a predetermined formula, the average value of the overall respondents in Jetis District is classified into the less prepared level. The variables of the emergency plan and the ability to mobilize resources were the variables with the lowest level of readiness. Data from interviews with respondents on the emergency plan variable and the ability to mobilize resources are variables with a lower level of readiness than the other two variables.

The variable of resource mobilization ability is the variable with the least number of respondents who fall into the ready category. There are only 13.13% of respondents in the ready category while 41.41% of respondents are in the ready category. There are several indicators that are still weaknesses in efforts to support preparedness such as the participation and frequency of family heads and family members' participation in training and fire simulations, then funding and availability of logistics for emergencies, while indicators with low scores on the emergency response plan variable are the availability of goods. -goods for emergencies. These items include Hydrant extinguishers / light fire extinguishers, Handy Talky, First Aid Box, and alternative lighting tools.

3.3 Efforts to Improve Community Preparedness

The first attempt is related to knowledge and attitude variables. Efforts to improve preparedness are emphasized on indicators of knowledge of the causes of fire, knowledge of buildings with low risk when a fire occurs, and attitudes when a fire occurs. Activities carried out to increase the level of community knowledge include training and simulations to deal with disasters and fire hazards including attitudes when a fire occurs, how to extinguish fires independently, and family evacuation plans. The second effort carried out by the District Government in order to improve community preparedness is to coordinate with the Yogyakarta City Government for the provision of supporting facilities and infrastructure to deal with fire hazards.

The Bokomi community is here to provide breakthrough training so that the ability to handle a fire emergency can be comprehensive down to family members. Evaluating the training program held by the District Government, the Bokomi management made a regulation that training was held once a month with residents but carried out in rotation so that each family member was expected to participate. Public education related to disaster is also a concern of Bokomi. There were training and simulation activities with the community, then several times Bokomi also collaborated with universities and schools in order to educate and foster an attitude of disaster preparedness, especially fire. As a community that comes from residents, by residents, and for residents, of course, it always experiences ups and downs in its journey to increase community preparedness because it is only supported by minimal funds and resources. This is still homework for the bokomi administrators and the local government because until now there has been no intense coordination, especially to improve community preparedness in facing fire hazards in Jetis District.

The existence of bokomi, KTB, and Linmas should be able to be synergized so that they are able to work together mutually to further improve community preparedness. The bokomi management hopes that the local government can collaborate with KTB as an organization engaged in disaster that was formally formed by the government with bokomi.

4. Conclusion

From this research, we knew that the level of community preparedness in Jetis District in facing fire hazards. Jetis Subdistrict is classified as a fire incident red zone according to data from the Fire Service, which means that the fire incidence rate is still quite frequent. Efforts to increase preparedness in the face of fire hazards were carried out by the Jetis District Government as the policy maker and assisted by several organizations and communities, one of which was Bokomi. The first attempt is to conduct fire management training and simulation activities. Second, the procurement of supporting facilities for fire prevention by the Jetis District Government and the construction of a fire train by Bokomi. The third is the establishment of the Jetis District disaster post.

Acknowledgment

In this occassion, We thanks from Faculty of Social Sciences which has provided research opportunities until the author can write this article, support and facilities for us, Mrs. Hastuti and all in Geography Education, Faculty of Social Science, Yogyakarta State University lecturers who have always accompany, guide and motivate us. Parents, family, and friends, who always support us who cannot be mentioned one by one.

References

[1] Adhiarto, Yuniar. 2013. *Utilization of Quickbird Imagery and Geographical Information Systems for Mapping Fire Danger Levels in Parts of Surakarta City*. Yogyakarta: Gadjah Mada University

[2] BNPB. 2018. *Indonesian Disaster Data and Information*, http://dibi.bnpb.go.id/ accessed on 19 December 2018.

[3] Hidayati, et al. 2009. *Study on Community Preparedness in Anticipating Earthquake & Tsunami Disaster*. Jakarta: LIPI-UNESCO / ISDR.

[4] Law of the Republic of Indonesia No.24 of 2007. *Concerning Disaster Management*

[5] LIPI-UNESCO/ISDR. (2006). *Community Preparedness Study in Anticipating Earthquake and Tsunami Disasters*.

[6] Yogyakarta Fire Service 2019.

Assessment of human impact on flooding for Thai East coast - gulf basin

Norawit Suwannakarn*, Chanita Duangyiwa, Ekkamol Vannametee
Geography and Geoinformation Research Unit, Department of Geography, Faculty of Arts, Chulalongkorn University, Pathumwan, Bangkok, Thailand

Corresponding author e-mail: norawit.su1995@gmail.com

Abstract. Due to a rapid increase in urban and built-up areas, the East Coast – Gulf basin of Thailand faces flood hazards more frequently than in the past. In this study, we aim to assess the effects of building construction on flooding, and any link between them. The FloodMap model is used to simulate flooding in the study area in September 2015. Two flooding scenarios were designed; one based only on land surface elevation and the other one with building construction included on the land surface. According to the result, we found that human construction increases flood hazard in the study areas, particularly flood depth. Flood areas are also found to increase if the human factor is added into the model, but in a lesser extent. With human construction, paved road is found to have the highest flood potential compared to other types of road. Built up areas are more flooded, while flooding extent is almost similar to results from the scenario of no human construction in forest and agricultural areas.

1. Introduction

Flooding can be caused by heavy or prolonged precipitation. Flood could occur when the rainfall is beyond the average rate of the particular land that could accept. However, currently, human activities become one of the most influenced factors which cause flooding. To settle down, humans clear the land, removing native plants and land characteristics, which can increase the rainfall runoff [1]. Moreover, urbanization causes a decrease in land infiltration due to an increase in the number of impervious features such as roads and pathways [2].

East Coast - Gulf basin is located in the eastern region of Thailand. Most of the basin is mainly in four provinces, which are Chonburi, Rayong, Chanthaburi, and Trat, and some parts in Chachoengsao and Sakaeo [3]. The Department of Public Works and Town & Country Planning: DPT indicated that in 2013 there was an increase in the number of buildings around 575 km^2 or 129% rising in the basin [4]. The expansion of built-up areas may have resulted from the urbanization in the basin caused by the Eastern Seaboard economic region development program. Industrial support in the eastern region and growth in the tourism sector also caused the growth of building in the region. Flood events frequently occur in the basin. Major floods occurred three times in the last decade [4]. Furthermore, an ineffective drainage system due to inappropriate land use zoning could interrupt runoff and the settlement in the lower basin, therefore, causing the hazard of flooding in low areas of this basin.

This study aimed to investigate human influence on flooding in the East Coast - Gulf basin. The impact of building construction on the level of flood risk was analyzed. The assessment was done by the application of ArcMap program [5] and FloodMap-HydroInundation2D model [6]. The result of this study would help the stakeholders to gain better understanding of human impact on flooding and could enhance the effectiveness of urban planning for flood prevention.

Content from this work may be used under the terms of the Creative Commons Attribution 3.0 licence. Any further distribution of this work must maintain attribution to the author(s) and the title of the work, journal citation and DOI.
Published under licence by IOP Publishing Ltd

2. Methodology

2.1. Study area

The western sub-basin of the East Coast-Gulf basin is selected as a study area, as shown in Figure 1. This sub-basin is located in Chonburi and Rayong province. Many famous places are located in this sub-basin, including one of the most popular tourist attractions; Pattaya city, and several industrial estates. Moreover, DPT indicated that flooding in this sub-basin was caused by manmade-obstruction such as building and industrial estates, which would reduce the drainage capacity [4].

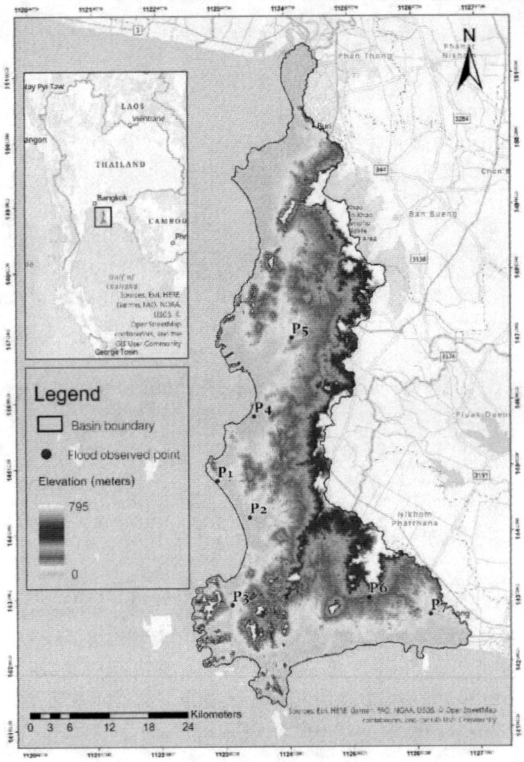

Figure 1. Elevation map of the western sub-basin with flood observed points

2.2. Model Description

FloodMap model is used to investigate the effects of building construction on urban flooding. FloodMap model is developed as a physically based two-dimensional flood inundation model on topographically complex floodplain [7-8]. The model has been applied under various conditions and proven as robust for different purposes, such as flooding prediction in the UK [9], sediment delivery problems in rivers [10].

The model simulates flooding in the urban areas in two stages. In the first stage, the model calculates the amount of surface runoff generated at each grid cell as an excess amount of effective rainfall after infiltration [11]. Infiltration capacity at time step, t, on the ponded surface is approximated using Green-Ampt infiltration equation [12].

$$f(t) = K_s\left(\frac{\varphi_f - h_o}{Z_f}\right) + 1 \tag{1}$$

Where K_s is the hydraulic conductivity of the soil at saturation area (m/h), φ_f is the capillary potential across the wetting front (m), h_0 is the ponding water on the soil surface (m), and Z_f is cumulative depth of infiltration (m).

In the second stage, surface runoff generated at grid cells is routed on the gridded surface. Surface flow is decoupled in x and y direction following the drainage pattern defined as the steepest topographical slope descent. Flow rate is modelled using the Saint-Venant equation. In case of surface runoff, the convective acceleration term in the full version of Saint-Venant equation can be neglected in the momentum equation with respect to the discretised time step of simulation becomes:

$$\frac{q_{t+\Delta t} - q_t}{\Delta t} + \frac{gh_t \partial(h+z)}{\partial x} + \frac{gn^2q^2}{R^{\frac{4}{3}}h} = 0 \qquad (2)$$

Where q is the flow per unit width, g is the acceleration due to gravity (m^2/s), R is the hydraulic radius (m), z is the bed elevation (m), h is the water depth and n is the Manning's roughness coefficient.

2.3. Model input and parameterization

The model requires topographical and meteorological inputs to simulation flooding. For topographical data, we use a digital elevation model (DEM) map derived from the ASTER Global Digital Elevation Model which is under the collaboration of the Ministry of Economy, Trade, and Industry of Japan (METI) and the National Aeronautics and Space Administration (NASA) [13]. DEM resolution was resampled to 50x50 m to reduce model run time.

Three-hour precipitation data were derived from the meteorological station located in Pattaya City. In this study, we chose the rainstorm event between 15th – 18th September 2015 that was caused by Tropical Depression Vamco and induced flash floods in the study area [14-19]. In this rain storm event, we only focused on the rainfall period over 15 hours between 16th to 17th September 2015, where the highest amount of rainfall in the study area was recorded. As there were no other rainfall records at three-hour resolution in the study area, uniform rainfall distribution was assumed in the model simulation.

Land use data was provided by the Land Development, which represents the land cover of the study areas in 2016. Infiltration parameters were assumed constant over the study area with K_s 0.015 m/h and φ_f 0.73 m. The evapotranspiration rate is set at 0.05 mm per day as descried in [11]. The drainage capacity for the flood modeling is set as 15 mm/h, which is the drainage capacity of Pattaya city reported by the director of Pattaya city sanitation division [20].

2.4. Model Evaluation

Performance of the FloodMap model is evaluated against validated using the observed flood levels (Table 1). Due to the scarcity of the observed flood levels in the study period, we used flood levels reported in the newspaper and websites measured by local news reporters. We also used online geolocated photos from news reporter websites [18-19], to estimate the flood depth at the particular areas by referring to nearby objects in the pictures, such as pathway and car. In total, seven estimated flood level data were used in the model evaluation. Table 1 summarized the collected flood level. Locations of the flood events are shown in Figure 1. Geolocated photos are presented in Figure 2.

IOP Conf. Series: Earth and Environmental Science **884** (2021) 012028 doi:10.1088/1755-1315/884/1/012028

Table 1. Estimated observed flood depth level at locations where online geo-located photos are available

Location	Flood depth (cm)	Sources of evidence
Jomtien Soi 7, Chonburi (P1)	30	car, pathway
Keha Eua Arthorn Na Jomtien, Chonburi (P2)	40	car, people
Sukhumvit road before J intersection, Chonburi (P3)	30	billboard
Pattaya Technical College, Chonburi (P4)	25	car
Jukkacher temple, Chonburi (P5)	40	news report
Noen Kra Prok Intersection, Rayong (P6)	100	news report
Maneewanwittaya school gate, Rayong (P7)	80	news report

P1 P2

P3 P4

Figure 2. Evidence of flood events during Tropical Depression Vamco

The root mean square error (RMSE) is used as an evaluation criteria on the model performance, which is widely and frequently used in the hydrological model [21 - 24]. And the definition of RMSE is in the following form:

$$RMSE = \sqrt{\frac{\sum_{i=1}^{N}(O_i - P_i)^2}{N}} \tag{3}$$

Where N is the number of samples of O_i and P_i which represent the observed data and estimated model result. RMSE can be from 0 to ∞, where 0 is the perfect value of RMSE [21].

2.5. Scenario design

In this study, we designed two flooding scenarios based on the elevation surface height. The first scenario, the original surface elevation was used in the model. The DEM was modified for the second scenario by incorporating building features on the elevation height. Building and road data in 2016 of the study area were derived from the Department of Public Works and Town & Country Planning. The resolution of these data is 1:4000. For building data, each grid cell was defined as 10 m for representing the height of the building in the grid. Then, it was overlaid on the elevation map to see the impact of human activity on flooding. It is to see where would be inundated if there were buildings on the surface. Then, we could see the difference between human-influenced flood and non-human.

As discussed in [13], the flood depth, which could hinder travel and affect transportation, is more than or equal to 30 cm. This depth threshold is based on the heights of car's inlets which many cities adopted to close the road if the water depth rises to 25 – 35 cm. In this study, we used this threshold to define the inundation depth. The depth of the inundated areas must be higher or equal to 30 cm. If, therefore, the water depth is lower than 30 cm, it will be defined as dry areas as it might not have much effect on road disruption

To identify the impact of building structures on the water depth, flood maps of each scenario were derived to compare the difference of water depth in each grid cell. If the difference of the water depth is positive, it means that there is an increase in the water depth, or the dry area becomes a wet area. For zero result, it means that there is no change in water depth or still dry. And for negative differences, it means that there is a decrease in the water depth or wet areas become dry areas.

After that, the rasterized road and land use would be overlaid on the flood maps derived from the model simulation for both scenarios. Then, it could be seen the impacts of flooding on road and land use from two different scenarios. For flooding impact on roads, the total length of inundated roads was estimated based on types of road, which are hard surface (paved road or asphalt road), soft surface (dirt road) and soil surface (pathway/ wagon track). For land use, flooding extent is calculated for each type of land use to evaluate the flooding impact separately, then summarized in total. There are four land use types in the analysis which are agriculture land, forestry land, build-up land, and miscellaneous land.

3. Results and discussion

3.1. Model performance

We compared the observed flood level with the model result. All are shown and summarized in Table 2. From the comparison, all observed places found affected by the flood in the flooding simulation. P6 is the only observed location in which the simulated flood level is equal to the observed level. There are four validated points, which the simulated level is higher than the observation (P1, P2, P4 and P5). Meanwhile the rest are lower than the observed level (P3 and P7). For the evaluation using RMSE, the error of flood depth from the model is 12.154 cm. However, it is important to note that there are many limitations which could cause any error such as, the wrong measurement in the observation. Because the flood level was referred to the nearest object, which is not directly measured.

Table 2. Model validation using observed flood level

	P1	P2	P3	P4	P5	P6	P7
Observed level (cm)	30	40	30	25	40	100	80
Simulated level (cm)	46	48	25	33	55	100	60
Difference (cm)	-16	-8	5	-8	-15	0	20

3.2. Effect of building construction on flooding

The total area of land in the western sub-basin using the hydrological tool in ArcMap to identify was 1,626 km^2. With a 30 cm water depth threshold, we found that 177.5 km^2 of the study areas are inundated under the scenario with raw elevation map. For the scenario with the building's height, the inundated areas increase to 178.6 km^2. It could be seen that human activity has an impact on flood events causing more hazard areas. However, the number of inundated areas in the second scenario might not reflect the total impact of human activity on flooding due to the different data scales. Some small building construction are omitted because of the difference in data resolution. The original resolution of the building features is 4x4 m (1:4000), while resolution of the DEM is coarser (50x50 m). When the rasterization of the building was done, most of the small polygons were not included in the data. So, the human influence on flooding may be underestimated. Table 3 breaks down the inundated areas into 5 intervals according to the flood depth.

Table 3. Inundated areas

Flood depth (m)	Without building construction	With building construction
	km^2	
0.30 – 0.50	41.9	43.2
0.50 – 1.00	66.0	66.4
1.00 – 2.00	56.3	55.9
2.00 – 3.00	10.0	9.8
3.00 – 14.0	3.4	3.4
Total	177.5	178.6

By water depth comparison, it is found that after including building construction in the simulation, there are 155 km^2 of the total study area where water depth increases. Only an area of 6 km^2 where flood depth decreases or becomes dry. And there is no change in water depth for the rest of the sub-basin; 1,464 km^2. Therefore, it could be implied that building construction has an impact on flood depth. The impacts of human construction on flood depth are more obvious than those on flood extent. Figure 3 illustrates (a) the inundated areas in the study area without building construction, (b) with building construction and (c) the change in flood depth.

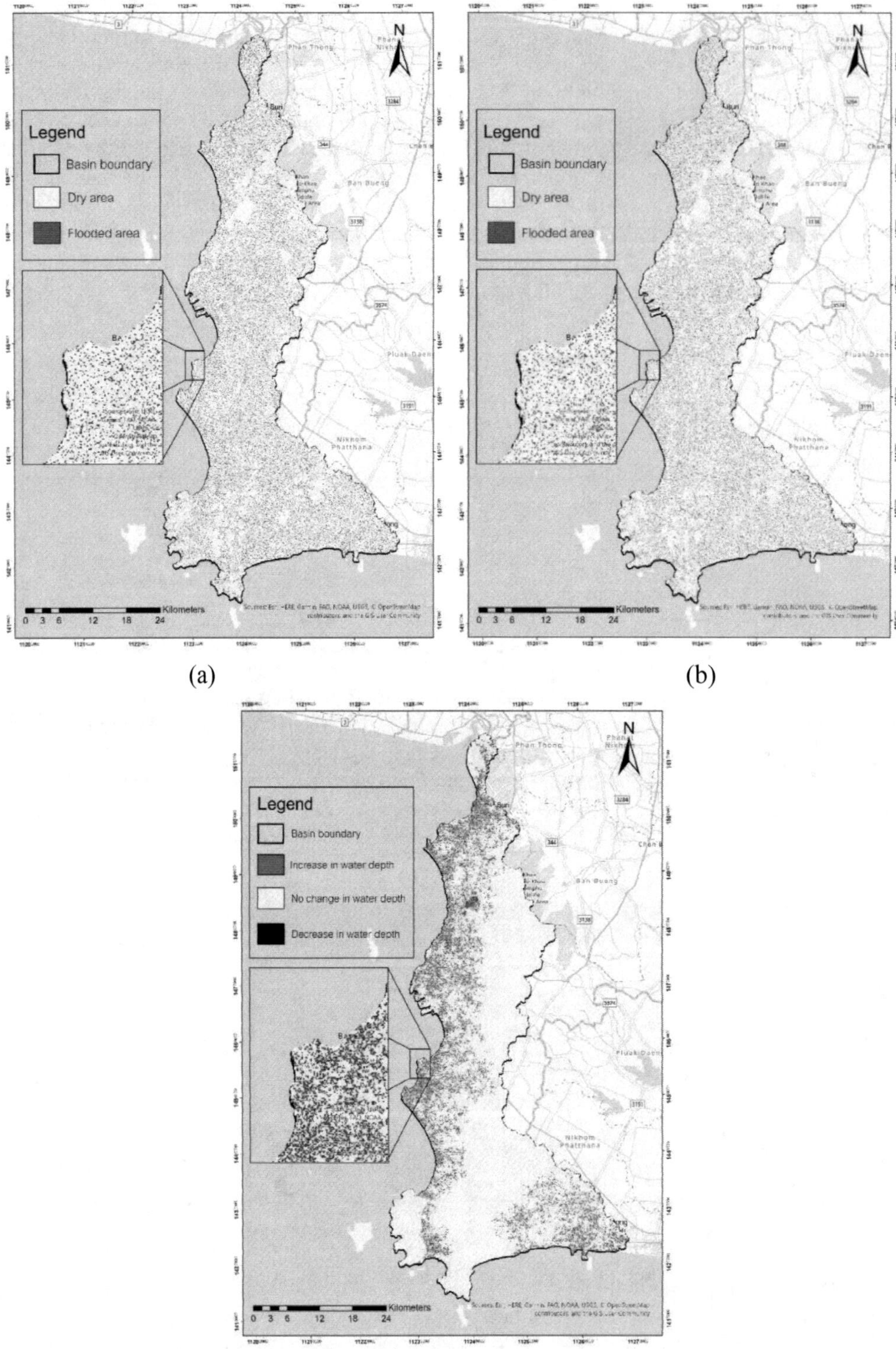

Figure 3. (a) Inundated areas without building construction; (b) Inundated areas with building construction; (c) Change in flood depth between (a) and (b)

3.2.1. Impact on road. Table 4 shows that, with incorporation of building construction, part of the roads that are inundated by flood slightly increases by 0.4% compared to the scenario of no building construction. Paved road is most affected by flooding when building construction is included as can be seen from the largest increase in percentage of flooded distance. Pathway or wagon track, on the other hand, is least affected. However, this road type is more flooded than the others for both scenarios.

Table 4. Inundated road length

Road type	Total Length (km)	Without building construction		With building construction	
		Inundated length		Inundated length	
		km	%	km	%
Paved/asphalt surface	3,720.9	412.5	11.09	431.6	11.60
Dirt surface	2,143.0	248.4	11.59	252.2	11.77
pathway/wagon track	539.6	66.9	12.40	67.2	12.45
Total	6,403.5	727.8	11.37	750.9	11.73

3.2.2. Impact on land use. Overall, all land use types get affected by flood for both flooding scenarios as shown in Table 5. For the most affected land use is miscellaneous land and the least affected land is forestry. Results suggest that the inundated areas increase for most land use types under the second flooding scenario, except forestry. It is because there is no human construction in the forest. It could be implied that there is no human impact on flooding in forest areas.

Table 5. Inundated areas based on land use

Land use	Total Area (km^2)	Without building construction		With building construction	
		Inundated areas		Inundated areas	
		km^2	%	km^2	%
Agriculture land	541.3	61.7	11.40	61.8	11.42
Forestry land	181.8	6.3	3.46	6.3	3.46
Build-up land	660.4	79.1	11.98	79.9	12.10
Miscellaneous land	176.0	21.3	12.10	21.5	12.22
Total	1,559.5	168.4	10.80	169.5	10.87

4. Conclusion

This study evaluates the effects of building construction on flooding in aspects of flood depth and flood extent in the west boundary of East Coast basin of Thailand. Application of the 2D flooding model to investigate the roles building construction in flooding shows that human constructions increase the flood depth up to 3 m in the study area. Flooding areas also expand, which mainly affects paved roads and built up areas. However, there are some limitations in this study, which may result in an underestimation of flood situation, such as the use of relatively coarse resolution topographical data and lacking detailed rainfall and land surface data. Future studies should focus on the use of high resolution of rainfall data and building construction. This can largely minimize model uncertainty and errors in the flood prediction

5. Acknowledgments

This research paper was funded by "the Chulalongkorn University-Faculty of Arts Research Fund".

References

[1] Doswell III C A 2003 Flooding *Encyclopedia of Atmospheric Sciences* 769-776
[2] Fitzpatrick F A, Diebel M W, Harris M A, Arnold T L, Lutz M A andd Richards K D 2005 Effects of Urbanization on the Geomorphology, Habitat, Hydrology, and Fish Index of Biotic Integrity of Streams in the Chicago Area, Illinois and Wisconsin *American Fisheries Society Symposium* **47** 87-115
[3] Water Crisis Prevention Center *Data Development Project for 25 Basins : East Coast - Gulf basin*
[4] Department of Public Works and Town & Country Planning 2015 *Flooding Mitigation Plaing for East Coast – Gulf Basin* (Bangkok: Progress Technology Consultant) chapter 1-2
[5] ESRI 2020 *ArcMap* Available at: https://desktop.arcgis.com/en/arcmap/
[6] Yu D 2005 *Diffusion-based flood inundation modelling over topographically complex floodplains* (Leeds: University of Leeds)
[7] Yu D and Lane S N 2006a Urban fluvial flood modelling using a two-dimensional diffusion-wave treatment, part 1: Mesh resolution effects *Hydrological Processes* **20**(7) 1541-1565
[8] Yu D and Lane S N 2006b Urban fluvial flood modelling using a two-dimensional diffusion-wave treatment, part 2: Development of a sub-grid-scale treatment *Hydrological Processes* **20**(7) 1567-1583
[9] Yu D 2010 Parallelization of a two-dimensional flood inundation model based on domain decomposition *Environmental Modelling and Software* **25**(8) 935–945
[10] Lane S N, Reid S C, Tayefi V, Yu D, and Hardy R J 2008 Reconceptualising coarse sediment delivery problems in rivers as catchment-scale and diffuse *Geomorphology* **98**(3-4) 227–249
[11] Yu D and Coulthard T J 2015 Evaluating the importance of catchment hydrological parameters for urban surface water flood modelling using a simple hydro-inundation model *Journal of Hydrology* **524** 385-400
[12] Yin J, Yu D, Yin Z, Liu M, and He Q 2016 Evaluating the impact and risk of pluvial flash flood on intra-urban road network: A case study in the city center of Shanghai, China *Journal of Hydrology* **537** 138–145
[13] Jet Propulsion Laboratory 2012 *ASTER Advanced Spaceborne Thermal Emission and Refection Radiometer* Available at: https://asterweb.jpl.nasa.gov/gdem.asp
[14] Thairat 2015 *Tropical Depression Vamco hits, putting Royong-Sattahip underwater* Available at: https://www.thairath.co.th/news/local/526088
[15] Morning News TV3 2015 *Heavy rain hit Map Ta Phut, induating markets-schools, students evacuating riotously* Available at: https://ch3thailandnews.bectero.com/news/54748
[16] Nation TV 2015 *Tropical Depression Vamco hit Pattaya, flood cuasing chaotic traffic congestion* Available at: https://www.nationtv.tv/main/content/378471636
[17] Manager Online 2015 *Flash flood hits serveral places in Rayong* Available at: https://mgronline.com/local/detail/9580000105192
[18] Voice TV 2015 *Pattaya hit by heavy and 1 – 2 meter flood, cuasing traffic congestion* Available at: https://www.voicetv.co.th/read/259569
[19] Kapook 2015 *Tropical Depression Vamco cusing severe flood, Pattaya declreaed disaster area* Available at: https://thaiflood.kapook.com/view129448.html
[20] Voice TV 2016 *Small storm sewers cuasing repeated flood in Pattay after rainfall* Available at: https://www.voicetv.co.th/read/365898
[21] Ritter A and Muñoz-Carpena R 2013 Performance evaluation of hydrological models: Statistical significance for reducing subjectivity in goodness-of-fit assessments *Journal of Hydrology* **480** 33–45
[22] Bui D T, Pradhan B, Nampak H, Bui Q-T, Tran Q-A, and Nguyen Q-P 2016 Hybrid artificial intelligence approach based on neural fuzzy inference model and metaheuristic optimization for flood susceptibilitgy modeling in a high-frequency tropical cyclone area using GIS *Journal of Hydrology* **540** 317–330
[23] Kim B and Sanders B F 2016 Dam-Break Flood Model Uncertainty Assessment: Case Study of

IOP Conf. Series: Earth and Environmental Science **884** (2021) 012028 doi:10.1088/1755-1315/884/1/012028

Extreme Flooding with Multiple Dam Failures in Gangneung, South Korea *Journal of Hydraulic Engineering* **142**

[24] Wang Y, Hong H, Chen W, Li S, Panahi M, Khosravi K, Shizadi A, Shahabi H, Panahi S, and Costache R 2019 Flood susceptibility mapping in Dingnan County (China) using adaptive neuro-fuzzy inference system with biogeography based optimization and imperialistic competitive algorithm *Journal of Environmental Management* **247** 712–729

The Local Wisdom of The Tenggerese People to Coexist with Disaster of Mount Bromo

P Yuanjaya[1] and G Meiwanda [2]
[1]Departement of Public Administration, Faculty of Social Science Universitas Negeri Yogyakarta, Jl. Colombo No.1 Yogyakarta, 55281, Indonesia
[2] Departement of Public Administration, Faculty of Social and Political Science, Universitas Riau, Kampus Bina Widya Km 12,5 Simpang Baru Pekanbaru, 28293, Indonesia

Email: pandhu@uny.ac.id

Abstract. The highest losses from global disasters are concentrated in poor countries. Disaster risk in poor countries is exacerbated by conditions such as weak government, vulnerable people especially in rural areas, and environmental degradation (ecosystems). The experience of the people who have long lived on the slopes of the mountain, providing community knowledge to learn about nature and adapt to the dangerous threats of Mount Bromo. Religious and customary institutions play an important role during eruption and post-corruption in collaboration with the Probolinggo District government and related institutions. Religious and customary institutions are very trusted by the community, therefore at the time of the eruption of religious and customary institutions became very important. The decision to evacuate or stay in the village will be fully implemented by the community. Religious institutions with customs, although they can read the activities of Mount Bromo with their knowledge and beliefs, also follow the development of volcano science and collaborate many institutions.

Keywords: local wisdom, disaster, tenggerese people

1. Introduction

The highest losses from global disasters are concentrated in poor countries. Worsens disaster risk in poor countries, is a weak government, vulnerable people, especially in rural areas, and environmental degradation (ecosystem) [1]. Disasters usually take the form of disturbances caused by nature (such as mountain eruptions, earthquakes and landslides) that cause damage to community functions [2,3]. This damage is due to human limitations in handling disasters so that the nature of the disaster becomes a danger to human life. Disasters arise when behavioral threats meet vulnerability (human limitations) [4,5]. Disaster management is a systematic process that involves the government and the community in using their strategies and survival capabilities to reduce the impact of disasters [6,7,8]; Peleg, 2018; Kong, et al, 2020) .

This research focused on the Tengger community has the capacity to deal with disasters through local knowledge, experience and wisdom that has existed for hundreds of years. Mount Bromo is one of the famous mountains in Indonesia in East Java. Mount Bromo is one of five mountains found in the

Content from this work may be used under the terms of the Creative Commons Attribution 3.0 licence. Any further distribution of this work must maintain attribution to the author(s) and the title of the work, journal citation and DOI.
Published under licence by IOP Publishing Ltd

Tengger Mountains complex. Mount Bromo is a volcano that has a long history, both in the natural process of its formation and its role in the spiritual life of the people of Tengger who live around. From historical records, the eruption of Mount Bromo was first recorded by humans in 1804 which informed that Mount Bromo was a very active mountain, with a deadline for resting from one eruption with another eruption of only a few months and a maximum of 16 years. The eruption period can last for one day and a maximum of nine months [9]. Actually the mountain area is a distinct advantage for an area. Fertile land to be processed into agricultural land, beautiful natural scenery and cool air are an inseparable part of the mountain region. The high potential of the area that can be developed makes this area attractive to residents. Behind the good potential, the mountain area is a dangerous area for humans to live in, because mountains are like time bombs that can erupt violently at any time.

Mount Bromo is a potential source of danger that threatens the safety of humans around it (Center for Volcanology and Geological Disaster Mitigation (PVMBG), 2011). The 2010-2011 eruption was the longest eruption ever recorded in the history of Mount Bromo. Phreatomagmatic eruption which produces material in the form of dust to fine sand is a characteristic of Mount Bromo [9]. Fine dust and sand material spreads in the direction of the changing winds from north, east, northeast to southeast. Incandescent eruption material flew as far as 1.5-2 km in the desert area. Rain of ash and fine sand from the 2010-2011 Bromo eruption caused several areas on the slopes of the Tengger Caldera to fail to harvest. The main agricultural products such as potatoes, leeks, and cabbage cannot be harvested. Plants wither and eventually die due to volcanic ash covered. From the data held by the East Java BPBD, agricultural losses caused by the eruption of Bromo reached 2,440 hectares with a value of losses reaching Rp 28 billion. The largest loss occurred in Ngadirejo Village, the area of damaged land reached 304 hectares with a total loss of Rp 8.6 billion. Some villages that were also damaged due to the eruption of Mount Bromo]include Ngadisari, Wonokitri, Wonoroto, Jetak, Ngadas, Wonokerto, Sambikerep, Sambikerep, Pakel, Kadiasari, Sariwangi, Sukapura, and Ngepung Villages [10].

Disasters are seen from three management typologies namely routine, non-routine and complex [11]. The eruption of Mount Bromo when seen from the typology is in the first type, routine, with low to moderate scale and with local impact (small) only in the villages around Bromo (but if the wind is strong enough it will have wider impact), it can be estimated / measured. The mundane approach to mountain and eruption is carried out by PVMBG and people's experience of seeing natural phenomena. This community experience is often used as a guideline in understanding the activities of Mount Bromo when an eruption occurs it is very difficult to evacuate people to distant areas.

Efforts to optimize the role of the community in disaster risk management, the various potential ones in the community must continue to be explored. Review in the form of research by taking an example of Tengger community life in Ngadirejo Village which is prone to the dangers of Mount Bromo. The review is linked to local wisdom, knowledge, and perceptions of the dangers of the eruption of Mount Bromo. Forms of local wisdom in society can be in the form of; values, norms, ethics, beliefs, customs, customary law and special rules. Local wisdom can be in the form of rules concerning; 1) institutional and social sanctions, 2) provisions on spatial use and estimated seasons for planting, 3) preservation and protection of sensitive areas, and 4) forms of adaptation and mitigation of dwellings to climate, disasters or other threats [7,8,11]. Shifting the paradigm of disaster management from being responsive to preventive, demands a variety of new ways in the mechanism of disaster risk management [12]. Involving the role of the community is an alternative model for disaster management, especially in Asia.

2. Theoretical Framework

Changes in disaster mitigation approaches need to reposition the role of local wisdom in disaster governance [12,13,14]. Local wisdom has an important role in creating disaster resilient communities [6]. Disaster management that has been carried out based on top down policies has made the community vulnerable because they are rarely involved in the process [3,15]. In fact, local people have long knowledge and experience of disasters in their area [16].

Local wisdom is a view of life, knowledge and life strategies in the form of activities carried out by local communities in responding to various problems in meeting their needs, including facing disasters [12]. Local wisdom is a custom and custom that has been traditionally carried out by a group of people from generation to generation which is still maintained by certain customary law communities in certain areas [16,17]. Local wisdom is understood as local ideas that are wise, full of wisdom, of good value, which are embedded and followed by members of the community [19,20,21]. Local wisdom is an idea that arises and develops continuously in a community in the form of customs, rules / norms, culture, language, beliefs, and daily habits. Local wisdom in the success of disaster risk reduction is very potential [22,23,24].

3. Meterial and Method

This research is qualitative research, to uncover and understand something behind the phenomenon [25,26]. Qualitative research is methods to explore and understand meaning by a number of individuals or groups of people ascribed to social problems [26]. The location of this research focuses on Ngadisari Village, Sukapura District, Probolinggo Regency, East Java Province, Indonesia. The location was chosen as the closest village to Mount Bromo. In addition, all community activities, both agriculture and tourism services, are closely related to the condition of Mount Bromo. The ties between the local community and Mount Bromo are very strong, both spiritual, cultural and economic. Data collection techniques using observation, in-depth interviews, and review of documentation [27]. Primary data obtained from in-depth observations at the research location by doing life-in and following traditional rituals. In-depth inquiries were carried out to several key informants who had been previously selected [28]. In addition, this study also conducted interviews with local communities using snowball sampling. Documentation review is carried out based on PVMBG data, BPBD and reports from local governments.

4. Result and Discussion

4.1 Potential for Mount Bromo Disaster

Mount Bromo is located about 42 km from the center of Probolinggo City, East Java Province, Indonesia. Apart from the beauty of the natural panorama of the mountains and the distinctive local people, good transportation access and accommodation are the reasons why Mount Bromo is visited by many tourists making it one of the favorite destinations in Indonesia. However, as natural tourism, Mount Bromo also holds potential dangers for human activities due to frequent eruptions.

Mount Bromo has a height of 2,329 meters above sea level. Mount Bromo is actually the youngest active volcanic cone after the formation of the Sand Ocean Caldera from the Mount Tengger Complex. Mount Bromo is one of the volcanoes that often experience eruptions from 129 active volcanoes in Indonesia. The eruption of Mount Bromo when seen from the typology from Handmer and Dovers [11] is routine typology with low to moderate scale and local impact only in the villages around Bromo. The time span between one eruption and another erupted from less than one year to 16 years since it was first recorded in 1804.

Mount Bromo's volcanic activity cannot be separated from its position in the mountainous complex. Ash fall deposits or pyroclastic falls, which are dominated by fine sand, are characteristic of this volcanic complex. Rocks around and within the Ocean Sand Caldera are formed from ash and sand deposits associated with the formation of the Sand Ocean Caldera and the crater of Mount Widodaren. Mount Bromo has an eruption period of 4-6 years. More than 50 eruptions have been recorded since being recorded 215 years ago in 1804, with the last eruption occurring in 2019 (PVMBG). Mount Bromo is a type A volcano which has the characteristic of suddenly causing an eruption accompanied by materials such as gravel and ash. Mount Bromo can cause earthquake tremors with normal amplitudes with amplitudes below 10 mm but can increase to 30 mm. The Center for Volcanology and Geological Disaster Mitigation (PVMBG) has mapped the hazard-prone areas for the eruption of Mount Bromo based on a disaster-prone map (See Figure 1).

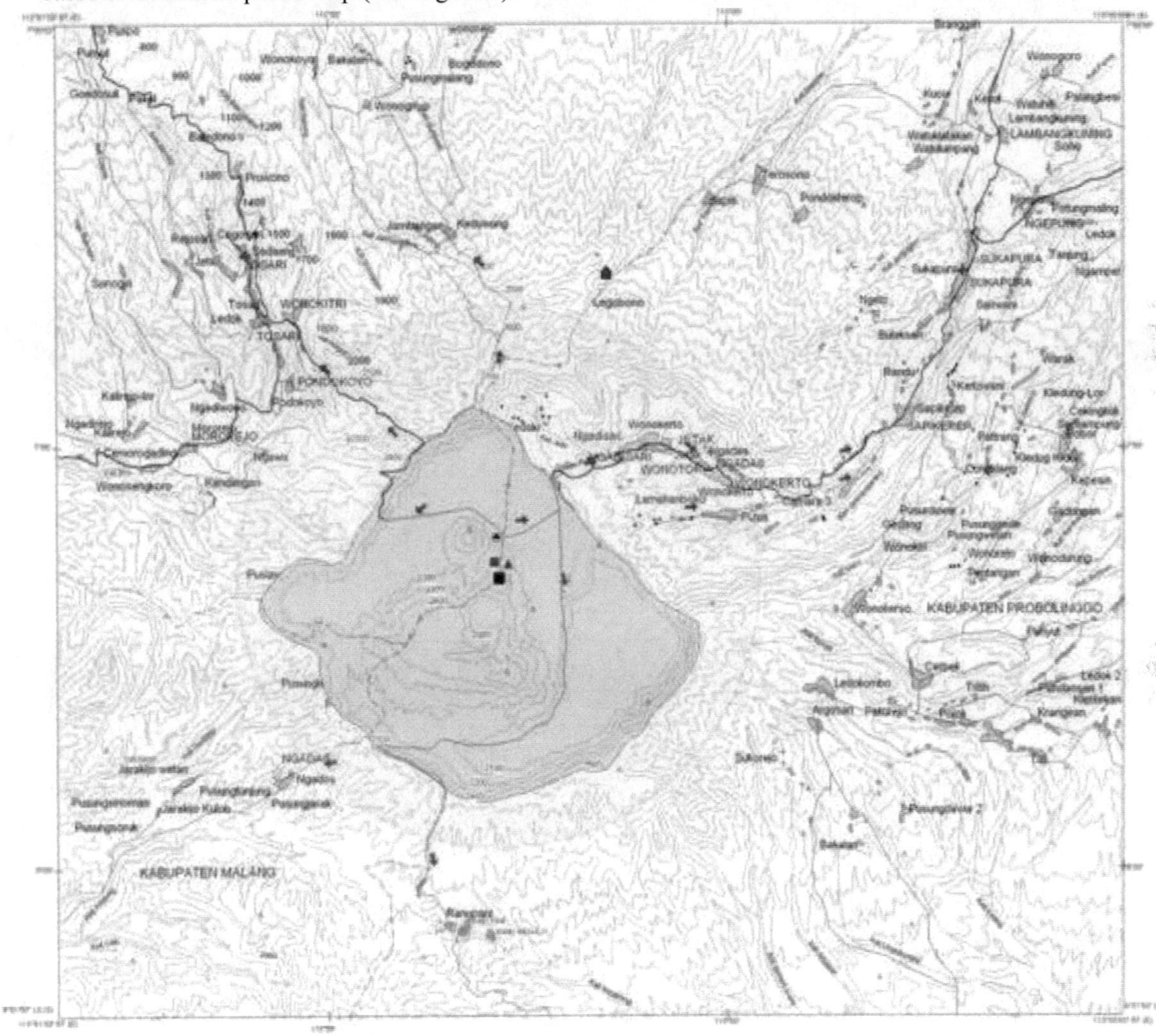

Figure 1. Volcanic Hazard map of Bromo Volcano (PVMBG), Scale 1:50.000

a. Hazard Zone I ( ), potentially threatened by ahs fals and possibly ajected (glowing) rock fragments. This area covers areas ranging from the caldera embankment of the sand sea to a radius of 6 km centered on the crater of Mount Bromo. Hazard Zone I is 63 km2. Apart from the Bromo-Tengger tourist complex in Cemorolawang, the villages included in Hazard Zone I are Ngadisari, Jetak, Wonotoro and Wonokerto. If there is an increase in the activity / eruption of Mount Bromo, residents, tourists and tourism managers in this Hazard Zone I must increase awareness by paying attention to and obeying orders from the authorities. In circumstances of forcing residents in this Hazard Zone I can be ordered to evacuate by the authorities. For this reason, residents must know and pay attention to the direction / route of self-rescue and the location of evacuation.

b. Hazard Zone II ( ) , risk of disasters in this area is higher than Zone I, potentially threatened by ash fals and possibly flows, and ajected (glowing) rock fragments. This area covers the entire base and perch caldera cliffs with an area of 50 km2 and a radius of 2-5.5 km centered on the crater of Mount Bromo. So that the hazard zone for Mount Bromo also includes Mount Batok, Mount Segorowedi, Mount Kursi, and Mount Widodaren. If there is an increase in activity / eruption of Mount Bromo, residents and tourists are prohibited from descending to the bottom of the caldera, let alone approaching Mount Bromo and other volcanoes around it.

4.2 Local Wisdom of the Tengger Community

a. Demographics of Ngadisari Village

The inhabitants of Ngadisari Village are the Tengger Tribe, hereinafter referred to as the Tengger Ngadisari Tribe. If there are some residents who say that they are not descended from the original Tengger Tribe, they are outsiders who live in the Tengger tribe area due to a marriage relationship. According to data belonging to Ngadisari Village, the number of residents of Ngadisari village based on village monograph data until 2017 amounted to 1697 consisting of 511 family cards (KK), 21 Rukun Tetangga (RT), and 3 Rukun residents (RW). The total population of women is 731 and men are 791. The majority of the population who are Hindus are indigenous people of Tengger and a minority of the Muslim population are immigrants or non-native Tengger residents. In terms of education, at first the level of education of the population in Ngadisari Village was still low. Where the majority of the population only graduated from elementary school (SD), but it began to change slowly when village leaders made village government policies requiring residents to study up to high school level.

The type of fertile soil in Ngadisari Village is very supportive for agricultural activities, so that in Ngadisari Village the majority of the population has a livelihood as farmers. The number of people who work as farmers in Ngadisari Village is 916 residents. The large number of residents whose livelihoods are farmers make farming activities a routine and the responsibility of family members ranging from husbands, wives to their children on agricultural land belonging to previous families.

Success in agriculture has made the people of Ngadisari Village prefer to continue or work on family-owned agricultural land instead of attending high school. In addition to their main livelihoods as farmers, residents in this area also work as traders, shop owners, kiosk owners, selling souvenirs, camping rentals, shop owners, hotel service owners, villa / homestay owners, guest house owners, tailors, barbers. , carpenters, builders, carpenters, batik craftsmen, civil servants, TNI, POLRI, tour guides, hardtop (jeep) drivers and horse guides. Currently, most of the population in this region also has a lodging business with the types of guest houses, homestays and villas. This relatively easy business opportunity to run is accompanied by an increasing tourism industry and the development of tourism products. Hotels or villas in the Ngadisari Village area are owned by private communities because only the people of Ngadisari Village have land rights in the village.

b. Values, Traditions and Institutions in the Tengger Tribe Society

Tengger tribe people believe in and apply what is called *Cattur guru bhakti* in their daily life. The concept of community trust perched on Mount Bromo, they divide the pillars of belief in *Catur guru Bhakti* into four parts which they apply in daily life and influence their belief in the existence of Mount Bromo, the four pillars of belief are:

1) *Guru Rupaka*, namely trusting and respecting one's own parents, because for the perch society, *Guru Rupaka* is taken from the word Ngereka from something that does not exist, because for that service, respect for parents is a must because of their great service

2) *Guru Pengajian* namely teachers who teach at schools or places to gain knowledge in the beliefs of the Tengger community, also uphold teachers because for them teachers have taught many things that they cannot get in their environment, such as reading, writing and so on.
3) *Guru Wisesa* namely belief in the government, is called *guru wisesa* because it is taken from the word "*ngawisesa*" or to rule, which creates tranquility and peace
4) *Guru Swadhaya* namely the belief in *Sang Hyang Widhi*, *Sang Hyang Widhi* is the creator of all the contents of the world, so people really respect things related to nature, because for them *Sang Hyang Widhi* also represents the power of nature

Catur Guru Bhakti is a concept of trust and public respect for what is considered Guru and trusted from generation to generation, the concept of trust is not only applied in daily life but also has an impact when people are perched on the eruption of Mount Bromo because their belief in *Sang Hyang Widhi* is a representation of Their natural forces consider the eruption to be a blessing given by *Sang Hyang Widhi* through the eruption of Mount Bromo to the Tengger community. Then the concept of the perch community trust in the government or in Guru Wisesa is also influential when the local government wants to evacuate and establish a safe radius distance from the eruption of Mount Bromo because the perch community trusts and respects all decisions made by the local government for them.

The Tengger Ngadisari Village perform several ritual traditions based on their customs and beliefs, namely performing ceremonies *Leliwet, Pujan, Munggah Sigiran* (*Among-among/ ngamongi* jagung), *Wiwit, Kasada, Mayu* (*Mahayu*) village, *Mayu Banyu* and *Pujan Mubeng*. Almost all rituals have a strong relationship with the closeness of the community to Mount Bromo. *Kasada Day* which is held every year on the full moon in December or January is also the same. *Kasada day* has a story, starting in the 15th century, starting with a story about a daughter *Roro Anteng* who led the Tengger kingdom with her husband *Joko Seger*. This couple did not have any children, so they prayed and asked the mountain gods to give them a child. From their request they were granted 25 children, then the god ordered the sacrifice of the last child (*Raden Kusuma*) to be thrown into the volcano and if the request was not done there would be a big disaster that befell him. The god's request was carried out, so it has become a tradition to this day by throwing crops and livestock. Tengger community believe in this mountains will not be angry with them. Therefore, the Tengger people gather, pray and perform ritual offerings from their agricultural products while still carrying out the cultural values of their ancestors.

Local wisdom in the Ngadisari Tengger community is also manifested in community institutions. Institutions play a role as a social system that regulates the hierarchical structure of social and community groups, which can be in the form of customary organizations consisting of several customary groups. This is the case with the Tengger Tribe of Ngadisari Village, where there are traditional organizations in charge of managing community life, namely religious leaders and traditional shaman institutions. The Hindu Tengger concept is that there is a grouping between the religious system which is derived from the teachings of divinity based on Hinduism and the traditional system which is derived from the beliefs and traditions passed down from generation to generation from the ancestors of the Tengger Tribe. However, in the implementation stage, the assimilation of Hindu religious teachings with the teachings of the customs / beliefs of the Tengger Tribe is carried out. This is reflected in the fact that apart from carrying out religious activities based on the teachings of Hinduism, the people of the Tengger tribe also obediently carry out various traditional ceremonies. The existence of groupings of religious and customary activities affects the division of tasks and functions of each religious leader and traditional dukun institutions.

Religious leader institutions, are religious institutions that accommodate the chairman and administrators of religious activities in Ngadisari Village. The management structure of Religious Leaders Institutions consists of: 1) *Mangku Gedhe*; 2) *Mangku Gelar*; dan 3) *Mangku Alit*. The area that is sanctified and becomes the area of responsibility of the Mangku is the Dhang Kahyangan Kerti Jaya Buana Temple which is a place of worship for Hindus. The traditional dukun institution, functions as a customary institution that accommodates the chairman and administrators of the custom. The management structure of the traditional dukun institute consists of: 1) *Dukun*

Adat; 2) *Legen*; 3) *Sanggar*; and 4) *Sepuh*. *Dukun, Legen, Sanggar* and *Sepuh* are responsible for *padhanyangan* which are customarily sanctified areas. In addition, the role of a traditional shaman is to oversee the implementation of customary rules and customary law. Like other villages in general, the administrative area of Ngadisari Village is headed by a village head, but what distinguishes it from most villages is the dukun / adat elder who plays an important role in leading the customary territory as a customary head.

The Tengger community is divided into two customary areas, namely sabrang kulon (represented by Tosari Village, Tosari District, Pasuruan Regency) and sabrang wetan (represented by Ngadisari, Wanantara, Jetak, Sukapura District, Probolinggo District) consisting of village groups each led by a customary chief. Thus, the boundaries of the work area of the traditional dukun are the customary areas and communities in the village where they serve as traditional shaman. In each district there is a regional dukun coordinator who is in charge of coordinating traditional shaman in their area. The traditional shaman who is in each village area of the Tengger Tribe community is generally respected and highly trusted because of his very influential role in the life of the Tengger Tribe community. Structurally, the traditional shaman in the life of the Tengger Tribe is classified as a respected person who becomes a role model for the community and is more respected than the village apparatus. As a customary head, the traditional dukun has both a spiritual and social function. The spiritual function of the traditional shaman is to lead traditional ceremonies. Meanwhile, its social function is to act as a mediator between society and affairs related to government. In addition, traditional healers also have certain powers in making decisions, rules, sanctions or social fines for violators of customary laws and regulations. For example, the dukun's authority in decision-making is that when a disaster occurs, the dukun has the right to determine when the community should evacuate or remain in the village.

c. Village Layout

The concept of space based on customary territory and administrative territory can be explained through two aspects, namely territorial boundaries based on physical markers and non-physical markers. Administrative boundaries based on physical markers can be clearly stated, for example roads and rivers. Likewise with physical markers on the boundaries of customary territories in the form of locations or buildings that are ritualistic such as temples, padhanyangan (dhanyang), and tombs with the orientation of Mount Bromo which function as the center of ritual activities (pancer). Non-physical markers at the boundaries of customary territories can be observed from the behavior settings of the community, for example still following beliefs, laws, customary rules, language, the nature and attitude of life of the Tengger Tribe. In connection with this, Pangarsa, Pamungkas and Subekti (1992) reveal that space in the Tengger architectural tradition can be explained through the concept of space boundaries, at the macro-spatial scale (area, village and environment) to the micro space scale (residence). On a regional scale, there are two conceptions of space that occur; customary territory and administrative area. The boundaries of customary territories are not as strict as the administrative areas of the village, and these two boundaries do not always coincide. The Tengger Ngare conception, which passes through the mountainous area of the local dialect as well as the work area of the dukun, is an indication of the boundaries of customary territories.

1) Settlement spatial orientation

Identifying the value of local wisdom in settlements can be seen from certain customary rules / regulations governing the orientation of the placement of the elements that form the settlement. The concept of the direction that has developed and has become a hereditary belief in the Tengger tribe has a philosophical meaning and is symbolized by certain color elements. The meaning contained in the concept of this direction is then interpreted in the *Pujan Mubeng* (*Nrundhung*) ritual which aims to ask for the safety of the village and to clean the village from disturbances and disasters. The form of application of the philosophical meaning contained in the concept of this direction based on the customs and beliefs of the Tengger tribe is in the form of various color offerings (red, white, yellow, black) which have a philosophical meaning symbolizing the four corners of the village.

Ngadisari Village has the application of a customary rule which is the basis for the concept of direction in laying the elements that form the settlement, including: 1) Tombs consist of sacred graves and ordinary tombs. The provision for laying sacred graves is to the north of the village and far from the location of residential areas. The belief that is believed by the Tengger tribe regarding the laying of tombs is that it is preferably outside the residential area and placed in the north. From the past until now the location of the sacred grave has remained in the same place with the area of land that cannot be increased or decreased. Until whenever the sacred tomb must remain in that place with a fixed area; 2) The temple as a place of worship is placed in a sacred place in the middle of the settlement, which is the place where there are the most pamujan studios in the vicinity. The location of the *Dhang Kahyangan Kerti Jaya Buana* Temple in Ngadisari Village is in the Northeast of the residential area. The philosophical meaning contained in the provision of laying the temple to the East is that it faces the sun. As a sacred place, the temple is placed on the highest contour of the land; 3) *Padhanyangan (dhanyang)* is a sacred place for the Tengger Tribe of Ngadisari Village. *Padhanyangan (dhanyang)* is located in the south of the village and is in one orientation with the sacred tomb (towards Mount Bromo); 4) Residential buildings (micro) are divided into several rooms, namely the pamujan (place of worship), patamon (living room), paturon (bedroom), pagenen (kitchen), pedaringan (storage room), pakiwan (bathroom), and pekayon (a place to store wood). The terms of the placement of each room are; a) *Sanggar pamujan* is placed in front of the house, must face East or South, must not face West and North, b) the *patamon* is placed at the front of the house, c) the *paturon* must be to the right of the direction of the *pelawangan* (door), d) the *pagenen* and pedaringan are placed in behind the patamon and it can be combined, e) the placement of clerks must be in the back of the house, f) *pekayon* is an additional space, placed at the back of the house. In addition there is a customary provision which states that children who are married are not allowed to build a house to the right of the house his parents; 5) The fields / moor used for agriculture are located in the South, North and East of the village; and 6) Mount Bromo, which is located in the south, is believed to be the axis (pancer) of spiritual activity for the entire Tengger tribe. There is a sacred axis that leads to Mount Bromo (South) which connects between the sacred tomb and the *padhanyangan (dhanyang)*.

Regarding the concept of Mount Bromo, which is located in the South as the axis of spiritual activity for the entire Tengger tribe, it has been stated by Hefner (1985) that in Tengger, the presumption of a Southern orientation was followed on the basis of a belief which states that the South is identified as the throne of Bromo or Dewa. Brahma. This can be seen not only at the position of the burial, praying, chanting facing south, or in the orientation of the door (the main door or the shrine), but also in the orientation of the holy places. With Mount Bromo as the spiritual center / axis, it means that people always pay attention to Mount Bromo at all times in order to anticipate all possibilities (mitigation)

2) Land tenure and ownership system

The life attitudes of the Tengger Tribe that are important are the tenure system (not many risks), *ojo jowal-jawil* (do not like to disturb others), work hard and maintain their own land from generation to generation [29]. As in other Tengger Tribe communities, the land tenure and ownership system is regulated by customary rules which state the prohibition or prohibition against selling land outside the Tengger Tribe community. Whatever the reason, the sale of inherited land or land can only be carried out among the Tenggerese people, usually the sale of inherited land or land is prioritized to close relatives. Land owned by the Tengger tribe in Ngadisari Village is generally obtained from the legacy of their parents. The inherited land distribution system has also been maintained since now with the provision of equal distribution between boys and girls. With customary regulations like this, land management will never change and the community will be at risk of carrying out dangerous activities.

3) Adaptation of residence to climate

The application of the value of local wisdom is the adaptation of a place to live to the climate. The construction of the traditional house of the Tengger Tribe in Ngadisari Village has the ability to adapt to the local climate. The existing condition of the appearance of the traditional house building of the

Tengger Tribe in Ngadisari Village is as follows: 1) Roof, in the form of a pyramid resembling an upright triangular pyramid with a slope of ± 45 °, made of zinc; 2) Wall, in the form of a wall or half wall, the wall is made of brick while what is meant by half wall is a combination of wall and wooden planks; 3) Doors, made of wood; and 4) Windows, made of wood and glass frames.

The characteristics of the adaptation form of the traditional house of the Tengger Tribe in Ngadisari Village to the local climate can be explained based on the functions of several components contained in the construction of residential buildings and the surrounding environment. The climatic conditions in Ngadisari Village include tropical climates with average daily temperatures between 16 ° -23 ° C. Changes in the use of roofing materials in the traditional houses of the Tengger Tribe in Ngadisari Village are caused by adaptation factors to the local climate. Around the 1950's the roof used zinc, but then in the 1970's it was replaced by tile. However, it turns out that the house whose roof uses tile causes the temperature inside the house to get colder. Therefore, since the 1980s, the roof of the traditional house of the Tengger Tribe in Ngadisari Village has been replaced again by using zinc which is more suitable if applied to cold climates such as in Ngadisari Village. From the changes in the use of materials for the roof, it can be identified the form of the value of local wisdom, namely through a process of trial and error then the most suitable and adaptive results are obtained if applied. With a triangular upright roof reduces the risk of collapse during volcanic ash rain.

4) Environmental Management

Local wisdom in environmental management can be seen in various economic activities of the community. The use of technology in the management of fields / moor in Ngadisari Village is limited to simple and environmentally friendly agricultural technology. The planting system uses an intercropping system. Because the land contours are quite steep, to avoid landslides and erosion, a terracing system is created by making plots of land called beds. Strong rooted plants, for example, cypresses, are planted in the fields of Ngadisari Village to prevent landslides and erosion, apart from having strong roots, the wood can also be used as building materials. The type of fertilizer used prioritizes the use of manure / compost which according to the Tengger Tribe of Ngadisari Village is environmentally friendly and does not damage the texture and soil fertility.

The farms cultivated in Ngadisari Village are cattle and pig farms. The form of application of local wisdom of the Ngadisari Village Tengger tribe community in the livestock raising system, namely by placing cages that are far from residential areas. The placement of the cattle sheds far from the settlement is a form of environmental wise action. Apart from this, the people of the Tengger Tribe in Ngadisari Village also use livestock manure to make manure which is able to fertilize without damaging the soil texture but is also environmentally friendly.

The value of local wisdom in the Tengger tribe of Ngadisari Village regarding the forest management and protection system is to classify the forest and use it. The Tengger tribe of Ngadisari Village has a high awareness in managing the forest. The proof of the concern of the Tengger tribe of Ngadisari Village in participating in maintaining the forest is by not cutting the forest carelessly. This attitude in forest management and protection is based on the slogan that is adhered to, which reads "cut one plant two" which means that if you cut one tree, you must plant at least two trees of the same type. This water resources management activity is also one of the social activities which is a weekly activity for the Tengger Tribe of Ngadisari Village, namely working together to clean public water reservoirs.

4.3 Reflection on local wisdom in disaster mitigation

The people of Ngadirejo Village have been living on the slopes of Mount Bromo eruptions often have an impact, but the people of Ngadirejo Village still inhabit the slopes of Mount Tengger. Mount Bormo, which is one of the threats living on the slopes of Mount Tengger, is considered a source of life by the community. The community knows that Ngadirejo Village is the "route" through which the eruption material passes every time Mount Bromo erupts. However, it is considered a blessing in itself. Eruption material such as ash is considered as fertilizer for agriculture. Agricultural yields usually increase after an eruption.

The way people read nature's signs also affects their perceptions [30,31]. Natural signs such as abundant yields are a sign for them that Mount Bromo will erupt again. There are signs that are around the environment, people use it as a danger alarm, it is believed by the community even though it cannot be studied from a scientific perspective [31,32]. Eruption materials such as ash and sand are not considered a hazard by the community compared to the hot cloud eruption material released by Mount Merapi. The community can say they are lucky, even though the agricultural land was totally damaged by the eruption of Mount Bromo.

The local government has prepared evacuation routes and evacuation tents in collaboration with community institutions, so that when Mount Bromo erupts, it is easy for people to be directed to save themselves. The government cannot force because of the belief in the community that Mount Bromo will not cause casualties. Steps taken by the government were monitoring the situation, alerting TNI and Polri officers at the scene of the incident. Transportation is provided whenever an evacuation must be carried out. Coordinate with PVMBG to find out the latest status during the eruption of Mount Bromo. The village head must report the condition of his citizens. Probolinggo's BPBD priority is saving human lives.

Perceptions of hazards are built by the knowledge of a person / community about these hazards [33]. The people of Ngadirejo village who have lived on the slopes of Mount Tengger for hundreds of years have their own view of Mount Bromo. Mount Bromo is considered the center of cosmology, their beliefs, their mother, and their brothers who inhabit the mountain. Perceptions formed in society are influenced by a strong cultural sector and have been passed down from generation to generation [33]. The Tengger community no longer felt that Mount Bromo was dangerous, so no response was given when Mount Bromo erupted. This is a form of public agreement with nature that people accept the risk of living in a mountain area by considering threats as a blessing.

5. Conclusion

The experience of the people who have lived on the slopes of the mountain for a long time provides the community with knowledge to learn about nature and adapt to the dangers of Mount Bromo. Traditional ceremonies that are often carried out can maintain the togetherness and mutual cooperation of the Tengger community. The local wisdom of the Tengger community in a disaster is as follows:

a) The community places Bromo as the axis and all face the mountain, indicating that the community is always monitoring mountain activities which can erupt at any time.
b) Customary institutions are highly obeyed and are very open to knowledge of disasters.
c) Togetherness in facing disaster
d) Environmental preservation by the community through customary regulations as a mitigation effort to reduce landslides, pests, famine, and social conflicts

The local values of the Tengger people are transformed into economic life, farming systems, land management, respect for nature and living side by side with Mount Bromo. This has had a very big impact on the Tengger community who live together with volcanoes and efforts to minimize the risk of disasters.

6. Acknowledgments

The authors would like to thank the Ministry of Environment and Forestry Republic Indonesia, The Center of Bromo Tengger Semeru National Park, and The Tengger Community for supporting this research.

References

[1] Jayasuriya, S and McCawley, P. 2010. *The Asian Tsunamy*. UK: Edwar Elgar
[2] UNDP, 2004. *Reducing Disaster Risk: A Challenger for Development*. New York: UNDP
[3] Oktari, R S et.al. 2020. *International Journal of Disaster Risk Reduction*. Vol 51
[4] Utami, Sri. 2011. "*Bertani Selaras Alam di Lereng Merapi*". Yogyakarta: Sekolah Pascasarjana UGM (CRCS).

[5] Lane M B 2003 *J. of Plann. Edu and Re.* **22** 360-373
[6] Kusumasari, B. and Alam, Q. (2012), *Disaster Prevention and Management*, Vol. 21 No. 3, pp. 351-369.
[7] Peleg, K et.al. 2018. *International Journal of Disaster Risk Reduction*. Vol 31, 750-757
[8] Kong, T M et. al. 2020. *International Journal of Disaster Risk Reduction*. Vol. 59
[9] Zaennudin, A.. 2011. *Jurnal Lingkungan dan Bencana Geologi*, Vol.2, No. 1, hal. 21-37.
[10] Rachmawati, T A, Rahmawati, D, and Rachmansyah, A. 2018. Disaster risk analysis of Mount Bromo eruption after the 2015 eruption in Sukapura District. MATEC Web of Conferences 229
[11] Handmer, J and Dovers, S. 2007. The Handbook of Disaster and Emergency Policies and Institutions. London: Earthscan
[12] Lin, P S and Chang, K M. 2020. *International Journal of Disaster Risk Reduction.* Vol 42.
[13] Douglas, E M, Reardon, K M, and Tager, M C. 2018. *Journal of Urban Management*. Vol 7, Issue 3, 152-160.
[14] TO, K and Gerrard, V 2013. ISCGP. Colombo: Sri Lanka
[15] Smith E A 2010. *Evolution and Human Behavior* **31** 231-245
[16] Kankanamge, N et al. 2020. . *International Journal of Disaster Risk Reduction*. Vol 42
[17] Gyau, et al 2014. Current Opinion in Environmental Sustainability **6** Pages 68-72
[18] Okubo, T. 2016. *Journal of Cultural Heritage* Vol 20, 715-724
[19] Yuanjaya, P and Fajri, H. 2020. IOP Conference Series: Earth and Environmental Science 448
[20] Pretty J and Ward H 2001 *World Development* **29** 209-227
[21] Poteete R and Ostrom E 2004 *J. Dev. and Change* **35** 435–461
[22] Coulson, N E, McCoy, S J, and McDonough, I K. 2020. *Regional Science and Urban Economics*. Vol 85
[23] Olson M. 1965. *The Logic of Collective Action*. Harvard University Press: Cambridge
[24] Karlsson B M and Hovelsrud G K 2015 *J. Glob. Env. Change* 96–107
[25] Yin R K. 2003. *Case Study Research*. Jakarta: PT. Raja Grafindo Persada.
[26] Creswell J W. 2010. *Research Design*. Yogyakarya: Pustaka Pelajar
[27] Sultana P and Thompson P. 2003. *Methods for consensus building for community based fisheries management in Bangladesh and the Mekong*. CAPRi working paper 30. Washington DC
[28] Strauss A. and Corbin J. 2007. *Basics of Qualitative Research*. Yogyakarta: Pustaka Pelajar.
[29] Sukari, et al. 2004. *Kerifan Lokal di Lingkungan Masyarakat Tengger Pasuruan Jawa Timur*. Yogyakarta: Kementrian Kebudayaan dan Pariwisata.
[30] Badstute L B et al., 2002. Working Paper 02-02. CIMMYT
[31] Meinzen-Dick R, Di Gregorio M and McCarthy N 2004 *J Agric Syst* **82** 197–214
[32] Bhatti, Amjad. 2007. *Risk Perception, Culture and Communication: A South Asian Experience*
[33] Laurens J M 2012 Procedia – Social and Behavioral Sciences. **36** 372-382

IOP Conf. Series: Earth and Environmental Science **884** (2021) 012030 doi:10.1088/1755-1315/884/1/012030

Disaster Preparedness in Physical Aspect of students in Bantul and Balikpapan Indonesia

Saryono[1], Ika Setya Pratiwi[1], Soni Nopembri[1], and Diana Septi Purnama[2]

[1] Departement of Sport Education, Universitas Negeri Yogyakarta, Jl. Colombo No. 1 Yogyakarta, Indonesia
[2] Department of Guidance and Counseling, Universitas Negeri Yogyakarta, Jl. Colombo No. 1 Yogyakarta, Indonesia

Email: saryono@uny.ac.id

Abstract. This study aims at identifying difference in physical fitness level between students in disaster-safe school in Bantul and in Balikpapan, Indonesia. This study is a quantitative descriptive study that adopted a comparative method. Data were collected through tests and measurement. Population of the study was all upper-grade students (105 students) of State Madrassa *Ibtidaiyah* (MI *Negeri*) 2 Bantul and all upper-grade students (102 students) of State Primary School (SD *Negeri*) 003 Balikpapan. Samples were selected using a purposive sampling technique with as many as 65 students of MI *Negeri* 2 Bantul and 87 students of SD *Negeri* 003 Balikpapan involved. Instrument of the study was the Indonesian Physical Fitness Test (TKJI) for children aged 10-12 years. Data were analyzed using a 2-sample t-test preceded by normality and homoscedasticity tests. Results indicate significant difference in physical fitness level between upper-grade students at MI *Negeri* 2 Bantul disaster-safe school and those at SD *Negeri* 003 Balikpapan. The t-test generated a t-value of 7.664 and t-distribution of 1.655 (df 150;5%) with significance level (p) of 0.000. With t-value being greater than t-distribution ($7.664 > 1.655$) and p-value being less than 0.05 ($0.000 < 0.05$), it is clear that physical fitness level of students in both schools differs significantly. It can also be concluded that physical fitness of students in Bantul (a mean of 12.40) is higher than that of students in Balikpapan (a mean of 9.93). Hence, it can be said that students in Bantul are more physically prepared for disaster than students in Balikpapan.

1. Introduction

Geographically speaking, Indonesia is situated in the Pacific Ring of Fire and is highly prone to disaster such as earthquake, volcanic eruption, tsunami and landslide. In addition, as a tropical country that is made up of forested areas and lives along with encroachment culture, the country is constantly faced with forest fires. Indonesia's geographical background and natural potential are among serious threats to Indonesians. One of the ways to cope with this threat is to make use of education to help people prepare for natural disasters, which can be done through, for example, Physical Education (PE). This subject provides students with opportunities to learn to move and to learn through movement as well as to create strong and healthy (both physically and mentally) characters. PE integrates motor skill, physical ability, knowledge and logic and teaches appreciation of attitudinal values and habituation of healthy lifestyle to stimulate balance growth and development. PE aims to help students improve their physical fitness and health. According to Suharjana, physical fitness indicates one's ability to perform daily activities without experiencing too much exhaustion and with some spare of energy to do other minor

Content from this work may be used under the terms of the Creative Commons Attribution 3.0 licence. Any further distribution of this work must maintain attribution to the author(s) and the title of the work, journal citation and DOI.
Published under licence by IOP Publishing Ltd

tasks [1]. Physical fitness is crucial for human and therefore needs to be built since early age, through activities in both formal and informal settings. With physical fitness, it is expected that students can study well and live more healthily, which can eventually improve the quality of human resource.

Good physical fitness will affect students' learning activities as they become more excited in participating in learning process and more ready for receiving any given materials. Physical fitness or excellent physical conditions improve students' ability and willingness to learn. In other words, physical fitness will influence one's physical and mental readiness and ability to handle certain workload (learning activity) that is part of students' responsibility. Besides, this study specifically aims to build physical preparedness for emergencies, or in this case, disaster. It is incredibly important to be physically prepared for disaster. Being mentally ready is not enough; one should also maintain physical preparedness for disaster. It is expected that a person with excellent physical fitness and physical ability can survive longer than other people with poor physical fitness. Physical fitness can be improved through PE subject at school or other types of training.

Primary school is embodiment of government's effort to improve physical fitness since early age through formal education. According to data presented on InaRisk website compiled by the National Agency for Disaster Management (BNPB), Bantul is a regency with high potential for earthquake (94%) and has been nationally identified as red zone [2] while Balikpapan is constantly exposed with forest fires every year [3] Balikpapan is a city with fairly unique characteristics. Topographically, Balikpapan is a city whose altitude ranges from 0 to 80 meters above sea level, situated mostly along the coast and made up of hills (85%). Such conditions cause erratic rainfall and temperature in the city, making it prone to forest fires. With such circumstances, disaster-safe schools in Bantul and Balikpapan should ensure that students are physically prepared and fit for responding to emergencies [4].

Physical preparedness can be built through PE subject taught at school. Data from a research on physical fitness level among students in volcanic-disaster-prone zone indicated the lack of preparedness [5]. There is a test namely Tes Kesegaran Jasmani Indonesia (TKJI) as an Indonesian Physical Fitness Test to measure students' physical fitness. In a physical fitness workshop conducted in 1984, TKJI was appointed an instrument that is valid for nationwide use as it was arranged and tailored to suit the conditions of Indonesian minors. TKJI is divided into four age groups including 6-9 years, 10-12 years, 13-15 years, and 16-19 years. This test is to measure and determine physical fitness level among Indonesian minors (based on their own age groups). This study involved students aged 10-12 years old and collected data from upper-grade primary school students. Involvement of primary school aged respondents was due to factors such as low activity level of children in Bantul[6] and an indication of poor physical fitness in Martapura and Banjar Baru, areas with similar geographic characteristics to Balikpapan [7], [8]. Prior to this study, MI Negeri 2 Bantul and SD Negeri 003 Balikpapan had never used TKJI to measure physical fitness of their students. Therefore, results of this study can be source of data/information for PE teachers. In addition, from the results, physical fitness of students in disaster-prone zone can be identified to prepare for emergencies that might occur. Referring to the problem stated, the objective of this study is to identify difference in physical fitness levels between students in disaster-safe school in Bantul and students in disaster-safe school in Balikpapan.

2. Method

2.1. Participants

This quantitative study adopted a comparative descriptive method and compared physical fitness level among upper-grade students in disaster-safe schools. This study took place in two different disaster-safe schools namely MI *Negeri* 2 Bantul and SD *Negeri* 003 Balikpapan in November 2019. Population of this study was all upper-grade students (grade IV, V and VI) in both schools, totaling 207 students. Meanwhile, samples of the study were 152 students selected using purposive sampling technique with such criteria as: primary schools in earthquake-prone zone in Bantul regency and fire-prone zone in Balikpapan municipality, upper-grade students (grade IV, V and VI), age ranging from 10 to 12 years old, being present during TKJI and being physically and mentally healthy, willing to or allowed by the schools to participate in the study.

Table 1. Total number of students in disaster-safe schools in Bantul and Balikpapan involved in this study

No.	School	Grade	Total
1.	MI N 2 Bantul	IV	15
		V	22
		VI	28
	Total		65
2.	SD *Negeri* 003 Balikpapan	IV	22
		V	34
		VI	31
	Total		87

2.2. Data Collection

Instrument used to measure physical fitness in this study was the TKJI 2010 for children aged 10 to 12 years old [9] TKJI was chosen as it has been appointed as a reference and designed to suit the conditions of children in Indonesia. Besides, this test is not only relatively easy to use but has also been proven valid and reliable for data collection. This study that adopted TKJI conducted five different tests, which included 40-meter sprint, pull up, sit up, vertical jump, and 600-meter running. Data were collected using test and measurement using the Indonesian Physical Fitness Test (TKJI). The initial stage of this study included test preparation that ranged from providing students with understanding about the test, preparing equipment and other necessities for TKJI, preparing track by dividing it into five sequential posts to assigning personnel to conduct TKJI. This stage was followed by execution of the physical fitness test using TKJI that comprised of some items including (1) 40-meter sprint, (2) pull-up, (3) sit-up, (4) vertical jump, and (5) 600-meter running. Following this, the test began by summoning students in the field, preparing all the necessities, conducting warming-up session, and providing students with explanation related to TKJI. Afterwards, students were called one by one based on their number and class to do all tests in sequence while being guided by two officers in each post. Each student took turn to do the test sequentially, starting from 40-meter sprint, pull-up, sit-up, vertical jump to 600-meter running. Data collected from each student were then recorded and converted in TKJI table as presented in Table 2 and 3.

Table 2. Indonesian Physical Fitness Test (TKJI) Score for Male Students Aged 10 to 12 Years Old

40-meter Sprint	Pull-up	30-second Sit-up	Vertical Jump	600-meter Running	Score
Up to – 6.3"	≥ 51"	≥ 23	≥ 46	Up to – 2'09"	5
6.4" – 6.9"	31" – 50"	18 – 19	38 – 45	2'10" – 2'30"	4
7.0" – 7.7"	15" – 30"	12 – 17	31 – 37	2'31" – 2'45"	3
7.8" – 8.8"	05" – 14"	04 – 11	24 – 30	2'46" – 3'44"	2
8.9" – etc.	04" etc.	0 – 03	23 etc.	3'45" – etc.	1

Source: [9] Kemendiknas (2010)

Table 3. Indonesian Physical Fitness Test (TKJI) Score for Female Students Aged 10 to 12 Years Old

40-meter Sprint	Pull-up	30-second Sit-up	Vertical Jump	600-meter Running	Score
Up to – 6.7"	≥ 40"	≥ 20	≥ 42	Up to – 2'32"	5
6.8" – 7.5"	20" – 39"	14 – 19	34 – 41	2'33" – 2'54"	4
7.5" – 8.3"	08" – 19"	07 – 13	28 – 33	2'55" – 3'28"	3
8.4" – 9.6"	02" – 07"	02 – 06	21 – 27	3'29" – 4'22"	2
9.7" – etc.	0" – 0.1"	0 – 01	20 etc.	4'23" – etc.	1

Source: [9] Kemendiknas (2010)

2.3. Data Analysis

Data in this study were analyzed using 2-sample t-test. t-test is a parametric statistical technique to compare ratio or interval of data [10]. Before conducting a t-test, basic assumption test, including normality and homoscedasticity tests, was conducted.

3. Result

This study used TKJI for students aged 10-12 years old, which involved: (a) 40-meter sprint, (b) pull-up, (c) sit-up, (d) vertical jump, and (e) 600-meter running. Each result was then converted into T-score and summed. Results of analysis of difference in physical fitness level between upper-grade students at MI *Negeri* 2 Bantul and upper-grade students at SD *Negeri* 003 Balikpapan are presented in Table 4.

Table 4. Data on Physical Fitness Level of Upper-Grade Students in Disaster-Safe Schools in Bantul and Balikpapan

Statistics	**MI *Negeri* 2 Bantul**	**SD *Negeri* 003 Balikpapan**
N	65	87
Mean	12.40	9.93
Median	12.00	10.00
Mode	12.00	10.00
Std. Deviation	2.01	1.93
Minimum	9.00	5.00
Maximum	17.00	15.00
Sum	806.00	864.00

Difference in physical fitness level between students at MI *Negeri* 2 Bantul and SD *Negeri* 003 Balikpapan is considered significant if t-value is greater than t-distribution (t-value > t-distribution) and if p-value is less than 0.05 (Sig < 0.05). From results of the analysis, it can be seen that fitness level of students at MI *Negeri* 2 Bantul and students of SD *Negeri* 003 Balikpapan differs significantly. In addition, it is also found that the means of physical fitness of students in Bantul and in Balikpapan are 12.40 and 9.93 respectively. These numbers indicate that physical fitness of students in Bantul is higher than that of students in Balikpapan and therefore answers Hypothesis 2 of the study. The results are presented in the Table 5.

Table 5. Summary of Analysis of Difference in the Means of Physical Fitness Level between Students in Disaster-Safe Schools in Bantul and Balikpapan

Disaster-Safe Schools	Mean	t-test for Equality of Means			
		t-value	t-distribution	*p*	Difference
MI *Negeri* 02 Bantul	12.40	7.664	1.655	0.000	2.47
SD *Negeri* 003 Balikpapan	9.93				

4. Discussion

Results demonstrate significant difference in the level of physical fitness between students at MI *Negeri* 2 Bantul and students at SD *Negeri* 003 Balikpapan. Referring to difference in mean (2.47), students at MI *Negeri* 2 Bantul demonstrate better physical fitness than students at SD *Negeri* 003 Balikpapan.

According to statistics, of 65 students from MI *Negeri* 2 Bantul and 87 students from SD *Negeri* 003 Balikpapan, 5 MI *Negeri* 2 Bantul students (7.69%) of and 31 SD *Negeri* 003 Balikpapan students (35.63%) fall under “very poor” category. It shows a significant difference (27.94%) in the number of students falling under “very poor” category between two schools: MI *Negeri* 2 Bantul and SD *Negeri* 003 Balikpapan. One factor that influenced SD *Negeri* 003 Balikpapan students’ poor performance is their lack of seriousness and enthusiasm while doing each of the tests.

In the meantime, 36 (55.38%) and 51 (58.62%) students of MI *Negeri* 2 Bantul and SD *Negeri* 003 Balikpapan fall under “poor” category respectively. According to statistics, there is a difference of 3.24% in the number of students under “poor” category with SD *Negeri* 003 Balikpapan being higher than its counterpart. One factor that contributed to the high number of students under “poor” category is fatigue. It is because students are not used to performing activities nonstop.

Results also indicate that 24 MI *Negeri* 2 Bantul students (36.92%) and 5 SD *Negeri* 003 Balikpapan students (5.75%) fall under “fair” category. A significant difference of 31.17% between MI *Negeri* 2 Bantul and SD *Negeri* 003 Balikpapan in this category demonstrates that physical fitness level of students of MI *Negeri* 2 Bantul is higher than their counterpart. One contributing factor to higher number of MI *Negeri* 2 Bantul students under “fair” category is related to environmental factor. Living in village area, students of MI *Negeri* 2 Bantul are allowed to engage in daily activities that require active physical moves such as playing traditional games or helping parents at home or in the field after school.

Surprisingly, 0% of students in both schools fall under “good” and “very good” categories. This is mainly caused by students’ internal factor that includes failure to perform at satisfactory level during physical fitness test. Besides, it is also caused by the lack of regular sport practices outside PE subject and the fact students had never known TKJI before the test was conducted, making this test unfamiliar to them. In general, data suggest that physical fitness level of MI *Negeri* 2 Bantul students is higher than their counterpart due to factors that influence their physical fitness. According to Suharjana (2013: 9-10), physical fitness is influenced by factors including: (1) age, (2) sex, (3) genes, (4) diet, (5) regular sleeping schedule, and (6) exercise. MI *Negeri* 2 Bantul and SD *Negeri* 003 Balikpapan differ in environmental conditions. For this reason, students of both schools may demonstrate different physical fitness levels.

MI *Negeri* 2 Bantul is located at Jl. Imogiri-Siluk Km 3, Kebonagung, Imogiri, Bantul. Most students of MI *Negeri* 2 Bantul live near the school and most of them prefer to walk or ride bicycle to get to school. Nevertheless, with the advancement of technology and various characteristics among students, parents of MI *Negeri* 2 Bantul often use motor vehicles to take their children to school. These children, however, will walk home together after school as their houses are fairly close to school. Students of MI *Negeri* 2 Bantul are used to playing traditional games together after school. They also help their parents at home or in the field.

SD *Negeri* 003 Balikpapan is located at Jl. Jend. A. Yani RT 49, Gunungsari Ilir, Central Balikpapan Subdistrict, Balikpapan Municipality, East Kalimantan Province. This school is situated in urban area with most parents working in formal sectors. However, as the city is located near the beach, some other parents work as fishermen. Most SD *Negeri* 003 Balikpapan students live pretty far away from school while some others live nearby. Most students take motor vehicles, together with their parents, to get to school while some other students living nearby prefer to bike. Students of SD *Negeri* 003 Balikpapan do not get used to playing together at school. They go home and play, watch TV, or play with their gadget at home after school. However, there are also students who do the homework right after school. This finding is in line with results of the study conducted by Fikri [11] arguing that physical fitness is influenced by a number of factors including: (1) physical activities and sports in daily basis, (2) food intake, and (3) technological advancement. In addition, uninteresting physical fitness test conducted by teachers also contributes to the lack of interest in physical activities. This is in line with results of previous study where students expressed some bad experiences with physical fitness test caused by poor practices [12].

It is clear that difference in characteristics between students of MI *Negeri* 2 Bantul and students of SD *Negeri* 003 Balikpapan affects their physical fitness. Irianto defined environment as a place where someone lives for a long time. It is related to not only physical but also socioeconomic conditions [13]. Environmental condition, job, habit, and economic condition are all contributing factors to one's physical fitness. Gill et al argued that children in rural areas were physically fitter than those in urban areas [14]. Joens-Marten et al claimed that physical fitness level among subjects in rural areas was higher than that in urban areas with mean of VO2 max of 28.54 ± 1.79 ml/kg/minute compared with 21.57 ± 1.79 ml/kg/minute [15] . Nevertheless, school-aged children in both rural and urban areas demonstrated poor physical fitness with VO2 max of 30 ml/kg/minute based on sex and weight. Mexitalia claimed that environment is one of factors contributing to children's health [16]. Differences in geographical location, socioeconomic condition, and lifestyle between people in rural and urban areas can influence children's health level. A dichotomy between rural and urban also relates to education, income, and health levels, with people in rural areas living with lower socioeconomic and health levels than those in urban areas. It was reported that poverty and health level in rural areas in the US was lower than those in the cities in 2015.

Physical fitness is influenced by a number of factors such as environment. In other words, schools' environmental condition, such as being located in disaster-safe zone, plays important role in students' physical fitness. As can be seen from the bar chart in Figure 1, many students from both MI *Negeri* 2 Bantul and SD *Negeri* 003 Balikpapan fall under "poor" category. According to statistics, physical fitness of students from both schools remains low. Therefore, it is important to practice sport regularly through PE subject at schools and to engage in other physical activities. Physical education can be a mean for physical activities and sports to develop preparedness for disaster, particularly physical preparedness through development of motor skills, physical abilities, knowledge and logic as well as habituation of healthy lifestyle to stimulate balance growth and development [17]. Physical Education that refers to collective practice of physical activities at school is a strong predictor of physical fitness, particularly cardiovascular fitness, and by implementing this subject, schools can play role in boosting habituation of physical activities that are favored and last until adulthood [18]. Good physical fitness in disaster-safe schools is important to prepare school personnel physically for emergencies, particularly disaster. With excellent physical abilities obtained through physical preparedness, one can survive longer in the event of emergency. In other words, one can use his physical ability to save himself by heading towards safer zone. In contrast, a person with poor physical fitness will lack of physical preparedness and therefore lack of ability to save himself and evacuate to safer area in case of emergency or disaster. This is in line with results of the study conducted by Nopembri, Saryono & Rithaudin stating that engagement in physical activities is the most effective as well as the safest way to obtain physical fitness as it helps to improve fitness and leads to better concentration level [19]. In general, results indicate that physical fitness level of MI *Negeri* 2 Bantul students is higher than that of SD *Negeri* 003 Balikpapan students. Hence, if disasters (with the same scale) occur in both disaster-prone areas, students in Bantul will survive better as they are equipped with better physical preparedness (indicated by TKJI results) than students in Balikpapan.

5. Conclusion

Referring to data analysis results, description, result tests, and discussion, it can be concluded that physical fitness level of upper-grade students in disaster-safe school in Bantul differs significantly from that of upper-grade students in disaster-safe school in Balikpapan with upper-grade students in disaster-safe school in Bantul demonstrating better physical fitness than students in disaster-safe school in Balikpapan. Referring to this finding, it can be said that students of MI Negeri 2 Bantul are more physically prepared for natural disasters than students of SD Negeri 003 Balikpapan.

Acknowledgments

The authors would like to thank the Ministry of Research, Technology, and Higher Education for the provision of PDUPT research grant in 2019. This writing is also part of second author's undergraduate thesis.

References

[1] Suharjana. 2013. *Kebugaran Jasmani* Cetakan Pertama (Yogyakarta: Jogja Global Media)
[2] BNPB (Indonesian National Board for Disaster Management) n.d. *Indeks Risko Bencana*, inaRISK viewed 03 September 2019 from http://inarisk.bnpb.go.id/ irbi
[3] Syarifah H., Poli D. T., Ali M., Rahmat H. K., & Widana I. D. K. K. 2020. Kapabilitas Badan Penanggulangan Bencana Daerah Kota Balikpapan dalam Penanggulangan Bencana Kebakaran Hutan dan Lahan *NUSANTARA: Jurnal Ilmu Pengetahuan Sosial 7*(2) 398-407
[4] PTS, A. Yushpie, Putri, A. P., Robiandi, F., & Endrawati, B. F. 2018. Agito (Agility Non-Thermal Wind Offensive) Sebagai Pendeteksi Cuaca Kota Balikpapan Menggunakan Gnss Wind Turbine. *JTT (Jurnal Teknologi Terpadu)*, *6*(1), 44-53.
[5] Triyono A 2018 Tingkat Kesegaran Jasmani Siswa Kelas IV dan V Daerah Rawan Bencana Gunung Merapi di Kabupaten Sleman *Skripsi* Universitas Negeri Yogyakarta
[6] Rahayuningsih W. S., & Khusnal E. 2010 *Faktor-Faktor yang Berhubungan dengan Pola Aktivitas Fisik pada Anak Kegemukan Usia Sekolah di Kecamatan Bambanglipuro Bantul Yogyakarta* (Doctoral dissertation, STIKES'Aisyiyah Yogyakarta)
[7] Maisaroh P., Athar A., & Kahri M. 2020. Analisis Komponen Kebugaran Jasmani Peserta Didik Sekolah Dasar Negeri Dilihat Dari Sarana Dan Prasarana Pendidikan Jasmani Kelas V Usia 10–12 Tahun Kota Banjarbaru *Stabilitas: Jurnal Pendidikan Jasmani dan Olahraga*, *1* (1) 80-83
[8] Mita E. 2020 *Status Kesegaran Jasmani Peserta Didik Sekolah Dasar Negeri Se Kabupaten Banjar Tahun 2016* https://repo-dosen.ulm.ac.id//handle/123456789/10858
[9] Kemendiknas 2010 *Tes Kesegaran Jasmani Indonesia,* (Jakarta: Pusat Pengembangan Kualitas Jasmani)
[10] Sugiyono 2007 *Statistika untuk Penelitian* Cetakan Ke Dua Belas (Bandung: Alfabeta)
[11] Fikri A. 2017 Meningkatkan Kesegaran Jasmani Melalui Metode LatihanSirkuit Dalam Pembelajaran Pendidikan Jasmani Olahraga Dan Kesehatan Di SMA N 1 Lubuklinggau. *Jurnal Pembelajaran Olahraga* 3(1) 95
[12] Wiersma L. D., & Sherman C. P. 2008 The responsible use of youth fitness testing to enhance student motivation, enjoyment, and performance *Measurement in Physical Education and Exercise Science 12* (3) 167-183
[13] Irianto D.P. 2018 *Pedoman praktis berolahraga* (Yogyakarta: UNY Press)
[14] Gill M, Deol N.S, & Kaur R. 2010 Comparative study of physical fitness components of rural and urban female students of Punjabi University, Patiala *Anthropologis*;12 (1):17-21
[15] Joens-Matre R elk GJ, Calabro MA, Russell DW, Nicklay E, Hensley LD. 2008 Rural–urban differences in physical activity, physical fitness, and overweight prevalence of children. *J Rural Health;* 24(1):49-54
[16] Mexitalia, M, Selina, H, Anam, M.S, Yoshimura, A, Yamauchi, Y, Hariyana, B.H. 2012 Perbedaan status gizi, kesegaran jasmani, dan kualitas hidup anak sekolah di pedesaan dan perkotaan *Jurnal Gizi Klinik Indonesia* Vol 8 No 4
[17] Sesfao A. 2019 Upaya Meningkatkan Kesegaran Jasmani Melalui Pendekatan Bermain

Dalam Pembelajaran Pendidikan Jasmani (PenelitianTindakan Kelas di kelas VII-G SMP Negeri 14 Tasikmalaya) *Jurnal Wahana Pendidikan* 5(3) 31-36
[18] Kelly, I. R., Phillips, M. A., Revels, M., & Ujamaa, D. 2010 Contribution of the school environment to physical fitness in children and youth *Journal of Physical Activity and Health* 7(3) 333-342
[19] Nopembri, S., Saryono. & Rhitaudin, A. 2019 *Aktivitas-aktivitas Jasmani Berbasis Psikososial* (Yogyakarta: Istana Agency)

Commercial Multirotor UAV Campaign on Data Acquisition for Disaster Management

S T Putro[1], **F Nucifera**[2], **E Febriarta**[3]
[1]Department of Geography Education, Faculty of Social Science. Universitas Negeri Yogyakarta, Indonesia.
[2]Department of Geography, Faculty of Science and Technology. Universitas AMIKOM Yogyakarta, Indonesia.
[3]Palawa Karya, Consultant.

E-mail:
[1]sutanto.tp@uny.ac.id; [2]fnucifera@amikom.ac.id; [3]e.febriarta@gmail.com

Abstract. UAV usage and interest have shown great popularity recently, especially for commercial multirotor. It is cheap, easy to use, flexible, and has many capabilities. That is why multirotor is widely chosen in disaster management. As a technology, it possesses certain aspects. Review of how the implementation of multirotor UAV in disaster management is important to gain improvement in the technology and evaluate at how we use it. This paper provides insight into commercial multirotor UAV campaign in disaster management through literature study and personal experience in a related theme. Similar ideas on the development of UAV and its application in disaster management then combine with personal experience was used to identify the advantage, disadvantages, challenges, and opportunities of multirotor drone applications. The result suggested that multirotor UAV has limited capability in conducting disaster management campaigns. The weakness in deploying multirotor is related to the lack of technology which may be overcome in the future as the development in hardware and software technology.

Keywords: Quadcopter, disaster management, data acquisition

1. Introduction

Rapid data assessment in every fieldwork related to any field is preferable. Since a long time ago, methods on data collection has developed to have adequate data with less time, effort, and capital. In the earlier times, data assessment for post-disaster relief was done manually by laboring surveyor to collect a large amount of data. This traditional method is time consuming, painstaking, and expensive. Another method comprising satellite imagery and aerial photograph has long been used for disaster management comprising pre- and post-disastrous campaign [1], [2].

Earth`s surface data acquisition or earth observation by satellite image for disaster management rapidly increases because of its rapidness and detail [3]. Technological advance in sensors and satellite is the factor of the increasing demand for satellite imagery. Satellite as part of aerospace technology, plays an important role in disaster risk reduction, thus rapid data acquisition, high temporal resolution, and all-weather capability is a must [4]. Satellite imagery and its wide options of sensors have the requirement mentioned by Verstappen. The satellite also has its advantages by its wide-area coverage, and the selection of sensors which some area weatherproof. Despite the advantages, satellite imagery is lacking on temporal resolution (only some have high temporal resolution) For non-radar sensors, most of the sensors are depended on the weather, which means that the cloud will be a big interference for

Content from this work may be used under the terms of the Creative Commons Attribution 3.0 licence. Any further distribution of this work must maintain attribution to the author(s) and the title of the work, journal citation and DOI.
Published under licence by IOP Publishing Ltd

IOP Conf. Series: Earth and Environmental Science **884** (2021) 012031 doi:10.1088/1755-1315/884/1/012031

them. The high-quality satellite imagery is costly, and it becomes one of its drawbacks, especially for NGOs or private sectors with limited resources.

Other than satellite imagery, surface imagery can be obtained through aerial surveys utilize planes or high-altitude aircraft. This method was popular before the advancement of satellite imagery and advanced Unmanned Aerial Vehicle (UAV). An aerial photograph provides a detail image but has several limitations. Weather conditions could be the biggest hindrance to the data acquisition. The complicated Resource management for flight surveys also must be taken into consideration. Compared to any other system, aerial photography (manned system) has many constraints and requirements which is difficult to obtain, handle, store, calibrate and interpret [5].

UAV along with UAS (Unmanned Aerial System) was originated for military and intelligence concerns with their specific technologies [6], [7]. Not only UAV technology, many technologies we use daily were derived from the military, for example is internet [8]. The application of UAV has reached many aspects of life both civil and military. Various classification and types of drones utilized for different missions [9]. Drones also catch the curiosity of the enthusiast. They try to make their drones for many purposes. Aside from custom drones made by an enthusiast, commercial drones made by industries come in many type and application. Even, some are customizable following the mission.

Multirotor is the type of several commercial drones available. Several types in multirotor has 3, 4, 6, and 8 rotors called tricopter, quadcopter, hexacopter, and octacopter respectively. The main attraction of these types of drones is on its practicality, price, and data accuracy [9]. As for its benefit, many parties tend to have this type of drone. Mostly this type of drone especially used for videography or documentation. For the scientific purposes, this type is largely used for mapping and area identification. The same idea is used for disaster management as a drone is mainly used for data acquisition [10], [11], [12], [13].

As for scientific purpose, some type of drones provides customizable function to match its mission. The ability to change the sensor/camera is beneficial. The changing of sensor demonstrated in thermal imaging for eruption monitoring in Stromboli Volcano, Italy [14]. Wakeford and his team change the camera installed in their DJI Inspire using a thermal camera designed for DJI Inspire. Another use for multirotor drones aside from image capturing is for water sampling as demonstrated by several works [15], [16], [17], [18].

Easiness, reliability, and flexibility made commercial multirotor is the logical choice for most people. With its properties, the performance of a multirotor drone for fieldwork, mapping, and its application on disaster management needs to be explicitly described. Therefore, review on the application of multirotor UAV, especially for disaster management is important, so, many users and people who interested in using drones in the future can evaluate their experience and use of drones. For enthusiasts and industry, the result of this paper can be used to improve and develop drone technology.

2. Methods

Commercial drone experience and its related data on disaster management were obtained not only from authors experience and previous research but also depends on the literature studies. Previous research selected in this study were disaster management covering multi-type of hazard which utilize drone/UAV/UAS for their data acquisition.

As literature study in this study is part of a narrative study which represent a generalization of ideas or cumulative knowledge [19], [20], [21] on the topic of disaster management with the utilization of multirotor drone for data acquisition. Cumulative ideas presented in a table to be easily recognized. Conclusion on the application of multirotor UAV synthesized from the shared ideas among the previous studies to find the benefit, disadvantages, challenges, and opportunities in the future.

This method was chosen as much data are available outside of the author`s possession. They came in books and journals. The only way to reach and collect the information was through the literature study as presented by several works in the various theme of drone [22], [23], [24]. There are several steps to obtaining the information and drawing conclusions. The first step was literature retrieval in the form of journals and books. The second step was information extraction. The information that was gathered including the development of drones and the application of multirotor on the field of disaster management.

3. Result and Discussion

3.1. Result

3.1.1. UAV development. The earliest application on the unmanned aerial vehicles was in 1849 during the military campaign of the Austrian Empire by using unmanned blimp to air raided Venice. The next unmanned aerial vehicle application was used in the American Civil War in 1861-1865. It presents the same system as used by the Austrian Empire in 1849. Non-military application of aerial photographs was demonstrated by British meteorologists by attaching a camera to a kite in 1887. In 1916, the first real unmanned vehicle was flown by Royal Flying Corps (recently called Royal Air Force), British. The rising application and development of UAS were conducted during World War 1 and 2, for combat and aerial survey. Modern use and technology on UAS were developed in the post-war era, but mostly still serving a military campaign. Since the 2000s, a vast application of commercial drones for recreational use emerged, a sign of the blooming of drone usage for civilians [6].

The expansion of UAV application from military operation in the beginning, into much more vast utilization, makes human-drone interaction is more complex [25]. From the study conducted by Tezza and Andujar in 2019 [25], the profile of the most used commercial drone and their application, their growth, and favourable model information were obtained. **Figure 1** shows the highest estimation of drones registered to the FAA in the US. That number can demonstrate that the growth of interest and application in drones is high. Unfortunately, Indonesia has no published data on the registered drone yet as the regulation on the UAS deployment in Indonesia has not evenly known to the public yet. But, from many sources, the prospect of the commercial drone in Indonesia is high, manifested from the drone sales figure (Hussein, 2019) and the increasing number of drone hobbyist communities, which is now globally increasing rapidly in number [26].

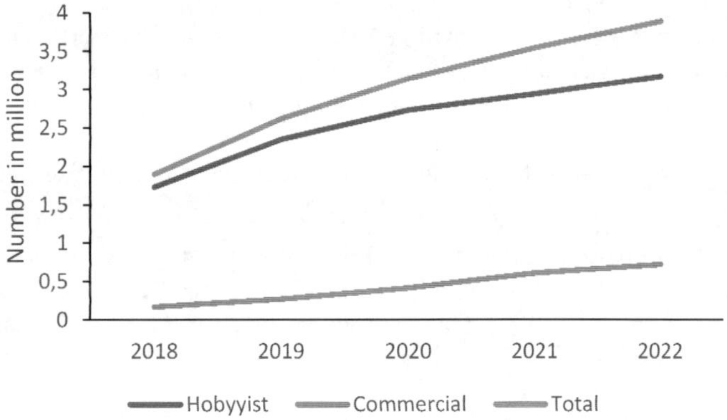

Figure 1. High estimation of drone growth by FFA [25], [27]

In the US, multirotor, especially quadcopter is the most favourable among the hobbyist. DJI, one of many drone corporations still becomes the ruler of drone market share globally (Mordor Intelegence, 2020) with multirotor as the biggest type of drones in the market. Multirotor drone tends to have a more stable image capturing which increase the accuracy and number of the image due to the low-speed flight coupled with the ease of vertical take-off and landing [28]. A comparison of the basics of technical properties of multirotor drones that commonly used for survey and mapping is sown in **Table 1**. The table show variety of multirotor drone, from the small recreational drone, up to the big-heavy-duty-industrial purposed drone. The price also varies between hundreds to thousands of dollars. Some are originated for recreation purposes but can be enhanced into survey and mapping with the addition of mapping and surveying software.

Table 1. Comparison of several Multirotor UAV commonly used in mapping

Brand	Model	Flight time (min)	Speed (Km/h)	Weight (gr)	Range (Km)	Picture resolution	Control Interface
DJI	Mavic Pro	27	64	734	7	12 MP	RC, phone
DJI	Mavic 2 Pro	31	70	907	8	20 MP	RC, phone
DJI	Phantom 3 Pro	23	57	1,280	5	12 MP	RC, phone
DJI	Phantom 4 Pro V.2.0	30	72	1,375	9	20 MP	RC, phone
DJI	Phantom 4 RTK	30	58	1,391	5	20 MP	RC, phone
DJI	Inspire 2	27	93	3,440	7	Varies by camera	RC, phone
DJI	Matrice 300 RTK	55	82	6,400	15	Varies by camera	RC
DJI	Matrice 200 series	38	80	4,672	8	Varies by camera	RC, phone
Parrot	Anafi	25	53	320	4	21 MP	RC, phone
Parrot	Bebop2	25	59	500	0.3	14 MP	RC, phone
Yuneec	H520	28	61	1,995	1.6	Varies by camera	RC
Yuneec	Tornado H920 Plus	24	50	4,990	1.6	16 MP	RC
Vulcan	Raven	12	28	7,250	8	Varies by camera	RC

The recreational purpose for several drones is mainly for image capturing and video recording with the enhance of the capability of the drone such as following objects, following points or tracks, fly in circles, or following several rules. Enhanced software for mapping is introduced by several developers. The software has two main functions, first is to plan the flight track and properties such as speed, altitude, overlapped size, type of orthophoto, and camera angle. Second, the software is used to aid the image processing process and analysis, for example, to produce the mosaic, DEM, or analysis on the vegetation. Some software has multi-platform use, desktop (pc-based), and mobile (installed the mobile device). To have better assistance from the software developer, the user must make a payment subscription, unless only basic or time-limited service can be provided by the developer. Several software developers provide free service for flight planner in a mobile platform, and it can be used only for several drone manufacturers. Some software for flight planning and analysis are available such as Pix4D, dronedeploy, DJI ground station pro, Map pilot, etc. That software can enable the autonomous and semi-autonomous flight over planned track/points with the ability to override the control during the flight. In mobile platforms, some software can be run under android or iOS which is come handy for most users.

3.1.2. UAV Operation on disaster management. Data demand right after the occurrence of a disaster covering the abrupt changes of the affected area is one of the top priorities in disaster management, so rapid data assessment is crucial [29] to provide data to improve the response and mitigation of the disaster [30]. Rapid data assessment after the onset of disaster can be covering a large sector including access and security, demographics, community resources, health, water, sanitation, food and non-food items, and shelters [31], and most of the data can be obtained from remote sensing survey [29], including the deployment of UAV. Aside from rapid data assessment, a UAV survey in a disaster can be applied to mitigation as a pre-disaster event. This application is widely used and reported by many researchers (see **Table 2**).

Table 2. Publication on disaster management with the implementation of multitirotor UAV

Publication		Stages			Application			Type of Disaster
Author	Year	Pre-disaster	Disaster assessment	Post-disaster	Monitoring, planning	Damage assessment	Cargo, SAR	
Wulan et al.	2017	√	-	-	-	√	-	Landslide, Flood
Baker, Rapp, Elwakil, & Zhang	2020	-	-	√	-	√	-	Hurricane

Publication		Stages			Application			Type of Disaster
Author	Year	Pre-disaster	Disaster assessment	Post-disaster	Monitoring, planning	Damage assessment	Cargo, SAR	
Leizer & Tokody	2017	√	√	√	√	√	-	Engineering
Gupta & Shukla	2018	√	-	-	√	-	-	Landslide
Lucieer et al.	2014	√	-	-	√	-	-	Landslide
Calantropio, Chiabrando, Sammartano, Spanò, & Losè	2018	-	-	√	-	√	-	Earthquake
Wakeford et al.	2019	√	-	-	√	-	-	Eruption
Estrada & Ndoma, 2019	2019	-	-	√	-	-	√	Multi
Aiello, Hopps, Santisi, & Venticinque	2020	√	-	-	√	-	-	Engineering
Mikami et al.	2019	-	-	√	-	√	-	Tsunami
Ononiwu, Okoye, Onojo, & Onuekwusi	2016	-	√	-	-	-	√	Multi
Qi et al.	2015	-	√	-	-	-	√	Earthquake
Zenkin, Berman, Pachkouski, Pantiukhin, & Rzhevskiy	2020	-	√	-	√	√	-	Forest fire

A previous study inferred that UAV campaign for disaster management is a sophisticated method that enables users to achieve data easily and rapidly with less effort and resources. Its ability to take off and land in limited space is also a key factor for the vast campaign of multirotor in disaster management. Various method of hazard management application, algorithm, and modelling has been developed to enhance the success of disaster management. Modelling and monitoring regarding mass movement have been conducted, so any changes in the environment that can trigger the landslide could be detected early to minimize the damage and loss [12]. The same concept also applied in mitigation of volcanic eruption of Stromboli Volcano by utilizing thermal images to produce 3 dimensional model of the volcano [14].

Frequently, UAV is mobilized due to assessing the surface conditions in the post-disaster event. Multirotor can also be deploy in this kind of event due to its flexibility. A study on the utilization of commercial quadcopter for mapping in the impact of landslide onset was carried in Banyumanik [42] and in Bangli, Bali as the result of flood and landslide a day before [32]. Based on the study carried by Azeriansyah, Prasetyo, & Yuwono, they can identify the affected area with good accuracy compared to the standard issued by Geospatial Information Agency (BIG/*Badan Informasi Geospasial*) under the regulation of the head of Geospatial Information Agency Number 15/2014. While based on the study by Wulan et al., they did not specify the technical result related to the data accuracy as they did not focus on that issue. Their study mainly focused on the mapping of the area affected by the flood and landslide. Another utilization of UAV and on landslide data measurement in India resulted in 5% difference between measurement using data assessed form UAV and data obtained from field measurement using a total station [13].

Disaster comes in many forms and triggering factor. Man-made structure and technology also exposed to a certain possibility of failure due to natural and artificial events. UAV mission for monitoring and assessing data on technology/engineering failure is common nowadays. Certain UAV deployment on an artificial structure is regularly made to investigate damage and verdict future possibilities and steps in disaster management. The sample of this work presented by Leizer & Tokody in 2017, when the utilize drone in the railway accident and disaster. Their work was motivated by the large amount of hazardous substances being transported using trains, and often accidents occur in the process of transporting the substance. They develop a method to monitor and assess the data utilizing RFID (Radio Frequency Identification) to identify the transported goods, their company, and any other data. They also revealed that deployment of drones, especially multirotor gave a huge benefit as their sensors are interchangeable and cheap. identification of hazardous radioactive waste or substance also can be monitored using a thermal sensor-equipped drone. Assessment on a structure which has high potential of danger requires the assistance of drone. If the result of drone assessment suggested detail manual observation, then human intervention is needed. The example of this work was done by Aiello et al. in 2020, they promote the use of mini-UAV in an inspection as to its participation in mitigation. Based on their study, mini-UAV with MTOW (Maximum Taking off Wight) is less than 300 grams, has a high potential in the monitoring stage of mitigating disaster risk in an industrial complex.

Even after several weeks of disaster occurrence, UAV participation is still possible. Works on tsunami impact and its inundation distribution and run-up heights did 1 month after the onset was conducted in Sulawesi as the response of landslide-generated tsunami triggered by an earthquake [38]. Based on the study, Mikami and his team able to identified the run-up height, inundation distribution, damage pattern. They conclude that the topographical factor plays an important role in the distribution of inundation. Another example of UAV deployment after the earthquake presented by Qi et al in 2015 after the Lushan earthquake in china. He deployed rotary-wing UAV to assist in SAR (search and rescue) process by capturing and processing data through low altitude flying. The mission proves its applicability and saves time by utilizing the rotary-wing UAV.

UAV comes in many sizes, capability, and prices. Each drone is unique due to its characteristics. Large size UAV with high capability and endurance may be used as a logistic carrier during the post-disaster campaign. This idea is brought by Estrada & Ndoma in 2019 in their study. They emphasize that the success of UAV campaign depends on the development of technology and training for pilots to have efficient and effective missions in different stages. Technological development and balance mentioned by Estrada & Ndoma comprise of hardware and software. This development of technology demonstrated by Zenkin et al in 2020 by promoting the UAV software and hardware for s forest fire.

Figure 2. Samples of DSM and aerial photograph from UAV deployment for Tsunami Mitigation in Sadeng, Gunungkidul Indonesia, and Geodetic GPS to obtain GCP data. Source: Authors fieldwork in 2019

Some disaster requires height data, such as for tsunami, mass movement, flood/inundation, liquefaction, etc. Height data is obtained by photogrammetry of the overlapping photograph from an aerial acquisition. Aerial image captured by UAV can be processed using various software developed by various developers such as global mapper, esri, Agisoft, dronedeploy, pix4d, etc. Many analyses and properties regarding the process can be adjusted. More accuracy in georeferencing can be enhanced using GCP (Ground control point) obtained by applying geodetic GPS during the survey. The example of the result is shown in **Figure 2**.

Height data extracted from the photogrammetric analysis will be in a DSM (*digital surface model*). DSM still including the height of land covers such as trees and buildings, while DTM only represents the bare earth [43]. Based on DSM dan DTM, some analysis and modelling can be conducted for mitigation and damage assessment. Meanwhile, an aerial photograph is used to interpret the surface properties such as land use and land cover, structural damage, affected area, and resources identification.

3.2. Discussion

Based on the result derived from the literature study and personal experience in utilizing commercial multirotor UAV in disaster management, several aspects can be identified respectively as seen in **Table 3** with the following discussion.

Table 3. Benefit, disadvantage, challenge, and opportunity of multirotor UAV

Benefit	Disadvantage
• VTOL (Vertical Takeoff and Landing) • Low-cost • Easy to control • Hovering maneuver • Play autonomous or semi-autonomous • Interchangeable parts • Some are small enough, flexible and easy to carry • Stable • Better accuracy	• Limited flight range • Short flight time • Small coverage area • Some small multirotor have problems in their camera • Wind-resistant • Small device means the small payload
Challenge	**Opportunity**
• Flight safety • Flight regulation • Pilot qualification, certification, and skills • Issue of invasion to a privacy	• A vast field of application • Favoured by market • Development of technology to meet the needs

3.2.1. Benefit.

The good thing coming from multirotor UAV is its many benefits. Many commercial multirotor available in the market are small, even foldable, and some can fit in a pocket. Its practicality will come handy during fieldwork where we have to carry so many additional equipments. We often use small size foldable UAVs such as DJI Mavic Pro. Its often used during our research and fieldwork to assess the environmental condition or reach inaccessible points from a safe place [44] and conduct aerial mapping to obtain a surface elevation model. Commercial multirotor can be controlled easily using a remote control or remote control with a mobile device connected to the controller. Some models provide double remote control as pilot control the movements and cameraman control the camera/sensor. To have consistent speed and direction, a flight plan for fly autonomously/semi is made. From the software flight pattern and properties can be adjusted to ensure the best overlap and result of the aerial photograph [45].

The many reasons why multirotor is chosen is for its price, easy to control, and maneuverability. Its VTOL (Vertical take-off-landing) capability assures the flexibility in any field and terrain of operation. Some small multirotor UAV can take -and land in the face of a palm. Movement of the UAV during flight contributes to the accuracy of the data it obtained. One movement that can only be done by multirotor than fix wing is hover maneuver. It is a still flight condition mid-air. This maneuver is beneficial as we can concentrate and examine objects without moving. The movement of multirotor UAV ensures the stability of data capturing [28], [46] which resulted in a better aerial image and accuracy [28]. Several missions require a custom-made apparatus to be mounted on a drone. Such as thermal, visible light sensor, lidar, radiation detector, sampler, etc. Multirotor is can provide stability to grab the sample, such as water sample [15], [16], [17], [18] and radiation monitoring [47], [48], [49].

3.2.2. Disadvantage.

Despite multirotor advantage, as opposites, it possesses several weaknesses that in the future may be overcome with the introduction of new technology. As for today, many commercial drones are lack of flight time, range, covered area, wind-resistance, payloads, and sensor. In several occasion, mapping mission using our quadcopter create a horrifying moment. As we have several mappings done in the windy coastal area, our quadcopter cannot meet the maximum flight duration. Once, we experienced only less than 5% battery remained with flight time less than the maximum duration. Based on our colleague`s experience, they have lost drones due to a signal lost on the hilly and forested area. It is not only happened in multirotor, fix wing also experienced similar events sometimes. This event demonstrates the deployment of multirotor UAV carrying risks. Topographic barrier, high and dense vegetation cover may disrupt the signal. The maximum range and flight time may be reached under preferable conditions. Problems in several cameras found in several commercial multirotor. The camera tends to condensate and produce blurry and misty images due to changes in atmospheric pressure and temperature during flight. This mist will be gone as the UAV stay longer in the air. The example of the blurry and misty image can be seen in **Figure 3**.

Figure 3. Blurry image as the result of condensation in the camera
Source: Author fieldwork in 2018

3.2.3. Challenge.

The introduction of UAV is a revolutionary thing [6]. Introduction of a drone to the rapidly changing society carries a great challenge. Issues on terrorism, crime, and invasion of privacy emerge related to drone deployment. To prevent misuse and protect the community, several regulations are made due to the application of UAV and the use of air space to ensure flight safety. Pilot as the controller and decision-maker in the drone deployment needs to be well educated on flight safety, regulation, and understand the use of air space as well as a responsible individual. Every UAV deployment is using air space that must be shared for some interest such as commercial, recreation, and military. Each air space has its regulation that needs to be followed by the user and the UAV pilot. Besides that, drones development is facing a level where a new novel drone is expected with the ability to fly autonomously in every condition for many purposes [9]. Related to disaster, UAS need to be developed to enhance the effectivity, flexibility, and the rate of success of UAV deployment in the field to support disaster management campaign. Several cases mention the technical issue as a problem need to be overcome for the future supports in disaster relief programs. The issues mention here are data transfer [50] and data safety [51].

3.2.4. Opportunity.
Technological development has created a vast possibility for multirotor drone to advance. From sensors, flight systems, power sources, and capability [52] that will open new opportunities of UAV, especially multirotor new mission and field. Also transformable multiotor UAV has been developed to increase its manuverable, movement, and flexibility [53]. Based on current technology, UAV has been applied in the field of disaster management, crime prevention, law enforcement, border police, life safety and healthcare service, military service, commercial, construction, private uses, education, research, entertainment, sport, logistic, firefighter, agriculture, geology, astronomy, meteorology, and environment [34], [36]. Commercial drone, especially for multirotor, has shown a huge interest in the market and provide a huge amount of profit. Business on drone become a new prospect. The huge drone technology will be achieved for the next decade as computer technology is greatly advancing [54].

4. Conclusion

Multirotor technology provides new insight into the possibilities of UAV deployment in many fields of interest. Furthermore, the large usage of multirotor UAV has risen alarms on privacy, safety, air space usage, and regulatory system. In needs time for UAV to be fully accepted in daily life. Multirotor is widely used in daily activity because of its advantages, from price, flexibility, and capability. The weakness of multirotor for a disaster management survey is related to the technological aspect related to the software and hardware. Those weaknesses can be overcome with the future advance of technology. As the technology in multirotor UAV is advancing, market interest will grow in its resemblance. Overall, according to the literature study and personal experience, commercial multirotor UAV has a proper but limited capability in disaster management campaigns due to its weakness. The high usage of multirotor UAV for disaster management represents its benefit as a system, as it provides many benefits and promises a high possibility in future technological development.

5. Acknowledment

This paper is a generalization and summarize of our several previous researches and projects for the previous 5 years. The researches and projects were done with the supports of many sources of funding and backed by student from Institut Teknologi Yogyakarta (ITY), Amikom University, and surveyors from many places for.

References

[1] K. Kaku, "Satellite remote sensing for disaster management support: A holistic and staged approach based on case studies in Sentinel Asia," *Int. J. Disaster Risk Reduct.*, vol. 33, no. April 2018, pp. 417–432, 2019.
[2] C. Thanawood, C. Yongchalermchai, and O. Densrisereekul, "Effects of The December 2004 Tsunami and disaster management in Southern Thailand," *Sci. Tsunami Hazards*, vol. 24, no. 3, pp. 206–217, 2004.
[3] S. Voigt *et al.*, "Global trends in satellite-based emergency mapping," *Science (80-.).*, vol. 353, no. 6296, pp. 247–252, 2016.
[4] H. T. Verstappen, "Aerospace technology and natural disaster reduction," *Adv. Sp. Res.*, vol. 15, no. 11, pp. 3–15, 1995.
[5] T. M. Lillesand, R. W. Kiefer, and J. W. Chipman, *Remote Sensing and Image Interpretation*, Fith., vol. 53, no. 9. New Jersey: Wiley, 2004.
[6] R. Bartsch, J. Coyne, and K. Gray, *Drones in Society. Exploring the strange new world of*

unmaned aircraft. New York: Routledge. Taylor & Francis Group, 2017.

[7] R. Luppicini and A. So, "A technoethical review of commercial drone use in the context of governance, ethics, and privacy," *Technol. Soc.*, vol. 46, pp. 109–119, 2016.

[8] B.-K. Kim, *Internationalizing the Internet. The Co-evolution of Influence and Technology*, vol. 53, no. 9. Northampton, Massachussets, USA: Edward Elgar, 2005.

[9] M. Hassanalian and A. Abdelkefi, "Classifications, applications, and design challenges of drones: A review," *Prog. Aerosp. Sci.*, vol. 91, no. November 2016, pp. 99–131, 2017.

[10] M. Aris Marfai, H. Fatchurohman, and A. Cahyadi, "An Evaluation of Tsunami Hazard Modeling in Gunungkidul Coastal Area using UAV Photogrammetry and GIS. Case Study: Drini Coastal Area," *E3S Web Conf.*, vol. 125, no. 201 9, 2019.

[11] S. Kashihara *et al.*, "Towards Practical Utilization of Unmanned Aerial Vehicle in Disaster Mitigation - UAV Operation Drill," *2019 IEEE Glob. Humanit. Technol. Conf. GHTC 2019*, pp. 2019–2020, 2019.

[12] A. Lucieer, S. M. de Jong, and D. Turner, "Mapping landslide displacements using Structure from Motion (SfM) and image correlation of multi-temporal UAV photography," *Prog. Phys. Geogr.*, vol. 38, no. 1, pp. 97–116, 2014.

[13] S. K. Gupta and D. P. Shukla, "Application of drone for landslide mapping, dimension estimation and its 3D reconstruction," *J. Indian Soc. Remote Sens.*, vol. 46, no. 6, pp. 903–914, 2018.

[14] Z. E. Wakeford, M. Chmielewska, M. J. Hole, J. A. Howell, and D. A. Jerram, "Combining thermal imaging with photogrammetry of an active volcano using UAV: an example from Stromboli, Italy," *Photogramm. Rec.*, vol. 34, no. 168, pp. 445–466, 2019.

[15] C. Koparan, A. B. Koc, C. V. Privette, and C. B. Sawyer, "Adaptive Water Sampling Device for Aerial Robots," *Drones*, vol. 4, no. 1, p. 5, 2020.

[16] S. Olsson, "Drones in Arctic Environments: Development of Automatic Water Sampler for Aerial Drones," Stockholm, 2018.

[17] H. T. Lally, I. O'Connor, O. P. Jensen, and C. T. Graham, "Can drones be used to conduct water sampling in aquatic environments? A review," *Sci. Total Environ.*, vol. 670, pp. 569–575, 2019.

[18] D. Castendyk, J. Voorhis, and B. Kucera, "A Validated Method for Pit Lake Water Sampling Using Aerial Drones and Sampling Devices," *Mine Water Environ.*, 2020.

[19] F. Lau and C. Kuziemsky, Eds., *Handbook of eHealth Evaluation: An Evidence-based Approach*. Victoria, Canada: University of Victoria, 2016.

[20] G. Paré and S. Kitsiou, "Methods for Literature Reviews," in *Handbook of eHealth Evaluation: An Evidence-based Approach*, F. Lau and C. Kuziemsky, Eds. Victoria: University of Victoria, 2015, pp. 157–178.

[21] G. Paré, M. C. Trudel, M. Jaana, and S. Kitsiou, "Synthesizing information systems knowledge: A typology of literature reviews," *Inf. Manag.*, vol. 52, no. 2, pp. 183–199, 2015.

[22] J. Seo, L. Duque, and J. Wacker, "Drone-enabled bridge inspection methodology and application," *Autom. Constr.*, vol. 94, no. October 2017, pp. 112–126, 2018.

[23] M. A. Khan, W. Ectors, T. Bellemans, D. Janssens, and G. Wets, "UAV-Based Traffic Analysis: A Universal Guiding Framework Based on Literature Survey," *Transp. Res. Procedia*, vol. 22, no. 2016, pp. 541–550, 2017.

[24] F. Klauser and S. Pedrozo, "Power and space in the drone age: A literature review and politico-geographical research agenda," *Geogr. Helv.*, vol. 70, no. 4, pp. 285–293, 2015.

[25] D. Tezza and M. Andujar, "The State-of-the-Art of Human-Drone Interaction: A Survey," *IEEE Access*, vol. 7, pp. 167438–167454, 2019.

[26] N. Vargas-Ramírez and J. Paneque-Gálvez, "The Global Emergence of Community Drones (2012–2017)," *Drones*, vol. 3, no. 4, p. 76, 2019.

[27] FAA, "FAA Aerospace Forecast Fiscal Years 2020-2040 Full Forecast Document and Tables," 2019.

[28] M. A. Boon, A. P. Drijfhout, and S. Tesfamichael, "Comparison of a fixed-wing and multi-rotor UAV for environmental mapping applications: A case study," in *International Archives of the*

Photogrammetry, Remote Sensing and Spatial Information Sciences - ISPRS Archives, 2017, vol. 42, no. 2W6, pp. 47–54.
[29] X. Han and J. Wu, "Disaster Emergency Rapid Assessment Based on Remote Sensing and Background Data," in *The International Archives of the Photogrammetry, Remote Sensing and Spatial Information Sciences, ISPRS TC III Mid-term Symposium "Developments, Technologies and Applications in Remote Sensing"*, 2018, vol. 42, pp. 481–483.
[30] D. Pastor-Escuredo, Y. Torres, M. Martínez-Torres, and P. J. Zufiria, "Rapid multi-dimensional impact assessment of floods," *Sustain.*, vol. 12, no. 10, 2020.
[31] M. Arii, "Rapid assessment in disasters," *Japan Med. Assoc. J.*, vol. 56, no. 1, pp. 19–24, 2013.
[32] T. R. Wulan *et al.*, "Pemetaan Cepat Kawasan Terdampak Bencana Longsor dan Banjir di Kabupaten Bangli, Provinsi Bali," *Maj. Geogr. Indones.*, vol. 31, no. 2, p. 44, 2017.
[33] C. A. Baker, R. R. Rapp, E. Elwakil, and J. Zhang, "Infrastructure assessment post-disaster: Remotely sensing bridge structural damage by unmanned aerial vehicle in low-light conditions," *J. Emerg. Manag.*, vol. 8, no. 1, pp. 27–41, 2020.
[34] G. K. K. Leizer and D. Tokody, "Radiofrequency Identification by using Drones in Railway Accidents and Disaster Situations," *Interdiscip. Descr. Complex Syst.*, vol. 15, no. 1, pp. 114–132, 2017.
[35] A. Calantropio, F. Chiabrando, G. Sammartano, A. Spanò, and L. T. Losè, "UAV strategies validation and remote sensing data for damage assessment in post-disaster scenarios," *Int. Arch. Photogramm. Remote Sens. Spat. Inf. Sci. - ISPRS Arch.*, vol. 42, no. 3W4, pp. 121–128, 2018.
[36] M. A. R. Estrada and A. Ndoma, "The uses of unmanned aerial vehicles -UAV's- (or drones) in social logistic: Natural disasters response and humanitarian relief aid," *Procedia Comput. Sci.*, vol. 149, pp. 375–383, 2019.
[37] G. Aiello, F. Hopps, D. Santisi, and M. Venticinque, "The Employment of Unmanned Aerial Vehicles for Analyzing and Mitigating Disaster Risks in Industrial Sites," *IEEE Trans. Eng. Manag.*, vol. PP, no. 99, pp. 1–12, 2020.
[38] T. Mikami *et al.*, "Field Survey of the 2018 Sulawesi Tsunami: Inundation and Run-up Heights and Damage to Coastal Communities," *Pure Appl. Geophys.*, vol. 176, no. 8, pp. 3291–3304, 2019.
[39] G. Ononiwu, A. Okoye, J. Onojo, and N. Onuekwusi, "Design and Implementation of a Real Time Wireless Quadcopter for Rescue Operations," *Am. J. Eng. Res.*, vol. 5, no. 9, pp. 130–138, 2016.
[40] J. Qi *et al.*, "Search and Rescue Rotary-Wing UAV and Its Aplication to the Lushan Ms 7.0 Earthquake," *J. F. Robot.*, vol. 00, no. 0, pp. 1–32, 2015.
[41] A. Zenkin, I. Berman, K. Pachkouski, I. Pantiukhin, and V. Rzhevskiy, "Quadcopter Simulation Model for Research of Monitoring Tasks," *Conf. Open Innov. Assoc. Fruct*, vol. 2020-April, pp. 449–457, 2020.
[42] R. Azeriansyah, Y. Prasetyo, and B. D. Yuwono, "Analisis Identifikasi Dampak Bencana Tanah Longsor dengan Menggunakan Unmanned Aerial Vehicle (UAV). Studi Kasus: Kelurahan Ngesrep, Kecamatan Banyumanik," *J. Geod. UNDIP*, vol. 6, no. 4, pp. 474–484, 2017.
[43] A. Nemmaoui, F. J. Aguilar, M. A. Aguilar, and R. Qin, "DSM and DTM generation from VHR satellite stereo imagery over plastic covered greenhouse areas," *Comput. Electron. Agric.*, vol. 164, no. July, p. 104903, 2019.
[44] S. Mintchev and D. Floreano, "A pocket sized foldable quadcopter for situational awareness and reconnaissance," *SSRR 2016 - Int. Symp. Safety, Secur. Rescue Robot.*, pp. 396–401, 2016.
[45] T. N. Tonkin, N. G. Midgley, D. J. Graham, and J. C. Labadz, "The potential of small unmanned aircraft systems and structure-from-motion for topographic surveys: A test of emerging integrated approaches at Cwm Idwal, North Wales," *Geomorphology*, vol. 226, pp. 35–43, 2014.
[46] D. C. Patel, G. S. Gabhawala, A. K. Kapadia, N. H. Desai, and S. M. Sheth, "Design of Quadcopter in Reconnaissance," in *International Conference on Innovations in Automation and Mechatronics Engineering 2013 (ICIAME2013)*, 2013, no. February 2013.

[47] S. Mochizuki *et al.*, "First demonstration of aerial gamma-ray imaging using drone for prompt radiation survey in Fukushima," *J. Instrum.*, vol. 12, no. 11, 2017.
[48] F. Marturano *et al.*, "Enhancing Radiation Detection by Drones through Numerical Fluid Dynamics Simulations," *sensors*, vol. 20, no. 6, p. 1770, 2020.
[49] Y. Sato *et al.*, "Remote radiation imaging system using a compact gamma-ray imager mounted on a multicopter drone," *J. Nucl. Sci. Technol.*, vol. 55, no. 1, pp. 90–96, 2018.
[50] F. Greenwood, E. L. Nelson, and P. Gregg Greenough, "Flying into the hurricane: A case study of UAV use in damage assessment during the 2017 hurricanes in Texas and Florida," *PLoS One*, vol. 15, no. 2, pp. 1–30, 2020.
[51] N. A. Khan, N. Z. Jhanjhi, S. N. Brohi, and A. Nayyar, *Emerging use of UAV's: secure communication protocol issues and challenges*. Elsevier Inc., 2020.
[52] N. Kerle, F. Nex, M. Gerke, D. Duarte, and A. Vetrivel, "UAV-based structural damage mapping: A review," 2019.
[53] D. Ruben Raphael, "Design of a Transformable Unmanned Aerial Vehicle," University of Minnesota, 2020.
[54] K. Nonami, "Drone technology, cutting-edge drone business, and future prospects," *J. Robot. Mechatronics*, vol. 28, no. 3, pp. 262–272, 2016.

IOP Conf. Series: Earth and Environmental Science **884** (2021) 012032 doi:10.1088/1755-1315/884/1/012032

Strengthening Disaster Literacy as An Effort to Reduce The Risk of Disaster in D.I. Yogyakarta Society

Triyanto[1], D A Agustinova[2], K Syamsi[3]
Post Graduate Program, Universitas Negeri Yogyakarta

Email: [1]yantotri49@gmail.com, [2]danangad96@gmail.com, [3]kastam@uny.ac.id

Abstract: This research aims to find out how to optimize GLS in building awareness of disaster literacy, the role of families in GLS to encourage the disaster literacy movement, and opportunities to strengthen disaster literacy in the digital era. The research subjects are the society and schools in D.I. Yogyakarta area. The research method used is literature study, data collection techniques are carried out by reviewing several books, literature, and other documents that are appropriate to disaster literacy studies. The results showed that optimizing the school literacy movement directed towards disaster literacy awareness can increase public understanding, disaster preparedness and can reduce the risk of disasters that will cause material and non-material losses. The strategy and pattern of implementing the school literacy movement includes three stages, namely: habituation, development, and learning activities. The role of the family will contribute to disaster literacy that can provide experience and knowledge of the society according to the conditions of the residence. Technological advances are also able to contribute to the society as literacy materials and provide information quickly and accurately related to natural disasters.

1. Introduction

The threat of disasters in Yogyakarta is still frequent, especially geological disasters such as floods, volcanic eruptions, landslides, and earthquakes. Yogyakarta is located between Mount Merapi in the north and the beach on the south side. Thus, the occurrence level of a disaster may be possible. Earthquakes often occur in Indonesia, especially in the Special Region of Yogyakarta (DIY). Seen from the map of earthquake areas in Indonesia, Yogyakarta is in region 4. This area is an area prone to earthquakes [6]. The last major earthquake occurred in Yogyakarta was on 27 May 2006 with a magnitude of 5.9 on the Richter scale. The earthquake centered on the bottom of the Indian Ocean with a depth of 33 km at the coordinates 8.26 ° latitude and 110.31 ° east longitude [9] . Based on BNPB data, the number of victims due to the earthquake reached 5,716 people died and 37,927 people were injured [3]. The earthquake caused many people being trapped in their homes, thus claiming more victims that were children and the elderly [15]. BMKG data published on the website [2] bmkg.go.id recorded that the last earthquake occurred in Yogyakarta on July 13, 2020 with a magnitude of 5.2 on the Ritcher scale centered at a depth of 10 km southwest of Bantul-DIY. Apart from the earthquake, Yogyakarta is also one of the cities affected by the existence of Mount Merapi.

Mount Merapi is one of the most active volcanoes in the world. Mount Merapi erupts in a cycle of 4 - 6 years [2]. The eruption cycle of Mount Merapi has changed. Data from the Geological Agency, Ministry of Energy and Mineral Resources accessed on August 5, 2020 [2], noted that the status of Mount Merapi is at level 2 (alert) since June 21, 2018. The largest eruption in the last ten years occurred on November 5, 2010 that resulted 346 people were died. Besides causing victims, the eruption of Mount Merapi had an impact on the housing, infrastructure, telecommunications, electricity and clean water sectors [14].

The disaster that occurred in Yogyakarta took a lot of losses from various aspects. Children are vulnerable to disasters due to their limited knowledge and understanding of disaster risks, besides that,

Content from this work may be used under the terms of the Creative Commons Attribution 3.0 licence. Any further distribution of this work must maintain attribution to the author(s) and the title of the work, journal citation and DOI.
Published under licence by IOP Publishing Ltd

the lack of understanding regarding the actions that must be taken when a disaster occurs is also one of the causes. Based on data about the incidence of natural disasters in several areas, the average number of victims is children, both during school hours and outside school hours [21].

There is still a potential for disaster in the Special Region of Yogyakarta, especially erupting volcanoes with alert status for more than 2 years without any decline in status, so all parties must pay serious attention to minimize the impact of losses due to the disaster. The facts described above show that there are still many victims, both materially and non-materially, so it can be seen that disaster preparedness has not been prepared and understanding of disaster risk is still low. One of the efforts to reduce the risk of community disasters is to increase the capacity of disaster literacy to the community [16]. Currently, the disaster education is very needed. Research show that 79.43% of it stated that their families were not ready to face the disasters. They got 88.64% of the disaster information through their TV or the internet, and 12.89% stated that they got information through the learning was suitable with the scientific discipline [27].

The important thing that must be owned by the community is disaster literacy or community awareness efforts in dealing with a disaster. Disaster literacy is a key that must be possessed by the community so that they can fully understand the geographical location of their area in a disaster-prone area. Disaster education is one that can encourage the realization of a disaster resilient generation for the younger generation [7]. Disaster literacy is an effort to make people aware of a disaster that is very important to have. This is done in order to reduce the risk of disasters causing losses [16].

Based on the problems above, this study aims to solve problems related to the form of strengthening literacy as an effort to reduce disaster risk carried out through schools and communities in Yogyakarta. The initial assumption in this study is a form of strengthening disaster literacy resulting in increased efforts to minimize material and non-material risks and losses as well as technological opportunities in encouraging awareness of disaster literacy in D. I. Yogyakarta community.

2. Method

The research method used in this research is literature study. Literature-based research is a form of research that uses literature as the object of study. The subjects of this study include the community and schools in the Yogyakarta area. There are several main features that need to be considered in the literature study research method, namely: First, the author is faced directly with text/numerical data. Second, books and articles are interpreted as second sources, meaning that the writer obtains information from the second party, not original from the first party in the field. Third, the data/information obtained is "ready to use". Fourth, books and articles obtained are not limited by time and space [26].Data collection techniques are carried out by reviewing several books, literature and other documents considered appropriate to disaster literacy studies.

3. Results and Discussion

Optimizing the School Literacy Movement in Building Disaster Literacy Awareness

The School Literacy Program can be an alternative in developing disaster literacy awareness. When referring to the objectives of GLS, both general and specific goals, it is clear that the school literacy movement provides space in developing the character of students through cultivating the school literacy ecosystem to become lifelong learners. More specifically, GLS is also a forum for increasing the capacity of residents and the school environment so that they are able to process various kinds of information / knowledge that can support the students' lives in a sustainable manner [24]. At this point, it is very possible that the School Literacy Movement can become a reinforcer in building awareness of disaster literacy through the implementation of existing GLS strategies and models in schools.

Ideally, the practice of the School Literacy Movement is not limited to reading and writing activities. GLS must be an activity in the form of a combination of developing receptive and productive skills so that competences can be maximized. There are several gaps that can be optimized in developing awareness of disaster literacy through the implementation of GLS including habituation, development and learning activities. The results of the research done by Suraya, 2019 reveal that the implementation of the School Literacy Movement has three stages, namely habituation, development, and learning activities. At the habituation stage, schools can design several strategies to create a literacy culture that can be integrated with disaster information content. For example, schools allocate15 minutes each day to read about disaster literature that is relevant to the potential threat of disasters in the school location and in the surrounding area. Schools can schedule reading time at the beginning, middle, or end of lessons, depending on the schedule and conditions of each school. This activity can be followed by an informal discussion about the references that are read so that there is a variety of communication related to the perception of disaster.

At the development stage, there is a great opportunity for schools to improve literacy skills related to disaster information by showing the involvement of students' thoughts and emotions through productive activities, both orally and in writing or as part of curricular activities. For example, holding a mini workshop related to disaster simulations that could potentially occur in the school area. This activity can be followed by short presentation assignments, simple writing, simple presentations, crafts, or acting to respond to reading, those are adjusted to the level and abilities of students so that the same perception will be built in the community in an effort to reduce the risk of disaster. Positive culture for critical, analytical, creative, and innovative thinking in reducing disaster risk can be transformed in a sustainable manner.

At the learning activities stage, it provides strategic opportunities for schools to integrate disaster literacy skills in all subjects, either directly or indirectly. Direct literacy activities can be carried out through integration into certain subjects relevant to disaster backgrounds. For example, the subjects of Social Studies and Sciences are very free to provide disaster content related to current natural conditions and phenomena through various relevant sources, such as maps of disaster-prone areas, weather forecasts from BMKG and others. For indirect literacy activities, it is possible to integrate disaster content in a broader perspective. For example, Indonesian Language subjects contain disaster content from folk stories, fairy tales, and novels so that they can provide positive values that can be applied in an effort to deal with disasters. Islamic Education subjects also have the same opportunity to integrate religious values in supporting human life, especially in dealing with various natural phenomena or disasters that are parts of human life. Through the pattern of the school literacy movement, optimization in building awareness of disaster literacy can be maximized.

The active role of academics can be the key to successful disaster mitigation. The active role of academics to go directly to society is one of the most effective efforts to strengthen disaster literacy, both for teachers and students [11]. Schools can transfer knowledge to students, while students can continue the information obtained in the world of education to their families or people around them [12]. The connectivity of various elements, both schools and the community can be an asset in increasing the capacity of the community in dealing with disasters. Furthermore, one of the important variables in disaster management is increasing the capacity of people living in disaster-prone areas [5].

Geologically, the special region of Yogyakarta is limited by two great potential disasters, namely Mount Merapi in the north and the Subduction Zone in the south, therefore integrated mitigation measures need to be taken. Understanding the characteristics of disaster sources is also very important in order to estimate the potential for disasters that might arise and to reduce the impact of disasters on the surrounding life and public facilities around the location. For example, the Yogyakarta earthquake occurred on 27 May 2006 can be used as a case study to conduct disaster mitigation studies due to the earthquake. This incident caused several educational facilities, social facilities, villages and other infrastructure (roads, mosques, bridges, electricity and water networks) estimated to be damaged by the earthquake [8]. Knowledge related to disasters obtained from literacy that has become a habit will help reduce the risk of disasters that causes losses. This is in line with a research by [1] showing that increasing disaster literacy can raise public understanding and disaster preparedness. This can be proved the increased awareness of the community accompanied by increased ability to identify and evaluate the factors and signs of disaster. In addition, the community can also understand well what to do according to the type of disaster being experienced. Disaster preparedness will help reduce risks and losses due to disasters.

The Role Of the Family in the School Literacy Movement to Encourage the Disaster Literacy Movement

The first and foremost environment in instilling education, personality, and moral values is family. Parents and families around the house will be examples as well as role models for children in the development process [10]. The family is the first institution that is recognized as circulation in the development of an individual's life. A person can get to know the world starting from family education, therefore the family is commonly referred to as the first and foremost educational institution. Family creativity at home will influence in building a literacy culture. A special attention is needed in developing children's literacy skills. For example, a collection of books at home can encourage children to have a literacy spirit [25]. Home is the best place to foster interest and develop reading habits, especially because of the family atmosphere that has been built since childhood. Through a comfortable and quiet place, children will easily gain knowledges and experiences quickly that can be remembered for a long time [23].

Literacy is not only taught through school programs but can be taught outside formal educationlike family. Most of the time the child has is with his family, so the family will be more able to control the child for activities that are more beneficial for them than school hours that on average only spend a quarter of the child's time. This is also supported by [18] in his research suggesting that the basic foundation of children is family. So, it is necessary to develop awareness of parents and people around them to understand and introduce literacy culture since childhood. This is stated on the basis of literacy habits that cannot be cultivated in a short time. Loving reading behavior is influenced by habits developed by the family. Argued that literacy can be started through the family. Family literacy practices depend on individual activities, which are influenced by the language used at home, household activities, values, and family aspirations. Literacy practice is also related to learning activities at school. Other findings show that family literacy practices are not hindered by economic conditions [19].

Reading habits built within the family will influence children to make reading as a necessity. Another thing families can do is to create a private library in the house. The creation of a library that is made attractively will help children to enjoy reading activities, even children can invite their friends to play in the home library. This is supported by a research [25] that the literacy culture of children will be influenced by the creativity of parents at home. Children's literacy skills will also be able to develop with a collection of children's books in the house. Parents who have the habit of reading fairy tales to children will provide a stimulus for children's development. This habit will increase the child's curiosity through the habit of asking questions. The type of story that parents choose will also affect children's knowledge, for example by telling history and events around the house, including disasters that have occurred and how to deal with them.

Children's reading habits that have been formed from an early age can also be directed into disaster literacy. Children's reading books can be added with knowledge or related to natural disaster mitigation. Disaster literacy is the capacity of each individual to read, understand, and use the information received and then follow instructions in the context of disaster mitigation, preparedness, response and recovery [4]. Literacy refers to the ability to read and understand information. Argues that disaster literacy is basically related to how living people in disaster-prone zones are able to access information, then they are able to interpret this information in the form of disaster mitigation practices. The habit of disaster literacy can be started or continued with literacy habits that children have had since in the family environment [8].

Knowledge related to disasters obtained from literacy that has become a habit will help reduce the risk of disasters that cause losses. This is in line with a research conducted by [1] showing that increasing disaster literacy can improve public understanding and disaster preparedness. This is proven in the increased awareness of the community accompanied by the increased ability to identify and evaluate the factors and signs of disaster. In addition, the community can also understand well what to do according to the type of disaster being experienced. Disaster preparedness will help reduce risks and losses due to disasters.

Opportunities for Strengthening Disaster Literacy in the digital era

The development of technology in the 21st century has grown rapidly and the progress that has been made is irresistible. Nearly all users can reach the benefits of current technological developments. In the context of education, technological advances can be used as a learning resource with unlimited access like literacy. Suggests that technological advances in education are factors in changing behavior in learning activities, learning resources, and learning technology. So, technological advances can be maximally utilized to launch the school literacy movement program by using applications that have been developed. The success of disaster literacy will be significantly influenced by the sustainability of the school literacy movement [17].

As a form of positive benefits that can be used in disaster literacy is disaster information that can be accessed more quickly and easily. The existence of applications from BNPB, BMKG, BPPTKG, and social media such as Twitter, Instagram, Facebook, can provide fast and accurate information. The use of technology as disaster literacy can reduce anxiety, unpreparedness, losses, and other disaster risks. Extensive and easily accessible literacy sources will help the community to obtain information related to disaster mitigation. The results of a research conducted by Kamil et al. [12] showed that the geographic literacy can increase people's understanding and knowledge of disasters by 91.6%. Encouraging students to be able to increase awareness and individual skills about caring for disasters in the surrounding environment can be started through disaster literacy for children [20].

Yogyakarta as a city that is prone to and has the potential to be hit by disasters such as earthquakes, volcanic eruptions, and tsunami must have the ability and knowledge in disaster mitigation. The application of disaster literacy within the family, community, and school will be a factor that affects disaster risk. Based on the results of the observations, it is still in a low level that teachers and students implement the application of disaster literacy. The current use of information technology can also be functioned as a means of providing learning experiences related to disaster literacy to students in order to have life skills in dealing with disasters. The use of information technology can be integrated with the required information literacy. This is in accordance with the preparedness put forward by Kent [13] that will help minimize the adverse impacts of a disaster through effective prevention, rehabilitation and recovery measures due to a disaster.

4. Conclusion

Based on research conducted the research results showed that optimizing the school literacy movement directed towards disaster literacy awareness can increase public understanding, disaster preparedness and can reduce the risk of disasters that will cause material and non-material losses. The strategy and pattern of implementing the school literacy movement includes three stages, namely: habituation, development, and learning activities. The role of the family will contribute to disaster literacy that can provide experience and knowledge of the society according to the conditions of the residence. Technological advances are also able to contribute to the society as literacy materials and provide information quickly and accurately related to natural disasters.

Acknowledgments

Through this paper, the authors would like to thank all those who have contributed to the completion of this scientific work, so that it can be completed properly and on time.

References

[1] Afrian R., & Islami Z R 2019 Peningkatan potensi mitigasi bencana dengan penguatan kemampuan literasi kebencanaan pada masyarakat Kota Langsa. *Jurnal Pendidikan Geografi. 9251*, 132–143. http://journal2.um.ac.id/index.php/jpg/article/view/5313

[2] BMKG 2020 *No Title* (Badan Meteorologi Klimatologi Dan Geologi). https://www.bmkg.go.id/gempabumi/gempabumi-terkini.bmkg

[3] BNPB 2014 *Data dan Informasi Bencana Indonesia*. Badan Nasional Penanggulangan Bencana. http://dibi.bnpb.go.id/DesInventar/simple_data.jsp

[4] Brown L M, Haun J N & Peterson L 2014 A proposed disaster literacy model. *Disaster Medicine and Public Health Preparedness*, *8*(3), 267–275. https://doi.org/10.1017/dmp.2014.43

[5] Chieh Lee H & Chen H 2019 Implementing the Sendai Framework for disaster risk reduction 2015–2030: Disaster governance strategies for persons with disabilities in Taiwan. *International Journal of Disaster Risk Reduction*, *41*. https://doi.org/10.1016/j.ijdrr.2019.101284

[6] Dwisiwi R S, Surachman, Sudomo J & Wiyatmo Y 2012 Pengembangan Teknik Mitigasi Dan Manajemen Bencana Alam Gempabumi Bagi Komunitas SMP DI Kabupaten Bantul Yogyakarta. *Prosiding Seminar Nasional Penelitian. Pendidikan Dan Penerapan MIPA, Fakultas MIPA, Universitas Negeri Yogyakarta.*

[7] Hadiyati S & Hafida N 2018 Urgensi pendidikan kebencanaan bagi siswa sebagai upaya mewujudkan generasi tangguh bencana. *Jurnal Pendidikan Ilmu Sosial. 28*(2), 1–10.

[8] Haifani A M 2006 Manajemen Risiko Bencana Gempa Bumi (Studi Kasus Gempa Bumi Yogyakarta 27 Mei 2006). *Seminar Nasional IV SDM Teknologi Nuklir.*

[9] Hardjono I 2006 Hirarki Gempa Bumi Dan Tsunami (Aceh, Nias, Bantul, Pangandaran, dan Selat Sunda) The Hierarchy of Earthquake and Tsunami: The Cases of Aceh, Nias, Bantul, Pangandaran, and Sunda Staits. *Jurnal, Fakultas Geografi, Universitas Muhammadiyah Surakarta*, *Vol. 20*, *N*, 135–141. http://hdl.handle.net/11617/258

[10] Inten D N 2017 Peran Keluarga dalam Menanamkan Literasi Dini pada Anak Role of the FamilyToward Early Literacy of the Children. *Jurnal Pendidikan Anak Usia Dini. 1*(1), 23–32. https://doi.org/10.29313/ga.v1i1.2689

[11] Izumi T, Shaw R, Djalante R, Ishiwatari M & Komino T 2019 Disaster risk reduction and innovations. *Progress in Disaster Science*, *2*. https://doi.org/10.1016/j.pdisas.2019.100033

[12] Kamil, Puspita A, Utaya S, Sumarmi & Utomo D H 2020) Improving disaster knowledge within high school students through geographic literacy. *International Journal of Disaster Risk Reduction*, *43*. https://doi.org/10.1016/j.ijdrr.2019.101411

[13] Kent R 1994 *Kesiapan bencana II. Program Pelatihan Manajement Bencana*. Banyuwangi:DHA-UNDIP.

[14] Marhaento H & Kurnia A N 2015 Refleksi 5 Tahun Paska Erupsi Gunung Merapi 2010: Menaksir Kerugian Ekologis Di Kawasan Taman Nasional Gunung Merapi. *Geoplanning: Journal of Geomatics and Planning*, *2*(2), 69–81. https://doi.org/10.14710/geoplanning.2.2.69-81

[15] Maidaneli 2019 *Kesiapsiagaan Sekolah Dasar dalam Menghadapi Bencana Gempa Bumi dan Tsunami Di Kecamatan Pariaman Tengah Kota Pariaman*. Skripsi Thesis, Universitas Negeri Padaang.

[16] Marlyono S G & Pasya G K 2016 Peranan Literasi Informasi Bencana Terhadap Kesiapsiagaan Bencana Masyarakat Jawa Barat. *Jurnal Geografi.* 16, 116–123. https://doi.org/10.17509/gea.v16i2.4491.g9968

[17] Masitoh S 2018 Blended Learning Berwawasan Literasi Digital Suatu Upaya Meningkatkan Kualitas Pembelajaran dan Membangun Generasi Emas 2045. *Proceedings of The ICECRS*, Volume 1 No 3 (2018) 13-34. https://doi.org/10.21070/picecrs.v1i3.1377

[18] Prasetyo E 2016 Keluarga Literat dan Orangtua Mendongeng. *Majalah UNESA*.

[19] Suatengco R T & Florida J S 2018 Family literacy in a low-income urban community in the Philippines. Journal of Early Childhood Literacy, 20(2), 327–355. https://doi.org/https://doi.org/10.1177%2F1468798418766604

[20] Suharini E, Meliana D, Sanjoto T & Kurniawan E 2020 The Strategy of Disaster Mitigation Literacy through Problem-Based Learning (PBL) in the School Prone to Tidal Floods. https://doi.org/10.4108/eai.18-7-2019.2290156

[21] Sunarto N 2012 *Edukasi Penanggulangan Bencana Lewat Sekolah*. BPBD. http://bpbd.banjarkab.go.id/?p=75

[22] Surono, Jousset P & Pallister J 2012 The 2010 explosive eruption of Java's Merapi volcano—A '100-year' event. *Journal of Volcanology and Geothermal Research*, *241–242*, 121–135. https://doi.org/10.1016/j.jvolgeores.2012.06.018

[23] Tampubolon 1993 *Mengembangkan Minat dan Kebiasaan Membaca Pada Anak*. Angkasa.

[24] Wiedarti P, Laksono K & Retnaningdyah P 2016 *Desain Induk Gerakan Literasi Sekolah. Menengah, Direktorat Jenderal Pendidikan Dasar Dan Kebudayaan, Kementerian Pendidikan Dan Kebudayaan.*

[25] Yulianingsih W 2016 Peran Pendidikan Keluarga dalam Literasi Anak.*Jurnal Pendidikan Anak Usia Dini*. 1(1). 23-32. https://doi.org/10.29313/ga.v1i1.2689

[26] Zed M 2003 *Metode Penelitian Kepustakaan*. Jakarta:Yayasan Obor Indonesia.

[27] Zhang D, Zhu X, Zhou Z, Xu X, Ji X & Gong A 2020 Research on Disaster Literacy and Affecting Factors of College Students in Central China. Disaster Medicine and Public Health Preparedness, 1-7. doi:10.1017/dmp.2020.33

Participatory Mapping of Tsunami Evacuation Routes (Case Study of Karangbenda Village Cilacap Regency)

W Sabani[1,2], Juhadi[1] and E Trihatmoko[1]
[1]Geography Department, Faculty of Social Science, Universitas Negeri Semarang, Semarang 50229, Indonesia

[2]e-mail : wahyusabani@students.unnes.ac.id

Abstract. Indonesia is one of the countries with a high tsunami risk, in this case the community plays an important role in disaster management, one of which can be done through the preparation of a tsunami evacuation route map which can be done through participatory mapping. This research was conducted in Karangbenda Village, Adipala District, Cilacap Regency, the purpose of this study was to determine how the skills of the community in determining tsunami evacuation routes were compared with the criteria for evacuation locations and effective evacuation routes and to test the effectiveness of tsunami evacuation routes prepared by the community. Data collection in this study was carried out through focus group discussions (FGD) and field surveys. The results show that the community can arrange a tsunami evacuation route properly, the evacuation route consists of the evacuation location and evacuation route, the designated evacuation location is in accordance with the evacuation location criteria and the evacuation route is in accordance with the route obtained through network analysis. The effectiveness level of the tsunami evacuation route was 98.82% and there were some people who were unable to reach the evacuation location within the specified time, totaling 36 people who were in the area of Dusun Congot to be precise in RT 02 / RW 02 because of the long distance from the evacuation location.

Keyword: Participatory Mapping, Disaster Mitigation, Network Analysis.

1.Introduction

Indonesian territory is influenced by the movement of the Eurasian, Australian and Pacific tectonic plates [1] thus making Indonesia often experience tsunami disasters [2], most of the tsunami cases in Indonesia were caused by earthquakes and the rest were due to volcanic eruptions and landslides [3]. Compared to other disasters, the number of incidents of tsunami disasters is relatively low, but the impact is very destructive and causes high casualties. [4], One of the causes of this is the waves of sea water coming quickly, in some cases the tsunami disaster in Indonesia only left a limited time for evacuation, which was about 10-30 minutes after the earthquake

Content from this work may be used under the terms of the Creative Commons Attribution 3.0 licence. Any further distribution of this work must maintain attribution to the author(s) and the title of the work, journal citation and DOI.
Published under licence by IOP Publishing Ltd

IOP Conf. Series: Earth and Environmental Science **884** (2021) 012033 doi:10.1088/1755-1315/884/1/012033

occurred. [5], so the community had to evacuate in less than 30 minutes [6]. Responding to this risk, the Indonesian government developed a tsunami early warning system known as InaTEWS (Indonesia Tsunami Early Warning System) which was inaugurated on 11 November 2008 which provides early warning at the national level and is one of the 3 providers of early information for the Indian Ocean region together with Australia and India [7].

Even though an early warning system has been built, in reality, the tsunami disaster management efforts in Indonesia still have weaknesses, one of which is the absence of any evacuation instructions resulting in congestion. field of geographic information systems [8], This condition is also influenced by the low level of disaster literacy in Indonesian community [9]. With the high risk posed, the community must be aware of the threat of a tsunami disaster, one of which is by knowing the correct tsunami evacuation route to reduce the risk when a tsunami occurs. [10]. For the community to know the evacuation route for their area, it is necessary to involve the community to play an active role in determining the tsunami evacuation route through participatory mapping of the tsunami disaster evacuation route.

Participatory mapping is a way for community to interact to create and communicate knowledge, experiences, and aspirations about the world on a map [11], Participatory mapping in this study uses a participatory rural appraisal approach which is intended to provide flexibility for the community to determine tsunami evacuation routes for their communities, participatory mapping can integrate local knowledge and scientific knowledge which is the result of top-down and bottom-up approaches that enable disaster risk assessment in an area, identify local potential and scientific strategies for disaster risk reduction, and integrate these strategies into comprehensive action plans [12]. The purpose of this study was to determine how the skills of the community in determining tsunami evacuation routes were prepared through participatory mapping, then the evacuation routes determined by the community were compared with the criteria for the evacuation location and the shortest evacuation route, and to analyze the effectiveness of tsunami evacuation routes that had been prepared by the community.

2. Methods

The study is done in the village karangbenda , kecamatan adipala , kabupaten cilacap located at 109° 9' 42,26" E – 109° 11' 29,49" E and 7° 39' 52,94" S – 7° 41' 34,5" S. This village is one of the villages directly facing the Indian Ocean so that Karangbenda Village has a high risk of a tsunami disaster. This is based on the tsunami risk map published by GITEWS (German-Indonesia Cooperation for Tsunami Early Warning System) as shown in the Figure 1.

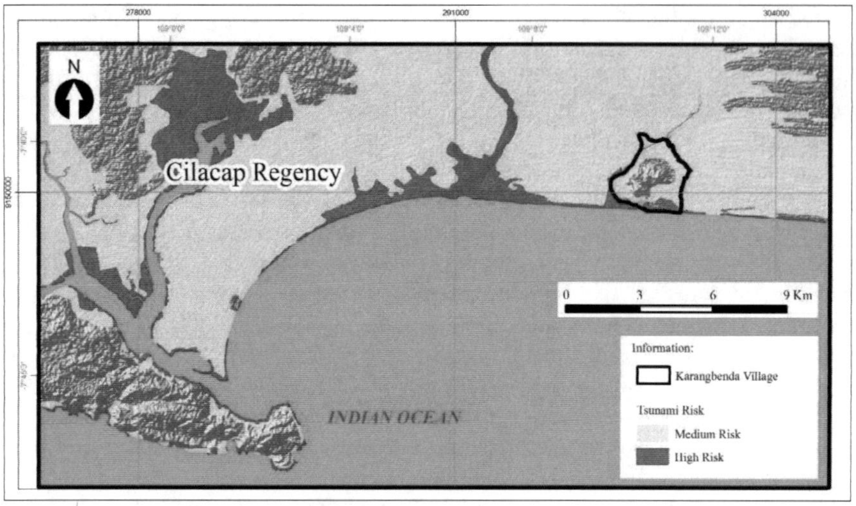

Figure 1. Research sites

This research was conducted in July 2019, primary data in this study were obtained through focus group discussions (FGD) and field surveys, while secondary data was obtained from related institutions, namely the Geospatial Information Agency (BIG). The population in this study were the people of Karangbenda Village, the samples were determined using a purposive sampling technique, namely the Karangbenda Village Government, Karangtaruna, PKK, and Community-Based Disaster Preparedness Groups (SIBAT). The series of research activities are described in the following table:

Table 1. Schedule of Participatory Mapping Activities for Tsunami Disaster Evacuation Routes

No.	Activity	Date
1	Introducing mapping ideas	July 4-6, 2019
2	Mapping training for the community (mapping team)	July 10, 2019
3	Field survey	Juli, 12-22 2019
4	focus group discussion (FGD) determination of tsunami evacuation routes	July 23, 2019
5	Make the final map	July 23, 2019
7	focus group discussion (FGD) for map validation	July 30, 2019

The FGD process is intended to discuss tsunami evacuation routes where the evacuation movement includes at least the destination or location of the evacuation, route of travel, and evacuation mode [13]. The variables in this study are the suitability of the evacuation route and the effectiveness of the evacuation route. The suitability of the evacuation route consists of the suitability of the tsunami evacuation location and the tsunami evacuation route, in this case, the evacuation routes prepared by the community are compared with the evacuation location criteria shown in Table 2 as follows:

Table 2. Tsunami Evacuation Location Criteria

Criteria	Parameter
Distance to River	Not in the tsunami-affected area based on the tsunami model
Distance to beach	Not in the tsunami-affected area based on the tsunami model
Elevation	Not in the tsunami-affected area based on the tsunami model
Slope	Less than 20°
Extensive	Adequate to accommodate the number of refugees assuming the need for refugee space, namely 1 m^2.

Sources: [6], [7]

Whereas for the evacuation route compared to the shortest evacuation route obtained using geographic information system (GIS) technology, the geographic information system in recent years has grown to become increasingly important with its varied uses, especially for evacuation [16], The information system technology used in this research is network analysis, namely the route function in the ArcGIS application which uses the Dijkstra algorithm to determine the best route from one point to another based on travel time. [17]. Then to test the effectiveness level of the tsunami evacuation route using network analysis, namely the service area function. The service area is an analysis to determine the area that can be served by a location in the road network [18], Service area analysis in this study is used to model the evacuation time with 5-minute intervals, tsunami early warning takes 5 minutes. [19], then the tsunami evacuation route is considered safe if it does not exceed 5 - 25 minutes after the tsunami occurs. Assuming the speed for a tsunami evacuation is 1 meter/second [20], then used modeling the maximum distance from the evacuation site, namely 1,400 meters.

3. Result and Discussion

The determination of the tsunami evacuation route was accommodated in the FGD activity on July 23, 2019, which took place at the Karangbenda Village Hall. This activity was attended by community components consisting of the Village Government, PKK, Karangtaruna, SIBAT, and involved related agencies, namely BPBD Cilacap Regency and PMI Cilacap Regency. The results of the FGDs are the tsunami evacuation locations for each neighborhood (RT) that shown in Table 2.

Figure 2. Focus Group Discussion on Determining Tsunami Evacuation Route in Karangbenda Village

Table 3. The location for the Evacuation of Karangbenda Village agreed by the community Through the Activities of Focus Group Discussion

Dusun	RT	Evacuation Location	Dusun	RT	Evacuation Location
Karangbenda	01	Sumur Windu	Babakan	01	Selok Hill (Bridge 1)
	02	Sumur Windu		02	Selok Hill (Bridge 1)
	03	Gili Jaran		03	Selok Hill (Bridge 1)
	04	Gili Jaran		04	Selok Hill (Bridge 1)
	05	Sumur Windu	Sodong	01	Mandara Giri Temple courtyard
Congot	01	Gili Jaran		02	Mandara Giri Temple courtyard
	02	Gili Jaran		03	Mandara Giri Temple courtyard
	03	Gili Jaran		04	Mandara Giri Temple courtyard
	04	SD 1 Karangbenda			
	05	SD 1 Karangbenda			

Then the evacuation location is compared with the indicators and parameters of the tsunami evacuation location as follows:

Tabel 4. Suitability of the Evacuation Location Determined by the Community with Evacuation Location Criteria

Evacuation Location	Evacuation Location Criteria							
	(1) Distance to River (m)	(2) Distance to Beach (m)	(3) Elevation (m)	(4) Slope (°)	Capacity (1 m2/person)	(5) Capacity (0,25 m2/person)	Number of Refugee (jiwa)	Information
Gili Jaran	520	1.628	16,8	1	972	3.890	930	suitable
Sumur Windu	636	1.687	19,1	2	3.126	12.507	594	suitable
SD N Karangbenda 1	997	1.654	11	4	298	1.195	320	suitable
Mandara Giri Temple courtyard	1.453	1.324	15	8	6.053	24.213	1.751	suitable
Bridge 1	1.320	1.300	34,3	9	475	1.901	739	suitable

The table above shows that the tsunami evacuation location determined by the community is following the tsunami evacuation location criteria. The community has also determined an evacuation route that is described in paper media, the route is then compared with the results of the analysis of the shortest route to the evacuation location obtained through the route function in the network analysis which is shown in Figure 3 as follows:

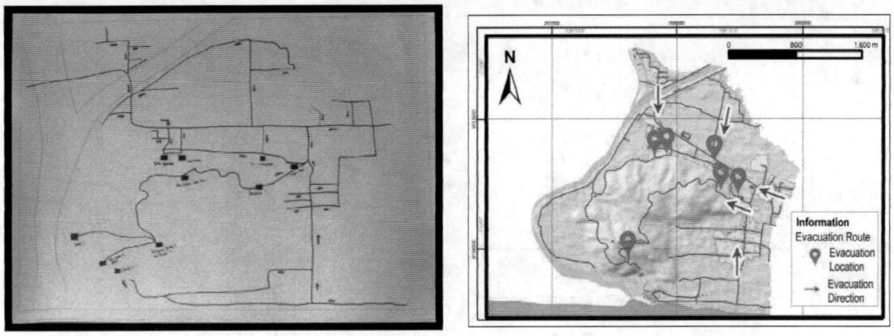

(a) (b)

Figure 3. Sketches of Tsunami Evacuation Routes Prepared by Communities (a) and Evacuation Route Map obtained through network analysis (b)

The image above shows the same evacuation route which shows the same result which shows that the community can determine the evacuation route well. Then for the analysis of the effectiveness level of the tsunami evacuation route is shown in Figure 4.From Figure 4 it can be seen that the number of people who can evacuate with the following evacuation time:

Table 5. Total Population Based on Evacuation Time

Total Population Based on Evacuation Time Intervals		Evacuation Time	Accumulated Population by Time of Evacuation	Efectivity
time interval	Number			
0 - 5	226 person	5 minutes	226 person	7,41 %
5 - 10	778 person	10 minutes	1.004 person	32,91 %
10 – 15	1.417 person	15 minutes	2.421 person	79,35 %
15 – 20	439 person	20 minutes	2.860 person	93,74 %
20 – 25	155 person	25 minutes	3.015 person	98,82 %
> 25	36 person	> 25 minutes	3.051 person	100 %

From the data above, it can be seen that the effectiveness level of the tsunami evacuation route in a period of 25 minutes is 98.82% where there are some people who cannot reach the evacuation location within the specified period of time totaling 36 people, which are known to be in the area of Dusun Congot precisely in RT 02 / RW 02.

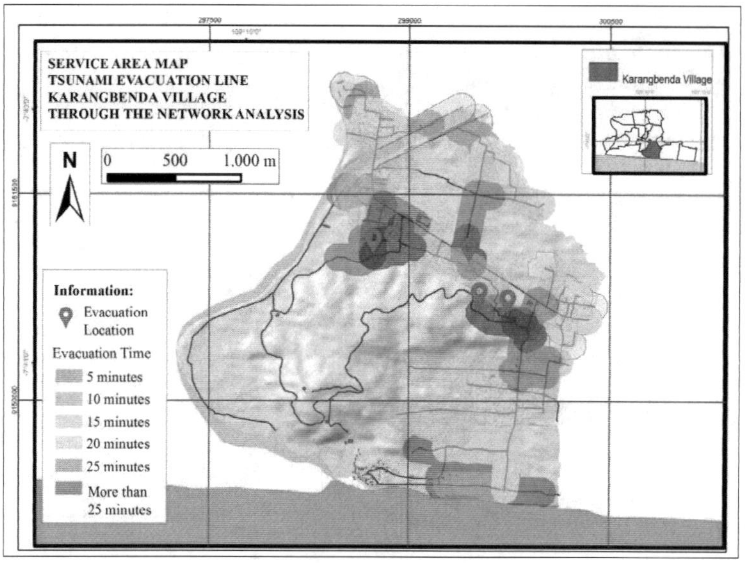

Figure 4. Tsunami Evacuation Route Map from Karangbenda Village Network Analysis

From the results above, it can be seen that participatory mapping activities through FGD activities can be an effective method for extracting information from village communities, information stored in community mental maps can be written in the form of informative maps, besides that FGD activities can be a forum to bring stakeholders together in the community. this is the government and local communities, so that disaster management policies do not only focus on top-down policies, as explained by Gaillard et. al [21] that participatory mapping can also be used to integrate the capacities of people of different castes, government, scientists, and community organizations in disaster risk reduction.

4. Conclusion

The conclusion obtained from this research is that the people of Karangbenda Village can develop a good tsunami evacuation route, seen from its suitability with the criteria for the evacuation location and evacuation route using network analysis. The evacuation route that has been agreed upon by the community has an effective level of 98.82%, this is obtained through service area analysis, although there are people who cannot go through the tsunami evacuation route within the specified time.

Acknowledgement

The author gives thanks to the people of Karangbenda Village who have been willing to play an active role in research activities and also to the Cilacap Regency BPBD and Cilacap Regency PMI for their cooperation in preparing the evacuation route map for Karangbenda Village.

Reference

[1] A. Hartoko, M. Helmi, M. Sukarno, and Hariyadi, "Spatial Tsunami Wave Modelling for the South Java Coastal Area, Indonesia," *Int. J. Geomate*, vol. 11, pp. 2455–2460, Jan. 2016.

[2] Juhadi and M. Herlina, *Pendidikan Literasi Mitigasi Bencana Di Sekolah*. Kudus: Parist Penerbit, 2019.

[3] H. Latief, N. Puspito, and F. Imamura, "Tsunami Catalog and Zones in Indonesia," *J. Nat. Disaster Sci.*, vol. 22, pp. 25–43, Jan. 2000.

[4] N. Parwanto and T. Oyama, "A Statistical Analysis and Comparison of Historical Earthquake and Tsunami Disasters in Japan and Indonesia," *Int. J. Disaster Risk Reduct.*, vol. 7, Nov. 2013.

[5] F. Imamura *et al.*, "Tsunami Disaster Mitigation by Integrating Comprehensive Countermeasures in Padang City, Indonesia," *J. Disaster Res.*, vol. 7, Dec. 2011.

[6] Y. Yunarto and A. Sari, "Analysis of Community Tsunami Evacuation Time: An Overview," *IOP Conf. Ser. Earth Environ. Sci.*, vol. 118, p. 12033, Feb. 2018.

[7] S. Harig *et al.*, "The Tsunami Scenario Database of the Indonesia Tsunami Early Warning System (InaTEWS): Evolution of the Coverage and the Involved Modeling Approaches," *Pure Appl. Geophys.*, vol. 177, no. 3, pp. 1379–1401, 2020.

[8] D. Mardiatno, M. Malawani, D. Wacano, and D. Ahmad, "Review on Tsunami Risk Reduction in Indonesia Based on Coastal and Settlement Typology," *Indones. J. Geogr.*, vol. 49: 2, Dec. 2017.

[9] Juhadi, M. Hernila, E. Kurniawan, and E. Tryhatmoko, "Disaster Mitigation Learning Literacy Towards Earthquake And Tsunami Based On Lamban Langgakh Local Wisdom For Student At School," *Int. J. GEOMATE*, 2019.

[10] A. Putra and H. Mutmainah, "The Mapping of Temporary Evacuation Site (TES) and Tsunami Evacuation Route in North Pagai Island, Mentawai Islands Regency - Indonesia," *IOP Conf. Ser. Earth Environ. Sci.*, vol. 47, p. 12020, 2016.

[11] G. Brown and M. Kyttä, "Key Issues and Priorities in Participatory Mapping: Toward Integration or Increased Specialization?," *Appl. Geogr.*, vol. 95, pp. 1–8, 2018.

[12] J. R. D. Cadag and J. C. Gaillard, "Integrating Knowledge and Actions in Disaster Risk Reduction: The Contribution of Participatory Mapping," *Area*, vol. 44, no. 1, pp. 100–109, 2012.

[13] F. Makinoshima, F. Imamura, and Y. Oishi, "Tsunami Evacuation Processes Based on Human Behaviour in Past Earthquakes and Tsunamis: A Literature Review," *Prog. Disaster Sci.*, vol. 7, p. 100113, 2020.

[14] M. Mück, "Tsunami Evacuation Modelling. Development and Application of A Spatial Information System Supporting Tsunami Evacuation Planning in South-West Bali," *Inst. für Geogr.*, vol. Degree in, p. 131, 2008.

[15] R. Fernando, B. Sujatmo, and A. Hendri, "Perencanaan Tempat Evakuasi Bencana Banjir Berbasis Teknologu Sistem Informasi Geografis (SIG)," *J. Online Mhs. FT UNRI*, pp. 1–9, 2017.

[16] A. Trindade, P. Teves-Costa, and C. Catita, "A GIS-based analysis of constraints on pedestrian tsunami evacuation routes: Cascais case study (Portugal)," *Nat. Hazards*, vol. 93, no. 1, pp. 169–185, 2018.

[17] S. Ahmed, R. Ibrahim, and H. Hefny, *GIS-Based Network Analysis for the Roads Network of the Greater Cairo Area*. 2017.

[18] S. M. H. Erfani, S. Danesh, S. M. Karrabi, R. Shad, and S. Nemati, "Using applied operations research and geographical information systems to evaluate effective factors in storage service of municipal solid waste management systems," *Waste Manag.*, vol. 79, pp. 346–355, 2018.

[19] Masturyono, T. Hardy, P. Susilanto, W. Setyonegoro, D. Ngadmanto, and M. Madlazim, "DESAIN SISTEM PENENTUAN POTENSI TSUNAMI MENGGUNAKAN : RUPTURE DURATION (Tdur), TIME DOMINAN (Td) DAN T50EX," *Pros. BMKG*, May 2013.

[20] Sea Defence Consultan, *Pedoman Perencanaan Pengungsian Tsunami (Tsunami Refuge Planning)*. Nias: Sea Defence Consultan, 2007.

[21] J. C. Gaillard *et al.*, "Participatory 3-Dimension Mapping: A Tool for Encouraging Multi-caste Collaboration to Climate Change Adaptation and Disaster Risk Reduction," *Appl. Geogr.*, vol. 45, no. December, pp. 158–166, 2013.

We are getting stronger as we stay longer: the influence of place attachment to urban community resilience

M Indayani[1,4], S Trisutomo[2], M Ramli[3]

[1]Graduate School, Hasanuddin University, Makassar, South Sulawesi 90245, Indonesia
[2]Department of Urban and Regional Planning, Hasanuddin University, Gowa, South Sulawesi 92171, Indonesia
[3]Department of Sociology, Hasanuddin University, Makassar, South Sulawesi 90245, Indonesia

E-mail: marinda.arin@gmail.com

Abstract. The increasing number of disasters happening in cities has put the sustainability of urban life at stake. One of the measures to strengthen the community's resilience is by understanding how they got attached to the place they live. However, the findings of the previous research regarding the influence of place attachment to community resilience show dissimilarities. This research aims to further reveal the influence of place attachment to communities' resilience in urban areas with a variety of socio-demographic and social dynamics. The study was conducted to 323 households that have frequently been affected by flood, at an urban-village, Tallo sub district, of Makassar City, Indonesia. With simple linear regression analysis, the aspects of place attachment which are personal, social, physical, and cultural are measured to assess its influence on each of their resilience perceptions. The research revealed that there is a positive influence of place attachment to a communities' resilience. Based on the communities' response, it can be seen that resilience is more likely to be formed by social-community aspects.

Keywords: Place attachment, community resilience, disaster.

1. Introduction

The number of the world population inhabiting urban areas has shown a continuous increase. The UN Population Division DESA's data shows that currently 56.2% of the world populations live in the city, and is forecasted to reach 68.4% at 2050[1]. In line with it, the number of global disasters has also increased, putting cities and urban areas at high risk, due to the population posture, density, and economic activity that are concentrated in urban areas [2]. The concept of city resilience is meant to reduce the risk of disasters in urban areas, and support essential systems (including the urban community) to be flexible in maintaining its function, before, during, and after the disaster.

Community resilience refers to the capacity of the social system to respond, adapt, and recover from disaster [3]. The capacity is determined by the characteristic of a community, one of which is being attached to the place they live [4], [5]. Emotion, cognition, and behavior reflected by

Content from this work may be used under the terms of the Creative Commons Attribution 3.0 licence. Any further distribution of this work must maintain attribution to the author(s) and the title of the work, journal citation and DOI.
Published under licence by IOP Publishing Ltd

individuals or groups to a certain place, is becoming important to understand psychological factor related to disaster, including the risk perception, and environmentally related behaviors [6].

A number of empirical studies have been performed to investigate how far place attachment contributes to community resilience. These researches are implemented to communities with different sizes, and with a variety of threat potentials [7], [8]. Despite the findings of a correlation, studies cannot yet firmly conclude how place attachment influences towards a stronger or weaker resilience of the community, especially those who have socio-economic and cultural complexity like in the cities. The inconsistent findings from previous research has inspired further studies in regards to how those two constructs influence each other. This research takes the community of Tallo sub-district, of Makassar city, Indonesia, who live at the frequently flood affected area due to the geographical position (at the coastal line), as the study case. Beside the influence of place attachment to the communities' resilience, this research also attempts to reveal the most influential factor that forms place attachment and community resilience by observing how the respondents react to the questions.

2. Literature Review

2.1. Place Attachment

The concept of place attachment in the scope of environmental-psychology refers to a bond that is formed between humans and the place they live. The word 'attachment' explains the aspect of affection, while place is an environmental setting where humans are attached to [9]. Attachment emerges when an individual attaches a value, or gives a meaning to a space, either personal (psychological), social, or cultural, by a process that involves feeling (emotion), cognition (knowledge), and action (behavior) [10].

The forming of place attachment is influenced by various aspects, which are personal, social, physical, and cultural. Personal aspects are connected to the socio-economic condition of the person in the community, and usually become a predictor of the presence of place attachment. One of the indicators of personal aspect which is the strongest and most consistent is the length of residence/ association at a place [11]. The longer someone stays at the same place, the stronger the person builds a value towards the place [12]. The home ownership and education level is another important predictor of the personal aspect. While length of stay and ownership is a positive predictor, education level which is related to the mobility of a person is a negative predictor to place attachment [10]. The social aspect of place attachment is often linked to the community attachment. This aspect is formed as a result of meaningful interactions among individuals in that community [13]. Attachment to community arises the feeling of security equipped by community members. This aspect is even assumed more meaningful compared to satisfaction of environment-physical attributes at the community with middle-low level income [9]. A number of researches put density, community size, and home types as indicators at physical aspect of place attachment [10], [11]. However, at the type of community that lives in the middle to lower level of residential area, the condition of the houses becomes the most dominant influencing factor[9]. The cultural aspect of place attachment is formed when the place is historically connected to a particular ethnic or group. To maintain the value of a place, rituals ceremonies are held to unite the group in the community.

2.2. Community Resilience

The concept of resilience, which is initially utilized in ecology, has now been applied extensively to describe the ability of a system, not only to absorb disturbance, but also to preserve the previous condition [14]. In social science, resilience refers to the ability of an individual to respond to changes, and recover from any kind of disturbance, and be prepared to anticipate any challenging conditions in the future [4]. Community resilience focuses on the capacity possessed by a group of people that share the same place, purpose, value, and even the same destiny, to survive and adapt with any kind of threats and obstruction to their living environment.

Community resilience is formed by the interaction between internal factor (existing characteristics in a community), and the external factors that influence their ability in surviving and adapting. It becomes the basis of the development of methods in measuring communities' resilience by experts. Disaster Resilience of Place (DROP) measures the community's resilience based on the exogenous factors (ecology, social, economy, infrastructure) and endogenous factors (community competence) [3]. Aharanson and Lahad developed The Conjoint Community Resilience Assessment Measure (CCRAM) that asses the objective aspect and community resilience perception [15]. The Urban Community Resilience Assessment (UCRA), a collaboration project between World Resources Institute (WRI) and Cities Alliance has comprehensively assessed the place vulnerability, social cohesion, and individual competence in community [16]. This research aims to assess the community resilience related to place attachment, thus the assessed resilience is the one that is subjective (resilience perception). Resilience perception is used to understand affection and community cognition elements in evaluating everyone's adaptive capability [17].

The socio-demographic aspect (age, education, and profession) are the internal factors that influence community resilience[18], in addition to the competency/ characteristics that exist within the community. Local knowledge and past experience in encountering disaster, perception of disaster risk, preparedness, efficacy, and participation are the keys of community resilience [3], [16], [18], [19]. While for the external factors, social context in a community (social cohesion and leadership in a community) will ensure aid availability and support among each other member of the community crucial moments [5], [15]. The physical aspect is also an important factor that is strongly related to the perception of the safety of the neighborhood/living place towards disasters [20].

2.3. *Place Attachment and Community Resilience*

Place attachment is commonly used at the attempts to understand resilience to environmental changes and disasters. Attachment to places contributes to enhance the community participation to the environmental change related issues by arises the environmental preservation behaviors [21]. Place attachment is also proven to enhance the adaptive capacity of the community, especially those who are economically dependent on a place [8]. Nevertheless, a higher level of a place attachment might hinder the community's preventive behavior to disaster risk [22], when presented with a high risk threat and high risk-perception. High risk-perception to disasters are shown by those who have high level of place attachment, but at the same time they also tend to ignore other effects that might occur [7]. This is a form of coping mechanism to handle fear of disaster risk potentials.

3. Research Methods

This research uses quantitative-deductive approach, with a survey method. The variables researched are the place attachment (independent variable), and community resilience (dependent variable).
The survey is performed using a questionnaire that consists of 18 questions, each of which is provided with answers on a scale from 1 to 4. The place attachment variable is assessed with 8 indicators, which are rootedness, social bonds, social interaction, home quality, sense of security, environment responsible behavior, cultural characteristics, and cultural similarity. Whilst the community resilience variable is assessed with 9 indicators which are the social cohesion, leadership within the community, perception of home safety, disaster experience, disaster risk perception, preparedness, planning, efficacy, and participation.

Tallo sub district, which is located at northern part of Makassar City, Indonesia, was selected as location of study, focused at four neighborhoods (RW A, B, C, and D) that are frequently affected by flood due to its geographical position at the coastal line. Tallo sub district is categorized as a flood-affected zone with a surface area of 51,431sqm, and flood level reaching 50cm [23]. The social life of community at Tallo sub-district reflects a village characteristic even though it administratively located at urban area.

Data were collected from 323 respondents, assisted by 3 enumerators. The questionnaires were not self-administered due to the low level of community's literacy and education. Instead, they were asked directly to the respondents. The age of respondents in this research was selected to at least 20 years old, and only one respondent at each of household was chosen as representative. The data was collected at daytime to evening for five consecutive days. Other than questionnaires, observation was also conducted to get the overall picture of the settlement's condition objectively. The observation focused on the community's behavior in adapt to disaster risk regarding their settlement, and the government's intervention as a mitigation action to flooding event in this neighborhood as well.

Data analysis was conducted with simple linear regression analysis, assisted by the SPSS version 22 application. This kind of analysis is commonly used in studies that aim to discover the influence between 1 independent variable and 1 dependent variable [24]. It requires two classical assumption tests to be performed as condition before the analysis is run, which are (1) Data validity and reliability test, and (2) Data normality and linearity test. Validity test was conducted using The Pearson product-moment correlation analysis, where the correlation coefficient value, r_{count}, should greater than r_{table}'s. value. Whilst the reliability test requires Cronbach's alpha greater than 0.600. Kolmogorov-smirnov test is used for the normality test, with the requirement of the residual value more than 0.05. The same applies for the data linearity test; the value of the sig.deviation accepted is more than 0.05. The results for Pearson's correlation showed that coefficient value of r_{count} at both variables were greater than r_{table}'s. The Cronbach's alpha for place attachment variable was at 0.785, while for community resilience was at 0.655. Those results indicate that the instrument used for this study has met the condition for validity and reliability tests. The Kolmogorov-smirnov test came with residual value at 0.08. Sig.dev value from linearity test was at 0.06. Both were greater than 0.05, which means that the data used in this study is normally distributes and has linearity.

4. Results and Discussions

4.1. Respondent's Response Frequency

Respondents at the research location are dominated by women (53.6%) with the dominant range of age (63.47%) at 30-49 years old. Averagely, the education level of the head of the household is Elementary School (32.8%), and Senior High School or higher (35%), where 52.3% are working in the informal sector as casual laborers. Most of the respondents (57.9%) have been staying at this place for more than 15 years, and holding ownership status of house and land (94.7%). The average value of household's dependency ratio is at the range of less than 46%.

Determining the average value of respondents' answers of each question in the questionnaires is intended to understand the respondents' response to each of the variables [24]. By doing so, it can be comprehended which of the indicators of the variables gains the highest or the lowest scores, thus the overall picture of the variable-related condition/phenomenon of the study can be discovered. On the place attachment variable, the respondents mostly gave the highest scores (3.8 out of 4) at the question of "Do you know the people in this neighborhood well?" which is the indicator of social bond. This condition is in-line with the theory that, social aspect (community attachment) has predominantly influenced place attachment compared to satisfaction towards place's physical attributes (community satisfaction) in community with middle-low level of income. The social bonds of the community residing in Tallo sub-district are merely formed because of the intensive interaction within the community, either by formal events (social organizations/activities), and also non-formal gatherings (to get together just to mingle and have light conversation). While the lowest score (2.04 out of 4) comes from the question of "Do you have religious celebration/cultural rituals held on a regular basis in this neighborhood?" which indicates a common cultural value. This shows the cultural aspect has not become the important factor of forming place attachment of the community in Tallo sub-district. The scores of the respondents' responses at place attachment variable are provided at the following figure 1.

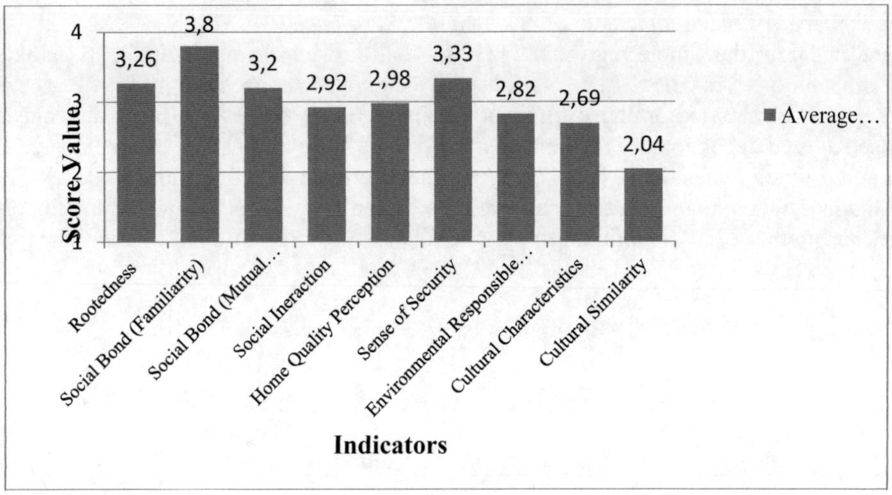

Figure 1. Graph of the respondent's score on the place attachment variable.

The respondents showed highest agreement (score 3.43 out of 4) to the question of the social cohesion indicator, which is "Do you reach out to your neighbors when you are in need of assistance (i.e. accident, grief)?". This is in-line with the research conducted by Patel & Gleason [25] which found out that a high score of social cohesion in the society at a slum area in Haiti, was correlated with the strong resilience of that community in dealing with disaster. The social cohesion in a community ensures the availability of aid and assistance when required during hurdle/emergency times [16]. The lowest score given by the respondents is at the question "Do you have enough savings and/or (any kind of) insurance to back up your state of emergency?" This question originated from the indicator of preparedness which is one of the aspects in community competence. Community in Tallo sub-district that is frequently affected by flood mostly do not have long term planning regarding mitigation action towards flood disaster in financial terms, which is merely caused by their economic condition with low income at most of the households, and overall low community literacy as an addition. The scores of the respondents' responses at community resilience variable are provided at the following figure 2.

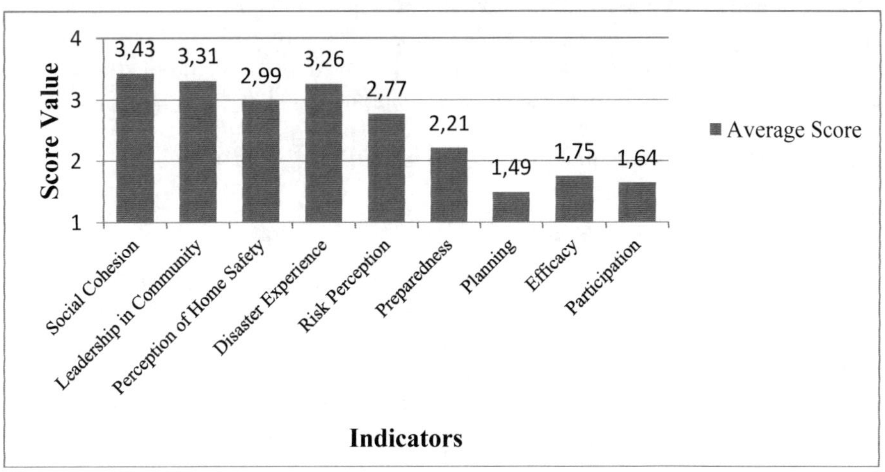

Figure 2. Graph of the respondent's score on the community resilience variable.

4.2. The Influence of Place Attachment to Community Resilience

It is found out from the simple regression analysis that the significance value (Sig.) is 0.000, lower than the probability score 0.05. The t_{count} of the regression also shows value of 9.678, greater than the t_{table}'s value of 7.79. Based on those findings it can be concluded that the place attachment variable can be used to predict community resilience variable. The correlation score between place attachment variable and community resilience is 0.475. While the determination coefficient scores (R Square) is at 0.226. This shows that the place attachment variable has influencing factors to community resilience at 22.6%, while another 77.4% is influenced by other factors. The SPSS output of coefficient can be seen at table 1.

Table 1. SPSS output for simple linear regression analysis.

coefficients [a]

Model		Unstandardized Coefficients B	Unstandardized Coefficients Std. Error	Standardized Coefficients Beta	t	Sig.
1	(Constant)	11,750	1,427		8,233	,000
	Place Attachment	,485	,050	,475	9,678	,000

a. Dependent Variable: Community Resilience

The equation of the regression is as follows:

$$Y = 11.75 + 0.485X \quad (1)$$

The constant value of 11.75 shows that the community resilience variable's consistency level is at 11.75. The X regression coefficient of 0.485 indicates that for each 1% increase in place attachment level, community resilience grows stronger 0.485. Since the regression coefficient shows positive value, the direction of the place attachment variable to community resilience is positive.

The findings in this research confirm the theory which claims that place attachment is a catalyst that can strengthen the community's resilience [8], [26]. In the particular location of this research, the strong attachment to the place has encouraged the local community to carry out adaptive behaviors by reinforcing their building structure which was initially made of wood, gradually being replaced or being added with brick structure. A strong social cohesion within the community of Tallo sub-district facilitates and enables that movement to be self-subsistent. This is in line with theory states that resilience in a community with rural characteristic is likely to emerge due to the high social context amongst the community members [27]. Nevertheless, the ability of place attachment variable to predict the community resilience is considerably small. The lack of government involvement and other related parties in the activities of disaster risk mitigation actions in Tallo sub-district is the cause of the low level of the community's knowledge on the importance to have long-term planning related to disasters. Revitalization of the environment by the government is limited only to the improvement of roads and supply of clean water piping network from PDAM, which unfortunately often does not work, thus makes no point for the people.

5. Conclusions

In conclusion, built upon the feedback of the respondents at Tallo sub-district of Makassar City, Indonesia, place attachment and community resilience are mostly emerged by social aspects (social bond and social cohesion). Not only increase their attachment to the place they live, the existence of social context within a community can also ensure the availability of support when needed, which eventually also strengthens the community's resilience collectively. Place attachment gives positive influence to the attempt of strengthening the community's resilience. At Tallo sub-district, the communities carry-out self-subsistent mitigation movements as part of their behavior of protecting their meaningful living place.

References

[1] United Nations, Department of Economic and Social Affairs, Population Division 2018. World Urbanization Prospects: The 2018 Revision, Online Edition. [Accessed: 14-Aug-2020].
[2] Amaratunga D Sridarran P and Haigh R 2019 *Making Cities Resilient Report 2019* (United Kingdom: United Nations Office for Disaster Risk Reduction)
[3] Cutter L Susan, Barnes L, Berry M, Burton C, Evans E and Jennifer W 2008 *Glob. Environ. Chang* **18** 598–606
[4] Chapman A Daniel, Trott D Charlie, Silka L, Lickel B and Clayton S 2018 *Psychol. Clim. Chang* 0–22
[5] Faulkner L, Brown K and Quinn T 2018 *Ecol. Soc.* **23**
[6] Scannell L and Gifford R 2010 *J. Environ. Psychol.* **30** 1–10
[7] Bonaiuto M, Alves S, De Dominicis S and Petruccelli I 2016 *J. Environ. Psychol.*
[8] Guo C Yongrui, Jie Z, Yuling Z and Zheng C 2018 *Sustainability* **2347** 1–14
[9] Altman I and Low M Setha 1992 *Place Attachment* (New York: Plenum Press)
[10] Gifford R 2013 *Environmental Psychology: Principles and Practices* (Canada: Optimal Books)
[11] Lewicka M 2011 *J. Environ. Psychol.* **31** 207–230
[12] Smaldone D 2006 *Proc.of the 2006 Northeastern Recreation Research Symposium* pp. 47–56
[13] Brown B, Perkins D and Brown G 2003 *J. Environ. Psychol.* **23** 259–271
[14] Holling S 1973 *Annu. Rev. Ecol. Syst.* **4** 1–23
[15] Cohen O, Leykin D and Lahad M 2013 *Technol. Forecast. Soc. Chang.*
[16] Rangwala L Elias-trostmann K Wihanesta R Chandra M and W. R. I. Org 2018, *Prepared Communities: Implementing the Urban Community Resilience Assesment in Vulnerable Neighborhoods of Three Cities* (Washington: World Resource Institute)
[17] Jones L and Tanner T 2016 *Reg. Environ. Chang.*
[18] Lessy M Bemba J and Nagu N 2018 *MATEC Web of Conferences, ICDM 2018* pp. 0–11
[19] Grothmann A, and Patt T 2005 *Glob. Environ. Chang.* **15** 199–213
[20] Domingues R B, Santos M C, Neves S, Jesus D and Ferreira O 2018 *Ocean Coast. Manag.* **157** 248–256
[21] Vaske J Jand Kobrin K C 2010 *J. Environ. Educ.* **32** 16–21
[22] De Dominicis S, Fornara F and Ganucci U 2015 *J. Environ. Psychol.***43** 66–78
[23] Mahardy A I 2014 Skripsi (Unpublished)
[24] Sugiyono 2017, *Metode Penelitian Kombinasi (Mixed Methods)*, Cetakan ke-9 (Bandung: Alfabeta)
[25] Patel R B and Gleason K M 2018 *Int. J. Disaster Risk Reduct.* **27** 161–167
[26] Butu H M, Hariza A, Hashim B and Ahmad N 2018 *IOSR J. Env. Sci. Tox and Food Tech.* **12** 12–19
[27] Rapaport C, Hornik-Lurie T, Cohen O, Lahad M, Leykin D and Aharonson-Daniel L 2018 *Int. J. Disaster Risk Reduct.* **31** 470–477

The Utilization of Sentinel-2A and ASTER Imagery for Monitoring the Changes of Public Green Open Space and Oxygen Needs in Sukoharjo Regency in 2004-2019

Agnes Putri Devinta[1*], Prima Widayani[2]
[1,2]Departement Sains Informasi Geografi, Fakultas Geografi, Universitas Gadjah Mada, D.I. Yogyakarta, Indonesia

Email: [1]agnesdevinta@gmail.com [2]primawidayani@ugm.ac.id

Abstract. High land requirements have an impact on land conversion. This study aims to calculate the accuracy of the results of mapping public green open space from ASTER and Sentinel-2A imagery, know the changes in green public space, calculate oxygen demand and the needs of green space in 2004 and 2019. The types of green open spaces that are interpreted visually include urban forests, river borders, cemeteries, fields, and city park. Oxygen demand is calculated by the gerrarkis method including livestock, industry, population, and motor vehicles. The mapping accuracy with the ASTER is 96% while the Sentinel-2A imagery is 90%. The mapping of changes in public green open space show that 17,62 km^2 public green open space has not changed, increased 1,15 km^2, and decreased 2,61 km^2. Oxygen demand in 2004 was 1053531,92 kg/day with green open space needs covering 10,41 km^2, while in 2019 it was 1923959,31 kg / day with Green Open Space needs covering 19 km^2. The need for green space in 2004 has been fulfilled from public green space of 20,22 km^2. In 2019 the area of public open green space is 18,77 km^2, so that public open green space has not been able to fulfill the needs of overall green open space.

Keyword: Green Open Space, ASTER Imagery, Sentinel-2A Imagery, Oxygen Needed

1. Introduction
Sukoharjo Regency as one of the pillars of economic growth in Central Java has a rapid economic development. This has an impact on increasing the population significantly. Data from BPS Sukoharjo Regency in 2019 states that the population of Sukoharjo Regency has continued to increase from 2004-2019. In 2018 the population density reached 1,897 people / km^2 with an area of 466.66 km^2 [1]. Economic development has an impact on increasing the need for space and land. High land requirements, not accompanied by adequate land availability. This has resulted in a tendency to change the function of green open space land into a built-up area. Change of land function is a change in land function from one function to another and generally occurs in the suburbs. Data from the Sukoharjo Regency Agriculture Office states that the area of paddy fields decreases every year while built-up land increases by 50-100 hectares per year [11].

The Minister of Public Works Regulation in 2008 states that every city area must provide a Green Open Space of 30% of the total area, namely 20% for Public Green Open Space and 10% for private Green Open Space. The main functions of green open space are as an ecological function, socio-cultural function, economic function, and aesthetic function. Green Open Space also plays a role in providing oxygen for living things and the combustion process [2]. The increase in population, livestock,

Content from this work may be used under the terms of the Creative Commons Attribution 3.0 licence. Any further distribution of this work must maintain attribution to the author(s) and the title of the work, journal citation and DOI.
Published under licence by IOP Publishing Ltd

motorized vehicles, and industry has an impact on increasing oxygen demand. Oxygen is very important to sustain ecological life. oxygen plays a fundamental role in preserving the life of living things in the world. With this very important role, the balance between consumption and production needs to be fulfilled [9].The increase in oxygen demand continues to increase while the availability of green open space is decreasing every year.

Previous research regarding the estimation of green open space needs has been carried out by D. Purba, S. Subiyanto, and Hani'ah, 2018 [3]. In this research, the calculation of the area of green open space was carried out with a visual interpretation of the 2015 Quickbird Image and updated using the Sentinel-2A image in 2018. Oxygen demand parameters based on this research used parameters of population, number of motorized vehicles, and number of livestock only. The research only focuses on estimating the need for green open space and its oxygen demand. While in this study the research included monitoring related to the availability of green open spaces and oxygen demand within 15 years. In addition, this study uses the parameters of the number of residents, the number of motorized vehicles, the number of livestock, and the number of industries for the calculation of oxygen demand.

Monitoring of green open space changes is carried out to determine the environmental degradation that has occurred. Remote sensing can be used as a tool in overcoming urban problems. This is because remote sensing has advantages in time, cost efficiency and is environmentally friendly. The availability of Sentinel-2A images and ASTER images which are obtained for free and have good temporal resolution can make changes and the extent of green open space availability can be monitored properly. The aim of this research is:

a. Calculating the accuracy of the mapping results of the public green open space fromthe ASTER image and the Sentinel-2A image
b. Knowing the changes in public green open space in Sukoharjo Regency from 2004-2019
c. Calculating the oxygen demand in Sukoharjo Regency in 2004 and 2019
d. Calculating the estimated green open space requirement based on oxygen demand in Sukoharjo Regency in 2004 and 2019.

2. Methos

2.1 Data source

Image processing was carried out on the ASTER image recorded on July 7, 2004 and the Sentinel-2A image recorded on September 26, 2019. Both images used were geometrically and radiometrically corrected. The area of green open space availability was obtained from on-screen digitization and visual interpretation of images in the two different years. The tools and materials used in this research are:

a. Handheld GPS; Camera; Laptop;
b. ArcGIS software; ENVI Software; RStudio software; Microsoft Office;
c. Stationary; Field checklist;
d. ASTER image recording year 2004 and Citra Sentinel-2A in 2019;
e. Graphic data (shp format) administrative boundaries of Sukoharjo Regency;
f. Map of the City Spatial Patterns of Sukoharjo Regency 2011-2031;
g. Secondary data on oxygen demand variables in 2004 and 2019 in Sukoharjo Regency:
   - Population Data
   - Data on the number of livestock
   - Data on the number of motorized vehicles
   - Data on the number of industries
h. Regulation of the Minister of Public Works Number 05 / PRT / M / 2008 concerning Guidelines for the Provision and Utilization of Green Open Space in Urban Areas.

2.2 Analysis

The two images have different resolutions. The ASTER image on the band appears to have a spatial resolution of 15 m while the Sentinel-2A image has a spatial resolution of 10 m. Resampling needs to be done on the image, to overcome the difference in spatial resolution. Image resampling was done using RStudio software. The resampling method used is bilinear interpolation. The advantage of this method is that the time required for processing tends to be fast with smooth image results [10]. The interpretation of Green Open Space is carried out by referring to the Regulation of the Minister of Public Works Number 05 / PRT / M / 2008. The classification of Green Open Space types can be seen in **Table 1**

Table 1. Type of Green Open Space

No	Type of Green Open Space
1.	Yard Green Open Space - Home yard - Office and shop yard - Rooftop garden
2.	Green Open Space Parks and Urban Forests - RT Park; RW - Taman kelurahan; sub-district - City Park - Urban Forest - Green Belt
3.	Green Line Green Open Space Road - Island roads and road medians - Pedestrian path - Space under the flyover
4.	Green Open Space Specific Functions - Railroad Green Open Space border - The green line of the power grid - Green Open Space for river borders - Green Open Space on the beach - Green Open Space safety source of air springs - funeral

Based on this classification, not all types of green open space can be interpreted. Interpretation is only carried out on the type of Public Green Open Space, while Private Green Open Space such as office pages is not interpreted. This is due to the spatial resolution factor of the two images used. The types of green open space that were interpreted were river borders, fields, city parks, funeral, and urban forests. Interpreted appearance is one with a size exceeding the Minimum Mapping Unit (MMU) used. The determination of the MMU size is adjusted from the appearance of the object, where the appearance of the expanding area uses MMU 3mm x 3mm, while in areas that tend to lengthen using a minimum width of 1 mm.

The two interpreted images have different spectral characteristics. The VNIR band on the ASTER image starts with the green band (0.556 μm), red band (0.661 μm), and near NIR / Infrared band (0.807 μm). The difference with the Sentinel-2A image is that the blue band has a wavelength of 0.49 μm, the green band is 0.56 μm, the red band is 0.665 μm, and the NIR band is 0.842 μm. The interpretation was carried out by utilizing a standard false color composite (NIR, red, and green bands), namely band 321 in the 2004 ASTER image and band 843 in the 2019 Sentinel-2A image.

Monitoring of changes in public green open space is carried out by comparing the results of reinterpretation of the ASTER image and the Sentinel-2A image. The area of change in public green open space in 2004 and 2019 was obtained from processing using a pivot table. Accuracy test is done by using confusion matrix to determine the ability of ASTER image and Sentinel-2A image in mapping public Green Open Space. Determination of field samples is to use the simple random sampling method with a total sample of 50 samples. Field checks are carried out in two different ways. Field checks in 2004 made use of Google Earth imagery, while field checks in 2019 were conducted live.

Oxygen demand is calculated using the gerrarkis method. The calculation of oxygen demand uses variables in the form of oxygen demand from the population, the number of livestock, the number of industries, and the number of motorized vehicles. The formula for calculating the need for green open space based on oxygen demand is:

$$Lt = \frac{Pt + Kt + Tt + It}{(54)\,(0{,}9375)\,(2)}$$

Lt	: City green open space in year t (m^2)
Pt	: The amount of oxygen needed by the population
Kt	: The amount of oxygen needed for motor vehicles
Tt	: The amount of oxygen needed for livestock
It	: Amount of industrial oxygen demand
54	: A determination which shows that 1 m^2 of land area produces 54 grams of plant dry weight per day.
0,9375	: The rule indicating that 1 gram of plant dry weight is equivalent to the production of oxygen 0.9375 grams
2	: Number of seasons in Indonesia

3. Result and Discussion

3.1 Preprocessing

In the resampling process, each band will experience a change in pixel size. The result of the resampling process is changing the Sentinel-2A image resolution from 10 m to 15 m. Image cutting is done on ENVI software using ROItool (Region of Interest). The study area includes 12 sub-districts in Sukoharjo Regency.

3.2 Interpretation

Types of public green open space that can be interpreted are the field, river border, and urban forest, while the funeral and city park only appear to have more than MMU size which are interpreted. The dominant elements of interpretation are form, color, pattern and site.

The type of public green open space in the form of a funeral can be identified by looking at the visible shapes and patterns. The pattern is grouped with features that tend to be unique and distinctive from a collection of tombstones. Cemeteries are social facilities where this object is able to provide green open space reserves, as a water catchment area, as well as a microclimate creator. There are two burial objects that have been successfully interpreted, namely the funeral in Kartasura District, where this funeral has a very large size of 0.13 km^2.

The field is built with its use that is open, public, and capable of being a water catchment area. The key interpretations of this type of green space are shape, color, and site. Most of the forms of Green Open Space are square or rectangular. The color that appears in both images is clearly visible from the reflection of the soil and the surrounding vegetation. In addition, field interpretation can be related to its location in the middle or near the settlement. This is because the social function of the field is to accommodate community activities.

City parks are in the form of parks open to the public that function as micro-climate creators. This type of green space is difficult to interpret without the help of local knowledge from the manufacturer. This is because most of the city parks are located in the middle of dense settlements with a park area that is not too large. City parks that can be interpreted are Satya Negara Square Sukoharjo Regency and Ananda City Park. The two city parks can be interpreted due to their size which is more than the specified MMU.

River border in the form of green lines on either side of the river to protect the river and maintain its sustainability. This green space interpretation uses key interpretations in the form of patterns, colors, and sites. This green space has an elongated, linear pattern along the river. In addition, their appearance is easily recognizable by the red, clustered reflection of the vegetation. This appearance can also be related to its location on the right and left of the river.

Based on the classification of urban forests listed in the Minister of Public Works Regulation Number: 05 / PRT / M / 2008, green open space in the form of urban forest can be in the form of clustered urban forest, urban forest spread, the area planted with an area of 90% - 100%, and urban forest in the form of a path. Urban forest is clustered in the form of clustered vegetation in one area, with a minimum of 100 trees and irregularly dense spacing. The urban forest spreads out without a pattern, and the growing vegetation spreads out in clumps with minimum area of 2500 meters. The urban forest in the form of a lane has a minimum width of 30 meters with a pattern following the formation of rivers, roads, channels and others.

IOP Conf. Series: Earth and Environmental Science **884** (2021) 012035 doi:10.1088/1755-1315/884/1/012035

The interpretation carried out in Sukoharjo Regency shows that the form of urban forest in this area tends to be clustered and spread. The clustered form of urban forest can be found in production forest in Polokarto District, while urban forest that spreads out is found in protected forest in Bulu District. The process of interpreting urban forest is a little difficult to interpret, so assistance is needed in the form of a map of the spatial pattern of Sukoharjo Regency 2011-2031 [7]. The map of the spatial pattern of Sukoharjo Regency 2011-2031 is shown in **Figure 1**

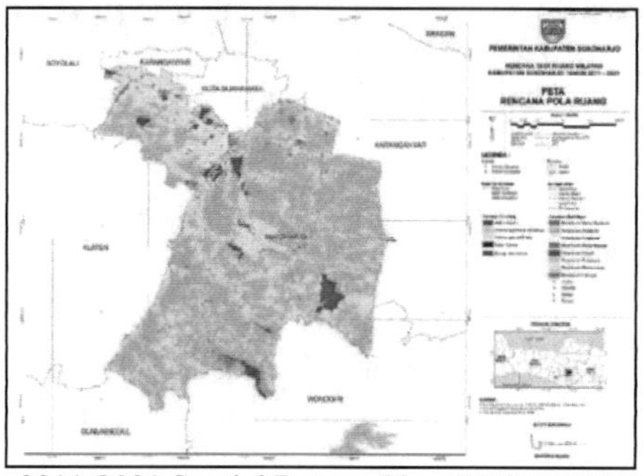

Figure 1. Map of the 2011-2031 Spatial Pattern Plan (PUSDATARU, 2017)

The urban forest spreads out in a protected area. The morphology tends to be unique and has certain ecological functions with an area of up to 3 km^2. In addition to being seen from the spatial pattern map of Sukoharjo Regency in 2011-2031, urban forest is clustered in shape, it can also be identified from the appearance of vegetation with clustering patterns and tree shapes that tend to be uniform. This type of urban forest that can be interpreted is rubber and teak production forest. The results of the map of public green open space in Sukoharjo Regency in 2004 and 2019 are shown in **Figure 2**

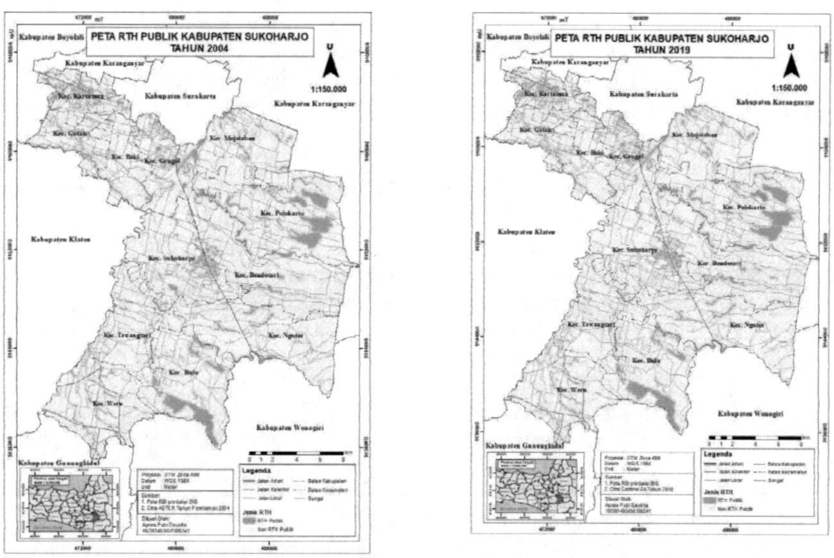

Figure 2. (a) Map of Public Green Open Space 2004 (b) Map of Public Green Open Space 2019

3.3. Determination of samples and field checks

The total sample used was 50 samples that were evenly distributed in each study area. The results of the field check in 2004 showed two wrong interpretation results, namely the type of green open space for city parks and the field, while in the field survey in 2019 it was found that there were five wrong interpretation results, namely on the river and field boundaries. Field green open space has been mistaken because this object is similar to unplanted agricultural land or dry agricultural land. In addition, this green open space is also similar to a garden owned by residents. Borders of rivers and city parks experience errors in interpretation because of their proximity to each other so that they are difficult to distinguish from one another. Urban forest samples located in mountainous areas are difficult to reach. This results in a shift in the sample point to a place that is more easily accessible, without changing the type of green space to be checked for conditions in the field.

3.4. Test accuracy

The accuracy test is carried out by comparing the interpretation of ASTER images with google earth. The overall accuracy results in 2004 can be seen in Table 2. Mapping public green open space using the ASTER image, the accuracy is 96% with a kappa coefficient value of 0.94. Based on the accuracy of the manufacturer and the accuracy of the user, all types of public green open space can be identified properly.

The accuracy test of public green space mapping in 2019 was carried out by comparing the interpretation of Sentinel-2A images with field surveys. The results of the creator's accuracy and user accuracy are more varied, as shown in Table 3. The overall mapping using the Sentinel-2A image produces an overall accuracy of 90% with a kappa coefficient value of 0.86. Based on the accuracy of the maker and the user, there are four types of public green open space that can be identified properly.

The accuracy produced by the two images exceeds 85%, which according to Sutanto (1986) states that if the accuracy rate exceeds 85% then it is feasible for further analysis [8]. The kappa value produced by the two images ranges from 0.8 to 1, where according to Altman (1991) this value is in the very strong or very good category [5]. Based on this, the mapping of public green open space with both images can be further analyzed because the higher the overall accuracy value and the kappa coefficient value, the better the mapping results can be.

Table 2 Confussion Matrix 2004

		Field							
		Urban Forest	Field	Non Public Green Open	Funeral	River Border	City Park	Total	User's Accuration
Interpretasion	Urban Forest	7	0	0	0	0	0	7	100%
	Field	0	18	1	0	0	0	19	95%
	Non Public Green Open Space	0	0	0	0	0	0	0	0%
	Funeral	0	0	0	2	0	0	2	100%
	River Border	0	0	0	0	20	0	20	100%
	City Park	0	0	1	0	0	1	2	50%
	Total	7	18	2	2	20	1	50	
	Producer's Accuration	100%	100%	0%	100%	100%	100%		96%
	Indeks Kappa				0,941961695				

Table 3 Confussion Matrix 2019

		Field							
		Urban Forest	Field	Non Public Green Open Space	Funeral	River Border	City Park	Total	User's Accuration
Interpretasi	Urban Forest	7	0	0	0	0	0	7	100%
	Field	0	17	2	0	0	0	19	89%
	Non Public Green (	0	0	0	0	0	0	0	0%
	Funeral	0	0	0	2	0	0	2	100%
	River Border	0	0	0	1	17	2	20	85%
	City Park	0	0	0	0	0	2	2	100%
	Total	7	17	2	3	17	4	50	
	Producer's Accuration	100%	100%	0%	67%	100%	50%		90%
	Kappa Indeks				0,8606466				

3.5. Monitoring of Public Green Open Space

a. Public Green Open Space in 2004

The results of the mapping of green open space in 2004 that have been reinterpreted indicate that the identified public open green space is a funeral, city park, field, river border, and urban forest. The area of public green open space in 2004 reached 20.22 km^2 where this area is 0.043 of the total area. If viewed from the Regulation of the Minister of Public Works Number: 05 / PRT / M / 2008, which states that the percentage of public Green Open Space is 20% of the area, then the area of public Green Open Space available is still inadequate.

Urban forest has the largest area, reaching 10.06 km^2. The river border has the second largest area, namely 9.43 km^2. These two types of public green open space are able to have the largest area because this type of green open space is the easiest to recognize and most of them have a sufficient size for interpretation. The lowest area of all types of green open space is a city park, which is only 0.02 km^2. This is because it is very rare for city parks to have an area that exceeds the smallest mapping unit used. The area of public green open space for each type of green open space in 2004 is shown in **Figure 3.**

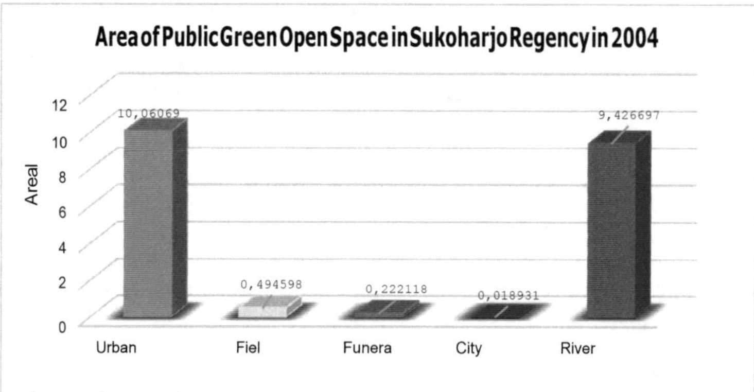

Figure 3. Area of Public Green Open Space in Sukoharjo Regency in 2019

Based on the distribution of public green open space in 2004, it was found that the sub-districts with the widest public green open space were Polokarto District with an area of 7.25 km^2 and Bulu District 4.82 km^2, while the sub-district with the lowest area of green open space was Tawangsari District with an area of 0.42 km^2. Polokarto and Bulu Subdistricts have a high area of public green open space because in these two sub-districts there are quite la rge urban forests in the form of protected forests and production forests. In addition, Polokarto District has the largest area among the 11 other sub-districts. Tawangsari sub-district has very few public green open spaces because this sub-district is dominated by agricultural land in the form of large irrigated rice fields.

b. Public Green Open Space in 2019

The public green space that has been identified in 2019 are cemeteries, city parks, fields, river borders, and urban forests. The area of this public green open space reaches 18.77 km^2. If this area is calculated based on the need for public green open space of 20% of the area, namely 93.33 km^2, then the area of public green open space available in 2019 is still inadequate. The area of public green open space for each type of green open space in 2019 can be seen at **Figure 4.**

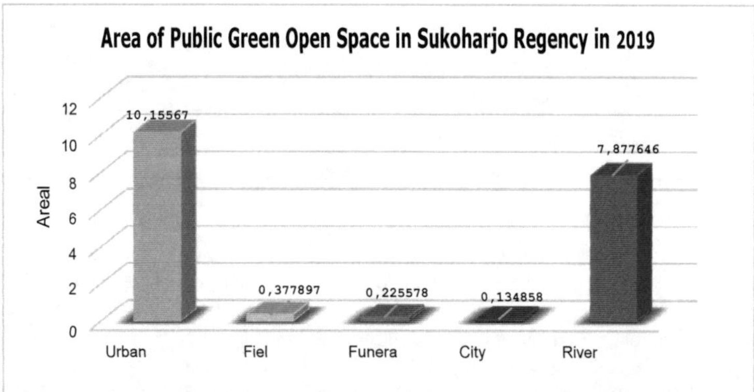

Figure 4 Area of Public Green Open Space in Sukoharjo Regency in 2019

Based on Figure 4, it is known that there is a change in the area of the river border and there is an additional area in the types of green open space for city parks and urban forests. This is because from 2004-2019 the construction of several city parks such as Ananda City Park and Grogol Indah City Park were carried out in order to meet the needs of the community's green open space. The sub-district with the largest public green open space is in Polokarto District, covering an area of 7.14 km^2, while the sub- district with the lowest public green open space is in Sukoharjo District with an area of 0.34 km^2. The two sub-districts both experienced a decrease in the amount of green open space when compared to 2004.

c. Monitoring Change in Public Green Open Space

Monitoring of changes in green open space that occurs is carried out by looking at the extent of changes using a pivot table. In this process, a comparison of the green open space map in 2004 and 2019 will be carried out where the results will be obtained from a matrix of changes in the area of green open space for each type of green open space. The results of the green open space change matrix show that the highest change in the area of public green open space occurred on river boundaries which turned into non-public green open space, namely an area of 2.48 km^2. The extent of changes in each type of public green open space can be seen in **Table 4**.

Table 4 Matrix Changes in Public Green Open Space

Green Open Space 2004 (km2)	Green Open Space 2019 (km2)						
	Urban Forest	Field	Non Public Green Open Space	Funeral	River Border	City Park	Total
Urban Forest	10,075158						10,075158
Field		0,348521	0,131963811		0,014112939		0,4945975
Non Public Green Open Space	0,0818241	0,029377	470,3443487		0,921728014	0,1158076	471,49309
Funeral				0,22211852			0,2221185
River Border			2,48176461	0,0034595	6,940471103	0,0001197	9,4258149
City Park						0,0189315	0,0189315
Total	10,156982	0,377898	472,9580771	0,22557802	7,876312056	0,1348588	491,72971

Changes in public green open space as a whole during 2004-2019, namely public green open space covering an area of 17.62 km^2 did not change or remain permanent. Over the past 15 years, there has been an increase in the area of public green open space to an area of 1.15 km^2, but during that period there was also a reduction in the area of public green open space by 2.61 km^2. The increase in the area of public green open space has occurred due to the construction of several city parks in order to meet the needs of public green open space, including the Ananda City Park, Masdulkabi City Park, and Grogol Indah City Park. The highest reduction in public green open space occurred at river boundaries. The reduction in the area of river borders has occurred in many river border areas adjacent to residential areas. This is due to the increasing population and increasingly varied community activities, so that more space is needed to live or do activities. Vegetation on the border of the river also turns into a dwelling for the community.

Monitoring can also be done by overlaying the map of green open space in 2019 and 2004. The result can be seen the addition or reduction of public green open space in Sukoharjo Regency, as shown in **Figure 5**. Most of the Sukoharjo area is a flat area consisting of Kartasura, Sukoharjo, Baki, Mojolaban, Tawangsari, Gatak, and Grogol, while Nguter, Polokarto, Bulu, Bendosari and Weru sub-districts tend to have a bumpy to mountainous topography. Flat areas are dominated by agricultural land in the form of irrigated rice fields and people's fields so that there are still few public green open spaces. In addition, kecamatan with flat relief tend to be used for industrial estates.

The western part of Sukoharjo Regency which includes Polokarto, Bendosari and Nguter Districts is bordered by Karanganyar Regency, so the relief tends to be bumpy to hilly. These three districts are still dominated by the use of agricultural land in the form of terraced rice fields and gardens owned by residents, and in some areas used for rubber plantations. This is different from the situation in Karanganyar Regency, which is dominated by standing trees.

Based on the map of changes in public green open space in Sukoharjo Regency, it can be seen that the area of change in public green open space per district is shown in **Table 5**. The most additional public green open spaces were added in Grogol District, but the highest reduction in public green open space was also found in Grogol District. This addition occurs because there is a development of urbanparks, while the reduction in the area of public green space occurs due to residential areas that are increasingly densely populated and also the development of industrial areas.

Table 5 Area of Change in Public Green Open Space

No	District	Area of Change in Public Green Open Space (km2)			
		Not change	Increase	Decrease	Non Public Green Open Space
1.	Weru	0,88	0,10	0,11	43,57
2.	Bulu	4,71	0,02	0,11	41,42
3.	Tawangsari	0,33	0,03	0,09	39,19
4.	Sukoharjo	0,28	0,06	0,26	46,05
5.	Nguter	0,72	0,14	0,10	54,64
6.	Bendosari	0,88	0,13	0,31	56,56
7.	Polokarto	7,00	0,14	0,26	59,51
8.	Mojolaban	0,70	0,15	0,27	37,10
9.	Grogol	0,96	0,23	0,47	29,93
10.	Baki	0,38	0,08	0,22	22,50
11.	Gatak	0,40	0,03	0,26	19,26
12.	Kartasura	0,37	0,04	0,15	20,50

The addition of public green open space indicates that at these locations there was an increase in the area of public green open space in 2019 compared to the area available in 2004. This could be in the form of the construction of new types of public green open space or an increase in the area of the existing public green open space. The reduction in public green open space illustrates that at these locations there has been a reduction in the area of public green open space compared to the area in 2004. The reduction in public green open space that occurs is only in the form of a reduction in the area of public green open space that has been there before.

Public green open space in the form of urban forest has experienced the least change, its area has even increased by 0.08 km^2 in Polokarto District. This is different from other types of public Green Open Space, such as river borders and fields. The two types of green open space had the most changes in area, either praying or not. In this type of public green open space, all sub-districts in Sukoharjo district experienced a reduction in their area compared to their area in 2004. However, at several points in all sub-districts there was also an increase in the area of this public green open space, although the additional area was not too large. This is due to land conversion due to human settlements. An additional area of green open space in the form of construction of new types of public green open space is available in Grogol and Baki Districts. The development of city parks is aimed at meeting the need for green open space, this is because this sub-district has a large industrial area.

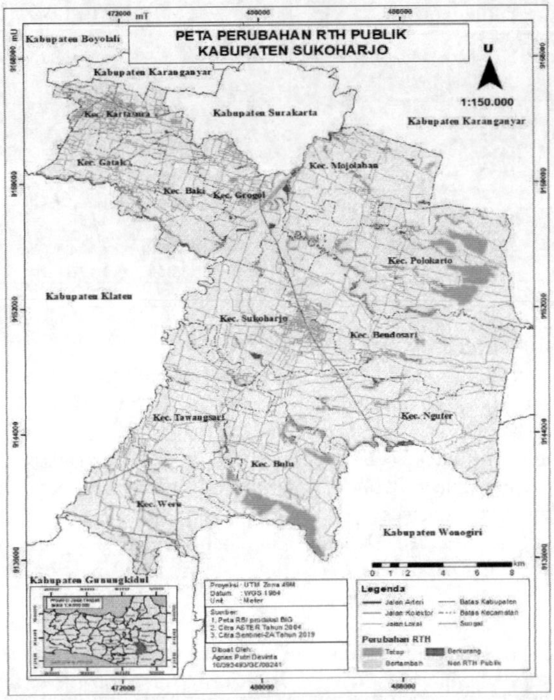

Figure 5 Map of Changes in Public Green Open Space for Sukoharjo Regency 2004-2019

3.6. Oxygen Needs

a. Oxygen Needs for Population

In 2004 the population in Sukoharjo Regency was 815,089 people with the largest population in Grogol District (97,273 people) and the lowest was in Gatak District (46,581 people). The population in 2019 has increased to reach 891,912 people with the highest population in Grogol District (141,407 people), while the lowest number is located in Bulu District (27,575 people). According to White et al. (1959) the need for oxygen per person in one day can reach 0.864 kg of oxygen / day [6]. Based on the calculation of oxygen demand, it is known that the total oxygen demand based on the population in 2004 was 704,236.90 kg / day, while in 2019 it reached 770,611.97 kg / day.

b. Oxygen Needs for Livestock

The difference in the number of livestock was clearly visible between 2004 and 2019, this was due to an increase in the livestock sector that was developing in Sukoharjo. In 2004, the number of livestock was 1,307,279, while in 2019 it was 5,604,339. The livestock sector is dominated by poultry, both in 2004 and 2019. During the past 15 years, there has been an increase in the number of livestock by 4,297,060 heads. In 2004, there were gaps in data on native chicken types, while in 2019 the number of dairy cows was unknown.

The total oxygen demand for livestock is not limited to the boundaries of each district, but is analyzed at the district level. This is because there is no specific data in each district. Oxygen requirements for large livestock have greater oxygen requirements than small livestock and poultry. According to Muis (2005) the oxygen demand for cows and buffaloes reaches 1.70 kg / day while 2.86 kg / day for horses. Small livestock types such as goats and sheep require oxygen of 0.314 kg / day while poultry is 0.17 kg / day [3]. The oxygen demand for pigs is based on the research of Hannon et al. (1989) which is equal to 1.24 kg / day [4]. The oxygen needs of each type of livestock and their numbers in 2004 and 2019 are detailed in **Table 6**. The increase in livestock numbers also has an impact on increasing the amount of oxygen demanded. This can be seen in the oxygen demand in 2019 which reached 1,014,770.48 kg / daywhile in 2004 it was only 279,325.70 kg / day.

Table 6 Oxygen requirements for each type of livestock

No	Types of Livestock	Need for Oxygen (kg/day)	Total		Total Need for Oxygen (kg/day)	
			2004	2019	2004	2019
1	Cow	1,7	25582	29745	43540,56	50625,99
2	Buffalo	1,7	2398	597	4081,4	1016,09
3	Horse	2,86	209	243	597,74	694,98
4	Goat	0,314	36076	49626	11327,86	15582,56
5	Sheep	0,314	33337	46247	10467,82	14521,56
6	Pig	1,24	6798	16331	8429,52	20250,44
7	Chicken	0,167	1117447	5224425	186613,65	872478,98
8	Duck	0,167	85432	237125	14267,14	39599,88
	TOTAL				279325,7	1014770,48

c. *Oxygen Needs for Industry*

The calculation of oxygen demand based on the number of industries is not limited to the boundaries of each district, but is analyzed based on district administrative boundaries. This is because there is no specific data in each district. The division of industrial types is classified into the agro and forest products industry, the textile and various industries, and the electro-machine metal chemical industry. The calculation of oxygen demand based on industry uses a large number of industries. This is based on the assumption that large industries use a large number of machines and for a longer period of time than small and medium industries. In 2004, the total number of large industries was 25 industries, while in 2019 it grew to 129 industries. This development mostly occurred in the electro-machine metal chemical industry.

The use of large industries for the calculation of oxygen demand is based on research by Ryadi (1984). The research states that the average fuel requirement for the engine is 185.76 kg / day, while each kg of diesel motor fuel requires 2.86 kg of oxygen [6]. Based on this formula, the oxygen demand in 2004 was 13,281.77 kg / day and in 2019 was 68,533.92 kg / day.

d. *Oxygen Needs for Motor Vehicle*

The number of motorized vehicles in 2004 was 169,968 units and in 2019 801,581 units. Over the past 15 years, there has been an increase in motor vehicle consumption by 631,613 units of vehicles. The number of motorized vehicles is obtained from BPS data from Sukoharjo Regency which is derived from vehicle tax revenue data collection in Sukoharjo Regency. The assumption used in calculating oxygen demand is the number of vehicles originating from Sukoharjo Regency, while vehicles originating from outside the area are ignored. The total oxygen demand is not limited to the boundaries of each sub-district, but is calculated at the district level.

The largest number of motorized vehicles is dominated by motorbikes. Bus cars have the smallest number and the number continues to decrease every year. This is followed by an increase in the number of passenger cars and motorbikes, indicating that the use of private vehicles is increasing every year. The fuel and oxygen requirements for each type of vehicle are detailed in **Table 7**.

Table 7. Oxygen Needs of Each Type of Vehicle

No	Motor vehicle	Fuel	Fuel Requirements (kg/PS)	Oxygen Needs Per Liter of Fuel (kg)	Oxygen Needs (kg/hours)
1	Motorcycle	gasoline	0,21	2,77	0,58
2	Passenger car	gasoline	0,21	2,77	11,63
3	Bus Car	diesel fuel	0,16	2,86	44,32
4	Load Car	diesel fuel	0,16	2,86	22,88

Source: (Wisesa, 1988 in Putrajaya, 2017)

The total oxygen demand of each type of motorized vehicle in each year can be seen in **Table 8**. The calculation results show that there is an increase in oxygen demand from motorized vehicles, namely in 2004 of 56,687.56 kg / day while in 2019 it was 70,042.94 kg / day.

Table 8 Total Oxygen Needs

No	Type of Vehicles	Oxygen Needs (kg/day)	Total of vehicles (unit)		Total Oxygen Needs (kg/day)	
			2004	2019	2004	2019
1	Motorcycle	13,92	152811	586447	1340,09	5142,91
2	Passenger car	279,12	10059	57557	1768,83	10121,14
3	Bus Car	1063,68	1066	820	714,35	549,5
4	Load Car	549,12	6032	156757	52864,29	54229,39
	TOTAL				56687,56	70042,94

3.7. Need for green open space

The need for green open space is calculated using the Gerrarchic approach. This method shows that green open space in every m^2 will produce 54 grams of dry weight, wherein one gram of dry weight is equivalent to oxygen production of 0.9375 grams / day [6]. The total oxygen demand of the population, livestock, industry and motor vehicles in 2004 was 1,053,531.92 kg / day, while in 2019 it was 1,923,959.31 kg / day.

This can be seen from the available open green space area of 20.22 km^2 from the need for green space that is only 10.41 km^2. The increase in population, industry, motorized vehicles, and livestock has an impact on increasing the need for green open space. This can be seen in the oxygen demand in 2019 which increased by 870427.39 kg / day from 2004. The green space available in 2019 was 18.77 km^2, where this area experienced a reduction of 1.45 km^2 from the previous 15 years. The reduction in the area of green open space is inversely proportional to the need for green open space in 2019 which has increased to 19 km^2. Based on the number of requirements, it can be said that the availability of green open space in 2019 is still not fulfilled by the available public open green space.

4. Conclusion

The conclusions obtained from this study are ASTER imagery and Sentinel-2A imagery can be used for visual interpretation in mapping public open space in the form of fields, river borders, and urban forests while for public open green space in the form of cemeteries and city parks can only be interpreted if the size is more than MMU. The overall accuracy value produced by the ASTER image reaches 96%, while the accuracy of the Sentinel-2A image is 90%. Mapping of changes in public green open space in 2004-2019 shows that 17.62 km^2 of public green open space has not changed. Changes in the area of public green open space, namely an addition of 1.15 km^2 of green open space and a reduction of 2.61 km^2 of green open space, which was dominated by changes in the area of the riverbank to non-public green open space of 2.48 km^2.

Based on the population, industry, livestock, and motorized vehicles, it was found that the estimated total oxygen demand in Sukoharjo Regency in 2004 was 105353192 kg / day while oxygen demand in 2019 reached 1923959.31 kg / day. Estimated need for green open space based on oxygen demand in Sukoharjo Regency in 2004 was 10.41 km^2 and in 2019 was 19 km^2. The availability of green open space in 2004 has been fulfilled from the available public green open space, which is 20.22 km^2, while in 2019 public green open space has not been able to meet the needs of green open space in 2019, which is only able to meet 18.77 km^2.

In this research, the interpretable type of green open space can only include public green open space. This is adjusted to the type of image used. The suggestion for further research is the use of remote sensing images with better spatial resolution and the same characteristics, so that it can simplify the process of interpreting types of public green open space in more detail and with more accurate results. In addition, data on the number of motorized vehicles in and out of Sukoharjo Regency can also be added so that the calculation of oxygen demand can be more accurate. Monitoring changes in green open space is very important to do. Future research is not only limited to monitoring changes that have occurred, but also estimating the need for green open space in relation to oxygen demand and the city's rapid development.

Acknowledgement

Thank you to the Department of Geographical Information, Gadjah Mada University for supporting this research. The author also thanks the related agencies in Sukoharjo Regency for the assistance provided with secondary data.

References

[1] BPS (Badan Pusat Statistik) Kabupaten Sukoharjo 2019, '*Kabupaten Sukoharjo Dalam Angka 2019*', accessed date 1 Oktober 2019, https://sukoharjokab.bps.go.id/publication/2019/08/16/45a43f981083d82ef531ecef/kabupaten-sukoharjo-dalam-angka-2019.html.

[2] Kementerian Pekerjaan Umum dan Perumahan Rakyat Republik Indonesia 2017, '*Peraturan Menteri Nomor 05/PRT/M/2008*' accessed date 5 January 2020, http://jdih.pu.go.id/produk-hukum-detail.html?id=1236.

[3] Purba, D, Subiyanto, S, & Hani'ah 2018, 'Analisis Kebutuhan Ruang Terbuka Hijau Berdasarkan Kebutuhan Oksigen di Kota Pekalongan dengan Menggunakan Penginderaan Jauh dan Sistem Informasi Geografis', *Jurnal Geodesi Undip*, Vol. 7, No. 4, pp. 264 - 273.

[4] Putra, EH 2012, 'Analisis Kebutuhan Ruang Terbuka Hijau Berdasarkan Kebutuhan Oksigen Menggunakan Citra Satelit EO-1 ALI (Earth Observer-1 Advanced Land Imager) di Kota Manado', *Jurnal Info BPK Manado*, Vol. 2, No. 1, pp. 41-54.

[5] Putra, IWGAE, Sutarga, IM, Kardiwinata, MP, Suariyani , NLP, Septarini, NW & Subrata, IM 2016, Modul Penelitian Uji Diagnostik Dan Skrining, Fakultas Kedokteran Universitas Udayana, accessed date 17 June 2020, https://simdos.unud.ac.id/uploads/file_pendidikan_1_dir/d204d4a5ad0870a0965416e671a38791.pdf.

[6] Putrajaya, IK 2017, 'Analisis Indeks Vegetasi Menggunakan Citra ALOS AVNIR-2 Untuk Estimasi Kebutuhan Ruang Terbuka Hijau Berdasarkan Kebutuhan Oksigen di Kota Denpasar, Provinsi Bali', *Jurnal Pendidikan Geografi*, Vol. 22, No. 1, pp. 49-59.

[7] PUSDATARU Jawa Tengah (Pekerjaan Umum Sumber Daya Air dan Penataan Ruang Provinsi Jawa Tengah) 2017, Dokumen RTRW Kabupaten Sukoharjo, accessed date 20 May 2020, http://pusdataru.jatengprov.go.id/dokumen/RTRW-Prov/14-Kab-Sukoharjo/Lampiran-Peta-Pola-Ruang-Wilayah.jpg.

[8] Ramadhani, F & Susilo, B 2016, 'Integrasi Penginderaan Jauh Dan Sistem Informasi Geografis Untuk Penentuan Prioritas Perbaikan Jalur Evakuasi Di Daerah Rawan Erupsi Merapi (Kecamatan Pakem Dan Cangkringan)', *Jurnal Bumi Indonesia,* vol. 5, no. 4, accessed date 16 June 2020, http://lib.geo.ugm.ac.id/ojs/index.php/jbi/article/download/860/833.

[9] Sahid, Rohmat ,D, Malik, Y & Wahyuni, L 2018, 'An Investigation of Oxygen Need Based on Green Open Space in Indonesia University of Education (UPI) Campus Bandung' Proceedings of 3rd Annual Applied Science and Engineering Conference Series: Materials Science and Engineering 434 (2018) 012234, pp 1-12, accessed date 13 January 2021, https://iopscience.iop.org/article/10.1088/1757-899X/434/1/012234/pdf.

[10] Yanuar, RC, Hanintya, R & Muzaki, AA 2018, 'Penentuan Jenis Citra Satelit Dalam Interpretasi Luasan Ekosistem Lamun Menggunakan Pengolahan Algoritma Cahaya Tampak Studi Kasus: Wilayah Pesisir Sanur'. *Geomatika*, vol. 23, no. 2, pp. 75-86, accessed date 14 May 2020, http://jurnal.big.go.id/index.php/GM/article/download/704/500.

[11] Yashinta, MR, & Utomo, IH 2017, 'Efektivitas Pengendalian Alih Fungsi Lahan Pertanian oleh Badan Koordinasi Penataan Ruang Daerah Kabupaten Sukoharjo', *Jurnal Wacana Publik,* Vol. 1, No. 2, pp. 1-10.

Physical vulnerability assessment of settlements in lahar hazard prone areas along the river originated from Merapi Volcano

A Z Syarafina[1], and D R S Sumunar[2]

[1]Departement of Geography Education, Universitas Negeri Yogyakarta, Indonesia.
[2]Faculty of Social Science, Yogyakarta State University, Indonesia.

Email: akalili0466pasca.2019@student.uny.ac.id, dyah_respati@uny.ac.id

Abstract: Merapi Vulcano is one of the most active volcanoes in the world. This volcano has a secondary hazard that is lahar flow. This research refers to the 2010 eruption of Merapi Vulcano in Indonesia. Eventhough there was a phreatic eruption in 2018, it did not increase the deposition of lahar material on its slopes. The communities living along the lahar prone rivers have various vulnerabilities including physical, social, economic and environmental vulnerability. This research aims to determine the physical vulnerability distribution of settlement in lahar hazard prone areas alongside Pabelan River. This study is a descriptive quantitative research using secondary data and spatial analysis. The results of this research show that the high vulnerability distribution covers the area of Mungkid Sub-district and Sawangan Sub-district as many as 5 clusters or equal to 7.49% of the research area. The medium level of physical vulnerability is Dukun Sub-district, Mungkid Sub-district and Muntilan Sub-district as many as 26 clusters or equal to 41.27% of the research area. The low physical vulnerability of settlements is scattered in Dukun Sub-district, Mungkid Sub-district, Muntilan Sub-district and Sawangan Sub-district as much as 32 clusters or equal to 50.79% of the research area. It can be concluded that the physical vulnerability of settlements in lahar hazar prone along the Pabelan river after the 2010 eruption of Merapi Vulcano is dominated by low vulnerability.
Keywords: physical vulnerability, lahar, Merapi Vulcano

1. Introduction

The events of post-volcano eruption in the form of lahar hazards caused various damages including properties, buildings and even fatalities. Therefore, lahar flow can be categorized as a disaster according to Law of the Republic of Indonesia Number 24 of 2007. This research refers to the 2010 eruption of Merapi Vulcano in Indonesia. Eventhough there was a phreatic eruption in 2018, it did not increase the deposition of lahar material on its slopes. Lahar flow once damaged 19 Sabo Dam which 3 of those were completely destroyed, more than 307 houses damaged and even the flow caused the need to closed down a national road access for 18 times in the span of 25 days [1]. Lahar flows consist mixture of water (can be rainwater or melted snow), sand, gravel and even massive rocks. They move in a rapid motion down the slopes while sweeping away all the things they pass through [2].

History once recorded that lahar coming from the eruption of Merapi Vulcano destroyed the hydrography system near Borobudur basin and created at least six paleolake in the last 119.000 year

Content from this work may be used under the terms of the Creative Commons Attribution 3.0 licence. Any further distribution of this work must maintain attribution to the author(s) and the title of the work, journal citation and DOI.
Published under licence by IOP Publishing Ltd

[3]. There are 12 rivers originated from Merapi Vulcano which becomes streams vulnerable to lahar [4]. One of those rivers is Pabelan River in which this river cut across highly populated settlements in Magelang District. Pabelan River crosses 4 Sub-districts in Magelang District including Mungkid Sub-district, Muntilan Sub-district, Dukun Sub-district and Sawangan Sub-district. The level of population density in the areas along the lahar prone river is quite high. The data can be seen in Table 1.

Table 1. Population density in each Sub-district directly bordering with Pabelan River in Magelang District

Sub-district Name	Total of Population (resident)	Sub-district Area (Km^2)	Population Density (resident/ Km^2)
Dukun	46.900	53,41	878
Mungkid	74.295	37,95	1.958
Muntilan	79.136	28,61	2.766
Sawangan	57.287	72,37	791

Source: [19], [20], [21], [22]

Based on Table 1, it can be concluded that the densest population is Muntilan Sub-district with the level of density 2.766 residents per Km^2, and the lowest population density is Sawangan Sub-district. This numbers, if based on the classification of SNI 03-1733-2004 about area classification, is still categorized as low density (less than 15.000 residents per km^2) but the varied characteristic of settlement distribution can increase the level of vulnerability especially physical vulnerability of buildings. According to Ikhsan [16], in the period between 2012 and 2015, in the headwater of Pabelan River there is an increased erosion of the riverbed which reached 30 – 45 meter in depth and lessen in the downstream to a depth of around 5 meter. This means that Pabelan River experiences normalization post mining of material deposit from lahar flow in 2010 – 2011.

The aim of this research is to determine the distribution of physical vulnerability of settlements in the lahar prone area alongside the Pabelan River. Many researches study the analysis of disaster risk of volcano eruption but there are not many research specifically discuss the vulnerability from disasters caused by lahar flow. The level and distribution of physical vulnerability of settlements along the Pabelan river is not yet known, so this research can increase the level of capacity of both the local community and also for the local government. It is expected that through a map of settlement vulnerability in this research there will be an awareness enhancement of the people who live alongside the rivers vulnerable to lahar. It can also be used by the local government regarding the policy of relocating settlements with high risk in the lahar prone areas alongside the Pabelan River. Taking into consideration that Merapi Vulcano is highly active, the results of this study can be used as one of considerations for the government to plan disaster management in increasing resilience and as a strategy in lowering disaster risk in order to achieve the goal of sustainable development.

2. Literatur Review

The hazard from post-eruption activities of Merapi Vulcano is the lahar flow. One of the rivers that is most at risk to lahar flow are Pabelan River, Gendol River, Krasak River, Boyong River and Kali Putih River [4]. According to UN-ISDR (www.unisdr.org), hazard is a dangerous phenomenon, substance, human activity or condition that may cause loss of life, injury or other health impacts, property damage, loss of livelihoods and services, social and economic disruption, or environmental damage. Rivers in the slope of Merapi Vulcano which become the streams vulnerable to lahar cut across farm fields, settlements and even schools which make those areas to be highly vulnerable to lahar flow. The enhancement of settlement vulnerability in the area near Merapi slope is caused by the shift of demography and land use in Merapi Vulcano especially in the mount ring plain [5]. The changes in terms of social-economy, lahar frequency, and the size of lahar-affected area are the impact caused by lahar flow; furthermore, the spatial expansion from natural hazards has a direct influence to vulnerability pattern [6].

There is a need to shift the paradigm from quantification and hazard analysis to identification, assessment and level of vulnerability in disaster risk management where the challenge of the complexity in measuring vulnerability quantification will always exist [7]. Physical vulnerability against a certain hazard can be induced by other hazard and in this case a lahar hazard triggered by volcano eruption and the lack of available data can increase the number of uncertainty involved in vulnerability assessment [6]. Observations on rainfall are needed in order to create an early warning for lahar flow from Merapi Vulcano [8], [9]. Tiltmeter and extensometer can also be used to predict the weight of the stored debris alongside the river segments by real time and the concept of debris flow potential can help someone to evaluate the possibilities of debris flow during the volcano eruption [9] [18].

Disaster vulnerability is dynamic including physical, social, economic or environment vulnerability because vulnerability depends on the process, total number, disaster location and also the damage and consequences emerged [6]. Vulnerability against disasters can be defined as a condition of a community which leads or causes the inability to overcome existing disaster hazards (Regulation of Head of Indonesian National Board for Disaster Management Number 02 of 2012). The indicators used in determining physical vulnerability are measured by settlement density (permanent, semi-permanent and non-permanent), availability of public buildings/facilities and availability of critical facilities. Settlement density is derived from dividing them over built area or village area and divided by area (in ha) times the price of each unit from each parameter. Physical vulnerability index is obtained from the average of the weight of settlement density (permanent, semi-permanent and non-permanent), the availability of public buildings/facilities and the availability of critical facilities. The large and large number of public facilities can increase the level of physical vulnerability, especially public facilities located in settlement areas [17].

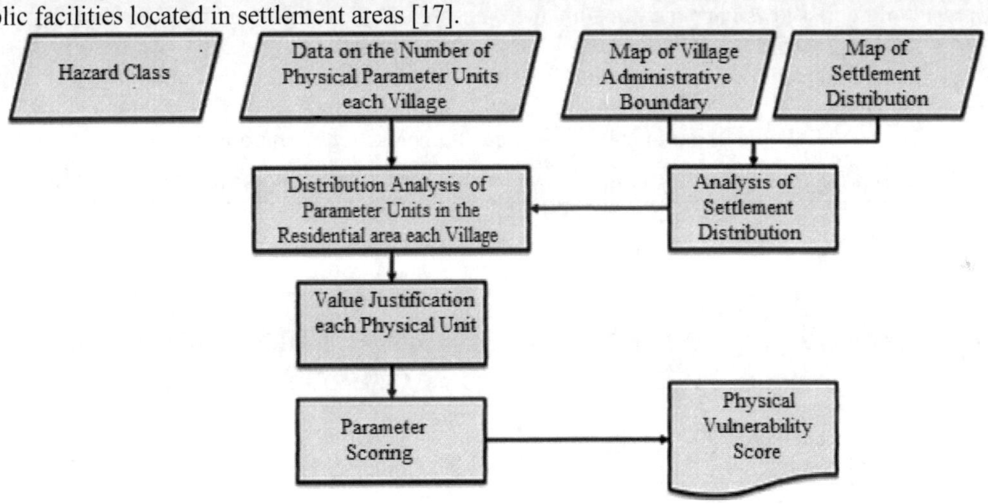

Figure 1. The Analysis of Physical Vulnerability according to Indonesian National Board for Disaster Management

The most fatal of secondary effect from volcanic activities is the debris flows and mudflows which collectively known as lahar [2]. Merapi Vulcano as one of the most active volcanoes in the world has the disaster potential in the form of pyroclastic and tephra. It was estimated that in the eruption of 2010, it produced more than 100 x 106 m^3 volume of pyroclastic debris flowed to lahar prone rivers [10]. The term lahar derived from one of the local languages in Indonesia (Javanese) refers to its emergence extensively in escarpments in wet tropical area [11]. Lahar is produced when a massive amount of volcanic ashes and other tephra saturate with water and become unstable and suddenly flow down the slope [2], [11]. Settlements located near the banks of rivers flowed with lahar have a various risk and hazard level to lahar but lahar flows tend to be more destructive on the areas alongside the streams [11], [12]. Unlike pyroclastic flows, lahar can happen without any eruption and generally a stream with low temperature [2]. The undecided rainfall intensity causes the difficulty to predict in which river the lahar will flow [4].

Lahar hazards are a result of the impact from the existence of a volcano in which this disaster can happen at the same time or after a short period of time and these events is not a rare thing [6]. The building vulnerability towards lahar flood is different from the building vulnerability towards regular flood. In the event of regular flood, a building of more than 20 years and built from natural materials tend to have higher damage level compared to the one that is built in the last 10 years and was built using materials that are more durable [13]. Debris flows are a combination of smooth sediments and huge rocks that have the consistency of a wet concrete; this flows can move for kilometers down the valleys from the part of the volcano where they were formed [2]. The size and level of the runoff from post-eruption volcano will gradually decrease after the eruption [14].

3. Method

The data collection technique used in this research is a secondary data collection. The data was taken from Citra Quickbird of Magelang District in 2017. The assessment of settlement physical vulnerability in the area alongside the lahar flow path from post-eruption of Merapi Vulcano in Pabelan River used the spatial analysis technique which measures the parameter of the distance of houses to the river (buffer), the number of the houses in each cluster and the building materials of the houses. According to Kang [15] who combined the probability of physical damage caused by debris flows and the criteria proposed by HAZUS and then studied the theory to a building damage which was a result of debris flows happened in July and August 2011. The formula that used to analyzed the data is weighting, according to the number of parameter that used and also the impact of each parameter so that the weights has different. The weight, parameter classification and score that used in this research can be seen in Table 2.

Table 2. Parameter of Physical Vulnerability of Buildings

Parameter	Weight	Parameter Classification	Score
The distance to the river	40 %	100-150 m	1
		50-100 m	2
		0-50 m	3
House	30 %	< 400 MIO	1
		400 – 800 MIO	2
		> 800 MIO	3
Building material	30 %	Wood	1
		Brick	2
		Concrete frame	3

Based on the Table 2 above, the parameter of the distance to the river has the highest weight because it has the highest influence to a threat of lahar hazards. The parameter of the distance of houses to the river (buffer) are divided into 50 meter, 100 meter and 150 meter; the close a house to the river the higher the settlement physical vulnerability. The method used in determining the vulnerability of a house unit is a classification of the number of houses by the Indonesian National Board for Disaster Management which in an area with a low disaster potential the total of the houses. The parameter of houses used in this study was obtained from totaling the houses in each cluster and then in the area with low disaster risk the number was multiplied by IDR. 5.000.000, in the area with medium disaster risk it was multiplied by IDR. 10.000.000 and in the area with high disaster risk the total number was multiplied by IDR. 15.000.000. The results from the calculation then were used to classify the above parameter. Moreover, for the building materials, there was only 1 building from all 4 Sub-districts built using concrete frame which was crossed by the lahar flow.

According to Kang [15] who combined the probability of physical damage caused by debris flows and the criteria proposed by HAZUS and then studied the theory to a building damage which was a result of debris flows happened in July and August 2011. From this research, it was obtained a building damage scale classified by the materials used to build the building. Wood materials are the most vulnerable, bricks have a medium level of vulnerability when hit by debris flows and steel frames are at lowest vulnerability level. In each parameter was given a score then later to be weight. The results from the analysis were then classified into three which are low, medium and high vulnerability level. The population in this study is the whole settlements in the radius of 0 – 150 meter from Pabelan River including Mungkid Sub-district, Sawangan Sub-district, Dukun Sub-district and Muntilan Sub-district. The flow process of this research can be seen on the Figure 2.

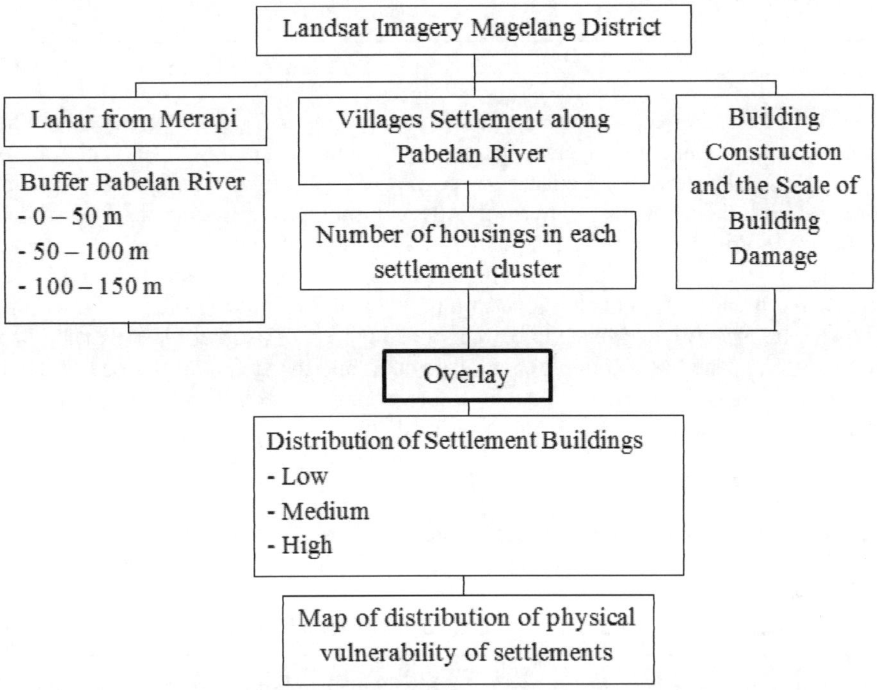

Figure 2. Flow Chart

4. Results

The results from the spatial analysis and imagery interpretation, it is obtained that in total of 1.419 houses are in the vulnerable areas. The areas prone to lahar hazards are scattered in four Sub-districts. A total of 522 houses are classified in low vulnerability level, a total of 661 houses are in medium vulnerability level and 236 houses have high vulnerability level. The vulnerability assessment starts from the parameter of the distance from the river to the houses using the buffer analysis which is 50 meter, 100 meter and 150 meter on the right and left side along the Pabelan River from the upstream to the downstream which is located in the intersection with Progo River. Furthermore, an overlay was done to the settlement polygon in order to count the number of the houses which overlaped with the results of the river buffer. The results from the calculation then were used to classify the above parameter. Moreover, for the building materials, there was only 1 building from all 4 Sub-districts built using concrete which was crossed by the lahar flow. There were a total of 1.419 houses all around the 63 clusters. The following are the discussion of each parameter of physical vulnerability of settlement buildings along the Pabelan River.

4.1. The Parameter of the Distance to the River

The areas along the riverbank of Pabelan River have different variation of vulnerability from low to high scattered in 63 clusters. The distribution of cluster settlements with the physical vulnerability from low to high can be seen in Table 3.

Table 3. The Distribution of the Settlement Distance along Pabelan River

Distance to the River	Score	Vulnerability Level	Total of Clusters	Percentage (%)
100-150 m	1	Low	36	57,14
50-100 m	2	Medium	18	28,57
0-50 m	3	High	9	14,29
Total			63	100

Based on Table 3, it can be seen that most of the housings are less than 100 meter to 150 meter from the riverbank hence the vulnerability classification is dominant in the low level. The total of clusters with the low vulnerability level is 36 clusters or equal to 57.14% from the research area. This finding is in accordance with the settlement pattern of village community in Indonesia which is centered in farm lands where the river function as the source of agriculture irrigation so the settlement are not in the riverbank area. The housing pattern of villages in Jawa Island is different of that in Kalimantan Island. The people living in Kalimantan tend to build their house close to the river since they use the river is a main transportation. Meanwhile, the housing in the riverbank of Pabelan River is an exception since it happened as a consequence of demography dynamic and the rise of need for land. The settlements which are less than or equal 50 meter from the river with a high level of vulnerability are in a total of 9 which equals to 14.29% from research area.

4.2. The Parameter of Housings

The imagery interpretation of the objects interpreted as housings resulted in 1.179 building suspected as houses. The house distribution with physical vulnerability level from low to high can be observed in Table 4.

Table 4. The Distribution of the Total Housing Loss in each Cluster

House	Score	Vulnerability Level	Total of Cluster	Percentage (%)
< Rp 400 mil	1	Low	42	66,67
Rp 400 – 800 mil	2	Medium	14	22,22
> 800 mil	3	High	7	11,11
Total			63	100

Based on the information in Table 4, it can be seen that most of the physical vulnerability of the house buildings are less than 400 million rupiahs in total of 42 clusters or equal 66% of the research area. The physical vulnerability of the house buildings is Rp. 400.000.000 to Rp. 800.000.000 in total of 14 clusters which equals to 22.22% of the research area. The clusters with high level of vulnerability are Rp. 800.000.000 in total of 7 clusters or equal to 11.11% of the research area.

4.3. The Parameter of Building Material

In General, the types of building along the Pabelan River are dominated with the ones which used bricks as the main material. House buildings have the variation of the materials used in building them.

The physical vulnerability of settlement buildings is based on the variation of strength scale of the house building types against the damage caused by lahar flows. The house distribution of varied building construction types can be seen in Table 5.

Table 5. Types of Building Construction and the Scale of Building Damage

Type of Construction	Score	Vulnerability Level	Total of Cluster	Persentase (%)
Wood	1	Low	0	0
Brick	2	Medium	62	98,41
Steel Frame	3	High	1	1,59
Total			63	100

Based on Table 5, it can be concluded that there is 1 cluster which has 1 building constructed using steel frame or equals with 1.58% form the research area. Referring to the above table, most of the buildings are constructed using bricks with the total of 62 clusters or equal to 98.41 %. Based on the results and imagery interpretations, there is no house building that uses woods as construction material.

4.4. The Distribution of Areas with Low Physical Vulnerability of Settlement Buildings in Lahar-Hazards Prone Area

The areas which have a low level of vulnerability against lahar flows along Pabelan River is varied from the ones with a low level to the highest ones. The settlement area distribution of the low physical vulnerability against the lahar flows in Pabelan River can be seen in Table 6.

Table 6. The Distribution of Settlement Buildings with Low Physical Vulnerability Level

Sub-district	Cluster	Total of House	Vulnerability	Level
Dukun	51	20	11	Low
Dukun	58	3	11	Low
Dukun	63	1	11	Low
Mungkid	2	12	11	Low
Mungkid	4	16	11	Low
Mungkid	6	30	14	Low
Mungkid	8	19	11	Low
Mungkid	10	2	11	Low
Mungkid	16	28	14	Low
Mungkid	19	3	11	Low
Mungkid	20	43	14	Low
Mungkid	31	5	11	Low
Mungkid	33	32	14	Low
Muntilan	1	6	11	Low
Muntilan	5	2	11	Low
Muntilan	11	15	11	Low
Muntilan	12	18	11	Low
Muntilan	14	7	11	Low
Muntilan	23	1	11	Low
Muntilan	29	6	11	Low

IOP Conf. Series: Earth and Environmental Science **884** (2021) 012036 doi:10.1088/1755-1315/884/1/012036

Sub-district	Cluster	Total of House	Vulnerability	Level
Muntilan	32	4	11	Low
Muntilan	38	24	11	Low
Muntilan	40	8	11	Low
Muntilan	55	7	11	Low
Sawangan	41	41	14	Low
Sawangan	46	38	14	Low
Sawangan	47	34	14	Low
Sawangan	50	34	14	Low
Sawangan	52	13	11	Low
Sawangan	53	9	11	Low
Sawangan	56	16	11	Low
Sawangan	59	25	14	Low
Total		522		

Based on Table 6, it can be concluded that the areas with low physical vulnerability of settlement buildings have the interval of 11 to 14. The Sub-district with the biggest number of low level settlement physical vulnerability is Sawangan Sub-district with the total of 11 clusters. The Sub-district with the smallest number of low level settlement physical vulnerability is Dukun Sub-district with the total of 3 clusters. Cluster 20 has the greatest number of house in total in which there are 43 house units including the ones classified under low vulnerability level. Sawangan Sub-district has te biggest number of houses in low level vulnerability with the total of 210 houses among all the subdistrict. This phenomenon is caused by the settlement is in lengthways pattern and its distance is more than 100 meter to 150 meter from the riverbank and the material use for the buildings is bricks.

4.5. The Distribution of Areas with Medium Physical Vulnerability of Settlement Buildings in Lahar-Hazards Prone Area

The distribution of areas with settlement physical vulnerability towards medium level lahar flows is mostly from the houses in which the distance classified as medium level vulnerability. However, there are some that come from the distance classified as low and high level vulnerability. The distribution of those with medium level vulnerability can be observed in Table 7.

Table 7. The Distribution of Settlement Buildings with Medium Physical Vulnerability Level

Sub-district	Cluster	Total of House	Vulnerability	Level
Dukun	49	49	5	Medium
Mungkid	7	3	17	Medium
Mungkid	9	5	17	Medium
Mungkid	34	2	17	Medium
Mungkid	17	17	29	Medium
Muntilan	3	54	17	Medium
Muntilan	15	1	17	Medium
Muntilan	24	1	17	Medium
Muntilan	26	56	17	Medium
Muntilan	30	2	17	Medium
Muntilan	39	4	17	Medium
Muntilan	44	1	17	Medium

Sub-district	Cluster	Total of House	Vulnerability	Level
Muntilan	60	10	17	Medium
Muntilan	27	27	48	Medium
Muntilan	13	13	2	Medium
Muntilan	25	25	1	Medium
Muntilan	28	28	2	Medium
Sawangan	35	68	17	Medium
Sawangan	45	8	17	Medium
Sawangan	48	20	17	Medium
Sawangan	57	5	17	Medium
Sawangan	61	61	55	Medium
Sawangan	62	62	1	Medium
Sawangan	54	54	1	Medium
Sawangan	42	42	28	Medium
Sawangan	43	43	9	Medium
Total		661		

Based on Table 7, it can be concluded that the areas with medium physical vulnerability level have the interval from 17 to 20. There are 26 settlement clusters which are included in the medium level. The Sub-district which has the most medium physical vulnerability for settlements is Muntilan Sub-district with 12 clusters. The Sub-district with the least medium level of physical vulnerability is Dukun Sub-district with only 1 cluster. Cluster 2, 26, 35 and 61 are the settlement clusters with the low level of distance vulnerability but the house classification is more than Rp. 800.000.000 hence the scores of the settlement vulnerability become the medium level. Sawangan Sub-district has te biggest number of houses in low level vulnerability with the total of 363 houses among all the subdistrict. Cluster 13, 25,28,43,49 and 50 have the distance vulnerability classified as high but the houses are Rp. 400.000.000 to Rp. 800.000.000 hence they are categorized as medium.

4.6. The Distribution of Areas with High Physical Vulnerability of Settlement Buildings in Lahar-Hazards Prone Area

There are two clusters classified as high physical vulnerability of settlement buildings in lahar-hazards prone area that come from the medium level of distance vulnerability. From the 4 Sub-districts bordering with Pabelan River, the clusters in Dukun Sub-district and Muntilan Sub-district are not classified as high. The cluster distribution with high vulnerability can be seen in Table 8.

Table 8. The Distribution of Settlement Buildings with High Physical Vulnerability Level

Sub-district	Cluster	Total of House	Vulnerability	Level
Mungkid	21	93	23	High
Mungkid	18	13	24	High
Mungkid	22	19	24	High
Sawangan	36	90	23	High
Sawangan	37	21	27	High
Total		236		

IOP Conf. Series: Earth and Environmental Science **884** (2021) 012036 doi:10.1088/1755-1315/884/1/012036

Based on Table 8, the information we can obtain is that the areas with high physical vulnerability of settlement buildings have the interval from 23 to 27. There are 5 settlement clusters classified as high. The Sub-district with the most number of high physical vulnerability of settlements is Mungkid Sub-district with 3 clusters. The Sub-district with the lowest number of high physical vulnerability of settlements is Sawangan Sub-district with 2 clusters. Cluster 21 and 36 are clusters from low level of distance vulnerability but the house is more than Rp. 800.000.000 so the vulnerability then classified as medium. Cluster 18, 22 and 37 are from the high level of distance vulnerability but classified as houses under Rp. 400.000.000 hence they are classified as high. Mungkid Sub-district has te biggest number of houses in low level vulnerability with the total of 125 houses among all the subdistrict.

The areas with physical vulnerability of settlement buildings in lahar-hazards prone area are scattered all over the research area. Each of the area has various vulnerability scores. The distribution of areas with physical vulnerability is arranged by the level of vulnerability of lahar-hazards prone area around Pabelan River. The vulnerability level of lahar flows is classified into three categories: low, medium and high. The higher the level of physical vulnerability the higher the chance to experience loss caused by lahar flows. That chance can be light damage, medium damage and severe damage and even completely collapse. The level of settlement physical vulnerability alongside Pabelan River from the lowest to the highest can be observed in Table 9.

Table 9. The Percentage of houses based on The Physical Vulnerability Level of Settlement Building along Pabelan River

Vulnerability Level	Total of Vulnerable House	Vulnerability Percentage (%)
Low	522	36,79
Medium	661	46,58
High	236	16,63
Total	1.419	100

Based on Table 9, it was found that the settlement buildings alongside Pabelan River are dominated by those with medium vulnerability level which is I total of 46,58% of all the areas. The high vulnerability levels are distributed in most of the research area. Sawangan and Mungkid Subdistrict had all of the variation risk level. The distribution of physical vulnerability level of lahar-hazards prone along Pabelan River can be observed in the Figure 3. In the figure 3a we can observe the cluster 1 until 6 which dominated by low vulnerability cluster in Mungkid and Muntilan Subdistrict. Cluster 8-16 shown in Figure 3b that variated form low untill high vulnerability. Cluster number 11 in Muntilan Subdistrict has a part or cluster that located near by the river, but in moderate vulnerability because of the low number of house and the material used for the building. Fom the Figure 3c we can informed that there was dominated by high vulnerability cluster. In map 3d we can only observed 5 cluster with various vulnerability. Cluster number 35 in Sawangan Subdistrict is the longest and the widest cluster with high vulnerability. It caused by the distance to the river and also affected by setllement density. In the Figure 3e we can conclude that it dominate with low vulnerability. In the figure 3f only 2 cluster shown, and it has low vulnerability. In Figure 3g we can observed that it has unique phenomenon. Cluster 61 has area which slightly far from the river but the higher vulnerability cluster reversed. In the Figure 3h only shown one cluster with low vulnerability density.

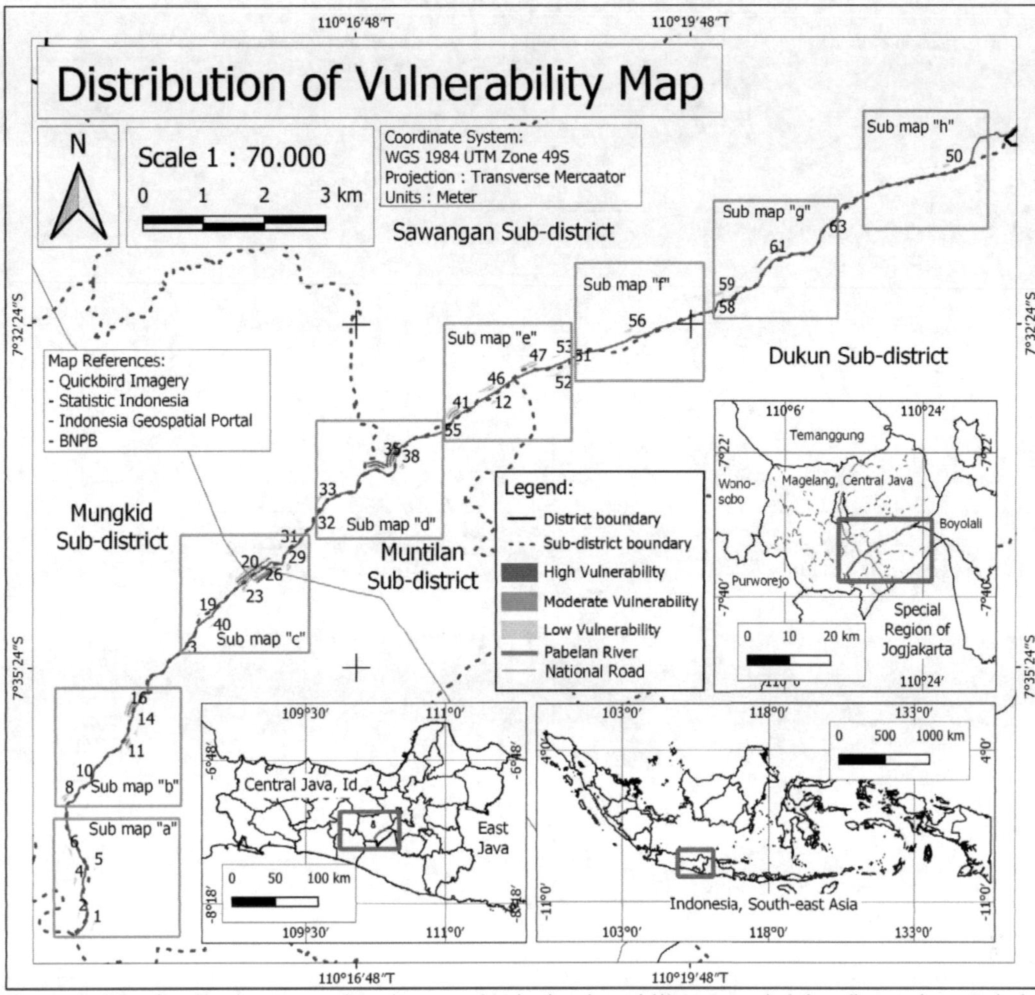

Figure 3. The distribution map of settlement physical vulnerability towards lahar flows along Pabelan River.

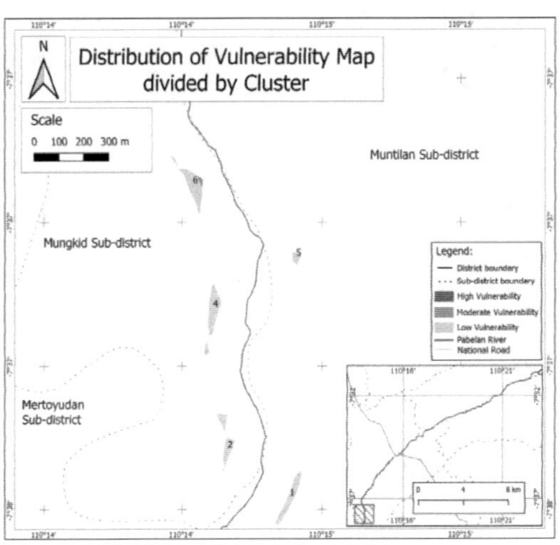

(3a)

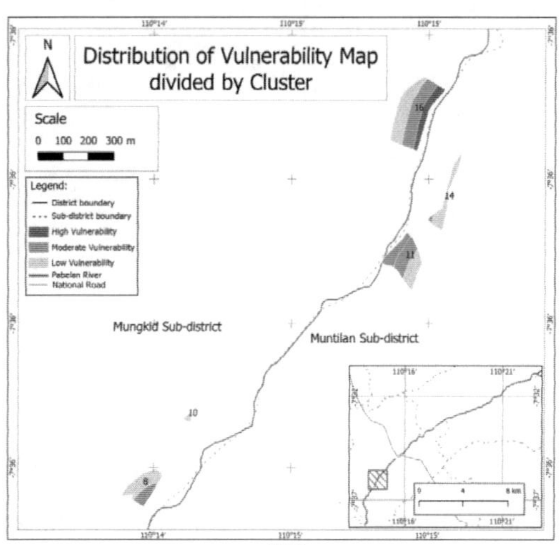

(3b)

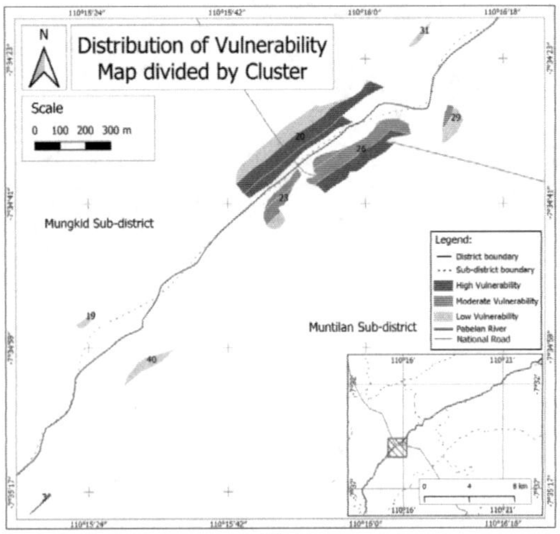

(3c)

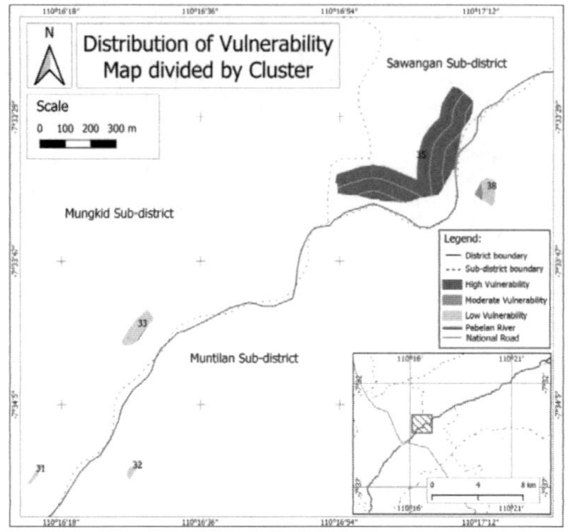

(3d)

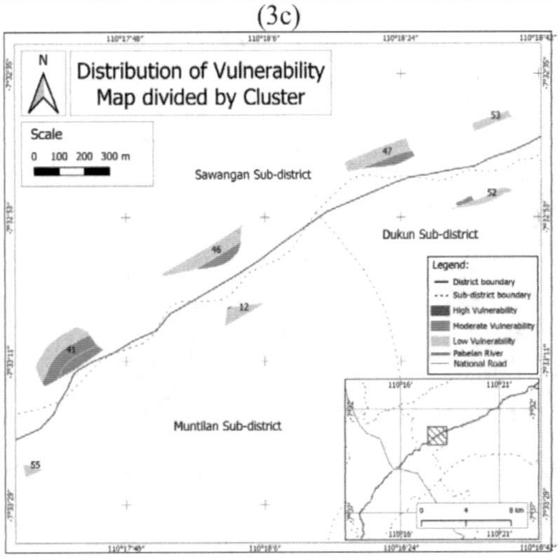

(3e)

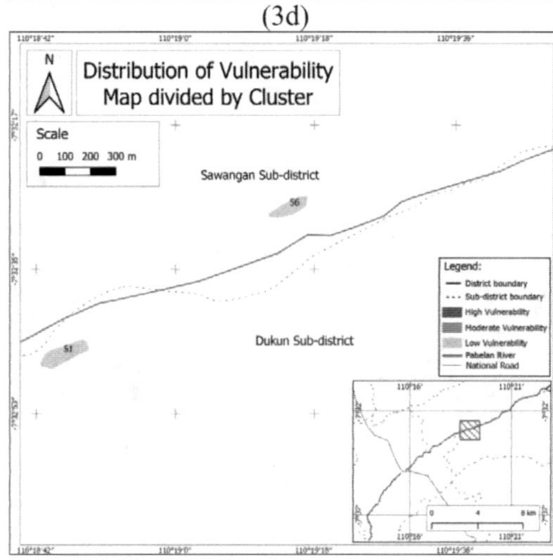

(3f)

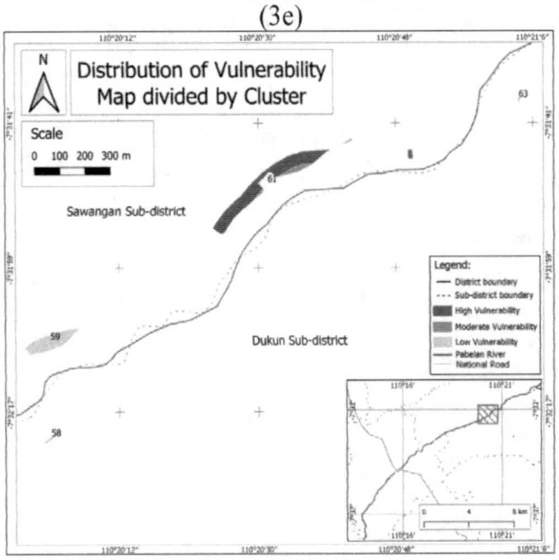

(3g)

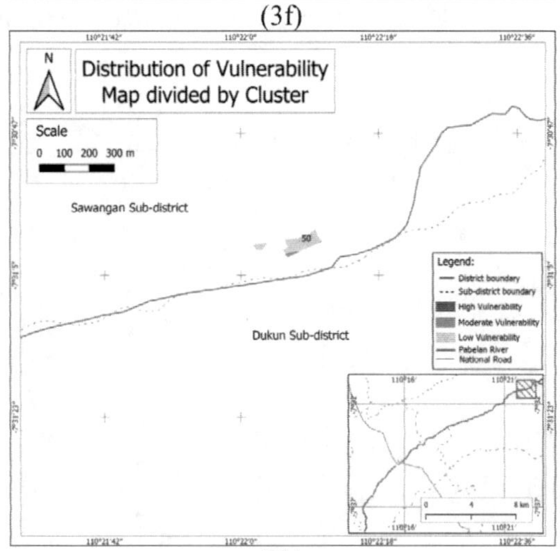

(3h)

Figure 3a-3h. settlement distribution

5. Discussion

Based on the results of this research, it can be concluded that the distribution of settlement physical vulnerability in lahar-hazards prone area along Pabelan River is affected by the distance of the settlement buildings to the river, the number if the buildings in each cluster and the type of the construction material. The closer the settlement buildings and the more buildings in each cluster, the wider the distribution of the physical vulnerability of settlement towards disasters. The highest distribution of high physical vulnerability is Mungkid Sub-district and Sawangan Sub-district with 5 clusters or equal to 7.49% of the research area. The medium level of physical vulnerability is Dukun Sub-district, Mungkid Sub-district and Muntilan Sub-district. The medium level is distributed to 26 clusters or equal to 41.27% of the research area. The low physical vulnerability of settlements is scattered in Dukun Sub-district, Mungkid Sub-district, Muntilan Sub-district and Sawangan Sub-district. The low level of physical vulnerability is distributed to 32 clusters or equal to 50.79% of the research area. The more we go to the upstream, the fewer houses we can found. Its also affected towards the result of this research. For the upcoming research, we suggested the reseacher to considering about the morphometry of the river. This research didn't used morphometry because of lack imagery before Pabelan river was hit by lahar. The benefits from this research can be used for the local community to be aware of their environtment, also for the local government to make the wisest policy. The advantages of this research can be considered as part of disaster risk analysis. This research can also be carried out in other rivers which originated in Merapi Volcano and becoming lahar prone rivers. We know that this research is far from perfection, but we also hope there will be more better research of this similar topic in the future.

Acknowledgements

The research is supported by data from BAPPEDA of Magelang District, Statistic Indonesia, and BNPB, and my colleagues (especially Ms. Assriyani, Ms. Aisah Puruhita and Ms. Ismi Sholihatun) that made priceless contributions to this research. Special thanks goes to Yogyakarta State University for supporting the first author to write this paper and the committee of ICMGEP which held this conference. Without them we could not have completed this research.

References

[1] Hadmoko D S *et al* Post-eruptive Lahars at Kali Putih following the 2010 eruption of Merapi volcano, Indonesia: occurrences and impacts *J Natural Hazards*, (German: Springer Nature)

[2] Keller E A and Blodgett R H 2008 Natural hazards : earth's processes as hazards, disasters, and catastrophes. (New Jersey: Pearson/Prentice Hall), pp. 85.

[3] Gomez C *et al*. Borobudur, a basin under volcanic influence: 361,000 years BP to present *Journal of Volcanology and Geothermal Research 196* pp. 245–264.

[4] Hapsari R I *et al*. 2019 X-MP Radar for Developing a Lahar Rainfall Threshold for the Merapi Volcano Using a Bayesian Approach *Journal of Disaster Research* Vol.14 No.5,

[5] Bélizal E de *et al*. 2013 Rain-triggered lahars following the 2010 eruption of Merapi volcano, Indonesia: a major risk", *Journal of Volcanology and Geothermal Research 261* pp. 330–347 .

[6] Papathoma-Köhle M, Gems B, Sturm M, Fuchs S 2016 Matrices, curves and indicators: A review of approaches to assess physical vulnerability to debris flows *J Earth-Science Reviews*

[7] Joseph J 2013 Measuring vulnerability to natural hazards: a macro framework *Journal Disaster* (UK: Blackwell Publishing) pp. 185-200.

[8] Hambali R, Legono D, Jayadi R, and Oishi S 2019 Improving Spatial Rainfall Estimates at Mt. Merapi Area Using Radar-Rain Gauge Conditional Merging *J of Disaster Research* Vol.14 No.1 pp. 69-79.

[9] Putra SS, Ridwan B W, Yamanoi K, Shimomura M, Sulistiyani, and Hadiyuwono D, "Point-Based Rainfall Intensity Information System in Mt. Merapi Area by X-Band Radar", *J of Disaster Research* Vol.14 No.1, pp. 80-89.

[10] Lavigne F et al. 2011 Lahar hazards and risks following the 2010 eruption of Merapi volcano, Indonesia". Geophysical Research Abstracts Vol. 13, EGU 2011-4400
[11] Smith K and Petley D N 2009 Environmental Hazard: Assesing risk and reducing disaster (Routledge)
[12] Franck Lavigne 1999 Lahar hazard micro-zonation and risk assessment in Yogyakarta city, Indonesia" in GeoJournal 49 (Kluwer), pp. 173–183.
[13] Fatemi M N, Shimoda M, Okyere S A, Diko S K, Kita M, and Matsubara S 2020 Physical Vulnerability and Local Responses to Flood Damage in Peri-Urban Areas of Dhaka, Bangladesh *J Sustainability* (MDPI) pp 12.
[14] Gonda Y, Miyata S, Fujita M, Legono D, and Tsutsumi D 2019 Temporal Changes in Runoff Characteristics of Lahars After the 1984 Eruption of Mt. Merapi, Indonesia". *JDR* Vol.14 No.1, (Japan : Fujipress) pp. 61-68.
[15] Kang H S and Kim Y T 2015 The physical vulnerability of different types of building structure to debris flow events *J Nat Hazards* 80 (Springer) pp. 1475–1493
[16] Ikhsan J et al. 2020 Dynamics of lahar-affected river tributaries of the Progo river after the 2010 Mt. Merapi eruption Proc. Int. Conf. Of Water Resources Engineering Department IOP Conf. Ser.: Earth Environ. Sci. 437 012009
[17] Fitria L M, Ni'mah N M, and Danu L K 2019 "Kerentanan Fisik Terhadap Bencana Banjir di Kawasan Perkotaan Yogyakarta". *J Reka Ruang* Vol.2, No.1, pp.1-9.
[18] Iguchi M 2019 Proposal of Estimation Method for Debris Flow Potential Considering Eruptive Activity, *J of Disaster Research* Vol.14 No.1 pp. 126-134.
[19] Badan Pusat Statistik 2018 "Kecamatan Dukun Dalam Angka 2018", ISSN: 2406-9949.
[20] Badan Pusat Statistik 2018 "Kecamatan Mungkid Dalam Angka 2018", ISSN: 2442-5753.
[21] Badan Pusat Statistik 2018 "Kecamatan Muntilan Dalam Angka 2018", ISSN: 2356-2331.
[22] Badan Pusat Statistik 2018 "Kecamatan Sawangan Dalam Angka 2018", ISSN: 2442-529X.

IOP Conf. Series: Earth and Environmental Science **884** (2021) 012037 doi:10.1088/1755-1315/884/1/012037

Impact of Optical Aerosol Depth (AOD) on Light Pollution Level: a spatio-temporal analysis

Amalia Gita Ayudyanti[1*], Iswari Nur Hidayati[2]
[1,2]Departement Sains Informasi Geografi, Fakultas Geografi, Universitas Gadjah Mada, D.I. Yogyakarta, Indonesia

Email: [1]amalia.gita.ayudyanti@mail.ugm.ac.id [2]iswari@ugm.ac.id

Abstract. Population dynamics and economical, also industrial, in Indonesia are growth as it increasing every years. It brings an environmental problems unconsciously. Increasing the amount of aerosol optical depth (AOD) has a considerable impact, one of them is light pollution increase due to uncontrolled scattering of particles in urban areas. Light pollution is a new problem, it can disturb the balance of the ecosystem. Increasing light pollution on trophosperic surface on the earth can cause the functional damage of natural light in the sky in providing guidance for animal noctural animals and human health. The consequence are emergencing problems with ecosystem imbalances that can be a new disasters in the future. Identification of the impact of increasing AOD with light pollution needs to be known to maintain the balance of the ecosystem in the future. Remote sensing has an effective, efficient and able to provide actual data on this problem. MODIS imagery was used as AOD and VIIRS DNB data sources were used for multi-temporal light pollution data in 2014, 2016 and 2018. Both data were then performed correlation tests using Rank Spearment methods and obtained very strong results, 0.8-1 . The higher AOD in urban areas, the higher the level of light pollution in the region. Java and Bali have high AOD levels accompanied by high light pollution.

Keyword: VIIRS DNB, MODIS, AOD, light pollution, rank spearment

1. Introduction

The development of remote sensing technology bring a posibility for us to understand the spatial distribution and intensity of artificial night time light (NTL) in various regions of the earth's surface. The two most widely used data sources are DMSP OLS and VIIRS DNB. VIIRS DNB which was released in 2013, as a newer satellite, it shows more improvement and calibration with an automatic filter system that can eliminated stray lights, natural lights, such as fire sources, snow and cloud disruptions automatically. The results of the VIIRS data delivered as daily, monthly or annual data on a large scale (742 x 742 m^2). The use of VIIRS data has been widely used in various fields, including demographic analysis, socioeconomic, settlement dynamics, nighttime atmospheric properties, multitemporal changes, and other topics such as light pollution and CO_2 emissions [1].

VIIRS is one of the key environmental remote-sensing instruments onboard the Suomi NPP satellite [2]. The VIIRS instrument observes and collects global satellite observations that span the visible and infrared wavelengths across land, ocean, and atmosphere. A whiskbroom radiometer by design, it has 22 channels ranging from 0.41 µm to 12.01 µm. Five of these channels are high-resolution image bands or I-bands, and sixteen serve as moderate-resolution bands or M-bands. VIIRS also hosts a unique

Content from this work may be used under the terms of the Creative Commons Attribution 3.0 licence. Any further distribution of this work must maintain attribution to the author(s) and the title of the work, journal citation and DOI.
Published under licence by IOP Publishing Ltd

panchromatic Day/Night band (DNB), which is ultra-sensitive in low-light conditions that allows us to observe nighttime lights with better spatial and temporal resolutions compared to previously provided nighttime lights data by the Defense Meteorological Satellite Program [3].
The VIIRS DNB nightly images are affected by a number of factors including the atmospheric condition, moonlight, snow and clouds [4]. Atmospheric scattering [5] and absorption [6] (especially with high-concentration aerosols), the atmosphere can affect the results of the analysis of the light value of the DNB VIIRS imagery [1]. An aerosol optical depth (AOD) is an important parameter in measuring the degree of atmospheric extinction that can also be used as an effective substitute to track the evolution of the aerosol-induced airpollution [7]. Aerosols in the atmosphere have a significant effect on environmental pollution [8]. Different AOD quality can affect different sky conditions, especially the influence of the seasons in the Wuhan region [9]. Therefore, it is important to determine the atmospheric effect and its extent on the light pollution data.

It should be understood that there are still various factors that cause errors in the process of data recording by the NPP-VIIRS Suomi, one of that is an atmospheric problems such as the presence of aerosols in an area. Aerosol itself can play a role in the absorption and scatter of energy at certain sizes and densities, so that, there can be a decrease or increase of the value of artificial light in VIIRS. MODIS Optical Depth Aerosols (AOD) are used in understanding the global distribution and intensity of aerosols. It should be known that the VIIRS DNB sensor is very sensitive to aerosol dynamics in an area, even more sensitive than water vapor. Consideration of aerosol concentrations is especially important in studies that use artificial light data. Moreover, the highest contributor to artificial light radiation is in urban areas with high socioeconomic activity, similar things occur in the existence of air pollution which is also high in urban areas so that it plays a role in causing high aerosol thickness in the same urban area.

Previous research conducted by X. Wang, XihanMu and G. Yan, 2020 [1] has conducted quantitative tests on AOD and light pollution (NTL) in four cities (Beijing, Wuhan, Chengdu, and Meishan) of China. The research was only carried out by using one sample point in each city and using the Linear and log-linear models test to generate similar coefficients of determination (R2) for AOD and NTL data. While in this study the test sample totaled 46 points with spearment rank test model. The test samples were not only placed in the city center but were scattered randomly throughout Java and Bali. The test model is also adapted to the type of data that is non-parametric.This study focuses on understanding that problem through analysis of the relationship between NTL and aerosols, especially those that occurred on Java and Bali, Indonesia in 2014, 2016, and 2018 through the distribution of random sample points that cover various regional conditions with varying levels of development.

2. Details Experimental

2.1 Research area

The measurements presented in this study were located at Jawa and Bali Island , Indonesia (7°30′ S, 115°42′ E). As a preliminary note, the study take an area in the tropical area with a fairly high level of cloud formation with a dry season in April - September and a rainy season in October - March. Java and Bali are islands with varying levels of development in various regions, as the most developed islands in Indonesia, those are functionally used as the center of government and the country's economy. The island has many urban area as a “core” that take a role in the development of the surrounding area, such as Jakarta, Bandung, Surabaya and Semarang, with varying levels of socio-economic activity resulting in varying levels of artificial ngiht time light (NTL) illumination and the formation of diverse air pollution in various regions. In its own territory, Java-Bali itself has varied reliefs, with the altitude range of 0 - 3,676 m asl. The socio-economics centers are in the lowlands in the central part of the island and along the coast of Jawa-Bali.

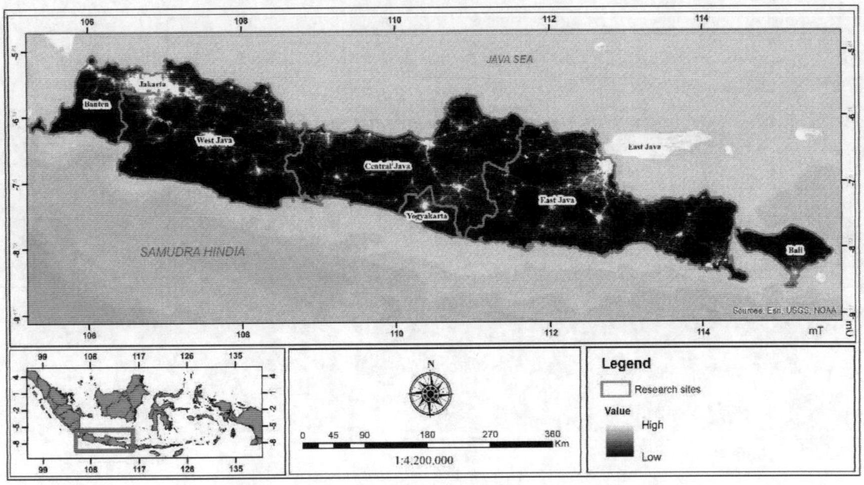

Figure 1. Study Area in Jawa – Bali Island

2.2. Data source

This study uses MODIS AOD (Aerosol of Depthness) and VIIRS DNB (Day-Night-Band) Data obtained through Google Earth Engine (GEE). Both took time in 2014, 2016, 2018 and produced 3 data by the monthly average data of VIIRS and MODIS AOD at 2014, 2016, and 2018. Note that the monthly average used is August because it has clear sky conditions. The resolution of the two image types are different so it should be resampled. Cloud masking on AOD and VIIRS is done to avoid the effect of impairment due to cloud cover that is too thick in the data used. More higher the VIIRS values indicate the higher light pollution. AOD are unitless, higher AOD means those areas are more polluted.

The Earth Observation Group in the National Oceanic and Atmospheric Administration's National Geophysical Data Center (NOAA / NGDC) of the United States released the annual NTL data. Each file consists of 2 data, namely NTL radiation data and average number of cloud-free (NCF). Both are presented in 742x742 m^2 resolution. Light pollution data (NTL) is presented in units of nW /cm^2 sr^{-1} which represents the value of artificial light radiation illumination at night. Pre-processing is not done on VIIRS data because it has been filtered by the system for the presence of data near the edges of the swath and data impacted by twilight, stray light, lightning, lunar illumination, and cloud cover [10].

The AOD data used from the recording of the Terra and Aqua MODIS sensor combined the Multi-angle Implementation of Atmospheric Optical Depth (MAIAC) of Land Aerosol Optical Depth (AOD) at a resolution of 1 km. So it is used as a reference of resample process of the VIIRS NTL data to produce uniformity in the area and location of the pixel grid. It's done to avoid decreasing of correlation value. The AOD average data has been automatically entered from the annual average cloud cover at GEE.

2.3. Sample and method

As explained before, the pre-processing data is done by cloud masking on AOD data, while VIIRS has been corrected by the system. Resample is also done referring to the AOD data (1000 x 1000 m^2). Then the two data were masked on the administrative boundary vector of Java and Bali in accordance with the experimental area. Data extraction is done by determining the sample points that presented the experimental area. The number of samples was determined as 46 sample points and randomly distributed to cover areas with different levels of development, such as urban, sub-urban, and rural area on the island of Java-Bali. Each sample point consists of AOD and VIIRS data which will be tested for correlation using Spearman's Rho. Classification of AOD and VIIRS values were not used in this study. Correlation test was conducted in this study using Spearman's Rho through one-tailed test with a significance level of relationship of 0.01. Furthermore, 46 sample points were also visualized in the form of scatter plot graphs to determine the pattern of relationships formed of VIIRS-AOD. Same method done for each 3 data in 2014, 2016, and 2018 to find out the dynamics that occur in the period of 2014

– 2018. It also can present the socio-economic development of study area. Multitemporal data is also used to make a spatial distribution model of AOD and VIIRS each year before spatial relationship analysis is performed.

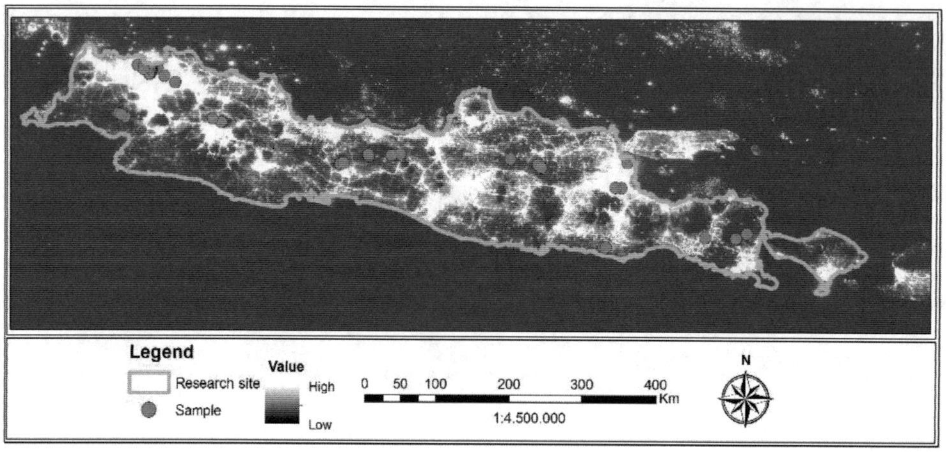

Figure 2. Sample points distribution.

3. Result and Discussion

The results of the average AOD and VIIRS-NTL distribution model for Java-Bali Island in 2014, 2016 and 2018 can be seen in Figure 3 (a) and (b). Visualized data has been sampled before. Dark brown warma represents a high level of aerosol density referring to areas with high air pollution, while dark blue indicates low aerosol density. In VIIRS, yellow represents high artificial light illumination, referring to areas with few sources of illumination, while blue represents low artificially illuminated light.

Based on its visualization, the AOD data in 2014 still has a location with a blue indicator that shows a low AOD value and the VIIRS image shows that light pollution in the islands of Java and Bali is still clustered in major cities only. In 2016 the distribution of AOD in the north coast increased and was followed by widespread light pollution in the 2016 VIIRS image on Java and Bali. In 2018 the Jakarta, Banten and West Java regions have very high AOD levels with dark brown color indicators and have an even distribution. The island of Bali is also experiencing widespread AOD conditions. In 2018 the existing light pollution conditions on these two islands will increase in almost all regions and have the effect of the distribution of light pollution to adjacent regions.

IOP Conf. Series: Earth and Environmental Science **884** (2021) 012037 doi:10.1088/1755-1315/884/1/012037

2014

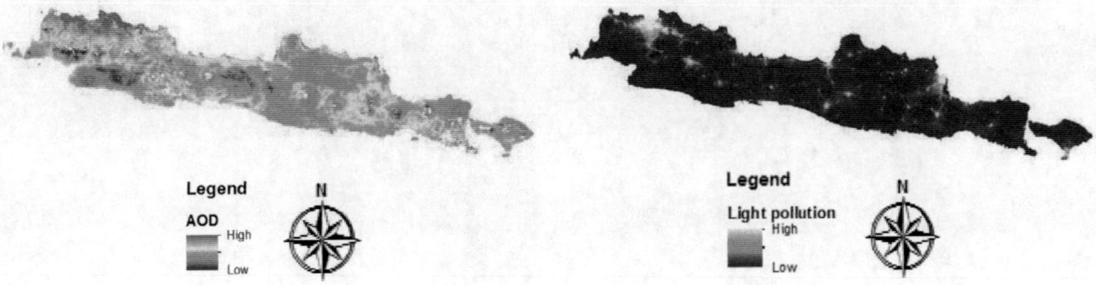

(a) (b)

Figure 3. (a) Distribution of AOD values in Java and Bali in 2014, 2016 and 2018 (b) Distribution of light pollution values in Java and Bali in 2014, 2016 and 2018

Areas with high aerosol values refer to the high density of material in the air that can be caused by various things, but most contributed factor are the presence of air pollution in an area. This also represents the high dynamics of human activity in these areas. While the value of high light pollution represents the level of socio-economic activity in an area at night, the higher the value means the area more developed and has been touched by humans.

The increase of light pollution in 2016 occurred because in 2016 the condition of dissolved air particles in Java and Bali experienced a significant increase compared to 2014. Dissolved particles in the air can increase light scattering, so that light pollution in the region is high. According to Lewin (2000) [11] the external lighting system still leaves more than 50% of the waste light emitted into the sky. If no light is scattered into the atmosphere, the problem is solved. However, the lighting system, however, emits light above the horizontal beam project so that it radiates directly into the night sky. This effect is known as sky glow which is a scattering of orange colors seen over urban areas caused by the emission of light through the atmosphere which is refracted and scattered by water or particle content (aerosols) in the form of dust, plant dust, bacteria, spores, salt content of bursts of sea water, light mineral particles from factory waste [12].

Based on the data obtained the average value of light pollution from the highest to the lowest in the Java-Bali Province sequentially is 2018> 2016> 2014. While on the AOD data, the highest to lowest aerosol thickness is respectively owned by 2018> 2016> 2014. This sequence is influenced by the condition of the Suspended Particulate Matter in each province in Java and Bali as shown in Figure 4.. [13]. The average of Suspended Particulate Matter has increased every year. At the same time the distribution of light pollution on the two islands has also increased.

Annual dynamics shows the increasing AOD and VIIRS values. Spearman; s Rho correlation test results in 2014, 2016, and 2018 can be seen in Figure 5.

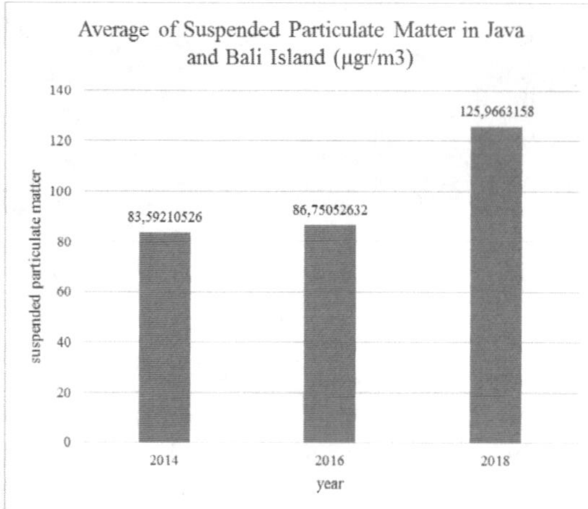

Figure 4. Value Graph of Suspended Particulate Matter in 2014, 2016, and 2018

Correlations

			AOD 2018	VIIRS 2018
Spearman's rho	AOD 2018	Correlation Coefficient	1,000	0,802**
		Sig. (1-tailed)	.	0,000
		N	46	46
	VIIRS 2018	Correlation Coefficient	0,802**	1,000
		Sig. (1-tailed)	0,000	.
		N	46	46

**. Correlation is significant at the 0.01 level (1-tailed).

(a)

Correlations

			AOD 2016	VIIRS 2016
Spearman's rho	AOD 2016	Correlation Coefficient	1,000	0,807**
		Sig. (1-tailed)	.	0,000
		N	46	46
	VIIRS 2016	Correlation Coefficient	0,807**	1,000
		Sig. (1-tailed)	0,000	.
		N	46	46

**. Correlation is significant at the 0.01 level (1-tailed).

(b)

Correlations

			AOD 2014	VIIRS 2014
Spearman's rho	AOD 2014	Correlation Coefficient	1,000	0,819**
		Sig. (1-tailed)	.	0,000
		N	46	46
	VIIRS 2014	Correlation Coefficient	0,819**	1,000
		Sig. (1-tailed)	0,000	.
		N	46	46

**. Correlation is significant at the 0.01 level (1-tailed).

(c)

Figure 5. (a) Spearman's Correlation Test rho for 2018 VIIRS Data and AOD Data (b) Spearman's rho Correlation Test for 2016 VIIS and AOD Data (c) Spearman's rho Correlation Test for 2014 VIIRS Data and AOD Data.

IOP Conf. Series: Earth and Environmental Science **884** (2021) 012037 doi:10.1088/1755-1315/884/1/012037

Based on the correlation test with the Spearman Rank Correlation method, the AOD-VIIRS correlation produce the R2 of 0.819 in 2014; 0.807 in 2016; and 0.802 in 2018. This correlation value looks increasingly declining. Based on Table 1., the R2 value of the VIIRS-AOD test shows a range of 0.8 - 1, indicating a very strong level of correlation at these different recording times. A positive ("+") R2 indicates a change in direction, or in other words, when the value of VIIRS rises it will be accompanied by an AOD value which also rises. This happened because of the highest contributor of artificial light illumination is an urban area with a variety of high socio-economic activities. In that same region, high levels of air pollution can also be found as a consequence of high levels of human activity.

The test results in this study are more complex considering the sample used covers the downtown area to the forest area without buildings, so the results obtained are different from the modeling carried out in China. Given the sample in the previous study was only one point with a different correlation model. Padalah particles in the air affects the level of light scattering in the city skyline [11].

Tabel 1. Spearman's rho Correlation Level

R^2	*Strength of Correlation*
0 – 0,19	*Very Weak*
0,2 – 0,39	*Weak*
0,4 – 0,59	*Moderate*
0,6 – 0,79	*Strong*
0,8 – 1	*Very Strong*

Spearman Rank Correlation Test is conducted in one-tailed relationship or one direction test with a significance level of 0.01, showing the fluctuation in the value of artificial light is influenced by an increase in aerosol thickness in the region. Fluctuations in the value of artificial light and aerosols can be understood by the phenomenon of decreasing the value of energy by absorption and adding value by scattering artificial light on the surface of the earth recorded by satellite sensors. Scatter plots are used to look at the distribution of samples and determine the pattern of relationships formed between artificial night light illuminations and aerosol thickness. Scatter plots in 2014, 2016 and 2018 can be seen in Figure 6.

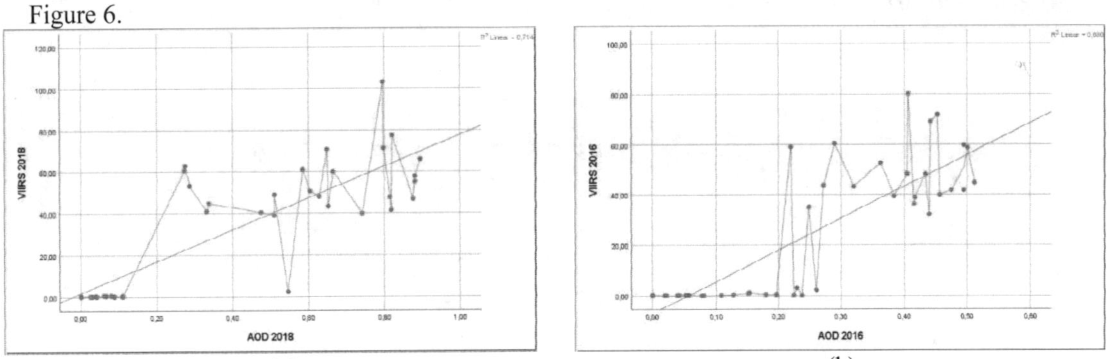

(a) (b)

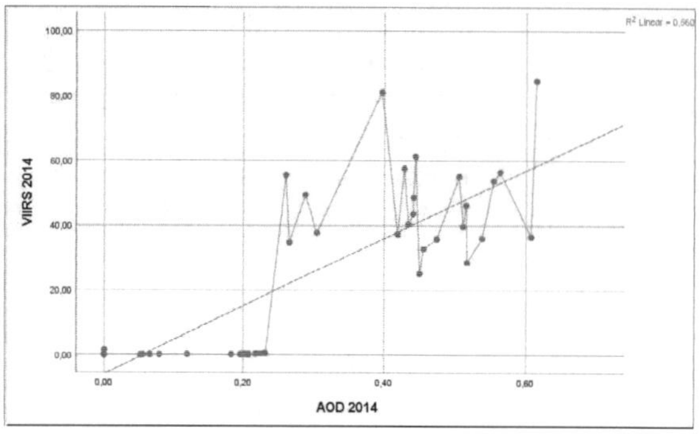

(c)

Figure 6. (a) Scatter plot in 2018 (b) Scatter plot in 2016 (c) Scatter plot in 2014

Flat responses are shown at aerosol values <0.2 in the data of 2014, 2016, and 2018, then begin to show fluctuations up and down in the range of values 0.4 - 0.8. The 2016 data shows that there are bring the different responses compared to other years in the range of aerosol values of 0.2 - 0.3. It's assumed that these values are outliers data. This fluctuating response can occur due to variations in composition that make up aerosols, both in terms of size and the particles. The operational limits in this study only examine the vertical column of the atmosphere that is realized in AOD, while the influence of the aerosol light extinction coefficient (the sum of the aerosol light scattering and absorption coefficients), aerosol single scattering albedo (the ratio of scattering to extinction coefficient), the aerosol upscatter fraction (the fraction of the incident solar radiation that is scattered upward to space), and aerosol size distribution [14] those are the outside of the study.

Schema of the effect of the presence of aerosols on artificial light energy propagation can be seen in Figure 7. Aerosol itself is a mixture of solid liquid particles suspended in the air, and affects other objects in the range of 0.001 - 100 μm [15].

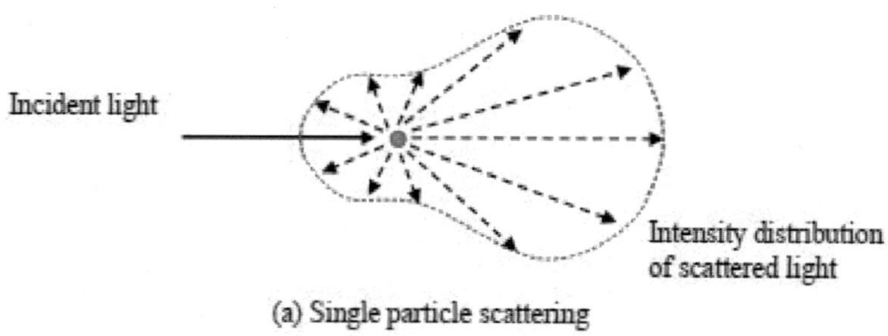

Figure 7. Schematic aerosol scattering of artificial light
(Lee dan Chao, 2016)

In this study, we assume single scattering albedo (w) = 0.9, and scattering asymmetry factor (g) = 0.61, which are both representative of values typically observed for tropospheric aerosols [16]. the optical depths of aerosols and clouds probably exhibit some covariance. This covariance can arise from (1) the aforementioned indirect effect atmospheric aerosols on clouds, (2) dependence of both aerosol scattering efficiencies and cloud formation on relativehumidity [17] and (3) the fact that the formation, accumulation, and removal of aerosols are affected by a host of meteorological conditions, including cloudiness. However, proper simulation of this covariance.

The occurrence of light scattering caused by a single particle in the air causes an increase of the NTL value, it cause the graph soaring up at a certain aerosol value. Aerosol play a role in interfering the recording of data by optical sensors which are characterized by the appearance of bright objects that are similar to the artificial nighttime glow in the VIIRS image. This phenomenon is known as 'ghost light' or in the perspective of remote sensing known as 'pixel blooming') (Carey et al, 2004. Pixel blooming is a phenomenon commonly found in artificial night light data, as happened in DMSP-OLS data [18].

4. Conclusion

The distribution of Aerosol Optical Depth (AOD) in Java and Bali in 2014, 2016, and 2018 experienced an increase and widespread distribution. This can be seen through the AOD distribution map which shows that in 2018 there are no areas with low AOD classification, while in 2014 the central and southern parts of Java and the central part of Bali still have low aerosol conditions. Even in 2018 the distribution of AOD is getting more massive. This one was relevant to the existing light pollution conditions in the regions of Java and Bali which are also higher by the yellow indicator becoming wider from the position of the city centers, following the distribution of aerosols which is increasing. The relationship between these two variables is even very strong at 0.8-1 using the Spearman Rank correlation test. Based on the results of the correlation test the value of AOD and light pollution is directly proportional and strong.

This research is a correlation models that exist in Java and Bali, so as to be different values for different regions. This research is highly dependent on the quality of satellite image data and is carried out on a wide scale. It would be better if the research was carried out using field data with a more detailed scale. This study provides an overview of the snowball effect of increased air pollution which can affect light pollution levels and disrupt the brightness of the night sky. It can interfere with astronomical observations and even affect the activity of noctural animals and result in extinction. Based on this research, the government can find out the very large impact of pollution, and make the right policies to immediately solve the problem.

Acknowledgement

Thank you to the Department of Geographical Information, Gadjah Mada University for supporting this research. The authors also grateful for the monthly global NTL composite data from the VIIRS DNB provided by the Earth Observation Group of Payne Institute, VIIRS data provided by NOAA , dan MODIS data provided by Google Earth Engine.

References

[1] X. Wang, XihanMu and G. Yan, "Quantitative Analysis of Aerosol Influence on Suomi-NPP VIIRS Nighttime Light in China," *IEEE Journal of Selected Topics in Applied Earth Observation and Remote Sensing,* vol. 13, 2020.

[2] R. C. Levy, L. A. Munchak, S. Mattoo, F. Patadia, L. A. Remer4 and R. E. Holz, "Towards a long-term global aerosol optical depth record: applying a consistent aerosol retrieval algorithm to MODIS and VIIRS-observed reflectance," *Atmospheric Measurement Techniques,* no. 8, p. 4083–4110, 2015.

[3] R. Wolfe, "Earth Data," NASA, 6 August 2020. [Online]. Available: https://ladsweb.modaps.eosdis.nasa.gov/missions-and-measurements/viirs/. [Accessed 14 August 2020].

[4] C. D. Elvidge, K. Baugh, M. Zhizhin, F. C. Hsu and T. Ghosh, "VIIRS night-time lights," *International Journal of Remote Sensing,* pp. 1366-5901, 2017.

[5] R. J. C. e. al, "Climate forcing by anthropogenic aerosols," *Science,* vol. 225, p. 423–430, 1992.

[6] M. Z. Jacobson, "Strong radiative heating due to the mixing state of black carbon in atmospheric aerosols," *Nature,* vol. 409, p. 695–697, 2001.

[7] G. d. L. e. al, "Two decades of satellite observations of AOD over mainland China using ATSR-2, AATSR and MODIS/Terra: Data set evaluation and large-scale patterns," *Atmos. Chem. Phys,* vol. 18, p. 1573–1592, 2018.

[8] Z. Pan, F. Mao, W. Gong, Q. Min and W. Wang, "The warming of Tibetan Plateau enhanced by 3D variation of low-level clouds during daytime," *Remote Sensing of Environment,* vol. 198, pp. 363-368, 2017.

[9] Y. Ma, B. Liu, W. Gong, Y. Shi and S. Jin, "Different AOD quality can affect different sky conditions, especially the influence of the seasons in the Wuhan region.," *Atmospheric Pollution Research,* vol. 10, pp. 2063-2071, 2019.

[10] K. Baugh, F.-C. Hsu, C. D. Elvidge and M. Zhizhin, "Nighttime lights compositing using theVIIRS day-night band: Preliminary results," *Proc.Asia-Pacific Adv. Netw. vol. 35,* p. 70–86, 2013.

[11] I. Lewin, "Light Trespass and Light Pollution- Practical Approaches to Dealing with Problems," Minneapolis, USA, 2000.

[12] House of Commoncs Science and Technology Committee, Light Pollution and Astronomy, London: The Stationery Office Limited, 2003.

[13] BPS- Statistics Indonesia, "Envinronment Statistic of Indonesia," 2019, Jakarta, 2019.

[14] J. Haywood and K. Shine, "The effect of anthropogenic sulphate and soot aerosol on the clear sky planetary radiation budget.," *Geophysical Research Letter 22,* p. 603–606, 1995.

[15] S. Hamdl, S. Saloka, L. Sofflat and A. Budtyono, "Aerosol Background Lapisan Stratosfer di Atas Bandung Berdasarkan Penelitian Tahun 1997 - 2000 Menggunakan LiDAR," *Jurnal LAPAN,* pp. 12 -23, 2010.

[16] O. Dubovik, B. Holben, T. F. Eck, A. Smirnov, Y. J. Kaufman, M. D. King, D. Tanre´ and I. Slutsker, "Variability of absorption and optical properties of key aerosol types observed in worldwide locations," *J. Atmos. Sci,* no. 59, p. 590–608, 2002.

[17] J. Li, J. G. D. Wong, J. S. Dobbie and P. Chylek, "Parameterization of the optical properties of sulfate aerosols," *J. Atmos. Sci,* no. 58, p. 193–209, 2001.

[18] L. Liu and Y. Leung., "A Study of Urban Expansion of Prefectural-Level Cities in South China Using Night-Time Light Images," *International Journal of Remote Sensing,* vol. 36, no. 22, p. 5557–5575, 2015.

IOP Conf. Series: Earth and Environmental Science **884** (2021) 012038 doi:10.1088/1755-1315/884/1/012038

Community Preparedness for Tsunami Hazard in Galur District, Kulon Progo Regency, Special Region of Yogyakarta

A B Surindar, Nurhadi

Department of Geography Education, Faculty of Social Sciences, Universitas Negeri Yogyakarta, Yogyakarta, Indonesia

anggarani.berliana2016@student.uny.ac.id; nurhadi_fise@uny.ac.id

Abstract. Kulon Progo Regency is a tsunami-prone area, including on the coast of the Galur District, because it coordinates with the Indian Ocean, the meeting point for two active tectonic plates. This study aims to see: (1) the level of community preparedness in facing the tsunami hazard in Galur District, and (2) the community's efforts to improve preparedness to face the tsunami hazard in Galur District. This research is a quantitative descriptive. The population in this study are all head of family in the three villages, which are Karang Sewu Village, Banaran Village, and Kranggan Village, totaling 4483 head of family. The sample in this study is calculated using the Slovin formula with an error rate of 10% resulting in a total sample of 98 head of family. The sampling technique used is proportional random sampling. Data collection is carried out by observation, interviews, and documentation. The collected data is then presented in a frequency distribution table. The data analysis used in this research is quantitative and qualitative descriptive. The results showed: (1) the level of community preparedness in Galur District to face a tsunami hazard was in the "ready" category (53.1%). Meanwhile, the category for the parameter of preparedness, namely parameter of disaster knowledge, was in the "ready" category (46.9%). The emergency response plan parameter was in the “full ready” category (40.8%). The early warning system parameter was in the “full ready” category (61.2%). The parameter of the ability to mobilize resources was in the "ready enough" category (34.7%). (2) Efforts that have been made by the community in Galur District in improving preparedness to face tsunami hazards, include: holding outreach and disaster simulations; establishment of the Forum of Disaster Risk Mitigation (FPRB); installation of Early Warning System (EWS) tower; installation of evacuation routes and determination of temporary and final gathering points.

1. Introduction

Indonesia is in a region that has high potential and varies greatly in terms of disaster types. This is because Indonesian region is located on three lines where the technical plates meet. The three plates are the Eurasian plate, the Indo-Australian plate, and the Pacific plate which is commonly called the Triple Junction. The movement of the three plates has a different direction of motion, due to the movement of the three plates causing the islands in Indonesia to event frequent earthquake vibrations, especially in areas where the Eurasian and Indo-Australian plates meet along with the islands of Sumatra, Southern Java, and Lesser Sunda Islands. Earthquakes that are caused by plate collisions are called tectonic earthquakes. As the result of plate collision, a subduction zone is formed [11]. Earthquake that occurs in subduction zones is one of the main causes of a local tsunami, a very short time that has the potential to large enough impact. According to BNPB, the cause of the tsunami in Indonesia was 90% due to

Content from this work may be used under the terms of the Creative Commons Attribution 3.0 licence. Any further distribution of this work must maintain attribution to the author(s) and the title of the work, journal citation and DOI.
Published under licence by IOP Publishing Ltd

tectonic earthquakes, 9% due to volcanic activity, and 1% due to landslides that occurred in bodies of water (lakes or sea) or landslides from land that enters the body of water [3].

The Tsunami that has occurred in Indonesia, the Aceh tsunami on December 26, 2004. Based on data from National Geophysical Data Centre in 2005, it is stated that tsunami disaster on 2004 impacted Aceh and North Sumatra caused approximately 300,000 deaths and losses of up to trillions rupiah [9]. Tsunami in Pangandaran is one of the tsunamis that occurred in the southern coastal area of Java Island. The tsunami affected until Kebumen Regency. Based on the 2013-2017 Disaster Risk Assessment Documents for Kebumen Regency, the tsunami disaster that had occurred caused 10 people to die, 24 people were injured, 8 people were missing, 581 were displaced, and 18 houses were severely damaged. Meanwhile, the tsunami that occurred in Banten in 2018 caused 437 fatalities [8]. The lack of community preparedness and knowledge of tsunami disaster was a major factor in the large number of fatalities when the tsunami occurred.

The southern area of Java Island, which is in danger of being struck by a tsunami as a result of a Megathrust earthquake in southern Java, along with the number of people affected. The following areas were affected by the Southern Java tsunami [2]

Tabel 1. Areas Affected by the Tsunami in Southern Java

No.	District/City	Province	Souls Affected
1	Cilacap	Jateng	629891
2	Kebumen	Jateng	220822
3	Purwokerto	Jateng	91943
4	Banyumas	Jateng	689
5	Wonogiri	Jateng	52
6	Kulon Progo	D.I.Y	60607
7	Bantul	D.I.Y	31369
8	Gunung Kidul	D.I.Y	366
9	Jember	Jatim	134207
10	Lumajang	Jatim	27706
11	Banyuwangi	Jatim	17107
12	Pacitan	Jatim	13188
13	Malang	Jatim	2144
14	Tulungagung	Jatim	297
	Total		1230388

Source: BNPB [2]

Based on table 1, one of the areas on the southern coast of Java Island with a possible tsunami is Kulon Progo Regency with 60,607 people affected. Kulon Progo Regency is a district located in the Special Region of Yogyakarta, which has a coastal area directly opposite the Indian Ocean. Kulon Progo Regency has 12 sub-districts, 4 of which are located in the southern part directly adjacent to the Indian Ocean. The Galur District is one of the four sub-districts.

German-Indonesian Cooperation for a Tsunami Early Warning System (GITEWS) is an Indonesia-Germany collaboration to support the development of a tsunami early warning system in the Indian Ocean. GITEWS has classified the tsunami hazard prone areas in Kulon Progo Regency through a tsunami hazard map as follows:

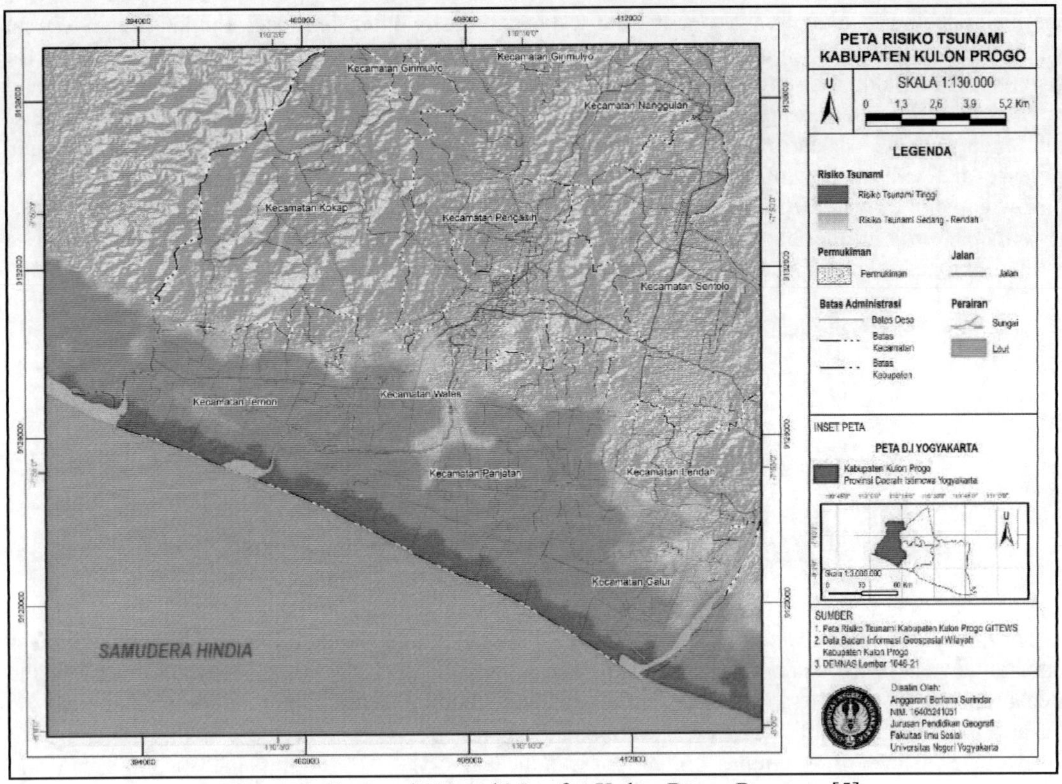

Figure 1. Tsunami Hazard Map for Kulon Progo Regency [5]

Based on figure 1, the tsunami hazard zone is divided into 3, which are: the red and orange zone is the warning zone and the yellow zone is the watch zone. Galur is one of the districts included in the tsunami danger zone, which is marked by red, orange and yellow zone. Tsunami hazard zone with an estimated water level >3 meters (red zone) and <3 meters (orange and yellow zone).

Tabel 2. Potential Tsunami Hazard in Kulon Progo Regency 2016-2020

Sub-district	Hazard Area Index (Ha)	Tsunami-Affected Population (soul)	Potential Loss (billion rupiah)
Temon	1020	3762	286.48
Wates	596	3220	147.33
Panjatan	959	3683	505.54
Galur	1682	13257	264.41

Source: BPBD DIY 2018

Based on table 2, Galur District has the highest potential hazard area index and tsunami-affected population compared to the other three districts which is 1682 hectares hazard area index and 13257 tsunami-affected population. It is also strengthened by the results of observations that have been carried out by researcher that the condition of land use in the coastal of Galur District is currently dominated by settlements and agricultural land in the form of rice fields and fishery ponds. Apart from land use in the form of settlements and agriculture, there is also very minimal vegetation in coastal areas. The lack of vegetation will increase the degree of damage because there are no barriers when the waves hit the coastal areas. In terms of morphogenesis, the relief of Galur District is located in an alluvial plain which extends from west to east of Kulon Progo Regency. This causes the topography in Galur District to be relatively sloping, if a tidal wave occurs, the water will easily enter and reach the land relatively far makes the water overflow area is very wide [12]. In addition, the existence of the Progo River in the eastern part of Galur District will also increase the tsunami hazard. The water carried by the tsunami waves will easily enter through the downdtream and reach further and wider plains, thus it will cause quite high damage. The narrower the river, the higher the speed and condition of the water level would increase, and with the flow of the same mass of water as it enters the shore, it must disperse across the narrow gap of the river [10].

Villages/wards in the Galur District with a possible tsunami hazard can be found in the data of Tsunami Prone Village/Ward, Galur District, Kulon Progo Regency follows:

Tabel 3. Tsunami-prone Villages/Wards, Galur District, Kulon Progo Regency

No.	Village/Sub-district	Hazard Class
1	Karang Sewu	Medium
2	Banaran	Medium
3	Kranggan	Medium
4	Nomporejo	Medium
5	Brosot	Medium

Source: BNPB [4]

This situation means that the population needs a special strategy so that they can survive in places that are vulnerable to tsunami hazards. The community in the Galur District needs to know the preparedness and awareness of the measures or actions required quickly and precisely to address the tsunami danger that is part of the preparedness. Based on this report, the purpose of this research is to (1) evaluate the level of community preparedness in the Galur District to address the tsunami hazard and (2) the community's efforts to improve preparedness to address the tsunami hazard in the Galur District.

2. Research Methods

This research is quantitative descriptive. This study aims to explain the community's preparedness to deal with the tsunami hazard in the Galur District and to identify the community's efforts to enhance preparedness for this danger. This study was carried out for seven months, from December 2019 to June 2020. The location chosen for this study was the Galur District, Kulon Progo Regency, Special Region of Yogyakarta. It focuses on the research area of three villages which are Karang Sewu Village, Banaran Village and Kranggan Village.

The variables in this study are: (1) community preparedness in the case of a tsunami hazard, as seen from the indicators: (a) knowledge, (b) emergency response plan, (c) early warning systems, and (d) resource mobilization, and (2)) efforts to improve preparedness, as seen from the following indicators: (a) planning and organization, (b) resources, (c) coordination, (d) readiness, (e) training and public awareness [7].

The population in this study are all heads of family in the three villages, namely Karang Sewu Village, Banaran Village, and Kranggan Village, totaling 4483 heads of family. The sample in this study is calculated using the Slovin formula with an error rate of 10% resulting in a total sample of 98 heads of family. The sampling technique used is proportional random sampling.

Data collection techniques in this study are: (1) observation, carried out by analyzing the conditions around the research location to gain an overview of the research location, (2) interviews, conducted with respondents (the community) using a questionnaire to obtain information about community preparedness seen from knowledge parameters, emergency response plans, early warning systems, and resource mobilization. Interviews with main informants, i.e. representative of Kulon Progo Regional Disaster Management Agency (BPBD), Forum of Disaster Risk Mitigation (FPRB), representatives of Galur District government using a questionnaire to gather information on the community's efforts to enhance preparedness in the face of tsunami hazards, and (3) documentation, carried out by taking data, such as maps, population data, other data, and photos of activities in the field that can support the research activities.

The data analysis used in this research is quantitative and qualitative descriptive. A description of the preparedness of the community in Galur District in the face of tsunami hazard is obtained from the scoring of the preparedness parameters. Scoring is performed to turn the information that has been gathered into data in the form of numbers. "yes" answer to each research instrument will be given 2 score, the "no" answer will be given a score 1 score, and the "do not know" answer will be given 0 score.The result of the sum of all preparedness parameter scores will obtain the minimum score and the maximum score. Furthermore, the score was categorized based on the Mean Ideal (Mi) and Standard Deviation (SDi) values obtained by calculations using the following formula [1] :

Mean Ideal (Mi) = 1/2 (maximum score + minimum score)
Standard Deviation (SDi) = 1/6 (maximum score - minimum score)

The results of the Mi and SDi calculations are then inserted into the trend distribution formula to determine the level of preparedness category with the criteria "full ready", "ready", "ready enough", "less ready", and "not ready". The collected data is then presented in a frequency distribution table and analyzed descriptively.

Tabel 4. Trend Distribution Formula

Criteria	Formula
Full Ready	X ≥ Mi + 1,8 (SDi)
Ready	Mi + 0.6 (SDi) ≤ X < Mi + 1,8 (SDi)
Ready Enough	Mi – 0,6 (SDi) ≤ X < Mi + 0,6 (SDi)
Less Ready	Mi – 1,8 (SDi) ≤ X < - 0,6 (SDi)
Not Ready	X < Mi – 1,8 (SDi)

Source: Azwar [1]

The efforts made by the community in Galur District to improve preparedness for tsunami hazards were obtained from the results of qualitative data analysis based on information from main informants.

3. Results & Discussion

3.1. Study Area

Galur District is one of the Districts located in Kulon Progo Regency, Special Region of Yogyakarta. Astronomically Galur District is located 7°55'30 " South Latitude - 7°58'30" South Latitude and 110°10'0 "East Longitude – 110°15'0" East Longitude. The administrative boundaries of Galur District are as follows north boundary is Lendah District, south boundary is Indian Ocean, west boundary is Panjatan District, and east boundary is Bantul Regency. The area of Galur Sub-district is 3,291.24 ha. The land area of each village, namely: Karangsewu (9,2624 km2), Banaran (9,0725 km2), Kranggan (2,3875 km2), Nomporejo (1,9091 km2), Brosot (3,2254 km2), Pandowan (1,4036 km2) and Tirtorahayu (5,6519 km2). Galur district has a tropical climate with two seasons, namely the dry season and the rainy season. Rainfall is 120mm / year with 6 months of rainy months. Daily average temperature of 30° C.

3.2. Community Preparedness to Face Tsunami Hazards in the Galur District

The community preparedness study to resolve the tsunami hazard in Galur District used a framework developed by LIPI in collaboration with UNESCO / ISDR in 2006. The community preparedness study is based on 4 (four) parameters: (1) disasters knowledge, (2) emergency response plans, (3) early warning systems, and (4) resource mobilization. Data regarding community preparedness was obtained through questionnaires distributed to the community, particularly the heads of family spread across three

IOP Conf. Series: Earth and Environmental Science **884** (2021) 012038 doi:10.1088/1755-1315/884/1/012038

villages, which are Karangsewu Village, Banaran Village, and Kranggan Village. The number of samples used as respondents was 98 heads of family spread across three villages. The following is a table of the results of the level of community preparedness in facing the tsunami hazard in Galur District, Kulon Progo Regency, Special Region of Yogyakarta.

Tabel 5. Community Preparedness Level to Face Tsunami Hazard in Galur District, Kulon Progo Regency, Special Region of Yogyakarta

No.	Indicator	Category (%)					
		Full Ready	Ready	Ready Enough	Less Ready	Not Ready	Total
1	Knowledge	24.5	**46.9**	22.4	3.1	3.1	100
2	Emergency response plan	**40.8**	26.5	19.4	9.2	4.1	100
3	Disaster warning system	**61.2**	23.5	11.2	2.0	2.0	100
4	Resources mobilization	8.2	24.5	**34.7**	26.5	6.1	100
5	All indicators	15.3	**53.1**	22.4	8.2	1.0	100

Source: Primary Data, 2020

Based on table 5, it shows the percentage of all community preparedness parameters in the face of a tsunami hazard in Galur District. The results of the data analysis showed that the community's preparedness to face the tsunami hazard in Galur District was in the "ready" category with a percentage of 53.1%. The greatest influence in determining the level of community preparedness in facing tsunami hazards comes from the parameters of the emergency response plan and early warning system which are categorized as “full ready”. Moreover, it also supported by the parameters of knowledge on disasters which have the "ready" category. Meanwhile, the parameter of resource mobilization is in the "ready enough" category which will affect the level of community preparedness to face the tsunami hazard in Galur District.

3.2.1. Community Preparedness for Knowledge Parameters of Tsunami Disaster

Public knowledge of tsunami hazard is the community's understanding of natural disasters. The results of the data analysis show that the level of community preparedness for the knowledge parameter of the tsunami hazard in Galur District can be categorized as "ready" with a percentage of 46.9%. This data can be interpreted that the people in Galur District have a good understanding of the tsunami hazard. Knowledge is the main factor for community preparedness [7]. The knowledge possessed by a community will influence how actions need to be taken in the preparation of disasters. The importance of knowledge about disasters must be possessed by every individual who is in a disaster-prone area such as in Galur District, Kulon Progo Regency, Special Region of Yogyakarta

Galur District's community stated that the definition of a natural disaster is a natural event that disturbs human life (92.9%), for example, tsunamis, earthquakes, floods, landslides, volcanic eruptions, and storms. While the cause of the tsunami was an underwater earthquake (93.9%) the community in Galur District agreed that not all earthquakes could cause a tsunami (55.1%). One of the signs of a tsunami disaster is the sudden receding of seawater (88.8%). The characteristics of tsunami resistant buildings/houses are buildings that have empty spaces for waterways (49.0%) based on this data, there are still many people who do not understand the knowledge of tsunami resistant buildings because socialization is more focused on knowledge of disasters, signs, and self-rescue in the event of a disaster, while the reconstruction effort was less emphasized during outreach. The majority of people in Galur District received information about the tsunami disaster from relatives, friends, and neighbours (80.6%).

In addition, the community received information about the tsunami disaster from socialization, seminars and meetings held by related parties (77.6%).

3.2.2. Preparedness Level for Emergency Response Plan Parameters
The emergency response plan for tsunami hazards in Galur District is a plan created by the community to anticipate hazards and understand what actions to take when a tsunami occurs. The results of the data analysis show that the level of community preparedness for the parameters of the emergency response plan against the tsunami hazard in Galur District can be categorized as "full ready" with a percentage of 40.8%. This means that the community in Galur District has the ability to have a good understanding of how to plan and act to deal with a tsunami disaster. The indicators contained in the parameters of the emergency response plan include the initial plan of what to do before a disaster occurs, the actions are taken when a disaster occurs, and a safe place for family in the event of a tsunami disaster.

The people of Galur District already have a plan to be alert to the possibility of a tsunami hazard, which is to prepare actions that must be taken by household members in the event of a tsunami (84.7%). Agree on evacuation place for family (86.7%) the community chooses a disaster tent/post which is provided as a place to save their families in the event of a tsunami disaster (94.9%). Prepare a first aid box (medicine box) (69.4%). Actions that must be taken in the event of the next tsunami disaster are to prepare important and valuable documents (91.8%) for important documents such as diplomas, birth certificates, family cards and other important documents that have been prepared in one bag, the aim is to make them easy to carry. At any time a tsunami disaster occurs so do not have to look for documents one by one, which of course will take a long time and endanger lives.

Another action is preparing alternative communication tools (Handy Talkie/ Radio/Phone) (83.7%). People choose cellphones as an alternative means of communication because the majority of family members communicate long distances using cellphones. Following evacuation drills/simulations (98.0%), in this case, the community is aware of the importance of evacuation drills/simulations against tsunami hazards. Evacuation drills/simulations are held to provide experience in the field to the community in the event of a tsunami disaster so that it will facilitate the evacuation process during an emergency and can minimize fatalities and property.

3.2.3. Community Preparedness for Disaster Warning system Parameters
The early warning system against tsunami hazards in Galur District is a warning sign and distributing information about a disaster. The results of data analysis show that the level of community preparedness for the parameters of the disaster warning system against the tsunami hazard in Galur District can be categorized as "full ready" with a percentage of 61.2%. This means that the people of Galur District already have a good awareness and understand what actions to take when they hear the warning sounds/signs of a tsunami disaster. Early warning is the main factor linking the stages of disaster preparedness and emergency response [6]. If the early warning system is sounded at the right time when the signs of a disaster are felt, the negative impact, both fatalilties and property losses, can be minimized. Indicators contained in the parameters of a disaster warning system include the presence of a tsunami warning sign/method, and actions that must be taken when hearing a tsunami disaster warning.

The people of Galur District already know about the signs of tsunami hazard warning, by using the traditional warning system in the form of kentongan (64.3%) and a technology-based tsunami warning system such as the Early Warning System (EWS) siren tower (85.7%). An early warning system by using kentongan is sounded with a long bead tone so that the community is alert and immediately evacuates to the agreed gathering point. People will stay away from the beach and run to high places/buildings (96.9%) in the event of a tsunami and also rush to rescue/evacuation place (98.0%). The chosen evacuation place or final gathering point is a place that is safe from the reach of the tsunami waves, wich is Banasara Elementary School and Cubung Jatirejo field, Lendah Sub-district.

3.2.4. Community Preparedness for Resource Mobilizing Parameters

The ability to mobilize resources is the ability of the community to reduce risks when a disaster occurs. The results of data analysis show that the level of community preparedness for resource mobilizing parameters against the tsunami hazard in Galur District can be categorized as "ready enough" with a percentage of 34.7%.

As many as 59.2% of family heads have attended training, seminars, or meetings related to tsunami hazard preparedness. However, apart from the family heads no one else has participated in tsunami hazard preparedness training. As a result, this is not maximized due to emergencies such as warning signs of a tsunami that can come at any time if the head of the family is not at home. It will greatly slow down the evacuation process because only the head of the family understands and know what kind of attitudes and actions should be taken when a tsunami warning occurs. Many factors influence this, such as the low level of community awareness to attend training. As many as (40.8%) of respondents stated that none of the household members had ever attended training, seminars, or meetings related to tsunami disaster preparedness. According to the respondent's statement, this can happen because people who take part in the training or seminar are representatives of several villages that make none can participate in these activities. As many as (86.7%) of respondents had relatives/friends who were ready to help in the event of a tsunami hazard

The ability to mobilize resources also includes the financial readiness of the community to face emergencies when a tsunami occurs. A total of (52.0%) respondents have assets/investments in the form of savings that can be used for family awareness of the possibility of a tsunami disaster. Meanwhile, 66.3% of respondents do not have life insurance/wealth/property insurance and also do not own another houses/place that are relatively safe from disasters. Public awareness for supporting preparedness from a financial perspective is still quite low because the economic factor of the community is still classified as middle to lower class, this is in line with the work of the majority of Galur District people, which is farmers. Not all people in Galur District have prepared measures to deal with the tsunami hazard. As much as (79.6%) of respondents did not make buildings/houses from light materials (eg wood, bamboo, zinc). Not building temporary huts to evacuate in safe places/hilly areas/high places and far from the coast (71.4%). Not preparing sufficient food and clothing supplies (75.5%). Not preparing a reserve stock (money, capital, land) to be stored in a safe place to deal with possible job losses due to the tsunami (65.3%).

3.3. Efforts to Improve Community Preparadness

Preparedness efforts are needed by the community to face the dangers in the area. In the disaster management cycle, preparedness efforts are included in the risk reduction phase before a disaster occurs. Efforts from the government and the community to increase preparedness in facing the tsunami hazard in Galur District is something that must be done because Galur District is an area that is prone to tsunami hazards. The following are some of the efforts made by the community in Galur District to improve preparedness in the face of a tsunami hazard:

a. There is a policy that regulates tsunami hazard management activities, the existence of Regional Regulation Number 8 of 2015 concerning Disaster Management and also the Regent Regulation Number 67 of 2018 concerning the use of village funds for disaster management. The villages which are directly adjacent to the Indian Ocean and the Progo River is very prone to tsunami hazards so that a disaster-prone villages have emerged, so the Disaster Resilient Villages (DESTANA) were formed especially for tsunamis. There are 5 villages that were formed as Disaster Resilient Villages especially for the tsunami, namely Banaran, Karangsewu, Kranggan, Brosot, and Pandowan villages. With the formation of Destana, it is hoped that the community can increase awareness of the dangers of a tsunami and can increase preparedness independently
b. Organizations are formed which are Forum of Disaster Risk Mitigation (FPRB) in each Destana. This organization is official and formed by BPBD Kulon Progo. The Forum of Disaster Risk Mitigation (FPRB) is a forum for elements of society or organizations in the field of disaster that are

formed to study, provide input, advice, and/or provide recommendations for consideration in the context of disaster management

c. Conduct socialization and simulations regarding disaster risks and the threat of tsunami hazards. Socialization and simulation activities regarding the tsunami hazard in each village of Galur District have been carried out, with 12 meetings, 9 meetings for socialization and counselling, while the 10th-12th meeting was conducted with simulations. Tsunami hazard simulation activities in Galur District have been held at the village level and usually take place in the village field or village meeting hall and also on the coastal of Trisik Beach. Participants are randomly chosen and the quota to carry out simulation activities is approximately 250 people per village. Earthquake simulation activities accompanied by a tsunami have been carried out 2 times in 2019 and 2018 with the theme Tsunami Disaster Response (The ability of the community to save themselves).
d. The availability of facilities and infrastructure to improve preparedness efforts in the form of an early warning system in the coastal of Galur District by establishing an Early Warning System (EWS) tower. The EWS tower is a siren that will sound as a warning sign of a tsunami hazard.
e. Availability of evacuation routes and final gathering points. The installation of signs for the evacuation route has been installed by the Regional Disaster Management Agency (BPBD) of Kulon Progo Regency since 2014. The final gathering point for the Galur District is at Banasara Elementary School, Lendah Sub-district and also at Cubung Field, Jatirejo, Lendah Sub-district, Kulon Progo Regency. Meanwhile, the temporary gathering point is at the Nomporejo Village Hall, Tirtorahayu Village Hall, Pandowan Village Hall. If the condition is too productive, they will immediately be directed to the final gathering point, which are Bonosoro Elementary School and Cubung Field.

4. Conclusion

a. The level of community preparedness in Galur Subdistrict to face tsunami hazard is in the "Ready" category (53.1%). Meanwhile, the level of preparedness for the knowledge parameter for disaster is in the "ready" category (46.9%). The emergency response plan parameter is in the "full ready" category (40.8%). The early warning system parameter is in the "full ready" category (61.2%). The parameter of the ability to mobilize resources is in the category of "ready enough" (34.7%).
b. Efforts to enhance tsunami hazard preparedness carried out by the government and the community include outreach, counselling and simulation by both the Kulon Progo Regency BPBD and the FPRB Village, the creation of the Forum of Disaster Risk Mitigation (FPRB) organization, the establishment of the Early Warning System (EWS) tower, located on Trisik Coastal of Galur District, as well as the installation of evacuation routes and determining the temporary and final gathering points.

5. Recommendation

a. For the Community

1) The public should understand of the importance of being aware of the dangers of a tsunami which can occur at any time by diligently participating in training and simulation activities held by the government and related organizations.
2) For people who have participated in training and simulations, it is better if they share the materials and skills obtained from these activities. So that other people can also understand how to evacuate themselves and their families in the case of a tsunami disaster.
3) The community should have prepared measures to deal with the possibility of a tsunami hazard.

b. For the Organization

There is a need for more structured and clear guidance for cadres who are prepared to assist in socializing and simulating the tsunami hazard.

Acknowledgments
Gratitude was expressed to the lecturers of geography education department for their many helpful suggestions which improved the original manuscript.

References
[1] Azwar S 2003 *Metode Penelitian* (Yogyakarta: Pustaka Belajar)
[2] BNPB 2012 *Masterplan Pengurangan Risiko Bencana Tsunami* (Jakarta: Badan Nasional Penanggulangan Bencana)
[3] BNPB 2012 *Menuju Indonesia Tangguh Menghadapi Tsunami* (Jakarta: Badan Nasional Penanggulangan Bencana)
[4] BNPB 2019 *Katalog Desa/Kelurahan Rawan Tsunami Kelas Bahaya Tinggi dan Sedang* (Jakarta: BNPB)
[5] German-Indonesia Tsunami Early Warning System 2012 *Peta Bahaya Tsunami DIY* https://www.gitews.org/tsunamikit/id/E1/sumber_lainnya/peta_bahaya/diy/Tsunami%20Hazard%20Map% 0Kulon%20Progo.jpg
[6] Hidayati D dkk 2006 *Kajian Kesiapsiagaan Masyarakat Dalam Mengantisipasi Bencana Gempa Bumi dan Tsunami* (Jakarta: LIPI Press)
[7] LIPI-UNESCO/ISDR 2006 *Kajian Kesiapsiagaan Masyarakat dalam Mengantisipasi Bencana Gempa Bumi dan Tsunami* (Jakarta: LIPIUNESCO/ISDR)
[8] Muhari A dkk 2019 The December 2018 Anak Krakatau Volcano Tsunami as Inferred from Post-Tsunai Field Surveys and Spectral Analysis. *Pure and Applied Geographysics* Vol 176
[9] Kodoatie dan Roestam S 2006 *Pengelolaan Bencana Terpadu* (Jakarta: Yarsif Watampone)
[10] Setyadi dkk 2015 Potensi Kerawanan Gelombang Tsunami di Pesisir Bantul Daerah Istimewa Yogyakarta *Jurnal Oseanografi* Vol 4 No 4 691-699
[11] Sukandarrumidi 2010 *Bencana Alam & Bencana Anthropagene* (Yogyakarta: Penerbit Kanisius)
[12] Widyawati A dkk 2013 Kajian Kerentanan Bencana Tsunami di Pesisir Kabupaten Kulon Progo Provinsi D.I.Yogyakarta *Jurnal of Marine Research* Vol 2 No 2 103-110

The roles of human geography in disaster studies

Assriyani[1] and Hastuti[2]

[1] Master Program of Geography Education, Graduate School of Yogyakarta State University, 55281, Indonesia
[2] Departement of Geography Education, Faculty of Social Science, Yogyakarta State University, 55281, Indonesia

Email: assriyani.2018@student.uny.ac.id

Abstract. A disaster is an event or series of events which threaten and disrupt the human lives. The studies of disaster are constantly considered related to humans because humans act as the causative factor, victim and at the same time the executant of the effort in disaster handling. Pioneered by Gilbert White and pursued by other experts, human geography later shows the roles in examining issues in disaster studies especially in disaster mitigation. Based on several research and notions, many experts in human geography agree that there is nothing natural in natural disaster. Although the disasters are often affected by geosphere physical phenomenon but various human actions and activities on the face of the earth cause certain population to be more prone to natural disaster. If the risks toward disaster need to be reduced, bigger attention needs to be given to minimize population vulnerability, increase people capacity to overcome disaster and strengthen people capability to adapt to disaster area in a long term. Currently, UN as the responsible party in reducing risk of international disaster has been doing community resilience effort adopting Hyogo Framework for Action (HFA) 2005–2015: building the nation and community resilience towards disaster, followed by Sendai Framework for Disaster Risk Reduction (SFDR) 2015-2030: reducing disaster risk and loss.

1. Introduction

Humans and nature live in harmony from time to time. Along with the development of human civilization, the nature keeps leaving their traces in the community, one of them has the quality to destroy and we know it as a disaster. Various disasters, such as flood, earthquake, landslide, tsunami, drought and many others keep coming and threatening the life of humans [1]. These disasters caused various and significant loss and damage in many sectors in human lives [2]. Even though they are often linked by geological, meteorological and hydrological process which underlie but an event can be called a disaster is not based on the cause but the consequences arise because of the interaction between the danger and community vulnerability [3], [4]. The Cambridge dictionary defines a disaster as an event that results in great harm, damage, or death, or serious difficulty which means that an event cannot be classified as a disaster unless human is involved in it [5].

In the beginning, a disaster was considered a punishment sent by God or other supernatural power for the mistake or sin committed by humans on earth [6]. However since the Lisbon earthquake in 1755, the paradigm of supernatural force of disaster began to shift into new paradigm that a disaster is not an interpretation but a phenomenon which can be controlled and predicted [7] [8]. Dranseika (2017) [6] in his writing emphasizes that a disaster is a series of events which can be worsen or lessen by human actions. A disaster can be massive or small depends on how humans prepare themselves through the

Content from this work may be used under the terms of the Creative Commons Attribution 3.0 licence. Any further distribution of this work must maintain attribution to the author(s) and the title of the work, journal citation and DOI.
Published under licence by IOP Publishing Ltd

choices they make. Kulatunga (2010) [2] states that several natural disasters have sharply increased all over the world and the studies on them are currently a global attention (Figure 1).

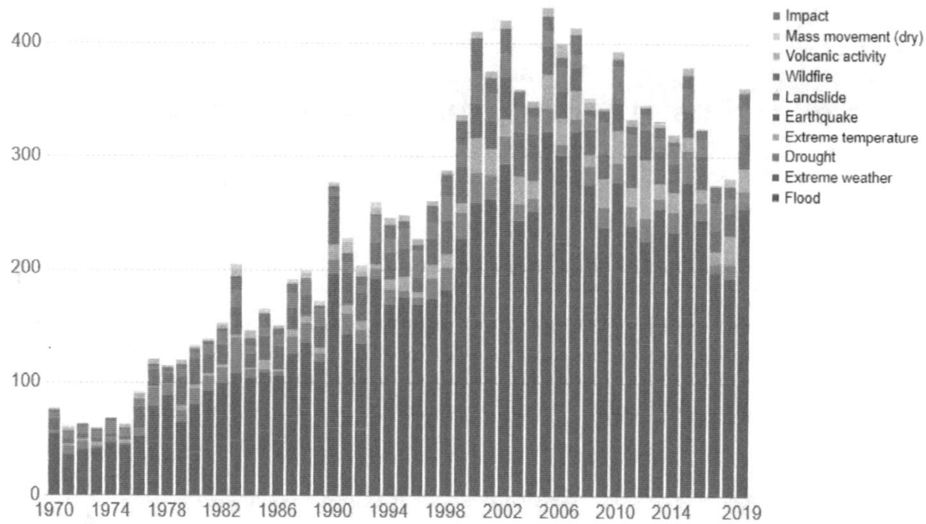

Figure 1. Data of the World Disasters Based on the Types (1970 -2020)
Source: EMDAT (2020): OFDA/CRED International Disaster Database, Universite catholique de Louvain – Brussels – Belgium [9]

Humans are the subject and the object of disaster and at the same time the main target of the countermeasures [6], [10]. Many human ideas, behavior and actions have affected how and what disasters that happened in human civilization from time to time. Therefore, human geography, as a branch of geography which focuses on the existence of humans on earth, is considered need to contribute its role in disaster studies [11], [12].

Human geography progressed from determinism and possibilism towards the appearance of probabilism as a compromise from both [13] [14]. This understanding emphasizes in reciprocal relationship between humans and nature in which the environment can affect the human lives and human have rooms to do some actions that can affect the environment [12], [14], [15]. Probabilism paradigm shows that based on their character, humans are more likely to make a certain decision compared to the environment. Hence, even both influence one another but the human roles as objects are more dynamic and can be seen as more dominant [16]. The emergence of probabilism in human geography has significantly bought out various new interests and studies of nature and humans such as the interest in cultural ecology and risk and hazard study [17]. Entering the twentieth century the human geography study becomes increasingly diverse, interesting and challenging. Research of the relationship of nature and humans in this discipline keep developing through new perspectives [17], [18]. Related to stated development, Montz and Tobin (2011) [19] agree that geographical research of nature and its danger which at first only focused on the physical process is now moving towards the recognition of interaction between physical environment and humans.

Through studies about various related literature, this article will discuss further about the linkages of humans and natural disaster, the standing of human geography in disaster studies, the experts who appeared as pioneers and contribute their ideas and the roles of human geography in disaster management especially in the effort of reducing disaster risk locally or globally.

2. Result and Discussion

2.1. What is the definition of "Human Geography"?

The term human geography started becoming known in the late 19th century when experts eventually agreed to break geography into two branches which are physical geography and human geography [12] [20]. While physical geography studies the physical geography environment for the natural elements especially geomorphology, climatology and hierology, human geography studies the relationships between the natural environment and sociocultural environment created by humans along with all spatial distribution and the impact caused by the social, economic culture and politic phenomenon all over the world [21] [22]. We cannot rely on explanations which only based on one of the disciplines. Both of them are related because socio-biophysics landscapes are diverse and not equal so in studying a phenomenon, it needs to combine explanations and perspectives of human geography and physical geography [4] [23]. That idea is strengthen by Goudie (2017) [24] who argues that currently, the interests towards both as disciplines related to one another keep growing. Some of them include the studies of disaster, hazard, vulnerability, resilience and the global change, reasons, mitigations and adaptations.

Boyle (2015) [12] defines human geography as a discipline which the effort is to give pictures and explanations about the difference of human activities on earth. Meanwhile, Knox and Marston (2004) [25] define human geography as a science that studies the spatial organization in various human activities along with their relationships with the environment. Human geography can also be defined as a study about when, where, why and how humans evolve, develop strategies to adapt to the environment, process the natural resources and transmigrate to other parts of the world [20] [26]. Based on the said definitions, it can be concluded that human geography is a study about the dynamic relationship of humans and nature in building and maintaining the civilization.

Human geography is a vast field which includes various sub-disciplines of geography such as agriculture, culture, economy, politic, social, tourism, population and many more [5] [18] [20] [27]. At the beginning of its development, human geography only focused on analyzing the population trend, social and cultural change, interpreting geo-political conflict and trying to explain economic activity all around the world in the perspective of geography [26]. At this time, the people and human geography experts start to develop analysis related to the relationship of humans and physical environment for example how humans can utilize the nature, how humans manage their activities locally or globally, how humans can make a place and how that place in certain time starts to turn and shape human lives [26] [27] [28]. The results of the analysis from human geography will always change over time. This is because humans and nature are dynamic and easily change by certain factors. Hastuti (2017) [29] states the behavior dynamic and human activities on earth are affected by the factors of resources availability, technology mastery and social culture behavior change.

Human geography can be used to study the phenomenon in the past, see the spatial changes related to human activities on earth, help overcome problems which are happening and prepare ourselves in facing the possibilities in the future. Robinson (2003) [30] argues that human geography can contribute through alternative approach to the future such as the social learning process to increase the analytical capability and participatory involvement. The statement is supported by O'Brien (2010) [31] who states that with human involvement as agent of change, human geography is able to give contributions for a better world in the future.

2.2. Gilbert White: The Pioneer of Disaster Studies in Human Geography

An American Geography expert emerged as the pioneer and realized that the natural disaster and extreme climate needed to be the main focus of experts in human geography and not only physical geography. White developed disaster and hazard study in human geography perspective and open the knowledge of the significance role of social, economy, culture and politic process in lowering the vulnerability and increasing the humans' capacity towards possible danger [12]. According to Hinshaw (2006) [32], White poured his approach and knowledge to his study about the hazard of disaster profoundly so he is then considered the founding father of hazard study.

White is famous as 'the founding father of floodplain management' because of his cognition and contribution to flood management [33]. Since he was given the responsibility to increase the floodplain management at the Missouri River Valley in the 1930s, White began to develop various ideas about predicting, preventing, mitigating, responding and adapting to disaster [12]. White's ideas about flood significantly contribute to human civilization as massive floods in the last few decades happened all around the world and caused high death rate and also severe damage in varied sectors.

The average of flood in many continents continues to rise every decade (Figure 2). Until now, there are many people who live in floodplain area. The data from UNEP/GRID-Arendal (2007) [34] states that 17% of the city land in the US is flood zone, 50% of Japan's citizen live in floodplain area and even Bangladesh has a high flood risk level that flood often happens over half of the country. Although from the beginning White is known for his ideas of flood management but most of his career had also been spent to study many types of hazards on the face of the earth; moreover, his comprehensive approach in lowering flood risk can also be applied vastly to other dangerous situation [12] [35]. White states that there are numerous adaptations that can be done by each individual or community when they are faced with danger in nature; this construct is then known as the 'dominant' approach [36] [37]. In 1976, White established Natural Hazards Center (NHC) in Colorado University as a well-known public policy authority in the world related to natural disaster [12].

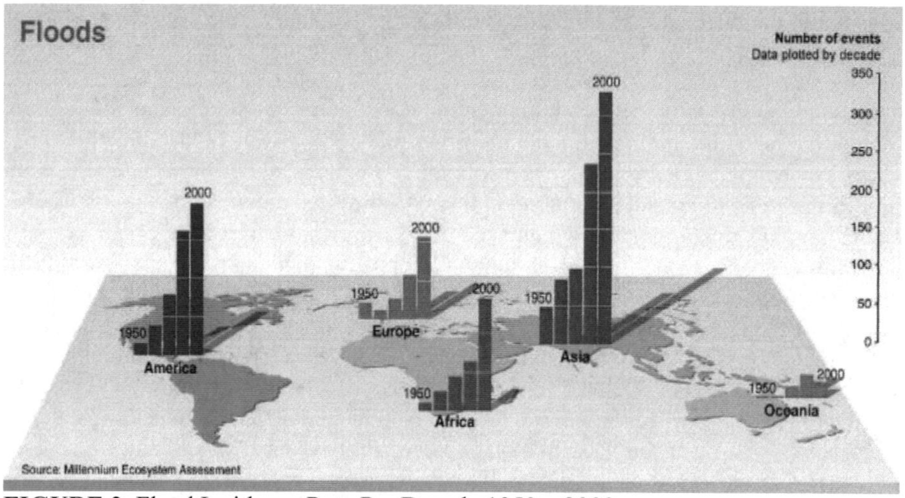

FIGURE 2. Flood Incidence Rate Per Decade 1950 – 2000
Source: Rekacewicz and Bournay, UNEP/GRID-Arendal, 2007
(https://www.grida.no/resources/6062) [34]

Gilbert White's footprints in disaster studies were followed by Robert W. Kates. Several Kates's essays and research in disaster field take up about the form of human adaptation to earthquake in Alaska 1964 (1970), the concept of the natural hazards in the perspective of human ecology (1971), the landscape planning to face hazards (1975), the risk study of technology (1983), the assessment method of hazard and its risk (1985), and the adaptation of global climate change (2012) [38], [39], [40], [41] [42], [43]. Based on his writings and concepts, it can be seen that Kates mainly focuses on the study of, hazard, adaptation and new concepts about natural hazards through other perspective that is more specific. Furthermore, many other names came up such as Kenneth Hewitt, Keith Smith, David N. Petley, J. C. Gaillard and Jonathan Rigg.

Human geography expert who is still actively contributing in disaster studies to this day is Jonathan Rigg. Rigg is a professor and also the chair of Human Geography in University of Bristol in which one of the focuses in his work is disaster and resilience. Rigg has studied issues related to human vulnerability and exposure towards various disasters such as tsunami, earthquake and climate change since 1980 covering area of Laos, Nepal, Thailand, Vietnam and Sri Lanka [44].

2.3. The Place of Human Geography in Disaster Studies

Natural hazards as earthquake, tornado, volcano eruption and tsunami appear as the word 'disaster' is mentioned. In fact, disaster needs to be defined based on the consequences received by humans and not based on the cause. Earthquake for example, it is merely a natural phenomenon. Even a powerful earthquake cannot be called as a disaster unless it causes casualty or damage of property. So an earthquake happened in an unpopulated area, like a huge quake every month, is solely a natural phenomenon and is not considered as disaster [45]. Apart from causing injuries, death and physical damage, big disasters also have long term social and political impact such as the tsunami happened in Indian Ocean on December 24, 2004, the monstrous earthquake happened in Kashmir on October 8, 2005 and hurricane Katrina which swept away New Orleans on August 25, 2005 [5]. Therefore, it can be said that when disaster happens, the position of human is to be the victim whose life is affected materially, physically and psychologically [46].

Further research shows that often times human are not only stand as the affected victim but also as one of the factors that triggers various hazards. Disaster caused by human actions is known as social disaster which includes war, terrorism, repressive regime and colonialism [47], [48]. Humans are also responsible for non-natural disaster including forest fires, famine, the lack of clean water, technology failures and the spread of disease outbreaks [31], [41], [49]. O'Brien (2010) [31] argues that resource scarcity, global climate change, shift in land utilization, reduction of water quality and quantity and many other issues which trigger disaster are the impact of human actions and responds towards the environment. The high rate of human mobility is directly proportional with the rising frequency and intensity of disaster on earth. Even the massive city structure expansion and the high number of urbanization have increased the exposure of infectious disease like COVID-19 which is now being a global pandemic [50], [51], [52].

Figure 3. Exposure of Population to Natural Hazards
Source: United Nations University – Institute for Environment and Human Security, 2016 [53]

Aroud 75% of the world's population live in the area which at least was once exposed to a disaster [12]. Most of the people exposed by a disaster live in a developing country with a low human development rate (Figure 3). The development of human resource needs to be level up in order to lower the disaster risk but rushed development can only be resulted in a new disaster risk because most of the policies and massive scale development programs cannot be explicitly configured for reducing risk [16]. In this stage, humans, who formerly solely stand as victim and trigger factor, then appear and play a part in disaster management. One of the indicators for the quality of human development is the level of knowledge and in this context focuses on the knowledge related to disaster specifically vulnerability and resilience [54]. Human geography then emerged as two main constructs. The first one is the science which examines the relationship of humans and environment that can affect the happening of a disaster. The second is a discipline which equips humans in the effort of lowering the disaster risk.

2.4. There is Nothing Natural about the Natural Disasters

Are natural disasters purely caused by nature? In the previous section, it has been discussed about how humans play a role as the causing factor of social and non-natural disaster. In this section, referring to the argument of Kenneth Hewitt (1983) [36], humans are also pictured as the cause of natural disaster. Flood for example cannot be considered that the cause of it is heavy rains. Heavy rains will never be a disaster if the region has a well water absorption area and a good conditioned river. Flood happens because of river siltation caused by the actions of dumping trash and industrial solid waste into the river and also the lack of water absorption area caused by the fact that the soil is covered with concrete and asphalt. Along with that Hewitt's idea, Henderson [55] states his notion that disaster risk, especially the ones in the city, is now more likely come from factors such as pre-disaster building constructions, environment topography and economic human resource in certain areas caused by the rapid development.

Another misunderstanding about natural disasters that often happens is that the population most at risk is the ones who unfortunate enough to be born in an exposed area. However, natural hazards are only a function from uneven distribution on earth so in reality, it is humans who put the population at a certain risk level by using the ability to manage vulnerability [12]. Meanwhile, there are various types of disasters happened to human race and most of the damage which follows the hazards might not be the direct effect of the hazards but it comes from the varied vulnerability which is not supported by capacity.

Kenneth Hewitt, a Canadian geographer, seriously considers the roles of the community in increasing and worsening the deathly hazards from natural disasters. In his book, Interpretations of Calamity: From the Viewpoint of Human Ecology (1983) [36], Hewitt shows that natural hazards are always threatening but they are only becoming disaster when humans chase after development without realizing that they are raising their vulnerability. In his other book, Regions of Risk: A Geographical Introduction to Hazards (1997) [56] Hewitt explains the claim by showing that a disaster is caused by three factors: hazard, vulnerability and capacity.

$$\mathbf{R} = \frac{\mathbf{H \times V}}{\mathbf{C}}$$

R = *Risk*
H = *Hazard*
V = *Vulnerability*
C = Capacity

There is nothing natural about natural disasters, even if the natural phenomenon such as earthquake, tsunami and volcano eruption affects humans' habitation, the impact might not be considered as massive calamity depends on the level of human vulnerability and capacity in the impacted area [12]. A natural disaster is a product built socially hence the trend depends on how big the interaction between the physical and social environment [3], [58]. Instead of understanding it as purely natural, disasters are more like a complex mosaic created by numerous factors of politics, technology, nature and moral [4].

IOP Conf. Series: Earth and Environmental Science **884** (2021) 012039 doi:10.1088/1755-1315/884/1/012039

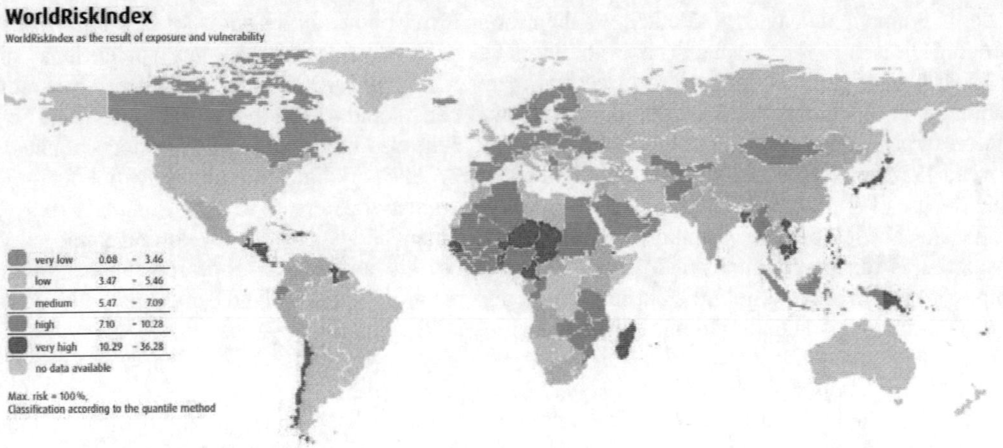

Figure 4. Risk Index of Global Disaster
Source: United Nations University – Institute for Environment and Human Security, 2016 [53]

Most countries around the world who have their disaster risk index in 'medium' to 'very high' are poor or developing countries with high vulnerability or low capacity (Figure 4). The vulnerability against hazards is affected by several social and demographical factors such as race, class, gender, age and education [11], [57]. Hazards create higher vulnerability for certain communities for example women, children, poor, minorities, refugees, people with low education experience and those who have no family [12], [59], [60], [61], [62]. Boyle (2015) [12] identifies six ways in which social, political and cultural process enhance population vulnerability exposed by natural hazards which are poverty, social exclusion, poor governance, war and violence, rapid urbanization and environment degradation. Experts agree that the most vulnerable zones against disasters are those inhabited by communities of poor families, immigrants and marginal people [5], [63]. In the disaster prevention and response, it is important to consider those vulnerabilities and how we can minimize them [64].

When the vulnerability cannot be minimized, building capacity or resilience locally, nationally or even globally is the effort that can be done to lower the disaster risk index. The capacity of a community to respond to hazards is influenced by the physical force and the economic and social relationships which underlie to act critically, respond to hazards positively and make a quick recovery from the impact [57], [65]. Emergency management plan in reducing disaster risk have to be able to build community capacity for resilience and in this case the resilience which involved several serious potential conflict such as stability and dynamics [65], [66].

The sustainably physical and human development plays a significant role in increasing the vulnerability through disaster adaptation and mitigation [67]. When two areas are exposed by hazards, the area with higher physical and human development will have a lower risk and impact compared to the other area with lower development index (Figure 4). For example, the earthquake happened in San Fernando, California in 1971. This earthquake has the magnitude of 6.4 Richter scale but the area around San Fernando Valley (with the population of more than seven million people) was only suffers small damage and 58 deaths. Two years later, an earthquake of 6.2 Richter scale hit Managua, Nikaragua and ravaged the city into ruins and killed 6.000 people [45]. Another case, for example, is the drought study done by Kates (2012) to compare the danger of drought for agriculture in Australia and Tanzania. The drought happened in Australia, from around 13 million populations, only one million were at risk with almost no record of death. Meanwhile in Tanzania, the number of deaths caused by the drought hit approximately 40 million from the total 12 million at risk. These two examples show that San Fernando and Australia have better human development compared to Managua and Tanzania in terms of disaster mitigation hence with the same disaster, Managua and Tanzania were calculated to have bigger disaster but not with San Fernando and Australia. This human development index is one of the capacities in decreasing the disaster risk.

IOP Conf. Series: Earth and Environmental Science **884** (2021) 012039 doi:10.1088/1755-1315/884/1/012039

Keith Smith and David N. Petley, in their book Environmental Hazards: Assessing Risk and Reducing Disaster [68], state that disaster studies have develop four main paradigms with each paradigm see hazards and approach mitigation in certain manners. (1) Before 1950, the engineering paradigm dominated. In this paradigm, the nature condition was considered as the root cause of disasters so the solution through engineering and technology development was recommended to protect humans. (2) Between 1950 and 1970, the behavior paradigm emerged. This paradigm focused on spatial planning to direct the use of land in the future to be a safe location against disasters. (3) The period of 1970 – 1990 was the era of development paradigm when underdevelopment and poverty in developing countries were considered as the main source of vulnerability. (4) In 1990 and so on, appeared the paradigm of complexity which aims to help the community in disaster prone areas to better manage the interaction between the people and nature to minimize their vulnerability against hazards.

2.5. Human Geography, GIS and Big Data in Disaster Management

Disasters have a long term impact for humans and spaces. Therefore, the emergence of human geography in disaster management often helps with the statistics, mathematics and space mapping technology such as GIS (Geographic Information System) who play significant role in spatial analysis [69] [70]. As one of the characters in geographical studies, spatial approach appears in human geography to show how each aspect of disasters can be mapped to understand the process on site. Because it is under the branch of human geography, the mapping done is focused on distributing the social process rather than the physical process [71].

The use of GIS in analyzing disaster studies cannot be separated from the needs of data which posse comprehensiveness, variation, resolution, indexation, accuracy and flexibility. Those needs are no longer a difficult matter to fulfill because right now we are in the era of big data where we can obtain information and use most of the data to create better and accurate understanding about the complexity of physical and social process [72], [73]. Most of the big data has spatial and temporal attributes so it gives a big potency for human geography including analyzing disaster [72]. Some experts agree that big data era can open numerous new possibilities for disaster management such as proposing and testing frameworks, planning physical recoveries and visualizing, analyzing and predicting disasters [74], [75] [76].

Hualou (2011) [16] states the roles of human geography that helps GIS and big data in disaster management:

1. Assessing Human-created Vulnerability: To a great extent, disasters are resulted from vulnerability as a consequence of various human activities concerning industrial revolution, natural resources exploitation and concentration of population. He methods of human geography and GIS are widely applied to vulnerability assessment aiming at disaster prevention and disaster management.
2. Assessing Human-created Vulnerability: The development of rapid industrialization and urbanization in developing countries often bring technological disaster due to the lack of adequate or effective countermeasures for risk reduction. In this field, geographical research mainly focuses on how to deal with the distribution, removing poisonous waste and their proximity to other land uses such as residential areas and farmland.
3. Developing Disaster Monitoring and Simulation Systems: Remote sensing, GIS and related technologies have been utilized to provide real-time value-added data and information to authorities in areas of natural resources and environmental management and the its role in detecting, modeling and monitoring natural hazards. Using GIS and remote sensing, natural disaster monitoring, situations assessment system and regional disasters reduction capacity assessment system can be established to prevent and manage diversified disasters.
4. Post-Disaster Recovery and Reconstruction: It is necessary for local post-disaster recovery plans to eliciting positive recovery outcomes that the plans include a sound participatory process and establish a nexus between local needs and policy objectives. The characteristics of human geography equipped by GIS with the quantitative statistic method have special advantages in this kind of plan such as the restoration and reconstruction post-earthquake.

2.6. The Roles of United Nations in Disaster Risk Reduction

Boyle [12] says his responsibility to promote and coordinate the international effort to minimize disaster risk falls under the shoulder of United Nations (UN). Since the first founded in 1945, the UN adopted at least four different approaches for disaster risk reduction:

1. 1st Phase (1946–1970): In this phase, the focus is to coordinate the stakeholders and partners after the natural hazards happened to give assistance and to help the countries to get back to their normal situation.
2. 2nd Phase (1971-1999): In the second phase, a series of formal policy frameworks and institutional arrangements were pioneered to help increase the countries' readiness, ability to countermeasure and capacity to lower the damage.
3. 3rd Phase (2000 – now): In the third phase, the focus is on disaster risk reduction. In 2005, the UN adopted Hyogo Framework for Action (HFA) 2005–2015: Building the Resilience of Nations and Communities to Disasters (UNISDR, 2005).
4. 4th Phase (the future): According to some experts, we are now standing in the verge of a new approach of disaster risk reduction which emphasize on integration of disaster risk management to social, political and economic policies.

The central concept promoted in Hyogo Framework for Action 2005–2015 is the concept of "resilience". Although it is possible to reduce the amount ad frequency of the disasters, it is possible that the strategies of disaster management in the future will focus on lowering the population vulnerability to natural hazards. These strategies will forty the vulnerable community from hazards, strengthen the capacity of numerous community to cope better to hazards and make it possible for all community members in establishing and conducting long term strategies especially adaptation enhancement [12]. Previously, Hyogo Framework for Action (HFA) 2005–2015: Building the Resilience of Nations and Communities to Disasters [77] gave the most authoritative frameworks in which the countries were rising up to confront the challenge of public policies faced with natural hazards. Approving with the International Conference on Disaster Reduction held in Kobe, Japan in 2005, Hyogo Framework for Action (HFA) had signed more than 168 countries and had a function as 10 years strategy which the damage caused by hazards can be minimized [12].

Sendai Framework for Disaster Risk Reduction (2015-2030) emerged as the next step of Hyogo Framework for Action (2005-2015). SFDR is a voluntary non-binding agreement in the span of 15 years which acknowledge that a country holds a significant role in overcoming disaster risk. That role can be distributed to local governments, private parties and others. These frameworks aim to achieve outcome in the substantial reduction of disaster risk and losses in lives, livelihoods and health and in the economic, physical, social, cultural and environmental assets of persons, businesses, communities and countries [78]. The framework for disaster risk reduction adopted on 14 – 18 of March, 2015 in Sendai, Miyagi, Japan established the Seven Global Targets which were expected to be reached in the span of 15 years [78]. The seven targets are:

1. Substantially reduce global disaster mortality by 2030.
2. Substantially reduce the number of affected people globally by 2030.
3. Reduce direct disaster economic loss in relation to global gross domestic product (GDP) by 2030.
4. Substantially reduce disaster damage to critical infrastructure and disruption of basic services.
5. Substantially increase the number of countries with national and local disaster risk reduction strategies by 2020.
6. Substantially enhance international cooperation to developing countries through adequate and sustainable support to complement their national actions for implementation of the present Framework by 2030.
7. Substantially increase the availability of and access to multi-hazard early warning systems and disaster risk information and assessments to people by 2030.

In the mission to implement disaster risk reduction globally, Sendai Framework for Disaster Risk Reduction (2015-2030) referring to the four priorities for action which are:

1. Understanding disaster risk.
2. Strengthening disaster risk governance to manage disaster risk.
3. Investing in disaster risk reduction for resilience.

IOP Conf. Series: Earth and Environmental Science **884** (2021) 012039 doi:10.1088/1755-1315/884/1/012039

4. Enhancing disaster preparedness for effective response and to "Build Back Better" in recovery, rehabilitation and reconstruction.

3. Conclusion

Human geography, as a big sub-discipline in geography, is currently developing its analysis related to physical environment specifically about human position in disaster studies. The studies of disasters are always considered related to humans because humans hold the position as the causing factor, victim and at the same time the one who put on the effort of disaster management. Gilbert White, an American geographer emerged as the disaster studies pioneer in human geography. White developed hazard and disaster studies in the perspective of human geography and also open the idea of the important role of social, economic, cultural and political process in minimizing the consequences from disasters. Other experts who later follow White are Robert W. Kates, Kenneth Hewitt, Keith Smith, David N. Petley, J. C. Gaillard and Jonathan Rigg.

Human geography experts agree that there is nothing natural about natural disasters because there is always human involvement in it. Although the global variation in the risk caused by natural disasters are partly the result of exogenous and endogenous force but numerous human behavior and activities on earth caused certain population to be more vulnerable to disasters. Population vulnerability on a specific area is affected by the factors of poverty, social exclusion, poor governance, war and violence, rapid urbanization and environment degradation. If the risk related to natural hazards need to be lowered, bigger attention needs to be given to minimize the population vulnerability, increase their capacity to cope with disasters and strengthen their ability to adapt in a long term period.

Human geography equipped with GIS and big data has significant roles related to disaster management; they are assessing human-created vulnerability, assessing human-created vulnerability, developing disaster monitoring and simulation systems, post-disaster recovery and reconstruction. In the real action in a global scale, the United Nations have promoted and coordinated the efforts in disaster risk reduction in which one of them is adopting Hyogo Framework for Action (2005-2015) which focuses on building the resilience of nations and communities to disasters and followed by Sendai Framework for Disaster Risk Reduction (2015-2030) which emphasize on reduction of disaster risk and losses in lives, livelihoods and health and in the economic, physical, social, cultural and environmental assets of persons, businesses, communities and countries.

Acknowledgments

Gratitude was expressed to the lecturers of master program geography education department for their many helpful suggestions which improved the original manuscript.

References

[1] Hastuti and Widyastuti M 2018 Happiness behind the threat of disaster: women in the south merapi slope *Karsa: Journal of Social and Islamic Culture* **26** 129-49
[2] Kulatunga U 2010 Impact of culture towards disaster risk reduction *International Journal of Strategic Property Management* **14** 304-13
[3] Britton N R 1986 Developing an understanding of disaster *The Australian and New Zealand Journal of Sociology* **22** 254–71
[4] Massazza A, Brewin C R and Joffe H 2019 The nature of ''natural disasters'': survivors' explanations of earthquake damage *International Journal of Disaster Risk Science* **10** 293–305
[5] Daniels P, Bradshaw M, Shaw D, and Sidaway J 2008 *An Introduction to Human Geography: Issues for* the *21st Century (4th edition)* (New Jersey: Pearson,).
[6] Dranseika V 2016 Moral responsibility for natural disasters *Human Affairs* **26** 73-79
[7] Rousseau J J 1756 *Correspondence complète de Jean Jacques Rousseau* ed J A Leigh **4** pp 37-50 (Translated by R Spang)
[8] Neiman S 2004 *Evil in Modern Thought: an Alternative History of Philosophy* (Princeton: Princeton University Press)

[9] EMDAT 2020 OFDA/CRED International Disaster Database, Universite catholique de Louvain – Brussels – Belgium (https://ourworldindata.org/exports/natural-disasters-by-type_v9_850x600.svg)
[10] Zein A M 2010 Community Based Approach to Flood Hazard and Vulnerability Assessment in Flood Prone Area: A Case Study in Kelurahan Sewu, Surakarta City, Indonesia (thesis: ITC, The Netherland)
[11] Fraser J C 2006 The relevance of human geography for studying urban disasters *Space and Culture* **9** 14 –19
[12] Boyle M 2015 *Human Geography : A Concise Introduction* (UK: Wiley Blackwell)
[13] Johnston R J 1983 *Philosophy and Human Geography : An Introduction To Comtemporary Approach* (London: Edward Arnold)
[14] Flowerdew R 2009 Probabilism *International Encyclopedia of Human Geography* ed R Kitchin and N *Thrift* (Amsterdam - Elsevier Science) pp 448–450
[15] Fekadu K 2014 The paradox in environmental determinism and possibilism: a literature review *Journal of Geography and Regional Planning* **7** 132–39
[16] Hualou L 2011 Disaster prevention and management: a geographical perspective *Disaster Advances* **4** 3-5
[17] Turner II B L 2015 Nature–society in geography *International Encyclopedia of the Social & Behavioral Sciences* **2** 334-339
[18] Kobayashi A 2017 Human Geography *International Encyclopedia of Geography: People, the Earth, Environment and Technology* ed D Richardson, N Castree, M F Goodchild, A Kobayashi, W Liu and R A Marston (New Jersey: John Wiley & Sons) pp 1–18
[19] Montz B E and Tobin G A 2011 Natural hazard: an evolving tradition in applied geography *Applied Geography* **31** 1–4
[20] Pitzl G R 2004 *Encyclopedia of human geography* (London: Greenwood Publishing)
[21] NCERT (National Council of Educational Research and Training) 2006 *Funfamentals of Human Geography*. (New Delhi: Government of India)
[22] Zhipeng T, Zhang J, Liu W and Wu H 2012 Differences between physical and human process simulation in geography *Journal of Geographical Sciences* **22** 497-508
[23] Lave R, Wilson M W, Barron E S, Biermann C, Carey M A, Duvall C S and Dyke C V 2014 Intervention: critical physical geography *The Canadian Geographer* **58** 1—10
[24] Goudie A S 2016 The integration of human and physical geography revisited *The Canadian Geographer* **61** 19–27
[25] Knox P and Marston S A 2004 *Human Geography: Places and Regions in Global Context - 3rd Edition* (New Jerseyy: Prentice Hall)
[26] Gibson C 2009 *Human Geography* (Australia: University of Wollongong)
[27] Fouberg E H and Murphy A B 2020 *Human Geography: People Place and Culture* (New York: John Wiley and Sons Inc)
[28] Tadaki M J, Salmond R, Heron L and Brierley G 2012 Nature, culture, and the work of physical geography *Transactions of the Institute of British Geographers* **37** 547—62
[29] Hastuti 2017 Memahami pariwisata melalui pendekatan geografi manusia *Geomedia* **15** pp 20
[30] Robinson J 2003 Future subjunctive: backcasting as social learning *Futures* **35** 839–56
[31] O'Brien K 2010 Responding to environmental change: a new age for human geography? *Progress in Human Geography* **35** 542–49
[32] Hinshaw R E 2006 *Living with Nature's Extremes: The Life of Gilbert Fowler White* (Colorado: Johnson Books)
[33] Macdonald N, Chester D, Sangster H, Todd B and Hooke J 2011 The significance of Gilbert F. White's 1945 paper 'human adjustment to floods' in the development of risk and hazard management *Progress in Physical Geography* **36** 125–33
[34] Rekacewicz and Bournay 2007 UNEP GRID-Arendal (https://www.grida.no/resources/6062)
[35] Wescoat J L 2006 Gilbert Fowler White (1911–2006) wisdom in environmental geography *Geographical Review* **96** 700–10

[36] Hewitt K 1983 *Interpretations of Calamity: From the Viewpoint of Human Ecology* (New York: Routledge)
[37] Burton I, Kates R W and White G F 1993 *The Environment as Hazard - second edition* (London: Guilford Press)
[38] Kates R W 1970 Human Adjustment to Earthquake Hazard *The Great Alaska Earthquake of 1964: Human Ecology, National Academy of Sciences-National Research Council Publication No 1607* pp. 7-31.
[39] Kates R W 1971 Natural hazard in human ecological perspective: hypotheses and models *Economic Geography* **47** 438-51
[40] Kates R W 1975 Planning for hazards: in everyday landscapes *Landscape Architecture Magazine* **62** pp. 165-168 (Published by: American Society of Landscape Architects)
[41] Kates R W and Kasperson J X 1983 Comparative risk analysis of technological hazards (a review) *Proc. of the Nat. Academy of Sciences* Vol 80 pp 7027–38
[42] Kates R W, Hohenemser C and Kasperson J X 1985 *Perilous Progress: Managing the Hazards of Technology* (Colorado: Westview Press) pp 251-64
[43] Kates R W, Travis W R and Wilbanks T J 2012 Transformational adaptation when incremental adaptations to climate change are insufficient *Proc. of the Nat. Academy of Sciences* Vol 109 pp 7156–61
[44] Kinvall C and Rydstrom H 2019 *Climate Hazards, Disasters, and Gender Ramifications* (New York: Rouletde Studies in Hazard, Disaster Risk and Climate Changes)
[45] Perez E and Thompson P 1986 *Natural Hazards: Causes and Effects* (US: Disaster Management Center at the University of Wisconsin-Madison)
[46] Wesnawa I G A and Christiawan P I 2014 *Geografi Bencana* (Yogyakarta: Graha Ilmu)
[47] Teich J M 2002 The informatics response in disaster, terrorism, and war *Journal of the American Medical Informatics Association* **9** 97–104.
[48] Harding S 2015 Man-made disaster and development *International Social Work* **50** 295–306
[49] Rose-Redwood R, Kitchin R, Apostolopoulou E, Rickards L, Blackman T, Crampton J and Buckley M 2020. Geographies of the COVID-19 pandemic *Dialogues in Human Geography* 1-10
[50] Song X, Zhang Q, Sekimoto Y and Shibasaki R 2014 Prediction of human emergency behavior and their mobility following large-scale disaster *Proc. of the 20th ACM SIGKDD International Conference on Knowledge Discovery and Data Mining - KDD '14*
[51] Acuto M 2020 COVID-19: lessons for an urban(izing) world *One Earth* **2** 317 -19
[52] Connolly C, Ali S H and Keil R 2020 On the relationships between COVID-19 and extended urbanization *Dialogues in Human Geography* pp 1-4
[53] Comes M et al. 2016 *World Risk Report 2016* (Published by: Bündnis Entwicklung Hilft and United Nations University – Institute for Environment and Human Security)
[54] Donovan A 2014 Geopower *Progress in Human Geography* **41** 44–67
[55] Henderson A 2006 The human geography of catastrophe: family bonds, community ties, and disaster relief after the 1906 San Francisco earthquake and fire *University of California* **88** pp 37-70
[56] Hewitt K 1997 *Regions of Risk: A Geographical Introduction to Hazards* (New York: Routledge)
[57] Hutton D and Haque C E 2004 Human vulnerability, dislocation and resettlement: adaptation processes of river-bank erosion-induced displacees in Bangladesh *Disasters* **28** 41–62
[58] Weichselgartner J 2001 Disaster mitigation: the concept of vulnerability revisited *Disaster Prevention and Management* **10** 85–95
[59] Cupples J 2007 Gender and hurricane mitch: reconstructing subjectivities after disaster *Disasters* **31** 155–75
[60] Hastuti 2016 Peran perempuan dalam menghadapi bencana di Indonesia *Geomedia* **14** 13-21
[61] Moreno-Walton L and Koenig K 2016 Disaster resilience: addressing gender disparities *World Medical & Health Policy* **8** 46–57
[62] Gaillard J C, Gorman-Murray A and Fordham M 2017 Sexual and gender minorities in disaster *Gender, Place & Culture* **24** 18–26

[63] Rigg J, Law L, Tan-Mullins M and Grundy-Warr C 2005 The Indian Ocean tsunami: socio-economic impacts in Thailand *The Geographical Journal* **171** 374–79
[64] Hanson R 2007 *Catastrophe, Social Collapse, and Human Extinction* (Virginia: Department of Economics, George Mason University)
[65] Maguire B and Hagan P 2007 Disasters and communities: understanding social resilience *The Australian Journal of Emergency Management* **22** 16-20
[66] Alexander D E 2013 Resilience and disaster risk reduction: an etymological journey *Natural Hazards and Earth System Sciences* **13** 2707–16
[67] Kelman I, Gaillard J C and Mercer J 2015 Climate change's role in disaster risk reduction's future: beyond vulnerability and resilience *International Journal of Disaster Risk Science* **6** 21–27
[68] Smith K and Petley D N 1991 *Environmental Hazards: Assessing Risk and Reducing Disaster* (London: Routledge)
[69] Kitchin R and Tate N J 2000 *Conducting Research in Human Geography : Theory, Methodology and Practice* (Singapore: Addison Wesley Longman)
[70] Komac B 2009 Social memory and geographical memory of natural Disasters *Acta Geographica Slovenica* **49** 199–226
[71] Curti A and Jacqueline W M 2010 *GIS, Human Geography, and Disasters* (California: Cognella Academic Publishing)
[72] Kitchin R 2013 Big data and human geography *Dialogues in Human Geography* **3** 262–67
[73] Graham M and Shelton T 2013 Geography and the future of big data, big data and the future of geography *Dialogues in Human Geography* **3** 255–61
[74] Papadopoulos T, Gunasekaran A, Dubey R, Altay N, Childe S J and Fosso-Wamba S 2017 The role of Big Data in explaining disaster resilience in supply chains for sustainability *Journal of Cleaner Production* **142** 1108–18
[75] Akter S and Wamba S F 2017 Big data and disaster management: a systematic review and agenda for future research *Annals of Operations Research*
[76] Yu M, Yang C and Li Y 2018 Big data in natural disaster management: a review *Geosciences* **8** pp. 165
[77] UNISDR 2005 *Hyogo Framework for Action 2005-2015: Building the Resilience of Nations and Communities to Disasters* (Japan: World Conference on Disaster Reduction)
[78] UNISDR 2015 *Sendai Framework for Disaster Risk Reduction 2015-2030* (Japan: World Conference on Disaster Reduction)

Description of the dynamics of the resilience of people victims of the 2018 North Lombok earthquake

Dewi Febriani[1] dan Siti Rohmah Nurhayati[2]
[1]Department of Psychology, Yogyakarta State University, Indonesia.
E-mail: dewifebriani.2018@student.uny.ac.id
[2] Department of Psychology, Yogyakarta State University, Indonesia.
E-mail: siti_rohmah@uny.ac.id

Abstract. Before the earthquake struck lombok island on July 26, 2018, with a magnitude of 6.4 SR and on August 5, 2018 with a magnitude of 7.0 SR, it turned out that the history of earthquakes in Lombok had occurred 7 times from 1856 to 2013 with different strengths. So it can be concluded that lombok island is classified as a disaster prone area. Earthquakes include geological disasters that leave major impacts on lives, economies, infrastructure, the environment as well as psychological impacts. This research aims to describe the dynamics of the resilience of the victims of the 2018 North Lombok earthquake. This research aims to describe the dynamics of the resilience of the victims of the 2018 North Lombok earthquake. This article was compiled using two sections using literature studies and empirical research with a qualitative approach through in-depth interview motode. The results showed that the dynamics of resilience formation in earthquake victims had a profound effect on the success of synergy holistically both individual strength factors in mitigation implementation, family support factors and community factors both in a broad and narrow environment. The aspects of resilience in lombok society are the dominant aspects of optimism, empathy, self-efficacy and reaching out.

1. Introduction

The territory of the Republic of Indonesia is a disaster-prone area. In accordance with Law No. 24/2007 at least Indonesia has 12 types of high-risk disaster threats. Be it in geological disaster groups (earthquakes, tsunamis, volcanoes, landslide movements), hydrometeorological disasters (floods, flash floods, extreme weather droughts, extreme waves, forest fires, and land), and anthropogenic disasters (epidemic disease outbreaks and failed industrial accident technologies). The cause of natural disasters in Indonesia is due to Indonesia's geographical position flanked by two large oceans. Indonesia's geological position at the meeting of the world's three main plates (Indo-Australia, Eurasia, and pacific). As well as being included in the Pacific Ring of Fire volcano. The condition generally has many active faults and often causes earthquakes. This contributes to nearly 90% of earthquakes and almost all of them are major earthquakes in the world. From the period 2005 to 2015 more than 78% (11,648) catastrophic events were hydrometeorological disasters and only about 22% (3,810) were geological disasters. Although the number of disasters caused by geological factors is not significant but has an impact on the death toll and the greater economy (BNPB, 2014) [1].

The recent Lombok earthquake on July 29, 2018 with a magnitude of 6.4 SR was followed by a more devastating earthquake on August 5, 2018 with a strength of 7.0 SR as one example of a geological disaster that left a major impact on the death toll, economy, infrastructure, environment and psychological impact. According to the National Earthquake Study Center Team (2018) that two

Content from this work may be used under the terms of the Creative Commons Attribution 3.0 licence. Any further distribution of this work must maintain attribution to the author(s) and the title of the work, journal citation and DOI.
Published under licence by IOP Publishing Ltd

major earthquakes occurred in Lombok in just one week, sourced from the same area of target. Occurs because one field of defense with a slope of 30 degrees moves 2-3 Meters. The location of the fault or fault is about 1 KM off the coast of North Lombok. BMKG analysis showed that this earthquake was a shallow earthquake caused by a sesar riding the arc behind Flores while the epicenter was located at coordinates 8.28 LS and 115.99 BT. Precisely located in the sea at a distance of 20 KM Northwest of North Lombok, at a depth of 10 KM. According to bmkg records tectonic lombok is indeed an active seismic area. Lombok is potentially shaken by earthquakes from the South and North. To the South there is an Indo-Australian plate subduction zone, while from the North there is a geological structure rising up Flores.

The geological conditions and geographical location of North Lombok Regency have different levels of vulnerability. This is evidenced in research conducted in three sub-districts in Pemenang, Tanjung and Ganges sub-districts of North Lombok Regency. The research used microtremor data based on the HVSR (Horizontal to vertical Spectrum Ratio) method to obtain GSS (Group Support System) and earthquake intensity values. The results of his research obtained the distribution of GSS scores that stated the highest level of inewanan in Pemenang Sub-district, while in Tanjung sub-district has the lowest level. As for the intensity level of the earthquake is on a scale iii up to VI MMI (Modified Mercalli Intensity) (Ningsih et al., 2019) [3].

Before the earthquake hit lombok island on July 26, 2018. In fact, the history of earthquakes records that the island has often experienced destructive earthquakes. Namely on July 25, 1856 and April 10, 1978 on a scale of 6.7 SR. On 21 May 1979 a scale of 5.7 SR, on 20 October 1979 a scale of 6.0 SR, on 30 May 1979 and on 1 January 2000 a scale of 6.1 SR, on 22 June 2013 a scale of 5.4 SR. So it can be concluded that lombok island is prone to disaster. The earthquake that occurred in August 2018 was destructive, came suddenly and in a short time was able to damage various facilities and infrastructure, infrastructure, housing, and take the lives of both the injured and the dead. In the event of lombok earthquake 2018 recorded total damage estimated losses of 7.7 trillion including 73,843 damaged homes, education facilities 671 units, bridges damaged 6 units, health facilities 52 units, worship facilities 128 units, offices 25 units and 515 people died, injured 7,145 people, and refugees 431,416 people died [4].

The potential threat of natural disasters such as earthquakes and tsunamis on the island of Lombok still exists due to the volcano Rinjani. As we know that lombok island is recorded as an earthquake prone area from 1856 to 2018 coupled with aftershocks that occur many times. For that, everyone must understand how to save themselves in the event of an earthquake or other natural disaster. Of course, the common hope of such a thing does not happen again even if it happens but the public knows how to deal with disaster emergencies. Risks are studied in order to mitigate. Mitigation that is able to lead to development. The effect of this disaster is more on mitigation failures. So it is considered the need to provide information about the knowledge of handling that can be used as a recommendation for disaster mitigation in Lombok. The success of the mitigation program will have an impact on improving the resilience of earthquake victims and being able to reduce the risk of disasters. The people will be resilient to disasters not separated from the role of factors that affect resilience itself. The factors that affect resilience are factors in themselves, community synergy (policy determinants of both central, local and rural governments), outside factors such as strengthening mitigation education and the role of knowledge observers about disasters.

2. Method

This article was compiled using two sections using literature and empirical research. In literature studies researchers collect library data, read and record various information from various sources both from textbooks, research report results, articles, internet, journals and index documents from related agencies. While in empirical research, researchers used a qualitative approach. Researchers excavated information from the field using in-depth interview methods to several informants who were considered to have reliable information about him and the people of North Lombok who were victims of the earthquake.

3. Results and Discussions

3.1 Research contribution to the Lombok earthquake

After the Lombok earthquake which occurred on 5 August 2018 attracted the attention of disaster observers at the regional, national and international levels. No less important is the role of academics, social communities, health agencies (association of doctors, nurses and midwives) to the role of individuals who take part in contributing according to their respective expertise. This is done as a form of caring for others, a form of solidarity of human beings and a sense of own and responsibility towards others. The research conducted by academics and disaster observers is a reference for the people of Lombok in particular and the wider community in general know the impact in the event of a disaster, the post-disaster impact, what risks will be faced, as well as what form of solution or mitigation should be planned in the future to the stage of rehabilitation and reconstruction.

The discovery of the impact of mental trauma experienced by children becomes the impact of the disaster itself. The mental impact is greater than the physical impact. Children lose their parents, lose their education, friends, relatives, lose fun, lose their environment and community (Suryatiningsih, 2018) [5]. The different symptoms of trauma in the reality faced by humans need to be handled wisely by experts or society as a whole. Therefore, with the detection of the initial symptoms of a traumatic event, it will facilitate in the efforts of providing assistance (counseling) properly and continuously. One solution given the efforts of psychosocial application in the form of fun-educational games and games is needed to help the mental recovery of children (Tanjung, 2018) [6]. As for the general public trauma healing with the approach of Islamic da'wah. Trauma healing program with Islamic da'wah approach is able to raise awareness of good religious practice so that it has the spirit to bounce back in living its life (Sa'i & Acim, 2018) [7].

It is not just the psychological mental problems of the child as a result of this catastrophic event. Results showed victims experienced neurosis symptoms (85.2%), psychotic symptoms (25.9%), PTSD symptoms (64.7%) and did not experience the risk of mental health disorders of 13 people (Dwidiyanti et al., 2018) [8]. Other risk assessments are medical facilities and infrastructure to deal with earthquake victims hampered because access to earthquake areas is damaged. so that ground transportation access is not optimally operated. The solution to that, medical volunteers with the community of emergency hospitals, sparked the idea of Ksatria Airlangga Floating Hospital (RSTKA). RSTKA is a pinisi ship with surgical room facilities, and a treatment room. The earthquake was not accompanied by a tsunami, so the pier is still worth leaning on. The research aims to analyze patient data in RSTKA, as well as as an evaluation and is expected to be a model of field hospitals (Airlangga et al., 2019) [9].

As presented in the introduction of the trigger for the earthquake disaster is the geological and demographic form of lombok island itself. Research from the Bandung Institute of Technology's Global Geophysical Expertise Group, together with the Earth Observatory of Singapore (EOS) and the Bureau of Meteorology and Geophysics (BMKG), studied in more detail about earthquake shock patterns in Lombok, West Nusa Tenggara. One way is to record aftershocks that occurred by installing 13 seismograph units in North Lombok, West, East, Central, and Mataram City for three months, from August 3 to October 20, 2018. Similar research obtained microzoning maps of earthquakes in Medana and Jenggala Villages, Tanjung District, North Lombok Regency. Based on GSS (Ground Shear Strain) values and earthquake intensity on an MMI (Modified Mercalli Intensity) scale. Microtremor data is analyzed using the HVSR (Horizontal to Vertical Spectral Ratio) method to obtain the dominant frequency value and amplification factor at each measurement point, from the value used as the initial data to obtain the dominant period value, seismic vulnerability index (Kg), maximum ground vibration acceleration using kanai method, GSS value and Earthquake intensity obtained using Wald method (Ningsih et al., 2019) [3].

This is used to predict whether there is a possibility of aftershocks that include tsunamis. The effect of landslides accompanied by the eruption of Mount Rinjani was greatly avoided. There are some researchers such as Kusdia et al., (2013) who took part in researching geomagnetic signals in Lombok

IOP Conf. Series: Earth and Environmental Science **884** (2021) 012040 doi:10.1088/1755-1315/884/1/012040

region using kupang observatory (KPG), Kakadu (KDU), Guam (GUA) with fractal method to determine anomalies that can be precursor in the event of earthquake in 2013. The earthquake has caused landslides and landslides in several places around Mount Rinjani. The earthquake that accompanied the tsunami wave that has edested tens of thousands of lives in Aceh is the earthquake with the highest number of casualties in the last 45 years. The largest earthquake ever occurred in Peru, in 1970 with a death toll of 120,000 (Kusdia et al., 2013) [11].

The concern of lombok people about the earthquake and tsunami with much greater strength will make it difficult for the people to bounce back immediately after the great earthquake that has occurred on Lombok Island. The tsunami event in the southern Lunyuk Sumbawa area in 1979 as a trigger for the concern of the people of Lombok island because it impacted the southern areas of Lombok Island namely the tanjung luar bay area in Keruak Sub-district and the Teluk Awang area in Central Lombok. Therefore, socialization efforts are needed that will make the public more aware of the risk of disasters in the event of a natural disaster and also reduce excessive concern due to a lack of understanding of disaster problems. Therefore, it is necessary to solve this problem, through socialization about disaster mitigation. The results of this socialization have made people understand about the risk of disaster, how to deal with it and eliminate excessive concerns. Hope with socialization and strengthening mitigation will make lombok community resilien against the pressures faced. Resilience is essential for post-earthquake rehabilitation and reconstruction.

3.2 Community-based synergies

Research that has been done from various parties is expected to help in disaster risk mitigation efforts in the future, especially related to earthquake problems. An important study is to continue, although sometimes researchers acknowledge the constraints on the situation and the condition of the limitations of existing human resources and the amount of cost budget required in the study. Literacy on disaster risk mitigation must be continuously improved in order to convey well to the community. The role of local government is indispensable to the program. Moreover, when referring to law number 24 of 2007, the authority, policies, and evacuation orders in the event of a disaster are in the local government not at the authority of BMKG. The role of the school is also very influential in the success of disaster mitigation. The role of schools in disaster risk reduction efforts through the education sector is an important part of efforts to improve children's resilience to disasters. The child will later become a change agent who can teach his community to form a culture ready for disaster (Rahma, 2018) [12].

In order to implement the mandate of Law No. 24 of 2007 on Disaster Management (Law No. 24/2007), especially Article 36, the Government through BNPB prepared the document of disaster management plan (RENAS PB) 2015-2019. This RENAS PB document is compiled based on achievements, challenges and opportunities in implementing disaster management in 2010-2014. Based on rpjmn evaluation results 2010-2014, review and evaluation of RENAS PB 2010-2014 and National Disaster Risk Reduction Action Plan (RAN PRB) 2010-2012, then in general the achievements that have been made in disaster management efforts 2010-2014 are regulatory and institutional frameworks, disaster risk assessment and early warning determination, SCIENCE TO build a safe culture, fundamental risk reduction, and preparedness strengthening [1].

The implementation of the program and the results of policies that have been established as disaster response solutions must be synergized and synchronized with other government agencies and non-governmental institutions. Ministries and institutions jointly and individually run programs that have been planned to strengthen community capacity. Once again the role of local government is important. BMKG, BNPB and BPBD are always ready when involved in any disaster risk reduction efforts. So that the community will have the encouragement from inside and outside to immediately rise to the next stage of progress.

3.3 Socialization and strengthening of disaster mitigation implementation

Mitigation is a series of efforts to reduce the risk of disaster, both through physical development and awareness and improvement of the ability to deal with disaster threats (article No. 21 of 2008 1 paragraph 6 PP on disaster management) (National Earthquake Study Center Team, 2018) [2]. Mitigation aims to reduce the impact, especially on the population. Mitigation is also a guideline for development planning and improving people's knowledge in dealing with and reducing disaster impacts or risks, so that people can live and work safely.

Some of the general disaster mitigation activities that the Government of North Lombok has conducted in the lombok earthquake study[2] are the first, the introduction and monitoring of disaster risk. Second, participatory disaster management planning and disaster aware cultural development. Third, the application of physical, nonphysical, and disaster management arrangements. Fourth, identification and recognition of the source of the danger or threat of disaster. Fifth, monitoring of natural resource management. As for recommendations for earthquake disaster mitigation can be divided into three stages. That is the stage before the earthquake, during the earthquake and after the earthquake. This knowledge is given in the hope that the public has knowledge and skills when dealing with all three stages of disaster. So that the public will know how to survive, reduce the risk and how to grow after the disaster.

Table 1. Earthquake disaster mitigation.

Sebelum gempa	Saat gempa	Setelah gempa
Mendirikan bangunan sesuai aturan baku (tahan gempa).	Tetaplah tenang dan	Cepat keluar dari bangunan. Gunakan tangga biasa hindari menggunakan lift.
Mengenali lokasi bangunan tempat tinggal.	Hindarilah sesuatu yang kemungkinan akan roboh, berlindung di bawah meja serta jauhi jendela kaca.	Memeriksa sekitar jika ada yang terluka, lakukan pertolongan pertama.
Menempatkan prabotan pada tempat yang proposional.	Perhatikan tempat berdiri, kemungkinan ada retakan tanah danberusahalah bisa ke tanah lapang	Hindari bangunan yang berpotensi roboh.
Menyiapkan peralatan seperti senter, P3K, makanan instan, dll.	Ketika berada di dalam bangunan, segera evakuasi diri setelah gempa bumi berhenti. Saat keluar, perhatikan reruntuhan maupun benda-benda yang membahayakan dan bisa jatuh menimpa diri pada saat evakuasi.	
Memeriksa penggunaan listrik dan gas.	Ketika berada di dalam rumah, tetap berada di bawah meja yang dirasa kuat	
Mencatat nomor telepon penting.	Periksalah keberadaan api dan listrik di sekitar yang berpotensi menyebabkan kebakaran.	
Mengenali jalur evakuasi.	Apabila di luar bangunan berupa tebing di sekelilingnya, hindari daerah yang rawan longsor.	
Mengikuti kegiatan simulasi mitigasi bencana gempa. Yaitu latihan penyelamatan. Seperti merunduk, perlindungan terhadap kepala, berpegangan ataupun dengan bersembunyi dibawa meja.	Jika berada di dalam mobil, hindari berhenti di bawah atau di atas jembatan atau rambu-rambu lalu lintas, tiang listrik, pohon serta jauhi pantai.	

3.4 Resilience is formed in lombok earthquake victims

According to Reivich and Shatte in Dimascio (2018) Resilience is the capacity to respond healthily and productively when dealing with misery or trauma, which is necessary to manage the stresses of daily life. Resilience is the ability to bounce back from negative experiences that reflect the innate qualities of the individual or are the result of learning and experience. Resilience generally leads to positive adaptation patterns during or after facing difficulties or risks. Improving resilience is an important task because it can provide experience for humans in the face of life's challenges and difficulties.

There are seven aspects of resilience [13].

3.4.1 *Regulation of emotions* is the ability to remain calm under pressing conditions, be able to regulate emotions and maintain relationships with others and the environment.

3.4.2 *Impulse Control* i.e. the ability of the individual to control desires, urges of desire and pressure that arise from within. Individuals with low impulse control abilities will rapidly experience emotional changes when dealing with various stimulus from the environment. Individuals will tend to be reactive, display irritable behavior, lose patience, be impulsive, and be aggressive.
3.4.3 *Optimism* i.e. believing that they have the ability to overcome adversity that may occur in the future. It also reflects his self-efficacy, namely the belief that he is able to solve existing problems and control his life.
3.4.4 *Causal Analysis* i.e. the ability of individuals to accurately identify the cause of the problem at hand. Individuals who are unable to identify the exact cause of the problem will constantly make the same mistake.
3.4.5 *Empathy* i.e. the individual's ability to read the signs of other people's emotional and psychological condition.
3.4.6 *Self-efficacy* i.e. a belief that individuals are able to solve problems experienced and achieve success.
3.4.7 *Achievements* i.e. the ability to overcome adversity and rise from the deterioration and the ability to achieve the positive aspects of life after adversity.

The result of the in-depth interview looks at the dynamics of the resilience of the earthquake victims community with the special meaning and concept of the people of North Lombok itself. means that the concept of resilience that is formed in earthquake victims communities in North Lombok cannot be genic to earthquake victims elsewhere. It is likely that there will be similarities in some aspects. The dominant aspects seen to form resilience in the community of lombok earthquake victims 2018 consist of optimistic aspects, empathy, self-efficacy and apprelication. The people of North Lombok have been hit by earthquakes from 1979 to 2013 before the 2018 earthquake. But the people of North Lombok are always optimistic that they can afford to go through this difficulty. This is evidenced by the strength of good self-efficacy and empathy towards others both as survivors and not so as to achieve success starting the recovery stage, rehabilitation and to the further stage of reconstruction.

Table 2. Transcripts of interviews according to aspects of respect.

optimis	Empati	Efikasi diri	pencapaian
pemuda-pemuda ini juga inisiatif mereka sendiri untuk saling membantu dalam membenah memoto semua rumah-rumah warga.	*Ada juga si yang peduli karena kita sama-sama merasakan kondisi seperti ini. Sama-sama kita laper. Jadi banyak yang memberikan barangnya dengan cuma-cuma.*	*Karena orang butuh tempat tinggal ini yang pertama. Butuh kepastian tidak pingin lama. sehingga mereka semaksimal mungkin berusaha masing-masing.*	*tadi keinginan yang cepat ingin kembali punya rumah.*
Awal gempa ada seribuan tenda yang tersebar. Ternyata dalam waktu sebulan terjadi penurunan yang drastis. Yang mereka tidak percaya masa' yang awalnya ribuan jadi puluhan.	*Ini naluri kita saja bertanggung jawab untuk membantu sesama.* *Kita akan saling membantu sebisa mungkin.*	*Ada yang rusak ringan, sedang dan ada yang rusak berat juga yang mengerjakan rumahnya dengan biayanya sendiri. Karena terlalu lama menunggu dari pemerintah akhirnya dilakukan sendiri.*	*pada saat terjadi gempa di tahun 2013. pernah terjadi gempa di tahun 2013 tapi tidak separah tahun ini. Yang pemukiman di atas gunung itu yang rusak tapi tidak separah yang saat ini. Tapi cepat pulihnya sama.*
Pemerintah kita sangat-sangat mendukung sekali misalkan dalam konteks pelayanan itu kesatu. Kedua, karena kita dari segi moral punya tanggung jawab untuk membangun kembali semangat warga ini supaya jangan terpuruk. Dari segi moral juga punya tanggung jawab supaya warga jangan terlalu larut.	*Siapapun tanpa pandang bulu mereka selalu saling membantu. Meski bukan keluarga.* *Sebenarnya faktor kekuatan dari keluarga itu sendiri. Dorongan dari keluarga malah di tengah kebencanaan yang harusnya memikirkan diri sendiri dulu kan. Tapi gotong royongnya masih kuat..*	*Pemerintah pernah memberikan pelatihan siap siaga bencana sehingga masyarakat mengerti apa yang harus dilakukan saat terjadi bencana gempa bumi, banjir, longsor dan masyarakat tau apa yang harus dilakukan pada tahap awal.*	*ada beberapa rumah yang sudah rampung disini. Relatif-relatif cepat.* *Alhamdulillah sudah membaik sudah normal semuanya. karena memang dari awal-awal banyak NJO yang datang dan sumbangan-sumbangan*

While the factors that influence resilience are formed by individual factors, family factors and community factors. These three factors are the most dominant in the community of lombok earthquake victims 2018 which makes the victims quickly rise from the fall is the ability that individuals have very well. Seen in cognitive, self-concept, self-esteem and social competence. The second is the important role of caring for each other and supporting each other's fellow earthquake victims so as not to make the people of lombok earthquake victims too late in the deterioration. Finally, the community factor is the ability to synergize holistically in providing support both moril and materil so that the people of North Lombok always do their best to move to make changes for the benefit and progress of the people of North Lombok. It is seen that indigenous stakeholders, religious leaders, local and village governments, communities and non-governmental agencies and institutions are taking part in the changes in order to realize the north Lombok community that is resilient to disaster. Good synergy has led to the success of mitigation programs that have been programmed by local governments. As well as the compliance of the people of North Lombok implement the mitigation program. So self-strength, strong encouragement from families and others who care for each other and holistic community support has led the people of North Lombok into a resilient society.

4 Conclusions

The resilience of earthquake victims' communities is able to be formed by the encouragement of holistic synergistic support factors namely self-strength, the strength of family support and the strength of community support whether it is in a narrow environment or a large environment.

References

[1] BNPB. (2014). Rencana Nasional Penanggulangan Bencana 2015-2019. *Rencana Nasional Penanggulangan Bencana 2015-2019 RINGKASAN*, 1–115. https://www.bnpb.go.id//uploads/renas/1/BUKU RENAS PB.pdf

[2] Tim Pusat Studi Gempa Nasional. (2018). *Kajian Rangkaian Gempa Lombok Provinsi Nusa Tenggara Barat.*

[3] Ningsih, F., Fitrianingsih, & Didik, L. A. (2019). Indonesian Physical Review. *Indonesian Physical Review*, *2*(3), 1–8.

[4] MDMC. (2018). Laporan Perkembangan Gempa Bumi Nusa Tenggara Barat, Yogyakarta.

[5] Suryatiningsih, H. (2018). Dampak Gempa Bumi Lombok Terhadap Prilaku Anak Hetty. *Prosiding Seminar Nasional Pengambdian 2018 Univeristas Muslim Nusantara Al-Washliyah, September*, 262–268.

[6] Tanjung, D. I. D. (2018). *Psikososial Sebagai Bentuk Trauma Healing Untuk Anak-Anak Terdampak Gempa*. *1*, 12–16.

[7] Sa'i, M., & Acim, S. A. (2018). Trauma healing bagi masyarakat terdampak gempa Desa Gumantar Kecamatan Kayangan Lombok Utara. *TRANSFORMASI: Jurnal Pengabdian Masyarakat*, *14*(1), 1–12. https://doi.org/10.20414/transformasi.v14i1.570

[8] Dwidiyanti, M., Hadi, I., Wiguna, R. I., & Ningsih, H. E. W. (2018). Gambaran Risiko Gangguan Jiwa pada Korban Bencana Alam Gempa di Lombok Nusa Tenggara Barat. *Holistic Nursing and Health Science*, *1*(2), 82. https://doi.org/10.14710/hnhs.1.2.2018.82-91

[9] Airlangga, P. S., Waloejo, C. S., Sulistiawan, S. S., Utariani, A., Hamzah, H., Semedi, B. P., Andriyanto, L., Manurung, P. P., Harjana, L. T., Putri, H. S., Husain, T. A., Perdhana, F., & Rahardjo, E. (2019). Post-Serial Earthquakes Health Problems in Lombok, Indonesia: Experience of "Ksatria Airlangga" Floating Hospital [Profil Masalah Kesehatan Pasca Gempa Serial di Lombok, Indonesia: Pengalaman Rumah Sakit Terapung Ksatria Airlangga]. *Proceeding of Community Development*, *2*, 368. https://doi.org/10.30874/comdev.2018.373

[10] Ningsih, F., Fitrianingsih, & Didik, L. A. (2019). Indonesian Physical Review. *Indonesian Physical Review*, *2*(3), 1–8.

[11] Kusdia, I. G. A., Kanata, B., & Zubaidah, T. (2013). *Studi Pengolahan Sinyal Geomagnetik Sebagai Precursor Gempa Bumi di regional Lombok Dengan Metode Fractal.*

[12] Rahma, A. (2018). Implementasi Program Pengurangan Risiko Bencana(PRB) Melalui Pendidikan Formal. *Jurnal VARIDIKA*, *30*(1), 1–11. https://doi.org/10.23917/varidika.v30i1.6537

[13] Dimascio, J. (2018). Building resilience. *Aviation Week and Space Technology*, *180*(15), 52–54. https://doi.org/10.7748/ns.26.32.16.s21

IOP Conf. Series: Earth and Environmental Science **884** (2021) 012041 doi:10.1088/1755-1315/884/1/012041

Spatial modelling of tsunami exposure areas at Ujung Genteng coastal, Sukabumi, West Java

Dinda Zannuba Arifah, S. Supriatna*

Department of Geography, Faculty Mathematic and Natural Sciences, Universitas Indonesia, Indonesia.

E-mail: dinda.zannuba@ui.ac.id, ysupri@sci.ui.ac.id*

Abstract. Indonesia is a country located between three tectonic plates. It is the Eurasian Plate, the Indo-Australian Plate, and the Pacific Plate. The location caused Indonesia to be prone to disasters caused by the movement of plates, one of which was a tsunami. Tsunamis are high waves that hit ports or beaches. Relatively sloping beaches with low beaches can cause larger tsunami waves. This study was carried out on the coastal of Ujung Genteng, Sukabumi. The purpose of this study is to analysis the area of tsunami exposure that occurred on the coast of Ujung Genteng. GIS-based spatial modelling is used to combine different types of data, both spatial and non-spatial data to be processed into tsunami exposure area information. Modelling of the inundation was carried out with mathematical calculations developed by Berryman (2006). By modeling the tsunami with a worst-case scenario of wave height of 20 meters, it was obtained that the tsunami exposed area reached 663.29 Ha. The area belongs to a low hazard class of 2.59 Ha, a medium hazard class of 29.05 Ha, and an area that belongs to a high hazard class of 631.65 Ha. The results of this modeling are expected to be a reference for tsunami disaster mitigation planning in Ujung Genteng.
Keyword. Tsunami, exposure area, spatial modelling, inundation

*To whom any correspondence should be addressed.

1. Introduction

Indonesia is a country located between three tectonic plates. The three plates are the Eurasian Plate, Indo-Australian Plate, and the Pacific Plate. This location puts Indonesia in the ring of fire [1]. This condition makes Indonesia a disaster-prone region mainly caused by the movement of tectonic plates that can cause collisions between plates. Collisions between these plates can cause tectonic earthquakes. An earthquake on the seabed triggers a tsunami if it has considerable strength. The term tsunami is derived from the Japanese word for "tsu" and "nami" means wave. A tsunami means a high wave that hits a port or beach. Tsunamis are often caused by tectonic earthquakes that occur on the seabed, although in other instances tsunami can be caused by volcanic eruptions, landslides on the seabed, and even the impact of falling meteors.

Tsunamis have occurred in various parts of the world for a long time but the impact of this disaster is still unavoidable. Therefore, research on tsunamis in various regions of the world is carried out to minimize the impact of the disaster. This disaster was also recorded several times in Indonesia. One of the tsunami disasters that had the biggest impact ever in Indonesia was the Aceh tsunami that occurred on December 26, 2004 [2]. The tsunami that occurred at that time took many lives and damaged the environment affected by the tsunami. The tsunami event at that time had become a national disaster even

Content from this work may be used under the terms of the Creative Commons Attribution 3.0 licence. Any further distribution of this work must maintain attribution to the author(s) and the title of the work, journal citation and DOI.
Published under licence by IOP Publishing Ltd

gained the attention of more than various countries. Tsunami events have also occurred in the southern sea of Java, which occurred in Banyuwangi in 1994 and Pangandaran in 2006 [3]. The most recent tsunami in Indonesia occurred in 2018 that hit Palu and Donggala. The tsunami was triggered by a magnitude tectonic earthquake with a magnitude of 7.4. The damage caused by the tsunami was quite severe and took many fatalities as well. Records of tsunami events in Indonesia have proven that the region in Indonesia is prone to tsunami disasters. The main determining factors tsunami size is characteristic of vertical seabed deformation resulting from the large earthquakes, epicentre depths (depths below the seabed where earthquakes occur) [4].

Tsunamis that occur on bay-shaped beaches produce waves larger than straight-lined beaches. The potential for tsunamis will increase if the morphology of beaches with the coastline of the bay ramps has no protectors such as mangroves, coconut trees, and other coastal forests [5]. There are previous studies that say that the southern waters of Java Island have the potential for tsunami waves up to 20 meters. One of the bays located in the southern part of Java Island is located on the coast of Ujung Genteng, Ciracap sub-district, Sukabumi. Ujung Genteng has the shape of the coastline in the form of a ramped bay. Ujung Genteng coastline is directly connected to the Indian Ocean. These conditions make the area vulnerable to tsunami hazards. This research was conducted to analyze the area of tsunami exposure that occurred on the coast of Ujung Genteng. Spatial modeling in GIS is chosen as the right method because it can aid decision making by integrating a variety of factors. This research is also expected to be information and reference for local governments in tsunami disaster mitigation along the coast of Ujunggenteng.

This research was conducted along the coast of Ujung Genteng. Ujung Genteng is one of the villages located in Ciracap sub-district, Sukabumi Regency, West Java. Ujung Genteng is one of the tourist destinations located on the south coast of Java Island. The south coast of Java island including Bayah Coast and surrounding areas (including Ujung Genteng) is one of the tsunami areas in Indonesia [6]. The Ujunggenteng coastal has a shallow beach and large waves [7]. Therefore, analysis of tsunami-prone areas needs to be done at this site for disaster mitigation planning to reduce the losses that will occur when a tsunami hits this area [8]. The study area can be seen in Figure 1.

Figure 1. The study area located in Ujung Genteng, West Java.

2. Methods

The method used in this study is spatial modeling based on Geographic Information Systems (GIS). Spatial modeling with GIS was chosen as a research method because it can combine different types of data, both spatial and non-spatial data to be processed into information. GIS can help to determine the distribution of tsunami exposure areas [9]. The use of GIS in natural disasters can be used to create mitigation planning and strategies to minimize disaster impacts. In making spatial layout of an area is not separated from the variable distance and height of the coastline [10].

The method used for tsunami modeling refers to the modeling techniques that have been used by the National Disaster Management Agency (BNPB). The method used is numerical modeling with mathematical calculations developed by Berryman (2006) [11]. This modeling analysis is based on the calculation of tsunami height loss per 1 m of inundating distance (inundation height) based on the value of distance to slope and surface roughness. The formula used is as follows:

$$H_{loss} = \left(\frac{167\, n^2}{{H_0}^{1/3}}\right) + 5\, Sin S$$

Description:

H_{loss} : loss of tsunami height per 1 m inundation distance
n : surface roughness index
H_0 : the height of tsunami waves on the coastline (m)
S : surface slope size (degrees)

Modeling tsunami inundation with mathematical equations require three variables. The variables used are slopes, the maximum wave potential of tsunamis, and the coefficient of surface roughness. The maximum tsunami wave potential is the maximum wave level of tsunami to the ground obtained from tsunami modeling by tsunami experts. The maximum wave height of tsunami in the Sukabumi Regency is based on the Regulation of the Head of National Disaster Management Agency No. 2 of 2012 which is 10 m. The coefficient index value of surface roughness comes from land-use data. The data is converted from vector to raster for the Hloss calculation process. The maximum wave potential of a tsunami is combined with the flow of water flowing on rough or irregular topographic surfaces to determine how far puddles might go. The coefficient value of surface roughness refers to the determination of Berryman (2006) can be seen in table 1.

Table 1. Surface Roughness Index

Land cover/ Land use	Surface Roughness Coefficient
Water	0.007
Shrubs / Bushes	0.040
Forest	0.070
Agricultural land	0.025
Fields	0.020
Mangrove	0.060
Built-in Settlements	0.050
Ponds	0.010
Plantation	0.035
Sand	0.018

3. Result and Discussion

The spread of tsunami-exposed areas in this study was obtained from spatial modeling which refers to tsunami modeling techniques used by the National Disaster Management Agency. The model used is a mathematical calculation model developed by Berryman (2006). This modeling is based on the calculation of tsunami height loss per 1 m of inundation height distance obtained from slope data processing, coefficient of surface roughness, and maximum wave height.

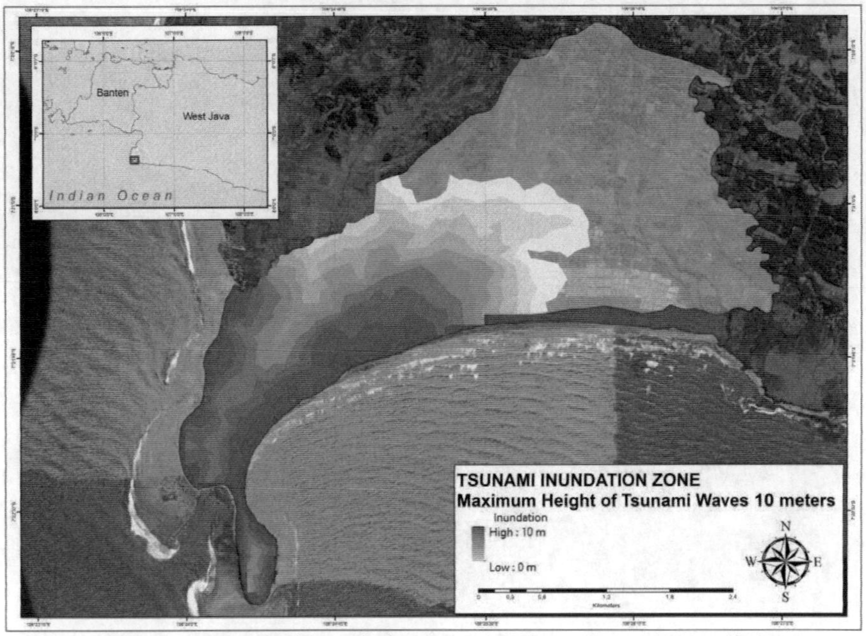

Figure 2. Tsunami inundation zone in Ujung Genteng coastline

The results of modeling the area of tsunami exposure can be seen in Figure 2. On the map we can know that along the coastline at Ujung Genteng has a maximum inundation height. Land use along the coastline in the form of fields, open land, and few coastal plants in the form of mangroves resulted in tsunamis reaching land with maximum inundation. In addition, the slopes that ramp in this region cause a wider range of tsunami inundation compared to areas that have steep slopes.

Based on the value of the inundation that has been obtained, we can then analyze the tsunami hazard index. Tsunami hazard index values are in the range of 0 to 1 by following a continuous classification pattern performed using fuzzy logic suppression. In accordance with National Agency Head Regulation 2/2012, there are three classifications of the tsunami hazard index. The three classifications are low hazard (inundation ≤ 1), moderate hazard ($1 <$ inundation ≤ 3), and high hazard (inundation > 3) [12]. The fuzzy membership of the inundation determined rule that the greater the inundation value (>3), then the value of the fuzzy membership inundation will be closer to the value of 1 or be at the limit of the value that can be referred to as the high danger class. Conversely, if the smaller the inundation value (≤ 1), then the value of the fuzzy inundation membership will be closer to the value of 0 or be at the limit of a value that can be referred to as a low hazard class.

Hazard analysis is carried out to obtain conclusions from the results of hazard index analysis (H) in the form of a hazard class. Hazard classes are classified based on the following grouping of hazard index values:

- Low ($H \leq 0.333$)
- Medium ($0.333 < H \leq 0.666$)
- Height ($H > 0.666$)

The tsunami hazard classification map on the Ujung Genteng coast can be seen in Figure 3. From the results of the processing of the data, it is known that along the coast of Ujung Genteng is a tsunami exposure area that belongs to the high hazard class. The farther from the coast, the safer the area is from the exposure of tsunamis that are classified in areas with low hazard classes. The total area exposed to the tsunami was 530.34 Ha. This area is half of the village area of 1,009.91 Ha. Areas included in the low hazard class include 54.71 Ha, areas with medium hazard classes covering 91.62 Ha, and areas with a high hazard class of 383.01 Ha.

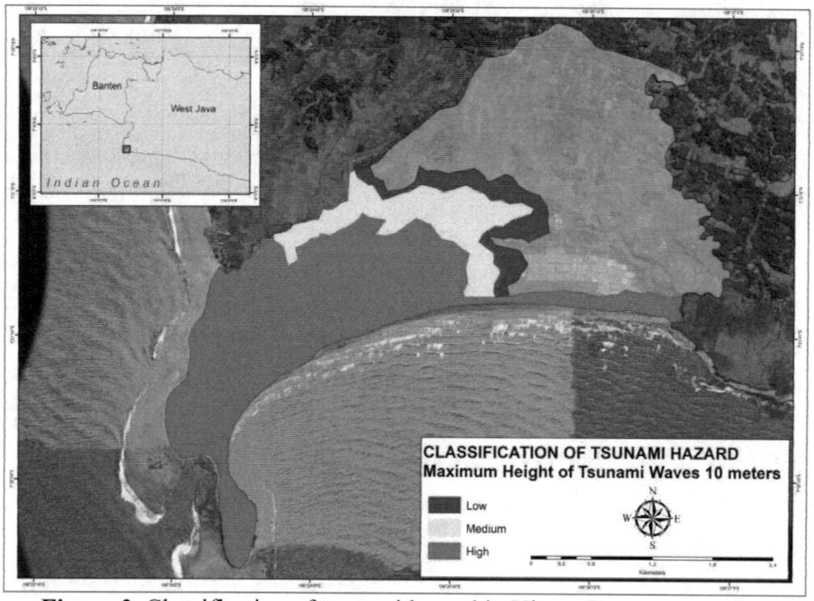

Figure 3. Classification of tsunami hazard in Ujung Genteng coastline

The tsunami modeling above uses a maximum tsunami wave estimate of 10 meters. Meanwhile, there are previous studies that say that the southern waters of Java Island have the potential for tsunami waves up to 20 meters [13]. In the study, it is said that judging from earthquake data and GPS data inversion, there is a gap seismic in the south of Java Island. The study used earthquake data, bathymetry, and data obtained from GPS. The worst-case scenario is that two megathrust segments in the south of Java Island can break simultaneously. If the earthquake occurs it could potentially trigger a tsunami with wave heights reaching 20 meters on the southern coast of West Java. From the research, on this occasion will also be done modeling tsunami area in Ujung Genteng with a maximum wave height of up to 20 meters to update the information of areas exposed to the tsunami.

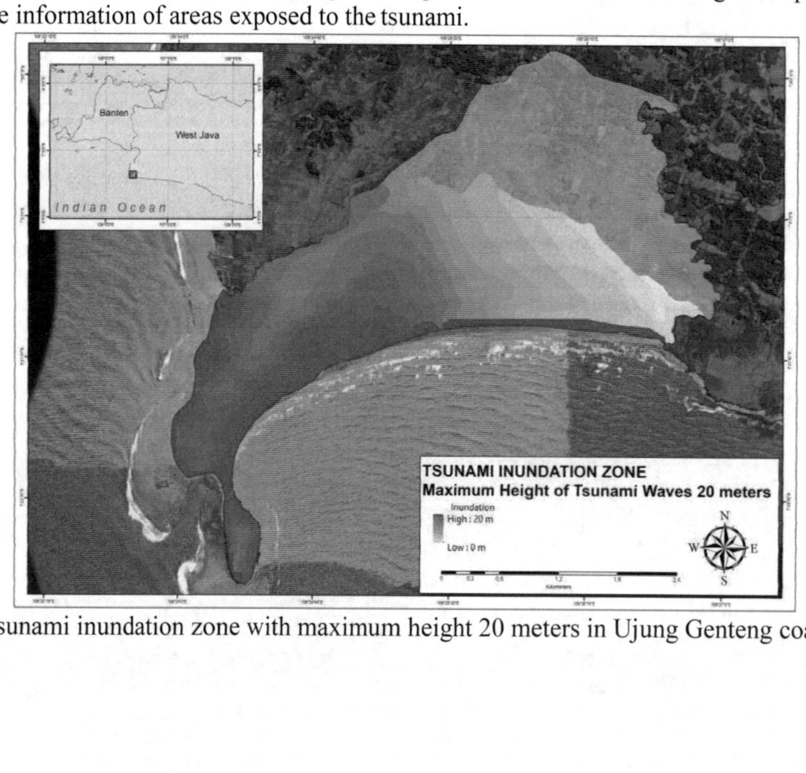

Figure 4. Tsunami inundation zone with maximum height 20 meters in Ujung Genteng coastline

Modeling of tsunami exposed areas with a wave height of 20 meters in Ujung Genteng can be seen in Figure 4. Tsunami-exposed areas cover almost the entire area of the village. The total area exposed to the tsunami reached 663.29 ha. While the total area of Ujunggenteng reached 1,009.91 ha. The tsunami-exposed area is divided into three classes. Areas included in the low hazard class include 2.59 Ha, areas with medium hazard classes covering 29.05 Ha, and areas with a high hazard class of 631.65 Ha. The tsunami hazard classification map on the Ujung Genteng coast can be seen in Figure 5.

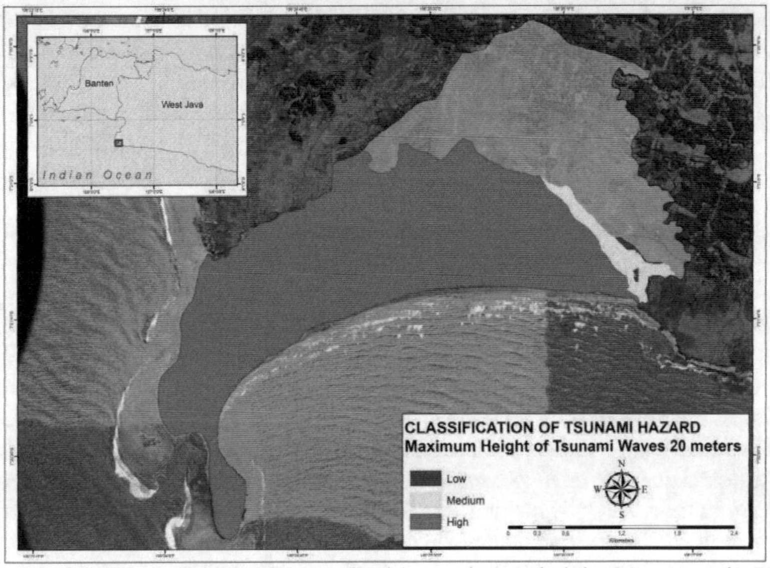

Figure 5. Classification of tsunami hazard with maximum height 20 meters in Ujung Genteng

From the results of the modeling, we can know the extent to which the tsunami will damage the Ujunggenteng area. The absence of coastal forests such as mangroves or coconuts along the coast causes the absence of barriers that can break the waves when a tsunami occurs. This resulted in the coverage of tsunami exposure areas in the region. As previously known, the shape of the coastline in the form of bays has a greater potential for tsunami height due to the accumulated waves of seawater to the coast. The results of the tsunami modeling can be used for tsunami disaster mitigation planning by local parties.

4. Conclusion

The exposure of the area along the Ujung Genteng coast to the tsunami is influenced by the condition of ramp slopes and land use on the coast. Along the coastline is classified as an area with a high danger class and increasingly avoiding the coast, the area is safe from tsunami exposure that belongs to low hazard class areas. By modeling the tsunami with a worst-case scenario of wave height of 20 meters, it was obtained that the tsunami exposed area reached 663.29 Ha. The area belongs to a low hazard class of 2.59 Ha, a medium hazard class of 29.05 Ha, and an area that belongs to a high hazard class of 631.65 Ha. The results of this modeling are expected to be a reference for tsunami disaster mitigation planning in Ujung Genteng.

Acknowledgement
Thanks to DRPM of Universitas Indonesia which has supported and funded this research grant PUTI 2020.

References

[1] Husna, A., S. Supriatna, Putut Ash Shiddiq, I., Dwi Mandini Manesa, M., & Ristya, Y. (2019). Environmental Vulnerability on Tsunami Hazard, Cisolok Village, Sukabumi. *KnE Engineering, 2019*, 9–19.

[2] Syamsidik, Tursina, Suppasri, A., Al'Ala, M., Luthfi, M., & Comfort, L. K. (2019). Assessing the tsunami mitigation effectiveness of the planned Banda Aceh Outer Ring Road (BORR), Indonesia. *Natural Hazards and Earth System Sciences*, *19*(1), 299–312.

[3] Aeda, S. A., Saputro, S., & Subardjo, P. (2017). Simulasi Penjalaran dan Penentuan Run-Up Gelombang Tsunami di Teluk Pangandaran, Jawa barat. *Jurnal Oseanografi*, *6*(1), 254–262.

[4] S. Supriatna; Pratiwi, Sri Fauza; Marko, Kuswantoro; Manessa, Masita Dwi Mandini; Ristya, Yoanna. (2020). Spatial Dynamics of Tsunami Prone Areas in Pariaman City, West Sumatera. *Journal of Computational and Theoretical Nanoscience* 17, 1474–1491 (2020).

[5] Diposaptono, S. (2005). Mitigasi Bencana Wilayah Pesisir Berbasis Ekosistem Mangrove. Ditjen Kelautan, Pesisir, dam Pulau-pulau Kecil, DKP.

[6] S Supriatna, S Sobirin, A N Pratiwi. (2018) Spatial model of vulnerability towards tsunami in Bayah coastal area, Banten, Indonesia. *AIP Conference Proceedings* 2023, 020178.

[7] E Ditriyani, Supriatna and M P Tambunan. (2020). Seagrass distribution using remote sensing along the Ujunggenteng Coastal, Sukabumi. *IOP Conf. Series: Earth and Environmental Science* 481, 012059.

[8] Utami, S. U., Muntasib, E. K. S. H., & Samosir, A. M. (2019). Hazard Management in Karang Hawu Beach, Sukabumi Distric, West Java. *Media Konservasi*, *24*(3), 322–333

[9] Sambah, A. B., & Miura, F. (2014). Integration of Spatial Analysis for Tsunami Inundation and Impact Assessment. *Journal of Geographic Information System*, *6*(1), 11–22.

[10] Natasya Syafira Putrie, Dewi Susiloningtyas, Supriatna, Mentari Pratami. (2019). Distribution pattern of Settlement in 2032 based on population density in Bengkulu City. *IOP Conference Series: Earth and Environmental Science*. 284, 012007.

[11] Berryman, Kelvin. (2006). Review of Tsunami hazard and risk in New Zealand. New Zealand: Institute of Geological and Nuclear Sciences.

[12] Badan Nasional Penanggulangan Bencana. (2018). Modul Penyusunan Kajian Risiko Bencana Tsunami.

[13] Widiyantoro, S., Gunawan, E., Muhari, A., Rawlinson, N., Mori, J., Hanifa, N. R., Susilo, S., Supendi, P., Shiddiqi, H. A., Nugraha, A. D., & Putra, H. E. (2020). Implications for megathrust earthquakes and tsunamis from seismic gaps south of Java Indonesia. *Scientific Reports*, *10*(1), 1–11.

Disaster Mitigation Snake and Ladder Game to Improve Earthquake Disaster Preparedness (A Case Study: Yogyakarta 5 Senior High School)

E Ersani and Mukminan

Postgraduate School, Yogyakarta State University, Indonesia

Email : errysani@gmail.com and mukminan@uny.ac.id

Abstract. Tectonic earthquakes are natural disasters that are difficult to predict when they occurs, therefore preparedness is needed to deal with them. This study aims to determine the improvement in student preparedness for tectonic earthquakes using Guncang Bumi snake and ladder learning media. This research is research and development using the ADDIE model (analysis, design, development, implementation, evaluation). The research subjects were 63 students of class XI of Science. Data collection techniques used were test questions. The data analysis technique used is descriptive analysis and independent *t*-test analysis. Based on the independent *t*-test analysis obtained sig <0,05, it can be concluded that there is a difference in preparedness between the experimental class and the control class. Based on the calculation of the improvement in preparedness score, the percentage of improvement in disaster preparedness for the experimental class is 67,65%. The improvement percentage in preparedness for the control class was 32,35%. Both classes have improved but the improvement over much experimental class. Therefore it can be concluded that the use of the Guncang Bumi snake and ladder can improve preparedness for a tectonic earthquake disaster.
Keywords: *snakes and ladders, earthquakes, disaster preparedness*

1. Introduction

The disaster laboratory is a term for the Indonesian state to illustrate that there are various natural hazards. The hazard can occur naturally or due to human activity and can be a disaster. An earthquake is an example of a natural disaster in Indonesia. Based on the cause, earthquakes can be divided into four types, namely tectonic earthquakes, volcanic earthquakes, ruins earthquakes, and meteoric earthquakes.

Earthquakes that often occur in Indonesia are tectonic. This is because geologically Indonesia is located at the confluence of three world plates. The Pacific Plate, the Eurasian Plate, and the Indo-Australian Plate continue to move. The consequence of tectonic plate movements is the existence of earthquake and volcanic pathways that are a hazard. According to [1] areas located around the Pacific Rim also have to compete with tsunamis, storms, typhoons, and floods.

Yogyakarta is one of the areas prone to tectonic earthquakes because it is located close to the Indo-Australian Plate and the Eurasian plate. This was evidenced by the earthquake on May 27, 2006, that caused casualties and damage to various facilities such as the education sector. The data in table 1 is data from the National Disaster Management Authority [2] about the number of fatalities caused by the 2006 Yogyakarta earthquake.

Content from this work may be used under the terms of the Creative Commons Attribution 3.0 licence. Any further distribution of this work must maintain attribution to the author(s) and the title of the work, journal citation and DOI.
Published under licence by IOP Publishing Ltd

Table 1. Total of Yogyakarta Earthquake Victims in 2006

Districts	Death (people)	Injuries (people)
Bantul	4141	12026
Sleman	232	3789
Yogyakarta	204	318
Kulonprogo	22	2678
Gunungkidul	81	1086
Total	**4.680**	**19.896**

Sources: [9]

Based on data from the National Development Planning Agency [3], most area severely affected due to the earthquake in Yogyakarta in 2006 was Bantul. The total of casualties died is 4.141 inhabitants. Damage to facilities and infrastructure also occurred such as houses and schools. A total of 208.991 houses were light to severe damage [4]. The Education Sector is also not spared from the impact of the disaster with the amount of damage as many as 197 schools destroyed and 765 schools in mild to severe damage. To reduce its impact, it is important to understand the many factors that influence people's ability to effectively respond to evacuation warnings and orders [5].

The education sector affected by the earthquake indicates that schools are vulnerable to the impact of earthquake disasters. The various impacts that have been caused by the tectonic earthquake in Yogyakarta are a lesson that the importance of tectonic earthquake disaster preparedness education. As a follow up of the most severely affected areas, schools in the districts of Bantul got the attention of the government. Education mitigation of earthquake tectonic continues such form disaster prepared school being planned.

The disaster prepared school being planned is directed to the most severely affected areas, but did not rule out the possibility for other area are also getting attention as the city of Yogyakarta. Yogyakarta City Region ranks third after Bantul and Sleman Regencies with 204 fatalities and 318 minor injuries. Therefore, tectonic earthquake disaster mitigation education is needed for schools in the city of Yogyakarta. This is because students who attend school in the city of Yogyakarta do not only come from within the city of Yogyakarta.

Disaster mitigation education for high school students is aligned with the Geography material in the 2013 curriculum. Therefore, disaster mitigation education can be given when learning Geography. This is consistent with the research [6] which states that there is no better way to manage unforeseen disasters than recognizing the potential for the worst events and taking active and cooperative preventive actions to minimize losses. According to [5] when a person knows the dangers it can be used to decide the steps and how that should be taken when a person must be personally involved in disaster preparedness. Accordingly, the research is done to develop learning media mitigation of earthquakes based games using snakes and ladders to improve disaster preparedness earthquake. To facilitate the students to understand the material on disaster mitigation then is required a medium of learning.

The snake and ladder game model was chosen because the snake and ladder game is a popular and popular game from the age of children, teenagers, and even adults. The game of snakes and ladders is very effective in the learning process [7]. Besides, playing snakes and ladders involves peer-to-peer interactions following the theory put forward by [8] that children's cognitive can develop better through interactions with peers compared with interactions with adults. Therefore, the author tries to develop learning media with the snake ladder game of disaster mitigation to improve tectonic earthquake disaster preparedness with a case study in Yogyakarta 5 Senior High School.

2. Methodology

Earthquakes are vibrations that occur and are felt on the surface of the earth from within the earth's structure [4]. The shift occurred as a result of the event of the release of seismic wave energy suddenly due to the deformation of tectonic plates in the earth's crust. Based on the cause, earthquakes are

divided into four, namely tectonic earthquakes, volcanic earthquakes, earthquake earthquakes, and meteoric earthquakes. According to [10], earthquake is a devastating but common natural disaster that results in billions of dollars in property damage and significant loss of life.
Preparedness is a series of activities carried out to anticipate disasters through organization and appropriate and efficient steps [11]. According to [6] Geohazards have increased globally. It is very important to manage natural hazards to minimize casualties and damage to various properties. Besides, there is a growing natural awareness that geological catastrophe events can adversely affect human groups. Enhancing natural disaster mitigation was identified as a priority issue by the island nation.

Table 2 . Indicators of Disaster Preparedness Assessment According to the Indonesian National Disaster Management Authority

Preparedness	Indicator	Aspect
Before a disaster	Knowledge and Attitude	Natural events and disasters Physical environmental vulnerability Attitudes to disaster risk Evacuation plan
Disaster response	Emergency	First aid, rescue, safety and security
After a disaster	Disaster Warning System	Traditional (local) and modern technology
	Resource Mobilization	Human resources and funding

Sources: [9]

According to [12] snakes and ladders games are a popular board game that has entrusted generations. This game is played on a grid board consisting of hundred squares in a ten by ten grid. It usually involves two or more players and they take turns to move by rolling a dice. On the way to the finishing line, the players meet some hurdles in the form of snakes and opportunities in the form of ladders. The player who reaches the finishing point (100th square) first wins the game. Snakes and ladders game is a famous board game that has captivated many generations. This game is played on a box in a game board consisting of a square box. Each grid can consist of ten squares. The game usually involves a minimum of two players. When traveling to the finish line, players will get obstacles depicted in the form of snakes and opportunities depicted in the form of stairs. The player who reaches the finish is first considered the winner of the game.

The research methodology is research and development (RnD). The development procedure used is adapted from the ADDIE model which consists of five stages [13]. The ADDIE model can be used to research the development of various learning media products, learning strategies, and learning methods [14]. The stages of the ADDIE model, namely analysis, design, development, implementation, and evaluation are presented in Figure 1:

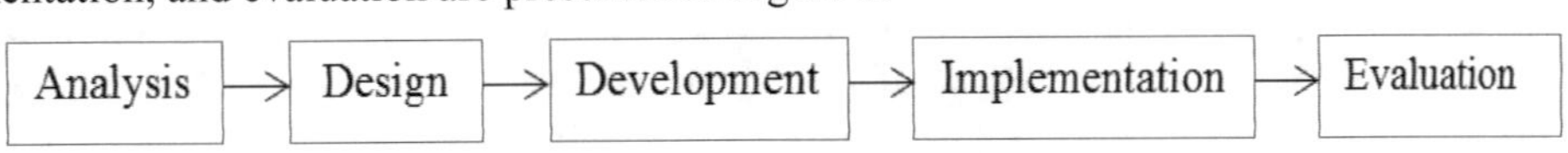

Figure 1. ADDIE model flow

The ADDIE model is the framework most widely used by instructional designers [15]. The ADDIE research procedure begins with analysis. The analysis conducted is an analysis of the problems that occur regarding the importance of earthquake disaster preparedness. The next step is to make a design in the form of a media design. The development phase includes validation by material experts and media experts. The assessment of the material and media experts aims to determine the feasibility, criticism, and suggestions. If there are criticisms and suggestions, a revision process will be carried out, but if there is no implementation process will be carried out immediately. The implementation process was carried out at Yogyakarta 5 Senior High School to find out the improvement for the earthquake disaster preparedness. Testing for increasing earthquake disaster preparedness is done by pre-test and post-test. Implementation involves the experimental class and the control class. The evaluation phase aims to find out the improvement in earthquake preparedness calculated using the independent t-test analysis.

3. Result and Discussions

Tectonic earthquakes are natural disasters whose timing is difficult to predict. This type of earthquake is more common than other earthquakes. The worst effects of an earthquake are fatalities. The school community is one that is vulnerable to the effects of the earthquake because there are activities that involve many people. Therefore, it is important to do an earthquake disaster preparedness education.

This tectonic earthquake disaster preparedness education is aligned with the Geography material in the 2013 curriculum. Following the 2013 curriculum, teachers are required to be creative and innovative in presenting learning in class. The form of a teacher's creativity can be reflected in efforts to develop learning media. The choice of media for snakes and ladders game is because the game is popular and is often played from the age of children, adolescents, and even adults.

The material in the snake and ladder game is tectonic earthquake disaster mitigation issued by the National Disaster Management Authority. The material is then arranged into a matter of pretest and posttest. Snakes and ladders that were develop were given the name Guncang Bumi snake and ladder. Components of the Guncang Bumi Snake and Ladder game that is: snakes and ladders board, cards, pawns, and dice. Snakes and ladders have the shape of a rectangle with a size of 42 x 29, 5 cm. This snake and ladder are packed using carton paper. Completeness in the Guncang Bumi snake ladder as follows:

- One snake and ladder board.
- Three pieces of different colors
- One dice
- A set of 10 answer cards for questions in red
- One set of motivation cards is 5 green cards
- A set of knowledge cards totaling 11 cards are yellow
- All cards have a size of 9 x 5 cm
- One sheet of game rules with a size of 21 x 15 cm.
- Snakes and ladders packaging has a length of 24 cm, a width of 15.5 cm, and a height of 3cm.

The number of boxes contained in the Guncang Bumi snake and ladder is 50 boxes. The rules of the game snakes and ladders Rake Earth the same as the rules of snakes and ladders in general. Players run pawns following the numbers that appear when throwing dice. If a player has a box with a ladder in it, the player must move his pawns up on the box that the ladder shows. If the player gets a box with a snake's tail, the player's pawn must go down on the box shown by the snake's head and must open the motivation box. The player must roll the dice twice if the dice shows the number six then add them up to find out the number of boxes to go.

Players in the knowledge box must take the knowledge card according to the number in the box and read it to other players. The player who gets the question box must answer the question and must open the answer card according to the box number is in the question box but does not answer the question then your next turn will be skipped once by another player. If the player has answered a question but your answer is wrong, then the turn will not be jumped over by other players. The game is considered finished if there are players who are in the finish box. Here are the components of the Guncang Bumi snake and ladder.

IOP Conf. Series: Earth and Environmental Science **884** (2021) 012042 doi:10.1088/1755-1315/884/1/012042

(a) (b) (c) (d) (e)

Figure 2. Components of the Guncang Bumi snake and ladder game. (a) packaging (b) board game (c) game instructions (d) cards (d) pawns and dice

Implementation in the experimental class is a class XI science one. The number of students who participated was 20 students. Learning activities carried out following the learning plan. The activity starts with a greeting, attendance, apperception, and pre-test. After that, it is continued by explaining the material to be studied briefly accompanied by a question and answer session.

Figure 3. Students play Guncang Bumi Snakes and Ladder game

The next activity is to bring students into seven groups. Each group consists of 2-3 students. When students have completed play Guncang Bumi snakes and ladders, learning terminated with post-test. The learning activities in the control class were carried out at XI science two. The number of students who participated was 19 students. The activity starts with greeting, attendance, apperception, and carrying out pre-test. After that, continued learning activities by explaining tectonic earthquake mitigation material. Explanation of earthquake disaster mitigation is interspersed with question and answer sessions. When the material that is delivered finished the students were given about the post-test. The thing that distinguishes the learning activities in the experimental class and the control class is the use of the Guncang Bumi snakes and ladder game. The experimental class used the Guncang Bumi snake and ladder while the control class did not use the Guncang Bumi snake and ladder.

Table 3 . Comparison of Tectonic Earthquake Disaster Preparedness for Experiment Class and Control Class

Class name	Average		Gain Score	Percentage Enhancement
	Pretest	Posttest		
Experiment	57	80	23	67.5%
Control	60	71	11	32.3%

Sources: Analysis or Primary Data (2020)

Based on the results of the average calculation of the pre-test and post-test scores of the experimental and control classes, it can be seen that the experimental class has a mean pre-test value of 57 and a post-test average value of 80. The control class has an average pre-test value of 60 and a post-test average value of 71. Each class experienced an improvement in tectonic disaster preparedness. The experimental class was 23 points and the control class was 11 points. The improvement in the average preparedness of the experimental group was higher than the control class.

Table 4. Analisis Independent T Test

		Levene's Test for Equality of Variances		t-test for Equality of Means						
		f	Sig	t	df	Sig (2-tailed)	Main Difference	Std. Error Difference	95% Confidence Interval of the Difference	
									Lower	Upper
GAIN	Equal variances assumed	1,752	,194	2,197	37	,034	11,43421	5,20469	,88851	21,97991
	Equal variances not assumed			2,215	34,631	,033	11,43421	5,16213	,95053	21,91789

Sources: Analysis or Primary Data (2020)

Based on the results of the independent t-test analysis it can be seen that the F value of 1.752 with a sig of 0.194. Therefore the sig value> 0,05 then the variance of the two groups is homogeneous. Therefore the interpretation that will be used is t which is at the top (equal variances assumed). T-test results found at a value of 2.197 with sig (2- tailed) 0.034. Therefore the value of sig <0,05 can be concluded that there is a difference in preparedness between the experimental class and the control class. Based on the calculation of improving preparedness score, an improvement in disaster preparedness score for the experimental class was 23 points with a percentage of 67.65% and an improvement in preparedness for the control class by 11 points with a percentage of 32.35%. The two classes have in common the improvement in preparedness but the increase is higher in the experimental class that uses Guncang Bumi snake and ladder. Therefore it can be concluded that the use of the Guncang Bumi snake and ladder can improve preparedness for a tectonic earthquake disaster.

4. Conclusion

Based on the independent t-test analysis obtained sig <0,05, it can be concluded that there is a difference in preparedness between the experimental class and the control class. Based on the calculation of the improvement in preparedness score, the percentage of improvement in disaster preparedness for the experimental class is 67,65 %. The percentage improvement in preparedness for the control class was 32,35%. Both classes have improved but the improvement over much experimental class. Therefore it can be concluded that the use of the Guncang Bumi snake and ladder can improve preparedness for a tectonic earthquake disaster.

5. Acknowledgments

This research was supported by the Geography Education Masters Program and Postgraduate of Yogyakarta State University.

6. References

[1] Paton, D. 2009. Living on the Ring of Fire: Perspectives on Managing Natural Hazard Risk in Pacific Rim Countries. *Journal of Pacific Rim Psychology, 3 (01), 1–3.* DOI: 10.1375 / prp.3.1.1.

[2] National Disaster Management Authority. 2010. *Indonesia Information Data Disaster: Natural Disaster in Yogyakarta province in 2010.* Retrieved from www.bnpb.dibi.go.id.

[3] National Planning and Development Agency. 2006. Public Disclosure Authorized: Initial Assessment of Damage and Disadvantage of Natural Disaster in Yogyakarta and Central Java. Jakarta: National Planning and Development Agency.

[4] National Planning and Development Agency. 2006. Earthquake Mitigation Action Plan in Central Java Province and Yogyakarta. Retrieved from http://bappenas.go.id.

[5] Bird, DK, Gisladottir, G., & Dominey-Howes, D. 2010. Volcanic risk and tourism in southern Iceland: Implications for hazard, risk and emergency response education and training. *Journal of Volcanology and Geothermal Research, 189 (1-2), 33–48.* DOI: 10.1016 / j.jvolgeores.2009.09.020.

[6] Bao, K., Wang, G., Lu, X., & McLaughlin, NB. 2011. A potential flood hazard is caused by Tianchi volcano eruption in Changbai Mountain, Northeast China. *Journal of Mountain Science, 8 (5), 677–681.* DOI: 10.1007 / s11629-011-2209-1.

[7] Mulyati, T. 2009. Snakes and Ladders Learning One Alternative to Increasing Results Studying History of Social Sciences Grade XI Students in SMA Negeri 1 Musuk Semester 2 Academic Year 2007-2008. *DIDAKTIKA Journal Vol 1 No 1.*

[8] Izzati, RE., Suardiman, SP., Ayriza, Y., et.al. 2013. *Student Development.* Yogyakarta: UNY Press.

[9] Christanto, J. (2011). Earthquakes, Environmental Damage, Management Policies and Strategies. Yogyakarta: Liberty.

[10] Xu, J., & Lu, Y. 2018. Towards an earthquake-resilient world: from post-disaster reconstruction

to pre-disaster prevention. *Environmental Hazards Journal, 17 (4), 269-275*. DOI: 10.1080 / 17477891.2018.1500878.

[11] National Disaster Management Authority. 2012. Resilient Pocket Book Tough Facing Disasters. Retrieved from http://www.bnpb.go.id.

[12] Shree, GGK, Premalatha, MR, & Pushpa, J. 2016. Snake and Ladder Nutritional Game for Enhanced Health Status of Obese Children. *Journal of Advance Research and Review Vol.1, No.6, 196–207.* Available at https: /ijjar.in/Admin/Pdf/103.pdf/.

[13] Molenda, M. 2003. In Search of the Elusive ADDIE Model. *Performance Improvement, 42 (5), 2003, 34-36*

[14] Ampera, D. 2017. ADDIE Model Through The Task Learning Approach In Textile Knowledge Course In Dress-Making Education Study Program Of State University Of Medan. *International Journal of GEOMATE 30 (12), 109 – 114.*

[15] Morrison, GR. 2010. *Designing Effective Instruction* (6th Edition.). John Wiley & Sons.

The Preparedness Level of Housewives in Dealing with the Earthquake Disaster in Tempel, Sidomulyo, Bambanglipuro, Bantul

F Latifah, S A Sutrisnowati

Departement of Geography Education, Faculty of Social Science
Universitas Negeri Yogyakarta, Jl. Colombo No.1 Yogyakarta, 55281, Indonesia

Email: fauziah.latifah2016@student.uny.ac.id

Abstract. One of geological disasters which cannot be predicted when it will occur is an earthquake. On May 27, 2006, an earthquake struck Bantul which cause many fatalities. The highest fatalities in Tempel were housewives. This study aims to know: (1) the preparedness levels of housewives in dealing with the dangers of earthquakes, and (2) the improvement efforts of housewives in dealing with the dangers of earthquakes in Tempel, Sidomulyo, Bambanglipuro, Bnatul. The method that used in this research is a qualitative method with a quantitative approach. The geographical approach used in this study is the environmental approach. This is a population study. The population of the study is all of the housewives in Tempel as many as 111 housewives. The data were obtained by using observations, interview, and documentations. The data analysis techniques used were a frequency table and descriptive analysis. The result of the study show that : (1) the preparedness levels of the housewives in dealing with earthquake dangers is on category of "full ready" (43,24%). Knowledge and attitude variables the housewives is on category of full ready (34,23%). Emergency plan variables the hosewives is on category ready (45,94%). Early warning system variables the housewives is on category ready (37,82%). Resource mobilization capabilities variables the housewives is on category ready (43,24%). (2) the improvement efforts to dealing with earthquake dangers are by conducting earthquake trainings and simulations as well as evacuation route installations.

1. Introduction

Indonesia is located at the meeting of three tectonic plates, the Indo-Australian plate in the south, the Eurasian plate in the north, and the Pacific Plate in the east. It is moved and collide with each other and caused an earthquake line and active volcanoes along Sumatra until South Maluku. This geological location also affect many disaster like earthquake, landslide, tsunamis, and volcanic erruption. From all of the geological disasters, earthquakes are the most dangerous disaster cause it cannot predicted when it will occur. Unpreparedness in dealing this disaster will cause even greater potential losses including from physical, phsycology, socio-economic, and cultural aspects. Seeing the impact of an earthquake disaster need to improve preparedness. Indonesia is the site of 12-15% of earthquakes on a world scale. Every year there are 6-12 earthquakes in Indonesia.

Special Region of Yogyakarta is one of the provinces in Indonesia which is prone to volcanic and tectonic earthquakes. Volcanic earthquakes are caused by the presence of an active volcano Mount Merapi and tectonic earthquakes caused by the geological position of the Special Region of Yogyakarta which is located on a plate subduction route. According to historical records there have been several large earthquakes that have occurred in this area. One of the earthquake disasters that caused DIY to

Content from this work may be used under the terms of the Creative Commons Attribution 3.0 licence. Any further distribution of this work must maintain attribution to the author(s) and the title of the work, journal citation and DOI.
Published under licence by IOP Publishing Ltd

experience a quite severe impact was an earthquake that occurred on May 27, 2006, at 05:54 WIB with the Ritcher scale 5.9, centered at 8° South Latitude and 110° East Longitude to be precise in the Opak Oyo River Fault around 25 km to the southwest of Yogyakarta City was destructive and caused the fatalities of more than 6000 people, tens of thousand were injured from bruises to fractures, and hundreds of rhousand buildings were severely damaged to totally destroyed [1]. The cause of the earthquake, which has a depth of 17 km below the ground surface, is the shift in the Opak Fault which stretches from the Bantul Coast to Prambanan along 40 km in a direction of 30° NE (North East). The death toll from this earthquake reached 5,738 people. The number of casualties in Yogyakarta was 4,143 people and 37,972 people was injured in this event. This earthquake destroyed 156,662 houses and destroyed historical sites both in Bantul and Yogyakarta. Bantul Regency is the district are the most seriously affected by this incident.

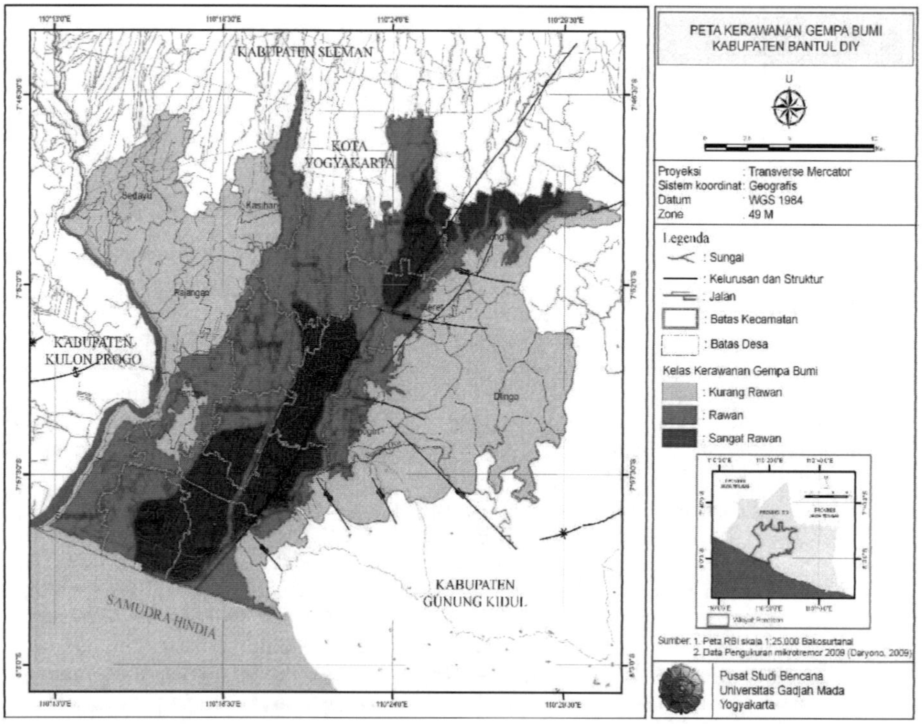

Figure 1. Eartquake Vulnerable Map of Bantul Regency

Of the seventeen districts in Bantul Region, there are five sub-districts that were most seriously affected by the 2006 earthquake, including Jetis District, Bambanglipuro District, Pleret District, Sewon District and Pundong District. According to data from BPBD Bantul, the number of casualties caused by the 2006 earthquake in Bambanglipuro District was 607 people. Of the many casualties in Bambanglipuro District, the victims in Sidomulyo Village were 78 people. The number of casualties and damaged buildings is because Sidomulyo Village is one of three villages in Bambanglipuro District which is directly adjacent to Pundong District, which is the area where the 2006 earthquake was located. The number of hamlets in Sidomulyo Village is 15 hamlets. Each hamlet has a different number of victims. One of the differences in the number of victims in each hamlet is influenced by the level of community preparedness.

Tempel is an area that has quite a lot of victims, totally 10 people. Of the ten victims, six of them were housewives, the rest were three toddlers and an elder. There are several reasons why housewives were the most victims of the 2006 earthquake. This is because housewives focus on taking care of domestic matters in the morning and preparing breakfast for their families. In addition, there were also those who ran into the house to save their child, but in the middle of the process a building fell down and resulted in the mother and child to dead. According to the research data, the number of housewives in Tempel Hamlet is 111 people spread over five RT (Rukun Tetangga) and most of the housewives are at home to take care of their domestic affairs. There are only a few housewives who have other jobs such as teachers, doctors, farmers, traders and factory workers.

At the time of the 2006 earthquake, the number of casualties indicated that the community, especially housewives, needed to know specific strategies to live in harmony with nature in earthquake-dangers areas. One of the efforts that can be done is to minimize losses due to the earthquake disaster with the provision of empirical experience of the 2006 earthquake with how to increase preparedness. Preparedness itself is a series of activities carried out to anticipate earthquake disasters through appropriate and efficient steps. This aims to monitor the preparedness of housewives to face earthquakes that can occur at any time as well as to find out efforts or efforts to increase the preparedness of housewives. The preparedness of a society or community can decrease over time according to changes and conditions, both social-economic, and cultural. Until now, there is still no research that examines the preparedness level of housewives in Tempel and the follow-up, namely efforts to improve the preparedness of these housewives. The level of vulnerability of women is higher than men. Women here are also one of the groups that vulnerable to earthquake disasters apart from the disabled, the elderly and children. Women also have a vital role so it is necessary to be directly involved in earthquake disaster preparedness because the mother is the de facto ruler of the household, has the nature of protecting her family, active in social groups, and a preparedness educator agent for herself, the family, and the environment. The existence of this dual role is the reason why it is important to measure the level of preparedness of housewives.

One of the impacts of knowing the level of preparedness of housewives in Tempel is to minimize the number of casualties if an earthquake occurs. Condition of preparedness housewives will greatly affect the level of preparedness of children and families. This because housewives become the first teaching agent for children as a vulnerable group. Based on this background, the authors feel he need to study and conduct relevant research with comunity preparedness, especially housewives.

2. Research Methods

The research design used in this research is descriptive analysis with quantitative approach. This study seeks to explain housewives preparedness in Tempel in dealing with earthquake disaster and the improvement efforts of housewives to dealing with a danger of earthquake in Tempel. This research conducted on Januari- Juni 2020. The research place is Tempel, Sidomulyo, Bambanglipuro, Bantul, D.I. Yogyakarta. Bantul Regency is a major area affected by the earthquake on May, 27th 2006. The variable in this research are (1) the preparedness level of housewive in dealing with the dangers of earthquakes, which can be seen from indicator: (a) knowledge and attitude, (b)emergency response plan, (c) disaster warning system, (d) resources mobilization, and (2) the improvement efforts of housewives to dealing with dangers of earthquake in Tempel, Sidomulyo, Bambanglipuro, Bantul.

Table 1. Framework for Housewives Preparedness Level on dealing with Earthquake disaster

Variable	Indicator	Sub Indicator
Knowledge and Attitude	Knowledge of Eathquake	• Understanding of natural disaster • Events that cause disaster • The cause of earthquake • Effect of earthquake • Actions taken in the event of an earthquake
Emergency Response	Related to evacuation and rescue.	• Evacuation Plan • Evacuation Transportation • Evacuation Equipment and supplies • Evacuation sites • Practice and simulation

Disaster Warning System	Warning signs and distribution of infiormation disaster to reduce casualities	• Knowledge of traditional warning system • Knowledge of modern warning system
Resource Mobilization	Knowledge of Eathquake	• Participating on training, workshop, or lecture • Frekuensi keikutsertaan pelatihan • Finance • Social network

The research is population study. The population in this study were all housewives in Tempel with total of 111 housewives. All members of the population is a housewives that lives in Dusun Tempel minimal since on 2006 and have earthquake's experience.

The data collected in this research are 2 (two) types, they are: (1) primary data, relating to the preparedness level of housewive in dealing with earthquake disaster and the improvement effort in dealing with earthquake disaster, and (2) secondary data, in the form of information about the physical, social, economic, and demographic conditions in the research location, which is expected to support the research findings.

Data collection techniques in this study are: (1) observation, carried out by observing the conditions around the research location to obtain an overview of the study area, (2) interviews, conducted to respondents using questionnaires to obtain information about housewives preparedness level that seen from indicator: knowledge and attitude, emergency response plans, disaster warning systems, and resource mobilization. Interviews with key informants using a questionnaire were used to obtain information about improvement effort in dealing with earthquake disaster, and (3) documentation, carried out by taking data, such as administrative maps, population data, and other data that can support research activities.

The data analysis used in this research is quantitative and qualitative descriptive analysis. The description of preparedness level of housewives in delaing with eartquake disaster is obtained from the results of preparedness parameter scoring. Scoring is done to change the information that has been obtained into data in the form numbers. The answer "Totally Agree" on each research instrument will be given a value of 5, the answer "agree" is given a score of 4, the answer "uncertain" is given a score 3, the answer "not agree" is given a score 2, and the answer "totally not agree" is given a score 1. Determination of the number of scores for each class is based on the range of minimum values up to the maximum range and the number of classes desired namely "full ready", "ready", "ready enough", "less ready", "not ready", so that interval determination class is obtained by formula:

$$\text{Length of class interval} = \frac{maximum\ value - minimum\ value}{number\ of\ classes\ (=5)}$$

3. Results

3.1. Decription area

Tempel Hamlet is located at 110°17'33'BT - 110°17'43'BT and 7°57'21'LS - 7°57'56'LS. , which is one of the hamlet in Bambanglipuro District, Bantul Regency, Daerah Istimewa Yogyakarta Province. Tempel hamlet has an area of 283,047 m^2 which is divided into 5 RT administrative areas. The population is 542 people with a population density of 1779 / km^2. The area of Tempel hamlet is located at an altitude of 140 meters above sea level. Tempel Hamlet temperature is 26,18 °C. The highest level of education in Dusun Tempel is high school graduates at 30.44% and the lowest is not attending school, which is 3.87%.

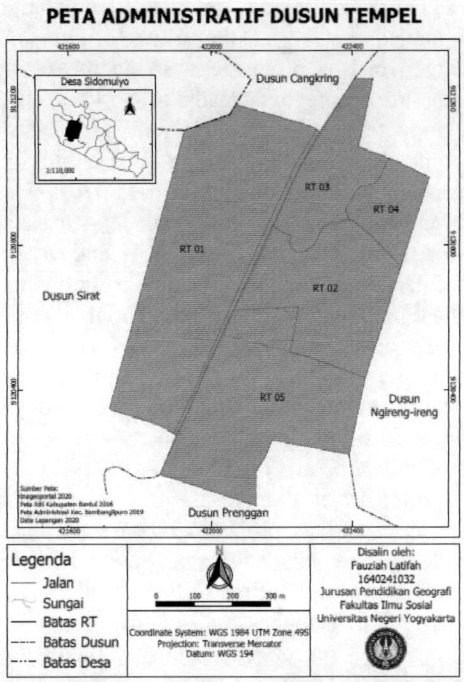

Figure 2. Administrative Map Tempel Hamlet

3.2 Preparedness Level of Housewives in Dealing with Earthquake Dangers in Tempel, Sidomulyo, Bambanglipuro, Bantul.

The study of housewive preparedness level in Tempel hamlet in dealing the earthquake disaster using a framework developed by LIPI in collaboration with UNESCO/ISDR in 2006. The housewives preparedness level is based on 4 (four) indicators, they are: (1) knowledge and attitude of phenomena earthquakes, (2) emergency response plans, (3) disaster warning systems, and (4) resource mobilization. Based on the 111 respondents surveyed, the level of housewives preparedness in Tempel hamlet was in the "full ready" category, which was 43% of all respondents. The following are four parameters used in measuring the level of housewives preparedness in dealing with earthquakes disaster.

Table 2. Preparedness Level of Housewives in Tempel on dealing with Earthquake disaster

No	Indicator	Category (%)					
		Full Ready	Ready	Ready Enough	Less Ready	Not Ready	Total
1	Knowledge and Attitude	**34,23**	30,63	27,02	08,10	00,00	100
2	Emergency Response Plan	21,62	**45,94**	24,32	08,10	00,00	100
3	Disaster Warning System	26,12	**37,83**	24,32	10,81	00,90	100
4	Resource Mobilization	27,02	**43,24**	00,18	09,90	00,90	100
5	All Indicator	**43,24**	28,82	25,22	6,30	00,00	100

Source: Primary Data through Processing (2020)

Based on table 2, it can be seen that the preparedness level of housewives in Tempel Hamlet in knowledge and attitude are in category "full ready" (34,23%). Knowledge and attitude parameters is a major factor in preparedness. Knowledge will affect the attitude of a person. This knowledge and attitude is very important for people living in disaster-prone areas. So that the resulting impact can be minimized.

The emergency response plan are in category "ready" (45,94%). This plan is an important part of preparedness, especially in relation to evacuation, help and rescue, so that disaster victims can be minimized. This effort is very crucial, especially during a disaster and in the first days after a disaster before assistance from the government and from outside parties arrives. The evaluation of the emergency response plan variables has four indicators, namely earthquake evacuation plans, earthquake evacuation transportation tools, earthquake disaster evacuation equipment, and earthquake training or simulations. Here are the levels earthquake emergency response plan in Tempel hamlet. The level of preparedness of housewives, especially in the emergency response plan variable in table 2 shows that as many as 45.94% of respondents are in the ready category. Respondents who already know the gathering point for all family members at home, the main priority of which group should be rescued immediately, transportation tools and evacuation equipment that need to be prepared, and knowing the importance of participating in earthquake training or simulations.

The disaster warning system indicator are in category "ready" (37,83%). This system includes warning signs and distribution of information about disasters. With this disaster warning, the community can take appropriate action to reduce casualties, property and environmental damage. Therefore, training and simulation are needed, what to do if you hear a warning, how to take steps to save yourself according to the location where the community was at the time of the warning. The following is the level of the early warning system for housewives in Tempel Hamlet. Early warning system parameters are in the ready category. This is based on 37.83 respondents who received information about the source of information in the form of two indicators, namely a traditional disaster warning system and a modern disaster warning system. Traditional disaster warning systems include kentongan, sirene, or bells. A modern disaster warning system can be done by following information from social media or television broadcasts. One example is following information from the BMKG's Instagram, Twitter, Facebook and another social media.

The resource mobilization indicator are in category "ready" (43,24%). Available resources, both human resources (HR), as well as funding and essential infrastructure for emergencies are potentials that can support or otherwise become obstacles in natural disaster preparedness. Therefore, resource mobilization is a crucial factor. The parameter of resource mobilization ability shows that, 70% of respondents is in the ready category. This is based on 60 respondents who already know what the potential supports and constraints, such as human resources, technical guidance such as training and provision of materials on preparedness, funding which can be in the form of savings, insurance or separate fund allocations, facilities and facilities. infrastructure, etc. The following is the level of ability to mobilize housewives' resources in Tempel Hamlet. The human resource that is meant is the presence of one of the family members who has received material, socialization or training at school or in the community regarding earthquakes. Support from the government in disaster preparedness, demonstrated by the provision of shelters and the installation of evacuation routes, greatly supports the readiness of the community in mobilizing resources.

Overall the preparedness level of housewives in dealing with the earthquake disasters are categorized as "full ready" (43,24%).

3.3. *Improvement Effort to increase the Preparedness Level of Housewives in Dealing with Earthquake Dangers in Tempel, Sidomulyo, Bambanglipuro, Bantul.*

Government and community efforts to increase preparedness in the face of earthquakes are something that must be done in disaster-prone areas. There is no specific policy regulating earthquake disasters. Government policy discusses disasters in general in Bantul Regency. There are nine disaster threats in Bantul Regency, one of which is an earthquake. This form of earthquake disaster response is carried out by socialization and training conducted at schools, villages, or village health centers where the BPBD will be the resource person. Every village in Bantul Regency has priority disasters that threaten the area

So that disaster-prone villages emerged. The village was named Desa Tangguh Bencana (DESTANA) and was fostered by the BPBD. In addition, efforts to improve disaster preparedness were carried out by holding comprehensive meetings between the village government, the community and the BPBD, which were held at the request of the village government.

The form of training held by the government at the village level is in the form of socialization, training and disaster simulation. Parties involved in training and simulations such as cadres of Puskesmas, representatives of housewives per hamlet and youth organizations.

The disaster organization that drives every disaster management activity at the Village level is the FPRB (Regional Disaster Management Forum). The FPRB is voluntary and is a partner of the BPBD. Until now, the FPRB of Sidomulyo Village is active. The activities of FPRB Sidomulyo include socialization and earthquake simulations in commemoration of Disaster Preparedness day with the target community representatives of each hamlet, as well as schools. The disaster simulation training involved 30 representatives of youth organizations, PKK, village officials and the Indonesian National Armed Forces (TNI). The group is divided into 6 small groups that are prone to disasters. Each group discussed the mitigation that will be carried out with the disaster. The results of the discussion were presented and evaluated. Then from the six disaster threats, priority was obtained for disasters that threaten the area. The disaster was simulated jointly by dividing the tasks. This simulation will involve and be evaluated by representatives of groups that are prone to earthquakes such as the diffable, the elderly, children, and women. From this small simulation, the follow-up was a field rehearsal which was attended by around 200 people. The duties of the PKK's representative following training is communicating to other housewives who did not attend the training in the hamlet. This educational process lasted quite a long time. This educational process is conveyed repeatedly and continuously in order to form a habit to be alert to the danger of earthquakes in the Tempel Hamlet environment.

Facilities and infrastructure used in supporting activities to improve community preparedness, including disaster-prone maps and evacuation route signposts, however, not all villages exist. The facilities and infrastructure used in the efforts to disseminate information included leaflets and billboards. The level of preparedness of housewives in facing earthquake hazards is measured using four variables, namely knowledge and attitudes, emergency response plans, disaster warning systems, and resource mobilization capabilities. The level of preparedness of housewives in Tempel Hamlet in the face of an earthquake is in the very ready category of 43.24%. The variable of knowledge and attitude of housewives was in the very ready category of 34.23%. This is supported by educational factors and economic factors. The emergency response plan variable is in the ready category of 45.94%. The early warning system variable was in the ready category at 37.83%. The variable of resource mobilization ability is in the ready category of 43.24%. Of the four variables, the variables that occupy the most prepared positions are knowledge and attitudes.

Efforts that have been made by the government in collaboration with related agencies include holding training or earthquake simulations. In addition, the provision of several facilities such as evacuation route boards, earthquake hazard maps at the village level.

4. Conclusion

From this research, we knew that earthquakes can occur anytime. Housewives need to follow information from either BMKG or local government and saving telephone numbers of relevant agencies such as, ambulances, BMKG, fire brigade, hospital, police office, etc. The government needs to add facilities and infrastructure such as evacuation route boards and assemble point, conduct even training and simulations for PKK members. The community must synergize with government to increase the preparedness level of the community.

5. Acknowledgment

In this occassion, We thanks from Faculty of Social Sciences which has provided research opportunities until the author can write this article, support and facilities for us, Mrs. Sri Agustin Sutrisnowati and all in Geography Education, Faculty of Social Science, Yogyakarta State University lecturers who have always accompany, guide and motivate us. Parents, family, and friends, who always support us who cannot be mentioned one by one.

References

[1] BPS Kabupaten Bantul. *Kecamatan Bambanglipuro Dalam Angka 2018.* Bantul: BPS Kabupaten Bantul

[2] D.I.Yogyakarta, B (2007) *Profil Kebencanaan DIY* Yogyakarta: BAPEDA.

[3] Hidayati, D (2006) *Kajian Kesiapsiagaan Masyarakat dalam mengantisipasi Bencana Gempa Bumi* Jakarta: LIPI-UNESCO/ISDR.

[4] Kumalasari (2014) *Manajemen Bencana dan Kapabilitas Pemerintah Lokal*. Yogyakarta: Gava Media

[5] Law of the Republic of Indonesia No. 24 of 2007 concerning *Disaster Management.*

[6] LIPI-UNESCO/ISDR (2006) *Community Preparedness Study in Anticipating Earthquake and Tsunami Disasters*

[7] Mineral, K.E. (2015) *Gempa Bumi dan Tsunami* Bandung: Kementrian Energi dan Sumber Daya Mineral.

[8] Purwantara, S., Khotimah, N., Dwiningrum, S.IA. (2016) Development of Disaster Mitigation Education Based on Disaster Preparedness Schools in D.I. Yogyakarta *Institutional Research.* Yogyakarta: Studies Center for Disaster Mitigation UNY

[9] Rakhman, A.N. and Kuswadarni, I (2012) Case Study of the Yogyakarta Earthquake 2006 Empowering Local Wisdom as a Resilient Community Capital to Disaster *Proceedings of National Conference* on Application of Science & Technology, third Period, Yogyakarta, November 3, 2012, ISSN: 1979-911X.

Assessing local community resilience to tsunami in a small port (Case Study: Sadeng Port, Indonesia)

F Nucifera[1], S T Putro[2], W D Setiawan[1]

[1]Department of Geography, Faculty of Science and Technology. Universitas AMIKOM Yogyakarta, Indonesia.
[2]Department of Geography Education, Faculty of Social Science. Universitas Negeri Yogyakarta, Indonesia.

E-mail: [1]fnucifera@amikom.ac.id

Abstract. Coastal community is vulnerable to hazards including tsunami hazard. Coastal communities suffer the most both physical and psychological impacts. The main purpose of this research to assess local community resilience to tsunami in small port which has a unique characteristic. Main data of this research was collected by field observation and interview. Interview process were addressed to key persons and households. USAID coastal community assessment is used as main method in this research. There are eight elements of resilience for coastal community resilience: governance, society and economy, coastal resource management, land use and structural design, risk knowledge, warning and evacuation, emergency response and disaster recovery. The results show that overall rating for eight elements of community resilience is 2,5 or 50 %. It is only halfway to reach an ideal criterion for community resilience. Strong elements in this community are risk knowledge and warning and evacuation, while weak elements are society and economy, coastal resource management, and disaster recovery. The community should gain more effort especially in developing a sustainable economy and strengthening integration among stakeholders for disaster recovery.

Keywords: resilience, local community, tsunami

1. Introduction

Building resilient communities have been recognized as a focus on disaster risk reduction. Improving institution structure and capabilities to build hazard resilience has been mentioned as one of the strategic goals in the Hyogo Framework for Action 2005-2015 [1]. Sendai Framework for Disaster Reduction 2015-2030 renewed commitment for building resilience in focused on sustainable development and poverty reduction [2]. Resilience is the capability of a system, community, or society threatened to hazard to survive, adapt and get back from the effects of hazard in a reliable condition [3]. Building community resilience requires an integration of several stakeholder elements.

Coastal community is vulnerable to hazards including tsunami hazard. Indian Ocean tsunami 2004 became a turning point in terms of disaster risk reduction, especially in Indonesia. Coastal communities suffer the most both physical and psychological impacts [4]. Local communities in Indonesia today are highly vulnerable to disaster and threatened by future disasters [5]. As many as

Content from this work may be used under the terms of the Creative Commons Attribution 3.0 licence. Any further distribution of this work must maintain attribution to the author(s) and the title of the work, journal citation and DOI.
Published under licence by IOP Publishing Ltd

tsunami events in the world occurred in Indonesia with the major events located in Aceh, Java, Flores, Mentawai, Halmahera, Sulawesi, and Papua [6]. Communities in those locations particularly live in high risk of tsunami hazard.

Resilient communities are required for enhancing disaster risk reduction. Resilient communities have six characteristics include: a) knowledgeable and healthy; b) well organized; c) well connected; d) good infrastructure and services; e) high economic opportunities; and 6) capability to manage its natural assets [7]. Also, resilience communities are capable to cope the damages resulted by hazard and recover by utilizing their own resources [8]. Tsunami-resilient communities have the following criteria: understanding tsunami hazard, having plans and tools for tsunami mitigation, sharing tsunami hazard information, understanding how to save themselves, having adaptation and recovery action plans to get back better [9],[10].

Port is one of the centers of activities in coastal area that constantly impacted by tsunami [11],[12]. Activities at the port can be transportations, trades, and fisheries. Port is prone to earthquake, tsunami, and liquefaction hazard due to its location close to sea level and built on unconsolidated material [13]. Improving tsunami resilience for port communities became main priority to reduce disaster impacts [14].

Sadeng is a small port in Gunungkidul coastal area which is directly facing Indian Ocean. The main function of this port is as a fishing port. Apart from having various infrastructures to support fishing port function, Sadeng port is inhabited by fishermen's group who are partly nomadic and come from various regions in Indonesia [15]. Villagra et al [16] mentioned that level of resilience may vary based on their social, politic, urban, rural, and native characteristics of certain area. Morin et al [17] stated that people who are well-educated about tsunami basic knowledge tend to be survived from tsunami events. Indigenous knowledge provided valuable knowledge for people to act during pre-tsunami, tsunami, and post-tsunami [18]. Currently, previous researches focused on the characteristics on tsunami-resilient communities whereas strength and weakness in local community to be tsunami-resilient still have deficient discussion.

The main purpose of this research to assess local community resilience to tsunami in small port which has a unique characteristic. Local community resilience assessment is aimed to identify strength and weakness of local community to cope tsunami hazard. The results provide: a) strength elements in community resilience which can be a lesson learned to other regions; and b) weak elements which should be improved to enhance community resilience.

2. Methods

Study area is in Sadeng port, a small port located in Songbanyu Village, Gunungkidul Regency, Yogyakarta. Sadeng port is situated in a bay and valley surrounded by a high and steep cliff on both sides (Fig 1). Land use types in Sadeng port includes port infrastructure, fish market, settlements, and dry agriculture land [15].

Figure 1. Research area location

USAID [4] has developed coastal community framework and benchmark for disaster management. There are eight elements of resilience for coastal community resilience: governance, society and economy, coastal resource management, land use and structural design, risk knowledge, warning and evacuation, emergency response and disaster recovery (Fig 2). Description for each elements of coastal community resilience is shown in Table 1. All elements in coastal community resilience are related and influence each other. The ratings in each element are manifested with a rating ranging from 0 to 5 (Table 2). This rating does not show the level of resilience but shows the current condition of resilience compared to ideal resilience conditions.

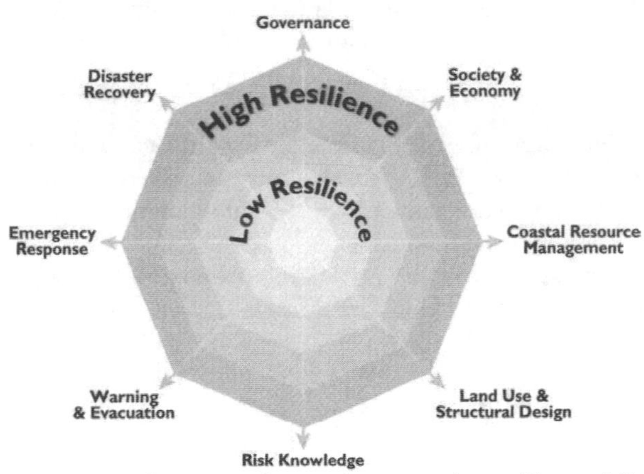

Figure 2. Elements of coastal community resilience [4]

Table 1. Description of elements in coastal community resilience [4]

Elements	Description
Governance	Government supports communities to adapt and get back better by providing policies, programs, leadership, and legal framework
Society and economy	Society and economy give direct influences on

	communities' livelihood in term of disaster preparedness, adaptive capacity, and recovery speed
Coastal resource management	Coastal resource management functions to manage and maintain valuable resources in coastal area to support community's live both pre-disaster and post-disaster
Land use and structural design	Useful land use management and well-structured infrastructure minimize potential impacts of disaster
Risk knowledge	Risk knowledge rise community awareness about potential hazard and negative impacts of disaster for them
Warning and evacuation	Warning and evacuation are a crucial part in disaster management. Effective warning and evacuation prepare communities to act faster during the disaster events
Emergency response	Adequate emergency response supports communities to protect their lives
Disaster recovery	Disaster recovery helps communities to recover and get back better from the disaster impacts in term of livelihood and basic services

Table 2. Rating for elements of coastal community resilience [4]

Rating	Description
0	Condition absent
1	Poor (1 to 20 percent fulfilled)
2	Fair (21 to 40 percent fulfilled)
3	Good (41 to 60 percent fulfilled)
4	Very Good (61 to 80 percent fulfilled)
5	Excellent (81 to 100 percent fulfilled, sustainable)

Main data of this research was collected by field observation and interview. Interview process were addressed to key persons and households. Key persons in this research were leader of fishermen group, leader of SAR (Search and Rescue) and local government officer. The key persons were determined by snowball sampling method. Sadeng port population consists of around 300 persons and 80 households. We interviewed all of households in the Sadeng port. Interviews on key persons and households were conducted to obtain more accurate data from two different perspectives for each element.

3. Results and Discussion

3.1. Governance

Good governance facilitates community to enhance their resilience through appropriate policies, programs, and leadership. There are several benchmarks for assessing governance elements that is showed in Table 3. Local government represented by BPBD (National Agency of Disaster Management) Gunungkidul has conducted regular tsunami risk assessment and evaluation with uncertain timing range. The tsunami risk assessment is held by BPBD in collaboration with NGOs and academics party. Tsunami risk assessment involves the community, especially DESTANA (Desa Tangguh Bencana) officials. Government launched DESTANA program to enhance community resilience to tsunami for coastal villages. DESTANA was established at the village level but in the

IOP Conf. Series: Earth and Environmental Science **884** (2021) 012044 doi:10.1088/1755-1315/884/1/012044

Sadeng port case, only few of fishermen community were involved in DESTANA. DESTANA is a program originating from the central government and implemented at the village level. The potential resources and culture of the fishermen community in Sadeng port have not been considered in the disaster risk management program.

The overall rating for governance elements is 2.5 or 50% of the ideal rating. The role of government in improving community resilience has been sensed by the community. However, it needs to be improved in involving local communities more in the disaster risk reduction program. The local community can be actively involved in the program so that the program can be implemented in line with the characteristics of community.

Table 3. Benchmark and rating of governance

Benchmark	Rating
Government assess and evaluate tsunami risk routinely	3
Government involves community to conduct tsunami risk assessment	2
Government provide plans and programs to enhance community resilience	3
Government provide basic needs (water, health, sanitation, education, security) efficiently	3
Public facilities and infrastructure remain operated during disaster events	2
National and local government policies related to tsunami risk management are well integrated	3
Disaster risk management programs consider potential resource and culture of the community	2
Average	2,5

3.2. Society and economy

Society and economy directly influence social vulnerability. People in poverty tend to be more difficult to bounce back when a disaster occurs. For marginalized communities, assistance from society will be valuable to restore their lives after a disaster [19]. Benchmarks on society and economy elements are described in Table 4. This element focuses on the strength of the organization in the community and the economic capacity to deal with the tsunami. The fishermen community in Sadeng port has a specific organization in tsunami disaster management consisting of community coordinators and a SAR team. This organization already has a good tsunami disaster mitigation program. Fishermen community has mutual funds, but the funds are not specifically intended for tsunami emergency response and recovery. However, it does not rule out if the funds are used during an emergency response.

All respondents in Sadeng port have their main livelihood as fishermen. As many as 51% of the respondents have a side job apart from fishermen include trading and farming. In terms of livelihood diversity, people in Sadeng port are very dependent on the fisheries sector. Dependence on one economic sector can lead to poor resilience. If a tsunami occurs, their livelihoods will be lost. The government through the Ministry of Maritime Affairs and Fisheries provides training programs for alternative livelihoods for fishermen and provides programs related to fishermen economic development. The training consists of industrialization of fisheries and quality life improvement. However, this program is not sustainable, so the effects of this program have not been seen.

As many as 91% of the people in Sadeng port do not have emergency funds for disasters. In addition, they also do not have sufficient assets to support their life post-tsunami. Most of the assets are movable assets such as motor vehicles and ships. Asset in property is only owned by 3% of all respondents. The existence of this asset is very important in relation to the ability to bounce back better after a disaster. If all the assets are lost by the tsunami, the community will have difficulty building their lives again. The elements of society and economy are classified as low with a rating of 2 or 40%. This element needs to be improved a lot, especially in the economic aspect, includes diversification of livelihoods and addition of economic assets.

Table 4. Benchmark and rating of society and economy

Benchmark	**Rating**
Community have a specific organization for tsunami risk reduction	3
Community provide tsunami mitigation program	3
Community have an insurance and assets for recovery process after tsunami	1
Community have a special fund collected among them for tsunami emergency response and recovery	1
Community have various livelihood, not only one type of livelihood	2
Existence of institution supported sustainable economic development	2
Existence of training program for alternative livelihood	2
Average	2

3.3. Coastal resource management

Coastal resource management planning and strategies are conducted by local governments (Table 5). Coastal resource assessments are conducted regularly but with uncertain timeframes. Community involvement in planning and implementation of coastal resource management is still low. The community has realized the importance of coastal resources for their lives, but the community has not had real action to implement coastal resource management. This element has a mean rating of 2 or 40%. This element needs to be improved by increasing community awareness to participate in coastal resource management.

Table 5. Benchmark and rating of coastal resource management

Benchmark	Rating
Existence of coastal resource management planning and strategies	2
Regular assessment of coastal resource potential	2
Existence of coastal resource zonation for conservation	3
Involvement of community in planning and implementing coastal resource management	2
Community is aware of the importance of coastal resource management and invest to sustain them	1
Average	2

3.4. Land use and structural design

Benchmark for land use and structural designed is shown in Table 6. Earthquake and tsunami resistant building standards are regulated by the central and local governments. This is implemented in government buildings and infrastructure located at Sadeng port. Regional spatial planning is regulated by the local government. Based on the spatial pattern plan for Gunungkidul Regency in 2010-2030, the Sadeng Port area is not designated as a settlement area but dry land agriculture. The fishermen community in Sadeng port has knowledge of tsunami resistant structures. However, most of the residential buildings in Sadeng port are vulnerable to tsunamis. The buildings are mostly built by wood with a weak structure. This is because the community does not have enough funds to build tsunami-resistant houses and considers the houses to be temporary residences. The land use and structural design elements have an overall rating of 2.2 or 44%. This element needs improvement, especially in the suitability of land use with regional spatial planning and improvement of building quality.

Table 6. Benchmark and rating of land use and structural design

Benchmark	Rating
Existence of standard and regulation about earthquake and tsunami resistant building	3
Appropriate land use planning	2
Design and structure of public infrastructures which consider tsunami hazard	3
Community have knowledge about tsunami resistant building	2
Community live in the tsunami resistant building	1
Average	2,2

3.5. Risk knowledge

Risk knowledge element have several benchmarks explained in Table 7. Tsunami risk assessment are regularly conducted by local government. Community participation in tsunami risk assessment needs to be improved. Fishermen community in Sadeng port is very aware about tsunami risk in their area. Training for disaster management are held regularly by BPBD Gunungkidul and SAR team. Overall for this element have good rating as many as 3,2 or 62 %.

Table 7. Benchmark and rating of risk knowledge

Benchmark	Rating
Regular tsunami risk assessment	2
Community participation in tsunami risk assessment	3
Community is well-understand about tsunami risk	4
Community know tsunami risk at household level in the surrounding area	3
Existence of training and information sharing about tsunami risk	4
Average	3,2

3.6. Warning and evacuation

Benchmark of warning and evacuation element is described in Table 7. Community receive earthquake and tsunami information from BMKG (Agency of Meteorology, Climatology, and Geophysics) server. Community share the information through SMS and social media network among them. Warning and evacuation procedure during tsunami are coordinated by trained volunteer. They ensure that all of people are safe. There is an early warning system for tsunami owned by BMKG and BIG (Geospatial Information Agency). Public information of hazard zone and evacuation route is displayed in strategic places (Fig 3). Warning and evacuation element have a good rating of 3,3 or 66 %. This element needs to be improved in coordination during evacuation process.

Table 8. Benchmark and rating of warning and evacuation

Benchmark	Rating
Community receive emergency information from national and regional system	4
Community share emergency information to others	3
Community have evacuation plan and procedures for tsunami	3
The early warning system is well-functioned and well-maintained	3
The existence of public information about hazard zone, evacuation route, shelters, and safe areas (map and sign)	4
The existence of volunteers in community who provide emergency information and coordinate during evacuation process	3
Average	3,3

Figure 3. Hazard map and early warning system

3.7. Emergency response

Emergency response elements have several benchmarks described in Table 9. Community provide emergency response plan which was coordinated by small group in their neighbourhood. But the emergency response team still do not have detailed and specific tasks so that sometimes they misunderstood. Basic needs include food, water and electricity was fulfilled by government. Community also share their own resource to be used together during emergency response. Government regularly conduct tsunami simulation training once in a month. Overall rating for this element is 2,6 or 56 %. Some improvement is required in detailing tasks for emergency response team.

Table 9. Benchmark and rating of emergency response

Benchmark	Rating
Community have specific emergency response plans	2
Existence of emergency response team with detailed and specific tasks	2
Basic needs (food, water, electricity) during disaster can be fulfilled	3
Community conduct regular training for tsunami emergency response	4
Identification resources owned by community that can be used together during emergency response	3
Average	2,8

3.8. Disaster recovery

Disaster recovery benchmark is explained in Table 10. Local community already have disaster recovery plan, but it is only for short-term goals. Disaster recovery process was conducted mostly by government. Government provide technical assistance for tsunami survivor. This element rating is valued 2 or 40 %. This low rating in disaster recovery elements due to low integration between stakeholders and there were no long-term recovery goals.

Table 10. Benchmark and rating of disaster recovery

Benchmark	Rating
Community have disaster recovery plans which provide long-term recovery goals	1
There is a coordination mechanism with stakeholders regarding disaster recovery	2
Existence of technical assistance for tsunami survivor	3
Average	2

3.9. Community resilience

Based on the analysis of eight elements, coastal community resilience in Sadeng port is described in Fig 4. The highest rating among elements are warning and evacuation and risk knowledge. Both elements become the strength for fishermen community in Sadeng port. Community in Sadeng port already have a good risk knowledge and appropriate program for warning and evacuation. Weakness in the community is shown at the lowest rating elements include society and economy, coastal resource management, and disaster recovery. Local community in Sadeng port need to improve these elements. Overall rating for eight elements is 2,5 or 50 %. It means community resilience in Sadeng port should be increased by another 50 %.

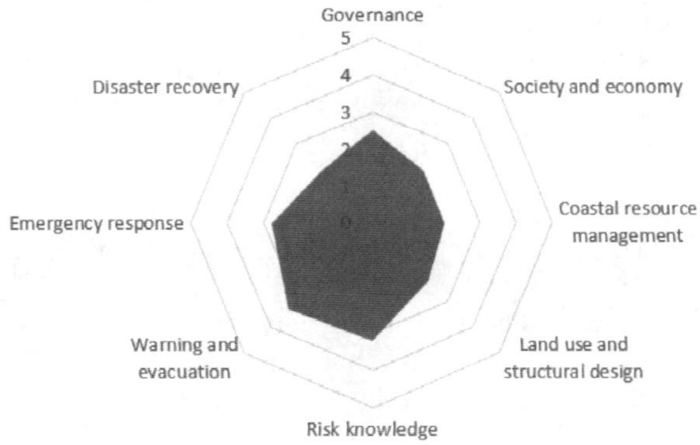

Figure 4. Results of coastal community resilience elements

4. Conclusion

Disaster risk reduction nowadays focus on building community resilience. Overall rating for eight elements of community resilience is 2,5 or 50 %. It is only halfway to reach an ideal criterion for community resilience. Strong elements in this community are risk knowledge and warning and evacuation, while weak elements are society and economy, coastal resource management, and disaster recovery. The community should gain more effort especially in developing a sustainable economy and strengthening integration among stakeholders for disaster recovery.

References

[1] UNISDR. (2005). Hyogo Framework For Action 2005-2015: Building the Resilience of Nations and Communities to Disasters. UNISDR: Geneva.
[2] UNISDR (2015) Sendai framework for disaster risk reduction 2015–2030. UNISDR, Sendai.
[3] UNISDR (2009). UNISDR Terminology on Disaster Risk Reduction. Geneva.
[4] USAID. (2007). How Resilient is Your Coastal Community?. Washington
[5] UNISDR. (2009). Building resilience to tsunami in the Indian Ocean. Kyoto.
[6] UNESCO and NOAA. 2018. Tsunami Sources 1610 B.C. to A.D. 2017. International Tsunami Information Center. Available at www.tsunamiwave.org
[7] Red Cross (2012). Understanding community resilience and program factors that strengthen them. Geneva.
[8] Gaillard, J.-C. (2007), "Resilience of traditional societies in facing natural hazards", Disaster Prevention and Management, Vol. 16 No. 4, pp. 522-44.
[9] Dengler, L. (2005), "The role of education in the national tsunami hazard mitigation program", Natural Hazards, Vol. 35, pp. 141-53.
[10] Jonientz-Trisler, C., Simmons, R.S., Yanagi, B.S., Crawford, G.L., Darienzo, M., Eisner, R.K., Petty, E.R. and Priest, G. (2005), "Planning for tsunami-resilient communities", Natural Hazards, Vol. 35 No. 1, pp. 121-39.
[11] Dengler, L., & Uslu, B. (2011). Effects of harbor modification on crescent city, California's Tsunami vulnerability. Pure and Applied Geophysics, 168(6–7). https://doi.org/10.1007/s00024-010-0224-8
[12] Wood, N. J., & Good, J. W. (2004). Vulnerability of port and harbor communities to earthquake and tsunami hazards: The use of GIS in community hazard planning. Coastal Management, 32(3). https://doi.org/10.1080/08920750490448622
[13] Wood, N. J., Good, J. W., & Goodwin, R. F. (2002). Vulnerability Assessment of a Port and Harbor Community to Earthquake and Tsunami Hazards: Integrating Technical Expert and Stakeholder Input. Natural Hazards Review, 3(4). https://doi.org/10.1061/(asce)1527-6988(2002)3:4(148)More references
[14] Goodwin, R.F. & Good, J.W. (2003). Making Ports and Harbors More Resilient to Earthquake and Tsunami Hazards. Proceedings of the 13th Biennial Coastal Zone Conference. Baltimore, MD. July 13–17, 2003
[15] Nucifera, F., Riasasi, W., Putro, ST., Marfai, MA. (2019). Penilaian Kerentanan dan Kesiapsiagaan Bencana Tsunami di Pesisir Sadeng, Gunungkidul. Jurnal Geografi Vol 11 No.2 (182-192). DOI: 10.24114/jg.v11i2.11475
[16] Villagra, P., Herrmann, M.G., Quintana, C., Sepulveda, R.D. (2017). Community resilience to tsunamis along the Southeastern Pacific: a multivariate approach incorporating physical, environmental, and social indicators. Nat Hazards (2017) 88:1087–1111. DOI 10.1007/s11069-017-2908-1.
[17] Morin, J., De Coster, B., Paris, R., Lavigne, F., Flohic, F., Le Floch, D. (2008). Tsunami-resilient communities' development in Indonesia through educative actions: Lessons from the 26 December 2004 tsunami. Disaster Prevention and Management, Vol. 17 No. 3, 2008, pp. 430-446. DOI 10.1108/09653560810887338.
[18] Rahman, A., A. Sakurai, K., Munadi. (2016). Indigenous knowledge management to enhance community resilience to tsunami risk: lessons learned from Smong traditions in Simeulue island, Indonesia. IOP Conf. Series: Earth and Environmental Science 56 (2017) 012018 doi:10.1088/1755-1315/56/1/012018.
[19] Rahman, MS. & Kausel, T. (2013). Coastal Community Resilience to Tsunami: A Study on Planning Capacity and Social Capacity, Dichato, Chile. IOSR Journal Of Humanities And Social Science (IOSR-JHSS). Volume 12, Issue 6 (Jul. - Aug. 2013), PP 55-63.

The Analysis of Dry Weather Aerial Photography in Sermo Kulonprogo Reservoir using DJI Spark Drone

Indreswari Suroso
Aerospace Technologycal College
indreswari.suroso@sttkd.ac.id

Abstract. Drones are a very important medium in the world of aerial photography to map military operations, agricultural areas, plantations, mining, urban planning, disaster mitigation, and outreach to remote areas and two types have been identified which are the winged and quadcopter. This research was conducted to map the drought-prone areas during the dry season in the Sermo Reservoir area, Kokap, Kulon Progo Regency between 14-15 August 2019. This involved the use of the DJI Spark quadcopter due to its more stability and wind resistance. The process involved connecting the DJI Spark drone remote controller to a smartphone, opening of DJI Go application, DJI Spark drone activation, smartphone installation on the remote controller holder, propeller installation, GPS settings in a locked position, drone flight and position adjustment, and capturing of the images on the smartphone screen. The results showed the water flow of the Sermo Reservoir has decreased by approximately 4 meters due to the dry season. This was evident in the changes from its original position of136.6 meters above sea level to
132.6 meters and this limits the quantity of water available to irrigate rice fields. The reservoir still has the capability to supply 800 liters per second in Pengasih District but the results of the aerial photographs showed the existence of drought and a decrease in the water discharge.

Keywords: drone; multicopter; aerial photography; Sermo reservoir; DJI Spark

1. Introduction

Drones are being used for several things in the world today and have also been discovered to be in two types which are the fixed-wing shown in Figure 1 and the quadcopter in Figure 2. The fixed-wing type resembles an insect or dragonfly, makes less noise while working, and produces a good image despite its low cost. Meanwhile, quadcopter type is usually used to capture aerial photographs in dangerous and disaster-prone areas as well as military operations. It has also been used to map aerial photographs at the entrance of Radar Congot, Temon Kulon Progo.

Drones have been applied for several reasons and these include its application in mapping aerial photographs for tourism, disaster mitigation, agriculture, mining, military, and the outer regions of Indonesia as well as the landslide areas in Somangari Village, Kaligesing District, Purworejo Regency. Moreover, the increased tourist visits in Malang Regency have encouraged the *Kelompok Sadar Wisata* (Pokdarwis) Gubugklakah Poncokusumo to improve their tourism services using aerial photographs and videos from drones. The studies in areas with potential iron sand such as Karangwuni, Temon, and

Content from this work may be used under the terms of the Creative Commons Attribution 3.0 licence. Any further distribution of this work must maintain attribution to the author(s) and the title of the work, journal citation and DOI.
Published under licence by IOP Publishing Ltd

Kulonprogo were also conducted using quadcopter drones while the Indonesian Center for Agriculture has also reported the importance of UAVs or drones in the agricultural mapping of West Java.

(a) (b)

Figure 1. (a) Fixed-wing Drone; (b). Quadcopter Drone

The research in Kotabaru, Lampung was conducted using fixed-wing drones as shown in Figure 2a, the mapping of the tourism area on the Congot Beach near the new Yogyakarta International Airport was also performed with drones as indicated in Figure 2b while the road to Yogyakarta International Airports on Glagah Beach was mapped by a quadcopter as presented in Figure 2c.

(a) (b) (c)

Figure 2. Mapping in (a) Kotabaru ; (b) Congot Beach (c) Glagah Lampung
(Source: Suroso, I., &Irmawan, E., 2018)

1.2. The Benefits of Utilizing Drones for Aerial Mapping

The flood and landslide mapping areas in Girimulyo, Kulonprogo were studied using a multicopter drone [1] due to its great stability and wind resistance [2]. Its application for the aerial mapping of sandy areas in Karangwuni, Temon, and Kulon Progo which are the producers of iron sand showed the presence of arid soil which is considered unsuitable for agriculture [3]. Moreover, a DJI Spark's drone with a camera was used to map the arid regions in the Hargowilis Kokap region, and good aerial photographs with clear quality were produced [4]. The quadcopter drone has been found to have the ability to hold up to 15 minutes in the air while producing clear pictures as observed from its application on the road to the new YIA Kulon Progo [5]. Furthermore, aerial mapping with multi-copter drones and Xiaomi cameras was observed to have lasted up to 30 minutes in the air with 1 km visible area [6] while a study conducted to map the development and agriculture in Sukadamai Village, Bogor Regency using drones is shown in Figure 3 [7].

IOP Conf. Series: Earth and Environmental Science **884** (2021) 012045 doi:10.1088/1755-1315/884/1/012045

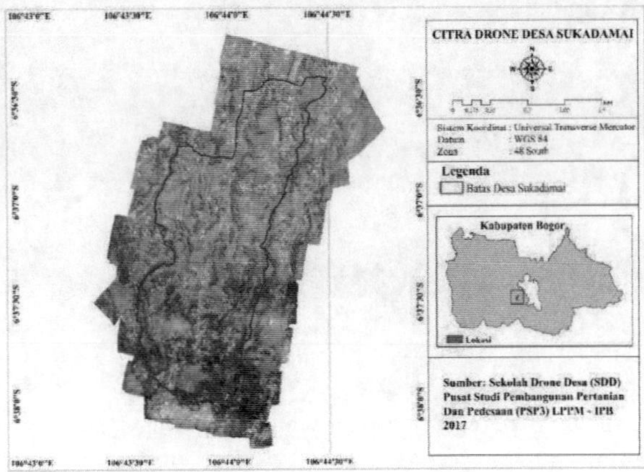

Figure 3. Aerial Photograph in Sukadamai Village, Bogor

Drones are also used as automatic hexacopter pesticide sprayers [8] and their application in spraying natural pesticides in Sungai Besar, Selangor, Malaysia was found to be twice as efficient as manual spraying as indicated in Figure 4 [9].

Figure 4. Aerial Photograph in Sungai Besar, Selangor, Malaysia

A study was also conducted using drones to capture shallow-water habitats in Lancang and Sebaru Besar Island. [10] and they were also used to have an overview of the administrative city in Langsa such as Sidodadi, Serambi, and Mulia Villages using aerial photographs. The region was selected due to its high population density [11]. Moreover, research has been conducted on the use of drones in mass media coverage by some journalists to facilitate reporting on television [12] while the report has also been made on its application in the surveillance of Indonesian territorial and patrol fishing areas [14]. There is, however, a Smart Flight in the country which is created as a form of drone piloting training by the Bogor Indonesia Drone Forces [13]. Meanwhile, the aerial photographs obtained with Geographic Information System (GIS) at the Muhammdiyah University of Gorontalo were used in imaging the Green Spatial Plan (RTH) [15] while the techniques involved in drone railway monitoring system are shown in Figure 5 [16].

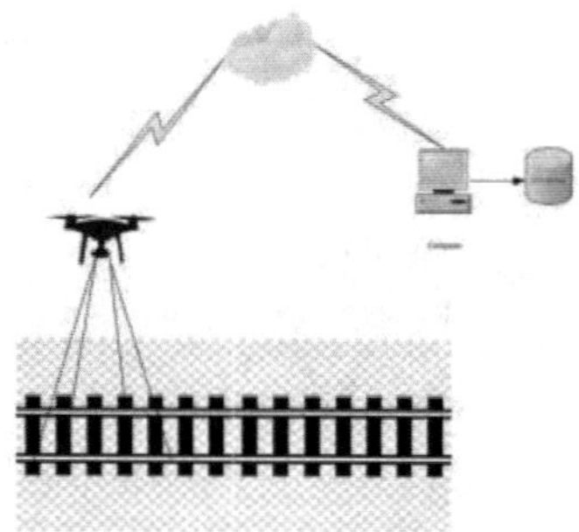

Figure 5. Drones Railway monitoring system

The quadcopter UAV platform is capable of spraying 2 m^2 agricultural land within 5 minutes with a liquid mass of 0.5 kg at a height 10 cm above the rice plant [17] and this indicates the use of drones in mapping paddy fields [18]. Moreover, drone media have been studied by students in SMAN 9 Tasikmalaya to help the learning process of remote sensing [19]. Another study was also conducted to eradicate rice pest attacks, especially White-capped munia, Pekingese, and sparrows, through the use of simple audio system drones [20]. They are also used with an exploder in the military as a target destroyer in an enemy area [21]. The use of drones in mapping Pancajaya District ecotourism areas was reported to have led to an improvement in the community's economy [23] while the flood mitigation efforts were discovered quickly using the drone rapid mapping method in Waygalih Village [24]. A quadcopter drone was also used to determine the landslide mitigation areas in Girimulyo, Kulon Progo Regency [25]. This study was, therefore, conducted to determine the effect of drought on Sermo Reservoir in Kokap, Kulon Progo regency using drones.

2. Method

A. **Instrument:** Drone Quadcopter DJI Spark
B. **Material:** Regional Areas of Sermo Reservoir, Kokap, Kulon Progo
C. **Location:** Drought areas in Sermo Reservoir, Kokap, Kulon Progo
D. **Time:** 14-15 August 2019
E. **Method:** A smartphone was connected to the DJI Spark drone remote controller, DJI Go application was opened to activate the DJI Spark drone after which the smartphone was installed on the remote controller holder, the propeller was installed, GPS position was locked, and the drone was flown and positioned to capture images on the smartphone screen.
F. **Condition of Sermo reservoir**
G. The Sermo Reservoir in the Kulon Progo area is located on the Kokap plateau, a winding road from the city center of Wates, about 3 km west of Wates, this sermo reservoir is the icon of Kulon Progo and the water in this Sermo reservoir for irrigation flows all the rice fields around Kokap, Pengasih, Ttemon, Panjatan and surroundings. If the Sermo Reservoir experiences a drought, it is certain that the rice fields will experience drough
H. **Map of Sermo Reservoir in Kokap, Kulon Progo**
Map of Waduk Sermo can be seen Figure 6.

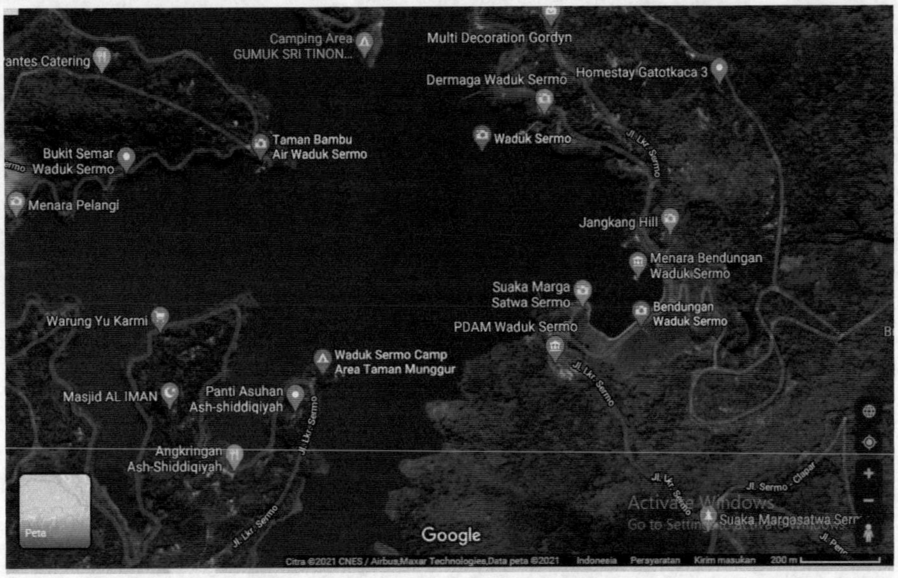

Figure 6. Map of Sermo in Kulon Progo

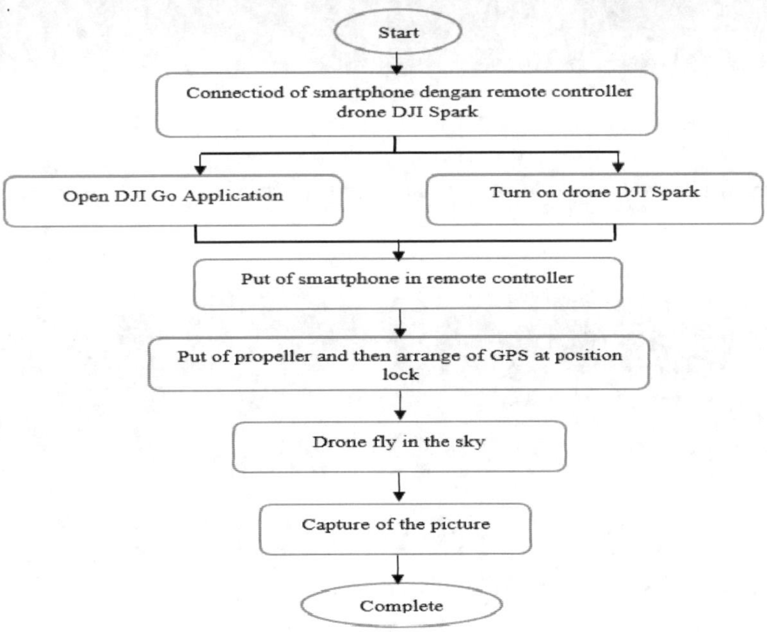

Figure 7. Flowchart

3. Results

The water flow in the Sermo Reservoir was observed to have decreased by approximately 4 meters due to the dry season. This is evident from the original position which was 136.6 to 132.6 meters above sea level. This reduction caused limitations to the irrigation for the paddy field even though the reservoir was found to be capable of providing 800 liters of supply per second in the Pengasih District. The description of the Sermo Reservoir at a height of 30 meters and from several cardinal directions is shown in Figures 8through 15. What distinguishes this research from others is that the research on the Sermo reservoir is highly anticipated by the Kulon Progo regional government because the local government wants to see the condition of the Sermo reservoir aerial photographs when the water flow decreases. This is because the Sermo Reservoir is a producer of irrigation for rice fields in Kulon Progo. If the discharge of the sermo reservoir water decreases, the government will prepare a way to overcome it by using irrigation from Kalibawang, Kulon Progo.

(a) (b)

Figure 8. (a) The northwest side aerial photograph of Sermo Reservoir, Kokap Kulon Progo at 30 meters height
and (b) DJI Spark quadcopter Drone

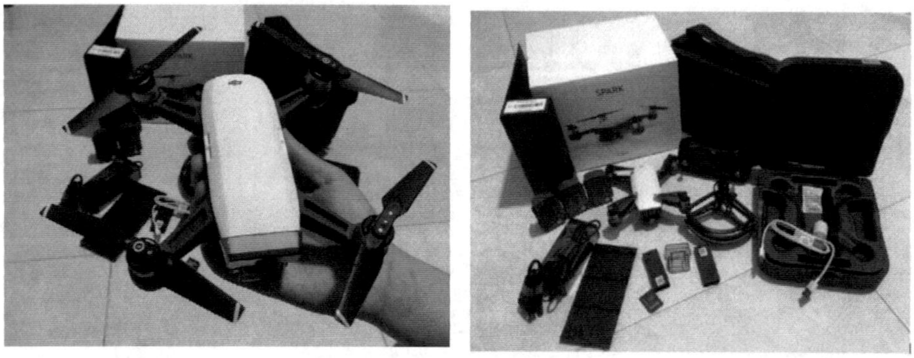

(a) (b)

Figure 9. (a) The southwest side aerial photograph of Sermo Reservoir, Kokap Kulon Progo at 30 meters
height and (b) DJI Spark Device

IOP Conf. Series: Earth and Environmental Science **884** (2021) 012045 doi:10.1088/1755-1315/884/1/012045

(a) (b)

Figure 10. (a) The west side Aerial photograph of Sermo Reservoir, Kokap Kulon Progo at a height of 30 meters and (b) DJI Spark camera device

(a) (b)

Figure 11. (a) The north side aerial photograph of Sermo Reservoir, Kokap Kulon Progo at 30 meters height and DJI Spark remote control

(a) (b)

Figure 12. a) The southern side aerial photograph of Sermo Reservoir, Kokap Kulon Progo at 30 meters height and (b) DJI Spark Drone

4. Conclusion

Drones are a very important medium in the world of aerial photography and useful in military operations, agricultural areas, plantations, mining, urban planning, disaster mitigation, and remote areas. This research was conducted on August 14-15, 2019 in Sermo Reservoir, Kokap, Kulon Progo Regency to map areas prone to drought during the dry season. The results showed the water flow of the Sermo Reservoir has decreased by approximately 4 meters due to the dry season. This was evident in the changes from its original position of136.6 meters above sea level to 132.6 meters and this limits the quantity of water available to irrigate rice fields. The reservoir still has the capability to supply 800 liters per second in Pengasih District but the results of the aerial photographs showed the existence of drought and a decrease in the water discharge.

5. Acknowledgments

The authors appreciate the Heads of STTKD Yogyakarta for providing adequate support for this research.

References

[1] Suroso, I 2018 Analisis Pemetaan Daerah Rawan Banjir Dan Longsor Dengan Drone Type Multicopter Di Girimulyo, Kabupaten Kulonprogo. Teknika STTKD: Jurnal Teknik, Elektronik, Engine. 5 1 34-43

[2] Suroso, I 2018 Analysis of Mapping Multicopter Drones in The Entrance Area of Prospective New Airports in Congot, Temon, Kulonprogo, Yogyakarta. JAGI. 2 2 130-134

[3] Suroso, I 2020 Analisis Foto Udara dengan Multicopter di Daerah Penghasil Pasir Besi Karangwuni, Temon,
Kulonprogo. REKAM. 16 1 29-35

[4] Suroso, I., & Pratama, H. H 2020 Analysis Mapping of Long Dry with Drone Type DJI Spark in Hargowilis,
Kokap, Kulon Progo. JAGI. 4 1 259-264

[5] Suroso, I., & Irmawan, E 2018 Analysis of Unmanned Aerial Vehicle Type Multicopter for Improving the Quality of Air Photography in Kulonprogo Airport. Journal of Applied Geospatial Information. 2 1 102-109

[6] Suroso, I., & Irmawan, E 2019 Analysis of UAV Multicopter of air photography in New Yogyakarta International Airports. TELKOMNIKA 17 1 521-528

[7] Arham, I., Sjaf, S., & Darusman, D 2019 Strategi Pembangunan Pertanian Berkelanjutan di Pedesaan Berbasis Citra Drone. Jurnal Ilmu Lingkungan. 17 2 245-255

[8] Hidayat, R 2020 Rancang Bangun Prototype Drone Penyemprot Pestisida Untuk Pertanian Padi Secara Otomatis. Jurnal Mahasiswa Teknik Elektro. 3 2 86-94

[9] Khoirunisa, H., and Kurniawati, F 2019 Penggunaan Drone dalam Mengaplikasikan Pestisida di Daerah Sungai Besar, Malaysia. Jurnal Pusat Inovasi Masyarakat 1 1 87-91

[10] Sugara, A., Siregar, V. P., & Agus, S. B 2020 Klasifikasi Habitat Bentik Perairan Dangkal Dari Citra Worldview-2 Menggunakan Data In-Situ Dan Drone. Jurnal Ilmu dan Teknologi Kelautan Tropis 12 1 135- 150

[11] Suciani, A., & Rahmadi, M. T 2019 Pemanfaatan Drone DJI Phantom 4 Untuk Identifikasi Batas Administrasi Wilayah. Jurnal Geografi. 11 2 218-223

[12] Yoedtadi, M. G 2019 Penggunaan Drone Pada Peliputan Berita Televisi (Perspektif Wartawan Televisi Terhadap Etika Peliputan Menggunakan Drone). Jurnal Muara Ilmu Sosial, Humaniora, dan Seni. 3 1 54-60

[13] Respati, R. E., & Irwansyah, I 2020 "Smart Flight" sebagai Bentuk Pelatihan Pilot Drone oleh Pasukan Drone Bogor Indonesia. Jurnal Komunikasi. 14 1 1-14

[14] Pratama, F. Y., Hendra, A. F., Maulana, F., & Yusuf, A 2020 Sistem Pengawas Kapal Penangkap Ikan Berbasis AI Menggunakan Drone. LKTI. 1 1 77-83

[15] Koto, A. G., & Taslim, I 2019 Kajian Ruang Terbuka Hijau Kampus Universitas Muhammadiyah Gorontalo Menggunakan Foto Udara Drone. Media Komunikasi Geografi. 19 2 153-164

[16] Mardiana, S., Hamdani, D., Chaniago, M. B., Wahyu, A. P., Heryono, H., & Suhendri, S 2020 Sistem Informasi Pemeriksaan Jalur Kereta Api Menggunakan Drone dan Teknik Image Processing. JOINT. 2 1 9-12

[17] Ade, V 2019 Analisa Pengembangan Drone Penyemprotan Hama Tanaman Dengan Jenis Nosel Dan Ketinggian Untuk Mengetahui Luas Semprotan. Engineering. 10 2 63-69

[18] Nurfaizah, D 2020 Analisis Pola Termal Citra Landsat 8 dan Drone Thermal pada Fase Pertumbuhan Padi Sawah di Kecamatan Bojongpicung, Kabupaten Cianjur.

[19] Rochaeni, E 2019 Penggunaan Media Pembelajaran Geografi Drone Melalui Metode Demonstrasi Pada Materi Penginderaan Jauh. Metaedukasi. 1 2 72-78

[20 Yani, I., Arifin, A., Santoso, D., Afrah, B. D., & Riady, M. I 2019 Implementasi Teknologi Drone Pada Pengendalian Hama Burung Di Persawahan. Prosiding Applicable Innovation of Engineering and Science Research. 1112-1116

[21] Priyono, E 2020 Pesawat Terbang Tanpa Awak (PTTA) Sebagai Salah Satu Komponen Kekuatan Udara. Jurnal Industri Elektro dan Penerbangan. 1 2 1-7

[22] Darmawan, I. G. B 2020 Pemanfaatan Drone Untuk Pemetaan Potensi Ekowisata di Kecamatan Panca Jaya,
Mesuji. Sakai Sambayan Jurnal Pengabdian kepada Masyarakat. 4 1 1-5

[23] Noor, M. S., Hafizianoor, H., & Suyanto, S 2020 Analisis Cadangan Karbon Pada Tanaman Reklamasi Lahan Bekas Pertambangan Batubara Di Pt. Borneo IndobarA. Jurnal Hutan Tropis. 8 1 99-108

[24] Rohman, A., & Prasetya, D. B 2019, August Rapid Mapping for Simple Flood Mitigation Using Commercial Drone at Way Galih Village, Lampung, Indonesia. In Forum Geografi. 33 1 101-113

[25] Suroso, I 2019 Analysis of Mapping Area of Flood with Drone Type Multicopter in Girimulyo, Kulonprogo. In IOP Conference Series: Earth and Environmental Science. 271 1 1-11

School Community Preparedness in Anticipation of Earthquake and Tsunami Threats in Temon Sub-District

A P Ira, S Purwantara

Department of Geography Education, Faculty of Social Sciences, Universitas Negeri Yogyakarta, Yogyakarta, Indonesia

Email: ira.putri2016@student.uny.ac.id ; suhadi_p@uny.ac.id

Abstract. The Kulon Progo Regency is a region prone to earthquake and tsunami, including on the coastal of Temon Sub-District. The study aims to determine: (1) school preparedness in Temon Sub-District in anticipating of the earthquake and tsunami threat, and (2) school efforts in Temon Sub-District in improving the preparedness of the earthquake and tsunami disaster. This study used descriptive research. The population of this study is all high schools in Temon Sub-District that enter the area prone to earthquake and tsunami disaster, which is used as a research analysis unit. The sampling technique used simple random sampling and calculated using the Slovin formula with a 10% error rate which results in a total sample of 6 principals, 84 teachers, and 177 students. Data collection techniques in this study are observation, interviews, and documentation. The research data presented in a frequency distribution table and analyzed descriptively. The results of the study showed: (1) school preparedness in Temon Sub-District from the schools aspect in "not ready" category (50.0%), the teachers aspect in "ready" category (42.9%), and the students aspect in "ready" category (44.1%). (2) Efforts to improve the preparedness of the earthquake and tsunami disaster carried out by schools in Temon Sub-District, including: (a) conducting disaster evacuation drills through scouting and PMR activities in schools, (b) holding regular disaster socialization and simulation routinely, (c) improving the quality and quantity of school facilities.

1. Introduction

Indonesia is one of the top ranked countries with significant potential for earthquake phenomena and is prone to tsunami disasters. It is because Indonesia is located at the confluence of world's tectonic plates, namely the Eurasian Plate in the north, the Indo-Australian Plate in the south, and the Philippine Plate and the Pacific Ocean in the east. The meeting between these plates in Indonesia creates a subduction zone that can cause earthquake vibrations with different strength intensities. In addition to the existence of a subduction zone, Indonesia is nicknamed the "Ring of Fire."

Yogyakarta Special Region is geologically located nearby the subduction zone between the Indo-Australian Plate and the Eurasian Plate. The subduction zone has a distance around 200-250 km from the Southern Coast of Java and has the potential to generate earthquakes. The risk of an earthquake in Yogyakarta is also influenced by the fault zone that can cause pressure, pull, and shear. Besides due to the plate collisions, local faults in Yogyakarta also contributed to tsunami. This local fault structure was formed from the pressure of the Indo-Australian Plate on the Land of Java Island. Some fault systems

Content from this work may be used under the terms of the Creative Commons Attribution 3.0 licence. Any further distribution of this work must maintain attribution to the author(s) and the title of the work, journal citation and DOI.
Published under licence by IOP Publishing Ltd

that are considered to be still active are the Opak Fault, Oya Fault, Dengkeng Fault, Progo Fault, and Micro Fault.

The earthquake on December 16, 2017 at 23.47.58 WIB (GMT+7) in the southeast of Bantarkalong, Tasikmalaya, West Java with a magnitude of 6.9 SR had also caused the coast in Ciamis and Tasikmalaya, West Java Province to have the potential for a tsunami with a "WARNING" level at an altitude of 0.5<3 meters, while in the coastal of Bantul, Kulonprogo, Cianjur, Garut, Sukabumi, Cilacap and Kebumen Districts, the level was "ADVISORY" at an altitude of <0.5 meters (detiknews, 2017) [4].

According to The National Disaster Mitigation Agency (BNPB) (2012: 15-17), the southern region of Java Island is highly active in terms of volcanism, seismicity, and vertical tectonic movements. Three major earthquakes were recorded in this area; they were in 1840, 1867, and 1875. The earthquakes that occurred in the subduction zone of the southern region of Java Island are feared to trigger a tsunami that could hit the Pangandaran Beach, Cilacap District with its oil refineries as well as other beaches in the southern province of Yogyakarta [2].

Table 1. Areas Affected by the Tsunami in Southern Java

No.	District/City	Province	Affected (People)
1.	Cilacap	Central Java	629.891
2.	Kebumen	Central Java	220.822
3.	Purwokerto	Central Java	91.943
4.	Banyumas	Central Java	689
5.	Wonogiri	Central Java	52
6.	Kulon Progo	Yogyakarta Special Region	60.607
7.	Bantul	Yogyakarta Special Region	31.362
8.	Gunung Kidul	Yogyakarta Special Region	366
9.	Jember	East Java	134.207
10.	Lumajang	East Java	27.706
11.	Banyuwangi	East Java	17.107
12.	Pacitan	East Java	13.188
13.	Malang	East Java	2.144
14.	Tulungagung	East Java	297
		Total	1.230.388

Source: Dewi (2010: 75) [5].

Table 1 shows districts/cities in the Southern Coast of Java that are susceptible to tsunami risk due to earthquakes from the Sunda Strait Megathrust and the South Java Megathrust with a total number of affected as many as 1.230.388 people. Three districts in Yogyakarta Special Region are also included in tsunami-prone areas. Kulon Progo Regency is one of the districts in Yogyakarta Province with the highest number of people affected as many as 60.607 people, followed by Bantul Regency with 31.362 people, and Gunung Kidul Regency with 366 people.

The relationship between earthquake and tsunami has been recorded in many historical events of disasters in Indonesia. Among 105 tsunami cases in Indonesia, 90.5% were caused by ocean floor earthquakes, 8.6% caused by volcanic eruptions, and 1% caused by landslides. The history recorded that earthquakes on the ocean floor and volcanic eruptions caused the occurrence of tsunamis along the southern coast of Java Island (Dewi, 2010: 75) [5].

Acknowledging the high level of vulnerability and susceptibility to tsunami, the government of Indonesia collaborated and worked with Germany, Australia, Japan and the United States to improve tsunami preparedness by building the Indonesia Tsunami Early Warning System (*InaTEWS*) [6].

IOP Conf. Series: Earth and Environmental Science **884** (2021) 012046 doi:10.1088/1755-1315/884/1/012046

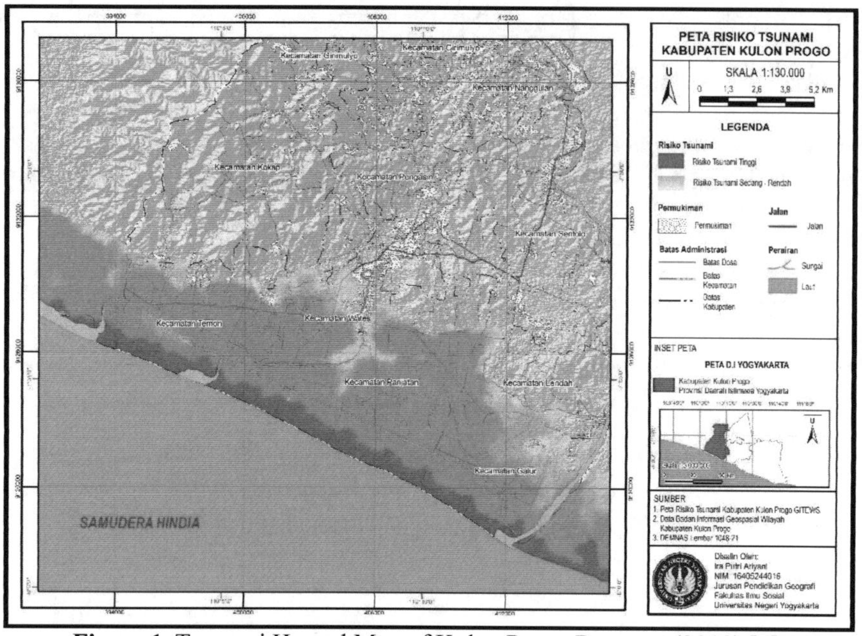

Figure 1. Tsunami Hazard Map of Kulon Progo Regency (2012) [6].

Figure 1 is a tsunami hazard map of Kulon Progo Regency, Yogyakarta. Based on this figure, the southern coast of Kulon Progo Regency which consists of four sub-districts, namely Galur sub-districts, Panjatan sub-districts, Wates sub-districts, and Temon sub-districts are included in the tsunami hazard zone with an estimated water level of >3 meters (red zone) and <3 meters (orange and yellow zones).

A school is considered to be the core of the community in many countries, and school plays a vital role in disaster issues, both in terms of preparedness and post-disaster recovery (Shiwaku et al., 2016:333) [12]. Preparing the school environment well is the key to "safety" (Khotimah, 2019:3) [7]. The survey of Chiu et al. (2015, pp. 175–196) shows that student, family, and school play an important role in building school resilience since school has a social function to develop cultural awareness on disaster and school resilience [3]. It is in accordance with Matsuura and Shaw (2015, pp. 613–633) who state that school has a role for disaster mitigation. In their research, it is shown that a school is very conducive to create a safe and comfortable environment in overcoming disasters when cultural awareness on disasters have been built by the school. Besides, in the recovery process, the school achievement in post-disaster rehabilitation is determined by the school resilience [10]. Safe schools are expected to save the lives of students, act as temporary shelters after a disaster occurs, and promote disaster prevention and mitigation efforts to the community [11].

Temon Sub-district is one of the four sub-districts in Kulon Progo Regency which is prone to earthquake and tsunami because it is located above an active fault and directly faces the Indian Ocean. Based on data from the Department of Primary and Secondary Education in Kulon Progo Regency, schools that have been based on disaster preparedness at the secondary level are mainly in Temon Sub-district. In contrast, the other three sub-districts do not yet have Disaster Preparedness Schools (Sekolah Siaga Bencana/SSB) for the secondary level. This condition indicates that there is a need for an assessment regarding the preparedness level of the school community elements to know the extent of school preparedness in Temon Sub-district in facing the earthquake and tsunami disasters as well as the school's efforts to improve preparedness.

Efforts to improve preparedness are everything that is done to improve the preparedness of the school community in facing the earthquake and tsunami disasters. The Law of the Republic of Indonesia No. 24/2007 Concerning Disaster Management, articles 33 and 34 explain that disaster management consists of three stages which includes:

a. Pre-disaster efforts. They are actions taken in the form of prevention of an earthquake or tsunami disaster in a situation without disaster and in a situation with potential disaster.
b. Emergency response. They are actions taken to rescue oneself to a secure place when an earthquake and tsunami occur.
c. Post-disaster. They are actions taken to recover from the impact of the earthquake and tsunami disasters, both in the form of physical and mental rehabilitation for the victims of earthquake and tsunami disasters [8].

2. Research Methods

The research method used in this research was descriptive research. This study attempted to explain the preparedness of schools in Temon Sub-district in facing the earthquake and tsunami disasters, and to describe what efforts that schools have made to improve preparedness for these disasters. This research was conducted in Temon Sub-district, Kulon Progo Regency, Yogyakarta in January-May 2020.

The population in this study were secondary schools in Temon Sub-district, an area which is prone to earthquakes and tsunamis. These secondary schools were used as a unit of research analysis. The researcher used simple random sampling to take the samples and then calculated using the Slovin formula with an error rate of 10% which resulted in a total sample of six principals, 84 teachers, and 177 students.

The variables in this study were: (1) school preparedness in facing earthquake and tsunami disasters, viewed from the indicators: (a) knowledge, (b) policies and guidelines, (c) emergency response plans, (d) disaster warning systems, and (e) resource mobilization, and (2) efforts made by schools in enhancing earthquake and tsunami disaster preparedness. Observation, interviews, and documentation was carried out as the data collection techniques. The collected data were then presented in a frequency distribution table and analyzed descriptively.

The description of school preparedness in Indonesia in the face of earthquakes and tsunamis is obtained from the results of preparedness parameter scoring. Scoring is done to change the information that has been obtained into data in the form numbers. The answer "yes" on each research instrument will be given a value of 2, the answer "no" is given a score of 1, and the answer "do not know" is given a score 0. Determination of the number of scores for each class is based on the Mean Ideal (Mi) and the Ideal Standard Deviation (SDi) obtained by calculations using formula as follows:

$$\text{Mean Ideal (Mi)} = \frac{1}{2}\,(\text{maximum value} + \text{minimum value})$$

$$\text{Standar Deviasi Ideal (SDi)} = \frac{1}{6}\,(\text{maximum value - minimum value})$$

The results of the Mi and SDi calculations are then categorized each according to the following provisions:

Table 2. Trend Distribution Formula

Category	Formula
Full Ready	$= X \geq Mi + 1{,}8\ (SDi)$
Ready	$= Mi + 0{,}6\ (SDi) \leq X < Mi + 1{,}8\ (SDi)$
Ready Enough	$= Mi - 0{,}6\ (SDi) \leq X < Mi + 0{,}6\ (SDi)$
Less Ready	$= Mi - 1{,}8\ (SDi) \leq X < Mi - 0{,}6\ (SDi)$
Not Ready	$= X < Mi - 1{,}8\ (SDi)$

Source: Azwar (2003: 163) [1].

Based on the calculation of the categorization of the trend distribution above then you will get the desired results, namely "full ready", "ready", "ready enough", "less ready", and "not ready".

3. Results & Discussion

School is one of the institutions that has a strategic position and plays an important role in efforts to prevent early disasters and mitigate them because various information, knowledge, and skills can be transmitted effective to the entire school community and society. Next, a study of school preparedness in Temon Sub-district will be presented in to face the earthquake and tsunami disasters using a framework developed by LIPI in collaboration with UNESCO/ISDR in 2006. The school preparedness study is based on 5 (five) indicators, they are: (a) knowledge, (b) policies and guidelines, (c) emergency response plans, (d) disaster warning systems, and (e) resource mobilization [9].

3.1. School Preparedness as an Institution for Earthquake Disaster and Tsunami Threat in Temon Sub-district

School preparedness as an institution in Temon Sub-district was measured using four parameters, namely school policies and guidelines, emergency response plans, disaster warning systems, and the ability to mobilize resources. The research subjects used to measure disaster preparedness viewed from the aspect of school as an institution were the principals of each school in Temon Sub-district, which consisted of six respondents.

Table 3. School Preparedness in Temon Sub-district in Facing Earthquake and Tsunami Disaster Viewed from the Aspect of Schools as Institutions

No	Parameter	Category (%)					
		Full Ready	Ready	Ready Enough	Less Ready	Not Ready	Total
1	Police and Guidelines	16,7	0,0	16,7	16,7	**50,0**	100
2	Emergency Response Plan	16,7	16,7	16,7	16,7	**33,3**	100
3	Disaster Warning System	16,7	0,0	0,0	33,3	**50,0**	100
4	Resources Mobilization	16,7	0,0	16,7	16,7	**50,0**	100
5	All Indicator	16,7	0,0	0,0	33,3	**50,0**	100

Source: Primary Data, 2020

Table 2 shows the percentage value of all school preparedness parameters in facing the earthquake and tsunami disasters in Temon Sub-district. The results of the data analysis indicated that the school preparedness as institutions in facing the earthquake and tsunami disaster in Temon Sub-district was in the "not ready" category (50.0%). This was inseparable from the four parameters of disaster preparedness. The first parameter to measure the preparedness of schools as institutions is policies/guidelines regarding earthquakes and tsunamis.

School policy/guidelines indicators would determine the emergency response plans, disaster warning systems, and resource mobilization carried out by the schools. The results of the data analysis showed that school policies/guidelines for dealing with earthquake and tsunami disasters fell into the category of “not ready” (50.0%). This means that the schools located in a disaster-prone area in Temon Sub-district did not have written policies related to the earthquake and tsunami disasters. The low level of school policies/guidelines was indicated by the absence of a disaster preparedness group (83.3%), most schools had not issued policies regarding regular or routine disaster practices/simulations (66.7%), and some schools had not issued policies of budget allocation for school preparedness (83.3%).

The second parameter concerning the school emergency response plan as an institution for earthquake and tsunami disasters was in the "not ready" category, with a percentage value of 33.3% and total frequency of two respondents. This means that schools in Temon Sub-district in making plans related to actions to be taken before, during, and after the disasters were still low. This was because most schools in Temon District did not have fixed procedures regarding evacuation, and there had been no evacuation simulation practice (83.3%), but some schools have made evacuation route signs placed on the side of the emergency staircase (66.7%).

The third parameter concerns the school warning system as an institution for earthquake and tsunami disasters that were in the "not ready" category, which has a percentage value of 50%. This was because some schools had not been prepared plans/steps to respond to disaster warnings (66.7%), tsunami warning information had not been disseminated (66.7%) and had not been tested in schools (83.3%). The fourth parameter is the last, namely the ability of schools as institutions to mobilize earthquake and tsunami disasters in the "not ready" category with a percentage value of 50% and a total frequency of three respondents.

The data from resource mobilization parameters showed that there were still many schools located in areas prone to earthquakes and tsunami hazards did not yet have disaster preparedness groups, which involved disaster warning groups, first aid groups, evacuation and rescue groups, and logistics group needed by the school community. The low parameter of resource mobilization was influenced by the absence of simulated emergency disaster evacuation for the school community as indicated by a percentage of (83.3%). Only one out of six schools that has conducted a disaster emergency evacuation simulation, namely SMK Negeri 1 Temon. Some schools also did not receive assistance from LSM (83.3%) and NGOs (50%).

3.2. Teacher Preparedness for Earthquake Disaster and Tsunami Threat in Temon Sub-District

Teacher preparedness in Temon Sub-district was measured using four parameters, i.e. knowledge, emergency response plans, disaster warning systems and resource mobilization abilities. The research subjects used to measure teacher disaster preparedness totalled 84 teachers, consisted of 26 junior high school teachers and 58 high school/vocational high school or the equivalent in Temon Sub-district.

Table 4. School Preparedness in Temon Sub-district in Facing Earthquake and Tsunami Disaster Viewed from Teacher Aspect

No	Parameter	**Category (%)**					Total
		Full Ready	Ready	Ready Enough	Less Ready	Not Ready	
1	Police and Guidelines	38,1	**50,0**	4,8	4,8	2,4	100
2	Emergency Response Plan	32,1	**42,9**	22,6	1,2	1,2	100
3	Disaster Warning System	15,5	21,4	**35,7**	19,0	8,3	100
4	Resources Mobilization	15,5	10,7	26,2	10,7	**36,9**	100
5	All Indicator	13,1	**42,9**	29,8	4,8	9,5	100

Source: Primary Data, 2020

Table 3 shows the percentage value of all teacher preparedness parameters in facing the earthquake and tsunami disaster in Temon Sub-district. The results of the data analysis showed that the school preparedness in Temon Sub-district in facing the earthquake and tsunami disasters, viewed from the teacher's aspect, fell into "ready" category with a percentage of 42.9%. This was inseparable from the four parameters of disaster preparedness.

The first parameter of disaster preparedness is knowledge. The data analysis results showed that the teachers' knowledge about earthquake and tsunami disasters was in the "ready" category with a percentage of 50%, and the number of respondents was 42 teachers. This data means that teachers in Temon Sub-district had good understanding skills related to the disaster. The indicators in the knowledge parameters are the common understandings of natural disasters, the definitions and causes of earthquake and tsunami, signs of strong earthquakes and tsunamis, earthquake and tsunami resistant buildings, and what actions to take if such disasters occur.

Most of the teachers in Temon Sub-district stated that the definition of natural disasters is human behaviour that causes natural damage (84.5%), for example, earthquakes. The earthquake's cause was a shift in the Earth's crust (96.4%), which was characterized by a strong earthquake causing dizziness (77.4%). When an earthquake occurs, the action to take is to seek shelter in a safe place (i.e., under a sturdy table) (88.1%). House buildings in earthquake-prone areas must also be considered, by making the parts of the building made of brick/concrete/wood which is strongly connected (88.1%).

The teachers in Temon Sub-district also believed that the cause of the tsunami was an earthquake under the sea (95.2%). One of the signs of a tsunami is the sudden receding of ocean water (70%). Areas that are prone to tsunami hazards must also pay attention to their house buildings. A tsunami-resistant building's characteristic is buildings that have gaps so that water can dissipate through it (64.3%). Regarding the ability to understand the knowledge of earthquakes and tsunamis, the teachers in Temon Sub-district are expected to be able to provide maximum knowledge or information to students regarding earthquakes and tsunamis, even though not included in particular disaster subjects.

The second parameter is the teacher's emergency response plan regarding earthquake and tsunami, which was in the "ready" category, with a percentage of 42.9% and having a frequency of 36 teachers. This means that the teachers in Temon Sub-district have a good understanding of how to plan and take action to deal with emergencies when an earthquake and tsunami occur. The analysis results showed that most of the teachers in Temon Sub-district had already prepared copies of essential documents and subject matter taught to computers or other safe places (73.8%). When a disaster occurs during teaching and learning hours, teachers in Temon District will guide students on the second floor, or more to use the emergency stairs and not use elevators/lifts (96.4%). Even though the teacher's emergency response plan parameter was in the "ready" category, there is still a need for improvement concerning plans carried out before, during and after a disaster.

The third parameter is the earthquake and tsunami disaster warning system in terms of teachers who was in "ready enough" category with 35.7% and a total frequency of 30 teachers. This means that the teachers in Temon Sub-district have a lack of understanding of the earthquake and tsunami early warning system. Based on the results of data analysis, most of the teachers in Temon Sub-district did not know the difference of the tsunami warning sounds (50%), the sounds of the tsunami cancellation (60.7%), and the sounds of safe conditions after the tsunami (61.9%). This was due to the low collaboration between the schools studied and the local The National Disaster Mitigation Agency (BPBD)/Satlak. Only one school intensively engages in disaster-related relations with the government because the school is already a disaster prepared school.

The fourth parameter is teachers' ability to mobilize the earthquake and tsunami disasters, which was in the "not ready" category (36.9%) and having a total frequency of 31 teachers. The data showed that the teacher's ability to mobilize disasters in Temon Sub-district was shallow due to the lack of seminars/workshops related to knowledge of disasters (59.5%), emergency response planning (66.7%), and disaster warning systems (67.9%) followed by teachers in Temon Sub-district. Only two out of six schools often received invitations and sent teacher representatives to disaster seminars organized by the government and the Islamic school centre. Teachers in Temon Sub-district also have never given lessons related to the earthquake and tsunami (58.3%), because most of the teachers who provided the lessons were only the social studies subject teachers, while at the SMK level there was currently no learning related to social and natural group.

3.3. Student Preparedness for Earthquake Disaster and Tsunami Threat in Temon Sub-district

The preparedness of students in Temon Sub-district was measured using four parameters: knowledge, emergency response plans, disaster warning systems, and the ability to mobilize resources. The research subjects used to measure disaster preparedness as viewed from the student aspect were 177 students consisted of 84 junior high school students and 93 high school/vocational high school students in Temon Sub-district. Junior high school students who were the research subjects consisted of class VIII and class XI students, while high school/vocational high school students have three classes: class X, class XI and class XII.

Table 5. School Preparedness in Temon Sub-district in Facing Earthquake Disaster and Tsunami Threat Viewed from Student Aspect

No	Parameter	Category (%)					
		Full Ready	Ready	Ready Enough	Less Ready	Not Ready	Total
1	Police and Guidelines	34,5	**40,7**	16,4	4,0	4,5	100
2	Emergency Response Plan	23,7	**43,5**	21,5	7,3	4,0	100
3	Disaster Warning System	**46,9**	39,0	12,4	1,1	0,6	100
4	Resources Mobilization	**46,9**	19,8	17,5	14,7	1,1	100
5	All Indicator	27,7	**44,1**	21,5	3,4	3,4	100

Source: Primary Data, 2020

Table 4 shows the percentage value of all students' preparedness parameters in facing the earthquake and tsunami disasters in Temon Sub-district. The data analysis results showed that the preparedness of schools in Temon Sub-district in facing the earthquake and tsunami disasters, viewed from the aspect of students, was in the "ready" category with a percentage of 44.1%. This was inseparable from the four parameters of disaster preparedness. The first parameter to measure students 'preparedness was knowledge about earthquake and tsunami. The results of the data analysis showed that students' knowledge of the earthquake and tsunami disaster was in the "ready" category with a percentage of 40.7%.

The high level of student preparedness was indicated by the scores of several sub-indicators that have been answered by 177 students in Temon Sub-district. The sub-indicator regarding the definition of natural disasters has a percentage value of 87.6%, with the statement that natural disasters are natural events that disturb human life. Most of the students in Temon Sub-district already understood natural events that can cause disasters, including earthquake and tsunami. The earthquake's cause was a shift in the earth's crust (96.4%) and volcanic eruptions (88.1%). The occurrence of a strong earthquake was characterized by vigorous shaking and caused people to be unable to stand (83.6%). Actions that must be taken when an earthquake occurs while at school are taking shelter in a safe place (for example under a study table) (83.1%), leaving the room regularly (not overcrowding) (93.8), and running towards the open field (96%).

Students in Temon Sub-district stated that the causes of the tsunami were earthquakes under the sea (94.4%), volcanoes under the sea (87%), and underwater landslides (48.6%). One of the signs of a tsunami is the sudden receding of the ocean water (89.3%), followed by the appearance of large waves on the horizon (67.2%). Students in Temon Sub-district also stated that earthquakes could not be predicted when they occur (41.2%), and not all earthquakes can cause tsunamis (73.4%). Even though the parameter of student knowledge was in the "ready" category, it is still necessary to increase the understanding of the causes and actions taken when earthquake and tsunami occur.

The second parameter is the student's emergency response plan regarding the earthquake and tsunami, which was in the "ready" category with a percentage of 43.5% and a total frequency of 77 students. This means that students in the Temon Sub-district have a good understanding regarding the action to take when dealing with emergencies when earthquake and tsunami occur, either by increasing knowledge from various media or participating in self-rescue exercises. The data analysis results showed that most students in Temon Sub-district already knew ideal places for evacuation when a disaster occurs (93.8%). Furthermore, they also knew essential places such as hospitals, fire-fighters, police stations, and the Red Cross (67.2%).

When a disaster occurred, some students chose to save securities and important items (71.8%) and ignore various school needs (61.6%). Students in Temon Sub-district got much information about disaster through disaster books (84.7%), posters, leaflets, and comics (66.7%), and only 42.9% of students get information through VCD or tapes about earthquakes and tsunamis. Easy access to information via the internet in the era of globalization also supported students in obtaining the latest news about disasters. Even though the students' emergency response plan parameter was in the "ready" category, efforts still need to be made to improve students' understanding of emergency response plans.

The third parameter is the student disaster warning system regarding earthquake and tsunami, which was in the "full ready" category. The percentage showed 46.9% and had a total frequency of 83 students. This means that students in Temon Sub-district had an excellent understanding and understood what actions to take when they hear traditional warning sounds/signs such as *kentongan* or modern warning signs, such as sirens.

The data analysis results showed that most students in Temon Sub-district already understood about self-evacuation when they heard a tsunami hazard sign, they would move away from the beach and run to high places (98.9%), go to the evacuation places immediately (94.9%), and calm themselves down or not be panic (91.5%). The information distribution to schools in Temon Sub-district used local agreement tools as disaster warning signs such as *kentongan*, bell, and drum (71.8%), but most students did not know the differences of sounds for warnings, cancellations, and safe conditions (74.6%). Even though it was in the "ready" category, there was still a low score on the parameters of the disaster warning system, about the sub-indicators related to differences in agreed signs, so it is still necessary to improve the understanding, especially on the agreed disaster warning signs.

Based on the results of data analysis, it is known that the students' ability of mobilization regarding earthquake disasters was in the "full ready" category with a percentage of 46.9%. This means that students in Temon Sub-district were already good at participating in disaster-related school activities. Most of the students had participated in scouting activities related to ropes, making stretchers, and setting up emergency tents (92.7%), but only 50.8% had participated in disaster evacuation practices and simulations. Schools that have involved students in disaster simulations are SMK Negeri 1 Temon.

4. The Efforts of Schools in Temon Sub-district in Improving Earthquake and Tsunami Disaster Preparedness

Schools in Temon Sub-district have made various pre-disaster, during, and post-disaster efforts to improve the preparedness of the school community, which consisted of:

a. Schools as institutions, teachers, and students with disaster simulation/practices activities, conduct *sharing session* related to disasters both through school activities at the time of giving mandates at flag ceremonies, providing disaster materials in social studies subjects, and supporting from extracurricular activities in the form of scouting and School's Health Clinic/Junior Red Cross.
b. Increasing the quality and quantity of facilities by building earthquake-resistant school buildings, increasing the number of stairs as accessibility equipped with evacuation route signs, providing medicines, providing School's Health Clinic, preparing disaster warning devices (bell, microphones, and *kentongan*), making posters related to disasters, as well as making a gathering point.
c. Ring the disaster warning signs and ask all school members to gather at the gathering point. If the disaster has been stopped and it is confirmed that there will be no further disaster, teaching and learning activities will continue.
d. Assist in healing trauma to students in the aftermath of a disaster.
e. Hold an intense and scheduled counselling and disaster simulations, by involving parties who are knowledgeable about disasters such as the Kulon Progo District National Disaster Mitigation Agency and the District Disaster Risk Reduction Forum (*FPRB*).

5. Conclusion

a. The preparedness of the school community in Temon Sub-district in facing the earthquake and tsunami threat viewed from the school aspect as an institution was categorized as "not ready" (50%), viewed from the aspect of teachers was in the "ready" category (42.9%), and from the aspect of the students was in the "ready" category (44.1%). The lack of emphasis on policies and guidelines, emergency response plans, early warning systems and mobilization of resources in schools means that the preparedness of schools as institutions was in the "not ready" category. The high percentage of teacher and student knowledge parameters causes preparedness to be in the "ready" category, because the knowledge parameter has the largest portion compared to the other three parameters.

b. Efforts to improve earthquake and tsunami disaster preparedness carried out by schools in Temon Sub-district, included:
   1) Equipping students and teachers with disaster simulation/practices, *sharing session* related to disasters through school activities (flag ceremonies), subjects, and extracurricular activities (scouting and Junior Red Cross).
   2) Providing a disaster socialization and simulation on a regular basis by involving parties who understand disasters such as The Disaster Mitigation Agency of Kulon Progo Regency and the Risk Reduction Forum.
   3) Improve the quality and quantity of facilities and infrastructure in schools.

Acknowledgments

The authors would like to thank Yogyakarta State University, which has facilitated excellent international cooperation research activities, the research team – Dr. Suhadi Purwantara who helped her during the research, and the schools, for their willingness to be a research setting, as well as teachers and students who became the research subjects.

References

[1] Azwar, S. (2003). *Research Method.* Yogyakarta: Pustaka Belajar.
[2] BNPB. (2012*). Regulation of the Head of The National Disaster Management Agency No. 04 of 2012 concerning Guidelines for The Implementation of Safe Schools / Madrasahs from Disaster* Jakarta: BNPB.
[3] Chiu, M. M., Chow, B. W.-Y., McBride, C. and Mol, S. T. (2015). Students sense of belonging at school in 41 countries: Cross-cultural variability. *Journal of Cross-Cultural Psychology, Vol. 47 No. 2, pp. 175–196.*
[4] Detiknews. (2017). *6.9 SR Earthquake in Tasikmalaya, This is House Damage Data.* Detiknews, Saturday, December 07, 2019, 09:08 WIB. https://news.detik.com/berita/d-3772078/gempa-69-sr-di-tasikmalaya-inidata-kerusakan-rumah.
[5] Dewi, R.S. dan Dulbahri. (2010). *"Tsunami", on Multirisk Assessment of Disasters in Parangtritis Coastal Area.* Yogyakarta: Gadjah Mada University Press.
[6] German-Indonesia Tsunami Early Warning System. (2012). *Peta Bahaya Tsunami DIY.* Monday, December 02, 2019, 10:10 WIB. https://www.gitews.org/tsunamikit/id/E1/sumber_lainnya/peta_bahaya/diy/Tsunami%20Hazard%20Map%0Kulon%20Progo.jpg
[7] Khotimah, N, dkk. (2019). School Preparedness in Anticipating the Threat of Earthquake and Tsunami in Bantul Regency. *IOP Conference Series: Earth and Environmental Science, Vol. 271, pp. 1-10.*
[8] Law of the Republic of Indonesia No. 24/2007. *Disaster Management.* Jakarta.
[9] LIPI-UNESCO/ISDR. (2006). *Community Preparedness Study in Anticipating Earthquake and Tsunami Disasters*.
[10] Matsuura, S. and Shaw, R. (2015). Exploring the possibilities of school-based recovery and community building in Toni District, Kamaishi. *Natural Hazards, Vol. 75 No. 1, pp. 613–633.*
[11] Shaw, R. and Kobayashi, M. (2001). Role of Schools in Creating Earthquake-Safer Environment. *Disaster Management and Educational Facilities*, Greece, 7-9 November, 2001.
[12] Shiwaku, K., Ueda, Y., Oikawa, Y. and Shaw, R. (2016). School disaster resilience assessment in the affected areas of 2011 East Japan earthquake and tsunami. *Natural Hazards, Vol. 82 No.*

1, pp. 333–365.

IOP Conf. Series: Earth and Environmental Science **884** (2021) 012047 doi:10.1088/1755-1315/884/1/012047

Senior High School Students' Preparedness In Facing Earthquake Disaster

I Nurkholifah[1] and D R S Sumunar[2]

[1]Departement of Geography, Education Postgraduate Program, Yogyakarta State University, Indonesia
[2]Faculty of Social Science, Yogyakarta State University, Indonesia

[1]isnaninurkholifah.2019@student.uny.ac.id

Abstract. Earthquake was a disaster that could not be predicted. The threat of earthquake in Imogiri needes attention to reduce the risk because of the disaster impact. The socialization and disaster training are needed to increase the students' preparedness at school. The purpose of the research was to know the students' preparedness level in Senior High School 1 Imogiri in facing the earthquake disaster. This research was descriptive research using quantitative approach. The data collection techniques used in the research were interview, documentation and questionnaire, that was the instrument of students' preparedness in facing earthquake disaster in Senior High School 1 Imogiri. The data analysis technique used in the research was descriptive analysis with quantitative approach using frequency table. The results of the research showed that:Senior High School 1 Imogiri students' preparedness in facing earthquake disaster was in the Ready category with average score 44,8. The indicators in the Ready category were the knowledge about earthquake and the resource mobility, while the indicators in the Less Ready were the attitude toward the earthquake, early warning system and the actions when the earthquake occurred.

1. Introduction

Natural disaster is phenomena or events that cannot be avoided, their occurrence can increase from year to year. According to Law Number 24 of 2007 concerning disaster management, a disaster is an event or series of events that threatens and disrupts the life and livelihoods of the community caused by natural factors and/ or non-natural factors as well as human factors resulting human casualties, environment damage, losses property, and psychological impact. Disaster can also be defined as a crisis situation that is far beyond human capacity to save oneself, it means that a natural disaster will not be called a disaster if the impact or the loss is not felt by human.

Earthquake is events that release energy caused by a sudden shift in the movement of the earth's interior (the earth's crust). Earthquake can occur due to tectonic activity or plate movement, earth, fault activity and volcanic activity. Earthquake is disaster that cannot be predicted when and in terms of their names, so that if an earthquake occurs in an area of great strength, it often causes casualties and damage to infrastructure. Therefore, it is necessary to have knowledge and efforts to deal with the earthquake., as to reduce casualties and damage to facilities and infrastructure as a result of the earthquake disaster.

Content from this work may be used under the terms of the Creative Commons Attribution 3.0 licence. Any further distribution of this work must maintain attribution to the author(s) and the title of the work, journal citation and DOI.
Published under licence by IOP Publishing Ltd

Geologically, Indonesia is located on three tectonic plates, it makes Indonesia has many volcanoes. These tectonic plates are the Indo-Australian plate, the Eurasian plate and the Pacific plate. Indonesia's location which is on three tectonic plates makes Indonesia becomes a country prone to earthquakes, tsunamis and volcanic eruptions. This condition is a treat to natural disaster in Indonesia.

Java island is one of the islands in Indonesia, which is located on the meeting point of the Indo-Australian Plate and the Eurasian Plate, namely the Indo-Australian Plate that skins below the Eurasian Plate which creates a subduction zone. The meeting of the two plates produces earthquake vibrations with different strength intensities. In addition, with the subduction zone, Indonesia also has many volcanoes, especially along the island of Sumatera, Java and Nusa Tenggara. BNPB stated that the Special Region of Yogyakarta had experienced earthquakes with a magnitude of 5 in 1867, 1943, 1976 and 2006. Bantul Regency is one of the areas in the Special Region of Yogyakarta that also felt the impact of earthquake on May 27, 2006 at 05:53 WIB with a magnitude of 5,9 SR (Zukhrufuddin Thaha,2007:67).

The earthquake on May 27, 2006 with a magnitude 0f 5,9 Richter Scale, killed 4.626 people and 19.202 people were injured. The most victims and the worst damage caused by the earthquake were in Bantul Regency which was the epicenter of the earthquake. In Bantul, 4.141 people died, 12.026 people were injured because of the earthquake. In Sleman Regency 232 people died and 3, 789 were injured. In Yogyakarta City, there were 204 people who died and 318 people were injured. 22 victims died in Kulon Progo Regency and 2. 678 people were injured, while in Gunung Kidul Regency, 81 victims died and 19.897 were injured (Akhmad Muktaf Haifani,2008:290).

Earthquake also had an impact to the public facilities damage, such as schools, hospitals, markets, and other public facilities. Areas of the Special Region of Yogyakarta province that were affected by the earthquake on May 27, 2006 included Bantul Regency, Sleman Regency, Kulon Progo Regency, Yogyakarta City and Gunung Kidul Regency. The number of houses damaged or collapsed in the villages in the Special Region of Yogyakarta province because of the tectonic earthquake which could be obtained from the Yogyakarta Earthquake Data Center dated June 12, 2006 was 96.360 houses, 117.182 severely damaged, and 156.568 slightly damaged. The worst damage was found in Bantul Regency which was assumed to reach 45% of the total number of damaged houses (Akhmad Muktaf Haifani, 2008: 291). Details of house damage due to the earthquake 0n May 27, 2006 are as follows:

Tabel 1: Data of damage house in Special Region of Yogyakarta

District	Damage (Residents' House)		
	Destroyed	Severely Damaged	Slightly Damaged
Bantul	71.683	70.796	66.512
Sleman	5.243	16.003	33.233
Yogyakarta	7.161	14.535	21.192
Kulon Progo	4.527	5.178	8.501
Gunung Kidul	7.746	10.670	27.130
Total	96.360	117.182	156.568

Source: Akhmad Muktaf Haifani, 2008: 291

Based on table 1, data of damaged house in Yogyakarta Province due to the earthquake on May 27, 2006 showed that Bantul Regency was the worst. There were 71.683 houses in the category of destroyed, 70.796 houses were severely damaged, while 66.512 houses were damage in the slightly category. The least damaged house was in Kulon Progo Regency. There was 4.527 houses that destroyed 5.178 houses were severely damage and 8.501 were slightly damaged. In Sleman Regency, there were 5.243 houses destroyed 16.0003 were severely damaged and 33.233 were slightly damaged.

Gunung Kidul Regency 7.746 houses destroyed 10.670 houses were severely damaged and 27.130 were slightly damaged. In Yogyakarta City, 7.161 houses were destroyed 14.535 houses were severely damaged and 21. 192 houses that were slightly damaged. The areas that were severely damaged during the earthquake in May 2006 were the areas that were transferred by the Opak Fault. The areas included Piyungan District, Pleret District, Imogiri District and Pundong District. One area that was categorized having a high danger was Imogiri district. The impact of the earthquake on May 2006 was also felt in the Imogiri District. The damages were casualties, wounded victims and damaged facilities.

The damage of facilities pr building will also have an impact to people live. For example: the victim caused by broken building. Schools are one of the most dangerous place when an earthquake occurs. It is because the school is one of the vital buildings as a gathering place for many individuals, especially during the school hours. School buildings are susceptible to the threat of earthquake with the impact of broken building and casualties. Senior High School 1 Imogiri is located on Imogiri Timur street number 14 Wukirsari Imogiri Bantul Yogyakarta. Seeing the location of the school, the existence of this school is in the area with high earthquake hazard. Imogiri subdistrict experienced on earthquake on May 27, 2006 had casualties and damaged infrastructure. It showed earthquake had a big impact. The impact of the earthquake could also disrupt learning activities at schools.

The threat of earthquake in SMA N 1 Imogiri which can occur anytime requires a response from the school both by teachers and students. This is done to obtain information regarding understanding of disasters, disaster concept, disaster insight, and disaster readiness. The information can be used to determine the efforts that can be done to reduce the risk of the earthquake. Therefore the teachers' and students' readiness is necessary to prevent casualties.

The efforts to reduce the impact of the natural disaster need the students' readiness to face the earthquake. Disaster readiness is defined as actions aimed at improving the safety of life. Readiness includes actions designed to increase the ability to take emergency measures to protect property from damage and chaos caused by disasters, as well as the ability to be involved in restoration and early post-disaster recovery activities (LIPI UNeSCO, 2006).

Preparedness in facing earthquake can be in the form of increasing the ability to handle hazards by participating in training, understanding evacuation routes, taking steps to safe one self, and others. In addition, students' understanding of earthquake disaster characteristics and disaster mitigation needs to be improved as an effort to reduce disaster risk. Increasing students' understanding of earthquake disasters can be done through education and dissemination of disasters from schools, government and related institutions. It needs to be improved with the aim of increasing students' knowledge about disaster management, especially earthquake disasters to reduce the impact of the disasters.

Building preparedness does not mean teaching students to resist earthquake in their area, but teaching students about the effort that can be made to prevent and anticipate the adverse effects of an earthquake that can occur any time. Preparedness at school is important, considering that schools are locations that have a high risk of casualties. Therefore, it is necessary to have information about the level of preparedness of students in facing disasters. It can be used to make efforts to reduce the impact of the earthquake.

Based on the background above, the researcher tried to know students' understanding and preparedness level in facing disaster. Therefore, the researcher was interested in conducting the research with title " The Preparedness of Senior High School Students in Facing Earthquake Disaster"

2. The Method of The Research

This research was a descriptive quantitative research using frequency tables to explain facts related to the preparedness of SMA N 1 Imogiri students in facing earthquakes. The population of this study was all 568 students. The number of sample was 221 samples, using Isaac & Michael formula with 5% error rate. The data collection techniques were questionnaire/close questionnaire, interview, and documentation. The data analysis technique used in this study was quantitative approach using non-statistical analysis and frequency tables.

IOP Conf. Series: Earth and Environmental Science **884** (2021) 012047 doi:10.1088/1755-1315/884/1/012047

3. Results and Discussion

Earthquake prepared is a series of activities to anticipate disasters. Disaster preparedness has five factors. The indicators of preparedness include: knowledge of earthquake disaster, attitude towards earthquake disaster, types of emergency action, level of early preparedness, and level of resource mobility.

Based on the five indicators above, it can be seen that students' preparedness in facing earthquake. After that The total score obtained by the students was assumed into five categories, very ready, ready, quite ready, less ready, not ready. The results obtained from the five indicators were as follows:

3.1. Students' knowledge of earthquake disaster

Students' knowledge of earthquake disaster is related to knowledge of earthquake characteristics. In addition, knowledge also includes students' understanding of the definition of disaster preparedness. The results of the research was as follows:

Tabel 2. Students' knowledge of earthquake disaster.

Indicator	Yes		Doubt		No	
	f	%	f	%	f	%
Sudden earthquake	190	86	28	12.7	3	1.3
Preparedness definition	213	96.4	7	3.2	0	0
Preparedness toward to earthquake	209	94.6	11	4.9	1	0.5

Based on table 2, students' knowledge of earthquake characteristics was good. It was shown from the results of the study that 86% of respondents or 190 students know the characteristics of earthquake. The earthquake characteristics in question was an earthquake that came suddenly. Students who answered doubtfully were 28 students or 12% of the respondents. While students who do not know was 3 students or 1,3% of the respondents.

Knowledge indicator was also related to the understanding of preparedness. Students know that preparedness was an activity carried out to prepare themselves for disaster quickly and and precisely. Students' knowledge of disaster preparedness was very good. It was indicated by the number of students who knew a number of 213 students or 96,4% of the respondents. Students who answered doubtfully were 7 students or 3,2 % of respondents, while students who did not know were 0% of respondents.

In addition, students understanding of the importance of earthquake preparedness was also very good. There were 209 students who answered yes , 11 students answered doubt and 1 student answered no. It showed that 94,6% of respondents knew the importance of being prepared for an earthquake. The percentage of respondents who answered doubt was 4,9%, while the respondents who answer no were 0,5%.

3.2. Students' attitude toward earthquake disaster

Attitude toward earthquake disaster was a response made by students to face earthquake. These attitudes included: providing first aid medicine, attitude when the earthquake occurs. knowledge before an earthquake occurs, and attitude after an earthquake happens. The results of the research were as follows:

Tabel 3. Students' attitude toward earthquake disaster

Indicator	Yes		Doubt		No	
	f	%	f	%	f	%
Providing first aid medicine	18	8.2	71	32.1	132	59.7
Stay calm and not panic	114	51.6	94	42.5	13	5.9
Knowing evacuation route	21	9.5	75	34	125	56.5
Provide evacuation route	159	72	48	21.7	14	6.3

Based on table 3, students' attitude in facing earthquake disaster were quite lacking. It was indicated by 132 students or 59,7% of the respondents who did not prepare first aid medicine (P3K) in their school bags. There were only 18 students who always brought first air medicine or 8,2% of respondents. 71 students or 32,1% of respondents answered doubtfully. When an earthquake occurred, students tried to stay calm and not panic. There were 114 students or 51,6% of respondents who were calm and did not panic when there was an echo of the earth. 94 students answered doubt, while 13 students or 5,9% of respondents answered no. It meant that 5,9 % of respondents still felt uneasy and panic when an earthquake occurred.

In addition, the attitude in facing the earthquake disaster was knowing the evacuation route and preparing the evacuation rout signs. There were 125 students or 56,5% of respondents who did not know the evacuation route signs at school. 75 students answered doubtfully, and 21 students answered yes. The answers of these respondents indicated that the students had never known the existence of an evacuation route at SMA N 1 Imogiri. There was no sign of an evacuation route at SMA N 1 Imogiri at the time of the observation. It was also made clear from the statement of the principal of SMA N 1 Imogiri. He confirmed that at SMA N 1 Imogiri no evacuation route had been installed.

Even though there were no signs of an evacuation route installed at school, the attitude of students about the importance of the evacuation route was very good. Based on the table 28, there were 159 students or 72% of respondents answered yes, 48 students or 21,7% of respondents answered doubtfully and 14 students or 6,3 of respondents answered no. It showed that 72% of respondents agreed that one of the actions to be prepared for disaster was to prepare an evacuation route.

3.3. Types of Emergency Response done by Students

Types of emergency response are actions that are taken when an earthquake occurs. These actions include actions during an earthquake and actions after an earthquake. The results obtained were as follows:

Tabel 4. Types of emergency response actions done by students

Indicator	Yes		Doubt		No	
	f	%	f	%	f	%
Go Out of the class when the earthquake occurs	172	77.8	33	15	16	7.2
Go to evacuation site	114	51.6	94	42.5	13	5.9
Refuge under the table	21	9.5	75	34	125	56.5
Preparing for the second earthquake	159	72	48	21.7	14	6.3
Preparing tent	34	15.3	70	31.7	117	53
Helping injured friends	92	41.7	79	35.7	50	22.6
Stock of first aid medicine in Health Care Unit	201	91	16	7.2	4	1.8

The actions done during the earthquake are effort to save oneself. Efforts to save yourself include rushing out of the classroom, refuge under the table, and gathering at the evacuation site. Based on the table 4, there were 172 students or 77% of respondent who chose to immediately leave the classroom when an earthquake occurs. 33 students or 15% of respondents answered doubtfully and 16 students or 7,2% of respondents answered no. Based on these data, students prefer to leave the classroom immediately when compared to refuge under the table. It was indicated by the very small percentage of respondents who chose to refuge under the table, 9,5%. There were 34% of respondents who answered doubtfully and 56,5% of respondents answered no.

Actions taken by the students during an earthquake were gathering at the evacuation site. There were 114 students or 51,6% of respondents chose to immediately gathered at the evacuation site, 94 students or 42,5% of respondents answered doubtfully and 13 students or 5,9% of the respondents answered no. The location of the evacuation points at SMA N 1 Imogiri is on the ceremony court, basketball court, and volleyball court.

Self-rescue measured after an earthquake included preparing for aftershock, setting up tents, and treating injured friends.159 students or 72% of respondents prepared themselves in case of aftershocks. However, 117 students or 53% of respondents chose not to help prepare the tents, 70 students or 31,7% answered doubtfully and 34 students or 41,7% answered yes. The preparedness of schools to handle victims when an earthquake occurred was good. It was shown by the presence of first aid medicine (P3K) at the Health Care Unit(UKS). UKS in SMA N 1 Imogiri has sufficient first aid kits.

3.4. Level of early warning preparedness

The level of early warning preparedness is an activity carried out to provide a warning about disaster. It was related to the disaster warning system in SMA N 1 Imogiri. The warning can be in the form of traditional warning tools, modern, or from related institutions. The data obtained were as follows:

Tabel 5. The level of early warning preparedness

Indicator	Yes		Doubt		No	
	f	%	f	%	f	%
Warning by siren	65	29.4	86	38.9	70	31.7
Warning by *kentongan*	52	23.5	90	40.7	79	35.8
Warning from *BPBD (Regional Disaster Management Agency)*	189	85.5	23	10.4	9	4.1

Based on table 5, the existing of early warning system for earthquake came from information from the Regional Disaster Management Agency (BPBD). It was indicated by a number of 189 students who answered yes, 23 students answered doubtfully, and 9 students answered no. A percentage of 85,5% of respondents received an early warning against an earthquake from the BPBD. Early warnings in the form of sirens and *kentongan* were not used for earthquake warnings. However, sirens and bells were less used as a warning when an earthquake occurred. It was also conveyed by the principal of SMA N 1 Imogiri. He said that school has bells. However, the bell is used to tell the change of hours. According to the principal, when an earthquake occurred, students spontaneously left the classroom.

3.5. Level of Resource Mobility

The level of resource mobility was the level of awareness of human resources, both students and institutions as well. The availability of infrastructure and facilities related to disaster preparedness. Efforts to prepare students in facing disasters were by participating in scout activities or Youth Red

Cross (PMR) to improve their self-healing skills. In addition, the availability of transportation facilities also played a role in evacuation of victims. The data obtained were as follows:

Tabel 6. The level of resource mobility

Indicator	Yes		Doubt		No	
	f	%	f	%	f	%
Extracurricular Scout and Youth Red Cross	206	93.2	12	5.4	3	1.4
Joining Youth Red Cross extracurricular	44	19.9	34	15.4	143	64.7
Joining scout extracurricular	210	95	8	3.6	3	1.4
Knowing how to set a tent	156	70.6	58	26.2	7	3.2
Students could make a sedan chair	136	61.5	71	32.1	14	6.4
Indonesian Red Cross played a role in victims evacuation	196	88.7	19	8.6	6	2.7

The level of human resource mobility was related to the participation of students in activities that supported their ability to save themselves. These activities were the Youth Red Cross (PMR) and scouts. Scouting and Youth Red Cross activities were already existed in SMA N 1 Imogiri as extracurricular activities. The participation of students in scout and Youth Red Cross had been good.

Based on the table above, there were 210 students or 95% pf respondents who had participation in scout activity. It was because all students had to follow the program. The participation of students in Youth Red Cross extracurricular was 19,9% of respondents or 44 students. There were 64,7% students who did not take extracurricular activities or 143 students. It was because PMR is an option of the extracurricular, so not all of the students took it.

The participation of students in scout and PMR could increase students' knowledge about how to make stretcher and tent. The table showed that there were 156 students or 70,6% of respondents who knew how to set up a tent. In addition, 136 students or 61,5% of respondents were able to make stretchers. The existence of these two extracurricular activities provided additional knowledge for the students, especially in action before it occurs, when it occurs and after it occurs.

3.6. Students' of SMA N 1 Imogiri preparedness level in facing earthquake disaster

Students' of SMA N 1 Imogiri prepareness level was the preparedness level in facing earthquake disaster quickly and precisely. Indicators of the preparedness level in this study was the students' of SMA N 1 Imogiri knowledge in facing earthquake disaster, students participation in earthquake disaster education and training activities, and students' preparedness in facing the earthquakes.

Based on those three indicators, it could be seen the level of students' preparedness in facing earthquakes. The total score obtained by students from the three indicators was assumed to break into five categories. The categories of preparedness were very ready, ready, quite ready, less ready, and not ready. The results obtained from this study were as follows:

Tabel 7. SMA N 1 Imogiri students' preparedness level in facing the earthquake disaster

Category	Frequency	%
Very Ready	1	0.5
Ready	115	52
Quite Ready	102	46.2
Less ready	3	1.3
Not ready	0	0

Based on table 34, there were 0,5% of respondents in the Very Ready category, 52% of the respondents were in the Ready category, 46,2% of respondents were in the Quite Ready category, 1,3% of respondents were in the less ready category and 0% of respondent was included in the Not Ready category. This data showed that the level of preparedness level of SMA N 1 Imogiri Students in facing earthquake disaster was in the Ready category. The preparedness of SMA N 1 Imogiri students who were included in the Ready category reached 52% of respondents. The percentage of preparedness was influenced by the level of preparedness of students' regarding knowledge about earthquake and students' preparedness to face it. Most students' knowledge about earthquake disaster was included in the Ready category, reached 105 students or 47,5% of respondents. Students who fall into the Very Ready category related to knowledge reached 102 students or 46,2% of respondents. In addition, this preparedness also related to the preparedness students to face the earthquake. Students' preparedness to face the earthquake was included in the Ready category with percentage of 68,3%. There were still some of students in SMA N 1 Imogiri who were in the Quite Ready and Less Ready category. The percentage of students who were Quite Ready reached 46,2%, while the Less Ready category reached 1,3% of respondents. The existence of students who were still in the Quite Ready Less Ready category influenced by students' participation in earthquake disaster counseling activities. The percentage of students' preparedness in participating in disaster counseling reached 47,10%, including in the Quite Ready category.

4. Conclusion

Based in the result and discussion of the research which was done, it can be concluded that the preparedness level of SMA N 1 Imogiri students in facing the earthquake disaster is in the Ready category, with average score 44,8. The indicator of students' knowledge about earthquake and resource mobility is in the Ready category, indicator of early warning is in the Less Ready category and attitude is in the not ready category.

Acknowledgments

The authors would like to thank all parties who helped in the smooth running of this research.

References

[1] Andespa, D., Fauzi, A. (2018). Analysis Of Senior High School Student Preparedness In Dealing With Earthquake Disaster In The Mentawai Island.. *Internasional Conference On Research And Learning Of Physics*. Doi:10.1088/1742-6596/1481/1/01205 doi:10.1088/1742-6596/1185/1/012081

[2] BAPPENAS 2006 Buku Utama Rencana Aksi Rehabilitasi dan Rekontruksi Wilayah Pasca Gempa Bumi di Provinsi Daerah Istimewa Yogyakarta dan Jawa Tengah Jakarta. BAPPENAS

[3] Dodon 2013 Indikator dan Perilaku Kesiapsiagaan Masyarakat di Permukiman Padat Penduduk dalam Antisipasi Berbagai Fase Bencana Banjir. *Jurnal Perencanaan Wilayah dan Kota, Vol 24 No 2, Agustus 2013*. Bandung: Kelompok Keahlian Perencanaan Wilayah dan Desa, Program Studi Perencanaan Wilayah dan Kota. Available at: http://www.sappk.itb.ac.id/jpwk/wp-content/uploads/2014/02/Jurnal-9-Dodon.pdf Accessed date : 11 Desember 2017.

[4] Dwiningrum, S.I.A, & Sudaryono 2010 Peran Sekolah dalam Pembelajaran Mitigasi Bencana. *Jurnal Dialog Penanggulangan Bencana Volume 1 Nomor 1, Tahun 2010.*

[5] Haifani, A.M. 2008 *Manajemen Resiko Bencana Gempa Bumi (Studi Kasus Gempa Bumi Yogyakarta 27 Mei* 2006) Jakarta. Pusat Pengkajian Sistem dan Teknologi Keselamatan, Instalasi, dan Bahan Nuklir, Bapeten. Available at: http://yuli.blog.uns.ac.id/files/2010/04/gempa-yuli.pdf Accessed date: 12 Februari 2018

[6] Hidayati, D. 2006 Kajian Kesiapsiagaan Masyarakat dalam Mengantisipasi Bencana Gempa Bumi Dan Tsunami. *LIPI-UNESCO. (Online)*. Available at:

http:///unesco.unesco.org/images/0015/001536/153617ind.pdf. Accessed date:1 Januari 2018.
[7] Hidayati, D. 2008 Kesiapsiagaan Masyarakat: paradigma baru pengelolaan bencana alam di Indonesia Vol. III, No. 1. Jakarta: *Pusat Penelitian Kependudukan-Lembaga Ilmu Pengetahuan Indonesia*. Available at: http://ejurnal.kependudukan.lipi.go.id/index.php/jki/article/view/164 Accessed date: 15 Januari 2018.
[8] Kusumasari, B 2014 Manajemen Bencana dan Kapabilitas Pemerintahan Lokal. Yogyakarta: Gava Media
[9] LIPI-UNESCO/ISDR 2006 Kajian Kesiapsiagaan Masyarakat dalam Mengantisipasi Bencana Gempa Bumi dan Tsunami, Meliputi Ilmu Pengetahuan Kebumian Lembaga Ilmu Pengetahuan Indonesia. Jakarta.
[10] Marlyono, S,G., Nandi, N. (2017). The Preparedness Level of Community in Facing Disaster in West Java Province. *UPI International Geography Seminar. IOP Conf.Series: Earth and Environmental Science.* doi :10.1088/1755-1315/145/1/012103
[11] Nur, A.M. 2010 Gempa bumi, tsunami, dan mitigasinya. *Jurnal Geografi Balai Informasi dan Konservasi Kebumian Karangsambung-Lipi.* Kebumen. *Vol 7 No 1* Available at: https://journal.unnes.ac.id/nju/index.php/JG/article/view/92/93 Accessed date: 15 Februari 2018.
[12] Nurjanah , et al. 2012 *Manajemen Bencana*. Bandung: Alfabeta
[13] Peraturan Kepala Badan Nasional Penanggulangan Bencana No.44 Tahun (2012) Tanggal 30 April 2012 Tentang Pedoman Penerapan Sekolah / Madrasah dari Bencana Aman
[14] Purnomo, H. & Sugiantoro, R. 2010 Manajemen Bencana Respond dan Tindakan Terhadap Bencana. Yogyakarta. Medpress
[15] Putri, G, F., Fauzi, A. (2019). Analysis of Student Preparedness Levels in Facing Flood Disasters for Development of High School Physics Digital Books (E-Books) With Flood Theme. *The 2nd International on Research and Learning of Physics. IOP Conf. Series: Journal of Physics.* Doi:10.1088/1742-6596/1481/1/012050
[16] Sari, S,A., Dirhamsyah, M., Finzia, P,Z. (2014). The Influence Of Earthquake Disaster Risk Reducation Simulation Towards The Preparedness Knowledge For Elementary School Student Of Elee Paya, Pulo Breuh, Aceh Besar, Indonesia. *Internationl Journal Of Education And Research Vol. 2 No. 9* . Https://Www.Ijern.Com/Journal/2014/September-2014/37.Pdf
[17] Syarif, H. & Mastura. 2015 Hubungan *Self Efficacy* gengan Kesiapsiagaan Bencana Gempa Bumi dan Tsunami Pada Siswa Sekolah Menengah Atas Negeri 2 dan 6 Banda Aceh.. *Idea Nursing Journal Vol. VI. No. 2.* Available at:http://journal2.um.ac.id/index.php./jpg/ Accessed date:13 Desember 2017.
[18] Tuladhar, G., Yatabe, R., dkk. (2013). Knowledge Of Disaster Risk Reduction Among School Student In Nepal. *Journal Geomatics, Natural Hazards and Risk.* DOI: 10.1080/19475705.2013.809556

Disaster Mitigation for Elderly: The Challenges of Social Welfare Services for Elderly in The COVID-19 Pandemic

K. N. Fitriana[1], P. W. Kuncorowati[2]
[1]Department of Public Administration, Faculty of Social Sciences, Yogyakarta State University

[1] Email: kurnianurfitriana@uny.ac.id
[2]Department of Citizenship Education and Law, Faculty of Social Sciences, Yogyakarta State University
[2] Email: puji_wulandari@uny.ac.id

Abstract. This study aims to determine the achievement of disaster mitigation in handling COVID-19 for the elderly through social welfare services. The elderly are the age group that has the highest risk of death due to the COVID-19 pandemic. This research is descriptive qualitative using in-depth interviews, observation, and documentation. Respondents selected by purposive sampling included government agencies, elderly families, social workers, nursing homes, senior regional commissions, the Yogyakarta Province COVID-19 task force, and the elderly. The study found that disaster mitigation in handling COVID-19 for the elderly was not optimal in lesson learn of understanding and knowledge to the elderly. The result of an emergency response plan has not been responsive because there is no accurate database and systematic and integrative standard procedures in handling the COVID-19. Thus, the assisted early warning system is already running well because of the integrated coordination from the family, local government to the national level in the authority of the COVID-19 Task Force. The mobilization of resources needs has not been able for the particular needs of the elderly during the COVID-19 pandemic in several areas in Yogyakarta Province, especially for neglected elderly. Therefore, the findings of this study recommend optimizing each stage in disaster mitigation to minimizing the risk of death in the elderly effectively.

1. Introduction

COVID-19 is a disaster that occurs in every country. COVID-19 outbreak considered a non-natural disaster. Due to caused by non-natural events or series of events, including technological failure, modernize failure, and disease outbreaks [1]. In general, the factors causing disasters are due to the interaction between threats and vulnerabilities. Currently, COVID-19 has become a pandemic due to the rapid increase in the number of cases and massive human-to-human spread, resulting in a global humanitarian and social disaster. WHO data (2020) shows that the number of COVID-19 sufferers in the worlds has reached 94,124,612 people, with 2,034,527 deaths [2]. In Indonesia, cases of COVID-19 as of March 31, 2020, has invected 1,528 people with 136 deaths and 81 recovered. A high risk of death globally occurs at the age of over 50 years, while in Indonesia at more than 40 years. The most deaths occurred in people with COVID-19 who were 80 years old. According to WHO and CDC report, the mortality rate of pre-elderly (50-59 years) is almost 2%, middle-elderly (60-69 years) 4%, and old-elderly (above 70 years) 15%. Most deaths occurred in people with COVID-19 aged 80 years and over, with a percentage reaching 21.9% [3]. Yogyakarta Province-as the highest number of elderly in Indonesia-has achievements as of April 2020 as many as 67 cases of

Content from this work may be used under the terms of the Creative Commons Attribution 3.0 licence. Any further distribution of this work must maintain attribution to the author(s) and the title of the work, journal citation and DOI.
Published under licence by IOP Publishing Ltd

patients being treated, with seven deaths, all of whom are elderly. Meanwhile, the total number of people under surveillance is 3,692 people spread across the province [4].

Elderly is one of the population age groups that have a high proportion of the population in Indonesia. Nowadays, Indonesia facing an ageing population due to the number of the elderly increases 10% in 2020 with a predominantly percentage of the number of young elderly (60-69 years) of 63.82% followed by middle elderly (70-79 years) of 17, 68%, and the elderly. old (80+ years) by 8.50% [5]. Meanwhile, the number of elderly in Indonesia counted at 25.7 million in 2019, or around 9.6% of the population. More specifically, in 2019 Yogyakarta Province had the highest proportion of elderly in Indonesia at 14.5%, followed by two other provinces in Java, namely Central Java Province and East Java Province, each with 13% [6]. Thus, the largest number of elderly in Indonesia increase the risk of COVID-19 in the elderly. This due to the elderly are vulnerable to infectious diseases and chronic diseases as the immune system decrease, all organ functions and movement functions are reduced [7]. However, Indonesia does not have a special procedure for handling elderly COVID-19 patients that can reduce the risk of death and reduce the burden of care costs for families. Therefore, disaster mitigation based on social welfare services needed for handling COVID-19 for the elderly.

In handling COVID-19, the most effective way to reduce the risk of death in the elderly is through family-based social welfare services. This is based on the fact that the majority of people in Indonesia prefer to take care of their parents themselves at home rather than send them to a nursing home. A recent report in 2019 shows that only less than 6% of respondents in Indonesia agree to send their parents to a care institution rather than taking care of them themselves [8]. This tendency makes the burden of elderly care costs under the full responsibility of the family. Elderly care should also be the responsibility of the state. Therefore, the government needs to provide a supporting health system, such as providing assistance to families and providing assistance in the form of medicines for elderly patients and assistance with personal protective equipment for families who care for them during this pandemic. This paper will discuss the challenges of disaster mitigation based on social welfare services for the elderly in handling the COVID-19 pandemic in Yogyakarta Province, Indonesia.

2. Literature Review

Disaster mitigation for the elderly can through social welfare services. Disaster mitigation is a series of efforts to reduce disaster risk, through physical development as awareness and increased capacity to face disaster threats. Mitigation activities carried out through (1) spatial planning implementation; (2) development policy, infrastructure development, building layout; and (3) providing education, counselling, and training both conventional and modern [9]. According to the Indonesian Institute of Sciences (LIPI) and the United Nations Educational, Scientific and Cultural Organization (UNESCO), four factors influence disaster preparedness, namely: (1) knowledge and attitudes towards disaster risk, (2) plans for disaster emergencies response, (3) early warning system, and (4) capacity for resource mobilization [10].

Social welfare is an institution or social welfare field that involves organized activities organized by both government and private institutions to prevent, overcome to solving social problems and improving the quality of life of individuals, groups and communities. Social welfare is a condition for the material, spiritual, and social needs fulfilment of citizens to live and develop themselves properly in social functions [11]. Thus, social welfare reflected as a condition of individuals, groups and communities with the fulfilment of life's needs such as physical, psychological, and social needs so everyone can carry out roles in society according to their duties, functions and basic needs.

Social welfare services are a form of disaster mitigation in handling COVID-19 for the elderly. This form can be conducted guidance, counselling, assistance, compensation and care a directed, planned and sustainable manner. Furthermore, social welfare services provided through:

a. Services based on nursing homes for the elderly and community-based. Elderly home-based services are a form of service that places service recipients in elderly homes. Meanwhile, community-based service is a social welfare service form that the recipient outside nursing home institutions (orphanages), for example, family, community and others;

b. Elderly social institution is the process of social welfare service activities for the elderly in coordination starting from the planning stage, that carried out by organizations/institutions, both formal and informal;
c. Social protection is an effort by the government and society to provide easy services for non-potential elderly to realize and enjoy a reasonable standard of living [12].

Elderly is someone aged 60 years and over who has specific physiological, social, spiritual, and psychological needs because they are degenerative. The elderly supposed to someone who reached the age of 60 years and over. The elderly can be classified into (1) Potential elderly as a productive elderly due to still able to do work and activities, (2) Bedridden elderly as an unable elderly to make a living resilience so that their lives depend on the help of others [13]. The more comprehensive characteristics of the elderly are as follows: (1) Over 60 years of age; (2) Has needs and problems that vary from healthy to sick, from psychosocial to spiritual needs, and from adaptive to mal-adaptive conditions; (3) Affected by varying living environmental conditions [14]. The characteristics of the elderly have an impact on changes that occur in the elderly include:

a. Physical changes: which include physical changes, including changes in cells, cardiovascular, respiration, breakfast, musculoskeletal, gastrointestinal, genitourinary, bladder, vagina, hearing, vision, endocrine, skin, learning and memory, intelligence, personality and organization, and achievement ;
b. Social change: which includes social change, including changing roles, family (emptiness), friends, abuse, legal issues, retirement, economy, recreation, security, transportation, politics, education, religion, nursing homes;
c. Psychological changes.

3. Methodology

The design of this study is research and development, which carried out continuously for two years. Data collection techniques in this study used the method of observation, in-depth interviews, focus group discussion, and documentation studies. Resource taking is done by purposive sampling method, namely by determining resource persons according to expertise and expertise in the field of disaster management and social services for the elderly, including bureaucrats, practitioners, social observers and public services, academics, and verifiers for the development and application of models. In this study, data analysis used a qualitative descriptive approach. The results of the first year research were intend to determine the results of the results of the implementation of social services for the elderly in Yogyakarta Special Region and to form a model of participatory social services for the elderly. In the second year it was intended to determine the development of factual data on the elderly and obtain the results of the development of a disaster management strategy oriented to social protection for the elderly in disaster emergency conditions in Yogyakarta Special Region. This research is very important because research studies on social protection for the elderly in disaster emergency conditions are still limited.

The credibility of the research was fulfilled through the use of the research data source triangulation technique, namely through three checking stages: First, triangulation of data sources, which is comparing the data from observations, in-depth interviews, and documentation. Second, peer review to find out the opinions of researchers and experts who conducted similar research. Third, theoretical triangulation, namely comparing empirical data with theoretical studies that have developed and are recognized as true. Fourth, researchers conducted data analysis to obtain valid and reliable data [15]. This research is focusedon Yogyakarta Special Region with considerate of: (1) Yogyakarta Special Region is an area that has the highest life expectancy in Indonesia; (2) The proportion of the elderly in Yogyakarta Special Region is the highest in Indonesia and the distribution ratio of the number of elderly people in the districts / cities is there; (3) Yogyakarta Special Region has the potential for disaster; and (4) fulfilling the need for social protection for the elderly becomes an important matter and must be done immediately.

IOP Conf. Series: Earth and Environmental Science **884** (2021) 012048 doi:10.1088/1755-1315/884/1/012048

4. Result and Discussion

4.1. The Vulnerability of the Elderly in the COVID-19 Pandemic

In the COVID-19 pandemic, the elderly have the highest vulnerability as victims. Even the death of the elderly due to COVID-19 has dominated the highest. Referring to WHO data, more than 95% of deaths from COVID-19 occur in people over 60 years of age. More than 50% of all deaths involve those aged 80 years or older. According to the WHO reports showed that 8 out of 10 deaths occur in individuals with at least one comorbidity, particularly those not only with cardiovascular disease, hypertension and diabetes, but also in various other chronic conditions. The cause is the elderly have experienced physiological, social, spiritual, and specific psychological degeneration. Physiologically, the elderly must face decreased organ function, body metabolism, dementia and limited physical mobility due to degenerative factors.

In the spiritual and psychological aspects, during the COVID-19 pandemic, the elderly often experience excessive spiritual and psychological anxiety because of the high risk of death in the elderly. The limitation on the accessibility of social activities during the COVID-19 pandemic has an impact on the fear and anxiety of the elderly. Emotional reactions can be observed in individuals who are victims. There are three stages of an emotional reaction that can be catastrophic:

a. Individual reactions immediately 24 hours after the disaster are: (1) Tense, anxious, panicked, (2) Stunned, shock, distrustful, (3) excited or euphoric, not feeling too much suffering, (4) tired, confused, (5) Restless, crying, withdrawing, (6) Feeling guilty;
b. Individual reactions from the first to the third week after the disaster: (1) fear, alertness, sensitivity, irritability, difficulty sleeping, (2) worry, very sadness, (3) reoccurring events, (4) sadness, (5) reactions positives that are still possessed: hope or think about the future, get involved in helping and saving activities, (6) Accepting disasters as destiny. This condition is still a response that requires minimal psychosocial action;
c. Individual reactions more than the third week after the disaster. The reactions shown can be applied and manifested by fatigue, panic, sadness continuously, pessimism, unrealistic thinking, inactivity, isolation, and withdrawal. Anxiety is manifested by palpitations, dizziness, fatigue, nausea, headaches, and other symptoms of health problems. Some of the survivors of the disaster may experience acute mental disorders that arise from several weeks to months after the disaster. Some of these disorders include acute reactions to stress, grieving and mourning, diagnosed mental disorders, adjustment disorders. This condition requires psychosocial assistance from health professionals [16].

Meanwhile, socially, the elderly also have social welfare vulnerabilities during the COVID-19 pandemic. This is related to the high socioeconomic status of the elderly who are categorized as poor. The elderly are faced with the fact of poverty due to limited accessibility of physical productive economic sources and accessibility of social welfare sources. The highest number of neglected elderly people in Yogyakarta Province in 2017 was in Gunungkidul Regency with a total of 18,420 people and the lowest was in Yogyakarta City with 1,549 people. Meanwhile, the number of neglected elderly people in Kulonprogo Regency is 9,336 people, Bantul Regency is 9,278 people and Sleman Regency is 7,128 people [17].

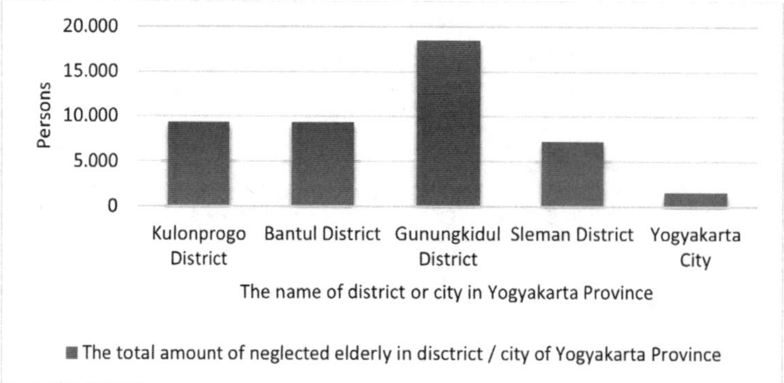

Figure 1. The total amount of neglected elderly in Yogyakarta Province on 2017

Source: Social Agency of Yogyakarta Province, 2018.

Based on the data analysis in the figure, it shows that the distribution of the number of neglected elderly people in Yogyakarta Province is more in districts that have rural characteristics than urban areas that have urban characteristics. This indicates that rural areas are highly prone to poverty compared to urban areas. Although on the other hand, rural areas actually have a higher potential source of social welfare than in urban areas because of the strong values of local wisdom and social capital possessed by local communities.

4.2. Disaster Mitigation Based on Social Welfare Services for the Elderly in Handling COVID-19

Efforts to minimize the social impact of the COVID-19 pandemic require disaster mitigation. Disaster mitigation is efforts to reduce disaster risk, both through physical development as awareness and increased capacity to face disaster threats. The scope of disaster mitigation is to reduce the risk of disaster impacts including health, safety and public welfare issues [18]. All stages in disaster mitigation reduce risks to human life sustainably now and then. Mitigation's activities carried out through (1) Spatial planning, (2) Development arrangements, infrastructure development, building layouts, (3) Education, counselling, and training both conventional and modern [19]. The principles of disaster management include: (1) Fast and precise. Disaster management should implement quickly and appropriately according to the circumstances; (2) Priority. In disaster management should be given priority to save human lives; (3) Coordination and Integrity. Disaster management constructed on good coordination and mutual support; (4) Efficient and effective. Disaster management activities must be implemented effectively, especially in overcoming community difficulties by not wasting excessive time, effort and cost; (5) Transparency and accountability. Each funding and implementation can be done collaboratively from various parties, including government, private sector, non-governmental organizations, non-governmental organizations, and donor agencies. The use of the budget must be accounted for through an audit as a form of accountability; (6) Partnership. There must be cooperation and partnership between the community and the government in handling disaster situations. This partnership is sustainable and requires a strong commitment; (7) Empowerment. Community empowerment is one of the most urgent parts of disaster management to optimize and minimize the impact of existing losses; (8) Fair and non-discriminatory. In providing disaster management, it does not give different treatment to gender, ethnicity, religion, race and any political ideology; (9) Nonprolition. In assisting, all parties are not allowed to impose their beliefs in times of disaster.

Disaster mitigation based on social welfare services for the elderly in handling COVID-19 carried out by the Yogyakarta Province COVID-19 Response Task Force with the following stages:

a. Preventive and promotive stages. The steps taken in the precondition stage are to provide information that is holistic, clear and easily understood by the elderly. This information includes the prevention and handling of COVID-19. Information conveyed through public communication effectively and intensively by various parties.

b. Delivery Stage. This stage includes the physical, health, psychosocial and psychosocial needs of the elderly during the COVID-19 pandemic.
c. Assessment stage. This stage includes identifying the physical, health, psychosocial and psychosocial needs of the elderly during the COVID-19 pandemic. The needs of the elderly during the COVID-19 pandemic include psychological guidance and spiritual guidance, healthy nutritious food, masks, face shields, hand sanitiser, hand washing soap, diapers, vitamins, and traditional medicine, light exercise at home, health care guidance based on COVID-19 health protocols. The regional COVID-19 Task Force to the village level conducts early detection and screening of the elderly through health centres and hospitals to reduce the high elderly death rate due to COVID-19. If the test results show positive, the elderly referred and treated to a referral hospital. COVID-19;
d. The Mentoring stage. Assistance is provided through strict COVID-19 health protocols when providing social welfare services by elderly social workers. The assistance is also carried out on the basis of elderly families and communities so that they can protect and educate each other as the main support system. Currently, assistance can also be provided through online health applications;
e. Monitoring and evaluation stage. The results of monitoring and evaluation are used as input in the formulation of policies and programs for handling COVID-19 in the elderly [20].

4.3. Opportunities and Challenges in Mitigation of Handling COVID-19 for the Elderly in Yogyakarta Province

In carrying out efforts to mitigate the handling of COVID-19 for the elderly in Yogyakarta Province, there are several opportunities. This opportunity can improve the handling of COVID-19 for the elderly and reduce the high risk of death in the elderly. These opportunities include:

a. Potential sources of social welfare in the community include the value of local wisdom and community social capital which is still strong;
b. Developments in information technology that have contributed to creating various online health applications and COVID-19 early detection applications so that they can provide education for the elderly;
c. Support policies for handling COVID-19 that integrated locally, nationally and internationally;
d. Engagement of stakeholders to handle the COVID-19 pandemic morally and socially.

However, some challenges have the potential to undermine the results achieved. The mapping of the challenges faced in handling the COVID 19 pandemic, namely:

a. The database does not review in properly and correct manner so that the information becomes less accurate;
b. Distortion of technology and information. The rapid development of technology has turned out to be an obstacle for the elderly in accessing it due to technological gaps. Also, the flow of information about the COVID-19 pandemic from various sources is difficult to identify whether valid or fake for the elderly;
c. Limitations of health workers, health facilities, and special assistants for the elderly in handling COVID-19;
d. Limited understanding and abilities of the elderly.

IOP Conf. Series: Earth and Environmental Science **884** (2021) 012048 doi:10.1088/1755-1315/884/1/012048

4.4. Discussion

Disasters can be particularly disruptive to the daily living of older adults and their caregivers. Chronic conditions that exist prior to an emergency can be exacerbated, equipment damaged or lost, and services or treatments interrupted, causing additional harm or stress [21]. Older persons are a very diverse group; not all older persons are equally vulnerable to hazards. Therefore, it is important to identify those who are vulnerable. The degree and severity to which older persons are affected in emergencies and disasters depend on the specific characteristics of the person, the type and severity of the hazard, the disaster management systems in place and the interactions among the three. Advanced age itself does not constitute vulnerability, but rather the problems common in old age often increase vulnerability. These can include deteriorating physical and mental ability, decreased strength, low tolerance for physical activity, functional limitations and decreased sensory awareness. Planners and policy makers should be aware of these factors and consider them when planning to meet the needs of older persons in disasters [22].

5. Conslusion

The results of disaster mitigation in handling COVID-19 for the elderly are still low. The failure in disaster mitigation showed from the high mortality rate in the elderly. This problem is due to the high dependence of the elderly on other people. Also, the elderly have less understanding and knowledge of COVID-19 and its health protocols. The emergency response plan that has been designed by the government has not been able to affirm the specific needs of the elderly. The best way to provide an emergency response is through family-based social welfare services and assistance from elderly social workers integratively. The logistics mobilization of the elderly needs can deliver by the closest chain system. Disaster mitigation results optimized by strengthening understanding and knowledge through socialization, internalization of the COVID-19 protocol, service assistance, and database integration. The results of disaster mitigation showed that knowledge about the dangers of COVID-19 by the elderly does not achieve to implement the COVID-19 protocol. There are due to (1) the COVID-19 health protocol is new to the elderly and can not adapt; (2) the elderly's motivation and willingness to comply with the COVID-19 health protocol is still low; (3) local thought patterns, habits and culture that hinder the internalization of life values during the COVID-19 pandemic. Furthermore, it is necessary to increase promotion and outreach to internalize new life values for the elderly. In the aspect of the emergency response plan carried out for handling COVID-19 for the elderly in Yogyakarta Province, it has not been responsive. After all, there is no accurate database and systematic and integrative standard procedures in handling the COVID-19 protocol for the elderly because there are no special procedures that applied to the elderly. Thus, the assisted early warning system is already running well because of the coordination from the family level, neighbourhood, sub-district / village, sub-district, district/city, provincial to national level integratively under the authority of the COVID-19 Task Force. The early warning system built is based on technology and information through applications, mass media and social media. This early warning made the public more aware of official information. Publication of this information can be done online and interactively by authorized parties to ensure that this is valid. The mobilization of resources needed by the elderly has not been able to meet all the specific needs of the elderly during the COVID-19 pandemic in several areas in Yogyakarta Province, especially for neglected elderly. The caused factors are location accessibility, limited human resource capacity, budget constraints, and logistics limited. Ideally, this resource mobilization should be able to increase understanding and skills in responding to emergencies. Thus, this study suggests increasing the achievement of each stage in disaster mitigation. Reducing the risk of death in the elderly can be through by staying healthy at home, community-based integrated care, and adapting to a new normal life.

Acknowledgement

This research funded by the Ministry of Education and Culture of the Republic of Indonesia in collaboration with Yogyakarta State University as a superior applied research grant for higher education in 2020. The author would like to thank the Ministry of Education and Culture of the Republic of Indonesia, Yogyakarta State University, Yogyakarta Provincial Social Service, COVID Task Force. -19 Yogyakarta Province, Yogyakarta Province Elderly Regional Commission, research assistants, and all parties who were respondents in this study.

References

[1] WHO (2020, Desember 12) *WHO Coronavirus Disease (COVID-19) Dashboard* Retrieved from https://covid19.who.int/?gclid=CjwKCAiAxp-ABhALEiwAXm6IycB33vP--rSS0JsvsEMfiGJ39cY0_RWcxFqorqKISLxh6WZEFk_mlRoCtLsQAvD_BwE.

[2] Ministry of Health of The Republic of Indonesia (2020, Desember 10) *Preparedness Facing COVID-19* Retrieved from https://www.kemkes.go.id/article/view/20012900002/Kesiapsiagaan-menghadapi-Infeksi-Novel-Coronavirus.html.

[3] Yogyakarta Provincial Health Office 2020 *Progress Report of COVID-19 in Yogyakarta Province* (Yogyakarta: Yogyakarta Provincial Health Offive).

[4] Katadata (2020, December 12) *Proportion of Elderly Population in Yogyakarta* Retrieved from (https://databoks.katadata.co.id/datapublish/2020/05/29/proporsi-penduduk-lansia-di-yogyakarta-tertinggi-di-indonesia#).

[5] COVID-19 Task Force of Republic of Indonesia (2020, December 6) *Analisis Data COVID-19 Indonesia* Retrieved from https://covid19.go.id/p/berita/analisis-data-covid-19-indonesia-update-6-desember-2020.

[6] Central Bureau of Statistics of the Republic of Indonesia (2019) *The Profile of Elderly in Indonesia* (Jakarta: Central Bureau of Statistics of the Republic of Indonesia).

[7] The National Team For The Acceleration of Poverty Reduction (TNP2K) of Republic of Indonesia (2020, September 18) *Vulnerability of the Elderly Amid the Covid-19 Outbreak: Efforts to Ensure Comprehensive Social Protection to Address Vulnerability* Retrivied from http://tnp2k.go.id/articles/vulnerability-of-the-elderly-amid-the-covid-19-outbreak:-efforts-to-ensure-comprehensive-social-protection-to-address-vulnerability.

[8] The Conversation (2020, December 12) *The Death Risk of the Elderly With Covid-19 Is High But Health Services Are Not Yet On Their Side What Should To Do* Retrievied from https://theconversation.com/risiko-kematian-lansia-dengan-covid-19-tinggi-tapi-pelayanan-kesehatan-belum-berpihak-pada-mereka-apa-yang-harus-dilakukan-138107).

[9] Law Number 24 of 2007 of Republic of Indonesia article 47 paragraph 2 concerning disaster management.

[10] LIPI - UNESCO/ISDR 2006 *Assessment of Community Preparedness in Anticipating Earthquake and Tsunami Disasters* (Jakarta: LIPI - UNESCO/ISDR).

[11] Fahrudin, A. 2014 *Pengantar Kesejahteraan Sosial* (Bandung: Rafika Aditama).

[12] Khotimah, N; Fitriana, K.N; and Pratiwi, P.H. 2019 *Handbook of The Implementation of Participatory Social Service Model for the Elderly* Yogyakarta: Yogyakarta State University Press.

[13] Law No. 13 of 2008 of Republic of Indonesia concerning Elderly Welfare.

[14] Keliat, B. A. 1999 *Gangguan Konsep Diri Edisi 1* (Jakarta: EGC).

[15] Miles M. B. and Huberman A. 1994 *Qualitative Data Analysis : An Expanded Sourcebook (*London: Sage Publications, Inc).

[16] Kristanti 2013 *Kesiapsiagaan Masyarakat terhadap Bencana Gempa Bumi di Dusun Piring Desa Shardono Kecamatan Pundong Kabupaten Bantul Yogyakarta (*Yogyakarta: Cipta Pustaka).

[17] Social Agency of Yogyakarta Province. (2018). *Profile of the Neglected Elderly in Yogyakarta Province.* Yogyakarta: Social Agency of Yogyakarta Province.

[18] Sylves, R. 2008 Disaster *Policy and Politics, Emergency Management and Homeland Security* (Washington DC: CQ Press).

[19] Law Number 24 of 2007 of Republic of Indonesia Article 47 concerning disaster management.

[20] Kunocorowati, P. W. and Fitriana, K.N. 2020 *Development of a Social Welfare Service Model for the Elderly in Handling COVID-19 in Yogyakarta Province.* (Yogyakarta: Yogyakarta State University).

[21] Department of Health and Human Services of United State of America (2020, December 12) *Disaster Preparedness Planning for Older Adults* Retrievied from https://www.phe.gov/Preparedness/planning/abc/Pages/older-adults.aspx.

[22] WHO 2012 *Pan American Health Organization Guidelines for Mainstreaming the Needs of Older Persons in Disaster Situations (*Washington D.C.: PAHO) Retrivied form https://www.who.int/hac/events/disaster_reduction/guide_for_older_persons_disasters_carib.pdf.

Student's Level of Understanding on Mobile Learning based Volcano Eruption

N F Wardaya[1], Pujianto[1], Jumadi[1]

[1]Science Education, Postgraduate Program, Yogyakarta State University, Sleman, Yogyakarta, Indonesia.

Email : nungkifebriliana.2019@student.uny.ac.id

Abstract. This study aims to analyze student's level of understanding on mobile learning based volcano eruption. This research is quantitative descriptive. The sample were 200 students who lives in the area that affected by the eruption of Merapi Volcano (Magelang Regency, Sleman Regency, and Yogyakarta City), recruited using a simple random sampling. The instruments were online survey questionnaire of Student's Level of Understanding on Mobile Learning based Volcano Eruption. The distribution of the data is normal, reliable and homogen based the analysis used IBM SPSS Statistics 22 software. The results of this study indicate that students who lives in Merapi Volcano prone area have a good understanding about mobile learning based volcano eruption. Level of student's understanding on mobile learning meet a good criteria with percentage 72,80%, level of student's understanding on volcano eruption meet a good criteria with percentage 73,40%, and student's understanding on disaster mitigation meet a good criteria with higher percentage 76,40%.
Keywords: *disaster mitigation,mobile learning, volcano eruption.*

1. Introduction

The number of crisis events caused by natural disaster such as volcano outbreaks, earthquakes, flood, and hurricane has dramatically increased over the past decades. Given its location in the Pacific Ring of Fire, volcanic eruptions pose a significant threat. 13% from the world's volcanoes are in Indonesia, which is 129 volcanoes are active. Merapi is a part of the most volcano in Indonesia. In October 2010, the biggest Merapi Volcano eruption ranked third in the world in terms of impact [8] and it claimed at least 386 people were killed and more than 300.000 people evacuated [15]. The loss of life was partly due to basalt landslides which generated pyroclastic flows that cascaded 16 km from the Merapi Volcano [6]. The ashes covered nearby towns completely, destroying infrastructure, houses, economic activities, and education activities.

Merapi eruptions has destroyed most of education facilities. Disaster looses data shows that damage ocurred to physical facilities such as classroom, teacher's room, libraries, and teaching-learning equipments [3]. The main damage is in school activities that experiencing a short-time vacuum because the building made into a refuge within 40 days. The disaster somehow tells us a value about the importance relations between education and the disaster itself.

Disaster related with education to ensure student has a good mental preparedness when disaster and after disaster events, having right attitude to face crucial moments and critical, minimize panic, reduce victims, and know the next danger in the environment [10]. One of the efforts made by the goverment

Content from this work may be used under the terms of the Creative Commons Attribution 3.0 licence. Any further distribution of this work must maintain attribution to the author(s) and the title of the work, journal citation and DOI.
Published under licence by IOP Publishing Ltd

especially in Central Java is to strengthen the disaster education and disaster mitigation in every aspects of life. The research by Sadewo shows that Central Java has a high level rank on efforts to anticipate / mitigate disasters based on aspects of early warning systems, safety equipment and evacuation routes.

A disaster is usually understood as a critical, widely shared, devastating event following unusual procedures. It makes the planning process for communities under continuous threat particularly challenging [14]. Ideally, those who are aware of the hazards and concerned should feel motivated to become better prepared; however, research repeatedly shows that an accurate risk perception often fails to generate adequate preparedness.

How these individuals respond to the threats posed by volcanic hazards influences the effectiveness of official hazard mitigation, response, and recovery efforts. Ideally, those who are aware of the hazards and concerned should feel motivated to become better prepared [5]. One way to increase student response to disasters is to strengthen preparedness behaviour and disaster understanding.

Preparedness is a crucial behaviour for people who live in disaster prone areas. Preparedness is an individual concept that preceived as part of adjustment behaviors in others [12]. Appraisals of the preparedness behavior and one's ability to enact the behavior, including both cognitive and social factors, were also found to influence preparedness for several behavioral theories and hazard types examined.

The factors that influence individual and household preparedness intentions and behaviours are complex and multifaceted [11]. As we know, the integration of disaster education and curriculum at school is one way to shape the preparedness behavior and strengthen the disaster understanding to students. Based on research that has been done by Tyas about student perception toward their preparedness for volcanic eruption disaster, the student from SMP Negeri 2 Cangkringan have good knowledge towards their preparedness for volcanic eruption disaster and their preparedness to face disasters is instinctively formed.

To save our generations of natural disaster requires two priorities; education on disaster risk and safety in the schools. Disaster education is very important for building an understanding of the teachers and students about the causes, nature, and effects of disaster. It also fosters the knowledge and behavior to contribute proactively for the prevention and mitigation of disasters. Based on research by Pamungkasih, student's knowledge and attitude facing disaster preparedness volcanic eruptions. The research result shows that the knowledge and attitude facing disaster preparedness volcano eruption of student class VIII SMP around Merapi Volcano had a good knowledge about knowing the danger sign of eruption with percentage 29% and had a ready for disaster preparedness with percentage 59%.

Education is a key mechanism which is student can participated directly in disaster risk reduction. When student are supported and provided with sufficient knowledge and skills, they can protect themselves, save others, and promote significant changes to reduce risk of disaster [2]. Disaster education aims to increase student's resiliency on disaster risks by solidifying knowledge about disasters, developing skills to prepare, adapt, mitigate, and respond to effect of disasters [13]. The importance of disaster education in schools implies disaster mitigation education needs to be introduced at the level of schooling in Indonesia [9]. As one of the efforts to prepare the people of Indonesia disaster preparedness. One of the innovations that can be applied is by integrating disaster education and technology.

Indonesia is currently entering the era of the Industrial Revolution 4.0, where technology, physical, digital and biological space combine into one. The impact is that there is less geographic locations activity because most of all are converting into digital. Mobile learning is an alternative learning media that integrated with technology. Mobile learning is a technology that able to change the way students communicate, interact, and behave with one another and shape their perceptions of learning [1].

4.0 urgently needed in the era of the successor generation of the Indonesian nation has a lot of ability. The positive impact of using mobile learning stated in research by Oyelere, especially for blended learning, social learning, student-centered learning, and project based learning, all of which are supported in mobile learning implementation. The research also indicates that developing mobile learning application through a well-established is expedient because its specific guidelines for evaluation and iteration.

Based on these problems, this study aims to analyze student's level of understanding on mobile learning based volcano eruption. This study focuses on student who lives in Merapi Volcano's prone areas. The third questions raised in this research are: how the understanding of mobile learning? How the understanding of volcano eruptions? And how the disaster mitigation that they already know? To address these questions, this study explores the understanding of student who lives in disaster-prone of Merapi Volcano.

2. Methods

These online survey questionnaire was administered to students who lives in the area that affected by the eruption of Merapi Volcano (Magelang Regency, Sleman Regency, and Yogyakarta City). Survey participants were shown a list of statments that could indicate the understanding about mobile learning based on volcano eruptions. These statements and items divided into three categories: understanding of mobile learning and its technological tools, volcanic eruption disaster, and disaster mitigation.

Respondents were recruited using a simple random sampling method. The simple random sample means that every case of the population has an equal probability of inclusion in sample [19]. The questionnaire was distributed through science teacher around in area that affected by the eruption of Merapi Volcano. Between 4 - 13 August 2020, 200 individuals participated in the survey (Table 1. survey demograpic data).

Table 1. Survey Participant Demographic

Demographic	Survey participants
Sex; n = 200	
Female	72%
Male	28%
Age; n = 200	
12	11%
13	18%
14	39%
15	22,5%
16	7,5%
17	2%
Origin; n= 200	
Magelang Regency	62,7%
Sleman Regency	21,3%
Yogyakarta City	16%
Parent's Education Qualification; n = 200	
No formal education	1%
Primary	34%
Secondary	48%
Diplomas	2%
Bachelor's/ Master's/ Doctorate	14%

The checklist understanding on the questionnaire included 33 supply items that develop from 3 categories: 12 items of understanding mobile learning (mobile learning definition, learning experience with mobile learning, mobile learning purposes, distance learning platform, science learning with mobile learning, the weakness of mobile learning, and mobile learning features needed in science learning); 13 items of volcano eruption understanding (characteristics of active volcano, the stages of volcano eruption warning, volcanic materials, volcanology information, self eruption experience, volcano eruption in science learning, and type of volcano eruption); and the last is 8 items of disaster mitigation (understanding disaster mitigation, disaster mitigation purposes, disaster risks, disaster-prone areas, volcano eruption sign, evacuation routes, pre-disaster mitigation, disaster mitigation when eruption ocuurs, and post-disaster mitigation).

This research is quantitative descriptive. About the questionnaire consists of 33 items that have been validated by expert lecturers. Data analysis techniques in this research namely (a) the qualitative data

IOP Conf. Series: Earth and Environmental Science **884** (2021) 012049 doi:10.1088/1755-1315/884/1/012049

analysis techniques, with some variations, asked participants to rate their agreement with statements on a scale of 1 (strongly disagree) to 5 (strongly agree). Where necessary, an "I don't know" (neutral) category was included [5], (b) qualitative data analysis techniques converted into quantitative data with measurements SS: 5, S: 4, N:3, KS: 2, TS: 1. Data from the analysis of the answers to the students and then calculating the average and standard deviation. The analysis result categorized into criteria knowledge and disaster preparedness for volcanic eruptions learners [17]. The category of criteria divided into two items, positive item and negative item. The criteria according to calculations using interval equations. The criteria can be seen in Table 2 and Table 3.

Table 2. Level of understanding for positive items.

Interval Score (%)	**Criteria**
80 – 100	Very good
60 – 79,99	Good
40 – 59,99	Enough
20 – 39,99	Low
0 – 19,99	Very low

Table 3. Level of understanding for negative items.

Avergae Interval Score	**Criteria**
80 – 100	Very low
60 – 79,99	Low
40 – 59,99	Enough
20 – 39,99	Good
0 – 19,99	Very good

Descriptive statistics recommended for ordinal measurement scale items include a mode or median for central tendency and frequencies for variability [4]. The distribution of the data is normal and homogen based the analysis used IBM SPSS Statistics 22 software. The Cronbach's alpha value is 0.75> 0.7 it means sufficient reliability, the normality value of kolmogorovsmirnov significicance equal to 0,128 > 0,05 and the homogenity significance equal to 0,293 > 0,05. Normal and homogeneous data allow these data to be analyzed statistically.

3. Results and Discussion

Disaster learning should be implemented in formal or non-formal education in an innovative way such as integrating with local potential of the volcano eruption . Disaster learning is rising with assumption that the focus of mitigation is not only about disaster management, but rather refers to the understanding and preparedness. Disaster preparedness and readiness refers to the knowledge and capacities of a person to efficiently anticipate and effectively respond to and recover from imminent and present disasters [13].

Hazard-related attributes represent a more detailed measure of perceived response-efficacy that includes efficacy for protecting people, efficacy for protecting property, and utility of the adjustment for other purposes. Respons related attributes measure characteristics of protective actions rather than characteristics of the people responding. People's responding depend on their understanding about disaster and mitigation.

The best way to educate the disaster and mitigation is through education. Revolution 4.0 encourage education to integrate with technology. Nowadays, smartphone is the most needed device for people. An innovative ideas to support learning disaster integrated with technology is mobile learning. Mobile learning is one of the answers to the demand the revolution 4.0. Mobile learning proven that learning can be more effective, can be done anywhere, and last a long time.

Science learning is one of subject that includes disaster knowledge in its basic competence. Spesifically, science learning in junior high school loads disaster material in the structure and dynamics of the earth.To know how student's understanding about integrating between mobile learning and volcano eruption. Then a survey research must be carried out for students who live in prone areas of volcanic disasters. This research divided into three categories; mobile learning understanding, volcano eruption understanding, and disaster mitigation.

3.1. Student's Understanding on Mobile Learning

Mobile learning understanding of students measured using indicators mobile learning definition, learning experience with mobile learning, mobile learning purposes, distance learning platform, science learning with mobile learning, the weakness of mobile learning, and mobile learning features needed in science learning. The results obtained to know how much student's understanding about mobile learning. The percentage data of student's understanding about mobile learning in Merapi Volcano-prone area can be seen in table 4.

Table 4. Student's Understanding on Mobile Learning

No	Items	Percentage (%)	Category
1.	Mobile learning is a technology in education by mobile devices that can only accessed at school.*	58	Enough
2.	I have experienced using mobile learning taught by family or relatives.	63	Good
3.	The aims of mobile learning is to make student easier to engage the instruction in anywhere and anytime, improving communication, interaction and collaboration skills.	78	Good
4.	I have participated online learning by using Google classroom, Edmodo, or whatsapp group.	84	Very good
5.	Using mobile learning make student easier to more active in science learning.	67	Good
6.	Using mobile learning make student easier to master the material.	64	Good
7.	Online mobile learning consumes a lot of internet data.	76	Good
8.	Learning with online mobile learning is not effective in areas with a signal which is 4G not available.	80	Very good
9.	Platforms such as google class, edmodo, and other web sites limit students to express their opinions.*	71	Low
10.	The less sophisticated mobile features cause students to be constrained from accessing the mobile learning.	77	Good
11.	Students prefer learning with mobile learning than face-to-face learning in class.	44	Enough
12.	Mobile learning equipped with audio or video features increases students' interest in learning science.	76	Good

*) *Negative items*

The development of new technologies and teaching-learning process have caused the emergence of new mobile devices strategic tools with the capacity to deliver such an instruction. Due to the relationship with technological innovation, mobile learning has been adapting to the new possibilities and demands, evolving into industrial revolution 4.0.

Understanding mobile learning means that learning can do in everytime and anywhere. Student's appreciation, skills, and motivation play significant roles in engaging and sustaining to apply mobile devices to meaningful learning. The survey result of student's understanding on mobile learning indicates that 2 items meet the criteria very good, 7 items meet criteria good, 2 items meet criteria enough, while 1 item meet criteria low. Based on result can be concluded that students already have a good understanding about mobile learning, but they still have a negative perception if their expression will be restricted when engage in online learning.

3.2. Student's understanding on volcano eruption.

The second question research is about student's understanding on volcano eruption. This research measure using indicators characteristics of active volcano, the stages of volcano eruption alert level, volcanic materials, volcanology information, self eruption experience, volcano eruption in science learning, and type of volcano eruption. The percentage data of student's understanding about volcano eruption in Merapi Volcano-prone area can be seen in table 5.

Table 5. Student's Understanding on Volcano Eruption

No	**Items**	**Percentage (%)**	**Category**
1.	I know the observable characteristics of an active volcano.	85	Very good
2.	Volcano with normal status means that volcanic activity increased. The threat is in the form of eruptions around the crater.*	64	Low
3.	The characteristics of volcano "*waspada*" alert level is when the activity of the volcano has increased due to magma, tectonic, and hydrothermal activity.*	81	Very low
4.	The characteristics of volcano "*awas*" alert level is when the activity of the volcano has increased due to magma, tectonic and hydrothermal activity.	79	Good
5.	The characteristics of volcano "*siaga*" alert level is when the activity of the volcano continues to experience an increasingly obvious increase or an eruption occurs.*	81	Very low
6.	Hazardous materials from volcanic eruptions include hot clouds, lava flows, poisonous gases, ash rain, and eruptive lava.	86	Very good
7.	I follow the latest information on Volcano activity at http://www.merapi.bgl.esdm.go.id/ and https://magma.esdm.go.id/ .	68	Good
8.	I was one of the people affected when the volcano erupted.	53	Enough
9.	I feel scared when volcano erupted because I didn't have the knowledge about self-evacuation efforts.*	63	Low
10.	I always feel anxious when a volcanic eruption suddenly occurs due to trauma in the past.*	66	Low
11.	The science lesson in junior high school discusses volcanoes and their eruptions deeply.	76	Good
12.	I understand the type of volcanic eruptions because it discussed circumstantially in science learning.	72	Good
13.	Hazardous materials from volcanic eruptions are discussed circumstantially in science learning.	79	Good

**) Negative items*

Disasters affect segments of the population in different ways while education provides protection in the face of a large-scale natural disaster is investigated. One of natural disaster that threaten Indonesia is volcano eruption. Volcano eruptions are one of the disasters that result in complex consequences. Volcano eruption uderstanding have an urgency to be owned by students who lives in merapi volcano prone area.

Understanding about volcano eruption for the surrounding communities is a knowledge to know and understand the dangers and preventive effort deal with the Merapi Volcano eruption. These research result indicates that 2 items meet the criteria very good where it's signifing student knows the characteristic, trigger and hazard of volcano eruption. 3 items meet the criteria good, 1 item meet criteria enough, 3 items meet the criteria low, and 2 items meet criteria very low. It might caused by student's lack of understanding to the negative items statement. Due to the statements relate to the 4 stages alert level of volcano when the eruption will occur.

Indonesia Volcano Alert Level consists of 4 stages, from the lowest to highest level of unrest and eruption there are: Normal, Waspada, Siaga, and Awas. Every stage indicates a particular level of activity and each include a recommendation for community response. When students have a limited knowledge to understanding volcano alert level, it will influence actions that will be taken in order to mitigate disasters.

3.3. Student's understanding on disaster mitigation

The third question research is about student's understanding toward disaster mitigation. This research measure using indicators understanding disaster mitigation, disaster mitigation purposes, disaster risks, disaster-prone areas, volcano eruption sign, evacuation routes, pre-disaster mitigation, disaster mitigation when eruption ocuurs, and post-disaster mitigation. The percentage data of about student's understanding toward disaster mitigation in Merapi Volcano-prone area can be seen in table 6.

Table 6. Student's understanding on disaster mitigation

No	**Categories**	**Percentage (%)**	**Category**
1.	Effort when a volcanic eruption occur is disaster mitigation. Disaster mitigation is an effort to reduce the risk of disaster through physical development and improving people's capabilities.	82	Very good
2.	Long-term disaster risk reduction, reducing casualties, and minimizing impacts are the goals of disaster mitigation.	81	Very good
3.	Understanding Disaster Prone Area is important to determine action in disaster mitigation efforts.	85	Very good
4.	As a resident of leren volcano, I understand the early warnings and signs of nature when there will be eruptions.	73	Good
5.	As a resident of the volcano slopes, I know the evacuation route and the closest gathering point from home or school.	69	Good
6.	Heeding the direction of the Center for Volcanology and Geological Disaster Mitigation (PVMBG) and the development of volcanic activity is not a mitigation effort before a disaster occurs.*	69	Low
7.	Not in valleys or watersheds, avoiding open spaces, and not in the recommended location to be emptied is	79	Good

	an attempt at disaster mitigation in the event of a volcanic eruption.		
8.	Avoid driving in areas affected by volcanic ash and being wary of potentially cold lava-stricken watersheds is not a mitigation effort after a disaster.*	73	Low

*) *Negative items*

Disaster mitigation can be implemented in Natural Science (IPA) subjects at junior high school level, in line with the contents of Law No. 24 of 2007 that disaster management should be integrated into development programs, including in education. Natural Science learning is very strategic to integrate with disaster mitigation. Understanding disaster mitigation through science perspective so as to better meet the challenges of natural disasters.

The result about disaster mitigation understanding indicates 3 items meet the criteria very good, 3 items meet the criteria good, and 2 items meet the criteria low. It's low criteria included 2 negative items about pre-disaster mitigation and post-disaster mitigation. There are two important lesson to effective mitigation measures against natural disaster. The first is disasters must be considered as unresolved development problems and that they are not unpredictable, isolated or independent events. Natural disaster must be increased the vulnerability. The second lesson is that mitigation of natural disasters cannot depend solely upon technological solutions and should be based on a wide range of measures including engineering devices, land management, social regulation and economic improvements [7].

In line with The Sendai Framework by United Nations (UN) have 4 priority due for action. This research include one of 4th priority. Understanding disaster mitigation belong to 1 priority : understanding disaster risk.This framework prevent new and reduce disaster risk through implementation of integrated and inclusive economic, structural, legal, social, health, cultural, educational, environmental, technological, political, and institutional measures that prevent and reduce hazard exposure and vulnerability to disaster, increase preparedness for response and recovery, and thus strengthen resilience. Substantially reduce disaster damage to critical infrastructure and disruption of basic services, among them health and educational facilities, including through developing their resilience by 2030.

Disaster education and disaster mitigation is gonna be forever concern for every research and focus in every aspects of living. It's not only understood and strengthened by people who lives in disaster prone areas but in every place. Disasters never know the time and place and it so importance to have an understanding about disaster and mitigation disaster. From this research can be inferred, there are great opportunities to integrate technologies such as mobile learning with volcanic eruptions. It's supported by level of student understanding that can be seen in the figure 1.

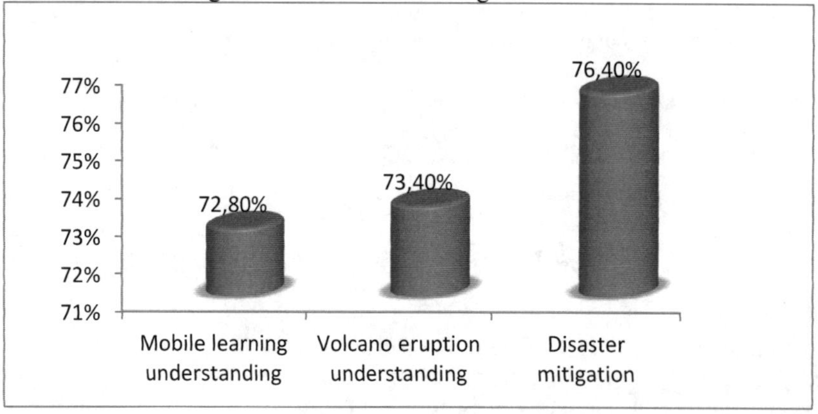

Figure 1. Student's understanding on mobile learning based volcano eruption.

Based on the summary of the figure 1,it can be known entirety that the level of student's understanding on mobile learning based volcano eruption meet the criteria good. Proven with percentage mobile learning understanding 72,80%, volcano eruption understanding 73,4%, and understanding on disaster mitigation 76,40%.

Furthermore, it takes a serious development of disaster education in junior high school that following technological advances. Technology and education integration such as mobile learning supports students who live on the slopes to still be able to learn anywhere and anytime. Undeniable, settling in volcanic regions is a challenge for most people to survive and live in disaster-prone areas. The threat of an eruption occurs at all times and is unpredictable. This causes people around the volcanic region to have knowledge and readiness at any time the disaster happens.

Knowledge is a key factor in understanding natural disasters. The inability of people to evacuate in mitigation when disasters occur will only create new problems. People's lack of understanding of volcanoes and dangers is key to dealing with the eruption crisis. This condition influenced by the role of Natural Science education in secondary schools.

In an effort to support improving people understanding on volcanic disaster and mitigation, there are things to do, among others: 1.) Strengthening the SSB (Sekolah Siaga Bencana), because it plays an optimal role in preparedness and response to volcanic eruptions; 2.) Continued socialization of BNPB to teachers about the importance of disaster knowledge and disaster mitigation capabilities in science learning; and 3.) Development of integrated disaster learning technology to realize a thriving mountainside community and live peacefully with the threat of catastrophic eruptions of Merapi Volcano.

4. Conclusion

Merapi Volcano will always be a threat to people who have an understanding of volcano eruptions and low disaster mitigation. To deal with disasters, there should be a harmonization and strengthening efforts from government, education, health and the community itself. In the aspect of education, disaster education can be synergized with learning technologies such as mobile learning. The findings of this study indicate that students who lives in Merapi Volcano prone area have a good understanding about mobile learning based volcano eruption. Level of student's understanding on mobile learning meet a good criteria with percentage 72,80%, level of student's understanding on volcano eruption meet a good criteria with percentage 73,40%, and student's understanding on disaster mitigation meet a good criteria with higher percentage 76,40%.

Disaster management involves all component of the community and stakeholders. Disaster management cycle explains 4 frame work (mitigation, preparedness, respon, and recovery) to control disasters and emergencies. Disaster management development efforts must be sustainable because disasters will continue to be a threat. This research is a preliminary study on the understanding of students living in disaster prone areas of Mount Merapi. Further research was conducted to validate these results as a reflection and introduction research of mobile learning based on volcanic eruptions.

Acknowledgment

The author wants to say thank you to the Department Science Education, Postgraduate Program, Yogyakarta State University that provided funding to the publication of this study. The author also want to give thanks to Mrs. Supiyah, S.Pd.SD for the biggest support for carrying out research to the preparation of articles.

References

[1] Al-Emran M, Elsherif HM, and Shaalan K 2016 Investigating attitudes towards the use of mobile learning in higher education. *Computer in Human Behavior***56** p93.

[2] Amri A, Bird DK, Ronan K, Haynes K and Towers B 2017 Disaster risk reduction education in Indonesia: challenges and recommendations for scaling up. *Natural Hazard Earth System Sciences***17** p 595.

[3] BAPPENAS 2010) *Laporan Kerusakan Letusan Gunung Merapi* (Jakarta: Badan Perencanaan Pembangunan Nasional).

[4] Boone HN and Boone DA 2012 Analyzing Likert Data.*Journal of Extension***50** p 3.

[5] Corwin KA, Brand BD, Hubbard ML and Johnston DM 2017 *Journal of Applied Volcanology***6** p 2.

[6] Cronin SJ, Lube G, Dayudi DS,Sumarti S, Subrandiyo and Surono 2013 *Journal of Volcanology and Geothermal Research* 261 p 258.

[7] El-Masri S and Tipple G 2002 Natural Disaster, Mitigation and Sustainability: The Case of Developing Countries. International Planning Studies 7 p 157.

[8] Guha-Sapir D,Hoyois P, Below R 2016 *Annual Disaster Statistical Review 2015* (Belgium: CRED).

[9] Kastolani W and Mainaki R 2018 Does Educational Disaster Mitigation Need To Be Introduced In School? *SHS Web of Conferences***42** p 1.

[10] Khairuddin, Zubir F, and Kismullah 2009 4th Annual International Workshop & Expo on Sumatra Tsunami Disaster & Recovery. p 105.

[11] Kohn S, Eaton JL, Feroz S, Bainbridge AA, Hoolachan J, and Barnett DJ 2012 Personal disaster preparedness: An integrative review of the literature. *Disaster Medicine and Public Health Preparedness*, **6** p 217.

[12] Lindell M 2013 *North American cities at risk: household responses to environmental hazards*. In: Joffe H, Rossetto T, Adams J (eds) Cities at risk (Dordrecht: Springer) p 109.

[13] Mamon MA, Suba RA, and Son IL.Disaster Risk Reduction Knowledge of Grade 11 Students: Impact of Senior High School Disaster Education in the Philippines. *International Journal of Health System and Disaster Management***5** p 69.

[14] Muñoz P, Kimmitt J, Kibler E, and Farny, S 2018 Living on the slopes: entrepreneurial preparedness in a context under continuous threat. *Journal Entrepreneurship & Regional Development* **31** p 413.

[15] Nakada S, Maeno F, Yoshimoto M,Hokanishi N, Shimano T, Zaennudin A,and Iguchi M 2019 Eruption Scenarios of Active Volcanoes in Indonesia. *Journal of Disaster Research***14** P 40.

[16] Oyelere SS, Suhonen J, Wajiga GM, and Sutinen E 2018 Design, development, and evaluation of a mobile learning application for computing education. *Educ Inf Technol***23** p 471.

[17] Pamungkasih W and Atun S 2020 Students' knowledge and attitudes facing disaster preparedness volcanic eruptions: A case study in Merapi Mt. areas. *Journal of Physics: Conference Series* doi:10.1088/1742-6596/1440/1/012099.

[18] Sadewo MG, Wanto A, and Windarto AP 2018 Penerapan Algoritma Clustering dalam Mengelompokkan Banyaknya Desa/Kelurahan menurut Upaya Antisipasi/ Mitigasi Bencana Alam menurut Provinsi dengan K-Means. *Jurnal STMIK Budidarma***2** p 314.

[19] Taherdoost H 2016 Sampling Methods in Research Methodology; How to Choose a Sampling Technique for Research. International *Journal of Academic Research in Management (IJARM)***5** p 21.

[20] Tyas RA and Pujianto 2020 Students Perception Toward Their Preparedness for Volcanic Eruption Disaster. *Journal of PhysiC***1440**p 4.

Image Pattern Recognition in Spatial Data using Artificial Neural Network

Nursida Arif[1*], Edi Nursantosa[2]

[1]Geography Education, Faculty of Social Science, Universitas Negeri Yogyakarta, Indonesia
[2]Directorate of Topography Army, Jakarta Pusat, Indonesia

nursida.arif@uny.ac.id

Abstract. This study predicts erosion based on the image patterns as the input data by using an ANN approach. Several simulations had been carried out to get the ANN parameter combination in producing the best accuracy through trials and errors. The results show that the accuracy of artificial neural network training is not influenced by the number of channels, namely the input dataset (erosion factors) and the dimensions of the data, but it is determined by changes in the network parameters. The best combination of parameters is 2 hidden layers, learning rate 0.001, Momentum 0.9, and RMS 0.0001 with an accuracy of 98.55%

1. Introduction

The development of remote sensing technology from aerial photography to satellite imagery in the 1980s makes the pixel-based spatial analysis widely used. Remote sensing has become a solution in the acquisition of spatial data, which is fast and efficient but requires proper analytical techniques, if it is integrated with non-spectral spatial data such as measurements in the field. At present, the development of studies on spatial data analysis techniques is increasing along with the technological advances that demand high accuracy and time efficiency both in the analysis and data collection. The pixel-based spatial data analysis is strongly influenced by the classification method used because each method has a different treatment for the input data [1], [2]. The classification of input data that relies on spectral can be done using conventional methods based on statistical principles such as maximum likelihood where the data are normally distributed [3]. The spatial data with mixed or combined spectral characteristics between spectral and non-spectral data require analysis techniques with machine learning including decision trees [4], [5], [6], [7], fuzzy logic [8], [9], [10], and artificial neural networks [2], [11], [12], [13].

Artificial Neural Network is an effective method with higher accuracy than that of the decision tree [14], while fuzzy logic is more appropriate to be used for data with probabilistic membership of a vague class [9]. Artificial neural network is an information processing system that mimics the work of the neural network system in the human brain that represents knowledge with learning ability [15]. The main problems of the spatial data processing are large data dimensions and data uncertainties [16], [17]. ANN can easily understand uncertain data through network training and the right combination of input parameters [12]. ANN is widely used for an image analysis including filtering, interpretation, and prediction. In this study, ANN was used to predict the level of erosion through pattern recognition of several input data combinations.

Content from this work may be used under the terms of the Creative Commons Attribution 3.0 licence. Any further distribution of this work must maintain attribution to the author(s) and the title of the work, journal citation and DOI.
Published under licence by IOP Publishing Ltd

The sample was taken from each input data as the training data on ANN representing each class of erosion. Input data are a combination of spectral images, namely SPOT imagery and non-spectral images. SPOT imagery is as the data source of vegetation cover data, while non-spectral images are obtained from the interpolation of data from the measurements in the field. In training data, the ANN architecture used is Multilayer Perceptron (MLP) with backpropagation algorithm. Some previous researchers proved that ANN with MLP is able to provide accurate predictions in cases with complex data [1], [18]. Backpropagation is a supervised learning algorithm and is usually used by MLP to change the weights associated with neurons in the hidden layer [15].

The aim of using backpropagation is to obtain a balance between correct training pattern recognition and good responses to other similar patterns (testing data). The network can be trained continuously until all training patterns are correctly identified. In this study, a testing was conducted through simulations of various combinations of ANN parameters to find the best accuracy in recognizing patterns. The ANN parameters include iteration, learning rate, momentum, hidden layer, and RMS error. Iteration (I) is the number of repetitions performed during data training. [19] produces better accuracy (> 60%) in iterations of 10000 to 25000. Learning rate (LR) is a constant speed of learning, where the selection of the right constant will speed up the network in learning to recognize patterns, so that only little repetition occurs [15], [20]. Momentum (M) is a network parameter that serves to accelerate convergence and prevent local optimum, with values ranging from 0 to 1. Hidden layer is the layer that receives responses in the form of weights from the input layer to be forwarded to the output layer. RMS error is the value used to determine the RMS value limit, so that the iteration is stopped. Iteration will stop if the RMS value is lower than the tolerance limit value set on the network. Basically, there are no provisions on the exact value of each network parameter to obtain high accuracy, because ANN is based on "trial and error". Therefore, it is necessary to do a lot of simulations to test the best combination of the parameter values. The simulation results become a reference for analyzing the ability of the network in the recognition of erosion spatial data.

2. Methods

2.1. ANN-Multi Layer Perceptron

Information processing on ANN is divided into three, namely the input layer, hidden layer, and the output layer. Information will be collected in layers called neuron layers. Information given to ANN will be propagated from layer to layer, starting from the input layer to the output layer through another layer, which is often known as the hidden layer.

The algorithm used in this study is backpropagation. Farrokhzad [21] made an ANN-based modeling with backpropagation algorithm to analyze certain elements causing landslides in the study area and soil profiles. The results of the modeling accuracy with ANN were compared to data which were not used in the ANN training in accordance with the analysis of field data; the accuracy of ANN training obtains > 93%.

The backpropagation algorithm uses an error output to change its weight values in the backward direction. To get this error, the forward propagation stage must be done first. At the time of forward propagation, neurons are activated by using an activation function that can be differentiated [15]. The activation function used in this study is the sigmoid function:

$$y = f(x) = \frac{1}{1+e^{-\sigma x}} \quad (1)$$

$$f'(x) = \sigma f(x)[1-f(x)] \quad (2)$$

Backpropagation Algorithm

- Initialization of weights (weights are taken with a fairly small random value)
- Set: iteration, target error, and learning rate (α)

Feedforward

a. Each input unit (Xi, i = 1,2,3, ..., n) receives the xi signal and forwards the signal to all units in the layer above it (hidden layer)
b. Each unit in the hidden layer (Zj, j = 1,2,3, ..., p) sums the weighted input signals:

$$z\_in_j = b1_j + \sum_{i=1}^{n} x_i v_{ij} \quad (3)$$

use the activation function to calculate the output signal:

$$z_j = f(z\_in_j) \quad (4)$$

and send the signal to all the units in the upper layer (output units)

zj = f (z_inj) (4)

c. Each unit of output (Yk, k = 1,2,3, ..., m) sums the weighted input signals.

$$z\_in_k = b2_k + \sum_{i=1}^{p} z_i w_{jk} \quad (5)$$

Use the activation function to calculate the output signal:

$y_k = f(y\_in_k)$(2.7)

and send the signal to all the units in the upper layer (output units)

note:
Step (b) is done as many as the number of hidden layers

Backpropagation

1. Each unit of output (Y_k, k = 1,2,3,....,m) receives a target pattern related to the learning input pattern. Calculate the error information:

$$\delta 2_k = (t_k - y_k)\, f'\,(y\_in_k) \quad (6)$$

$$\varphi 2_{jk} = \delta_k z_j \quad (7)$$

$$\beta 2_k = \delta_k \quad (8)$$

Then calculate the weight correction (which will be used to correct the w_{jk} value):

$$\Delta w_{jk} = \alpha\, \varphi_{jk} \quad (9)$$

Also calculate the correction bias (which will later be used to correct the $b2_k$ value):

$$\Delta b2_{jk} = \alpha\, \beta_{jk} \quad (10)$$

Step (a) is also done as many as the number of hidden layers, i.e. calculating error information from a hidden layer to the previous layer.

2. Each hidden unit (Zj, j = 1,2,3, ..., p) adds up the input delta (from the units in the upper layer):

$$\delta\_in_j = \sum_{k=1}^{m} \delta_i w_{jk} \quad (11)$$

multiply the value by δ_in_j with the derivative of the activation function to calculate the error information:

$$\delta 1_j = \delta\_in_j\, f'(z\_in_j) \quad (12)$$

$$\varphi 1_{ij} = \delta_j x_j \quad (13)$$

$$\beta 1_j = \delta 1_j \quad (14)$$

calculate the weight correction (which will be used later to correct the v_{ij} value):

$$\Delta v_{ij} = \alpha\, \varphi 1_{ij} \quad (15)$$

calculate the bias correction (which will later be used to correct the $b1_j$ value):

$$\Delta b1_j = \alpha\, \varphi 1_j \quad (16)$$

3. Each unit of output (Yk, k = 1,2,3, .., m) corrects the bias and weight (j = 0,1,2,...,p):

$$w_{jk}(newer) = w_{jk}(older) + \Delta w_{jk} \quad (17)$$

$$b2_k(newer) = b2_k(older) + \Delta b2_k \quad (18)$$

Each hidden unit (Zj, j = 1,2,3, ..., p) fixes its bias and weight (i = 0,1,2,...,n):

$$v_{ij}(newer) = v_{ij}(older) + \Delta v_{ij} \quad (19)$$

$$b1(newer) = b1_j(older) + \Delta b1_j \quad (20)$$

2.2. Input Dataset

The erosion control factor becomes the input and output layers, namely the class of erosion (very slight, slight, moderate, severe, and very severe). Artificial neural networks require two stages of analysis, namely stages of data training and data testing. The overview of the inputs and outputs that occur in this study is presented in Figure 1.

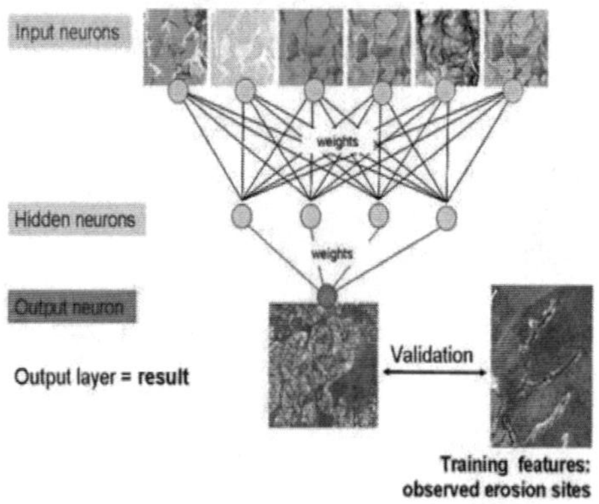

Figure 1. ANN structure of Erosion Model
Source: http://www.beak.de/advangeo/advangeo_prediction

The sample area is taken from each erosion class with a total pixel number of 2216 (Table 1) consisting of 30 samples. The determination of the sample area is based on the results of interpretations on SPOT 5 satellite imagery and the results of field analysis. There are 9 input images, consisting of 4 channels in Spot Image 5, Image R as the representations of climate factors, Image K as the representations of soil factors, Image LS as the representations of topographic factors, Image C as the representations of vegetation factors, and Image P as the representation of land management factors. Each input image is presented in Figure 2.

Table 1. Number of Training Data Pixels

Erosion Class	Total Pixel	Total (%)
Very slight	162	7.31
Slight	295	13.31
Moderate	359	16.20
Severe	425	19.17
Very severe	975	43.99
Total	**2216**	**100**

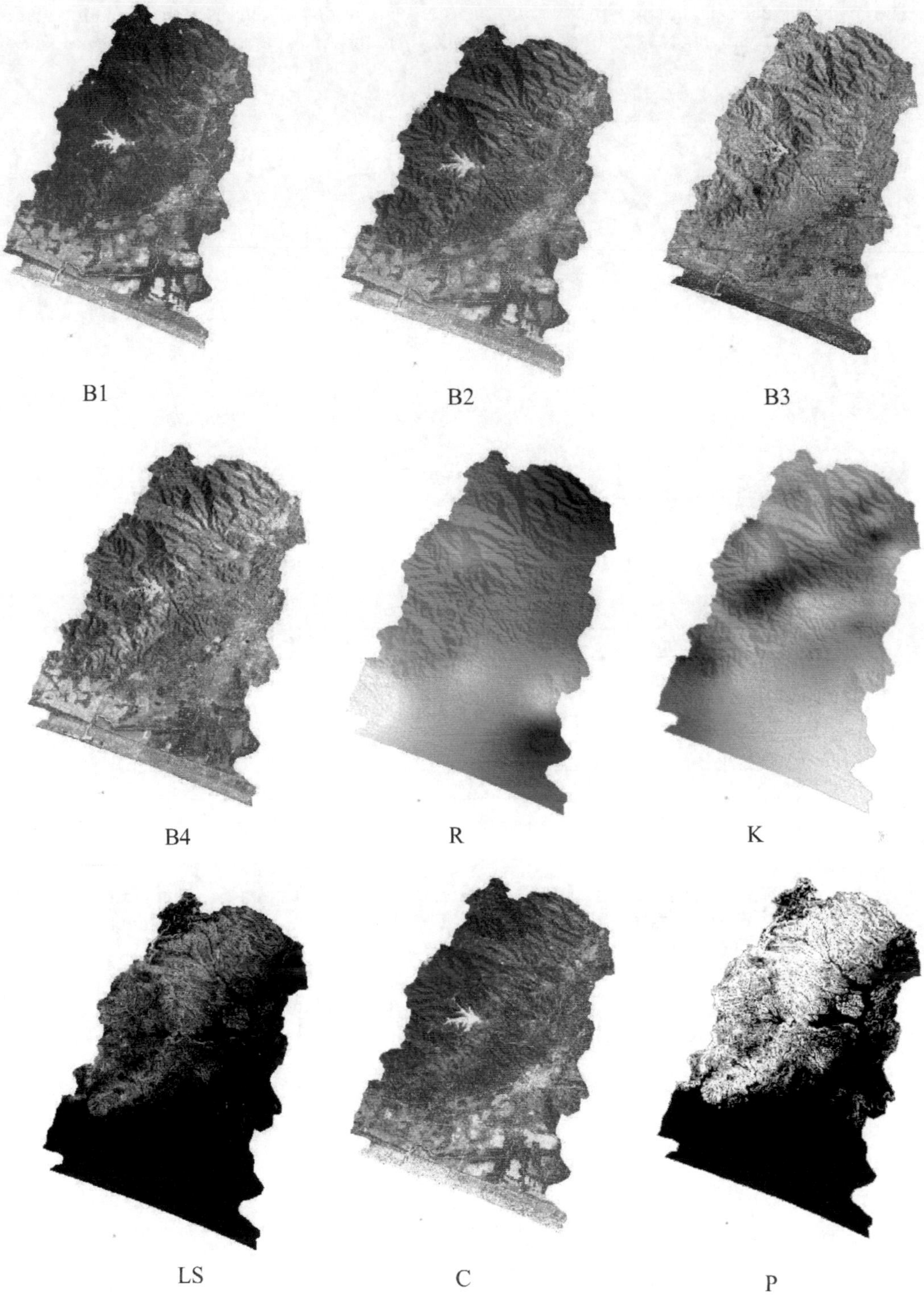

B1,B2,B3,B4 = SPOT Image Channels; R=climate factor; K=soil factor; LS=topography factor; C=vegetation factor; P=land management factor

Figure 2. Input images of ANN Model

The mathematical calculation of ANN from the image input data above uses the backpropagation algorithm done by following the illustration in Figure 3 where the ANN parameters used refer to Table1.

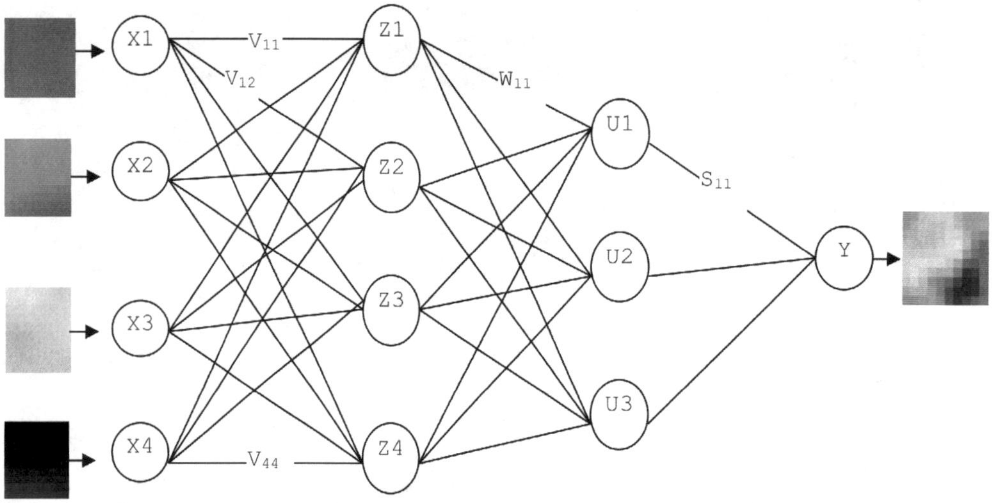

Figure 3. Architecture of ANN Model

Table 2. Network Parameters

Network Parameters	Values
Hidden layer	1 and 2 [15]
Learning rate	0.001, 0.01 [2] 0.4 [22]
Momentum	0.4, 0.5 and 0.9 [1], [23]
RMS	0.001; 0.0001 [1], [2]

3. Results and discussion

The recognition of input data patterns with ANN is done using 23 trials and error simulations with the different levels of accuracy as presented in Table 2. Arif and Danoedoro [2] conducted 20 ANN simulations to get the best classification results.

Table 3. ANN Simulation and Accuration

Simulation	HL	LR	M	RMS	I	Overall accuracy
ANN 1	1	0.001	0.5	0.0001	5000	11.68
ANN 2	1	0.001	0.5	0.0001	10000	62.54
ANN 3	1	0.001	0.5	0.0001	15000	85.65
ANN 4	2	0.001	0.5	0.0001	5000	*error*
ANN 5	2	0.001	0.5	0.0001	15000	*error*
ANN 6	1	0.001	0.9	0.0001	15000	97.65
ANN 7	2	0.4	0.5	0.0001	5000	*error*
ANN 8	1	0.01	0.5	0.001	5000	91.96
ANN 9	1	0.01	0.5	0.001	10000	90.7
ANN 10	2	0.01	0.5	0.001	5000	45.39
ANN 11	2	0.01	0.5	0.001	10000	92.28

ANN 12	2	0.01	0.4	0.001	10000	41.96
ANN 13	1	0.001	0.5	0.0001	25000	95.35
ANN 14	1	0.01	0.5	0.0001	15000	92.14
ANN 15	2	0.01	0.9	0.001	10000	*81.94*
ANN 16	2	0.001	0.9	0.0001	15000	98.15
ANN 17	1	0.001	0.9	0.0001	5000	5.09
ANN 18	1	0.001	0.5	0.0001	20000	93.54
ANN 19	2	0.001	0.5	0.0001	20000	8.70
ANN 20	1	0.001	0.9	0.0001	20000	97.65
ANN 21	2	0.001	0.9	0.0001	20000	**98.55**
ANN 22	2	0.001	0.9	0.0001	25000	98.51
ANN 23	1	0.001	0.9	0.0001	30000	98.42

Table 2 shows that the lowest accuracy occurs in ANN 4, ANN 5, and ANN 7 simulations. ANN 1 simulation only produces an accuracy of 11.68%; in ANN 4 the same parameters are tried using 2 HL, in which the results cannot be executed by the network (error) as happened in ANN 5 and ANN 7 simulations. The examples of outputs produced under error conditions are that among the 30 trained patterns, ANN usually only recognizes two patterns. The parameter on ANN 4 was tried on ANN 5 and ANN 19 by adding the number of iterations; low results were still obtained. This means that the addition of iteration to the momentum values of 0.5, 2 HL, LR 0.001 and RMS 0.0001 is less optimal. Most of the low accuracy occurs in the simulations with 2 HL. HL is used to analyze the impact of ANN training iterations [24]. However, if the simulation with HL 1 produces low accuracy, the number of iterations as in ANN 1 is added with the same parameters (momentum, HL, LR, RMS); the number of iterations is added to 10000 on ANN 2 which results in a higher accuracy of 62.54%. The results of the 23 simulations carried out based on the trials and errors obtain several simulations with an accuracy of >80% as presented in Figure 4.

(a) (b) (c)

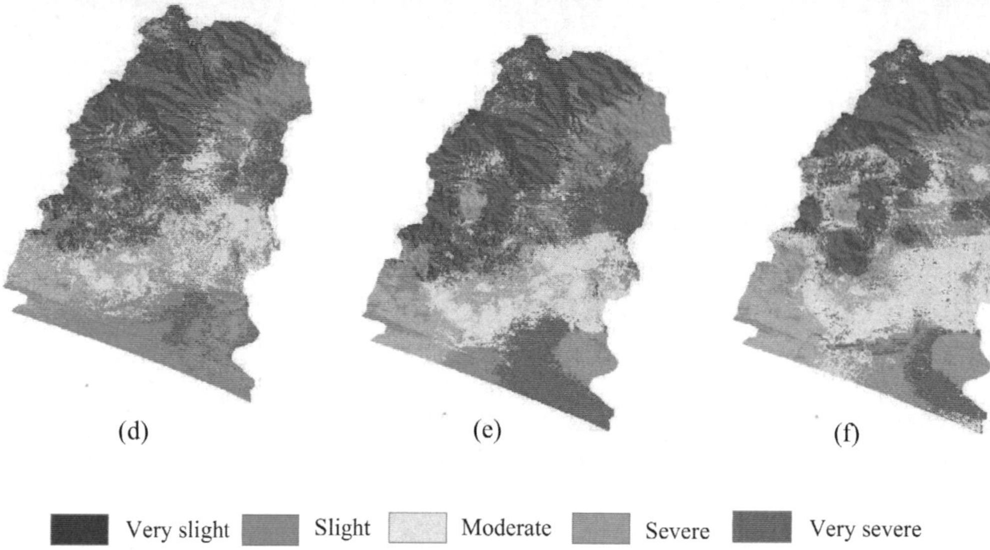

Figure 4. Some of classification results of ANN

Figure 4 shows several classification results with different patterns. Figure 4(e) is a classification result with higher accuracy than the overall simulation that has been done. The highest accuracy reaches 98.55% with 2 HL network parameters; LR 0.001; M 0.9; RMS 0.0001; iteration of 20000. The results were higher than those produced by [24], which reached 95% in the land use classification. The accuracy of ANN training is influenced by the input dataset (erosion factors) and changes in the network parameters [1][1]. These findings provide an opportunity for future research to test other ANN algorithms and validation methods for pattern recognition results to determine the most optimal method.

Acknowledgements
The authors would like to thank National Institute of Aeronautics and Space, Indonesia and Meteorological Climate and Geophysics Agency for providing the data.

References

[1] Arif N Danoedoro P and Hartono, 2017 Analysis of Artificial Neural Network in Erosion Modeling: A Case Study of Serang Watershed in *IOP Conference Series: Earth and Environmental Science* **98**, 1.
[2] Arif N and Danoedoro P, 2013 An analyze of a backpropagation neural network in the identification of critical land based on ALOS imagery in *34th Asian Conference on Remote Sensing 2013, ACRS 2013* **1**.
[3] Shivakumar B R and Rajashekararadhya S V., 2018 Investigation on land cover mapping capability of maximum likelihood classifier: A case study on North Canara, India *Procedia Comput. Sci.* **143** p. 579–586.
[4] Amirruddin A D Muharam F M Ismail M H Ismail M F Tan N P and Karam D S, 2020 Hyperspectral remote sensing for assessment of chlorophyll sufficiency levels in mature oil palm (Elaeis guineensis) based on frond numbers: Analysis of decision tree and random forest *Comput. Electron. Agric.* **169**, May 2019 p. 105221.
[5] Ghulam A Porton I and Freeman K, 2014 Detecting subcanopy invasive plant species in tropical rainforest by integrating optical and microwave (InSAR/PolInSAR) remote sensing data, and a decision tree algorithm *ISPRS J. Photogramm. Remote Sens.* **88** p. 174–192.
[6] Otukei J R and Blaschke T, 2010 Land cover change assessment using decision trees, support vector machines and maximum likelihood classification algorithms *Int. J. Appl. Earth Obs. Geoinf.* **12**, SUPPL. 1 p. 27–31.
[7] Yang C C *et al.*, 2003 Application of decision tree technology for image classification using remote sensing data *Agric. Syst.* **76**, 3 p. 1101–1117.

[8] Ahmad I *et al.*, 2020 GIS and fuzzy logic techniques-based demarcation of groundwater potential zones: A case study from Jemma River basin, Ethiopia *J. African Earth Sci.* **169**, April p. 103860.

[9] Saadoud D Hassani M Martin Peinado F J and Guettouche M S, 2018 Application of fuzzy logic approach for wind erosion hazard mapping in Laghouat region (Algeria) using remote sensing and GIS *Aeolian Res.* **32**, January p. 24–34.

[10] Silver M Svoray T Karnieli A and Fredj E, 2020 Improving weather radar precipitation maps: A fuzzy logic approach *Atmos. Res.* **234**, June 2019 p. 104710.

[11] Gholami V Booij M J Nikzad Tehrani E and Hadian M A, 2018 Spatial soil erosion estimation using an artificial neural network (ANN) and field plot data *Catena.*

[12] Jahangir M H Mousavi Reineh S M and Abolghasemi M, 2019 Spatial predication of flood zonation mapping in Kan River Basin, Iran, using artificial neural network algorithm *Weather Clim. Extrem.* **25**, June p. 100215.

[13] Philippopoulos K and Deligiorgi D, 2012 Application of artificial neural networks for the spatial estimation of wind speed in a coastal region with complex topography *Renew. Energy.*

[14] Xiong Y Z Wang R and Li Z, 2010 Extracting land use/cover of mountainous area from remote sensing images using artificial neural network and decision tree classifications: A case study of Meizhou, China *Proc. - 2010 Int. Symp. Intell. Inf. Process. Trust. Comput. IPTC 2010* November p. 133–136.

[15] Fausett L, 2005 Fundamentals of Neural Networks: Architectures, Algorithms and Applications Fundamentals of Neural Networks: Architectures, Algorithms and Applications *J. Chromatogr. B. Analyt. Technol. Biomed. Life Sci.* **814**, 2 p. 315–323.

[16] Li L Ban H Wechsler S P and Xu B, 2017 Spatial Data Uncertainty *Compr. Geogr. Inf. Syst.* **3** p. 313–340.

[17] Pogson M and Smith P, 2015 Effect of spatial data resolution on uncertainty *Environ. Model. Softw.* **63** p. 87–96.

[18] Yang S Feng Q Liang T Liu B Zhang W and Xie H, 2018 Modeling grassland above-ground biomass based on artificial neural network and remote sensing in the Three-River Headwaters Region *Remote Sens. Environ.*

[19] Arif N and Danoedoro P, 2014 Study on Experiments of Artificial Neural Network Using Spatial Data **4**, 8 p. 732–736.

[20] Mollalo A Mao L Rashidi P and Glass G E, 2019 A gis-based artificial neural network model for spatial distribution of tuberculosis across the continental united states *Int. J. Environ. Res. Public Health* **16**, 1 p. 1–17.

[21] Farrokhzad F Barari A Choobbasti A J and Ibsen L B, 2011 Neural network-based model for landslide susceptibility and soil longitudinal profile analyses: Two case studies *J. African Earth Sci.* **61**, 5 p. 349–357.

[22] Chen G and Zhai M, 2019 Quality assessment on remote sensing image based on neural networks *J. Vis. Commun. Image Represent.*

[23] Song H Xu R Ma Y and Li G, 2013 Classification of ETM+ remote sensing image based on hybrid algorithm of genetic algorithm and back propagation neural network *Math. Probl. Eng.* **2013**.

[24] Abbas Q Haider Bangyal W and Ahmad J, 2013 The Impact of Training Iterations on ANN Applications Using BPNN Algorithm *Int. J. Futur. Comput. Commun.* **2**, 6 p. 567–569.

Disaster Risk Analysis of Merapi Volcano Eruption in Cangkringan District Sleman Regency

M. Rani and N. Khotimah

Department of Geography Education, Faculty of Social Science, Universitas Negeri Yogyakarta, Yogyakarta, Indonesia

Email: rani.muliarti2016@student.uny.ac.id; nurulkhotimah@uny.ac.id

Abstrak. Cangkringan is located in the Merapi Volcano Disaster Prone Areas which has the potential to be affected by eruption. The eruption of Merapi Volcano is a consequence that must be faced by the local resident, so that the need for disaster risk analysis in the region through research is a must. This disaster risk analysis research aims to (1) Analyze the risk level of Merapi Volcano eruption in Cangkringan. (2) Analyze the risk distribution of Merapi Volcano eruption in Cangkringan.This research is a descriptive research with a quantitative approach conducted in Cangkringan District, Sleman Regency, Special Region of Yogyakarta. The population in this study is the entire village in Cangkringan. The entire area is the subject of this research. The variables of this reseach are hazard, vulnerability and capacity. This study used primary data and secondary data. Data collection techniques used are observation, interviews, and documentation. Data analysis techniques used are scoring, overlay and descriptive.The results of this study indicate: (1) The level of risk of Merapi Volcano Eruption in Cangkringan is divided into four levels which are high, medium, low and very low. The area of Cangkringan has a high level of risk covering an area of 19,00% of the total area, the medium-risk level is 38,38% of the total area, the low-risk level is 16,61% of the total area, the very low-risk level is 20,23% of the total area of Cangkringan District. The higher the level of disaster risk, the greater the potential loss due to the eruption of Merapi Volcano. (2) The distribution of disaster risk of Merapi Volcano Eruption in Cangkringan is in the entire village. The distribution of high-risk level is in part of Umbulharjo Village, part of Glagaharjo Village and part of Argomulyo Village. The distribution of medium-risk level is in part of Umbulharjo Village, part of Kepuharjo Village and part of Glagaharjo Village. The distribution of low-risk level is in part of Kepuharjo Village, part of Wukirsari Village and part of Argomulyo Village. The distribution of very low-risk level is in part of Wukirsari Village and part of Argomulyo Village.

1. Introduction

Indonesia is a country located between three active plates in the world which are the Eurasian Plate, the Indo-Australian Plate, and the Pacific Plate. The three plates collided with each other and caused an active subduction zone in Indonesian territory. The existence of this active subduction movement causes volcanic activity in Indonesian territory.

One of the active volcanoes in Indonesia is Mount Merapi. Merapi Volcano is located in two provinces which are Special Region of Yogyakarta and Central Java, and four districts including Sleman Regency, Magelang Regency, Klaten Regency and Boyolali Regency. Merapi Volcano is a Strato-type volcano with a height of 2.965 meters dpal. Volcanoes have two characteristics, on the one hand, volcanoes can provide the availability of natural resources, especially fertile land, so that it can be used by local resident who live around Merapi Volcano. The existence of these fertile land conditions makes the population on the slopes of Mount Merapi on average work in the agricultural sector, besides that mineral C is also abundant [3].

Content from this work may be used under the terms of the Creative Commons Attribution 3.0 licence. Any further distribution of this work must maintain attribution to the author(s) and the title of the work, journal citation and DOI.
Published under licence by IOP Publishing Ltd

The activity of Merapi Volcano tends to be high in eruptive activity. Since 1006, Merapi Volcano continues to experience eruptions regularly with a span of one to seven years and until has recorded eruptions of more than 80 times until now. On this basis, Mount Merapi is often referred to as a *never sleeps volcano* [1], but currently the volcanic activity of Mount Merapi is slightly different from before. Since the 2010 eruption, it has been counted more than seven years that Merapi Volcano has not yet experienced an eruption again. This indicates that a greater magnitude of potential eruptions is likely to occur and of course the magnitude of the risk is also greater. The last explosive eruption of Merapi Volcano occurred in 2010 and caused enormous damage and disaster.

The area of Cangkringan District is one of the areas affected by the eruption of Mount Merapi in 2010. In the eruption that occurred in 2010, precisely in October to November 2010, the direction of the eruption, especially the eruption material in the form of hot clouds, was issued several times to the west, southwest, south and southeast [4]. The location of Cangkringan District, which is on the southeast side of Merapi Volcano, is certainly affected by the eruption material that was generated when the eruption occurred. The Center for Volcanology and Geological Hazard Mitigation (CVGHM) showed that the Cangkringan District area is in a Disaster Prone Area where it has the potential to be affected by eruptions.

The eruption of Merapi Volcano on October 26, 2010 until it reached the peak of the biggest eruption on November 5, 2010 caused considerable damage and loss. Sleman Regency is the area most affected by the disaster where it is estimated that around 65% of the damage and losses [2]. Based on the data compiled by the Sleman Regency Government, there are 2.271 houses in the Cangkringan District suffered damage due to the eruption of Merapi Volcano.

Table 1. Data of Damaged Houses in Cangkringan District

Data on Damaged Houses Due to the Eruption of Mount Merapi 26 October – 5 November 2010					
No	District	Village	The Number of House		
			Total	Damaged	Percent (%)
1.	Cangkringan	Glagaharjo	855	625	73
2.		Argomulyo	862	297	34
3.		Kepuharjo	802	736	91
4.		Wukirsari	531	268	50
5.		Umbulharjo	345	345	100
		Total	**3385**	**2271**	**348**

Source : Sleman Regency Goverment Data, 2010

In addition to causing physical damage, the 2010 Merapi Volcano eruption also caused casualties. Based on the data from the Sleman Regency Health Office compiled by the Sleman District Health Office in 2010, the number of victims who died as a result of the 2010 Merapi eruption was 277 people on December 2, 2010. Most of the victims came from the Cangkringan District area.

Cangkringan District is on the southeast side of Mount Merapi which consists of 5 villages namely Wukirsari, Argomulyo, Glagaharjo, Kepuharjo and Umbulharjo. According to the data from the Central Bureau of Statistics in Sleman Regency 2017, the population in this district is 29,456 people in 2017, with a population density of 617 people per km2. Cangkringan District has abundant natural resources and also supported by fertile land. The land in this area is 47.99 km2 with various agricultural commodities in it. Therefore, people in this area are reluctant to leave this area. They still do some activities and lives in this area, even though this area is in Disaster Prone Area of Merapi Volcano.

Community activities in the Disaster Prone Area of Merapi Volcano have the potential to be exposed to primary and secondary hazards that can threaten human life, property and livelihoods there. From the conditions in the field, it is necessary to make efforts to identify and reduce the risk posed by the hazard of Merapi Volcano in the Cangkringan District by the disaster management agency and community.

Disaster management measures must require a strong basis in their implementation. Therefore, the need of a disaster risk assessment in the form of a tool is important to assess the possible magnitude of losses due to existing threats. By knowing the possible risk of a disaster that occurs, it can minimize the impact and loss that can occur someday. Based on the observations that have been made, Cangkringan

District area has not been maximized in disseminating information about the level and the distribution of the disaster risk of Merapi Volcano eruption in the Cangkringan District.

Because of the condition of Cangkringan District that has not been maximal in disseminating information about the level and the distribution of the disaster risk of eruption in the region, this research is carried out with the aim of analyzing the risk level of Merapi Volcano eruption and analyzing the risk distribution of Merapi Volcano eruption in Cangkringan District.

2. Theoritical Framework

Cangkringan District is located in the Merapi Volcano Disaster Prone Areas. In this way, it causes catastrophic disasters due to the eruption of Merapi Volcano. Disaster analysis is needed to measure the potential loss of both properties and casualties of volcanic eruption. The disaster analysis not only considers the level of hazard from volcanic eruptions, but also considers the level of vulnerability and capacity in the area.

The hazard level of Merapi Volcano eruption in Cangkringan District is affected by the location of the area towards Mount Merapi. Based on the Map of Merapi Volcano Disaster Prone Areas published by the Geological Agency, the level of hazard due to the eruption of Merapi Volcano is divided into KRB I, KRB II, and KRB III. The level of vulnerability is influenced by several variables, such as physical vulnerability using the parameter of buildings vulnerability (houses and public facilities), economic vulnerability using the parameter of productive land area (rice fields, moor, ponds and plantations) and Gross Regional Domestic Product (GRDP), environmental vulnerability using the parameter of land use (protected forest, plantations, shrubs, rice fields / moor and settlements) and social vulnerability using the parameter of population density and the ratio of vulnerable groups (ratio of age group, ratio of sex, ratio of disabled people and ratio of poverty). The level of capacity in the volcanic eruption disasters is influenced by five variables including the existence of an early warning system, the reduction of basic risk factors, the existence of disaster management organizations, the presence and type of disaster socialization, and the presence and type of disaster mitigation.

In this research, the disaster risk analysis used scoring and map overlay methods. The scoring method is done by giving a score to each of the hazard, vulnerability and capacity variables. The overlay method is done by overlaying all the variables of hazards, vulnerabilities and capacities that have been changed in the form of maps and which have been marked for each disaster in the study area. After scoring and overlaying all relevant variables, the distribution and level of risk in volcanic eruption disaster will be shown on a map. Later, the map of the risk level of Merapi Volcano eruption can be used by the residents and the government agencies in Cangkringan District as a guide for carrying out disaster management in Cangkringan District.

3. Research Methods

This research is based on the method and problem discussion and it is a descriptive research. This research will describe everything in the field, especially the risk of disaster that may occur when the eruption of Merapi Volcano occurs again in the Cangkringan District area. This study uses a quantitative approach as one of the supporting methods during the research activities. The data from the results of this study will be presented in the form of a frequency table for further scoring. The scoring result data is then used as a guide in the analysis of map overlay.

This research was conducted in Cangkringan District, Sleman Regency. The determination of the location of this research is based on the Map of the Volcano Hazard Prone Areas of Merapi. This research was conducted from October 2019 to March 2020. The population in this study was the entire area of Cangkringan District which consists of five villages namely Umbulharjo Village, Kepuharjo Village, Glagaharjo Village, Argomulyo Village, and Wukirsari Village. This research is a population study so that the entire area is the subject of the research.

Data collection methods in this study are observation, interviews, and documentation. The data analysis techniques used in this study were scoring analysis, map overlay analysis in geographic information systems, and descriptive analysis.

4. Result & Discussion

4.1 The Description of the Research Area

Cangkringan District is located in the northern part of Sleman Regency. Astronomically, Cangkringan District is located between 434000-444000mT and 9152000-9162000 mU, Cangkringan District has an area of 4,799 hectares (47.99 km2) with a population of 29,592 people, consisting of 14,673 male residents and 14,919 female residents. Cangkringan District has a population density of 617 people / km^2. Cangkringan District is divided into five villages including Wukirsari, Argomulyo, Glagaharjo, Kepuharjo and Umbulharjo. The following bellow is an area division of each village or sub-district in Cangkringan District.

Table. 2 The Allocation of the Research Area

No.	Village	Total (Soul)	Larege (Km^2)	Population Density Level ($Soul/Km^2$)
1.	Wukirsari	10. 291	14,56	707
2.	Argomulyo	6.905	8,47	815
3.	Glagaharjo	3.803	7,95	478
4.	Kepuharjo	3.297	8,75	377
5.	Umbulharjo	5.296	8,26	641
	Total	29.592	47,99	3.018

Source : Secondary Data Analysis, 2020

4.2 The Disaster Risk Level of Merapi Volcano Eruption in Cangkringan District

4.2.1 The Hazard Level of Merapi Volcano Eruption in Cangkringan District

The hazard level of Merapi Volcano eruption is divided into three levels which are low, medium and high. Cangkringan District has five villages, but not all of them are included in the Disaster Prone Areas map. The low hazard level does not exist in Cangkringan District. The moderate level of danger covers part of Argomulyo Village, part of Wukirsari Village, part of Kepuharjo Village and part of Umbulharjo Village. The high hazard level covers all of Glagaharjo Village and part of Argomulyo Village, part of Kepuharjo Village, part of Wukirsari Village and part of Umbulharjo Village.

Table. 3 The Hazard Level of Merapi Volcano Eruption Disaster in Cangkringan District

No	Village	Criteria	Large of KRB (Ha)	Large in percent %	Score	Hazard Level
1.	Wukirsari	KRB II	363,25	24,94	20	Medium
		KRBIII	105,25	7,22	30	High
2.	Argomulyo	KRB II	266,01	31,40	20	Medium
		KRB III	326,47	38,54	30	High
3.	Glagaharjo	KRB III	795,00	100	30	High
4.	Kepuharjo	KRB II	198,72	22,71	20	Medium
		KRB III	567,36	64,84	30	High
5.	Umbulharjo	KRB II	532,48	64,46	20	Medium
		KRB III	390,14	47,23	30	High

Source : Secondary Data Analysis, 2020

4.2.2 The Vulnerability Level of Merapi Volcano Eruption Disaster in Cangkringan District

The results of the assessment of the total vulnerability variable to the eruption of Merapi Volcano in Cangkringan District are included in the high and medium categories. The high category includes five villages, namely Wukirsari Village, Argomulyo Village, Umbulharjo Village and parts of Glagaharjo Village and Kepuharjo Village. The medium category includes two villages, namely Glagaharjo and Kepuharjo villages. The factors that cause high vulnerability in Cangkringan District are mainly economic and physical vulnerability variables.

Table 4. The Distribution of Total Vulnerability in Cangkringan District

No	Village	Level of Total Vulnerablity	Large (Ha)	%
1.	Wukirsari	High	1.456	30,33
2.	Argomulyo	High	847	15,65
3.	Glagaharjo	High	360,46	7,51
		Medium	537,54	11,20
4.	Kepuharjo	High	72,77	3,52
		Medium	699,23	14,57
5.	Umbulharjo	High	826	17,22
Total			4.799	100

Data Analysis, 2020

4.2.3 Capacity Level of Merapi Volcano Eruption Disaster in Cangkringan District

The existing capacity in the Cangkringan District area in facing the eruption of Merapi Volcano is included in one criterion, namely high. This high level of capacity can be found in all villages in the Cangkringan District. The factor affecting the high capacity level in the area is that all capacity nodes are functioning properly.

Table 5. The Distribution of Capacity in Cangkringan District

No	Village	Capacity Level	Large (Ha)	%
1.	Wukirsari	High	1.456	30,33
2.	Argomulyo	High	847	15,65
3.	Glagaharjo	High	795	16,56
4.	Kepuharjo	High	875	18,24
5.	Umbulharjo	High	826	17,22
Total			4.799	100

Data Analysis, 2020

4.3 The Distribution of the Risk of Merapi Volcano Eruption in Cangkringan District

4.3.1 The Distribution of the Hazard of Merapi Volcano Eruption in Cangkringan District

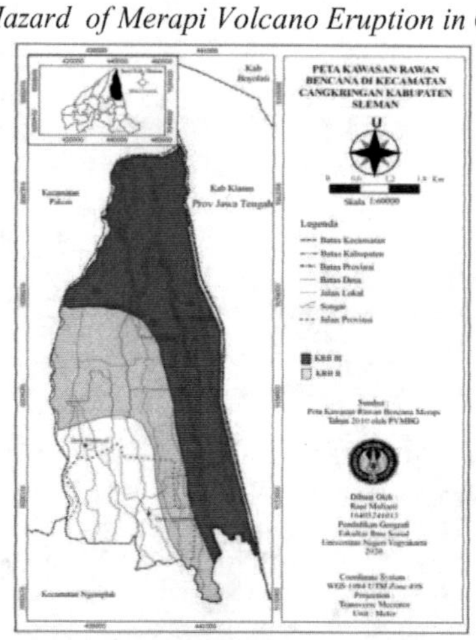

Figure 1. Map of Disaster Prone Area in Cangkringan District

The Disaster Prone Area III covers part of Umbulharjo Village, part of Kepuharjo Village, part of Wukirsari Village, whole of Glagaharjo Village and part of Argomulyo Village. The village in Disaster Prone Area III is a village that has threats in the form of hot clouds, lava flows and also threats of throwing incandescent rocks and dense volcanic ash. The Disaster Prone Area II is located in part of Umbulharjo Village, part of Kepuharjo Village, part Wukirsari Village and part Argomulyo Village.

The village in Disaster Prone Area II is a village that has threats in the form of mass flows of lava as well as throwing materials and heavy ash rain.

4.3.2 The Distribution of Vulnerability of Merapi Volcano Eruption in Cangkringan District

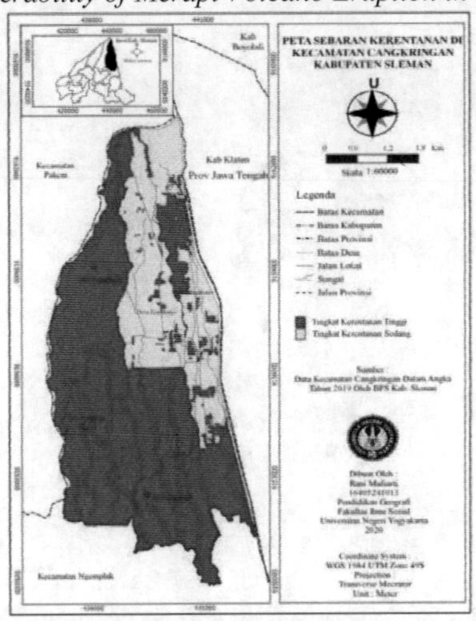

Figure 2. Vulnerability Distribution Map

The results of the assessment of total vulnerability to the eruption of Merapi Volcano in Cangkringan District are included in the high and medium vulnerability category. The high category includes five villages, namely Wukirsari Village, Argomulyo Village, Umbulharjo Village and parts of Glagaharjo Village and Kepuharjo Village. The medium category includes two villages which are part of Glagaharjo and part of Kepuharjo villages. The factors that cause high vulnerability in Cangkringan District are mainly caused by economic and physical vulnerability variable

4.3.3 Distribution of Capacity against the Eruption of Merapi Volcano in Cangkringan District

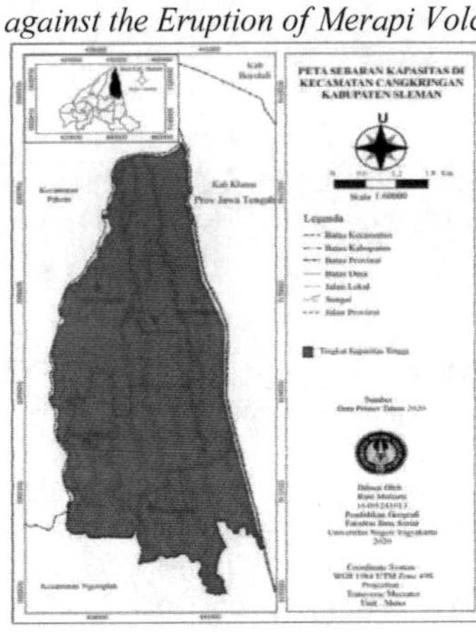

Figure 3. Capacity Distribution Map

The existing capacity in the Cangkringan Subdistrict area in facing the eruption of Merapi Volcano is included in one criterion, namely high. This high level of capacity can be found in all villages in the Cangkringan District. Factors affecting the high capacity level in the area are the presence of OPB, the existence of an early warning system, basic risk reduction factors and the existence of well-functioning disaster mitigation.

IOP Conf. Series: Earth and Environmental Science **884** (2021) 012051 doi:10.1088/1755-1315/884/1/012051

4.4 The Disaster Risk Level and Distribution of Merapi Volcano Eruption in Cangkringan District

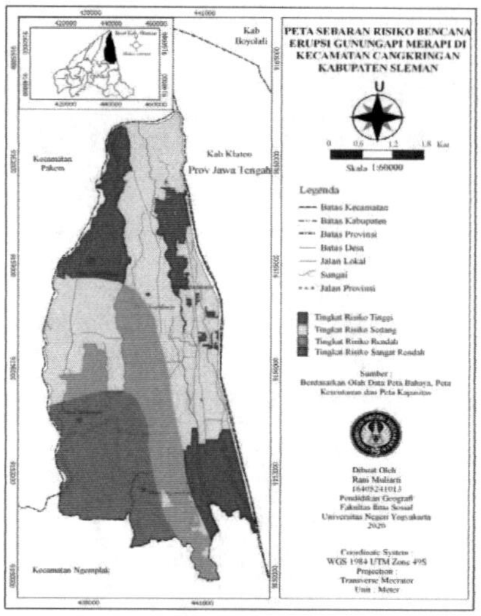

Figure 4. Map of Disaster Risk Level and Distribution

Disaster risk in Cangkringan District is in all villages. The distribution of high-risk level is in part of Umbulharjo Village, part of Glagaharjo Village and part of Argomulyo Village. The distribution of medium-risk level is in part of Umbulharjo Village, part of Kepuharjo Village and part of Glagaharjo Village. The distribution of low-risk level is in part of Kepuharjo village, part of Wukirsari village and part of Argomulyo Village. The distribution of very low-risk level is in part of Wukirsari Village and part of Argomulyo Village.

5. Conclusion

The risk level of Merapi Volcano eruption in Cangkringan District has four levels which are high, medium, low and very low disaster risk levels. The distribution of high-risk level is in part of Umbulharjo Village, part of Glagaharjo Village and part of Argomulyo Village. The distribution of medium-risk level is in part of Umbulharjo Village, part of Kepuharjo Village and part of Glagaharjo Village. The distribution of low-risk level is in part of Kepuharjo Village, part of Wukirsari Village and part of Argomulyo Village. The distribution of very low-risk level is in part of Wukirsari Village and part of Argomulyo Village.

Acknowledgments

Gratitude was expressed to the lecturers of geography education department for their many helpful suggestions which improved the original manuscript.

References

[1] Ashari dan Setyawati 2017 Geomorfologi Lereng Baratdaya Gunungapi Merapi Kaitannya Dengan Upaya Pengelolaan Lingkungan Dan Kebencanaan *Jurnal Geomedia* 15 1 45-60

[2] Muhamad 2015 Tingkat Kesiapan Masyarakat Pengelolaan Lingkungan dan Kepariwisataan Pasca-Erupsi 2010 *Jurnal Kawistara* 5 2 132-143

[3] Sumunar & Ardi 2017 Analisis Risiko Bencana Erupsi Gunung Merapi di Kecamatan Dukun Kabupaten Magelang. *Jurnal Geomedia* 15 1 101

[4] Wahyunto dan Wastio 2012 *Pengembangan Pertanian Berbasis Inovasi di Wilayah Bencana Erupsi Gunung Merapi* (Jakarta: Kementrian Pertanian)

IOP Conf. Series: Earth and Environmental Science **884** (2021) 012052 doi:10.1088/1755-1315/884/1/012052

The Role of Village Tourism in Environmental Education and Social Disaster Management

Saliman [1], Satriyo Wibowo [1], Anik Widiastuti [1], Raras Gistha Rosardi [12]

[1]Social Science Education, Faculty of Social Science, Yogyakarta State University

[2]Social Science Education Doctoral Program, Postgraduate Program Semarang State University

Email: salimanjaper@uny.ac.id, satriyo@uny.ac.id, anikwidiastuti@uny.ac.id, rarasgistha@uny.ac.id

Abstract. The development of the tourist village is very fast. People are aware of the need for tourism which tends to return to nature and offers a natural atmosphere. There have been many villages that have tried to develop their territory into a Tourism Village. However, this tourist village is not yet fully ready to provide for the various needs of tourists who come at any time. Some are only labeled as tourist villages, but have not maximally applied their tourism charm. The purpose of this study was to analyze the role of tourist villages in providing environmental education and social disaster management. This research uses descriptive qualitative. The use of qualitative methods is also considered because the research instrument is a human being or the researcher himself. The research location is in Grogol Sleman Tourism Village. Researchers conducted observations and interviews with village officials and the community in the tourist village of Grogol. The research subjects were determined by purposive sampling technique. Purposive sampling is a technique of taking data sources according to the special considerations of the researcher. Data collection techniques in this study using a questionnaire (questionnaire), interview (interview), observation (observation), and documentaries (secondary sources). The data analysis technique used in this study was the triangulation technique. The triangulation technique used in this study includes triangulation of methods to compare research results obtained from interviews, observations, and documentation. The results show that the development of the Grogol Tourism Village is one of the steps for entrepreneurial opportunities, environmental management and social disaster management.

Content from this work may be used under the terms of the Creative Commons Attribution 3.0 licence. Any further distribution of this work must maintain attribution to the author(s) and the title of the work, journal citation and DOI.
Published under licence by IOP Publishing Ltd

Grogol tourism village is an independent tourism village. Grogol tourism village initially relied on cultural potential in self-promotion. Over time, the management of the tourism village has innovated to make Grogol Tourism Village a nature-based tourism village, but without eliminating the cultural elements and local wisdom of the local community.

Keywords: Potential, tourism village, environmental education, social disaster

1. Introduction

Tourism is one of the largest sources of income in Indonesia. Tourism development is experiencing rapid dynamics and adaptation. Tourism as a catalyst for development [1]. The essence of tourism is an activity to provide comfort for the audience. Traveling is a basic human right to improve the quality of human life[2]. Tourism, with its various positive aspects, is seen as a passport for development, a new type of money, a means of regional development, invisible exports, non-polluting industries and others [3]. This sector is considered capable of being a strategic support for economic growth for the country [4]. This economic growth can have an impact on the level of welfare of the population and is expected to reduce economic problems such as unemployment and crime.

Local communities that can systematically build tourism activities, namely villages. Tourism experts say that a community-based tourism development approach is a prerequisite for sustainable tourism [5], [6]. Tourism must be developed at the community level to ensure this flow of money reaches local communities[7]. The community that builds eco-friendly tourism activities comes from the village. Nowadays, the development of tourist villages is experiencing rapid progress, especially in the province of Yogyakarta as a favorite tourist destination in Indonesia after Bali. Within 10 years the emergence of tourist villages has increased significantly.Nature-based tourism is a reference for developing tourism with high social values.Symptoms of a shift in tourism patterns that appear to Indonesia indicate that tourists (58%) want tourism activities to natural objects [8]. The community is aware of the need for tourism which tends to return to nature and offers a rural atmosphere for a moment to release mental and physical fatigue from routine activities.

BPS data shows that the number of tourist villages in 2018 reached 1,734 tourist villages and scattered where the island of Java - Bali occupies the highest position with 857 tourist villages. Then followed by Sumatra as many as 355 villages, Nusa Tenggara 189 villages, Kalimantan 117 villages. In addition, Sulawesi Island also recorded 119 tourist villages, Papua 74 villages, and Maluku 23 villages [9].The Ministry of Tourism is targeting Indonesia to have 2,000 tourist villages by the end of this year. The Head of the Tourism Ministry's Rural and Urban Tourism Acceleration Team Vitria Ariani said the government continues to encourage the potential of villages that can be used as tourist villages [10].

Tourism villages provide freedom for people to manage their hometowns in accordance with the authenticity of the village. This is in accordance with Republic of Indonesia Law Number 10 Year. 2009 concerning Tourism Article 4 that tourism aims to increase economic growth; improve people's welfare; eradicating poverty; overcoming unemployment; conserve nature, the environment and resources; promote culture [11]. The mutual concern between the government and parties involved in developing a tourist village is felt to be able to stimulate the community's economy. Environmental education is part of tourism destinations, especially in rural nature tourism. This type of tourism is called green tourism which pays special attention to tourism and nature[12]. However, natural tourism locations have not prioritized the values of environmental education, which are still limited to natural landscapes to be enjoyed and documented.

Nature tourism that is present in tourism villages actually provides benefits not only for relaxation but also for the value of environmental and social education for the community. Ministerial Regulation (Permen) of the Republic of Indonesia Number 14 of 2016 concerning Guidelines for Sustainable Tourism Destinations that is to create orchestration and ensure the quality of activities, facilities, services, and to create experiences and value benefits of tourism in order to provide benefits and benefits to society and the environment [13]. The social life of the community is full of dynamics and prone to conflict if there are differences and gaps, especially those related to welfare. To minimize negative things and bring positive things, tourism can use a special approach, namely the special approach, namelyThe Adaptancy approach as an approach in tourism states that the negative effects of tourism activities can be controlled by looking for other forms of tourism development than the ones that are already common. This approach proposes strategies such as small-scale development, controlled tourism, sustainable tourism, tourism by enjoying the local community and tourism related to ecology [14].

The People's Core Company is a Tourism Village, which is a rural area that offers an overall atmosphere that reflects the authenticity of the village both from socio-economic life, socio-culture, customs, daily life, has a unique building architecture and village spatial structure, or economic activities that are unique and attractive and have the potential to develop various components of tourism, for example: attractions, accommodation, food and beverages, and other tourism needs [15]. In the attraction component, the whole life of the population can be studied and used as an insight enhancer for tourists who live together. In this case, CBT (Community Based Tourism) is manifested in the implementation of tourism villages.

The interaction of tourists and communities produces contacts and exchanges of cultural values, generates new knowledge for the community and acceptance of modernity symbols from outside the community. Factors that influence the application of environmental principles. CBT is a condition of the global environment and local wisdom of the community [16]. It can also be said that tourism villages provide a harmonious life so that conflicts prone to local residents and tourists can be dammed because they interact with each other in the harmony of tourism activities. Community empowerment is a means to be able to create an independent tourism village. The local community is fully involved in the management of the tourist village, so that the income earned can be enjoyed by the local community. However, the community is not yet fully aware that the existence of a tourist village where they live brings opportunities to achieve economic independence. Tourism has a significant impact on the growth of creative entrepreneurs [17].

Yogyakarta Special Region has abundant tourism potential in the form of natural and cultural tourism. Its various tourist attractions make this area one of Indonesia's main tourist destinations. Yogyakarta has modern tourist attractions that are managed professionally and traditionally(Aji, 2020). Grogol Tourism Village is located in Margodadi, Seyegan, Sleman,Yogyakarta. This tourist village is a cultural village because Grogol Hamlet is a village for making wayang kulit and gamelan. Besides visitors can learn about the process of making leather puppets, visitors can also watch performances such as Karawitan, Wayang Kulit / Pedalangan, Campursari, Keroncong. Apart from the show there is a Tuk Bedhug ceremony which is a spring in the Grogol Tourism Village. There are also interesting tourist attractions such as the beauty of the village environment, ancient buildings, Tobong gendeng, and Petilasan Sunan Kalijogo: "KETANDAN". Culinary tours are also available in this tourist village, namely with foods such as Cethil, Sego Gudangan, Tumpeng.

IOP Conf. Series: Earth and Environmental Science **884** (2021) 012052 doi:10.1088/1755-1315/884/1/012052

2. Research Method

This research uses descriptive qualitative research. The use of qualitative methods is also considered because the research instrument is a human being or the researcher himself and reveals the noumena of an event in a place.Qualitative research is research that is used to examine the conditions of natural objects, (as opposed to experiments) where the researcher acts as a key instrument, sampling at the data source is done purposively [18]. The research method used in qualitative descriptive is a case study approach [19]. The research was carried out within 5 months (January - June 2018). The research is located in Grogol Tourism Village, Margodadi Village, Seyegan, Sleman, Yogyakarta Special Region, Indonesia.The informants in this study consisted of 5 people, namely the head of the Grogol hamlet, the originator of the Grogol tourism village, the manager of the Grogol tourism village, youth leaders, and business actors. The informants were determined by using purposive sampling technique. The informants were chosen because they are considered to know and can provide the data needed in the research. Data collection techniques using interviews, observation and documentation. The data validity technique used data triangulation. The data analysis technique in this research is descriptive, namely the data analysis technique using a simple explanation of the data obtained and then doing an in-depth analysis. The analysis technique used in this research is interactive analysis. In this analysis model, the three components of the analysis are data reduction, data presentation,

3. Finding

Grogol tourism village was launched in 2001 by being spearheaded by 10 residents of Grogol hamlet. Grogol tourism village was originally a culture-based tourism village, this is due to the many cultural potentials that can be highlighted in Grogol hamlet. The cultural potentials that exist in Grogol village, such as adat midang, wijayakusuma studio, and brahmaniten studio. The results achieved by the tourist village of Grogol have received appreciation from various parties, for example, the government and local communities. The tourist village of Grogol now has the status of an independent tourism village. This means that this village has been able to manage its potential well. The government's role is no longer intense because the government believes that independent tourism villages can manage and develop their own potential.

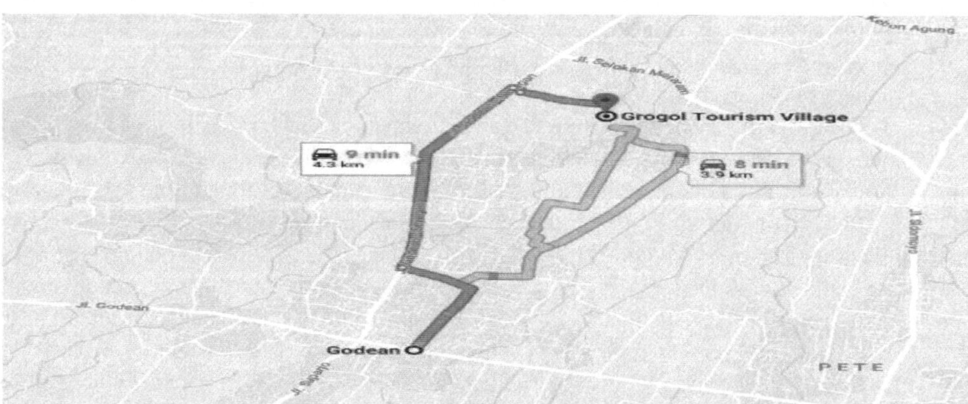

Figure 1. Map of Grogol Tourism Village

Grogol hamlet is located in Margodadi sub-district, Seyegan sub-district, Sleman. The border of Grogol hamlet in the north is kadipiro hamlet, in the east it is bordered by Mlati sub-district, in the south it is bordered by Nyangkringan, Klabgkapan, and in the west it is bordered by Gendengan hamlet. Most of the natural environmental conditions in Grogol are rice fields and settlements, the rest are gardens or people's yards. 70% of the residents of Grogol Hamlet work as farmers, both private rice owners and rice cultivators.

At first, the attractions in the tourist village of Grogol were in the form of arts and culture. At that time, the Grogol tourist village offered local cultural wisdom and arts in Grogol. In the past, the tourist village of Grogol relied on its cultural and artistic potential, such as midang, especially the custom of the midang and the local wisdom that exists here. Besides that, there is also a studio owned by Mr. Sancoko. There is also a lot of history in Grogol, there are Hindu Buddhist relics, there are stones and statues.

The life of the Grogol community is thick with Javanese culture. Based on the observations of researchers, it can be seen from the various community activities in Grogol hamlet. Community activities that reflect Javanese cultural wisdom are musical activities, dancing, upload-uploading or manners that are always upheld by the local community, which are unique in their own right which is rarely found in the modern era like today. In Grogol hamlet, we are rich in culture, about midang customs, about the culture of the studios in Grogol, such as the Wijayakusuma studio where Mr. Sancoko is, the brahmaniten studio owned by mbah kadi there, which has artistic potential such as leather puppet crafts, as well They also teach dance and gamelan for children and adults.

The tourist village manager feels that the tourist attraction in the form of local wisdom is lacking, therefore the tourism village manager has made an innovation to manage the green and beautiful natural environment of the Grogol village. The result is that the tourist village of Grogol is increasingly visited by tourists. Natural tourist attractions are developed and attract many tourists to visit the tourist village of Grogol. The rice fields and rivers that are managed are a vehicle for outbound activities and river tubing has become the prima donna for tourists visiting Grogol. The attractiveness of the tourist village of Grogol continues to develop until now. Based on the observations of researchers, in addition to a pleasant rice field and river play area, tourists can learn to do TABULAMPOT, which stands for planting fruit or flowers in pots.

Table 1. Potential of Grogol Tourism Village

No.	Potency	Activity	The value contained in the activity
1.	Natural	River tubing, outbound,	Train courage, confidence, team building, love nature, participate in protecting nature so that it remains sustainable
2.	Environmental education	Tabulampot (Planting Fruits and Flowers in Pot), Clean the river	participate in protecting the environment in order to remain sustainable,

			maintaining local wisdom, conservation insight,
3.	Art and culture	Midang custom, leather puppet craft,	Local wisdom, preserving indigenous culture, strengthening local and national cultural identities.

Natural tourism in the tourist village of Grogol is increasingly in demand by the wider community. The tourism village manager continues to strive so that tourists don't get bored and want to come back to visit the tourist village of Grogol, even though there are many natural tourism enthusiasts, elements of culture and art are still inserted in every tourist attraction. The tourism village manager realizes that culture is something important, as well as the local community always participates in cultural preservation. Indigenous knowledge refers to the methods and experiences developed by a group of people who understand the local environment, which has been formed from previous generations [21], [22], [23], [24]. The preservation of culture and the natural environment is important for the people of Grogol tourism village, because it is a tourist attraction, therefore the potential in the tourist village of Grogol continues to be developed so that it continues to attract tourists to visit the tourist village of Grogol.

4. Discussion

The environment is a place for life to take place and a place for living things to do activities. Environment is a condition of the ongoing relationship between humans and humans, between humans and animals, between humans and plants, and between humans and inanimate objects around them [25]. In theory that we are familiar with environmental ethics, both anthropocentrism, biocentrism and ecofenism, all environmental ethical theories equally recognize that the universe needs to be respected [26]. The form of respect for the universe is to maintain and care for it in order to remain sustainable. Together with the local community, maintain the existence of the environment so that it can provide balance to nature.

One model for implementing local communities is tourism that focuses on culture so that it can initiate sustainable tourism which has elements in the form of culture, community, economy and the environment [27]. The existence of the Grogol tourist village needs to be managed so that it is attractive and preserved its environment in order to maintain its existence. The form of environmental management in the tourist village of Grogol has been carried out. Management of the natural environment in the tourist village of Grogol includes the use of rivers for playing facilities or river tubing. There are two programs that must be carried out for environmental management, namely programs to improve quality and access to information on natural resources and the environment [28]. This program aims to improve the quality and access to information on natural resources and the environment in order to support planning for the use of natural resources and the environment.

The forms of environmental conservation activities carried out by the Grogol tourist village community vary. The form of environmental conservation activities that are often carried out by the community is community service. Community service includes cleaning the tourism village environment, facilities and infrastructure. The purpose of carrying out community service in addition to preserving the environment is to maintain the beauty of the Grogol environment. The entrepreneurial activities carried out are Home-based Entrepreneurs. Home-based entrepreneurs are entrepreneurs who are run at home. Fathers or housewives who start their business activities from the household, for example mothers who are good at making cakes and various dishes, then send the products to retail stores in the surrounding area.

Environmental management has a broad scope, stating that there are various ways of environmental management [25]. First, is routine environmental management. Second, is the early planning of the management of an area which becomes the basis and guidance for development planning. Third, is environmental management planning based on the estimated environmental impact that will occur as a result of the planned development project. The fourth is environmental management planning to repair damaged environments, either due to natural causes or due to human actions.

Environmental management in the tourist village of Grogol takes the form of preventing environmental damage, according to the theory above. The people of Grogol hamlet routinely carry out community service in order to maintain cleanliness and environmental sustainability. The implementation of community service programs is routine and incidental. Incidental community service activities are carried out when a tourist activity is over. When tourists have finished carrying out activities, officers consciously immediately carry out cleaning activities. Cleaning activities include collecting and collecting trash from tourists, and tidying up the place so that it is as usual.

This environmental conservation program is carried out so that the environment of the Grogol tourist village remains beautiful and comfortable to visit. Public awareness of the importance of preserving the environment has begun to appear, this is evidenced by solid community participation when there are environmental conservation activities. The importance of preserving the environment is a shared responsibility because the tourism village environment is an asset so that the Grogol tourist village can maintain its existence as an advanced tourist village.

The United Nation Environment Program (UNEP) defines green economy as an economic activity capable of improving welfare and social justice on the one hand. But on the other hand, it is able to eliminate the negative impact of economic growth on the environment and scarcity of natural resources. According to UNEP, a green economy is an economic activity that is low carbon, does not rely on fossil fuels, saves natural resources and is socially just(Asj'ari, 2018). This will support the realization of a Tourism Village with the principles of Sustainable Tourism. Sustainable tourism is a development model in which tourism, resources and the human environment come together, where tourism is well coordinated with socio-economic, natural resources and the environment.(Guo et al., 2019).

5. Conclusion

The development of the Grogol tourism village is one step for entrepreneurial opportunities, environmental management and social disaster management regarding the value of peace, cooperatives and preservation of local culture. Grogol tourist village is a tourism village that is classified as independent. This independent predicate is not obtained instantly. It takes struggle and perseverance in developing Grogol Tourism Village which was originally a developing tourist village. Tourism village managers and the surrounding community continue to strive to develop a tourist village into an independent tourism village. Grogol tourism village initially relied on cultural potential in self-promotion. Over time, the tourism village management innovated to make Grogol Tourism Village a nature-based tourism village, but without eliminating the cultural elements and local wisdom of the local community.

Acknowledgments
The author wishes to acknowledge the financial support from the Social Science Faculty, Yogyakarta State University

Reference

[1] Yoety, OA 2010. The economy of tourism: introduction, information, and applications. Kompas Book Publisher..
[2] IG Ardika. 2018. Sustainable Tourism; Pathfinding through the community. 2018
[3] Pitana. 2002. Introduction to Tourism Science. Yogyakarta PT. Andi.
[4] Virgie Baker. 2017. Tourism and People's Economy.https://mediaindonesia.com/read/detail/119411-pwisata-dan-ekonomi-rakyat. Saturday 26 August 2017
[5] Murphy, PE 1985. Tourism A Community Approach. London and New York: Longman
[6] Woodley, S., & Kay, J. 1993. Ecological integrity and the management of ecosystems. CRC Press.
[7] RR Aji ,. 2020. "Tourism social entrepreneurship in community-based tourism: A case study of Pentingsari tourism villageIOP Conf. Ser. Earth Environ. Sci, vol. 447, p. 12009, 2020.
[8] Sari, LIAK. 2012. Agro tourism area development center rod Pagilaran Central Java. Semarang: Diponegoro University Press
[9] Dwi Aditya Putra. Monday, 10 December 2018 13:53. BPS data: Indonesia Has 1,734 Tourism Villages.https://www.merdeka.com/uang/data-bps-indonesia-miliki-1734-desa-wisata.html
[10] Yanita Petriella. July 15, 2019. In 2019, Indonesia is targeted to have 2,000 tourism villageshttps://ekonomi.bisnis.com/read/20190715/12/1124347/2019-indonesia-ditarget-punya-2.000-desa-wisata
[11] Republic of Indonesia Law Number 10 Year. 2009 About Tourism
[12] F. D'Alessandro. 2016. "Green Building for a Green Tourism. A New Model of Eco-friendly Agritourism, "Agric. Agric. Sci. Procedia, vol. 8, pp. 201–210
[13] Aribowo, H., Wirapraja, A., & Putra, YD 2018. Implementation of the Pentahelix Collaboration Model in Developing Tourism Potentials in East Java and Improving the Domestic Economy. Journal of Mebis (Management and Business), 3 (1), 31–38. https://doi.org/10.33005/mebis.v3i1.21
[14] James J. Spillane. 1994. Indonesian Tourism; Economic Strategy and Cultural Engineering. Yogyakarta: Kanisius. Thing28-30
[15 Hadiwijoyo, SS 2012. Community-based rural tourism planning: A conceptual approach. Graha Ilmu
[16] Nurhidayati, SE 2012. Development of Community-Based Sustainable Agro-tourism in Batu City, East Java. text.
[17] Iwan Nugroho. 2015. Ecotourism and Sustainable Development. Yogyakarta: Student Library
[18] Sugiyono, 2016. Educational Research Methods. Bandung: Alfabeta
[19] Ghony, MD, & Almanshur, F. 2012. Qualitative research methodology. Jogjakarta: Ar-Ruzz Media.
[20] Huberman, Michaels. 2007. Qualitative Data Analysis. Jakarta: UI Press
[21 Shaw R, Uy N and Baumwoll J 2008 Indigenous knowledge for disaster risk reduction: Good practices and lessons learned from experiences in the Asia-Pacific Region United Nations Int. Strateg. Disaster Reduction, Bangkok
[22] Mavhura E, Manyena SB, Collins AE and Manatsa D 2013 Indigenous knowledge, coping strategies and resilience to floods in Muzarabani, Zimbabwe Int. J. Disaster Risk Reduct. 5 38–48
[23] Sariffuddin S and Wijaya AP 2014 Patterns of Community Adaptation to Environmental Degradation in Genuk Coastal Area, Semarang) Tataloka 16 245–253
[24] Pratama, AY, & Sariffuddin, S. (2018). Community-Based Disaster Management: A Lesson

Learned from Community Emergency Response Management in Banyumas, Indonesia. IOP Conference Series: Earth and Environmental Science, 123 (1), 012003. https://doi.org/10.1088/1755-1315/123/1/012003

[25] Soemarwoto, O. (2004). Ecology, Environment and Development 10th Edition. Jakarta: Djambat Publisher.

[26] Sony Keraf. 2010. Environmental Ethics. Jakarta: Kompas 167

[27] Dewi, NIK, Astawa, IP, Siwantara, IW, & Mataram, IGAB (2018). Exploring the potential of cultural villages as a model of community based tourism. Journal of Physics: Conference Series, 953 (1). https://doi.org/10.1088/1742-6596/953/1/012072

[28] Hardjasoemantri, Koesnadi, (2012). Environmental Law. Yogyakarta: Gadjah Mada University Press

[29] Asj'ari, FMS (2018). GREEN ECONOMY IN SUPPORTING MILLENNIUM DEVELOPMENT GOALS (MDGs). 3 (2), 139–157. https://doi.org/10.18041/2382-3240/saber.2010v5n1.2536

[30] Guo, Y., Jiang, J., & Li, S. (2019). A sustainable tourism policy research review. Sustainability (Switzerland), 11 (11). https://doi.org/10.3390/su11113187

Landslide Susceptibility in Majalengka Regency, West Java Province

S Selaby[1], E Kusratmoko[1], A Rustanto[1]

[1] Department of Geography, Faculty of Mathematics and Natural Sciences, University of Indonesia
*eko.kusratmoko@sci.ui.ac.id

Abstract. Majalengka is one of districts in Indonesia which is susceptible to landslides. Landslides in Majalengka caused enormous losses such as damage to infrastructure, loss of property, and even human fatalities. Seeing of the impact, mitigation efforts are needed to reduce risks and losses by making landslide susceptibility maps. This study aims to map areas landslide susceptibility and as a reference for the government and related agencies to reduce losses. The method used overlay using Spatial Multi-Criteria Evaluation (SMCE), using weighting values from the Minister Public Works Regulation NO.22/PRT/M/2007, Puslittanak Bogor (2014) and Directorate Volcanology and Disaster Mitigation (DVMBG) (2004). Then comparison of these sources is carried out to determine weighting value with the highest accuracy. The variables are slope, rainfall, soil type, lithology, and land use. The results of this study indicate that landslide susceptibility areas are divided into non-susceptible, low, moderate, and high areas. Where areas Majalengka Regency is dominated by moderate susceptibility level. For the accuracy value of the landslide susceptibility map produced by the weighted value source from the Minister of Public Works Regulation NO.22/PRT/M/2007 has the highest accuracy value of 76%. For weighting from the Bogor Puslittanak is 73%, while weighting source from DVMBG is 68%.

Keywords: Majalengka Regency, Landslide Susceptibility, *Overlay*, SMCE

1. Introduction

Landslides are one of the disasters in Indonesia, many factors affect the stability of the land making up the slopes, causing landslides in one area. Landslide is a type of soil movement, both rock, and mixture that moves down or out of the slope caused by disturbing the stability of the soil or rock making up the slope [1]. Landslide as a land movement and is a natural phenomenon in the form of soil mass moving rapidly following the slope [2]. The steeper the slope, the more likely the material will fall to a lower place [3]. Factors that cause landslides naturally include the morphology of the earth's surface, land use, lithology, geological structure, rainfall, and seismicity. Apart from natural factors, human activity factors also cause landslides such as agricultural activities, slope cutting, plantations, and mining [4].

The Regional Disaster Management Agency (BPBD) of Majalengka Regency, West Java, noted that the largest contributor to the disaster came from landslides, namely 82 incidents in 2018 - 2019. There are 19 sub-districts from 24 sub-districts that have the potential to experience landslides and land movements, especially in peak rainy season. Where the landslides that occur cause infrastructure damage such as cut off-road access, damaged houses, and even casualties.

Seeing the impact of the losses caused by landslides in Majalengka Regency, it is necessary to carry out disaster mitigation efforts. One of the mitigation efforts that can be done to reduce the

Content from this work may be used under the terms of the Creative Commons Attribution 3.0 licence. Any further distribution of this work must maintain attribution to the author(s) and the title of the work, journal citation and DOI.
Published under licence by IOP Publishing Ltd

risk and impact of landslides is by making a landslide susceptibility map. Making landslide susceptibility maps using a Geographic Information System (GIS) can provide convenience in providing information and perform repeated, continuous, fast, and accurate spatial analysis. Landslide susceptibility in an area can be identified through a Geographic Information System (GIS) overlay using the Spatial Multi-Criteria Evaluation (SMCE) method. SMCE is a qualitative and quantitative analysis approach to identify disaster zones and can facilitate a sustainable disaster mitigation process and assist decision-makers based on predetermined criteria based on priority values [5]. Spatial Multi-Criteria Evaluation is a process that can combine and converting some data (input) into a decision (output) generated [6]. The variables used in this study are slopes, rainfall, soil type, lithology, and land use which will be given a weighting value that comes from the Minister of Public Works Regulation NO.22 / PRT / M / 2007 [7], Puslittanak 2004 [8], and the Directorate of Volcanology and Geological Disaster Mitigation in 2004 [9].

This research was conducted in the Majalengka Regency, which is located in the eastern part of West Java Province with a total area of 134,585 hectares or about 2.71% of the area of West Java Province. There are 26 sub-districts, 13 wards, and 334 villages. Majalengka Regency has very diverse topographical conditions, in the southern part, the topography is hilly and has high rainfall. The purpose of this study is to map the distribution of landslide susceptibility areas in Majalengka Regency by comparing the sources used in determining the weighting value so that the most relevant weighting values can be found that can be applied to the same physical conditions so that they can be used as a reference in mitigation. landslide disasters to reduce the number of victims and losses caused by landslides.

2. Data and Method

2.1. Data

The data needed in this study are primary data and secondary data in the form of spatial data and other supporting data obtained from websites, institutions, related institutions, and direct observations in the field when conducting validation tests at the point of landslides. The data needed during the research process and analysis to get the results of landslide susceptibility areas in the Majalengka Regency are data on slopes, lithology, rainfall, soil types, land use, and landslide locations.

2.2. Method

In this study, the method used is the Spatial Multi-Criteria Evaluation (SMCE). SMCE is used to process physical factors in the form of drought, rainfall, soil type, lithology, and land use and conducts multi-criteria assessments using weighted values derived from the Minister of Public Works Regulation NO.22/PRT/M/2007, Puslittanak Bogor in 2004, and the Directorate of Volcanology and Geological Disaster Mitigation in 2004.

2.2.1. Model 1 is sourced from the Minister of Public Works Regulation NO.22/PRT/M/2007.

Model 1 which is used to analyze the susceptibility of landslides in Majalengka Regency is an estimator model which refers to the Minister of Public Works Regulation NO.22/PRT/M/2007 type A with the following values:

IOP Conf. Series: Earth and Environmental Science **884** (2021) 012053 doi:10.1088/1755-1315/884/1/012053

Table 1. The Weighted Value of Model 1 is sourced from the Minister of Public Works Regulation NO.22/PRT/M/2007.

Indicator	Weighted (%)	Scoring	Classification
Slope Inclination	30	5	> 40°
		4	35° - 40°
		3	20° - 30°
		2	15° - 20°
		1	0° - 15°
Rainfall	15	5	> 5.500 mm/year
		4	4.500 - 5.500 mm/year
		3	3.500 - 4.500 mm/year
		2	2.500 - 3.500 mm/year
		1	< 2.500
Lithology	20	3	Volcanic
		2	Sediment
		1	Limestone and igneous rock
Soil Type	15	5	Regosol
		4	Andosol, grumosol
		3	Podsolik
		2	Latosol
		1	Alluvial
Landuse	10	5	Irrigated rice fields
		4	Rain-fed rice fields
		3	Plantations, fields
		2	Forest
		1	Rocky ground, bush, savanna

IOP Conf. Series: Earth and Environmental Science **884** (2021) 012053 doi:10.1088/1755-1315/884/1/012053

2.2.2. Model 2 originated from the Puslittanak Bogor in 2004.

Model 2 used to analyze the susceptibility of landslides in Majalengka Regency is an estimator model that refers to the 2004 Puslittanak with the following values:

Table 2. The Weighted Value of Model 2 sourced from Puslittanak Bogor in 2004.

Indicator	Weighted (%)	Scoring	Classification
Slope Inclination	20	5	> 45°
		4	30° - 45°
		3	15° - 30°
		2	8° – 15°
		1	< 8°
Rainfall	30	5	> 2.500 mm/year
		4	2.000 – 2.500 mm/year
		3	1.500 – 2.000 mm/year
		2	1.000 – 1.500 mm/year
		1	< 1.000 mm/year
Lithology	20	3	Volcanic
		2	Sediment
		1	Limestone and igneous rock
Soil Type	10	5	Regosol
		4	Andosol, Podsolik
		3	Litosol
		2	Latosol
		1	Alluvial
Landuse	20	5	Fields, Rice fields
		4	Shrubs
		3	Forests, Plantations
		2	City/Settlement, Airport
		1	Water body

2.2.3. Model 3 is sourced from the Directorate Volcanology Geological Disaster Mitigation in 2004.
Model 3 which is used to analyze the susceptibility of landslides in Majalengka Regency is an estimator model that refers to the Directorate of Volcanology and Geological Disaster Mitigation in 2004 with the following values:

Table 3. The Weighted Value of Model 3 is sourced from the Directorate of Volcanology and Geological Disaster Mitigation in 2004.

Indicator	Weighted (%)	Scoring	Classification
Slope Inclination	15	5	> 45°
		4	30° - 45°
		3	15° - 30°
		2	8° – 15°
		1	< 8°
Rainfall	30	5	> 3000 mm/year
		4	2500 – 3000 mm/year
		3	2000 – 2500 mm/year
		2	1000 – 2000 mm/year
		1	< 1000 mm/year
Lithology	20	3	Sediment
		2	Volcanic
		1	Limestone and igneous rock
Soil Type	15	5	Regosol, Litosol
		4	Andosol, Laterik, Grumosol, Podsolik
		3	Mediteran
		2	Latosol
		1	Alluvial
Landuse	20	5	Empty land
		4	Industrial and residential areas
		3	Plantation and irrigated rice fields
		2	Garden and scrub mix
		1	Water Body, Forest

2.3. Accuracy Test
The results of the landslide susceptibility map model were tested using the accuracy assessment. Accuration assessment is a method used to validate data by comparing landslide susceptibility map data with field survey results [10]. The purpose of the accuration assessment is to determine the accuracy of the classification of landslide incident points with field data. Landslide susceptibility maps that have been overlaid with the distribution of landslide event points will be tested by comparing the actual landslide event data, with the following formula:

$$\text{Accuration Assesment} = \frac{\Sigma\text{Point In Accordance With Ground Check}}{\Sigma\text{ Observation Point}} \times 100\%$$

IOP Conf. Series: Earth and Environmental Science **884** (2021) 012053 doi:10.1088/1755-1315/884/1/012053

3. Result and Discussion

Landslide susceptibility maps processed using the Spatial Multi-Criteria Evaluation (SMCE) method will produce 4 classifications of landslide susceptibility in Majalengka Regency, namely areas not susceptible, low, medium, and high. Determinants of landslides such as slopes, rainfall, lithology, soil types, and land use that have been given weighted values will affect the area of each resulting landslide classification.

3.1. Landslide susceptibility Model 1 (Minister of Public Works NO.22/PRT/M/2007).

The landslide susceptibility area model 1 in this study is the result of an overlay of the ILWIS software SMCE method with a weighted value that has been determined by Minister of Public Works Regulation NO.22 PRT/M/2007 as follows: 30% slope, 15% rainfall, type soil 15%, 20% lithology, 10% land use. Based on the results of the data processing that has been carried out, it is known that in model 1, Majalengka Regency is dominated by the classification of moderate susceptibility which is found with an area of 57,666 hectares or about 42.85%.

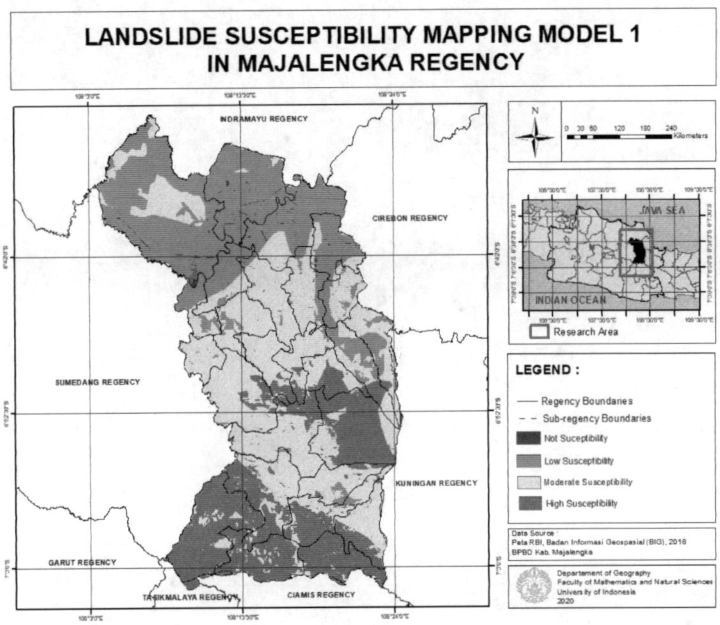

Figure 1. Landslide Susceptibility Map Model 1 Based on Minister of Public Works Regulation NO.22 PRT/M/2007.

Table 4. The total area in Majalengka Regency based on the landslide susceptibility class model 1

Classification	Area (Ha)	Percentage (%)
Not Susceptibility	774	0.58
Low	41.006	30.47
Moderate	57.666	42.85
High	35.139	26.11
Total	134.585	100

3.2. Landslide Susceptibility Model 2 (Puslittanak, 2004).

The susceptibility of the landslide area model 2 in this study is the result of an overlay of the ILWIS software SMCE method with the weighting value determined by Puslittanak in 2004 as follows: 20% slope, 30% rainfall, 10% soil type, 20% lithology, usage land 20%. Based on the results of the data processing that has been carried out, it is known that in model 2, Majalengka Regency is dominated by low susceptibility classifications which are found with an area of 63,737 Ha or around 47.4%.

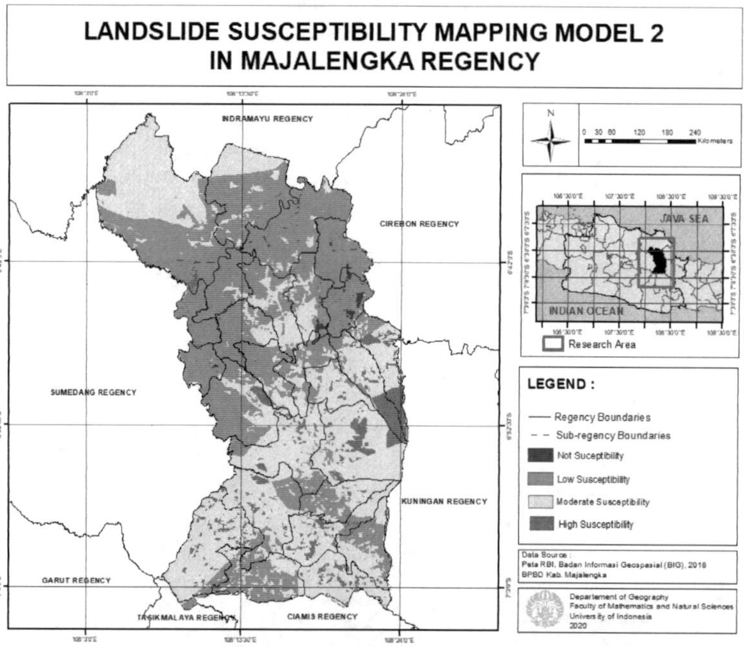

Figure 2. Landslide Susceptibility Map Model 2 Based on Puslittanak 2004.

Table 5. The total area in Majalengka Regency based on the landslide susceptibility class model 2

Classification	Area (Ha)	Percentage (%)
Not Susceptibility	890	0,7
Low	63.737	47,4
Moderate	62.120	46,2
High	7.600	5,7
Total	134.585	100

3.3. Landslide Susceptibility Area Model 3 (Directorate of Volcanology and Geological Disaster Mitigation, 2004).

The susceptibility of the landslide area model 3 in this study is the result of an overlay of the ILWIS software SMCE method with a weighted value that has been determined by the Directorate of Volcanology and Geological Disaster Mitigation in 2004 as follows: 15% slope, 30% rainfall, 20% soil type, lithology 20%, land use 15%. Based on the results of the data processing that has been carried out, it is known that in model 3, Majalengka Regency is dominated by the classification of moderate susceptibility which is found with an area of 85,055 hectares or about 63.23%.

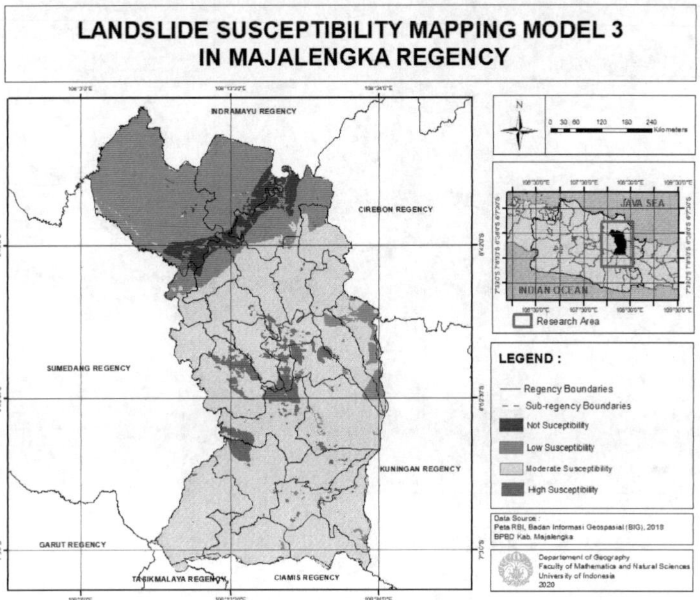

Figure 3. Landslide Susceptibility Map Model 3 Based on the Directorate of Volcanology and Geological Disaster Mitigation in 2004.

Table 6. The total area in the Regency of Magalengka is broken down based on the landslide Susceptibility class model 3

Classification	Area (Ha)	Percentage (%)
Not Susceptibility	5.361	3,99
Low	36.535	27,16
Moderate	85.055	63,23
High	7.555	5,62
Total	134.585	1000

IOP Conf. Series: Earth and Environmental Science **884** (2021) 012053 doi:10.1088/1755-1315/884/1/012053

3.4. Accuracy Test of Landslide Susceptibility Map in Majalengka Regency
Based on Figure 4, in Majalengka Regency in 2018-2019 there were 82 landslide incident points spread across 19 sub-districts. Argapura Kecamran had the most landslide incidents, namely 12 events.

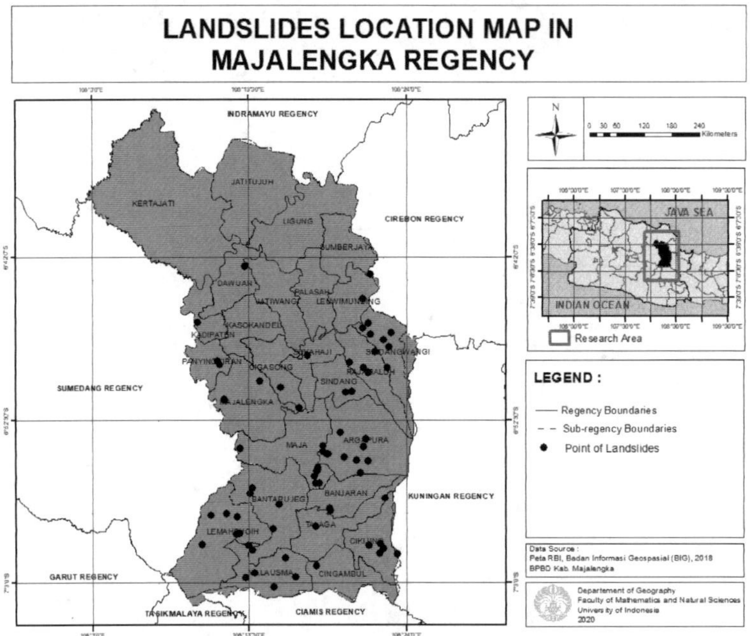

Figure 4. Landslide Location Map in Majalengka Regency

To find out the level of accuracy of the landslide-prone maps that have been generated from 3 sources, the Accurate Assessment method is used, namely by calculating the points that correspond to the ground check then divided by the number of landslide events and multiplied by 100.

The point that matches the ground check is known by looking at what variables are present at the landslide point, then calculating the standardization for each variable then matching it with the classification of susceptible areas from the map that has been generated, if it matches the standardization value with the classification of susceptible areas, then that point matches the ground check. The following is a table of the accuracy of the resulting landslide-prone area map:

Table 7. Results of Calculation of Landslide Susceptibility Map Accuracy Level

	Number of Points Corresponding	Accuracy Results
Landslide Susceptibility Map Model 1 (Minister of Public Works Regulation NO.22 / PRT / M / 2007)	64	76%
Landslide Susceptibility Map Model 2 (Puslittanak, 2004)	60	73%
Landslide Susceptibility Map Model 3 (DVMBG, 2004)	56	68%

There is a difference between accuracy result carried out in this study and the research of I Nengah Sinarta et all [11]. In research of I Nengah Sinarta, the accuracy result derived from Minister of Public Works Regulation NO.22 / PRT / M / 2007 produces an accuracy value of 71.4% while in this study the accuracy reaches 76%.

4. Conclusion

Based on the three landslide susceptibility maps produced using different sources, it can be seen that Majalengka Regency is dominated by moderate susceptibility areas that are scattered in the central to western parts of the Majalengka Regency. Increased incidence of landslides also caused by rainfall rainfall, steep slopes and utilization activities land above it [12]. This region has a slope of >30% with rainfall >1,500 mm / year. Slopes that have a high degree of steepness cause the location of rainwater is close to the maximum so that it can lead to possibilities landslides [13]. The types of soil contained in it are regosol, latosol, and andosol, where these soil types have a high erodibility level. Loose soil has high permeability so that water will be easy to pass, so that it will have more potential landslides occur compared to dense soils [14]. This area contains lithology in the form of volcanic and sedimentary rocks and uses of land in the form of rice fields, moor, and settlements. Based on the three sources used to determine the weighting value of the landslide susceptibility map in Majalengka Regency, the source of the Minister of Public Works Regulation NO.22 / PRT / M / 2007 model has a higher accuracy rate of 76% compared to the 2004 Puslittanak 2 model with a value of 73% and model 3 of the Directorate of Volcanology and Geological Disaster in 2004 with a value of 68%.

Evaluation in this study for further research is to determine the weighting value for each variable, you should pay attention to the physical condition of the research area, because each source of weighting values does not necessarily correspond to the conditions of the research area to be studied, and it is better to use the point of landslides 3 or 4 years earlier, so that the accuracy value is more accurate. Based on the research results, the SMCE method has a better accuracy value by using a weighted value than other weighting value methods[15].

Acknowledgments

The author would said thank you for the Universitas Indonesia who supported and funded this international conference, and thanks to Regional Disaster Management Agency Majalengka Regency which helps in providing information and data on landslides disaster

References

[1] Badan Nasional Penanggulangan Bencana 2012 *Resiko Bencana Indonesia* http://geospasial.bnpb.go.id/
[2] Suriadi A B Arsyad M & Riadi B 2013 *Potensi Resiko Bencana Alam Longsor (Potential Risk of Landslide Related to Extreme Weather in Ciamis Region, West Java)* Jurnal Ilmiah Geomatika, 19(1), 57–63.
[3] Karnawati D 2004 *Bencana Gerakan Massa Tanah/ Batuan di Indonesia; Evaluasi dan Rekomendasi, Dalam Permasalahan, Kebijakan dan Penanggulangan Bencana Tanah Longsor di Indonesia.* Jakarta: P3 - TPSLK BPPT dan HSF
[4] Karnawati D 2005 *Bencana alam gerakan massa tanah di Indonesia dan upaya penanggulangannya* (Yogyakarta: Tek.Geologi FT UGM)
[5] Castellanos Abella E, de Jong S, van Westen C, and van Asch W 2008 *Multi - scale landslide risk assessment in Cuba* (ITC Dissertation) p 154
[6] Malczewski, J 1999 *GIS and multicriteria decision analysis* John Wiley & Sons
[7] Peraturan Menteri PU NO.22/PRT/M/2007 2007 *Pedoman Penataan Ruang Kawasan Rawan Bencana Longsor*
[8] Pusat Penelitian dan Pengembangan Tanah dan Agroklimat 2004 *Laporan Akhir Pengkajian Potensi Bencana Kekeringan, Banjir dan Longsor di Kawasan Satuan Wilayah Sungai Citarum-Ciliwung, Jawa Barat Bagian Barat Berbasis Sistem Informasi Geografi* Bogor

[9] Direktorat Vulkanologi dan Mitigasi Bencana Geologi 2004 *Model Perhitungan Skor Kawasan Rawan Tanah Longsor, Bandung.*

[10] Bramantia S Aditya N P and Sudarto 2017 *Pemetaan Daerah Rawan Longsor Di Kecamatan Pujon Menggunakan Metode Analytic Hierarchy Process (AHP)* (Jurnal Tanah dan Sumberdaya Lahan) Vol 4 No 2 : 567-576, 2017 e-ISSN:2549-979

[11] I Nengah S Ahmad R Teuku F F and Wahyu W 2017 *Slope Stability Assessment Using Trigger Parameters and SINMAP Methods on Tamblingan-Buyan Ancient Mountain Area in Buleleng Regency, Bali* Geosciences 2017, 7, 110; doi:10.3390/geosciences7040110

[12] Hamza T & Raghuvanshi T K 2017 *GIS based landslide hazard evaluation and zonation – A case from Jeldu District, Central Ethiopia, GIS based landslide hazard 70 Universitas Indonesia evaluation and zonation* Journal of King Saud University - Science, 29(2), 151–165. https://doi.org/10.1016/j.jksus.2016.05.002

[13] Ristya Y 2018 *Perbandingan Wilayah Potensi Longsor Berdasarkan Metode Sinmap, Smorph, dan Index Storie di Kecamatan Pelabuhan Ratu dan Sekitarnya* Journal of 72 Universitas Indonesia Chemical Information and Modeling, 53(9), 1689–1699. https://doi.org/10.1017/CBO9781107415324.004

[14] Priyono K D & Priyana Y 2016 Kajian Tingkat Perkembangan Tanah pada Kejadian Bencana Longsor Lahan di Pegunungan Menoreh Kabupaten Kulonprogo Daerah Istimewa Yogyakarta, 489–495.

[15] Shahabi H Hashim M 2015 *Landslide susceptibility mapping using GIS-based statistical models and Remote sensing data in tropical environment* Sci Rep 5, 9899. https://doi.org/10.1038/srep09899

[16] Perera ENC, J. D. (2018). Spatial Multi Criteria Evaluation (SMCE) Model for Landslide Hazard Zonation in Tropical Hilly Environment: A Case Study from Kegalle. Geoinfor Geostat: An Overview, 10.4172/2327-4581.S3-004.

Characteristics of Landslide and Rainfall Areas in Majalengka Regency, West Java Province

S Fathaya[1], E Kusratmoko[1*], and R Saraswati[1]
[1]Department of Geography, Faculty of Mathematics and Natural Sciences, University of Indonesia, UI Campus, Depok, 16424, Indonesia

[*]E-mail: eko.kusratmoko@sci.ui.ac.id

Abstract. Majalengka Regency is one of the districts with a high landslide hazard in West Java Province. They are mostly affected by heavy rainfall or prolonged rain. This study aims to classify landslide events in 2018-2019 based on physical factors consisting of slopes, soil types, lithology, land use, and vegetation density using the K-Means Clustering analysis method. To analyze the characteristics of rainfall that triggered landslides in 2018-2019 using the Thiessen polygon method. The results showed that the clustering of landslide events in 2018-2019 in Majalengka Regency was formed five clusters with the highest rainfall on the D-Day average in cluster 5, which is 49 mm/day. The highest average cumulative rainfall 3 days before the landslide events was in cluster 4, which is 80 mm/day. The highest average cumulative rainfall 5 days before the landslide events was in cluster 3 is 112 mm/day. The highest average cumulative rainfall 10 days before the landslide events was in cluster 1, which is 174 mm/day.

Keywords: K-Means Clustering, Landslide, Majalengka Regency, Physical Factors, Rainfall

1. Introduction

West Java Province nationally ranks first in terms of landslides. The Regional Disaster Management Agency of West Java Province noted that the most frequent disaster events were landslides. During February 2019, there were 119 landslides in West Java [1]. The Disaster Risk Index of Majalengka Regency is ranked seventh in West Java Province, and at the national level, it is ranked 16th out of 497 districts and cities in Indonesia [2].

Besides, Majalengka Regency is dominated by hilly and mountainous topography, which covers 71.3% of the total area of this district [3]. The topography is spread across the central and southern parts of this district. In the Majalengka Regency south part, there are mountains, namely Mount Ciremai (3,076 masl), which is the highest mountain in West Java Province. This is related to landslides that usually occur in mountainous areas, hills, steep slopes, and cliffs, so it is straightforward to support landslides. However, in most of the events landslides have a significant impact, namely heavy rainfall or prolonged rain.

Throughout 2018, Majalengka Regency was subjected to rain with the highest rainfall that occurred in January, which reached 705.8 mm with the number of rainy days 28, and the lowest was in August, which was 20.0 mm with the number of rainy days 1. One of them was a landslide in the Gunung Herang Landeuh Block, Sadasari Village, Argapura District, Majalengka Regency, which allegedly occurred due to continuous rain [4]. The Meteorology, Climatology, and Geophysics Agency of the Jatiwangi meteorological station explained that the peak of the rainy season in January and February 2019 is expected to be precise mid-January to mid-February. This has contributed to the increasing incidence of landslides in January and February. This is because the intensity of rainfall that occurs is in the high category, which is above 300 millimeters per month [5].

Content from this work may be used under the terms of the Creative Commons Attribution 3.0 licence. Any further distribution of this work must maintain attribution to the author(s) and the title of the work, journal citation and DOI.
Published under licence by IOP Publishing Ltd

Thus, based on the occurrence of landslides that have occurred, this research is more emphasized to be able to classify landslide events based on physical factors such as soil type, lithology, slopes, land use, and vegetation density in 2018-2019 and can analyze the characteristics of trigger rainfall landslides are based on groups of landslides in the Majalengka Regency. This study analyzed during 2018-2019 in Majalengka Regency, due to a significant increase in the number of landslides.

2. Data and Method

2.1. Data

Data collection in this research consists of two stages: primary (field survey) and secondary (agency) data collection. The field survey in this study uses the application *Avenza*, work maps, digital cameras, and writing instruments. During the field survey, observations are made of the location of landslides in Majalengka Regency. The survey was conducted to determine the physical condition of the landslide events area. While other data are data on the distribution of landslides, distribution data of rainfall stations, daily rainfall data, National DEM (for slope maps), land use maps, vegetation density maps, soil type maps, and lithology maps.

2.2. *Method*

This study used the Thiessen Polygon method and the K-Means Clustering. For the characteristics of landslide-triggering rainfall using the Thiessen polygon method by applying the calculation of rainfall at the time of the landslide, three daily rainfall before the landslide, five daily rainfall before the landslide, and ten days of rainfall before the landslide based on the landslide event cluster in Majalengka Regency. While the characteristics of the landslide area use the K-Means Clustering method, which aims to classify landslide events based on physical factors consisting of soil type, lithology, slopes, land use, and vegetation density in Majalengka Regency to visualize the results of grouping landslide events.

Based on the data that has been collected, processed using ArcGIS 10.4 software, Microsoft Excel, and SPSS software. The stages of the processing process are as follows:

2.2.1. *Regional Physical Data Processing*

The physical factors of the area in this study consisted of soil type, lithology, slope, land use, and vegetation density. The stages in processing regional physical factor data are using ArcgGIS 10.4 software. Soil type maps, lithology maps, land use maps, and slope maps are categorized based on classifications in Majalengka Regency, meanwhile, for vegetation density maps using the results of Sentinel 2 Image processing which is processed using the NDVI algorithm which is adjusted to the results of the validation in the field and adjusted to Google Earth and land use maps from the National Defense Agency. Furthermore, each map of the physical factors is extracted to each landslide events point to produce a map of the distribution of landslide events based on physical factors.

2.2.2. *Landslide Events Data Processing*

Landslide location data for 2018-2019 is input into the physical factor map of the area consisting of soil type maps, lithology maps, slope maps, land use maps, and vegetation density maps in the Majalengka Regency. Furthermore, to find out the characteristics of the landslide area, the data on the location of landslide events for 2018-2019 in the tabular format is arranged based on the sequence of the date of the event according to the location and time of occurrence.

After that, the coding process was carried out with SPSS software using cluster analysis (K-Means Clustering) to classify the characteristics of landslide areas based on physical factors by scoring the five factors, according to their level of influence on the hazard of landslides. In this study, groups or clusters of landslide events were adjusted based on five physical factors, so that five groups or clusters of landslide events were formed in Majalengka Regency.

2.2.3. *Rainfall Data Processing*

Spatially and temporally, the rainfall station data is made by dividing the area by the Thiessen polygon. The rainfall data is divided into rainfall on the D-day landslide events, rainfall 3 days before the landslide events, rainfall 5 days before the landslide events, and rainfall 10 days before the landslide events. In determining the cumulative amount of rainfall that triggers landslides, it is done by summing the daily rainfall data at the nearest rain gauge station from the location of the landslide events. This is done through overlapping between the time of the landslide and the rain gauge station.

3. Result and Discussion

3.1. Distribution of Landslide Events in Majalengka Regency

3.1.1 Distribution of Landslide Events Based on Slope

Based on Figure 1, it is known that the occurrence of landslides in Majalengka Regency mostly occurred in the steep slope classification (8-16°), namely 29 events with a proportion of 35.37%. Then, the classification of slightly steep slopes (4-8°) has a total of 23 events with a proportion of 28.05%, the classification of ramps slopes (2-4°) has a total of 19 events with a proportion of 23.17%, the classification of slopes is flat (0-2°) has a total of 9 events with a proportion of 10.98%, the classification of slopes is slightly abrupt (16-35°) which is 2 events with a proportion of 2.44%. So it can be seen that Majalengka Regency is dominated by steep slopes and has the highest number of landslides by comparison to other slopes.

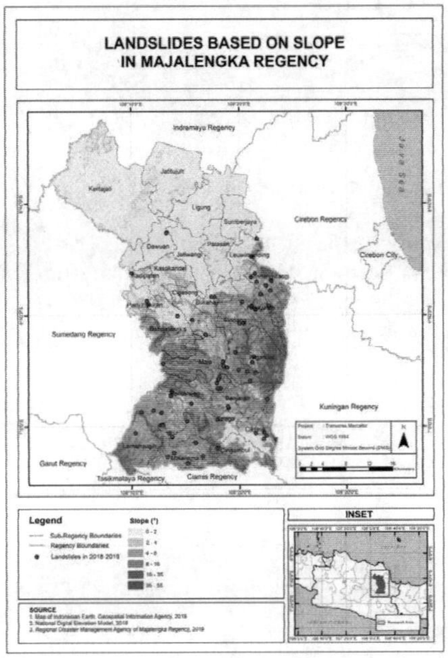

Figure1. Landslide distribution Map Based on Slope in Majalengka Regency

3.1.2 Distribution of Landslide Events Based on Type of Soil

Based on Figure 2, it can be seen that the frequency of landslides by soil type in the Majalengka Regency, West Java Province is in six classifications of soil types. Based on the six classifications of soil types, the most landslide occurrences are latosol soil types, which dominate the Majalengka Regency area with 41 landslides with a percentage of 50%. For the distribution of landslides according to other soil types, namely regosol soil types, there were 20 landslides with a percentage of 24.39%. Then, andosol soil types were 9 landslides with a percentage of 10.98%, grumusol soil types were 6 landslides with a percentage of 7.32%, litosol soil types were 5 landslides with a percentage of 6.10%, and 1 alluvial soil types with a percentage of 1.22%. This can be seen in the spatial distribution in Figure 2.

IOP Conf. Series: Earth and Environmental Science **884** (2021) 012054 doi:10.1088/1755-1315/884/1/012054

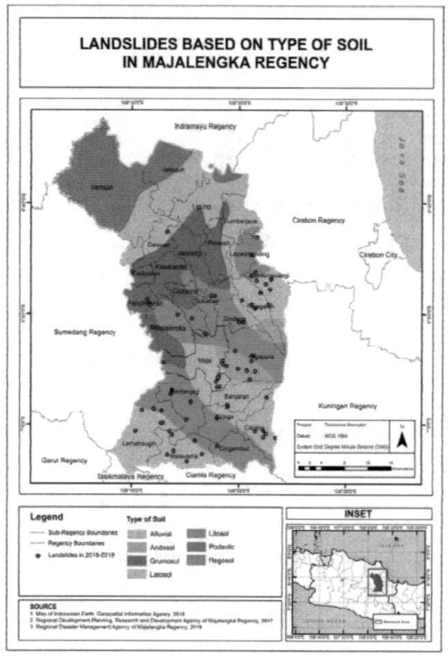

Figure 2. Landslide Distribution Map Based on Soil Types in Majalengka Regency

3.1.3 Distribution of Landslide Events Based on Lithology

Based on Figure 3 it can be seen that the frequency of landslides according to the lithology of Majalengka Regency, there are five lithological classifications. Of the five lithological classifications, landslides in Majalengka Regency are only in three lithological classifications, namely volcanic rocks, loose sediments, and solid sediments.

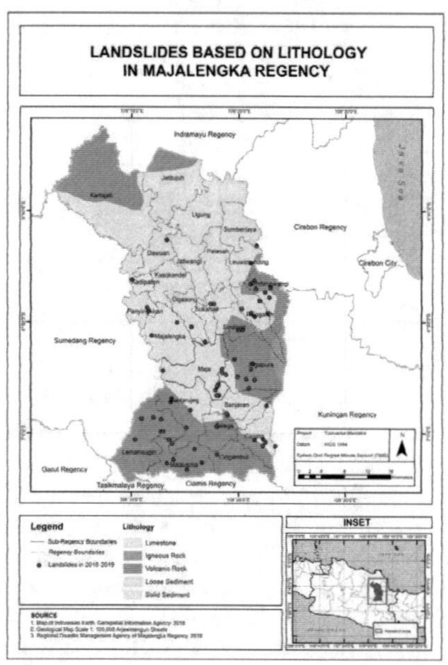

Figure3. Land slide Distribution MapBasedon Lithology in Majalengka Regency

Most landslides occurred in the Majalengka Regency, namely in volcanic rocks as many as 40 landslides with a percentage of 48.78% supported by the lithological area of the volcanic products in Majalengka Regency which dominates compared to other lithological classifications. Meanwhile, for the distribution of landslides, according to other lithologies, solid sediments were 26 landslides with a percentage of 31.71%. Furthermore, in loose sediments, there were 16 landslides with a percentage of 48.78%.

3.1.4 Distribution of Landslide Events Based on Land Use

The distribution of landslide events based on land use shows that residential land is a land-use classification with the highest number of landslide events, namely 39 events with a percentage of 47.56%. Then, shrubs and built land have the same number of landslide occurrences, namely 11 events with a percentage of 13.41%, moor land having 7 landslide occurrences (8.54% percentage), irrigation rice fields having 5 landslide events (percentage 6.10%), plantations had 4 landslide events (4.88% percentage), water bodies had 3 landslide occurrences (3.66% percentage), and forests had 2 landslide events (percentage 2.44%). The spatial distribution of landslides based on land use in Majalengka Regency can be seen in Figure 4.

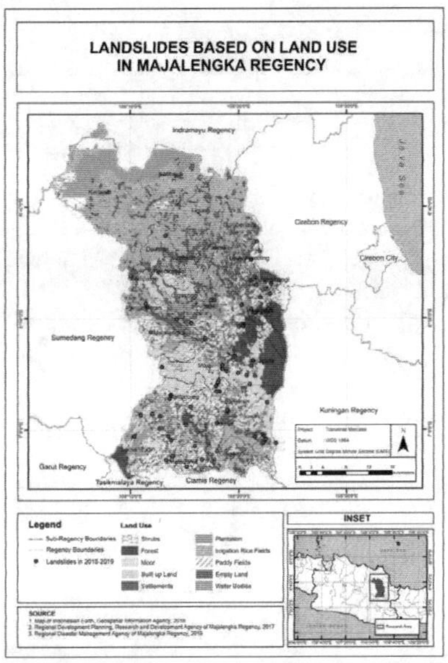

Figure 4. Landslide Distribution Map of Land Use in Majalengka Regency

3.1.5 Distribution of Landslide Events Based on Vegetation Density

Based on Figure 5a, it is known that the distribution of landslide events in 2018 with a value of> 0.5 (high density) has the largest number of landslides, namely 10 events with a percentage of 33.33%. Vegetation density with a value of <0.2 (non-vegetation) and a value of 0.3-0.5 (medium density) had the same number of landslide events, namely 8 events with a percentage of 26.67%. Vegetation density with a value of 0.2-0.3 (low density) has a number of landslide events of 4 incidents with a percentage of 13.33%. The distribution of landslide event points based on vegetation density in 2018 is centralized in the eastern part of Majalengka Regency which is an area with high density (Figure 5a).

Based on 5b, it is known that the distribution of landslides in 2019 vegetation density with a value of <0.2 (non-vegetation) has the highest number of landslide events, namely 17 events with a percentage of 32.69%. Low density with a value of 0.2-0.3 has a total of 16 landslide events with a percentage of 30.77%. Medium density with a value of 0.3-0.5 has the number of landslide events of 11 events with a percentage of 21.15%. The density is high with a value >0.5, the number of landslide events is 8 with a percentage of 15.38%. The distribution of landslide event points based on vegetation density in 2019 is scattered in the southern part of Majalengka Regency (Figure 5b).

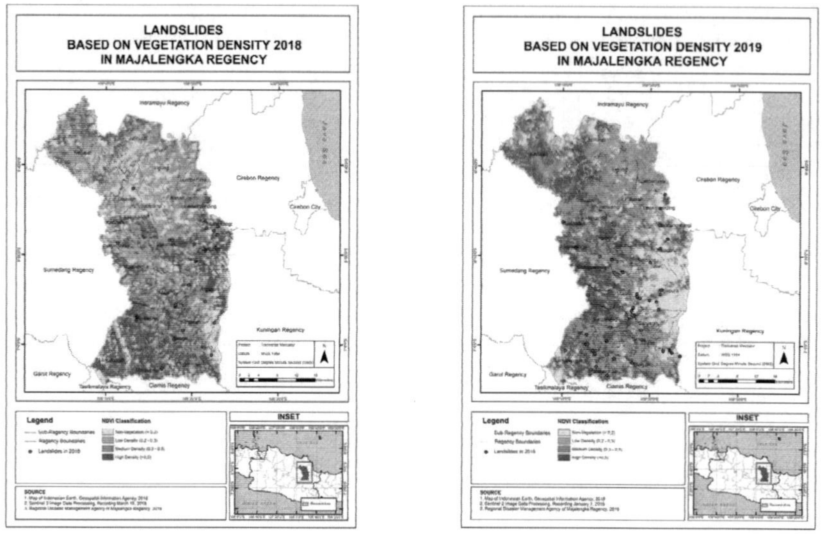

(a) (b)

Figure 5. (a) Landslide Distribution Map Based on Vegetation Density 2018 in Majalengka Regency; (b) Landslide Distribution Map Based on Vegetation Density 2019 in Majalengka Regency

3.2. Landslide Occurrences Classification Based on Physical Factors in Majalengka Regency

Table 1. Physical Characteristics of Each Cluster

Cluster	Slope	Type of Soil	Lithology	Land Use	Vegetation Density
1	Flat, Ramps, and Slightly Steep	Latosol, Litosol, and Grumosol	Volcanic Rock, Loose Sediment, and Solid Sediment	Water Bodies and Shrubs	Non-Vegetation and Low Density
2	Ramps, Slightly Steep, Steep, and Slightly Abrupt	Latosol, Regosol, Litosol, and Andosol	Volcanic Rock and Solid Sediment	Built Land, Settlements, and Moor	Non-Vegetation, Low Density, and Medium Density
3	Flat, Ramps, Slightly Steep, and Steep	Latosol, Regosol, Andosol, Grumosol, and Litosol	Volcanic Rock, Loose Sediment, and Solid Sediment	Built Land, Settlements Plantation, Irrigation Rice Fields, and Moor	Non-Vegetation, Low Density, Medium Density, and High Density
4	Flat	Alluvial	Loose Sediment	Irrigation Rice Fields	Non-Vegetation
5	Ramps and Steep	Latosol, Regosol, and Andosol	Loose Sedimet, Solid Sediment, and Volcanic Rock	Shrubs, Forest, Plantation, and Water Bodies	Non-Vegetation, Medium Density and High Density

In Table 1, it is known that the characteristics of landslide areas processed using cluster analysis (K-Means Clustering) in Majalengka Regency are divided into five clusters or groups of landslide events adjusted based on the five physical factors of the area. The clustering technique has wide use and tends to increase along with the growing amount of data. K-Means is a simple technique in clustering analysis that aims to find the best division of entities "n" into groups of "k" or what can be called clusters so that the total distance between group members and centroids is appropriate [6].

In cluster 1, the number of members is 5 landslide event points, cluster 2 has 28 landslide event points, and cluster 3 has 36 landslide event points, cluster 4 has 1 landslide event point members, and cluster 5 has a total membership of 12 landslide event points. The factors that cause the increasing incidence of landslides are soil type, slope morphology, rainfall, and human activities. Most landslides are triggered by high rainfall [7].

Based on Table 1, it can be seen that each cluster is based on physical factors consisting of slopes, soil types, lithology, land use, and vegetation density. The following is an explanation of the physical characteristics of the cluster, as follows:

1. Physical Characteristics of Cluster 1

In cluster 1, the dominant slope is a ramps slope (2-4°), which is 3 landslide events. Furthermore, soil type is dominated by latosol soil namely 2 events and lithosol soil namely 2 landslide events. For the lithology classification, it is dominated by volcanic rocks namely 2 events and loose sediments, namely 2 landslide events. If the land use was dominated by shrubs, namely 4 landslide events. Meanwhile, the vegetation density was dominated by non-vegetation with a value of <0.2, namely 4 landslide events.

2. Physical Characteristics of Cluster 2

In cluster 2, the predominant slope is steep (8-16°) with 14 occurrences. It is also known that the steeper the slopes will increase the potential for landslides [8]. The soil type in cluster 2 is dominated by latosol soil, which is 16 landslide events. For lithology, it is dominated by volcanic rocks, which are as many as 21 landslide events. For land use, it is dominated by settlements, namely 18 landslide events. Meanwhile, vegetation density was dominated by non-vegetation classification (value <0.2), which was 14 landslide events.

3. Physical Characteristics of Cluster 3

In cluster 3, the dominant slope is a slightly steep slope (4-8°), which is 13 landslide events. Soil type is dominated by latosol soil, which is 15 events. For lithology, it is dominated by volcanic rocks, which are 16 landslide events. Sedimentary rock types are rocks formed from the consolidation of sediment, as loose material, which is transported to the deposition site by wind, water, gravitational landslides, and landslides [9]. If land use is dominated by settlements, there are 21 landslide events. Furthermore, the vegetation density in cluster 3 is dominated by high density (value> 0.5), which is 12 landslide events.

4. Physical Characteristics of Cluster 4

In cluster 4, it is known that this cluster only has a number of members, namely 1 landslide event point in Majalengka Regency. Because the landslide events point is located in the northern part of Majalengka Regency, it is different from the other landslide event points which are scattered and close to each other, the distance between each landslide event point. Cluster 4 has physical characteristics only on flat slopes (0-2°). The soil type classification only consists of alluvial soils and the lithological classification only consists of solid or undifferentiated sediments. Meanwhile, the land use classification only consists of irrigation rice fields. Furthermore, for the classification of vegetation density, it only consists of non-vegetation which has a value of <0.2.

5. Physical Characteristics of Cluster 5

In cluster 5, the slopes are dominated by the steep classification (8–16°), which is 11 landslide events. If an area has an increasingly steep slope, the potential for landslides to occur in that area will be even higher. Slopes with a steep slope require soil cover in the form of plants that have strong roots so that the soil on the slope does not experience landslides. Furthermore, for the classification of soil types, latosol soil is dominated by as many as 8 landslide events. If the lithology classification is dominated by loose sediments, there are as many as 8 landslide events. For the classification of land use, it was dominated by shrubs, which were 6 landslide events. Meanwhile, the classification of vegetation density was dominated by high density (value> 0.5), which was 6 landslide events.

Based on Figure 6, it is known that cluster 1 is scattered in the West, East, and South parts of Panyingkiran District, Maja District, Cikijing District, Sindangwangi District, and Leuwimunding District. Cluster 2 is scattered in the East, Central, and South parts of Malausma District, Lemahsugih

District, Cikijing District, Bantarujeg District, Cingambul District, Sindang District, Sindangwangi District, and Argapura District. Cluster 3 is scattered in the West, Central, East, and Southern parts of Majalengka Regency, which are in Kadipaten District, Cigasong District, Argapura District, Rajagaluh District, Sindang District, Sindangwangi District, Leuwimunding District, Panyingkiran District, Bantarujeg District, Lemahsugih District, Banjaran District, Cikjing District, Dawuan District, Majalengka District, Maja District, Sumberjaya District, and Malausma District. Cluster 4 is located in the northern part of Dawuan District. Cluster 5 is scattered in the Central, Eastern, and Southern parts of Majalengka Regency, which are in Sukahaji District, Argapura District, Banjaran District, Sindangwangi District, and Rajagaluh District.

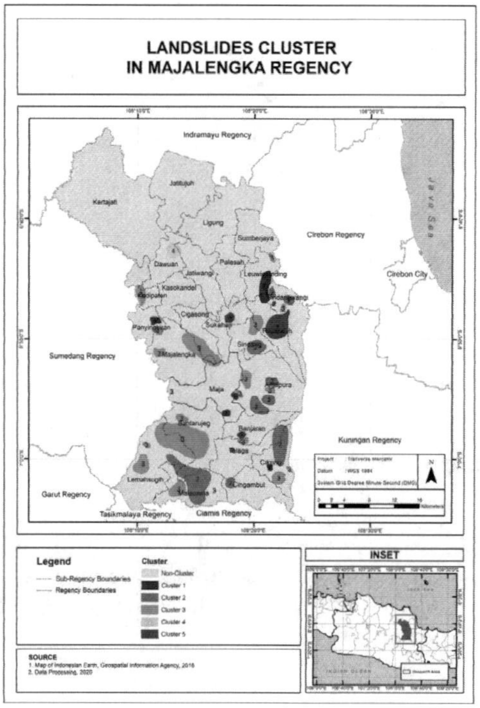

Figure 6. Landslide Events Cluster Map in Majalengka Regency

3.3 Characteristics of Landslide Trigger Rainfall Based on Landslide Events Groups

Landslides often occur in Majalengka Regency along with the arrival of the rainy season and occur after heavy rains or prolonged rains. When the rainy season arrives there is an increase in the amount of infiltration water which affects the soil, therefore soil pores are easily destroyed and soil aggregation becomes very weak so that the soil shear resistance decreases [10]. Data and information obtained from secondary data state that landslides that occurred in the 2018-2019 period in Majalengka Regency hit 20 sub-districts. As one of the significant natural hazards, landslides cause causalities and tremendous property losses in mountainous areas [11].

3.3.1 Characteristics of Rainfall During the Rainy Season Month

Based on the graph of the average monthly rainfall for 2018-2019 in Figure 7, it is known that the peak of the rainy season generally occurs in March, while the beginning of the rainy season occurs in November. Besides, the highest number of landslides also occurred in February to March as shown in Table 3. At the beginning and peak of the rainy season, the high intensity of rain causes the water content in the soil to fill in a short time and cause landslides through the cracked soil, water will enter and accumulate at the bottom of the slope. The rainfall-induced landslides rely principally on the increased pore pressure in the soil, which reduced effective normal stress to a critical level [12]. Landslides caused by heavy rainfall events and earthquakes are major natural disasters in many countries and on a global scale [13].

Then, in the January period, the rainfall in Majalengka Regency increases (rainy season period). Meanwhile, in the period from June to October, the rainfall was relatively low (dry season) so that there were no landslides. This situation is evidenced by the difference seen between the number of landslides occurring in the rainy and dry seasons. The number of landslides occurring at the beginning of the rainy season, from January to May, was 82 landslide events, while the number of landslides in the dry season did not have landslides.

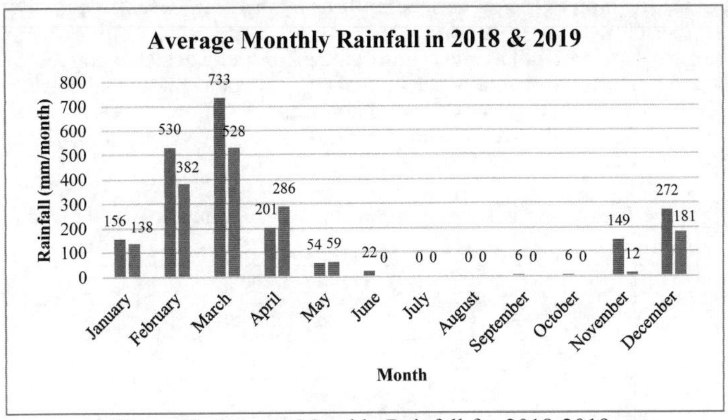

Figure 7. Average Monthly Rainfall for 2018-2019

3.3.2 Characteristics of Rainfall in the Landslide Event Groups

Based on Table 2, the characteristics of rainfall based on each cluster in Majalengka Regency, it can be seen that the highest average rainfall on the D-day landslide events is in cluster 5, which is 49 mm/day, while the lowest is in cluster 4, which is 5 mm/day. For the average rainfall for 3 days before the landslide events, the highest was in cluster 1 and cluster 3, which is 29 mm/day, while the average rainfall for 3 days before the lowest landslide events was in cluster 5, which was 25 mm/day.

Furthermore, the highest average rainfall for 5 days before the landslide events was in cluster 1, which was 31 mm/day, and the lowest average rainfall for 5 days before the landslide events was in cluster 5, which was 22 mm/day. Then, the average rainfall for 10 days before the occurrence of the highest landslide was in cluster 3, which was 26 mm/day, and the lowest average rainfall for 10 days before the landslide was in cluster 4, which was 14 mm/day.

At the average cumulative rainfall 3 days before the landslide, the highest was in cluster 4, which was 80 mm/day and the lowest was in cluster 5, which was 54 mm/day. The highest mean cumulative rainfall for 5 days before the landslide events was in cluster 3, which is 112 mm/day and the lowest was in cluster 5, which was 71 mm/day. The highest average cumulative rainfall for 10 days before the landslide events was in cluster 1, which is 174 mm/day and the lowest was in cluster 4, which was 115 mm/day.

For the average rainfall intensity on D-day, the highest landslide events were in cluster 5, which was 62 mm/hour and the lowest was in cluster 4, which was 2 mm/hour. The highest average rainfall intensity for 3 days before the landslide events was in cluster 3, which was 48 mm/hour and the lowest was in cluster 1, which was 7 mm/hour. The average rainfall intensity for 5 days before the landslide events was the highest in cluster 2, which is 28 mm/hour, and the lowest in cluster 4, which was no measurement at the rainfall station. Meanwhile, the highest average rainfall intensity for 10 days before the landslide events was in cluster 2 and cluster 3, which was 30 mm/hour, and the lowest was in cluster 4, which is 1 mm/hour.

In Table 2 it can be concluded that the average rainfall for each cluster with cumulative rainfall 3 days before the landslide events, rainfall 5 days before the landslide events, and rainfall 10 days before the landslide events always has rainfall of >50 mm/day, where the rainfall is included in the category heavy rain. Meanwhile, the average rainfall intensity based on each cluster is also included in the large rainfall category, because it has a rainfall intensity of >10 mm/hour. This is supported by the Meteorology, Climatology, and Geophysics Agency which states that indicators of heavy to very heavy rain can trigger landslides. The indicator of major rainfall is rainfall with an intensity of >50 mm/day and intensity of >10 mm/hour [14].

In general, there are two characteristics of landslide-triggering rain, which is the type of heavy rain with rainfall reaching 70 mm/hour or >100 mm/day and the normal type of rain with <20 mm/day but lasting a long time. The type of rain is normal but lasts a long time in Majalengka Regency, which is included in the five clusters of landslides. This type of heavy rain is only effective in triggering landslides on slopes where the soil has water-absorbing properties, such as sandy loam. Meanwhile, the normal type of rain if it lasts for several weeks can also trigger landslides on slopes composed of more impermeable soil, such as clay [15]. Besides, it is known that landslides do not only occur during heavy rain but when the rain stops or drizzle for some time (hours) it can cause landslide event in Majalengka Regency, West Java Province.

Table 2. Characteristics of Rainfall Based on Clusters in Majalengka Regency

	A	B	C	D	E	F	G	H	I	J	K
Cluster 1	27	29	31	25	70	97	174	23	7	16	8
Cluster 2	43	28	25	25	69	98	168	48	26	28	30
Cluster 3	37	29	30	26	74	112	170	48	48	23	30
Cluster 4	5	27	27	14	80	80	115	2	10	-	1
Cluster 5	49	25	22	23	54	71	146	62	25	10	7

Information :
A= Average rainfall on the day landslide events (mm/day)
B = Average rainfall for 3 days before the landslide events (mm/day)
C = Average rainfall for 5 days before the landslide events (mm/day)
D = Average rainfall for 10 days before the landslide events (mm/day)
E = Cumulative average rainfall 3 days before landslide events (mm/day)
F = Cumulative average rainfall 5 days before landslide events (mm/day)
G = Cumulative average rainfall 10 days before landslide events (mm/day)
H = Average rainfall intensity (mm/hour) on the day of the landslide events (mm/hour)
I = Average rainfall intensity (mm/hour) for 3 days before landslide events (mm/hour)
J = Average rainfall intensity (mm/hour) for 5 days before landslide events (mm/hour)
K = Average rainfall intensity (mm/hour) for 10 days before the landslide events (mm/hour)

4. Conclusion

Based on the physical factors that cause landslides in 2018-2019 in Majalengka Regency, the grouping of landslide events is formed into five clusters with physical characteristics in each cluster dominated by steep slopes (8-16°), latosol soil types, volcanic rock, settlements land use, and non-vegetation with NDVI value <0.2.

The rainfall characteristics that trigger landslides are based on the landslide events group in Majalengka Regency in 2018-2019 have the highest rainfall on the D-Day average in cluster 5, which is 49 mm/day. The highest average cumulative rainfall 3 days before the landslide events was in cluster 4, which is 80 mm/day. The highest average cumulative rainfall 5 days before the landslide events was in cluster 3 is 112 mm/day. The highest average cumulative rainfall 10 days before the landslide events was in cluster 1, which is 174 mm/day.

For research future, it is suggested to use the landslide event point data for a longer period so that the landslide event cluster is more varied and this research can help reduce the occurrence landslide events.

Acknowledgment
Thanks are due to the University of Indonesia which has supported and also thanks to Regional Disaster Management Agency Majalengka Regency which helps in providing information and data on lansdlides disaster.

References
[1] Warsudi, A (2019, November 1) Sepanjang Februari 2019 Jabar Diterjang 234 Bencana dan 9 Korban Tewas. Retrieved from https://daerah.sindonews.com/read/1383245/174/sepanjang-februari-2019-jabar-diterjang-234-bencana-dan-9-korban-tewas-1551460134
[2] Badan Penanggulangan Bencana Daerah Kabupaten Majalengka (2019, November 19) *BPBD Kabupaten Majalengka*. Retrieved from http://bpbd.majalengkakab.go.id/2019/05/
[3] BPS Kabupaten Majalengka 2017 *Kabupaten Majalengka dalam Angka 2017*. (Majalengka: BPS Kabupaten Majalengka)
[4] Baehaqi, A. I (2019, November 1) *BPBD Sebut 19 Kecamatan di Kabupaten Majalengka Rawan Bencana Alam*. Retrieved from https://jabar.tribunnews.com/2019/01/11/bpbd-sebut-19-kecamatan-di-kabupaten-majalengka-rawan-bencana-alam
[5] Ibnudin, A (2020, April 20) *Waspada BMKG Sebut Wilayah Majalengka Mulai Masuk Puncak Musim Hujan*. Retrieved from https://www.pikiran-rakyat.com/jawa-barat/pr-01305482/waspada-bmkg-sebut-wilayah-majalengka-mulai-masuk-puncak-musim-hujan
[6] Sharma, R., Alam, M. A., & Rani, A 2012 K-Means Clustering in Spatial Data Mining Using Weka Interface. International *Journal of Computer Applications pp 26-30*
[7] Lazzari, M., & Piccarreta, M 2018 Landslide Disasters Triggered by Extreme Rainfall Events: The case of Montescaglioso (Basilicata, Southern Italy) Journal of *Geosciences vol 8 chapter 10 pp 377*
[8] Buchori. I. dan Susilo. J 2012 Model Keruangan untuk Identifikasi Kawasan Rawan Longsor *TATALOKA vol 14* chapter 4 pp 282-294
[9] O'Dunn, S., & Sill, W. D 1988 *Exploring Geology: Introductory Laboratory Activities* TH Peek
[10] Abe, K., & Ziemer, R. R 1991 Effect of tree roots on shallow-seated landslides *USDA Forest Service Gen. Tech. Rep. PSW-GT chapter 130*, pp 11-20
[11] Zhao Y., Wang, R., Jiang, Y., Liu, H., & Wei, Z 2019 GIS-based logistic regression for rainfall-induced landslide susceptibility mapping under different grid sizes in Yueqing, Southeastern China *Journal of Engineering Geology pp 105147*
[12] Ran, Q., Hong, Y., Li, W., & Gao, J 2018 A modelling study of rainfall-induced shallow landslide mechanisms under different rainfall characteristics *Journal of Hydrology pp 790–801*
[13] Yano, A., Shinohara, Y., Tsunetaka, H., Mizuno, H., & Kubota, T 2019 Distribution of landslides caused by heavy rainfall events and an earthquake in northern Aso Volcano, Japan from 1955 to 2016 *Journal of Geomorphology pp 533-541*
[14] Badan Meteorologi Klimatologi dan Geofisika (2020, April 20) *Curah Hujan dan Potensi Gerakan Tanah*. Retrieved from http://www.hrdnetwork.com/pirba/content/e5781/e5795/e6331/e15201/eventReport15215/CurahHujan_PotensiGertan_BMKG.pdf
[15] Karnawati, D 2001 Bencana Alam Gerakan Tanah Indonesia Tahun 2000 Evaluasi dan Rekomendasi (Yogyakarta: Teknik Geologi Universitas Gadjah Mada)

IOP Conf. Series: Earth and Environmental Science **884** (2021) 012055 doi:10.1088/1755-1315/884/1/012055

Hazard Level of Slum Areas in Palembang City

Sukmaniar[1], Andri Kurniawan[2*], and Agus Joko Pitoyo[3]

[1]Student of Doctoral Program in Population Study of the Postgraduate School of Universitas Gadjah Mada and Lecturer of Geography Education Study Program FKIP University of PGRI Palembang
[2]Lecturer of Doctoral Program of Population Study Program at the Postgraduate School of Gadjah Mada University and Geography Study Program, Faculty of Geography, Gadjah Mada University and Researchers at the Center for Population and Policy Studies at Gadjah Mada University
[3]Lecturer of Doctoral Program of Population Study Program at the Postgraduate School of Gadjah Mada University and Geography Study Program, Faculty of Geography, Gadjah Mada University

*Corresponding Author: andri.kurniawan@ugm.ac.id

Abstract. The study aimed to identify the hazard level distribution in the slum areas in Palembang City. The research was a quantitative type with 382 families that live in the slums as the samples. The data were collected using proportional area random sampling and analyzed using the K-means Cluster test through SPSS 23. Spatial analysis through Arcgis was conducted to complete the understanding of the hazard distribution in the slum areas. The results showed that the hazard clusters formed by 64 areas could be categorized into three: low, medium, high. High level of hazards, reaching the one-third portion of the areas, dominated the slums in Palembang City. The cluster with a high hazard level was mostly in the city center, which was geographically located on the riverbanks, causing the areas to be vulnerable to flood, diseases, and crimes. The city center was known to offer high numbers of the job vacancy. People in the slums working in the city center were day laborers.

1. Introduction

Slums are areas with poor environmental quality, less qualified building materials, and low quality of life. On average, people living in the slums are under the poverty line. It has been a common phenomenon in cities of developed countries with poor quality of life and a high poverty rate [1]. People in the areas are from a low level of economy and education. The condition is continued from generation to generation. Slums quickly appear in the cities because of the high level of urbanization. The advances in technology and economic activities in the cities have triggered a population explosion. Unfortunately, technological advances were not followed by workforce absorption. Therefore, people who cannot compete and cannot find suitable occupations contribute to the increase of the new unemployment rate. Unemployment becomes one of the causes of slum emergence [2], [3].

Slum areas have provided affordable dwellings and living costs for people under the poverty line. However, they cannot prevent themselves from facing any kind of threat or hazard in the areas, such as natural disasters [4]. These people become vulnerable [5]. High vulnerability, especially in the economy, and the low capacity of the people have increased the disaster risk. For example, slum areas located on

Content from this work may be used under the terms of the Creative Commons Attribution 3.0 licence. Any further distribution of this work must maintain attribution to the author(s) and the title of the work, journal citation and DOI.
Published under licence by IOP Publishing Ltd

the riverbanks can face numerous problems in several aspects, such as environment, social, and economy [6]. The issues are the hazards for the slum areas. They take the form of building damage, diseases, crimes, domestic conflict, drug abuse, promiscuity, unemployment, brawl/dispute, and drowning. Other kinds of hazards that may occur in the slum areas are fire [7] and flood [8]. Based on the types, the distribution of risks in the slum areas can be determined through mapping [9].

The study was significant; it can prevent the increase of poverty level and the spread of slums. Besides, it can also reduce the hazard upon those who live in the slum areas in Palembang cities. Therefore, the present study aimed to observe the hazard level in the slum areas in Palembang City.

2. Method

The research employed a quantitative method. The variables include floods, building fire, damage, epidemic, crimes, drug abuse, domestic conflict, promiscuity, and unemployment. The samples were 382 families living in the slum areas in Palembang City. The data were gathered through proportional area random sampling, of which the results were analyzed to categorize the hazard levels. The categorization was carried out using the K-means Cluster test of SPSS.23. The clusters of the hazard level were low, medium, or high, with different characteristics. The data were analyzed spatially using the Arcgis program to observe the hazard distribution in the slum areas of Palembang City.

3. Results and Discussion

3.1. The Hazard Distribution in the Slum Areas in Palembang City

The data of threats or dangers were gathered directly from the field because they were not available in the relevant agencies. The hazard level was determined by using the K-Means Cluster, of which the data were normally distributed based on the Kolmogorov-Smirnov formula. The normality test results were presented below.

Table 1. Data Normality Test.

No.	Hazard Variable	Kolmogorov-Smirnov[a]		
		Statistic	df	Sig.
1.	Average number of flood cases per year	0,107	64	0,068
2.	The average number of building fire cases per year	0,110	64	0,053
3.	Average number of building damage cases per year	0,106	64	0,072
4.	Average number of epidemic cases per year	0,108	64	0,063
5.	Average number of crimes per year	0,077	64	0,200*
6.	Average number of domestic conflict cases per year	0,107	64	0,065
7.	The average number of drug abuse cases per year	0,106	64	0,072
8.	Average number of promiscuity cases per year	0,109	64	0,057
9.	Average number of unemployment case per year	0,107	64	0,068
10.	Average number of brawl/dispute cases per year	0,193	64	0,000
11.	Average number of drowning cases per year	0,534	64	0,000

IOP Conf. Series: Earth and Environmental Science **884** (2021) 012055 doi:10.1088/1755-1315/884/1/012055

Sources: Data analysis using *IBM SPSS 23*

Nine variables that were normally distributed were used, while the other two (dispute, drowning, and attacked by animals) were not. Based on the normality test, the research employed eleven variables with a significant value above 0.05 and the confidence rate of 5%. In other words, the nine variables were used for every 64 areas of the slums. Even so, data standardization was conducted to align the diverse data of each variable. To obtain accurate data, K-Means Cluster was employed to determine the area category based on the hazard level. From the calculation using IBM SPSS 23, the results were as follows.

Table 2. Iteration History[a].

Iteration	Change in Cluster Centers		
	1	2	3
1	2.860	2.628	2.626
2	.245	.351	.203
3	.136	.131	.169
4	.000	.000	.000

a. Convergence achieved due to no or small change in cluster centers. The maximum absolute coordinate change for any center is .000. The current iteration is 4. The minimum distance between the initial centers is 6.175.

Source: Data analysis using *IBM SPSS 23*

Based on the table, it needed five data iteration to obtain accurate data. Further, the final cluster centers' values were presented in the following table.

Table 3. Final Cluster Centers.

	Cluster		
	1	2	3
Zscore: Average number of flood cases per year	.48915	.02237	-.40832
Zscore: Average number of fire case per year	.66505	-.32214	-.28721
Zscore: Average number of damage cases per year	.86134	-.44301	-.35238
Zscore: Average number of epidemic cases per year	.88471	-.88041	-.03866
Zscore: Average numbers of crime cases per year	.72798	.21979	-.74943
Zscore: Average number of domestic conflict cases per year	.69245	-.20394	-.39897
Zscore: Average number of drug abuse cases per year	-.39426	1.14739	-.55661
Zscore: Average number of promiscuity cases per year	-.40450	1.12789	-.53360
Zscore: Average number of unemployment case per year	.79566	-.35380	-.36764

Source: Data analysis using *IBM SPSS 23*

IOP Conf. Series: Earth and Environmental Science **884** (2021) 012055 doi:10.1088/1755-1315/884/1/012055

The table showed the calculation results of Final Cluster Centers of the hazard variables with similar characteristics. The clusters were as below.

Table 4. Characteristics of Hazard Clusters.

Characteristics of Hazard Clusters		
Low Clusters (*Cluster 3)*	Medium (*Cluster 2)*	High (*Cluster 1)*
Low average number of flood cases per year	Medium average number of flood cases per year	**High average number of flood cases per year**
Medium average number of fire cases per year	Low average number of fire cases per year	**High average number of fire cases per year**
Medium average number of damage cases per year	Low average number of damage cases per year	**High average number of damage per year**
Medium average number of epidemic cases per year	Low average number of epidemic cases per year	**High average number of epidemic cases per year**
Low average number of crime cases per year	Medium average number of crime cases per year	**High average number of crime cases per year**
Low average number of domestic conflict cases per year	Medium average number of domestic conflict cases per year	**High average number of domestic conflict cases per year**
Low average number of drug abuse cases per year	**High average number of drug abuse cases per year**	Medium average number of drug abuse cases per year
Low average number of promiscuity cases per year	**High average number of promiscuity cases per year**	Medium average number of promiscuity cases per year
Low average number of unemployment cases per year	Medium average number of unemployment cases per year	Low average number of unemployment cases per year

Source: Research analysis results

Cluster 1: The variable hazard score was higher than cluster 2 and 3, or six out of nine variables. Those were flood, fire, damage, epidemic, crime, and domestic conflict. From the characteristics, it was assumed that people living in the slums were vulnerable to high hazard, evident in the highest score of the six variables. Two variables, which were drug abuse and promiscuity, were in the medium category. Meanwhile, only one division was in the low cluster (unemployment). The results were in line with [10], revealing that the slums along the riverbanks were highly vulnerable to danger because of poor health and occupation resulted in poor disaster risk management.

Cluster 2, The group consisted of two variables with a high score (drug abuse and promiscuity). Similar to [11], social hazards mostly occurred to those living on the riverbanks. The low significance value was occupied by three variables, which were fire, damage, and epidemic. The values were lowered compared to clusters 1 and 2 on the variables of the flood, crime, domestic dispute, and unemployment.

Cluster 3, The cluster did not contain the high value. Three variables were in the medium category (fire, damage, and epidemic). The flood had brought about diseases to humans and animals [12]. The characteristics indicated that people in the slums were vulnerable to flood, crime, domestic disputes, drug abuse, promiscuity, and unemployment. It can be concluded that the cluster has a low vulnerability.

Table 5. Data analysis results from ANOVA.

	Cluster	Error	F	Sig.

	Mean Square	df	Mean Square	df		
Zscore: Average number of flood cases per year	4.482	2	.886	61	5.059	.009
Zscore: Average number of fire cases per year	6.440	2	.822	61	7.838	.001
Zscore: Average number of damage cases per year	10.836	2	.678	61	15.993	.000
Zscore: Average number of epidemic cases per year	15.209	2	.534	61	28.476	.000
Zscore: Average number of crime cases per year	12.779	2	.614	61	20.819	.000
Zscore: Average number of domestic disputes per year	7.180	2	.797	61	9.004	.000
Zscore: Average number of drug abuse cases per year	17.934	2	.445	61	40.319	.000
Zscore: Average number of promiscuity cases per year	17.280	2	.466	61	37.066	.000
Zscore: Average number of unemployment cases per year	9.209	2	.731	61	12.601	.000

Source: Data analysis calculation using *IBM SPSS 23*

The table revealed that nine variables of the present study had a significance rate above 0.05. It can be concluded that there were differences between the three clusters. The number of cases identified using K-Means Cluster and tested using ANOVA was presented as follows.

Table 6. Number of Cases in each Cluster.

Cluster	1	20.000
	2	19.000
	3	25.000
Valid		64.000
Missing		.000

Source: Data analysis using *IBM SPSS 23*

The table showed that the highest number reached by cluster 3, which included 25 areas. It means that the areas had a low hazard level. Meanwhile, the lowest was in cluster 2, with 19 areas in the medium hazard category. Twenty areas in the high category showed that they were vulnerable to hazards. The results were presented in the following table.

Table 7. Slum Areas with Low Hazard Level.

No.	Name of the slums	Hazard Level	Notes
1.	1 Ilir A	1	Low
2.	2 Ulu & Tuan Kentang	1	Low
3.	5 Ulu B	1	Low
4.	9-10 Ulu	1	Low
5.	10 Ilir	1	Low
6.	Kampung Arab A	1	Low
7.	13 Ilir	1	Low
8.	16 Ulu A	1	Low
9.	16 Ulu B	1	Low

10.	16 Ulu C	1	Low
11.	29 Ilir	1	Low
12.	Bagus Kuning	1	Low
13.	Karang Anyar B	1	Low
14.	Karya Jaya A	1	Low
15.	Plaju	1	Low
16.	Plaju Darat	1	Low
17.	Keramasan B	1	Low
18.	Silaberanti	1	Low
19.	Pulokerto B	1	Low
20.	Pipareja	1	Low
21.	Sei Selayur	1	Low
22.	Sentosa	1	Low
23.	Tangga Takat	1	Low
24.	24 Ilir, 26 Ilir, & 26 Ilir DI	1	Low
25.	Karya Jaya D	1	Low

Source: Data analysis using Hasil perhitungan analisis data dari *IBM SPSS 23*

Twenty-five areas of the slums in Palembang City were in a low category. The rests were medium and high. In other words, there was only 39% showing a high vulnerability to hazard. Nevertheless, the people still need the government and Non-Government organizations to help them improve the physical and social conditions in the areas [13]. The medium category of the areas with hazard vulnerability was presented below.

Table 8. Slum Areas with Medium Hazard Level.

No.	Name	Hazard Level	Note
1.	1 Ilir B	2	Medium
2.	1 Ulu A	2	Medium
3.	2 Ilir A	2	Medium
4.	3 Ilir	2	Medium
5.	15 Ulu & Tuan Kentang	2	Medium
6.	32 Ilir B	2	Medium
7.	32 Ilir A	2	Medium
8.	Gandus B	2	Medium
9.	Gandus C	2	Medium
10.	Gandus D	2	Medium
11.	Gandus E	2	Medium
12.	Karang Jaya B	2	Medium
13.	Kebun Bunga D	2	Medium
14.	Kebun Bunga E	2	Medium
15.	Kemas Rindo Tepian Sungai	2	Medium
16.	Kemas Rindo Dataran Rendah	2	Medium
17.	Pulokerto A	2	Medium
18.	35 Ilir	2	Medium
19.	Sei Buah	2	Medium

Source: Data analysis using *IBM SPSS 23*

Nineteen areas, or 29,7%, were in the medium category. The results were different from [14], mentioning that people living on the riverbanks were highly vulnerable to hazards. However, the high hazard level was also spread in the slum areas in Palembang City, shown in the following table:

Table 9. Slum areas with high hazard level.

No.	Name	Hazard Level	Note
1.	2 Ilir B	3	High
2.	2 Ilir B 1	3	High
3.	Kampung Kapitan	3	High
4.	8 Ulu	3	High
5.	15 Ulu B	3	High
6.	36 Ilir A	3	High
7.	Karang Anyar A	3	High
8.	Karang Jaya A	3	High
9.	Karang Jaya C	3	High
10.	Karya Jaya B	3	High
11.	Kebun Bunga C	3	High
12.	Kemang Agung	3	High
13.	Kuto Batu & Lawang Kidul	3	High
14.	Keramasan A	3	High
15.	Ogan Baru A	3	High
16.	Kawasan Plaju Ulu	3	High
17.	Talang Putri	3	High
18.	23 Ilir & 18 Ilir	3	High
19.	30 Ilir	3	High
20.	Karya Jaya C	3	High

Source: Data analysis result using *IBM SPSS 23*

Twenty areas, or 31/3%, were highly vulnerable to hazard. The areas were dangerous to live in as the result of the illegal settlement, exposing the people to disasters. The condition was worsened by the poverty level [15]. Spatially, the hazard level map based on the slum area in Palembang City can be described through the map below:

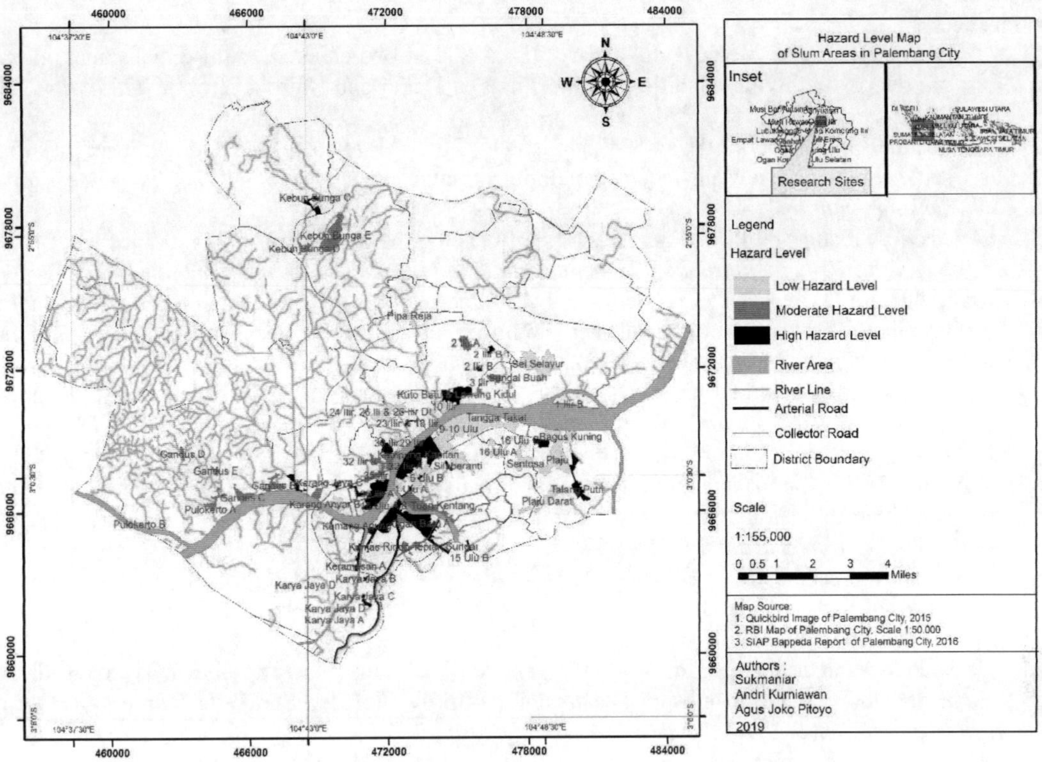

Figure 1. Hazard Level Map of Slum Areas in Palembang City.

The map showed that the slum areas with high hazard levels were in the middle of Palembang City. The areas were on the riverbanks, of which the waste management was poor. It worsened the slum areas. However, the strategic location with the economic center made the people stay for the sake of easy access to the market flow. Even so, the areas were prone to crimes. In total, almost all areas with a high level of hazard were on the main or branch riverbank depended on the tidal of the Musi River. The waste thrown to the rivers returned to the residence because of the tidal. Further, people were prone to floods and diseases. Slum areas were identical to the riverbank, where waste was thrown to the areas [16], which influenced the water quality [17], causing the areas vulnerable to flood [18], [19], [20], [21] and landslides [22]. Therefore, it is necessary to manage the waste in the area by establishing a proper waste management system and developing the people's awareness to process their household garbage [23].

4. Conclusion

The results and discussion revealed the clusters formed by the four areas of the slums. The areas were categorized into three groups: cluster 1 (low) consisted of 20 areas, cluster 2 (medium) 19 areas, and cluster 3 (high) 25 areas. Cluster 3 dominated the slum areas in Palembang City. Nearly one-third of the areas were highly vulnerable to hazard. The average danger or threat was the city center, which was located on the riverbanks. The geographical condition caused the areas susceptible to flood, disease, and crimes. The city center offered job opportunities for people. However, most of the people in the slum areas were day laborers.

Acknowledgments

The author would like to thank to Universitas Gadjah Mada for providing doctoral dissertation research funds in the 2019 fiscal year of RTA (Final Recognition) grant program.

References

[1] Gulyani S, Bassett E M and Talukdar D 2014 A tale of two cities: A multi-dimensional portrait of poverty and living conditions in the slums of Dakar and Nairobi *Habitat International* **43** 98-107

[2] Hofmann P, Taubenböck H and Werthmann C 2015 Monitoring and modelling of informal settlements-A review on recent developments and challenges *In 2015 joint urban remote sensing event* 1-4

[3] Adedayo A F and Malik N A 2015 Factors influencing the growth of slums in Lagos metropolis, Nigeria *Ethiopian Journal of Environmental Studies and Management* **8(2)** 113-119

[4] Satterthwaite D, Archer D, Colenbrander S, Dodman D, Hardoy J and Patel S 2018 Responding to climate change in cities and in their informal settlements and economies *International Institute for Environment and Development, Edmonton, Canada*

[5] Hossain M 2014 Pro-poor urban adaptation to climate change in Bangladesh: a study of urban extreme poverty, vulnerability and asset adaption *Doctoral dissertation, Heriot-Watt University*

[6] Rofiana V 2015 Dampak pemukiman kumuh terhadap kelestarian lingkungan Kota Malang (studi penelitian di jalan muharto kel jodipan kec blimbing, Kota Malang) *The Indonesian Journal of Public Administration* **1(1)** 40-57

[7] Rachman D N and Harison D 2018 Kajian kondisi persampahan dan rencana penanganannya di kawasan padat penduduk kelurahan talang bubuk plaju Palembang *Teknika: Jurnal Teknik* **5(1)** 66-80

[8] Jaswadi J, Rijanta R and Hadi M P 2012 Tingkat kerentanan dan kapasitas masyarakat dalam menghadapi risiko banjir di kecamatan pasarkliwon Kota Surakarta *Majalah Geografi Indonesia* **26(2)** 119-149

[9] Hizbaron D R, Hadmoko D S, Wibawa S, Dalimunthe S A and Sartohadi J 2010 Tinjauan kerentanan, risiko dan zonasi rawan bahaya rockfall di Kulonprogo, Yogyakarta *In Forum Geografi* **24 (2)** 119-136

[10] Ballesteros M M 2010 Linking poverty and the environment: Evidence from slums in Philippine cities *PIDS Discussion Paper Series* 2010 33.

[11] Rana M S and Nessa A M 2017 Impact of riverbank erosion on population migration and resettlement of Bangladesh *Science Journal of Applied Mathematics and Statistics* **5(2)** 60-69

[12] Sarker M N I, Ali M A and Islam M S 2015 Causes and possible solutions of poverty perceived by char dwellers in Bangladesh. *International Journal of Natural and Social Sciences* **2(1)** 37-41

[13] Hasan W and Mirwandono E 2007 Slum area on deli river bank. *Proc. Int. Conf. On The Knowledge City: Spirit, Character, and Manifestation (Medan, Indonesia)*

[14] Ghosh B K and Mahbub A Q M 2014 Riverbank erosion induced migration: a case study of Charbhadrasan Upazila, Faridpur *Oriental Geographer* **58(1)** 59-71

[15] Lein H 2000 Hazards and 'forced' migration in Bangladesh *Norsk Geografisk Tidsskrift* **54(3)** 122-127

[16] Brueckner J K, Mation L and Nadalin V G 2019 Slums in Brazil: Where are they located, who lives in them, and do they 'squeeze'the formal housing market? *Journal of Housing Economics* **44** 48-60

[17] Angriani P, Ruja I N and Bachri S 2018 River management: The importance of the roles of the public sector and community in river preservation in Banjarmasin (A case study of the Kuin River, Banjarmasin, South Kalimantan–Indonesia) *Sustainable cities and society* **43** 11-20

[18] Huq M E and Hossain M A 2012 Flood Hazard and Vulnerability of Slum Dwellers in Dhaka *Stamford Journal of Environment and Human Habitat* **1** 36-47

[19] Padawangi R, Turpin E, Prescott M F, Lee I and Shepherd A 2016 Mapping an alternative community river: The case of the Ciliwung *Sustainable Cities and Society* **20** 147-157

IOP Conf. Series: Earth and Environmental Science **884** (2021) 012055 doi:10.1088/1755-1315/884/1/012055

[20] Amoateng P, Finlayson C M, Howard J and Wilson B 2018. A multi-faceted analysis of annual flood incidences in Kumasi, Ghana *International journal of disaster risk reduction* **27** 105-117

[21] Tarigan A K, Sagala S, Samsura D A A, Fiisabiilillah D F, Simarmata H A and Nababan, M 2016 Bandung City, Indonesia *Cities* **50** 100-110

[22] Thorn J, Thornton, T F and Helfgott A 2015 Autonomous adaptation to global environmental change in peri-urban settlements: Evidence of a growing culture of innovation and revitalisation in Mathare Valley Slums, Nairobi *Global Environmental Change* 31 121-131

[23] Michiani M V and Asano J 2019 Physical upgrading plan for slum riverside settlement in traditional area: A case study in Kuin Utara, Banjarmasin, Indonesia *Frontiers of Architectural Research* **8(3)** 378-395

Settlement Suitability Analysis Based on the Catastrophic Eruption of Sinabung

W Utami, A Rahmat, BH Sialagan, S Exaudia, AL Turnip

National Land College, Indonesia

Corespondent author: westiutami@stpn.ac.id

Abstract. Mount Sinabung is an active volcano after being 'dormant' for a long time since the 1600s. Sinabung has begun to show its activities from 2010 to 2020, so that it has a long impact on community activities, damage to settlements, agricultural and threats to the lives of residents around Sinabung. This study aims to determine the conditions of land use in disaster-prone areas in relation to spatial patterns and levels of disaster prone. The research method is carried out through spatial analysis by overlaying the land use map with the spatial planning map to produce a land use suitability map. The spatial pattern suitability map is overlaid with a disaster hazard map to determine the level of security in settlements. The results showed that 5 affected Districts, namely in the Districts of Naman Teran, Payung, Simpang Empat, Tiganderket, Munte, the safe zone was 145.66 hectares, the less safe zone was 0.98 hectares, the unsafe zone was 36.91 hectares and the very insecure zone was 1.25 hectares. It is hoped that the spatial analysis regarding the distribution of safe locations for settlements is expected to become a policy material for the government to formulate arrangements for the use and utilization of space so that the spatial planning policies that are established are in line with efforts to reduce the risk of the eruption of Mount Sinabung disaster.

1. Introduction

Mount Sinabung is one of the active volcanoes in Indonesia [1], it was recorded that it had never erupted since 1600 (Strato volcano type B) [2], but this volcano showed eruptive activity on August 27, 2010 (changed to GA Strato Type A), November 2013, June 2015, May 2016, February - April 2018 and until now 2020 [3], [4]. This prolonged disaster began with an eruption in 2010 which activated a shallow active hydrothermal system [5]. The eruption then triggers and activates the deep hydrothermal system so that the eruption occurs in a prolonged manner [6]. The prolonged eruption with a very strong eruption force that produced lava and pyroclastic deposits from 2010 to 2015 amounting to approximately 2.89108 m^3 impacted damage inseveral areas including 5 (five) districts, namely Naman Teran, Payung, Tiganderket, Simpang Empat, and Munte in Karo, North Sumatra Province [7], [8], [9]. The extent of the impact of damage and losses caused by the eruption is due to the fact that the community considers Mount Sinabung to be inactive, so that residents live and depend on the agricultural sector with a radius very close to the peak of Sinabung [3]. As the most active mountain, the eruption has an impact on the area up to a radius of ± 7 kilometers from the peak causing residents to lose their homes and infrastructure, damage and disappearance of agricultural land and livestock, damage to plantation crops [10], [11], [12], decreased income in the tourism sector [13], health problems due to volcanic ash [14] and loss of other assets, which required residents to evacuate and be relocated to a safer place [15], [16].

Content from this work may be used under the terms of the Creative Commons Attribution 3.0 licence. Any further distribution of this work must maintain attribution to the author(s) and the title of the work, journal citation and DOI.
Published under licence by IOP Publishing Ltd

The main livelihood condition of the majority of the community as farmers where land and agricultural crops are damaged has implications for decreasing farmer income levels [17], [18].

The threat of the Sinabung eruption always threatens all the time, but the current condition is that there are still some residents who have not moved from their homes and their cultivated land, one of which is in part of the Guru Kinayan Village area. Socio-economic reasons make the community persist in the red zone radius so that there are still exposed people who are likely to be at risk if an eruption occurs [16]. As a disaster mitigation effort, the government has designated a radius of 3 km, 4 km to 5 km (red zone) Sinabung as the zone that must be vacated considering the threat of the impact of pyroclastic flows due to eruptions is very high [19], [20]. This relocation effort by the government was carried out through several stages, Phase I (residents of Sukameriah, Beekerah, Simacem Villages) with a total of 370 households for houses and 457 households in the form of farming land in Siosar Village [21], independent relocation Stage II (residents of Gurukinayan, Berastepu, Gamber, Kuta Tonggal Villages) with a total of 1,679 families and agricultural land have also been completed, while relocation Phase III (Sukanalu, Sigarang-garang, Mardinding, Lau Kawar) until 2019 is still incomplete for 892 residents or ± 3,500 people [22]. The difficulty of obtaining the availability of land that is safe from eruptions, land with clear and clean conditions for relocation is one of the obstacles to resettlement for victims of eruptions. In addition, high land prices and the unavailability of basic infrastructure for settlements make it difficult for residents to carry out independent relocation [16]. This condition results in the presence of people living in refugee camps and people who are still carrying out economic activities in the red zone [19].

The vulnerability of people living or farming in locations close to the center of the eruption requires the right solution [23], considering the number of victims due to hot cloud attacks that occurred in 2010 as many as 4 people died and 405 were injured, while in 2014 the number of victims increased to 17 people due to pyroclastic flows, and than in 2016 the Sinabung eruption caused 9 fatalities [24]. Regulating the functions and utilization of space through the determination of Spatial Planning Regulation (RTRW/RDTR) and proper zoning arrangements in disaster-prone areas is certainly one of the right solutions so that losses and casualties can be prevented [25], [26], [27]. In addition, monitoring efforts on the suitability of land use against predetermined spatial patterns and control of land use and utilization, especially in disaster-prone areas are the key so that the available space is able to support and create sustainable life [28], [29], [30]. The study aims to present a spatial analysis of the conditions of settlements around Mount Sinabung, to determine the level of suitability of settlements with spatial patterns, and to determine the level of security of settlements from disaster threats.

2. Method

This research was conducted through a spatial approach with an overlay technique to place one map graphic on top of another map graph and display the results in a new map. This method is able to combine digital data and its attributes so that it can produce a combined map of the two by having attribute information from the two maps. Some of the maps used in this study include land use maps, spatial planning maps and Mount Sinabung eruption hazard maps. The existing land use map is obtained through interpretation of satellite images from the SAS Planet providing. In this study, the land use classification is adjusted to the needs of the analysis so that land use is classified into residential and non-residential. Meanwhile, the spatial map was obtained from GISTARU data/information system about spatial planning from the Ministry of Agrarian Affairs and Spatial Planning/National Land Agency. This map provides spatial information regarding the distribution of directions for the use and utilization of space in the area around Mount Sinabung, especially directions for settlement locations. In order to provide direction for safe settlements from disasters in this study, a disaster-prone map determined by the Karo District Disaster Management Agency is used. The determination of safe areas is also based on several literature reviews by researchers related to the direction and impact of the Sinabung eruption that occurred from 2010 to 2021. The flow chart in this study is presented as follows:

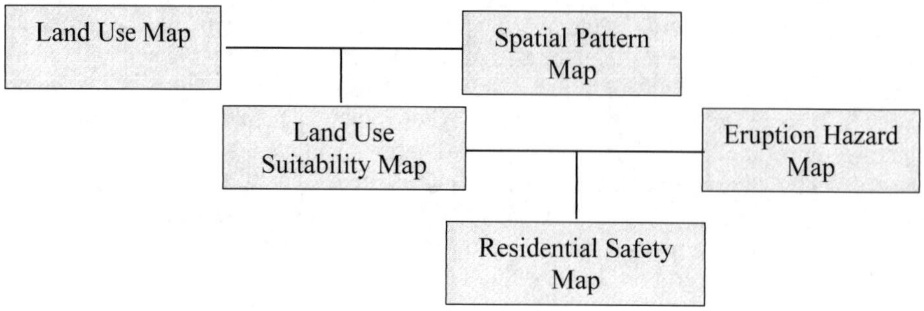

Figure 1. Research Flowchart

Based on the flow chart above, the overlay method is carried out between the land use map and the spatial plan map. From this overlay, it is obtained empty land that is still available for settlements where the location in the RTRW directions is used as a settlement. Furthermore, the overlay result in the form of a map of land availability for settlements is overlaid with a disaster-prone map to map the spatial distribution of the availability of settlements with the level of disaster security. From the results of this overlay, it is obtained the spatial distribution of settlement directions in accordance with the spatial layout and safe from the disaster location. And the final result of this analysis produces 4 criteria settlement security level, namely: Criteria 1). the location is available and safe from disasters; 2) The location is available but less safe; 3) the location is unsafe; 4) the location is very unsafe.

3. Result and Discussion

3.1. Mount Sinabung Eruption Prone Areas

Karo Regency has 16 (sixteen) sub-districts in total and there are 5 (five) sub-districts included in the eruption-prone area of Sinabung. We can observe this based on the map of the eruption-prone areas of Mount Sinabung which is overlaid with the administrative boundaries of Karo Regency as presented in Figure 2.

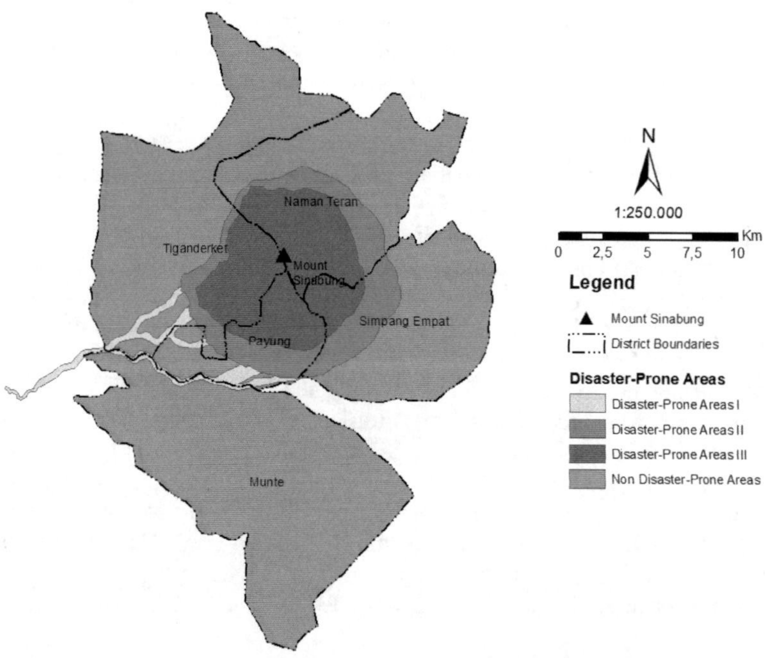

Figure 2. Map of Mount Sinabung Eruption Prone Areas in 5 Districts
(Source: Map of Mount Sinabung Eruption Prone Areas)

The following districts are included in the Disaster Prone Area (DPA) to the eruption of Mount Sinabung:

Table 1. Area of Sinabung Eruption Prone District

No	Sub-district	Large (M^2)				Total
		DPA I	DPA II	DPA III	Non DPA	
1	Munte	185,99	-	-	13494,36	13680,35
2	Naman Teran	-	1124,47	2383,29	4469,58	7977,34
3	Payung	388,41	779,61	1185,61	737,55	3091,17
4	Simpang Empat	-	1239,41	662,47	5393,87	7295,75
5	Tiganderket	383,00	663,25	1742,71	9536,71	12325,67
	Total	957,40	3806,73	5974,08	33632,07	44370,28

Source: Data Analysis 2020

As the most active volcano since 2010, the eruption that occurred has characteristics with different destructive forces and impacts. The characteristics and impacts of the Sinabung eruption can be presented as in table 2 below:

Table 2. Characteristics of Eruption and the Impact of the Sinabung Eruption

Year	Eruption Characteristics	Casualities and displacement	Damage
2010	Freactive eruptions, followed by magma activity in the form of explosive materials: ash /dust, bombs, lava flows to the east-southeast reaching a radius of up to 2 km. Affected areas (Sukameriah, Gungpitu, Sigarang-garang, Sukodebi and Susuk villages)	1 died ± 28.756 people evacuated	Damage to agricultural land and people's settlements
2013	Volcanic type magmatic eruption, by throwing volcanic material as high as 10,000 m, followed by hot clouds with a sliding distance of up to 1500 m, ash rain spreads up to 5 km	± 19.286 people evacuated	Damage to agricultural land (33.586.6 Ha) in 7 districts & damage to houses up to 5000 buildings.
2014	Magmatic eruptions of volcanic and merapi types, volcanic material as high as 5000 m, glide of hot clouds reaching 2 - 4 km to the south.	17 people died, 33.210 people evacuated	Damaged settlements, agriculture in 7 districts, loss of food and horticultural crops covering an area of 2.959 hectares, singed a forest out to at least 3.9 Km
2016	Radius of hot clouds reaches up to 4 km	9 people died (activities in the red zone); 11.113 Evacuated	Damage to agricultural
2017	Hot clouds bursting up to a radius of 4.6 km	± 7.255 people. Evacuated	Damage to community agricultural land
2018	Bursts of volcanic material as high as 5.000 m, rain of rocks to Kec. Munte	± 100 people Evacuated	Damage to community agricultural land

2019	The biggest eruption with volcanic material bursts as high as 7 km, hot cloud radius of up to 3.5 km. Areas affected by volcanic ash Mardinding, Naman Teran, Merdeka and Brastagi	± 7.256 people displaced	Damage to agricultural land is widespread because it is covered in ash
2020	The height of the ash column is ± 5000 m, the areas affected by the volcanic ash of Naman Teran, Berastagi, Simpang Empat and Merdeka District	-	± 1.483 Ha of crop failure in Namanteran, Merdeka, Berastagi and Dolat Raya

Source: Kusumayudha (2018), Indonesia Disaster Information Data (2020), Pallister (2019), Yulianto (2016)

Based on figure 2 shows that pyroclastic flow are the result of volcanic eruptions which are very dangerous and result in high casualties [32]. The Sinabung eruption that occurs quite often and it is not known when it will stop has had a wide impact on the villages around the dome of Mount Sinabung, the worst damage was in villages on the south, east and southeast sides which were influenced by the shape of the dome on the open side [33]. Data on villages that were affected by the Sinabung Eruption were quite severe, where some or even all areas were covered by volcanic material including:

Table 3. List of Villages That Have Been Damaged & Relocated Due to the Eruption

No	District	Village	Village with the worst impact & relocated
1	Naman Teran	Bekerah, Simacem, Kuta Tonggal, Sigarang-garang, Lau Kawar, Sukanalu, Kutarakyat, Kebayaken, Kutamblelin, Kutagugung.	Bekerah, Simacem, Kuta Tonggal, , Sigarang-garang, Lau Kawar, Sukanalu
2	Simpang Empat	Berastepu, Gamber, Kuta Tengah, Sibintun	Berastepu, Gamber,
3	Payung	Sukameriah, Guru Kinayan, Berastepu, Selandi,	Sukameriah, Guru Kinayan,
4	Tigan Derket	Mardinding, Perbaji,	Mardinding
5	Munte	Kutambaru, Munthe	

Sources: Several Online Media, Karo Regency Website, Interpretation of Google Earth Imagery (2020).

Naman Teran is an area that is experiencing the most severe conditions on the south, east and southeast side of the mountain, where the direction of the dome and pyroclastic flow directly leads to this area. Several villages that were severely damaged in this sub-district had been relocated in stage 1 and stage 2 considering that the residential conditions were already covered in very thick volcanic material and were always affected by the attack of pyroclastic flows. While the sub-district that experienced the least impact of damage was in Munte District, where this location is not directly facing Mount Sinabung and this location is in a fairly far radius from the peak of Sinabung, but in Kutambaru Village it is passed by a river that carries erupted pyroclastic material. Sinabung so that this village is quite vulnerable. In a large eruption with volcanic material bursts as high as 5000 m in 2018, rain of rocks and gravel reached up to Kutambaru and Kutarakyat villages, this shows that the impact of the Sinabung eruption was fluctuating.

3.2 Land Use in the Mount Sinabung Eruption Prone District

Land use has an influence on the high level of vulnerability and risk that can be caused by disasters. The more land use mismatches in disaster-prone areas, especially the red zone, the higher the level of vulnerability and risk that will be borne by the community. As happened in Sinabung, the improper use of land (settlements and agricultural/plantation land) in the red zone resulted in a high number of casualties, losses and damage as well as many people who had to flee due to the eruption. In this study, land use is determined based on the land cover obtained from the interpretation of the downloaded satellite images using the SAS Planet application with the selection of image sources from Bing Satellite. Based on image interpretation, land use classification is divided into two types, namely land use for residential and non-residential areas, land use maps are presented in Figure 3 below.

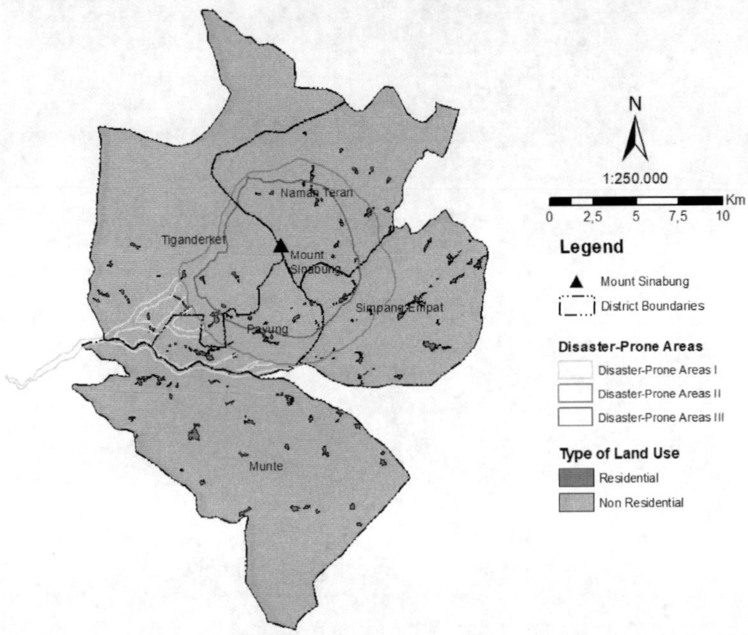

Figure 3. Land Use Map in Affected Districts. (Source: Data analysis 2020)

Based on the identification and data analysis carried out, the following results were obtained:

Table 4. Land Use

No	District	Land Use			
		Resettlement (Ha)	(%)	Non Resettlement (Ha)	(%)
1	Munte	177,68	1,30	13502,67	98,70
2	Naman Teran	74,96	0,94	7902,38	99,06
3	Payung	74,68	2,42	3016,49	97,58
4	Simpang Empat	131,76	1,81	7163,98	98,19
5	Tiganderket	94,11	0,76	12231,56	99,24
	Total	553,20	1,44	43817,08	98,56

Source: Data Analysis 2020

Based on image observations, it shows that land use for settlements in these five districts is still low, this is influenced by the level of population density in this area only in the range of 172.36/Km^2 to 266.89 / Km^2, this figure is much lower when compared to The population density in Kabanjahe District reached 1725.69 Km^2 [34].

3.3 Suitability of Land Use in Districts Affected by Mount Sinabung Eruption

Land use suitability analysis is an effort made to see the suitability of a land use with the spatial pattern plan determined by the Regional Government. This spatial pattern map functions to regulate the use and utilization of space including activities on it. Spatial data related to spatial patterns in these 5 districts is presented in the following figure:

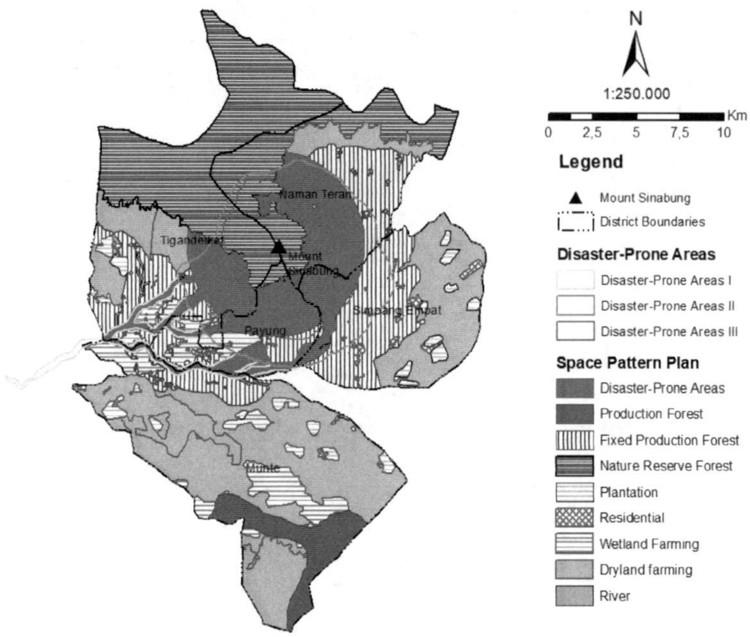

Figure 4. Karo Regency Spatial Plan Map
(Source: Karo District Government)

In the Karo Regency spatial pattern plan, it has been regulated that an area with a radius close enough to Sinabung is designated as Disaster-prone areas/DPA III, where this area has vulnerability and vulnerability to pyroclastic hazards. However, in this area there are still uses of land in the form of settlements or agricultural land cultivated by the community as a supplier of food/horticultural needs for Karo Regency. The spatial pattern as shown in Figure 4 shows that in the DPA III and DPA II areas the RTRW directions are designated as protected areas with the function of providing local protection against the impact of eruptions, while for residential areas, dry land agricultural land and wetlands the spatial pattern is set at a fairly far and safe from the impact of the eruption. This study of land use analysis focuses on settlements, by dividing the level of suitability of land use into 3 (three) categories, namely appropriate, unsuitable, and others.The explanation of the three categories is as follows:

(a). Appropriate means that the use of the land is the same as the spatial pattern plan (especially for settlements).

(b). Unsuitable means that the use of the land is not the same as the spatial pattern plan (especially for settlements).

(c). Miscellaneous means that land use is not assessed in this study, it may or may not be appropriate. This condition is because this research is limited/focused on the study of the suitability of land use in the form of settlements only.

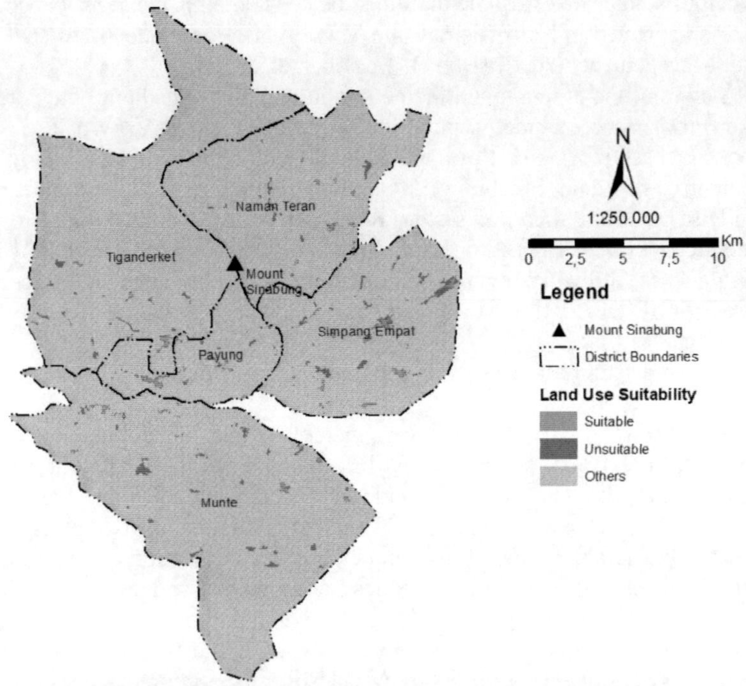

Figure 5. Suitability Map of Land Use for Settlement
(Source: Data Analysis 2020)

Based on the analysis of the suitability level of land use, the level of suitability of residential land use is obtained as follows:

Table 5. Suitability of Settlements with Spatial Patterns in Karo District

No	District	Settlements			
		Suitable (Ha)	(%)	Unsuitable (Ha)	(%)
1	Munte	34,96	18,3	190,94	84,5
2	Naman Teran	26,67	33,5	79,44	74,8
3	Payung	33,82	49,8	67,83	66,7
4	Simpang Empat	62,36	28,6	217,35	77,7
5	Tiganderket	26,98	28,06	96,12	78,1
	Total	**184,80**	**22 %**	**651,68**	**77,9%**

(Source: Data Analysis 2020)

The table above shows that more land uses for settlements are not in accordance with the predetermined spatial pattern. As in Naman Teran, Simpang Empat, and Payung Subdistrict as areas very prone to settlement mismatches reached 77,7 %. This condition has resulted in the post-eruption of the community having to do relocation because residents' settlements were destroyed by the eruption. The quite high mismatch of residential land use in this area can occur due to a lack of socialization in land use regulation or it can also occur due to low control and monitoring of spatial functions.

3.4 Safe Zone for Settlements in Affected Districts Prone to Eruption of Mount Sinabung

It should be noted that any land use for settlements that is in accordance with the spatial pattern plan is not necessarily safe for areas prone to eruption of Mount Sinabung. This condition could occur due to the preparation of the RTRW which probably paid less attention to disaster mitigation aspects or could also be caused by the impact of the damage caused by the Sinabung eruption that was expanding, resulting in changes in the level of disaster hazard. Therefore, in determining a safe and sustainable

IOP Conf. Series: Earth and Environmental Science **884** (2021) 012056 doi:10.1088/1755-1315/884/1/012056

settlement in previous studies, the variable that must be considered is the security aspect of the threat of disaster [35]. As happened in Karo District, the Map of Disaster Prone Areas (DPA) issued by the National Disaster Management Agency (BNPB) was updated in 2015.

Based on the nature of the eruption and the condition of Mount Sinabung, the potential dangers of an eruption that may occur include: pyroclastic flows (hot clouds), pyroclastic falls (throwing incandescent rocks and ash rain), lava flows and lava. Based on the potential hazards that may occur, the disaster-prone areas of Mount Sinabung can be divided into three levels of vulnerability from low to high, namely: Disaster Prone Area I, Disaster Prone Area II and Disaster Prone Area III.

a. Disaster-prone area III (DPA III) is an area that has the potential to be affected by hot clouds, lava flows and avalanches, throwing rocks (incandescent), heavy ash rain and poisonous gases. The disaster-prone area III (DPA III) of Mount Sinabung is located at a radius of ± 1.5 Km consisting of two parts, namely:
   1) Areas prone to disasters from mass flows (hot clouds, lava flows and avalanches), and poisonous gases.
   2) Areas prone to disasters due to rock throwing materials (incandescent) and heavy ash rain.
b. Disaster prone area II (DPA II) is an area that has the potential to be affected by hot clouds, lava flows, lava avalanches, throwing rocks (incandescent) and heavy ash rain, this area is located at a radius of about 5 km from the eruption center. This area is divided into two parts:
   1) Areas prone to disasters from mass flows (hot clouds, lava flows and avalanches).
   2) Areas prone to disasters due to throwing stones (incandescent) and heavy ash rain.
c. Disaster-prone area I (DPA I), is an area that has the potential to be affected by lava flow / flooding and may be exposed to expansion of lava / hot clouds. DPA I to flow is located along the river or valley. In DPA I there is still the possibility of pyroclastic falls ejecting incandescent stones or falling fragments < 2 cm [36].

To map the level of the safe zone for settlements, it is carried out through an overlay of land use maps for settlements that are in accordance with the spatial pattern plan against the map of the eruption-prone areas of Mount Sinabung. The zone classification of residential security levels in accordance with the spatial pattern plan is divided into 4 (four) categories, namely Safe, Less Safe, Unsafe, and Very Unsafe. The following is the classification of the four categories:

a. The safe category is obtained if the use of land for settlements that is in accordance with the spatial pattern plan does not overlap with the area prone to eruption of Mount Sinabung, be it DPA I, II, or III.
b. The less safe category is obtained if the use of land for settlements is in accordance with the spatial pattern plan that overlaps with the eruption-prone area of Mount Sinabung DPA I.
c. The unsafe category is obtained if the use of land for settlements is in accordance with the spatial pattern plan that overlaps with the eruption-prone area of Mount Sinabung DPA II.
d. The very unsafe category is obtained if the use of land for settlements is in accordance with the spatial pattern plan that overlaps with the eruption-prone area of Mount Sinabung DPA III.

The results of the mapping of settlement security levels against the threat of the Sinabung eruption based on the DPA map are presented as shown in map 6 below.

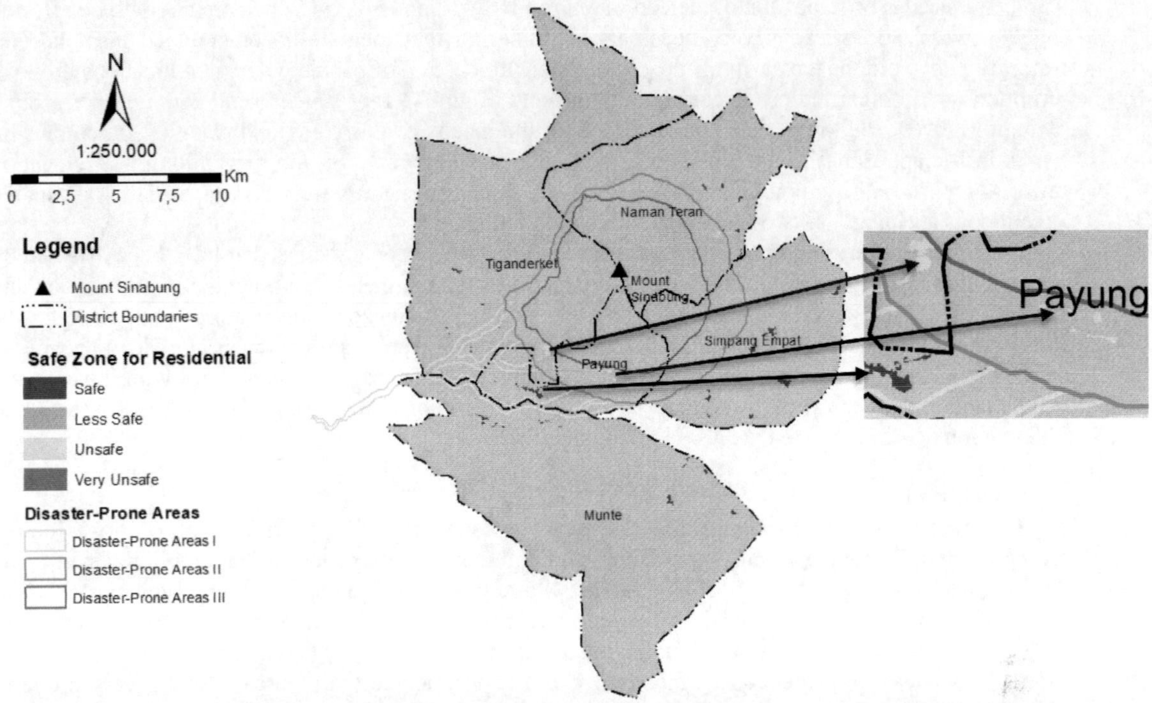

Figure 6. Residential Security Level Map
(Source: Data Analysis 2020)

Based on Figure 6 above, it shows that the distribution of settlements that are in accordance with the spatial plan but still in the DPA III and DPA II zones is still there, while the detailed data on the distribution of settlement security levels from disaster threats is presented in Table 6.

Table 6. Settlement Security Levels in Five Districts

No	District	Land Use Suitability							
		Safe (Ha)	(%)	Less Safe (Ha)	(%)	Not safe (Ha)	(%)	Very not safe (Ha)	(%)
1	Tiganderket	8,94	33,12	0,00	0,01	17,69	65,58	0,35	1,29
2	Simpang Empat	58,05	93,10	0,00	0,00	3,40	5,45	0,90	1,45
3	Payung	25,10	74,23	0,56	1,64	8,16	24,13	0,00	0,00
4	Naman Teran	19,02	71,30	0,00	0,00	7,66	28,70	0,00	0,00
5	Munte	34,54	98,79	0,42	1,21	0,00	0,00	0,00	0,00
	Jumlah	145,66	74,11	0,98	0,57	36,91	24,77	1,25	0,55

(Source: Data Analysis 2020)

The results of the analysis of the existing conditions in these 5 districts, there are still settlements that are currently in very unsafe condition, namely in the Simpang Empat and Tigan Derket Districts, although in small percentages. Meanwhile, the existence of people living in unsafe conditions, namely in the DPA II area, is also still found in 4 Districts, namely in Tiganderket, Naman Teran, Payung and Simpang Empat where the four Districts have previously carried out relocation of their residents. It is hoped that the distribution of security levels for residential settlements in 5 districts is expected to become a database to formulate appropriate disaster mitigation policies so that the level of disaster risk can be reduced. The relocation of the community in DPA III and DPA II which are likely to be affected by the pyroclastic threat is very necessary, considering that there is no other mechanism besides relocation to avoid the threat of this deadly disaster. Relocation of people who are in locations prone to eruptions are also carried out in communities around Mount Merapi Volcano and communities around Mount Manyon, Philippines [37], [38]. Based on the results of this study and resettlement studies on cases in Merapi, San Vicente Volcano El Salvador and Manyon Volcano the Philippines, relocation efforts emphasize the importance of community involvement, security from disaster threats, sustainable livelihoods and availability of public facilities [37], [38], [39].

Adaptation patterns and public awareness not to build resettlement areas that have been affected by the eruption need to be encouraged by enforcement of regulations through detailed spatial plans and zoning which function the disaster prone area as a non-residential zone and as a protection zone for the local area. Relocation efforts by providing land that is safe from disasters, clear and clean as well as efforts to restore community life by providing agricultural land that is in accordance with the capacity of the land as well as various assistance and assistance for the affected communities are needed so that the community is able to realize a safe and sustainable life.

4. Conclusion

The phenomenon of Sinabung as a 'sleeping' volcano for ± 400 years has implications for the community to live and cultivate agricultural land with a radius very close to the eruption center. This has resulted in a very high number of casualties, losses and damage due to the eruption from 2010 to 2020. This study shows that in 5 (five) districts affected by land use in the form of settlements, it reaches 553.20 Ha. Meanwhile, from this area, the mismatch level of settlements with spatial patterns reached up to 77,9 %, where this condition mostly occurs because people live in areas that are very prone to disasters/DPA III. Based on the data on the suitability level of settlements with the spatial plans overlaid with the DPA map, it shows that there are still settlements (0.5%) in very vulnerable areas, namely in Tiganderket and Simpang Empat Districts. Settlements that are in unsafe conditions (at DPA II) also still exist, covering an area of 36.91 hectares in 4 districts, namely in Tiganderket, Namanteran, Simpang Empat and Payung. Disaster mitigation efforts through relocation of people who are in very unsafe zones are the best alternative considering the ability of any technology has not been able to overcome the impact caused by eruptions in the form of pyroclastic flows, one of which is a hot cloud. In this context, disaster mitigation as outlined in the spatial plans regulation is very necessary so that the available space is able to support the security and sustainability of people's lives. Research development related to ideal spatial planning to support disaster mitigation efforts needs to be done so that the spatial plan that is prepared is capable of realizing sustainable development.

Acknowledgment

The author would like to thank the National Land College for conducting the research. The author thanks the Karo Regency government, the Ministry of Agrarian Affairs and Spatial Planning/National Land Agency for the data that has been provided.

References

[1] Mulyana A R 2010 *Pemetaan Kawasan Rawan Bencana Gunung Sinabung* Pusat Vulkanologi dan Mitigasi Bencana Geologi Kementerian Energi dan Sumber Daya Mineral.

[2] Hotta, K, Iguchi, M, Ohkura, T, Hendrasto, M, Gunawan, H , Rosadi, U, Kriswati, E 2018, Method for estimating the end of the defation initiated in 2014 at Sinabung volcano, Indonesia, under the assumption that the magma behaves as a Bingham fuid, *Earth, Planets and Space*, 70:107 https://doi.org/10.1186/s40623-018-0884-7.

[3] Kusumayudha, SB, Lestari, P, Paripurno, EK 2018 Eruption Characteristic of The Sleeping Volcano Sinabung, North Sumatera Indonesia And SMS Gateway For Disaster Early Warning System, *Indonesian Journal of Geography* **50** (1) pp 70-77.

[4] Pemerintah Kabupaten Karo 2020, Laporan aktivitas Gunungapi Sinabung Tahun 2020, https://www.karokab.go.id/

[5] Gunawan, H, Surono, Budianto, A, Kristianto, Prambada, O, McCausland, W, Pallister, J, Iguchi, M 2019, Overview of the eruptions of Sinabung Volcano, 2010 and 2013–present and details of the 2013 phreatomagmatic phase, Journal of Volcanology and Geothermal Research. **382**. 103-119, https://doi.org/10.1016/j.jvolgeores.2017.08.005

[6] Gonza´ Lez, PJ, Singh, KD, Tiampo, KF 2015 Shallow Hydrothermal Pressurization before the 2010 Eruption of Mount Sinabung Volcano, Indonesia, Observed by use of ALOS Satellite Radar Interferometry, Pure and Applied Geophysics. **172**, 3229–3245. Springer Basel DOI 10.1007/s00024-014-0915-7

[7] Afnimar, McCausland, WA, Hamidah, NN, Kristianto, Basuki, A, Indrastuti, N 2020 Local magnitude, coda magnitude, and radiated energy of volcanic tectonic earthquakes from October 2010 to December 2011 at Sinabung volcano, Indonesia, *Bulletin of Volcanology*, 82:45, Springer, https://doi.org/10.1007/s00445-020-01383-7

[8] Togatorop, Mouna, P 2016 Pemetaan Potensi Bencana Aliran Lava Gunung Sinabung Menggunakan Citra Aster GDEM *Jurnal Geodesi Undip* ***5*** *(2) pp 87-96.*

[9] Badan Penanggulangan Bencana Daerah (Local Agency for Disaster Management) Karo Regency 2020.

[10] Putri AC, Lubis SN, Sihombing L 2015 Dampak Erupsi Gunung Sinabung terhadap sosial ekonomi petani kopi (Studi kasus: Desa Guru Kinayan, Kecamatan Payung, Kabupaten karo) *Journal on Social Economic of Agriculture and Agribusiness* **4** (3).

[11] Hafni, Roswita, Lubis LS 2016 Dampak Erupsi Gunung Sinabung terhadap Kondisi Sosial Ekonomi Petani di Desa Suka Meriah Kecamatan Payung Kabupaten Karo *Jurnal Ekonomikawan* ***16*** *(1)*

[12] Tampubolon, J, Nainggolan HL, Ginting A, Aritonang, J 2018 Mount Sinabung Eruption: Impact on Local Economy and Smallholder Farming in Karo Regency, North Sumatera, *4th International Conference on Environmental System Research (ICESR), IOP Conf. Series: Earth and Environmental Science.*

[13] Febriaty, Hastina 2015 Dampak Erupsi Gunung Sinabung terhadap Pendapatan dari Sektor Pariwisata di Kabupaten Karo *Jurnal Ekonomikawan*

[14] Tarigan YG, Maimunah S 2019 Pengukuran paparan Particulate Matter Satu (PM1) dan CO2 serta keluhan kesehatan pada masyarakat pasca erupsi Gunung Sinabung *Jurnal Kesehatan Global* **2** (1) pp 28-36.

[15] Sidabutar AK, Yusak M, Salmiah 2016 Dampak Erupsi Gunung Sinabung terhadap Usaha tani Kentang (Kasus: Desa Kuta Rayat, Kecamatan Naman Teran, Kabupaten Karo) *Journal of Agriculture and Agribusiness Socioeconomics.*

[16] Maxriz J 2018 Analisis Faktor Penyebab Masyarakat Desa Guru Kinayan Tetap Bertahan di Zona Merah Gunung Sinabung *Jurnal Tunas Geografi 7 (2) pp 107- 114.*

[17] Sukarman & Suparto 2015 Sebaran dan Karakteristik Material Vulkanik Hasil Erupsi Gunung Sinabung Sumatera Utara *Jurnal Tanah dan Iklim* **39** (1)

[18] Nainggolan HL, Ginting A, Tampubolon J, Aritonang J, Hutagalung M 2019 Dampak erupsi Gunung Sinabung terhadap Kondisi Sosial Ekonomi Petani Hortikultura di Kabupaten Karo *Jurnal Ilmu-ilmu Sosial dan Humaniora* **21** (3) pp 287-295.

[19] Pandia SL, Rachmawati R, Mei ETW 2016 Relokasi permukiman Desa Suka Meriah akibat kejadian Erupsi Gunung Api Sinabung Kabupaten Karo *Jurnal Perencanaan Wilayah dan Kota* **27** (2) pp 137-150.

[20] Luqman A, Sulaksana N, Hamdani AH, Sulistri W 2017 Pemodelan Aliran Awan Panas (Aliran Piroklastik) Sebagai Data Pendukung Peta Kawasan Rawan Bencana Gunungapi (Studi kasus Gunung Api Sinabung Sumatera Utara) *Jurnal Lingkungan dan Bencana Geologi* ***8*** *(1) pp 1 - 12.*

[21] Lubis FW, Sabarina Y, Masril M 2019 Penanganan Bencana Erupsi Gunung Sinabung Ditinjau Dari Aspek Komunikasi dan Koordinasi *Jurnal Simbolika* ***5*** *(1) pp 1 – 22.*

[22] Alamudi AA 200 9, 10 Tahun Erupsi, Ribuan Pengungsi Sinabung Masih Menunggu Relokasi *IDN Times Sumut.*

[23] Smith K, Petley DN 2008 *Environmental Hazard: Assesing risk and reducing disaster*, Routledge Taylor & Francis Group, London and New York.
[24] Data Informasi Bencana Indonesia/DIBI (Indonesia Disaster Information Data) 2020, bnpb.cloud
[25] Korlena, Djunnaedi A, Probosubanu L, Ismail N 2011 Peraturan zonasi: Peran dalam pemanfaatan ruang dan pembangunan kembali di kawasan rawan bencana (Kasus: Arkadelphia city, Arkansas USA) *Forum Teknik 34 (1) pp 17 – 26.*
[26] Stephan C, Norf C, Fekete A 2017 How "Sustainable" are post disaster Measure? Lessons to be learned a decade after the 2004 tsunami in the Indian Ocean *International Journal Disaster Risk Science* Doi : 10.1007/s13753-017-0113-1, www. Springer.com
[27] Utami W, Wibowo YA, Afiq M 2019 Analisis Spasial Untuk Lokasi Relokasi Masyarakat Terdampak Tsunami Selat Banten *Bhumi Jurnal Agraria dan Pertanahan* 5 (1) pp 112-128.
[28] Wang C, Guldman J 2015 A Land Use Allocation Optimization Model To Mitigate Potential Seismic Damages *Environtment and Planning B: Planning and Design* **42** pp 730-753, doi: 10.1068/b130152p.
[29] Tiepolo M , Pezzoli A, Tarchiani V 2017 Renewing Climate Planning Locally to Attend the 11th Sustainable Development Goal in the Tropics, ©The Author(s) *Green Energy and Technology* DOI 10.1007/978-3-319-59096-7_1
[30] Utami W, Wibowo YA 2019 Pemanfaatan Data Spasial dan Data Kerawanan Bencana Sebagai Evaluasi Rencana Tata Ruang Wilayah (Studi kasus: pasca tsunami di Banten), *Prosiding 'Pendekatan Multidisiplin ilmu dalam Manajemen Bencana* Seminar Nasional Unriyo Yogyakarta
[31] Pallister, J, Wessels, R, Griswold, J, McCausland, W, Kartadinata, N, Gunawan, H, Budianto, A, Primulyana, S 2019 Monitoring, forecasting collapse events, and mapping pyroclastic deposits at Sinabung volcano with satellite imagery, *Journal of Volcanology and Geothermal Research* **382** pp. 149-163, Elsevier, https://doi.org/10.1016/j.jvolgeores.2018.05.012.
[32] Lube, G., Breard, E, Ongaro, TE, Dufek, J 2020 Multiphase flow behaviour and hazard prediction of pyroclastic density currents. *Nature Reviews Earth & Environment*, 348–365. https://doi.org.ezproxy.ugm.ac.id/10.1038/s43017-020-0064-8
[33] Yulianto, F, Suwarsono, Sofan, P 2016 The Utilization of Remotely Sensed Data to Analyze the Estimated Volume of Pyroclastic Deposits and Morphological Changes Caused by the 2010–2015 Eruption of Sinabung Volcano, North Sumatra, Indonesia, *Pure and Applied Geophysics* **173**, 2711–2725 Springer International Publishing, DOI 10.1007/s00024-016-1342-8
[34] Kabupaten Karo dalam Angka 2020, Tingkat Kepadatan Penduduk per Kecamatan, https://karokab.bps.go.id/publication/2020/05/14/20d25edf8b2b6a2c7d4402c9/kabupaten-karo-dalam-angka-2020.html
[35] Mardin R, Shen Z 2019 Integrated Criteria for Flood Disaster Mitigation in Indonesian Urban Masterplan; Housing and Settlement Suitability Case in Palu Urban Masterplan. In: Huang G., Shen Z. (eds) Urban Planning and Water-related Disaster Management. Strategies for Sustainability. Springer, Cham. https://doi.org.ezproxy.ugm.ac.id/10.1007/978-3-319-90173-2_8
[36] Badan Geologi Kementerian Energi dan Sumber Daya Mineral (Geological agency of The Ministry of Energy and Mineral Resources).
[37] Bawole, P 2015, Program Relokasi Permukiman Berbasis Masyarakat untuk Korban Bencana Alam Letusan Gunung Merapi Tahun 2010 *Jurnal Tesa Arsitektur* **13** (2)
[38] Usamah, M, Haynes, K 2012, An examination of the resettlement program at Mayon Volcano: what can we learn for sustainable volcanic risk reduction? *Bulleton of Volcanology* ©Springer-Verlag, 74:839–859 DOI 10.1007/s00445-011-0567-8
[39] Bowman, LJ, Henquinet, KB 2015, Disaster risk reduction and resettlement efforts at San Vicente (Chichontepec) Volcano, El Salvador: toward understanding social and geophysical vulnerability. *J Appl. Volcanol.* **4,** 14, https://doi.org.ezproxy.ugm.ac.id/10.1186/s13617-015-0031-0

Disaster mitigation and resiliency learning strategies of teachers during Covid-19 pandemic from several provinces in Indonesia

Rini Riris S[1], Alifi Nur P N [2], Satrio S P [3]

[1]Graduate School, Universitas Negeri Yogyakarta, Yogyakarta, Yogyakarta Special Province, Indonesia
[2] Department of Social Education, Universitas Negeri Yogyakarta, Yogyakarta, Yogyakarta Special Province, Indonesia
[3] Budi Utama School, Yogyakarta, Indonesia

Email: ririniris.2020@student.ac.id

Abstract. The Covid-19 disaster changed as well as limited students' learning space in carrying out learning activities at school. The face-to-face learning process was suspended and transformed into an online system. This education emergency demands rapid changes that must be made by all components of education, especially teachers. The enactment of the regulation on limiting the students' learning activities required teachers as educational facilitators to be ready and alert in changing the educational pattern. This study described the readiness of teachers and the resilience of teacher strategies in carrying out learning during the pandemic in accordance with teacher competency standards in Law Number 14 of 2005. The research data was collected through observation, questionnaires and interviews. Data analysis was performed using descriptive analysis and questionnaire data scoring. The results showed that according to the four teacher competencies, i.e. pedagogic, personality, social, and professional competences, only personality competency was in the good category range, while the other three components according to the average score are in the moderate category range. Therefore, some improvements on the the mitigation and resilience of teachers' strategies when facing emergency disaster are needed.

1. Introduction

Learning is a human process in adapting to the environment so that they can survive. Through the process of adapting to the environment when learning, someone can carry out social interactions [1]. The spread of SARS-CoV-2 or better known as the Novel Coronavirus (Covid-19) causes restrictions on human space for social interactions, and in the field of education this is not an exception [2]. The social distancing restriction imposed on the education sector forced the implementation of an educational emergency system that its mitigation had not been designed before. Face-to-face learning is suspended and it is temporary transformed into e-learning protocols to minimize the progression of the virus spread and avoid physical contact [3]. Indonesia followed this in mid-March 2020 through Ministerial Circular Letter Number 4 of 2020 concerning the emergency period of Covid-19 spread [4]. The adaptation to this pattern of emergency education has become something new in the field of education throughout history.

Content from this work may be used under the terms of the Creative Commons Attribution 3.0 licence. Any further distribution of this work must maintain attribution to the author(s) and the title of the work, journal citation and DOI.
Published under licence by IOP Publishing Ltd

The application of learning with an online system in education is one of the mitigation measures against the spread of Covid-19. Disaster mitigation has to be designed comprehensively for communities in dealing with the disaster [5]. At the educational level, the response to the Covid-19 pandemic related to the application of an online learning system that causes disruption in the cognitive, affective and psychomotor aspects of students as well as the competence of teachers as educators [6]. Education facilitators during the Covid-19 pandemic depended on IT infrastructure, the availability of learning technology support, digital literacy, and mental resilience [7]. Mitigation related to education should be an important issue for education practitioners. Therefore, it needs to be prepared in depth.

Knowledge of disasters by teachers is necessary to make sure that teacher competence and quality of education are well maintained. Teachers as the main agents in driving education must have the individual ability in reading, understanding, using and preparing mitigation when a disaster emerges [8]. The government and schools as legality holders also need to have emergency learning standards related to the models, strategies, methods and media that teachers must use [[9]. The relationship between teachers and other education stakeholders is an important key in mitigating the education sector, given the important role of teachers in education as a determinant of the direction of future generations [10].

This study serves as a reminder to all teachers in the world regarding the importance of mitigation and strategies resilience in a lesson in the realm of education when a disaster occurs. The descriptions of the results of this study were obtained from teachers in all subject teachers in Indonesia. First, the study describes four teacher competency standards including pedagogic, personality and social competences related to mitigation and teachers' strategy resilience in dealing with disaster emergencies. Second, the study describes the research method, the third explains the results of the research and discussion, and the fourth part is a description of the research conclusions.

2. Methods

This study is a qualitative study in the case of mitigation and teachers' resilience in implementing emergency learning during the Covid-19 pandemic in Indonesia. The research data was collected using Google Form that was distributed randomly via WhatsApp media by researchers. Respondents in this study were teachers of all subjects in senior high schools in 7 provinces in Indonesia: Bali, West Nusa Tenggara, East Java, Central Java, Banten, D.I Yogyakarta and Jambi. The teachers have implemented rules for transitioning the implementation of learning in schools to online learning from mid-March 2020 to September 2020. Online learning is expected to last until the end of 2020/2021. The decision to implementing online learning depends on government policy by taking into account the health risks resulting from the Covid-19 pandemic.

The data obtained from the results of this study were related to the mitigation and strategies resilience carried out by teachers in accordance with the four competencies that teachers must possess. Four teacher competencies as stated in Law Number 14 of 2005 including pedagogic, personal, social, and professional competencies. The competencies in the law was then developed by researchers in which there is implementation of learning, method selection, readiness of learning media, platforms, evaluation, teacher communication, and teacher professional attitudes. In addition, the data collected from observations were the difficulties experienced by teachers in implementing online learning. The data that researchers get was data from observations, discussions and filling out questionnaires by teachers of different subjects such as history, geography, Indonesian, Mathematics, English, Physics, Chemistry, and the principal.

Data analysis was carried out by combining the results of observation, scoring and interviews. The researchers' observations were made on several teachers either directly or through discussion to observe the difficulties or constraints that were felt during the implementation of online learning in schools. Researchers used descriptive analysis to discuss the mitigation and resilience of teacher strategies in dealing with the Covid-19 pandemic in accordance with four teacher competency standards. Descriptive analysis data were obtained from the results of a questionnaire on 57 respondents with 20 indicators representing the four teacher competency variables (Tables 1 to 5). The results of the questionnaire were analyzed using simple linear regression with the help of Microsoft Excel. On another hand, the data from the unstructured interview is used by the researchers as additional material as well as a supplement from the principal to confirm the readiness and resilience of the teacher's strategy in carrying out learning during the Covid-19 pandemic.

Table 1. Mitigation and Teachers' Strategy Resilience during Covid-19 Pandemic according to the Pedagogic Competence

No	Variables	Score				
		5 (Very Often)	4 (Often)	3 (Sometimes)	2 (Rarely)	1 (Never)
1	Designing learning by making a Learning Implementation Plan (RPP) in Distance Learning during the Covid-19 pandemic	3.5% (2)	1.8% (1)	15.8% (9)	57.9% (33)	21.1% (12)
2	Implementing various Information and Communication Technology (ICT)-based learning models prior to distance learning due to the Covid-19 pandemic	3.5% (2)	8.8% (5)	22.8% (13)	43.9% (25)	21.1% (12)
3	Implementing various Information and Communication Technology (ICT)-based learning models after distance learning is conducted due to the Covid-19 pandemic	26.3% (15)	54.4% (31)	17.5% (10)	1.8% (1)	0%
4	Applying a variety of learning media used in distance learning during the Covid-19 pandemic	0%	28.1% (16)	35.1% (20)	35.1% (20)	1.8% (1)
5	Conducting evaluation of learning in distance learning during the Covid-19 pandemic	0%	52.6% (30)	35.1% (20)	12.3% (7)	0%

Table 2. Mitigation and Teachers' Strategy Resilience during Covid-19 Pandemic according to the Personality Competence

No	Variables	Score				
		5 (Very Often)	4 (Often)	3 (Sometimes)	2 (Rarely)	1 (Never)
1	Facilitating students to develop their potential both academic and non-academic in implementing distance learning during the Covid-19 pandemic	3.5% (2)	24.6% (14)	19.3% (11)	52,6% (30)	0%
2	Providing motivation to students when implementing distance learning during the Covid-19 pandemic	56.1% (32)	40.4% (23)	3.5% (2)	0%	0%
3	Guiding students in completing tasks during the implementation of	42,1%	52,6%	5.3%	0%	0%

	distance learning during the Covid-19 pandemic	(24)	(30)	(3)		
4	Giving freedom to students in exploring learning strategies that students do during the implementation of distance learning during the Covid-19 pandemic	33.3% (19)	56.1% (32)	10.5% (6)	0%	0%
5	Conducting evaluation of learning in distance learning during the Covid-19 pandemic	36.8% (21)	56.1% (32)	7% (4)	0%	0%

Table 3. Mitigation and Teachers' Strategy Resilience during Covid-19 Pandemic according to the Social Competence

No	Variables	Score				
		5 (Very Often)	4 (Often)	3 (Sometimes)	2 (Rarely)	1 (Never)
1	Utilizing WA, Google classrooms, social media and other platforms to communicate with students before distance learning is implemented due to the Covid-19 pandemic	0%	14% (8)	21.1% (12)	61.4% (35)	3.5% (2)
2	Using the online lecturing method when delivering material during the implementation of distance learning during the Covid-19 pandemic	5.3% (3)	10.5% (6)	29.8% (17)	45.6% (26)	8.8% (5)
3	Teacher communication with students before the distance learning is implemented due to the Covid-19 pandemic	1.8% (1)	12,3% (7)	12,3% (7)	33.3% (19)	40.4% (23)
4	Communication with students during the implementation of distance learning during the Covid-19 pandemic	50.9% (29)	42,1% (24)	7% (4)	0%	0%
5	Inviting students to be active in discussions when distance learning is implemented during the Covid-19 pandemic	38.6% (22)	40.4% (23)	19.3% (11)	1.8% (1)	0%

Table 4. Mitigation and Teachers' Strategy Resilience during Covid-19 Pandemic according to the Professional Competence

No	Variables	Score				
		5 (Very Often)	4 (Often)	3 (Sometimes)	2 (Rarely)	1 (Never)
1	The quality of material comprehension by teachers decreased when carrying out distance learning during the Covid-19 pandemic	29.8% (17)	24.6% (14)	22,8% (13)	14% (8)	8.8% (5)

2	Designing various models, strategies, methods and learning media that have been prepared for use in emergency situations (such as during the Covid-19 pandemic)	1.8% (1)	3.5% (2)	19.3% (11)	33.3% (19)	42,1% (24)
3	Developing creative and innovative ICT-based learning materials for each material that will be taught during the implementation of distance learning during the Covid-19 pandemic	0%	0%	28.1% (16)	52,6% (30)	19.3% (11)
4	Adjusting quickly and easily when there are changes in policies, situations and conditions in the implementation of learning in schools during the covid-19 pandemic	3.5% (2)	17.5% (10)	29.8% (17)	49.1% (28)	0%
5	Understanding the concepts related to science, material comprehension and basic competencies of the taught subjects	56.1% (32)	38.1% (22)	5.3% (3)	0%	0%

3. Results and Discussions

3.1. *Mitigation and Teachers' Strategy Resilience during Covid-19 Pandemic according to the Pedagogic Competence*

Pedagogic competence requires teachers as educational facilitators to be able to manage learning in accordance with the applicable curriculum. The level of teacher pedagogic competence varies greatly depending on the level of professionalization at the school level and the educational background of the teacher [11]. The role of schools is of course very important for teachers in managing learning in schools, including support for school facilities. Thus, the educational background of teachers is also a factor of the broad scope of insights possessed by teachers in managing and providing variations in classroom learning.

Ideally, learning management that has to be done by the teacher includes designing learning, learning models, media and conducting evaluations. Through activities related to the implementation process of learning, the teacher can help students realize their own learning following certain methods to make learning more learner-centered [12]. The process of designing learning by making a Learning Implementation Plan (RPP) becomes an administrative completeness that must be carried out by the teacher. The fact is that some of the teachers often do not prepare the administrative requirements. From the results of research data obtained it shows that 57.9% of teachers rarely prepare lesson plans during the application of Distance learning during the Covid-19 pandemic. Based on the results of field observations, teachers still use the old lesson plans from the face-to-face learning lesson plans. This condition shows that teacher mitigation and resilience in the face of the Covid-19 pandemic is still minimal. The fundamental basis of the researchers' conclusions comes from the fact that the implementation of face-to-face activities with the implementation of online distance learning in terms of models, strategies and media are two different things.

The distance-learning model applied during the Covid-19 pandemic is based on the use of Information and Communication of Technology (ICT). ICT is currently a major component in implementing Learning Distance in Indonesia. This is because the ability of teachers to master ICT in Indonesia is not evenly distributed. Mastery of ICT and other supporting facilities often experiences various obstacles. Only 3.5% of teachers very often applied the ICT-based learning model even before the pandemic. The positive side of the covid-19 pandemic has an indirect impact on the distribution and mastery of ICT by teachers. Since the Covid-19 pandemic and the government implemented Learning Distance, 54.4% of teachers have used ICT in implementing Learning Distance. There is a movement for significant changes towards a positive direction from the covid-19 pandemic in terms of the use of ICT in learning.

The use of ICT in learning by teachers still needs to be developed and combined with learning media. The combination of ICT with learning media will make a big contribution in the field of education if it can continue to be developed. Based on the results of research, 28.1% of teachers often use varied learning media during the Covid-19 pandemic. This percentage is very low considering that now the world has entered into the revolutionary order 4.0 where everything related to ICT has been integrated with other media. Another interesting thing based on the research findings is that 52.6% of teachers often carry out learning evaluations during the Covid-19 pandemic. This relatively high percentage of evaluations shows that learning in Indonesia is still result-oriented, not process-oriented.

3.2. *Mitigation and Teachers' Strategy Resilience during Covid-19 Pandemic according to the Personality Competence*

The state of the Covid-19 disaster emergency demands the readiness and maturity of a teacher's personality in educating students. The psychological readiness factor of teachers greatly influences the learning motivation of students during the Covid-19 pandemic. Based on data compiled from WHO, most of the findings that have been published regarding the effects of a pandemic, the emergence of Covid-19 affect a fairly high increase in stress, PTSD, and psychological stress in most of the world's people [13]. The disturbances that arise will affect the quality of a person in carrying out their daily routine without exception to the teaching profession. Therefore, the teacher as a facilitator of students must have a strong personality readiness in order to continue to encourage students to develop.

In fact, the facilitation provided by teachers to students in developing academic and non-academic potential is still minimal, especially during the Covid-19 pandemic. The data from the research results prove that 57.9% of teachers said they rarely acted as facilitators in developing the potential of students in implementing distance learning during the Covid-19 pandemic. The limitations of space, time and teacher personnel are the reasons for the low role of teachers in developing the potential of students. In addition, the covid-19 pandemic made it difficult for teachers to adjust to changes in the direction of learning in schools. For example, it is difficult to transform face-to-face into online learning.

If course it is not only the teachers that feel the adjustment of the learning model during the Covid-19 pandemic, but also the students. Based on Circular Number 3 of 2020, the government advised the Department of Education to consider suspending the implementation of teaching and learning activities in schools in order to reduce the risk of spreading the covid-19 virus [14]. However, activities of staying at home and isolating oneself that last for a long time increase the risk of a depressed mood and lead to stress and anxiety in students [15]. In this condition, teacher motivation to students is very much needed in carrying out distance learning during the Covid-19 pandemic. The research data shows that 56.1% of teachers very often motivate students to remain enthusiastic in carrying out learning even in situations and conditions that are different from normal conditions. However, this does not rule out other factors that affect the low motivation of students to take part in learning during the pandemic.

The teacher's effort to guide students in completing tasks during the implementation of distance learning during the Covid-19 pandemic is also quite good. About 52.6% of the 57 respondents stated that they often provide assistance both individually and in groups to students to complete the tasks given by their respective teachers. The guidance carried out by the teacher is not conducted since it is important to consider the health protocols, therefore they use social media assistance such as WA, Instagram, Google Classroom, school's LMS and several other platforms.

Teacher control of students during the Covid-19 pandemic was very limited. Thus, a broad scope of leisure must be given to students related to their respective learning styles during the pandemic. From the respondents, 56.1% of teachers often gave freedom to students in exploring the learning strategies during the implementation of distance learning during the Covid-19 pandemic. Meanwhile, the questionnaire data showed that 56.1% of teachers often conducted learning evaluations. This situation shows that the freedom to explore the learning strategies of students given by the teacher makes the teacher integrated the learning style into teacher activities in evaluating the learning.

3.3. *Mitigation and Teachers' Strategy Resilience during Covid-19 Pandemic according to the Social Competence*

The implementation of education carried out by the teacher is divided into the delivery of theory and practice. A professional teacher needs social principles in carrying out learning including communication therein [16]. The communication aspect in the implementation of learning cannot be eliminated, thus the learning must be delivered with the right method to students. Changes in situations and conditions during the Covid-19 pandemic from delivering direct (face-to-face) learning to learning with virtual spaces encourage changes in teacher communication styles with students. Before the Covid-19 pandemic, 61.4% of teachers rarely used WhatsApp, Google classroom, social media and other platforms to communicate with students. Meanwhile, at this time almost all teachers in Indonesia depend on these media to establish communication with students. The communication that has been established due to Covid-19 has had a good impact on increasing the capacity of teachers in the development of the world of science and technology.

Another favorable condition is the limited space and time, causes learning to no longer be text-oriented. This is evidenced by the results of the study which show that only 5.3% of teachers still use the lecturing method when implementing Learning Distance during the Covid-19 pandemic. Then, in terms of teacher communication with students, it is getting more intense during the Covid-19 pandemic. Before the pandemic, 40.4% of teachers never communicated with students except when they were at school. The facts in the field today, 50.9% of teachers and students cultivate a quite loose communication by using various media platforms to establish communication. Even though there are differences when communicating directly through digital platforms, 40.4% of teachers still invite students to be active in discussions when the implementation of distance learning is implemented during the Covid-19 pandemic.

3.4. *Mitigation and Teachers' Strategy Resilience during Covid-19 Pandemic according to the Professional Competence*

One of the special requirements for teacher competence according to Law Number 14 is professionalism [17]. The quality of professional teachers also determines the quality of education in Indonesia. In any situation and condition, the professionalism of teachers has to be maintained. The results of the research questionnaire showed that 29.8% of the teachers experienced a decrease in quality of material comprehension due to the Covid-19 pandemic in implementing Learning Distance. After the researchers confirmed through closed interviews with several teachers, the decline in the quality of material comprehension by the teacher was due to the adaptation conditions of the teacher in carrying out online learning. The teachers are not familiar with the use of ICT, and thus the focus of the teacher has shifted to using Learning Distance supporting tools that are carried out online. The results of this research can also be used as a reference for evaluation that the government and education stakeholders need to provide assistance, development and widespread distribution of ICT-related facilities.

Teachers' readiness as professionals in dealing with disasters is still very low. From the respondents, 42.1% of the teachers have never designed various models, strategies, methods and learning media specifically for use in emergency situations (such as during the Covid-19 pandemic). As a result, the teachers are not ready to mitigate learning, and consequently they do not have a resilience strategy in dealing with sudden disasters. Another obstacle that arises later is the lack of development of ICT-based creative and innovative learning materials, in which there was only 28.1%. Teachers who should be able to adjust quickly and easily when there are changes in policies, situations and conditions in the implementation of learning in schools are difficult to implement. Only 3.5% of teachers were able to adjust quickly during the Covid-19 pandemic. Even though the Covid-19 pandemic has reduced the quality of teacher professionalism, at least 56.1% of teachers still understand concepts related to science, material comprehension and basic competencies of the taught subjects.

3.5. *Mitigation and Teachers' Strategy Resilience during Covid-19 Pandemic*

There are four competency standards that have to be used as guidelines for teachers in carrying out learning: pedagogic, personality, social and professionalism competence. During the Covid-19 pandemic that hit Indonesia, there were drastic changes in the education sector. The implementation of education that was originally carried out face-to-face is now carried out online. Changes in educational arrangements can be seen from the following picture of research result.

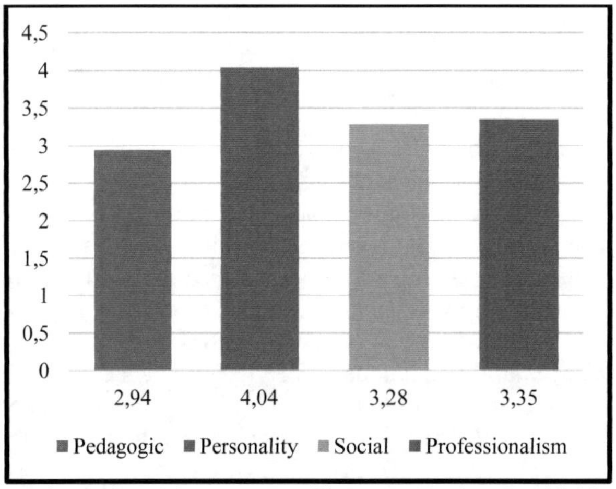

Figure 1. Mitigation and Teachers' Strategy Resilience In Facing Covid-19 Pandemic According to Teacher's Competency Standards

The bar chart image above explains that of the four teacher competency standards, the results of the research questionnaire show that pedagogic competence gets the lowest average score. The average pedagogic, social and professional scores, respectively, are 2.94, 3.28, and 3.35. They are in the range of $2.60 < X \leq 3.40$, therefore are included in the moderate category. Meanwhile, the average score for personality competence is 4.04 in the range of $3.40 < X \leq 4.20$ which is included in the good category.

4. Conclusion

The Covid-19 pandemic disaster changed the order of life in Indonesia, without exception in the field of education. Pandemic conditions and situations require education practitioners to make changes and adjustments quickly in the continuity of learning in schools. However, the adaptation to education disaster emergency conditions must still be implemented in accordance with teacher competency standards. Thus, the mitigation and teachers' strategies resilience as education implementers must be prepared in advance in order to avoid shocks or disturbances both mentally and physically when facing an emergency. Based on the results of observations, the researchers found constraints related to teacher readiness in carrying out distance learning, especially regarding the use of ICT-based tools. Furthermore, the researchers also observed that the teachers were less professional in their respective professions, as evidenced by their unpreparedness in making changes to learning models, strategies, and the use of learning media.

Based on the observational questionnaire, it was found that the teacher's pedagogic, social, and professional competence had decreased during the implementation of distance learning due to the Covid-19 pandemic. The three of them fall into the moderate category range according to the teacher competency standards in Law Number 14 of 2005. In contrast to the three competencies, the personal competence of teachers based on the research results is in the good category. This also shows that the situation is quite good where teachers can control their personalities to make sure they can continue to motivate students to learn during the Covid-19 pandemic.

5. Acknowledgement

The authors would like to thank all of subject teachers from various provinces who have shared their teaching experiences as materials for writing this paper. Also, the authors thank the lecturers and all of participant for discussion during data analysis.

References

[1] K. Khalil, N. M. Ardoin, and D. Wojcik, "Social learning within a community of practice: Investigating interactions about evaluation among zoo education professionals," *Eval. Program Plann.*, vol. 61, pp. 45–54, 2017.

[2] M. P. A. Murphy, "COVID-19 and emergency eLearning: Consequences of the securitization of higher education for post-pandemic pedagogy," *Contemp. Secur. Policy*, vol. 41, Number. 3, pp. 492–505, 2020.

[3] J. Dippong, "Status and vocal accommodation in small groups," *Sociol. Sci.*, vol. 7, pp. 222–241, 2020.

[4] *Surat Edaran Menteri Pendidikan dan Kebudayaan Nomor 4*. 2020, pp. 1–3.

[5] S. I. A. Dwiningrum, "Disaster Prevention and Management: An International Journal," *Emerald Insight*, vol. 26, Number. 4, pp. 437–451, 2017.

[6] M. P. A. Murphy, "COVID-19 and emergency eLearning : Consequences of the securitization of higher education for post-pandemic pedagogy," *Contemp. Secur. Policy*, Number. 30 April 2020, pp. 1–14, 2020.

[7] U. March *et al.*, "Considerations for Covid-19 Preparedness and Response To U.S. Schools of Nursing.," *Syllabus (American Assoc. Coll. Nursing)*, vol. 46, Number 2, pp. 1–3, 2020.

[8] O. Daramola, O. Odunsi, and O. Olowoporoku, "The corridor to survival: Assessment of disaster management literacy in a developing country," *Environ. Qual. Manag.*, vol. 27, Number. 2, pp. 15–24, 2017.

[9] E. Beaunoyer, S. Dupéré, and M. J. Guitton, "COVID-19 and digital inequalities: Reciprocal impacts and mitigation strategies," *Comput. Human Behav.*, vol. 111. April, 2020.

[10] A. Amri, D. K. Bird, K. Ronan, K. Haynes, and B. Towers, "Disaster Risk Reduction education in Indonesia: Challenges and Recommendations for Scaling up," *Nat. Hazards Earth Syst. Sci. Discuss.*, Number. February, pp. 1–28, 2016.

[11] L. Ezechil and P. Coman, "Analysis of didacticians' psycho-pedagogical competences," *Procedia - Soc. Behav. Sci.*, vol. 33, pp. 233–237, 2012.

[12] E. Topbaş and R. P. Şahbaz, "The views of tourism management teacher program students on the five step learning strategy: The example of the national and regional planning tourism course," *Procedia - Soc. Behav. Sci.*, vol. 2, Number 2, pp. 4643–4649, 2010.

[13] A. M. Lee *et al.*, "Stress and psychological distress among SARS survivors 1 year after the outbreak," *Can. J. Psychiatry*, vol. 52, Number. 4, pp. 233–240, 2007.

[14] *Surat Edaran Menteri Pendidikan Republik Indonesia Nomor 3 Tentang Pencegahan Corona Virus Disease (Covid-19) pada Satuan Pendidikan*. Indonesia, 2020, pp. 1–5.

[15] Z. H. Wang *et al.*, "Prevalence of anxiety and depression symptom, and the demands for psychological knowledge and interventions in college students during COVID-19 epidemic: A

large cross-sectional study," *J. Affect. Disord.*, vol. 275, Number. 1023, pp. 188–193, 2020.
[16] L. Mâţă, D. Cmeciu, and R. M. Ghiaţău, "A Reference Framework of Pedagogical Competences of Language Teachers in the Initial Training Programmes," *Procedia - Soc. Behav. Sci.*, vol. 93, pp. 648–653, 2013.
[17] *Undang-Undang Republik Indonesia Nomor 14 Tentang Guru dan Dosen*. Indonesia, 2005, pp. 1–84.

Identification of Spatial Clusters of COVID-19 in Yogyakarta using Moran's Index

Dyah Respati Suryo Sumunar[1], Nursida Arif[2], Nurul Khotimah[3]

[1,2,3] Department of Geography Education, Faculty of Social Science, Universitas Negeri Yogyakarta

dyah_respati@uny.ac.id

Abstract. Various studies on the spreading pattern of COVID-19 were carried out because, every day, there are always new progresses from scientists about this virus, including its unique characteristics in each region with different geographical conditions. The objective of this research is to find the spatial distribution pattern of COVID-19 and to identify the climate parameters influencing it in the Yogyakarta. This study used Moran's Index in analyzing COVID-19 distribution pattern. The results of the analysis show that out of 5 regencies/cities in Yogyakarta, Sleman Regency has the highest average spread of COVID-19 or an average infection rate of 1.09, followed by Bantul with 0.75. Kulonprogo is the regency with the lowest number of cases. The result of Moran's index of 0.32 means that there are correlation between cases.The correlation between temperature and infection rate is R2 = 0.05, while the correlation between humidity and infection rate is R2 = 0.003. The spread of COVID-19 in Yogyakarta is more influenced by mobility or interactions between infected and uninfected individuals.

1. Introduction

Corona Virus Disease 19 (COVID-19) is part of a type of coronavirus that attacks the immunity of the human body and can cause fatality. Preventive efforts were not only carried out by the affected areas, but also several other countries that were not affected, such as closing tourist access and closing several flight routes. It is due to the findings that the movement of humans from one place to another could trigger the transmission of COVID-19 [1,2]. In Indonesia, although several preventive steps have been taken by the government, such as quarantine at home, wearing masks, social distancing, physical distancing and Large-Scale Social Restrictions, the number of confirmed cases tends to increase [3]. Therefore, studies on the spread of COVID-19 continue to be carried out by scientists through various scientific disciplines. This prompted geographers to assess the ecological factors that influence the transmission and survival of the virus although there is no agreement among experts whether geographical factors, such as climate, affect the speed of transmission of COVID-19. There is no supporting evidence that if the weather is warmer, the number of cases will decrease, as happened in Brazil with tropical temperatures ranging from 16.8-27.4° C, which has a negative linear relationship with the number of confirmed cases [4]. In contrast to several previous researchers, some studies have shown that temperature has a positive relationship with the transmission of the coronavirus [5-7]. The risk of virus increase is greater at lower temperatures because it can increase suspended particles in the atmosphere [8]. An open environment that has high temperature and humidity is an environmental condition that is less ideal for the spread of coronavirus cases. On the other hand, low temperature and

Content from this work may be used under the terms of the Creative Commons Attribution 3.0 licence. Any further distribution of this work must maintain attribution to the author(s) and the title of the work, journal citation and DOI.
Published under licence by IOP Publishing Ltd

low humidity can facilitate the transmission of SARS coronavirus [9-11]. The aim of the study is to observe the COVID-19 spatial distribution pattern and to identify geographical parameters that influence the transmission of COVID-19. The case study was conducted in Yogyakarta, which is one of the provinces with fairly high mobility in Indonesia. Yogyakarta is a tourism and student city where migrants come from various regions [12-14].

This study uses geographic information system in the analysis of covid distribution patterns as well as analysis of influencing climatic factors. Currently, Geographic Information System (GIS) has become a vital tool in analyzing and visualizing the spread of COVID-19 [15].The development of GIS technology is able to present real time data related to the progress of the pandemic [16]. GIS plays an important role in the visualization and tracking of confirmed cases, prediction of regional transmission, as well as spatial segmentation of risk levels and prevention [15,16,17]. Ecological variables in various locations, such as surface temperature, often form patterns like gradients or clusters due to the reactions and interactions of geological, climatic, topographic and biological factors [20]. Therefore, these patterns can be analyzed using the spatial autocorrelation tool in ArcGIS.

Spatial autocorrelation is measured by scalar statistics, which are the function of the values observed in each plot and the values on neighboring plots [21]. Among several spatial autocorrelation methods, such as Getis-Ord General G [22,23], incremental spatial autocorrelation [24], multi-distance spatial cluster analysis [25,26], the Moran's Index is a widely used method to evaluate distribution patterns [20,27,28]. Moran's Index helped in validating statistically identified patterns [29]. This study is expected to support researchers and the government in identifying the pattern of the spread of COVID-19 in Yogyakarta from a geographic and environmental perspective.

2. Methodology

2.1. Study Area

Yogyakarta is located at 7°47′S and 110°22′E (Figure 1). It has an area of 3,133.15 km2 with a population of 3,594,290 [30]. Geographically, Yogyakarta is located in the middle of Java Island. There is Merapi Volcano in the north, which is one of the active volcanoes in Indonesia, and there are coastal plains in the south which located on the edge of the Indian Ocean. Administratively, Yogyakarta is located between Klaten Regency in the northeast, Wonogiri Regency in the southeast, Purworejo Regency in the west, Magelang Regency in the northwest and the Indian Ocean in the South. Yogyakarta consists of four regencies and 1 municipality, namely Kulonprogo, Bantul, Gunung Kidul, Sleman and Yogyakarta City.

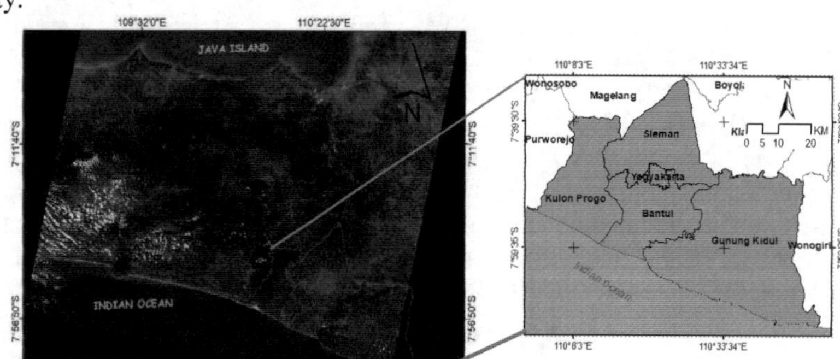

Figure 1. Study area

2.2. Moran's Index

Moran's I was used as the first measure of spatial autocorrelation. The values range from -1 to 1. A value of "1" means perfect positive spatial autocorrelation, while a value of "-1" indicates perfect negative spatial autocorrelation, and a value of "0" means a random spatial pattern [31]. A high positive Moran's I value implies the target value is similar to its environment and its location is spatially clustered, which includes the high-high clusters (high value in high-value environments) and low clusters (low value in low-value environments). If the value of Moran's I is negative, it means that it is a spatially potential outlier or different from the surrounding values [29]. Spatial autocorrelation means that there is a dependence between variable values at neighboring or proximal locations or a systematic pattern of variable values across locations on the map due to common underlying factors [32].

3. Results and Discussion

The first confirmed case in Yogyakarta was on March 13, 2020 in Sleman Regency. It then quickly spread to other areas in Yogyakarta. This research observed at the distribution of COVID-19 cases until June 16, 2020 throughout the regencies/cities in Yogyakarta (Table 1).

Tabel 1. Distribution of key variables by Sub-district

City	Latitude&longitude	Population	Study period (days)	Total cases	Total deaths	Average tenperature	Average humidity	Inf. Rate
Bantul	7.88461°S 110.33411°E	974.211	20/03-16/06 (88)	66	2	32,01	66,67	0.75
Gunung Kidul	7.96668°S 110.602561°E	768.215	25/03-16/06 (84)	50	1	31,24	61,18	0.59
Kulonprogo	7.645°S 110.0269°E	471.793	26/03-16/06 (85)	11	0	29,51	68,63	0.12
Sleman	7.68167°S 110.32333°E	1.068.370	13/03-16/06 (95)	104	4	30,87	69,12	1.09
Kota Yogyakarta	7° 48′5″ S and 110 ° 21′52″ E	412.487	18/03-16/06 (90)	33	1	32,72	69,41	0.36

*Inf. Rate = Number of infected/ Days of infection [32]

Table 1 shows the regency with the largest population, namely Sleman, is directly proportional to the highest number of COVID-19 cases, namely 104 cases. Kulonprogo Regency has the least number of cases and a smaller population than the other four districts. More people are likely to be infected with the virus in densely populated areas and it has the potential for active spread of COVID-19 to other areas [34]. In contrast, Yogyakarta City has more cases than Kulonprogo even though it has a smaller population. In other studies, population density is not an important factor in the spread of COVID-19 [17]. The climate data in Table 1, namely temperature and humidity, are the results of field measurements carried out in June 2019. The results of the analysis of the two climate parameters have no correlation with the number of infections or the number of the spread of COVID-19 in Yogyakarta. It is presented in Figure 2 and Figure 3. Several previous studies have shown the same thing where temperature and humidity have a negative correlation with the spread of COVID-19 [5,33,35].

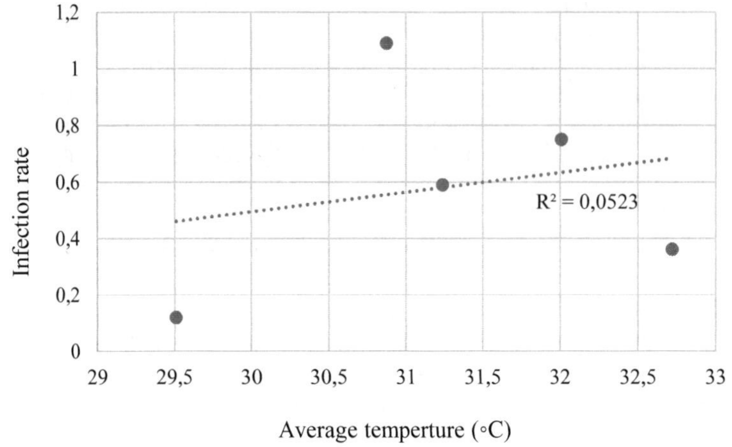

Figure 2. Correlation of average temperature and infection rate in regencies/cities in Yogyakarta

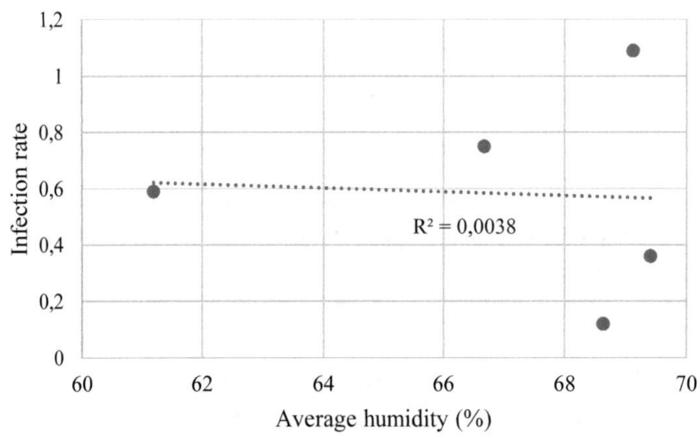

Figure 3. Correlation of average humidity and infection rate in regencies/cities in Yogyakarta

Figure 2 shows that the correlation between temperature and the infection rate is very weak (R2 = 0.05), similarly with the correlation for humidity and infection rate (R2 = 0.003). This means that the climate does not have a significant effect on the spread of COVID-19. However, ecological variables such as temperature and humidity which are measured in various locations can be considered as random processes with a certain degree of correlation with itself in space so that it becomes a challenge for conducting statistical tests for correlation with other variables [20]. Spatially, the humidity and temperature maps in Yogyakarta are presented in Figure 4 and Figure 5. When viewed from the COVID-19 distribution map (Figure 6), it shows that the distribution of COVID-19 is clustered in adjacent districts. In Gunungkidul Regency, the spread of COVID-19 is clustered in three adjacent sub-districts, namely Wonosari, Semanu and Karangmojo Districts. Likewise in other districts, it clustered in adjacent sub-districts.

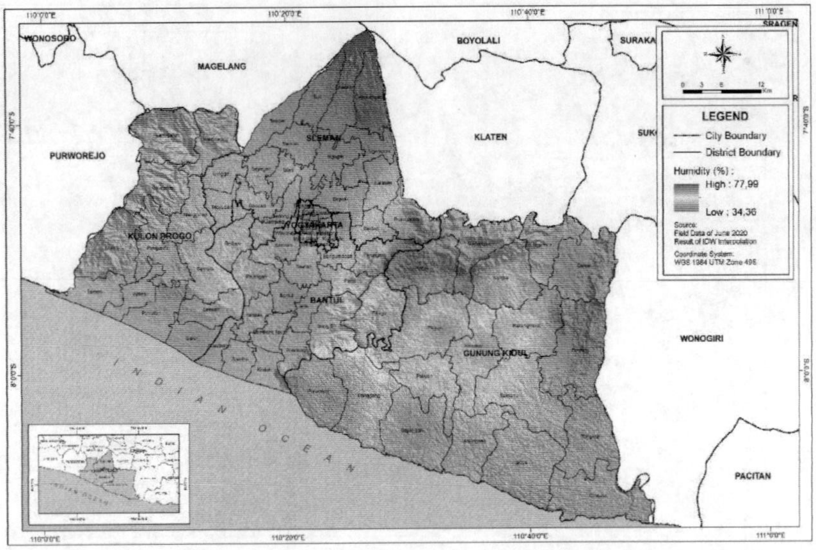

Figure 4. Map of Humidity in Yogyakarta

Figure 4 shows the highest humidity level, namely 77.99%, while the lowest is 34.36%. Gunung Kidul is the regency with the lowest average humidity compared to Bantul, Kulon Progo and Yogyakarta Regencies. However, on the map of COVID-19 distribution (Figure 6), Yogyakarta City, parts of Sleman and parts of Bantul, which tend to have high humidity, are the COVID-19 clusters.

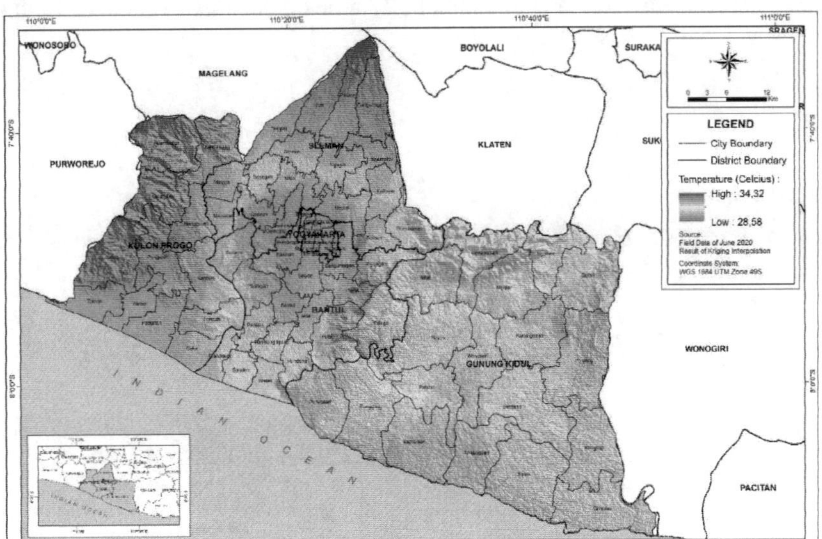

Figure 5. Map of Temperature in Yogyakarta

Figure 5 shows the high temperatures are distributed in Yogyakarta City, part of Sleman and parts of Bantul. Kulonprogo and northern Slemants have lower temperatures. Spatially, the temperature distribution pattern also has a reversed pattern to the spatial distribution of COVID-19 in Yogyakarta (Figure 6). The Yogyakarta City and Bantul, which have higher temperatures, are actually the COVID-19 clusters.

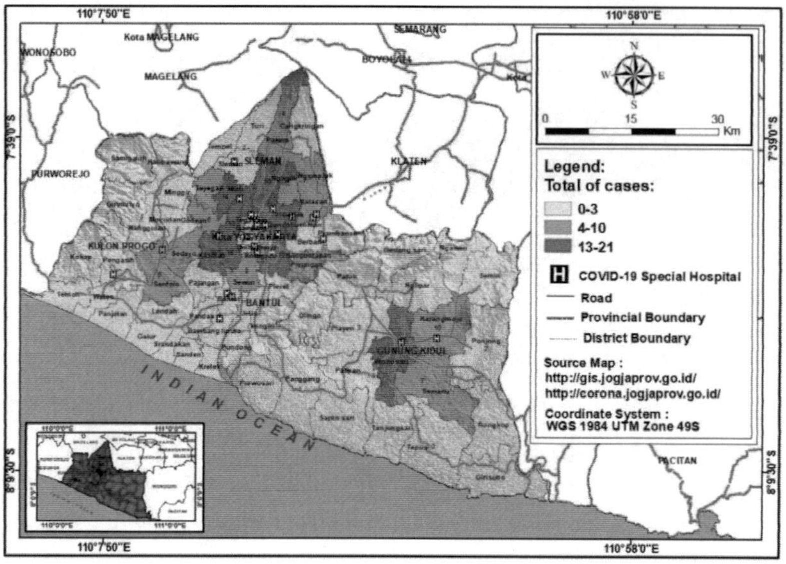

Figure 6. Map of COVID-19 Positive Patients in Yogykarta (Data 16 June 2020)

The results of Figure 6 were analyzed using Moran's Index (Figure 7) to find the spatial correlation between areas affected by COVID-19. The results show Moran's I index value of 0.32, which is in the range of 0<I≤ and indicates a positive spatial autocorrelation. It can be concluded that between regions affected by COVID-19, they have a similar or almost clustered data pattern [34]. This shows that the transmission of COVID-19 is more due to the mobility or movement of individuals from one place to another which causes interactions to form clusters.

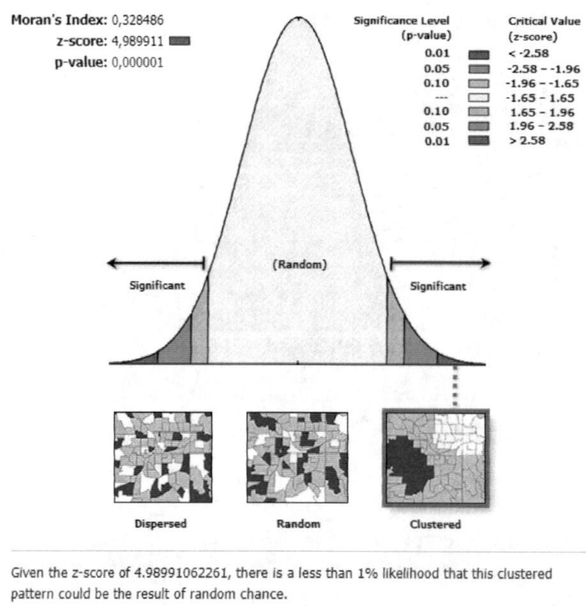

Figure 7. Curve of Moran's Index

Figure 7 shows the Moran's Index value of 0.328486, indicating the tendency for case clustering. The farther the statistic is from the 0 value, the stronger the spatial dependence [34]. The Z-score is 4.989911 and less than 1% which may indicate that the clustered pattern could be the result of random chance, so it has a confidence value of 99%.

4. Conclusion

The spatial distribution of COVID-19 in Yogyakarta, based on the results of Moran's Index analysis, is clustered, where the affected areas are inter-related. COVID-19 cases in Yogyakarta are not evenly distributed across regions but form clusters in adjacent areas. This shows that the transmission of COVID-19 is influenced by interactions between individuals in one affected area and then there is mobility so that the cases will form new clusters in other areas. The climatic factors, namely humidity and temperature, do not show a significant effect on the spread of COVID-19. In some areas, such as Yogyakarta City and parts of southern Sleman Regency which have high temperatures, the number of COVID-19 cases is higher than in areas with lower temperatures.

Acknowledgments

The authors would like to thank to the Institute for Research and Community Service, Yogyakarta State University for providing financial support for this research.

References

[1] Shereen M A, Khan S, Kazmi A, Bashir N, and Siddique R, COVID-19 infection: Origin, transmission, and characteristics of human coronaviruses, *J. Adv. Res.*, vol. **24**, pp. 91–98, 2020

[2] Sasidharan M, Singh A, Torbaghan M E, and Parlikad A K, A vulnerability-based approach to human-mobility reduction for countering COVID-19 transmission in London while considering local air quality, *Sci. Total Environ.*, vol. **741**, p. 140515, 2020.

[3] Djalante R *et al,* Review and analysis of current responses to COVID-19 in Indonesia: Period of January to March 2020, *Prog. Disaster Sci.*, vol. **6**, p. 100091, 2020.

[4] Prata D N, Rodrigues W, and Bermejo P H, Temperature significantly changes COVID-19 transmission in (sub)tropical cities of Brazil, *Sci. Total Environ.*, vol. **729**, p. 138862, 2020.

[5] Casanova L M, Jeon S, Rutala W A, Weber D J, and Sobsey M D, Effects of air temperature and relative humidity on coronavirus survival on surfaces, *Appl. Environ. Microbiol.*, vol. **76**, no. 9, pp. 2712–2717, 2010.

[6] van Doremalen N, Bushmaker T, and Munster V J, Stability of middle east respiratory syndrome coronavirus (MERS-CoV) under different environmental conditions, *Eurosurveillance*, vol. **18**, no. 38, pp. 1–4, 2013.

[7] Xie J and Zhu Y, Association between ambient temperature and COVID-19 infection in 122 cities from China, *Sci. Total Environ.*, vol. **724**, p. 138201, 2020.

[8] Qi H *et al.*, COVID-19 transmission in Mainland China is associated with temperature and humidity: A time-series analysis," *Sci. Total Environ.*, vol. 728, p. 138778, 2020.

[9] Sajadi M M, Habibzadeh P, Vintzileos A, Miralles-wilhelm F, and Amoroso A, This preprint research paper has not been peer reviewed. Electronic copy available at: https://ssrn.com/abstract=3550308," *Ssrn*, no. **410**, pp. 6–7, 1992.

[10] Bannister-Tyrrell M, Meyer A, Faverjon C, and Cameron A, Preliminary evidence that higher temperatures are associated with lower incidence of COVID-19, for cases reported globally up to 29th February 2020, *Nat. Sci.*, vol. **15**, no. 5, pp. 55–68, 2017.

[11] Chan K H, Peiris J S M, Lam S Y, Poon L L M, Yuen K Y, and Seto W H, The effects of temperature and relative humidity on the viability of the SARS coronavirus, *Adv. Virol.*, vol. **2011**.

[12] Cahya G A, Mahendra Y K D, and Damanik I I, Malioboro as a value of Special District of Yogyakarta City, *IOP Conf. Ser. Earth Environ. Sci.*, vol. **70**, no. 1, 2017.

[13] R. Abd Rahman and R. Ramli, "The History of Yogyakarta, and Education City," *Int. Proc.*

Econ. Dev. Res., vol. 63, no. **15**, pp. 73–77, 2013.

[14] Herliana E T, Preserving Javanese Culture through Retail Activities in Pasar Beringharjo, Yogyakarta, *Procedia - Soc. Behav. Sci.*, vol. **184**, pp. 206–213, 2015.

[15] Mollalo A, Vahedi B, and Rivera K M, GIS-based spatial modeling of COVID-19 incidence rate in the continental United States, *Sci. Total Environ.*, vol. **728**, p. 138884, 2020.

[16] Franch-Pardo L, Napoletano B M, Rosete-Verges F, and Billa L, Spatial analysis and GIS in the study of COVID-19. A review, *Sci. Total Environ.*, vol. **739**, p. 140033, 2020.

[17] Huang R, Liu M, and Ding Y, Spatial-temporal distribution of COVID-19 in China and its prediction: A data-driven modeling analysis, *J. Infect. Dev. Ctries.*, vol. **14**, no. 3, pp. 246–253, 2020.

[18] Kumari M, Sarma K, and Sharma R, Using Moran's I and GIS to study the spatial pattern of land surface temperature in relation to land use/cover around a thermal power plant in Singrauli district, Madhya Pradesh, India, *Remote Sens. Appl. Soc. Environ.*, vol. **15**, no. May, p. 100239, 2019.

[19] Czaplewski R L, Reich R M, and Bechtold W A, Spatial Autocorrelation in Growth of Undisturbed Natural Pine Stands Across Georgia, *For. Sci.*, vol. **40**, no. 2, pp. 314–328, 1994.

[20] Peeters A *et al.*, Getis-Ord's hot- and cold-spot statistics as a basis for multivariate spatial clustering of orchard tree data, *Comput. Electron. Agric.*, vol. **111**, pp. 140–150, 2015.

[21] Sarp G and Duzgun S, Morphometric evaluation of the Afşin-Elbistan lignite basin using kernel density estimation and Getis-Ord's statistics of DEM derived indices, SE Turkey, *J. Asian Earth Sci.*, vol. **111**, pp. 819–826, 2015.

[22] Zhang Y *et al.*, "Cluster of human infections with avian influenza a (H7N9) cases: A temporal and spatial analysis," *Int. J. Environ. Res. Public Health*, vol. **12**, no. 1, pp. 816–828, 2015.

[23] Khormi H and Kumar L, "Identifying and visualizing spatial patterns and hot spots of clinically-confirmed dengue fever cases and female Aedes aegypti mosquitoes in Jeddah, Saudi Arabia., *Dengue Bulletin* vol. **35**, pp. 15–34, 2011.

[24] Arroyo-Mora J P, Svob S, Kalacska M, and Chazdon R L, Historical patterns of natural forest management in Costa Rica: The good, the bad and the Ugly, *Forests*, vol. **5**, no. 7, pp. 1777–1797, 2014.

[25] Tepanosyan G, Sahakyan L, Zhang C, and Saghatelyan A, The application of Local Moran's I to identify spatial clusters and hot spots of Pb, Mo and Ti in urban soils of Yerevan, *Appl. Geochemistry*, vol. **104**, no. October 2018, pp. 116–123, 2019.

[26] Yuan Y, Cave M, and Zhang C, Using Local Moran's I to identify contamination hotspots of rare earth elements in urban soils of London, *Appl. Geochemistry*, vol. **88**, pp. 167–178, 2018.

[27] Fu W J, Jiang P K, Zhou G M, and Zhao K L, Using Moran's i and GIS to study the spatial pattern of forest litter carbon density in a subtropical region of southeastern China, *Biogeosciences*, vol. **11**, no. 8, pp. 2401–2409, 2014.

[28] BPS-Statistics of Daerah Istimewa Yogyakarta, Daerah Istimewa Yogyakarta Province in Figures 2020, Yogyakarta, **2020**.

[29] Tu J and Xia Z G, Examining spatially varying relationships between land use and water quality using geographically weighted regression I: Model design and evaluation, *Sci. Total Environ.*, vol. **407**, no. 1, pp. 358–378, 2008.

[30] Griffith D A, Spatial Autocorrelation, A. B. T.-I. E. of H. G. (Second E. Kobayashi, Ed. Oxford: Elsevier, **2020**, pp. 355–366.

[31] Ahmadi M, Sharifi A, Dorosti S, Jafarzadeh S, Ghoushchi, and Ghanbari N, Investigation of effective climatology parameters on COVID-19 outbreak in Iran, *Sci. Total Environ.*, vol. **729**, p. 138705, 2020.

[32] Kang D, Choi H, Kim J H, and Choi J, Spatial epidemic dynamics of the COVID-19 outbreak in China, *International Journal of Infectious Diseases*, vol. **94**. pp. 96–102, 2020.

[33] Wu Y *et al.*, Effects of temperature and humidity on the daily new cases and new deaths of COVID-19 in 166 countries, *Sci. Total Environ.*, vol. **729**, pp. 1–7, 2020.

Spatial Analysis of Covid-19 Distribution: case studies in Indonesia and Malaysia

N. Khotimah[1], DRS Sumunar[2], S. Purwantara[3], N. Arif[4], N. Nayan[5]

[1,2,3,4]Department of Geography Education, Faculty of Social Sciences, Universitas Negeri Yogyakarta, Yogyakarta, Indonesia
[5]Department of Geography, Faculty of Human Sciences, Universiti Pendidikan Sultan Idris, Tanjong Malim, Perak, Malaysia

Email: [1]nurulkhotimah@uny.ac.id

Abstract. Covid-19 has had an effect on various aspects of life. This study was to compare spatial distribution of covid-19 in Indonesia and Malaysia and its prevention efforts. The design of this study is a comparative study using descriptive analysis. Research was carry out in June through September 2020, taking location in Yogyakarta Special Region, Indonesia and Perak, Malaysia. Research focuses to data on infected positive patients with covid-19. Research data uses a secondary data, which is official statistics of covid-19. Statistical data then tabulated and mapping. The administrative map of both research sites used as a basis for the production spatial distribution map of positive patients infected with covid-19. Research shows that the number of positive patients infected by covid-19 June through September 2020 in five districts of Yogyakarta Special Region, Indonesia continued to increase significantly. Judging from time, spatial factors changes in covid-19 risk zones occurred in June to September 2020. Public movement and open access to Central Java Province causes 5 districts on Yogyakarta Special Region enter the high risk zone covid-19. Compared with Perak, Malaysia it different, spatially covid-19 infected number of positive patients in Perak, six district are relatively steady and four district are increasing. The increase occurred in Kinta District of July and August 2020, Kerian District of August 2020, Kuala Kangsar District of July 2020, and Perak Tengah District of September 2020. Judging from time, spatial factors do not changes the covid-19 risk zone from June to September 2020. Public movement and open access to Kedah, Kelantan, and Selangor caused two districts in Perak enter the high risk zone covid-19, five districts enter the medium risk zone covid-19, and three district enter the low risk zone covid-19. Indonesia and Malaysia's government efforts to prevent the spread of covid-19 are almost the same, which are restrictions on several public activities, clean and healthy socialization of living behaviour in all sectors, implementation of health discipline protocol in new normal, and provide of update covid-19 information data. The difference, lies in implementation lockdown, which some areas in Indonesia include Yogyakarta Special Region do not apply on Large Scale Social Restrictions (LSSR) like Jakarta, while a full implementation of lockdown in Malaysia through Movement Control Order (MCO) including Perak, so a growing number of covid-19 cases can be more pressure.
Keyword: Spatial distribution, Covid-19, comparative study

1. Introduction

In December 2019, reports of a virus outbreak in Wuhan, China, and spread rapidly across areas of the world, including Indonesia. The virus kills more than 800 people and infects more than 70,000 in the

Content from this work may be used under the terms of the Creative Commons Attribution 3.0 licence. Any further distribution of this work must maintain attribution to the author(s) and the title of the work, journal citation and DOI.
Published under licence by IOP Publishing Ltd

first 50 days of the epidemic [12]. On February 6, 2020, the World Health Organization (WHO) officially announced the name of the virus is Corona Virus Disease 19 (covid-19), a member of the coronavirus group that can cause death. Covid-19 resembles acute respiratory syndrome such as the Severe Acute Respiratory Syndrome (SARS) outbreak in China of 2002 and the Middle-East Middle-Relief Syndrome (MERS) that first appeared in the Middle East in 2012. While covid-19, SARS, and MERS are cause by a similar strain of the virus, coronavirus, there are differences in the incubation period, the rate of contagion, and the treatment. Experts continue to study the origins of covid-19 infections [15], which are genetically classified into four major generas, Alphacoronavirus, Betacoronavirus, Gammacoronavirus, and Deltacoronavirus [9].

Covid-19 is now a strategic issue for research in all fields of science, including geography. The first reason is that the covid-19 makes major circulation in geographic space automatically stop. The covid-19 pandemic has brought financial systems to the financial crisis of national governments, European agencies, and international systems [10]. Second, because the covid-19 occupying space is spreading in more than 200 countries with the highest number of cases in the America Continent and the European Continent. Then we might conclude that the covid-19 could spread over geographic space. Geographically, climate factors show an impact on amplitude and the speed of spreading covid-19 [13]. There is also possible a factor that rests on human activity so that the covid-19 can spread.

As the number of confirmed cases covid-19 in Indonesia continues to increase, the Indonesian government designated the covid-19 epidemic as a nonnatural national disaster. The covid-19 epidemic emergency status released by the Head of the National Disaster Management Agency on January 28, 2020, and could be extend to adjust existing conditions. By providing emergency response status with covid-19 means that provincial or district governments should be ready to work 24 hours a day and expend all available resources to save their citizens from covid-19 [4]. Malaysia is not unlike Indonesia, it is also affect by covid-19 and has imposed emergency status, but the number of confirmed deaths is far below Indonesia. According to data from April 20, 2020, covid-19 deaths in Malaysia total of 89, while in Indonesia the death toll is already 590. This is possible, because Indonesia's population is greater than Malaysia.

In the face of a covid-19 patient boom, each countries apply varying efforts to the country's conditions such as geographic conditions, population size, and technological preparedness. Prevention efforts are made by both the Indonesian and Malaysia governments, including hand washing habituation, using hand sanitizer, wearing mask, social distancing, physical distancing, closing schools and campus, and quarantine of citizens in affected areas. Nevertheless, the efforts have not been able to cope with the rapid spread of the covid-19. The movement of humans from place to place because of the daily activities involved in meeting the needs of the family's life is possible as a factor so that the spread of covid-19 can be faster and less controlled.

Spatial mapping using the Geographic Information System (GIS) today is especially important in providing information about affected areas of covid-19, which this study can help to see differences in the spread of covid-19 in Indonesia and Malaysia. GIS is a manual and or computer system that is use to collect, store, manage, and produce information that has spatial or geographic references [3]. Compared with other base data systems, GIS has the advantage of presenting spatial and nonspatial information [11]. GIS component consists of hardware, software, brainware, and input. The four components that make up a system for GIS function are data input, data management, data manipulation and analysis, and output. Through the help of GIS applications, regions affected by covid-19 can easily see in a map view [1].

Since the 1960s GIS has been use to analyze, visualize and detect disease spread, even now GIS technology development can present real time data from pandemic development [2]. GIS plays a major role in visualization and tracking confirmed cases, predictions of regional transmission, spatial segmentation of risk levels and prevention. Developed a GIS based risk assessment model using three variables that is the number of cases, population migration and transport networks [16]. A model result can be a recommendation to assess areas at the highest risk level. GIS can simulate predictions of transmission of disease and infectious disease using population mobility as important parameters [14].

With a systematic analysis using GIS it is expected to help the government with covid-19 treatment. Digital maps provide a communication tool for delivering spatial distribution of covid-19 to the public and to the authorities so that efforts to prevent the spread of covid-19 in the region can be do in time.

2. Methods

The design of this study is a comparative study using descriptive analysis. The study attempts to compare spatial distribution of covid-19 in Indonesia and Malaysia and describes what communities and governments have done to prevent its spread. Research carried out from June to September 2020, taking up a research site in Yogyakarta Special Region, Indonesia and Perak, Malaysia. Yogyakarta Special Region, Indonesia consists of five district, which are Bantul District, Sleman District, Kulon Progo District, Gunungkidul District, and Yogyakarta City. Perak, Malaysia consists of 10 district, which are Kinta District, Larut and Matang District, Manjung District, Perak Hilir District, Kerian District, Batang Padang District, Kuala Kangsar District, Perak Tengah District, Hulu Perak District and Kampar District. Research are carry out in Yogyakarta Special Region, Indonesia and Perak, Malaysia with a consideration of the educational center and cultural diversity that encourage public movement into the region and accelerate the spread of covid-19.

The study focused on mapmaking that was analyzed to evaluate the objects that were mapped. The mapped object is a covid-19 positive patient at two research sites. The distribution map of infected positive patients covid-19 is a spatial model used to describe conditions in the field at the risk of the spread of covid-19 in Indonesia and Malaysia. It would represent the spatial distribution of phenomena. Geographical production of spatial models for health studies is expected to explain where, why, and what are the subjects on a region's health issue [5].

Zoning mapping activity of covid-19 spread can be makes through secondary data study, which is the official statistical data of covid-19. Data is a basic and important ingredient in mapmaking because it determines the quality of the maps produced. Dickinson (1975) in [5] presents four reasons why a data is mapped, that is: (1) by map can create a greater attraction to the objects displayed, (2) by map can clarify, simplify, and explain a significant aspect, (3) by map can highlight subject points in writing or conversation, and (4) by map can be used as a source of data for the interest.

The official statistical data of covid-19 consisting collection of number then tabulated and mapping. The administrative map of both research sites is use as a basis for the production map of the spatial distribution of positive patients infected with covid-19. Spatial distribution information on positive patients infected with covid-19 according to place and time can be uses to study factors related to the covid-19 pandemic and be helpful to help establish a program's priority for covid-19 prevention efforts.

3. Result and Discussions

3.1. Spatial distribution of Covid-19 in Indonesia and Malaysia

In this study to see spatial distribution of covid-19 in Indonesia and Malaysia conducted case studies for Yogyakarta Special Region, Indonesia and Perak, Malaysia. Yogyakarta Special Region is one of the 34 provinces of Indonesia. Yogyakarta Special Region bordered the Indian Ocean in the south and bordered Central Java Province in the north, east, and west. Yogyakarta Special Region has an area of 3,185.8 km^2 and is divide into five district. Yogyakarta Special Region for the year 2020 has 3,882,288 people and is the fourth most populating province in Indonesia. Perak is one of 14 states in Malaysia. Perak borders on Kedah in the north, Yala and Narathiwat Provinces, Thailand in the northeast, Pinang Island in the northwest, Kelantan and Pahang in the east, Selangor in the south, and the Strait of Malacca in the west. Perak has an area of 20,976 km^2 and is divide into 10 district. Perak population in 2020 is 2,510,300 people and is the fifth most populous state in Malaysia. Given the geographic and demographic conditions in both areas of researched it is possible that there is spatial distribution differences in positive patients infected with covid-19.

3.1.1. Spatial distribution of covid-19 in Yogyakarta Special Region, Indonesia

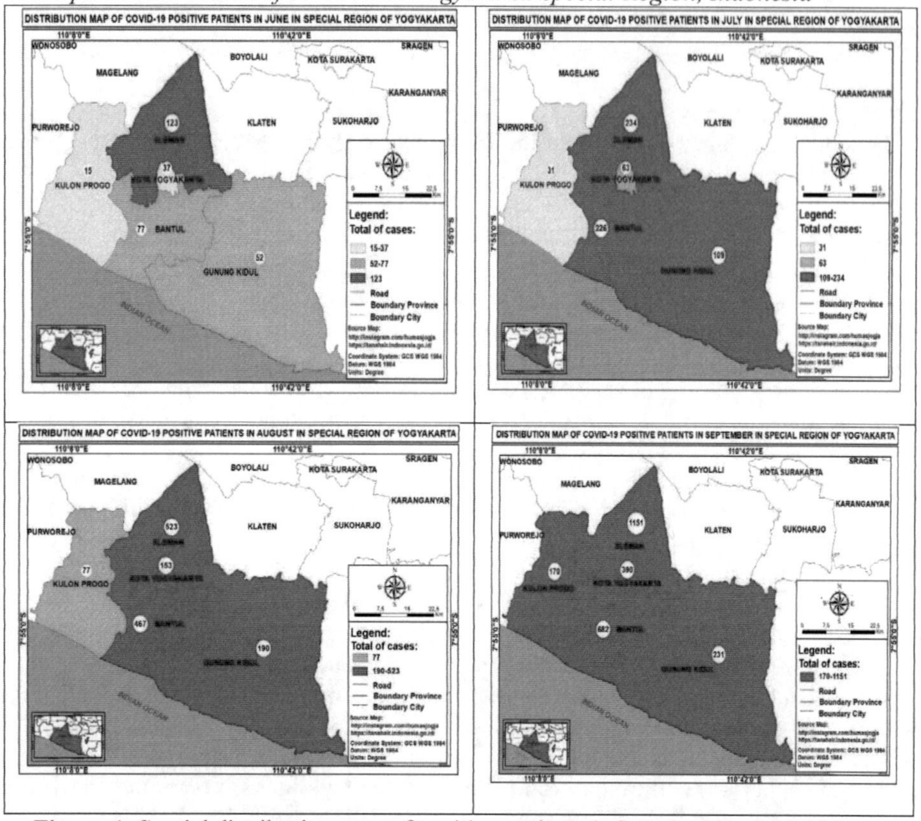

Figure 1. Spatial distribution map of positive patients infected with covid-19 in Yogyakarta Special Region, Indonesia, June to September 2020 [8]

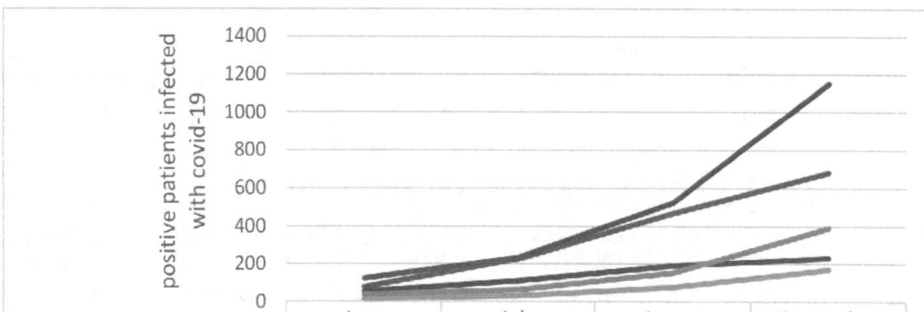

	June	July	August	September
Bantul District	77	226	467	682
Sleman District	123	234	523	1151
Kulon Progo District	15	31	77	170
Gunungkidul District	52	109	190	231
Yogyakarta City	37	63	153	390

Figure 2. Graph of development of positive patients infected with covid-19 in Yogyakarta Special Region, Indonesia, June to September 2020 [8]

From Figure 1, based on space or spatially, Yogyakarta Special Region has five districts, it is can be seen that the number of infected patients with covid-19 continues to experience significant increase in each district. According to time, a map that has been made by reference to the number of positive patients infected with covid-19 can be categorized into high risk (red zone), medium risk (orange zone), and low risk (yellow zone). Spatially, by June 2020, Sleman District entered the covid-19 high risk zone, Bantul District and Gunungkidul District enter the medium risk zone, while the Yogyakarta City and Kulon Progo District enter the low risk zone. In July 2020, Bantul District and Gunungkidul District followed Sleman District into the covid-19 high risk zone, Yogyakarta City entered the medium risk zone, and Kulonprogo District remained in the low risk zone. In August 2020, Yogyakarta City followed Sleman District, Bantul District, and Gunungkidul District entered high risk zone, while Kulonprogo District entered the medium risk zone. In September 2020, all district of Yogyakarta Special Region enter a high risk zone of covid-19. Judging from time, spatial factor changes in covid-19 risk zones occurred in June to September 2020. Public movement cause of all district of Yogyakarta Special Region to enter the high risk zone for covid-19, this is supported by open access to the Central Java Province.

From figure 2, in June to August 2020, it is can be seen that the highest number of positive patients infected with covid-19 in Yogyakarta Special Region was in Sleman District, followed by Bantul District, Gunungkidul District, Yogyakarta City, and the lowest in Kulonprogo District. However, in September 2020, the third position was replace by Yogyakarta City and Gunungkidul District in fourth position, while others district remained in their original position. The increase in the number of infected positive patients by covid-19 in all districts in Yogyakarta Special Region certainly requires a serious effort by local government to provide intervention in covid-19 prevention efforts in the region against existing health-care systems.

Each district in policy public activity intake is expect to take into account the covid-19 risk zone category. If areas enter high risk zones, most public activities should be stopped, such as the prohibiting of meetings, closing of some business activities, and making effort people to work from home and study from home. The spread rate of the virus is out of control in this zone as local transmission is happening fast and spreading widely. Divide into a medium category zone means there is a risk of transmission and potential of the virus in the direction of uncontrolled. What can be do is that the opening of limited business activities and the public are encouraged to study from home and to adopt strict health protocols when going out. For the region of a low risk zone, which means a controlled spread, despite the potential of transmission remains, so public activity may operate by applying health protocols. It is hope that a consideration of the covid-19 risk zone in policy to public activity would reduce a surge in the case. This is reinforce by the findings of [6], which suggests that the covid-19 epidemic situation in Hubei Province can be effectively controlled the basis of governmental already applied prevention. Strict quarantine has acted to prevent and control the virus.

3.1.2. Spatial distribution of covid-19 in Perak, Malaysia

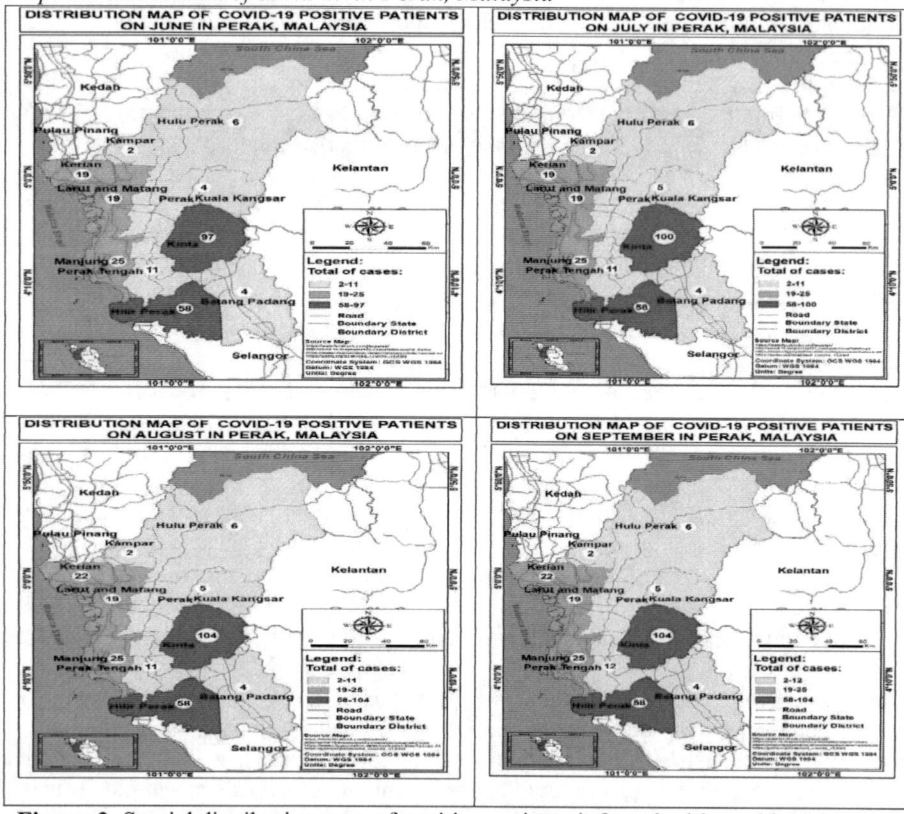

Figure 3. Spatial distribution map of positive patients infected with covid-19 in Perak, Malaysia, June to September 2020 [8]

	June	July	August	September
Kinta District	97	100	104	104
Larut and Matang District	19	19	19	19
Manjung District	25	25	25	25
Perak Hilir District	58	58	58	58
Kerian District	19	19	22	22
Batang Padang District	4	4	4	4
Kuala Kangsar District	4	5	5	5
Perak Tengah District	11	11	11	12
Hulu Perak District	6	6	6	6
Kampar District	2	2	2	2

Figure 4. Graph of development of positive patients infected with covid-19 in Perak, Malaysia, June to September 2020 [8]

From figure 3, spatially Perak consist of 10 districts, it is can be seen that the number of positive patients infected with covid-19 some are relatively steady and some are increasing. The number of positive patients infected with covid-19 is relatively steady in 6 district in Perak, which are Larut and Matang District, Manjung District, Perak Hilir District, Batang Padang District, Hulu Perak District, and Kampar District, while the increase in the number of infected patients with covid-19 occurs in 4 districts, wich are Kinta District, Kerian District, Kuala Kangsar District, and Perak Tengah District, although the increase in the number of infected patients is less significant. Public movement dominates the number of infected positive patients in Perak, supported by openness of access to the Kedah, Kelantan, and Selangor. Judging from time, spatial factors do not change the covid-19 risk zone from June to September 2020. Kinta District and Perak District both enter high risk zones; Batang Padang District, Hulu Perak District, Kampar District, Kuala Kangsar District, Perak Tengah District enters a medium risk zone; Larut and Matang District, Manjung District, Kerian District entered a low risk zone. The spatial analysis was helpful in understanding the spread of covid-19 in Perak, Malaysia. This is reinforce by the findings of [7] which suggests that spatial analysis is helpful in understanding the spread of infectious diseases, and spatial associations are key to spatial diffusion during the early stage of the covid-19 pandemic in China.

From figure 4, it is seen the increase in the number of infected positive patients in Kinta District in July and August 2020, Kerian District in August 2020, Kuala Kangsar District in July 2020, and Perak Tengah District in September 2020. Although the increase in the number of positive patients infected covid-19 in Perak is not as much in Yogyakarta Special Region, Indonesia, but the Malaysian government still has strict regulations to prevent local transmission quickly and widely spread, the one with lockdown policies. Some residents in Malaysia were arrest for conducting activities with large crowds, engaging in party arrangements, not wearing masks, not giving records to facilities, and violations because businesses operated outside the authorized hours.

According to data for development of covid-19 infections in Yogyakarta Special Region, Indonesia and Perak, Malaysia can be seen with the first case of positive people infected covid-19 in Perak on April 19, 2020, whereas Yogyakarta Special Region on March 15, 2020. The number of positive patients infected with covid-19 in Yogyakarta Special Region are 313 people (as of June 2020), 674 people (as of July 2020), 1,425 people (as of August 2020), 2,643 people (as of September 2020), while in Perak are 258 people (as of June 2020), 262 people (as of July 2020), 269 people (as of August 2020), 271 (as of September 2020). By dividing the number of infected covid-19 patients with the number of infected days since the first case [7], it is possible to report covid-19 infection rates in Yogyakarta Special Region, Indonesia and Perak, Malaysia in June through September, 2020 (Figure 5).

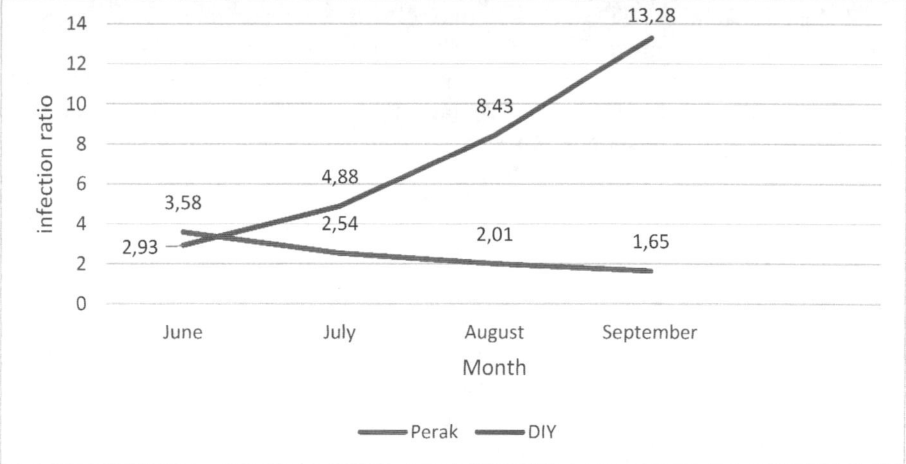

Figure 5. Graph of infection ratio of covid-19 transmission in Yogyakarta Special Region, Indonesia and Perak, Malaysia in June-September 2020 [8]

From Figure 5, it is can be seen that the infectious ratio of covid-19 in Yogyakarta Special Region, Indonesia from June through September 2020 continues to increase. The increase in the ratio of Covid-19 transmission in July 2020 by 66.6%, August 2020 by 72.7%, and September 2020 by 57.5%. It is can be concluded that in Yogyakarta Special Region during June to September 2020, there has been an increase in covid-19 transmission rate. This condition are different than in Perak, Malaysia, where covid-19 infection rates from June to September 2020 tend to decline. A decrease in covid-19 infection ratio in July 2020 by 29.1 percent, August 2020 by 20.9%, and by September 2020 by 17.9%. It is possible because lockdown policies in all of Malaysia, so the Perak is particularly effective in suppressing covid-19 case spread, in contrast Yogyakarta Special Region that does not implement a Large Scale Social Restrictions (LSSR) with some consideration so that the covid-19 case spread is significantly significant.

To make clear the conditions associated with covid-19 infections in Yogyakarta Special Region, Indonesia and Perak, Malaysia, the following is the data with a percentage of people infected by covid-19 in June through September 2020 (Figure 6). Percentage of people infected by covid-19 can be calculate by dividing the number of infected covid-19 patients with the total population [7]. The total population in Indonesia in 2020 was 3,882,288 people, while in Malaysia it was 2,510,300 people.

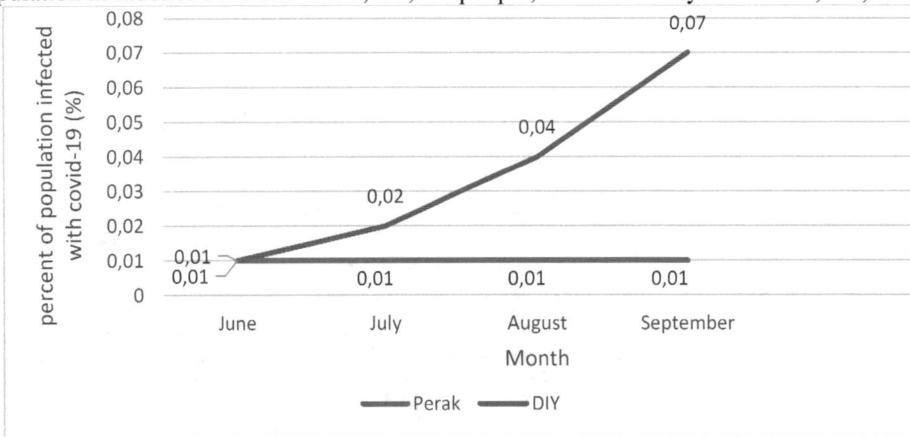

Figure 6. Percentage of population infected with covid-19 in Yogyakarta Special Region, Indonesia and Perak, Malaysia in June-September 2020 [8]

From figure 6 it is seen that the percentage of the population infected with covid-19 in Yogyakarta Special Region, Indonesia from June through September 2020 continues to increase. Increase in percentage of infected population covid-19 in July 2020 by 100.0%, August 2020 by 100.0%, and September 2020 by 75%. Different conditions occur in Perak, Malaysia, where the percentage of infected covid-19 is relatively constant. It shows setting off lockdown for all parts of the state in Malaysia, by prohibiting people out of doors unless there is urgent interest or some emergency, including in the Perak it is very effective to suppress covid-19 surmount in Malaysia, it is different from Yogyakarta Special Region for not applying the LSSR application with several considerations.

3.2. Efforts to Prevent the Spread of Covid-19 in Indonesia and in Malaysia

Similar efforts are made by the Indonesian and Malaysia governments on the prevention of the spread of covid-19, which are (1) restrictions on some public activities, such as the implementation of some work from office (wfo) and some work from home (wfh), study from home, and business space operating restrictions, (2) clean and healthy living behaviour socialization, both for family, school, environment, workplace, and health. Socialization by (a) regularly applying hand washing with soap and running water or by hand sanitizer, (b) always keep a safe distance from others who are coughing or sneezing, (c) masks are recommended if physical restrictions are not possible, (d) forbidden to touch the eyes, nose or mouth if you do not wash your hands, (e) coughing and sneezing please cover your mouth and nose with arms or tissues, (f) staying home when not feeling well, (g) if a fever, cough, or shortness of breath immediately seek medical attention, (h) eating foods is nutritious and balanced, (i) be diligent in exercising and sufficient rest, (3) implementation of disciplinary health protocols as key to new normal, the discipline of health protocols continues to be socialized so that people can always be productive and safe from the spread of covid-19, (4) provide information on the spread of covid-19 is updated with GIS help. GIS not only helps provide the covid-19 spread information but can be used to evaluate the quality, effectiveness, and accessibility of covid-19 treatment services for public health, such as the presence of hospitals. GIS can also provide a list of territory potential and demographic characteristics of the people, with a possible match between the number of people and existing health-care facilities.

The difference between Indonesian and Malaysian governments to prevent the spread of covid-19 follow the WHO's decision on January 30, 2020 regarding covid-19 as a global public health emergency, so with the first positive case on March 15, 2020, Yogyakarta Special Region will immediately establish the status of the covid-19 emergency response. The Government of Yogyakarta Special Region immediately works by mobilizing all available resources to save its people from the covid-19 disease, however, Yogyakarta Special Region does not carry out LSSR like Jakarta with several considerations. Unlike Yogyakarta Special Region, to prevent the spread of covid-19 where the first case occurred on April 19, 2020 in Perak, so Perak Government imposed a lockdown. In this case, there are restrictions on public movement to prevent the spread of covid-19, prohibit travel out of Perak or Malaysia and to close all business activities, except public services and grocery stores that remain in operation. Full lockdown is apply to areas that fall into the covid-19 category of high-risk zones so the area is temporarily locked down and society is not allowed in or out of the region. During a lockdown, all basic needs are guarantee by the kingdom. This is also applies to students living in campus dorms, both local and international students. Movement Control Order (MCO) became Malaysia's main weapon against covid-19. In this cased the MCO is supported with a clear legal basis, if public offense is punishable by 6 months imprisonment or a fine of RM 1000. Police and soldiers in Malaysia deployed to guard MCO policies.

4. Conclusion and Recommendation

Based on the results and discussions, the conclusions are as follows:

1. Spatially, the number of positive patients infected with covid-19 June to September 2020 in five district in Yogyakarta Special Province, Indonesia (Sleman District, Bantul District, Gunungkidul District, Kulonprogo District, Yogyakarta City) continued to increase significantly, resulting all districts entering the high risk zone covid-19. Judging from time, spatial factors changes in covid-19 risk zones occurred in June to September 2020. Public movement cause of all district of Yogyakarta Special Region to enter the high risk zone for covid-19, this is supported by open access to the Central Java Province. Compared with Perak, Malaysia it different, spatially covid-19 infected number of positive patients in Perak, six district are relatively steady which are Larut and Matang District, Manjung District, Perak Hilir District, Batang Padang District, Hulu Perak District, Kampar District while four district are increasing, which are Kinta District of July and August 2020, Kerian District of August 2020, Kuala Kangsar District of July 2020, Perak Tengah District of September 2020. Judging from time, spatial factors do not change the covid-19 risk zone from June to September 2020. Public movement and open access to Kedah, Kelantan, and Selangor causes two district in Perak which are Kinta District and Perak District enter the high risk zone covid-19, five district which are Batang Padang District, Hulu Perak District, Kampar District, Kuala Kangsar District, Perak Tengah District enter the medium risk zone covid-19, and three district which are Larut and Matang District, Manjung District, Kerian District enter the low risk zone covid-19.
2. Indonesia and Malaysia's government efforts to prevent the spread of covid-19 are almost the same, which are restrictions on several public activities, clean and healthy socialization of living behaviour in all sectors, implementation of health discipline protocol in new normal, and provide of update covid-19 information data. The difference, lies in implementation lockdown, which some areas in Indonesia include Yogyakarta Special Region do not apply on Large Scale Social Restrictions

(LSSR) like Jakarta, while a full implementation of lockdown in Malaysia through Movement Control Order (MCO) including Perak, so a growing number of covid-19 cases can be more pressure.

Recommendations in relation to the research findings are: a sense of crisis is needed by applying caution clean and healthy lifestyle, increasing public awareness through socialization, preparing additional health services for handling covid-19 actions, involving all elements in the region for cooperation in handling covid-19 down to the lowest government structure, paying special attention to the poor who are vulnerable to the economic impact and transmission of covid-19, provide information on the development of covid-19 by utilizing the official website and other social media, as well as providing support and appreciation for medical personnel.

Acknowledgments

The authors would like to thank Faculty of Social Science, Universitas Negeri Yogyakarta, which has facilitated excellent international cooperation research activities.

References

[1] Azhar, A. (2020). Sistem Informasi Geografis (SIG) dapat Memetakan Sebaran Covid-19. 17 April 2020. [Online]. Available: https://kptk.or.id/artikel/2020/04/17/1122-sistem-informasi-geografis-sig-dapat-memetakan-sebaran-covid19.html.

[2] Boulos, M. N. K and Geraghty, E. M. (2020). Geographical Tracking and Mapping Disease COVID-19/Severe Acute Respiratory Syndrome Coronavirus 2 (SARS-COV-2) Epidemic and Associated Events Around The World: How 21st Century GIS Technologies are Supporting The Global Fight Against Outbreaks and Epidemics. International Journal of Health Geographics 19: 8.

[3] Danoedoro, P. (2012). Pengantar Penginderaan Jauh Digital. Yogyakarta: Andi.

[4] Halidi, R., Varwati, L. (2020). Ini Alasan Mengapa Wabah Covid-19 Ditetapkan sebagai Bencana Nasional, 17 Maret 2020. [Online]. Available: https://www.suara.com/health/2020/03/17/210741/ini-alasan-mengapa-wabah-covid-19-ditetapkan-sebagai-bencana-nasional?page=all.

[5] Hanum, N.H., Saraswati, E., Widayani, P. (2015). Pemetaan Data Penyakit Menular di Kota Semarang (Studi Kasus : Penyakit DBD, Diare, Pneumonia, and TB Paru+). [Online]. Available: http://lib.geo.ugm.ac.id/ojs/index.php/jbi/article/download/176/173.

[6] Huang, R., Liu, M., Ding, Y. (2020). Spatial-Temporal Distribution of COVID-19 in China and its Prediction: a Data-Driven Modeling Analysis. Journal of Infection in Developing Countries, Volume 14(3), Pages 246-253.

[7] Kang, D., Choi, H. Kim, J.H., Choi, J. (2020). Spatial Epidemic Dynamics of the Covid-19 Outbreak in China. International Journal of Infectious Diseases, Volume 94, Pages 96-102.

[8] Khotimah, N., Sumunar, D.R.S., Purwantara, S., Arif, N., Nayan, N. (2020). Spatial Analysis of Covid-19 Distribution: a Comparative Study between Indonesia and Malaysia. International Cooperation Research Report. Yogyakarta: Faculty of Social Sciences, Universitas Negeri Yogyakarta.

[9] Li, F. (2016). Structure, Function, and Evolution of Coronavirus Spike Proteins. Annual Review of Virologi. 3:237-261.

[10] Lucchesse, M. and Pianta, M. (2020). The Coming Coronavirus Crisis: What Can We Learn? Intereconomics, Volume 55(2), pp. 98-104.

[11] Purwadhi, Hardiyanti S., Sanjoto T.B., (2009). Pengantar Interpretasi Citra Penginderaan Jauh. Semarang: Universitas Negeri Semarang.

[12] Shereen, M. A., Khan, S., Kazmi, A., Bashir, N., Siddique, R. (2020). COVID-19 Infection: Origin, Transmission, and Characteristics of Human Coronaviruses. Journal of Advanced Research, Volume 24, Pages 91-98.

[13] Tosepu, R., Gunawan, J., Effendy, D. S., Ahmad, L. A. I., Lestari, H., Bahar, H., Asfian, P. (2020). Correlation between Weather and Covid-19 Pandemic in Jakarta, Indonesia. Science

of The Total Environment, Volume 725, 138436.
[14] Wang, Y., Dong, L., Liu, Y., Huang, Z., Liu, Y. (2019). Migration Patterns in China Extracted from Mobile Positioning Data. Habitat International, 86, pp. 71-80.
[15] Wu, A., Peng, Y., Huang B., Ding, X., Wang, X., Niu, P., Meng, J., Zhu, Z., Zhang, Z., Wang, J., Sheng, J., Quan, L., Xia, Z., Tan, W., Cheng, G., Jiang, T. (2020). Genome Composition and Divergence of the Novel Coronavirus (2019-nCoV) Originating in China, Cell Host & Microbe, Volume 27, Issue 3, pp. 325-328.
[16] Zhou, C., Su, F., Pei, T., Zhang, A., Du, Y., Luo, B., Cao, Z., Wang, J., Yuan, W., Zhu, Y., Song, C., Chen, J., Xu, J., Li, F., Ma, T., Jiang, L., Yan, F., Yi, J., Hu, Y., Liao, Y., Xiao, H. (2020). COVID-19: Challenges to GIS with Big Data, Geography and Sustainability.

Seismicity map to analyze the depth and magnitude earthquake zone in Kwandang Area of North Gorontalo Regency

Salsabila Aulia Putri Zakaria, Intan Noviantari Manyoe, Alan Gani Abduh, Fadillah Budjang

Geological Engineering, Universitas Negeri Gorontalo, Jl. Prof. Dr. Ing. B. J. Habibie, Suawawa, Bone Bolango Regency, Gorontalo Province, 96119

Email corresponding author: Salsabila_s1geologi2017@mahasiswa.ung.ac.id

Abstract. Kwandang is a district located in the northern part of Gorontalo. The purpose of this research is to analyze the depth and magnitude earthquake zone that occurred in the district of Kwandang, Gorontalo Utara regency based on seismicity map. The astronomical research location is located at 0° 49′ 39″ S, 122° 55′ 8″ E. The method used in this research is seismicity map analysis. The earthquake that dominates in Kwandang, based on the value of its depth, namely shallow earthquake (0-70 km) and medium earthquake (70-300 km). This is caused by subduction activity in the direction of the subduction of the north arm of Sulawesi towards south of Tomini Bay. Whereas based on the strength of the earthquake in Kwandang sub-district is dominated by small earthquakes (-) with a light, mild earthquake. Based on the depth zonation, earthquakes mostly occurred in the west. Based on the magnitude, the earthquakes mostly occurred in the southwest.

1. Introduction

Indonesia is one of the regions in the Southeast Asia that lays in the equator and stretches between two continents of Asia and Australia. It is also surrounded by two oceans of the Pacific Ocean and Indian Ocean. Indonesia is one of the largest archipelago's country in the world with 17,504 islands. Indonesia is also known as Nusantara. The population is approximately 270, 054, 853 people. Indonesia is one of densely populated countries in the world and is the world largest Muslim population, with more than 230 million of its people are Moslem [1], [2].

According to [1], Gorontalo Province is categorized as an area with mild to moderate earthquake damage. The earthquake in Gorontalo occurred due to subduction of the Sulawesi Sea and active faults in Gorontalo. In addition, the maximum ground acceleration rate in Gorontalo area is 1,462 -99,714 gal [1], [7].

The seismicity of the Gorontalo region is made in the form of a seismicity map. Seismicity maps are created using QGIS software. Seismicity maps are divided into two maps, namely seismicity map based on earthquake depth and seismicity map based on earthquake magnitude [1].

The complex and highly diverse condition of Sulawesi tectonic makes it prone to earthquake. To describe the complexity of the tectonic of Sulawesi, several studies have carried out the Sulawesi suture [5], [6]. It characterizes by the active faults in Gorontalo, both strike slip faults and thrust faults. This

Content from this work may be used under the terms of the Creative Commons Attribution 3.0 licence. Any further distribution of this work must maintain attribution to the author(s) and the title of the work, journal citation and DOI.
Published under licence by IOP Publishing Ltd

study aims at finding out the degree of the earthquakes in Kwandang district of North Gorontalo regency based on its seismic data.

The north part of Sulawesi, especially Gorontalo city is the most complex area within the geological timeframe of the faults and tectonics. This area is the meeting point of three convergent plates, which interpreted as due to the interactions of three earth plates in Neogene area [5], [6].

The Gorontalo Fault, as a part of the Group B (the Gorontalo Fault Group), generally has been interpreted as dextral wrench fault. The present structural analysis suggests that this major fault was formerly developed as a dextral fault during Neogene, and later during Pleistocene it was reactivated in the sense of sinistral fault due to the changing of the stress system [4].

The Neogene stress orientation is supposed to be related to the subduction of the North Sulawesi Sea to the south, which during Pleistocene weakened as the Sangihe Subduction in the Mollucas Sea to the east commenced resulting in the change of field stress orientation [4].

Gorontalo is one of the areas that are vulnerable to earthquake disaster as it is located between two main faults of Pacific fault and Eurasian fault as well as other micro faults. Geological map of Gorontalo shows that the fault structure that crosses the Gorontalo city area. Gorontalo is one of the areas with complex tectonic condition as it is influenced by the subductions along the north and east of the north Sulawesi arm, and thus, critical to be properly mitigated. There are several local faults in Gorontalo that can move anytime and causes earthquakes, in areas such as Kwandang District of North Gorontalo regency. The purpose of this research is to find out the earthquake depth zonation and magnitude zonation based on the seismicity map.

2. Research Method

2.1. Research Site

The research site is in Kwandang Regency of North Gorontalo regency, of Gorontalo province with the total area of 1, 230, 07 km². administratively, this research site lays in the area of 0°49′ 39″ S,

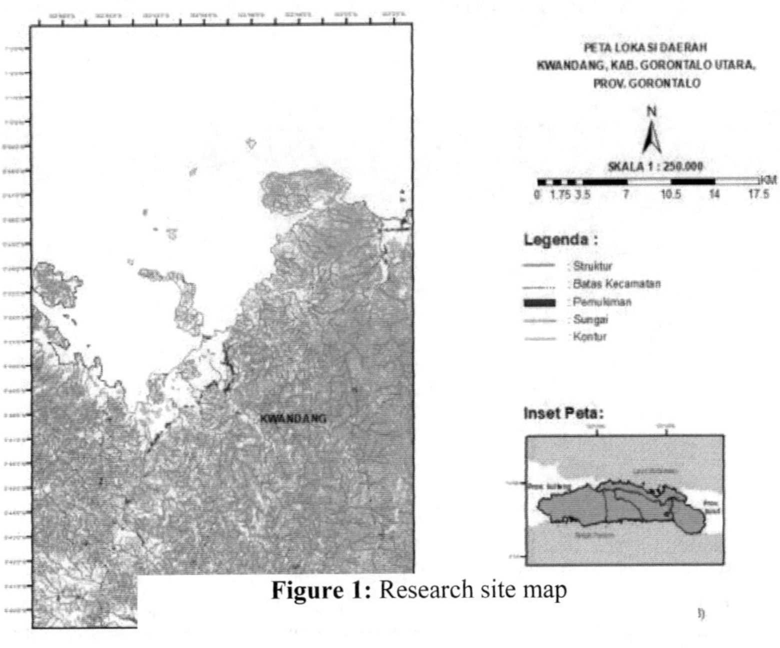

Figure 1: Research site map

122° 55′ 8″ E. This site can be reached using four or two-wheeled vehicles from Gorontalo city. The distance from the center of the city is ± 59 km and can be reached in 1 hour and 28 minutes.

IOP Conf. Series: Earth and Environmental Science **884** (2021) 012060 doi:10.1088/1755-1315/884/1/012060

2.2. Data Collection

The data used in this study are earthquake data taken from United States Geological Survey (USGS) which consisted of date, latitude, longitude, depth, magnitude, geological map, and topographic map obtained from RBI.

2.3. Data Analysis

The obtained-data are analyzed using seismic data analysis. The values are the depth magnitude taken from USGS. Further, the zonation is overlaid into the earthquake-depth map, and the zonation map of the magnitude of the earthquakes.

3. Results and Discussion

The figure below is the map of Kwandang district in North Gorontalo regency of Gorontalo province. The map has a scale of 1 : 250.000. Within this map, there are the geological structure, district borders, residence, rivers, and contour.

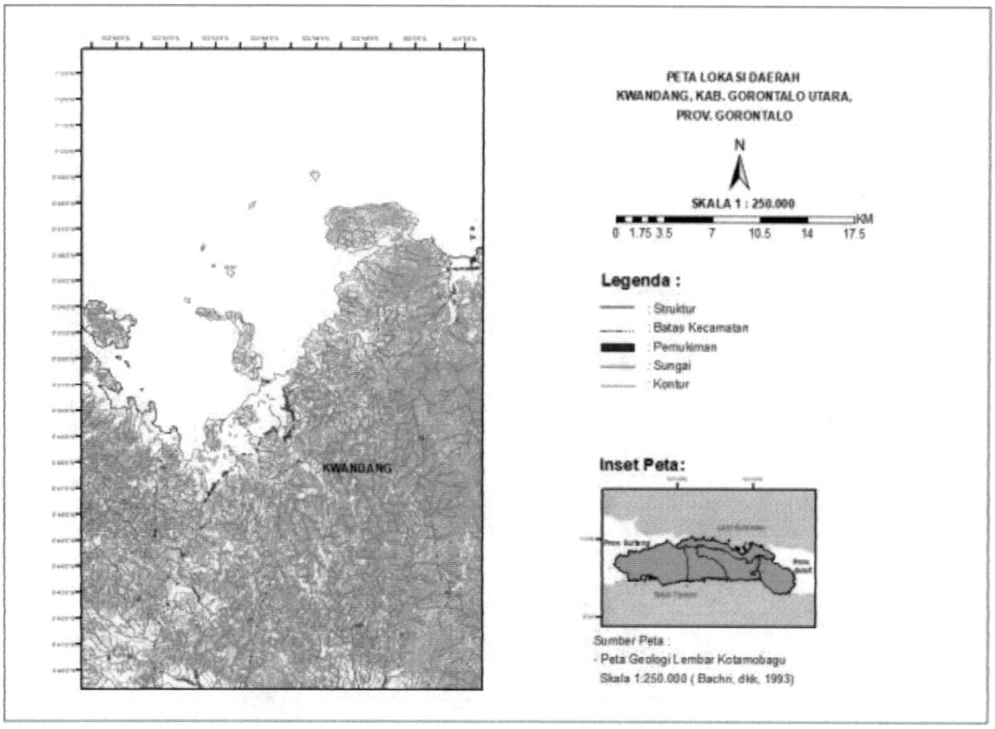

Figure 2: Research site map

The map below is the distribution of tectonic map points in the Kwandang district. Seismicity in the Gorontalo area and its surroundings is influenced by the Sulawesi Sea subduction. The direction of Sulawesi Sea subduction is north-south. Within the map, it describes the depth of the earthquake in the site is categorized as shallow and intermediate earthquakes. The shallow earthquakes happened in the depth between 0-70 kilometer from the surface, whereas the intermediate earthquake happen between 70-300 kilometer from the surface.

IOP Conf. Series: Earth and Environmental Science **884** (2021) 012060 doi:10.1088/1755-1315/884/1/012060

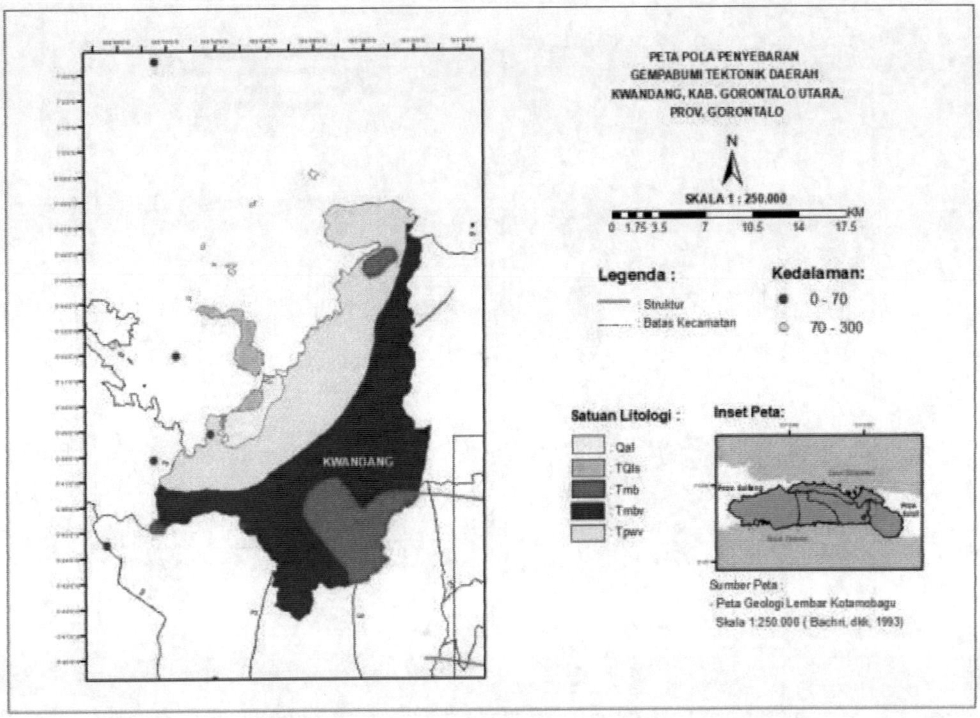

Figure 3: Seismicity map based on depth earthquake.

The figure below is the distribution of earthquakes based on their magnitudes in Kwandang district, of North Gorontalo regency using the scale of 1 : 250.000. Within the map, it is described that Kwandang experienced different magnitudes of earthquake, from shallow to deep magnitudes. The Kwandang regency is located into several soil formations such as, Qal, TQls, Tmb, Tmbv and Tpwv.

IOP Conf. Series: Earth and Environmental Science **884** (2021) 012060 doi:10.1088/1755-1315/884/1/012060

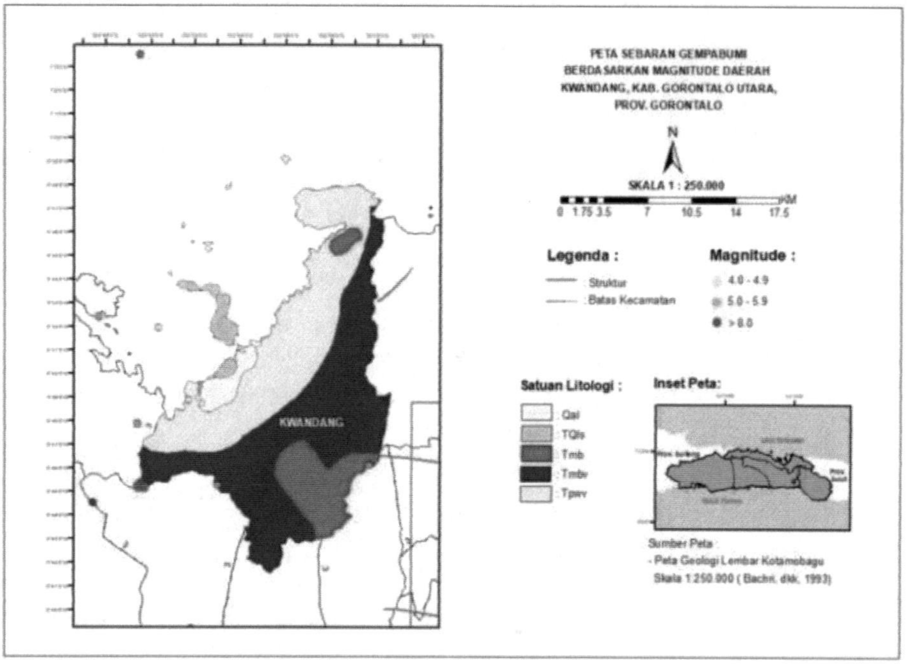

Figure 4: Magnitude-based earthquake distribution map

The map below is the zonation of the earthquake depth in Kwandang district. The zonation map shows that the depth of the earthquake varies from the shallowest of 45 km and the deepest is 90 km.

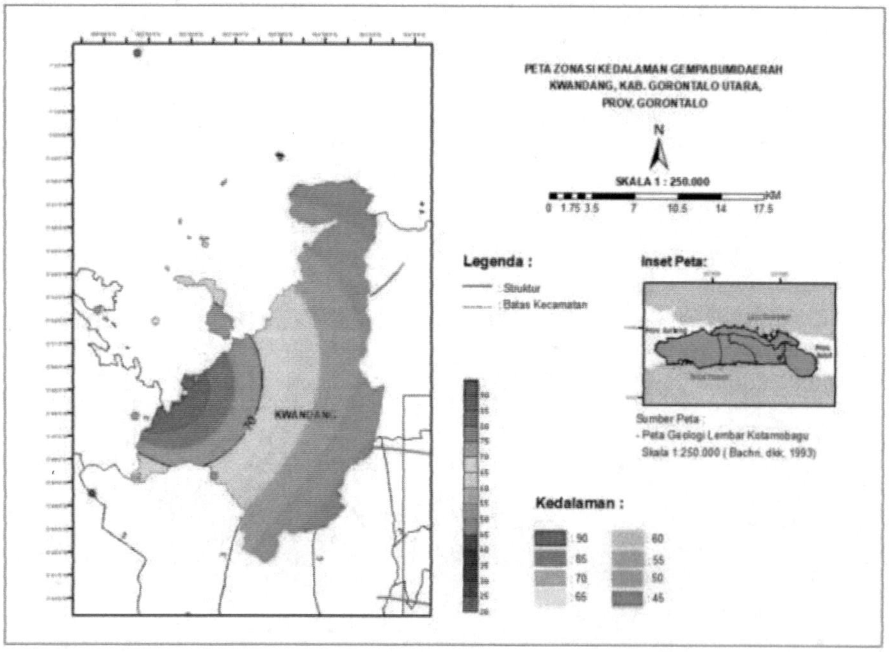

Figure 5: Earthquake depth zonation map

IOP Conf. Series: Earth and Environmental Science **884** (2021) 012060 doi:10.1088/1755-1315/884/1/012060

This map is the zonation of the earthquake magnitude in the Kwandang district with the scale of 1:250.000. The magnitude level of the earthquake is quite vary from the lowest of 3.8 in Richter Scale indicated by the purple color to the highest magnitude of 6 in Richter Scale indicated with the yellow color within the map.

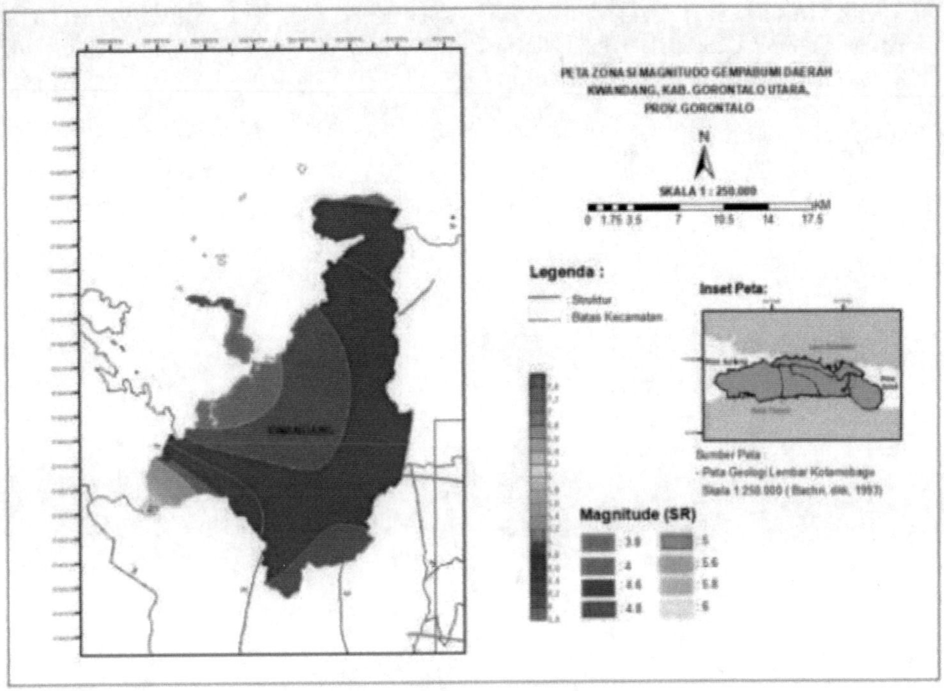

Figure 6: Earthquake magnitude zonation

4. Conclusion

Based on the map analysis, the earthquakes in Kwandang district are categorized into shallow earthquakes (0-70 km) to intermediate earthquakes (70-300 km). Whereas the deep earthquake (>300 km) is undetected in this area. The magnitude of the earthquakes in Kwandang district ranges from 4.0 - 4.9, 5.0 - 5.9, to > 80.

5. References

[1] Manyoe I.N., Lantu, Arif., Lahay R.J., 2019. Earthquake Damage Level Of Gorontalo Area Based On Seismicity And Peak Ground Acceleration. *Jambura Geoscience Review* (2019) Vol. 1 (1):7-1.

[2] Arifin, S.S., B.S. Mulyanto, Marjiyono, & R. Setianegara. 2012. Penentuan Zona Rawan Guncangan Bencana Gempa Bumi berdasarkan Analisis Nilai Amplifikasi HVSR Mikrotremor dan Analisis Periode Dominan Daerah Liwa dan Sekitarnya. Jurnal Geofisika Eksplorasi Vol. 2, No. 1.

[3] Bachri, S., Sukido, and Ratman, N. (1993). Peta Geologi Lembar Tilamuta, Sulawesi. Bandung: Pusat Penelitian dan Pengembangan Geologi.

[4] Bachri, S. (2011). STRUCTURAL PATTERN AND STRESS SYSTEM EVOLUTION DURING NEOGENE-PLEISTOCENE TIMES IN THE CENTRAL PART OF THE NORTH

ARM OF SULAWESI. *Jurnal Geologi dan Sumberdaya Mineral*, *21*(3), 127-135.

[5] Simandjuntak, T.O. 1992. An Outline of Tectonics of the Indonesian Region. Geological News Letter, 252(3), 4-6. Geological Research and Development Center, Bandung-Indonesia.

[6] Simanjuntak, T. O., & Barber, A. J. (1996). Contrasting tectonic styles in the Neogene Orogenic Belts of Indonesia. Dalam: Hall, R. & Blundell, D. (eds.): Tectonic Evolution of Southeast Asia. *Geological Society Special Publication, 106h*.

[7] Duwingik, Reski Fauzi, et al. "SURVEY AND EFFORTS TO IMPROVE MITIGATION ABILITY FOR DISABILITY STUDENTS IN GORONTALO CITY." Jurnal Sains Informasi Geografi 2.1 (2019): 40-43.

AUTHOR INDEX